TABLE OF ATOMIC MASSES AND NUMBERS

Based on the 1991 Report of the Commission on Atomic Weights and Isotopic Abundances of the International Union of Pure and Applied Chemistry and for the elements as they exist naturally on earth. Scaled to the relative atomic mass of carbon-12. The estimated uncertainties in values, between ±1 and ±9 units in the last digit of an atomic mass, are in parentheses after the atomic mass. (From *Journal of Physical and Chemical Reference Data*, Vol. 22(1993), pp. 1571-1584. © 1993 IUPAC)

Element	Symbol	Atomic Number	Atomic Mass		Element	Symbol	Atomic Number	Atomic Mass	
Actinium	Ac	89	227.0278	(L)	Molybdenum	Mo	42	95.94(1)	(g)
Aluminum	Al	13	26.981539(5)		Neodymium	Nd	60	144.24(3)	(g)
Americium	Am	95	243.0614	(L)	Neon	Ne	10	20.1797(6)	(g, m)
Antimony	Sb	51	121.757(3)		Neptunium	Np	93	237.0482	(L)
Argon	Ar	18	39.948(1)	(g,r)	Nickel	Ni	28	58.6934(2)	
Arsenic	As	33	74.92159(2)		Niobium	Nb	41	92.90638(2)	
Astatine	At	85	209.9871	(L)	Nitrogen	N	7	14.00674(7)	(g, r)
Barium	Ba	56	137.327(7)		Nobelium	No	102	259.1009	(L)
Berkelium	Bk	97	247.0703	(L)	Osmium	Os	76	190.23(3)	(g)
Beryllium	Be	4	9.012182(3)		Oxygen	O	8	15.9994(3)	(g, r)
Bismuth	Bi	83	208.98037(3)		Palladium	Pd	46	106.42(1)	(g)
Boron	B	5	10.811(5)	(g, m, r)	Phosphorus	P	15	30.973762(4)	
Bromine	Br	35	79.904(1)		Platinum	Pt	78	195.08(3)	
Cadmium	Cd	48	112.411(8)	(g)	Plutonium	Pu	94	244.0642	(L)
Calcium	Ca	20	40.078(4)	(g)	Polonium	Po	84	208.9824	(L)
Californium	Cf	98	251.0796	(L)	Potassium	K	19	39.0983(1)	(g)
Carbon	C	6	12.011(1)	(g, r)	Praseodymium	Pr	59	140.90765(3)	
Cerium	Ce	58	140.115(4)	(g)	Promethium	Pm	61	144.9127	(L)
Cesium	Cs	55	132.90543(5)		Protactinium	Pa	91	231.03588(2)	
Chlorine	Cl	17	35.4527(9)	(m)	Radium	Ra	88	226.0254	(L)
Chromium	Cr	24	51.9961(6)		Radon	Rn	86	222.0176	(L)
Cobalt	Co	27	58.93320(1)		Rhenium	Re	75	186.207(1)	
Copper	Cu	29	63.546(3)	(r)	Rhodium	Rh	45	102.90550(3)	
Curium	Cm	96	247.07003	(L)	Rubidium	Rb	37	85.4678(3)	(g)
Dysprosium	Dy	66	162.50(3)	(g)	Ruthenium	Ru	44	101.07(2)	(g)
Einsteinium	Es	99	252.083	(L)	Samarium	Sm	62	150.36(3)	(g)
Erbium	Er	68	167.26(3)	(g)	Scandium	Sc	21	44.955910(9)	
Europium	Eu	63	151.965(9)	(g)	Selenium	Se	34	78.96(3)	
Fermium	Fm	100	257.0951	(L)	Silicon	Si	14	28.0855(3)	(r)
Fluorine	F	9	18.9984032(9)		Silver	Ag	47	107.8682(2)	(g)
Francium	Fr	87	223.0197	(L)	Sodium	Na	11	22.989768(6)	
Gadolinium	Gd	64	157.25(3)	(g)	Strontium	Sr	38	87.62(1)	(g, r)
Gallium	Ga	31	69.723(4)		Sulfur	S	16	32.066(6)	(g, r)
Germanium	Ge	32	72.61(2)		Tantalum	Ta	73	180.9479(1)	
Gold	Au	79	196.96654(3)		Technetium	Tc	43	98.9072	(L)
Hafnium	Hf	72	178.49(2)		Tellurium	Te	52	127.60(3)	(g)
Helium	He	2	4.002602(2)	(g, r)	Terbium	Tb	65	158.92534(3)	
Holmium	Ho	67	164.93032(3)		Thallium	Tl	81	204.3833(2)	
Hydrogen	H	1	1.00794(7)	(g, m, r)	Thorium	Th	90	232.0381(1)	(g)
Indium	In	49	114.818(3)		Thulium	Tm	69	168.93421(3)	
Iodine	I	53	126.90447(3)		Tin	Sn	50	118.710(7)	(g)
Iridium	Ir	77	192.22(3)		Titanium	Ti	22	47.88(3)	
Iron	Fe	26	55.847(3)		Tungsten	W	74	183.84(1)	
Krypton	Kr	36	83.80(1)	(g, m)	Unnilhexium	Unh	106	263.118	(L, n)
Lanthanum	La	57	138.9055(2)	(g)	Unnilpentium	Unp	105	262.114	(L, n)
Lawrencium	Lr	103	262.11	(L)	Unnilquadium	Unq	104	261.11	(L, n)
Lead	Pb	82	207.2(1)	(g, r)	Unnilseptium	Uns	107	262.12	(L, n)
Lithium	Li	3	6.941(2)	(g, m, r)	Uranium	U	92	238.0289(1)	(g, m)
Lutetium	Lu	71	174.967(1)	(g)	Vanadium	V	23	50.9415(1)	
Magnesium	Mg	12	24.3050(6)		Xenon	Xe	54	131.29(2)	(g, m)
Manganese	Mn	25	54.93805(1)		Ytterbium	Yb	70	173.04(3)	(g)
Mendelevium	Md	101	258.10	(L)	Yttrium	Y	39	88.90585(2)	
Mercury	Hg	80	200.59(2)		Zinc	Zn	30	65.39(2)	
					Zirconium	Zr	40	91.224(2)	(g)

(g) Geologically exceptional specimens of this element are known that have different isotopic compositions. For such samples, the atomic mass given here may not apply as precisely as indicated.

(L) This atomic mass is the relative mass of the isotope of longest half-life. The element has no stable isotopes.

(m) Modified isotopic compositions can occur in commercially available materials that have been processed in undisclosed ways, and the atomic mass given here might be quite different for such samples.

(n) Name and symbol are assigned according to systematic rules developed by the IUPAC and used in the 1991 report. Since the 1991 report, the following names have been adopted: 104, rutherfordium; 105, hahnium; 106, seaborgium; 107, nielsbohrium; 108, hassium; 109, meitnerium.

(r) Ranges in isotopic compositions of normal samples obtained on earth do not permit a more precise atomic mass for this element, but the tabulated value should apply to any normal sample of the element.

ELEMENTS OF GENERAL, ORGANIC, AND BIOLOGICAL CHEMISTRY

NINTH EDITION

ELEMENTS OF GENERAL, ORGANIC, AND BIOLOGICAL CHEMISTRY

JOHN HOLUM
Augsburg College

JOHN WILEY & SONS, INC.
New York • Chichester • Brisbane • Toronto • Singapore

Cover Photo by Eric Kamp/Phototake NYC

Acquisitions Editor	Nedah Rose
Marketing Manager	Catherine Faduska
Senior Production Editor	Bonnie Cabot
Designer	Laura Nicholls
Manufacturing Manager	Susan Stetzer
Photo Researcher	Mary Ann Price
Illustration	Sigmund Malinowski

This book was set in 9.5/11.5 Caslon Book by The Clarinda Company and printed and bound by Von Hoffmann Press. The cover was printed by Phoenix Color.

Recognizing the importance of preserving what has been written, it is a policy of John Wiley & Sons, Inc. to have books of enduring value published in the United States printed on acid-free paper, and we exert our best efforts to that end.

The paper in this book was manufactured by a mill whose forest management programs include sustained yield harvesting of its timberlands. Sustained yield harvesting principles ensure that the number of trees cut each year does not exceed the amount of new growth.

John Wiley & Sons, Inc., uses recycled and recycled content papers in the manufacture of our books.

Library of Congress Cataloging in Publication Data:
Holum, John R.
    Elements of general, organic, and biological chemistry / John Holum. — 9th ed.
        p.   cm.
    Rev. ed. of: Elements of general and biological chemistry. 8th ed. c 1991.
    "A shortened version of Fundamentals of general, organic, and biological chemistry, fifth edition (1994)"—Pref.
    Includes bibliographical references.
    ISBN 0-471-31006-9
    1. Chemistry.   I. Holum, John R.   Elements of general and biological chemistry.   II. Holum, John R.   Fundamentals of general, organic, and biological chemistry. 5th ed.   III. Title.
QD31.2.H62   1995
540—dc20   94-22314
                                                                  CIP

Printed in the United States of America
10 9 8 7 6 5 4 3 2 1

PREFACE

This text is a shortened version of *Fundamentals of General, Organic, and Biological Chemistry,* fifth edition (1994), and a large number of the results of bringing that text up to date and making it pedagogically better have been incorporated into this ninth edition of *Elements of General, Organic, and Biological Chemistry.* Yet, since the fifth edition of *Fundamentals* "went to press," scientific advances have relentlessly driven on. Those relevant to the material herein have been incorporated. Such updating may seem of minor importance compared with the wealth of well-established principles that must be taught. However, having the most up-to-date version possible for any given edition has long been one of my major goals. Some of the most interesting applications of principles concern recent advances in environmental studies and in the molecular basis of life.

THE MOLECULAR BASIS OF LIFE AS A CENTRAL THEME

Two types of students study the material offered in this book. A number intend careers in the health sciences other than that of physician. Many others see their futures outside of science but are interested in how nature and human life work at their molecular levels. This book is suitable for both types of students because of its overarching theme, the molecular basis of life. I have allowed no topics that do not fit the theme or serve as background material for topics that do. At the same time, the book works for a one-term course, and high school chemistry is not a prerequisite.

Whether the book can be used in its entirety in one term depends somewhat, of course, on both admissions standards and the unique interests that teachers may have in particular areas. An appreciation of these facts has tempered the writing. I have placed many things in Special Topics that can be assigned or omitted as desired without worrying that their omission will complicate the study of later topics. Some whole chapters and some sections of chapters can be easily omitted, and teachers will have no trouble in identifying them as they tailor their courses to their own student populations. Many teachers, for example, find that Chapters 21 and 22 must be left out, which is why these chapters come at the end of the book. Chapters 19 and 20 begin with broad overviews of catabolism and anabolism, which may be as much of such material as time will allow in a one-term course.

STRUCTURE OF THE BOOK

Chapters 1−8: General Chemistry Topics.   We cannot go into the molecular basis of life without knowing about molecules as well as several fundamental concepts concerning the structure and properties of matter in general. I believe that as long as students retain confidence that a topic, no matter how seemingly remote, relates somehow to their interest in life and health, they will be motivated. Acids, bases, and buffers are studied, for example, because the acid–base status of the body is a matter of life and death. And, of course, any teacher realizes that acids, bases, and buffers cannot be studied without a good background in formulas, structures, equations, solutions, and equilibria.

Laboratory-Related Organization.   The vocabulary of terms and the background of concepts often required for good laboratory exercises develop too slowly under a traditional text organization. This is why I have *introduced* a number of topics early in the book, while knowing that their *full treatment* can only come later. Many teachers, for example, want to use solutions early in the first term. This is why I have defined a number of words concerning solutions in Chapter 4, although a fuller study comes in Chapter 6. The concept of molarity is presented in Chapter 4 because students will see it in the lab, but the other kinds of concentration expressions, like percentages, come later. The concepts of significant figures, accuracy, precision, uncertainty, and error (concepts much rewritten for this edition) are covered in the first chapter, although quantitative *chemical* reasoning (the mole concept) is discussed in Chapter 4. My willingness to split these kinds of topics is not only to facilitate the laboratory work. It stems also from my own experience that *the organization of knowledge for first-time learners must often be different from that for students who are reviewing what they have already studied in some depth.*

One change in this edition is in the treatment of acids and bases. The focus is now almost exclusively on the Brønsted concept so that we may routinely speak of something like sodium bicarbonate (or the bicarbonate ion) as a base. The Lewis concept is not mentioned because no use is made of it in the chapters on organic and biological chemistry. Credit, of course, is given to Arrhenius for starting acid–base theory.

Chapters 9–13: Topics of Organic Chemistry Essential to the Study of Biochemistry. The wedding between the theme of the course and the limitation of time results in a very abbreviated survey of organic chemistry. Some of the major topics developed in even a one-term course of "regular" organic chemistry, like the theory of resonance, nucleophilic substitution reactions, the Grignard synthesis, and many others, have had to be excluded. I have stressed only those functional groups that occur widely among the molecules of life and their reactions with four kinds of compounds: acids, bases, oxidizing agents, and reducing agents. I have provided some mechanisms because organic reactions otherwise seem too much like magic, and their learning becomes merely rote.

Chapters 14–22: Biochemistry, the Molecular Basis of Life.   Carbohydrates, lipids, and proteins begin this closing section of the book. Because of their importance to all that follows, I next take up enzymes, hormones, neurotransmitters, and the extracellular fluids of the body.

The citric acid cycle and the respiratory chain serve the metabolism of all types of biochemicals, so these pathways are studied next. Then come treatments of the metabolism of carbohydrates, lipids, and proteins. A study of nucleic acids completes the study of biochemistry. The role that nucleic acids have as genetic materials serves as a biochemical backdrop to a major threat to that role, ionizing radiation, the subject of the last chapter.

CHANGES TO EARLIER EDITIONS

Most of the Practice Exercises that require a calculation or a structure and many of those asking for an equation are new. Beyond that (and more importantly), there have been several changes in organization and content. In the general chemistry chapters (1–8), the following can be mentioned. Section 1.5 ("Accuracy, Error, Uncertainty, and Precision in Measurements") is revised thanks to a paper by Professor Charles Guare in the *Journal of Chemical Education* (August 1991, p. 649). The Brønsted concept of acids and bases, as mentioned earlier, now dominates Chapters 7 and 8.

New Special Topics Enriching the Organic Chemistry Chapters   New Special Topics such as "Organic Fuels," "Ozone in Smog," "The Ozone Shield," and "Ethyl Alcohol and Alcoholism" represent applications.

Considerable Updating in the Biochemistry Chapters.   Some reorganizations occur *within* chapters, such as the earlier placement of the study of fatty acids in Chapter 15 ("Lipids"). Similarly, the section in Chapter 19 on the citric acid cycle now (at the behest of many) comes *before* the section on the respiratory chain. Additional features include the following.

- Chapters 15 ("Lipids") and 16 ("Proteins") now offer a more extensive treatment of cell membranes than before. The lipid components are in Chapter 15 and are used to introduce membrane structure, but the equally important glycoprotein components are in Chapter 16, where cell membranes are revisited. The glycoproteins provide recognition sites for hormones, neurotransmitters, and certain drugs, so Chapter 16 has a new application in a Special Topic, "Mifepristone (RU 486)—Receptor Binding of a Synthetic Antipregnancy Compound." Another application, this one involving Chapter 15, is Special Topic 15.3, "Illicit Steroids—The Anabolic Steroids in Sports."

- Chapter 16 (Proteins), besides new material on cell membranes, also includes for the first time information on how gap junctions between cells enable the synchronized actions by many cells in certain tissues (e.g., the heart).

- Recent developments described in Chapter 17 include retrograde chemical messengers such as nitric oxide and carbon monoxide. A new Special Topic—"Molecular Complementarity and Immunity, AIDS, and the ABO Blood Groups"—applies the concepts of this chapter and earlier studies of cell membranes to topics of general interest.

- Chapter 18 ("Extracellular Fluids of the Body") remains about the same. *For future nurses, this chapter is the most important single chapter in the entire book.* Three sections concern the acid–base status of the blood. This topic is virtually the only topic studied in the course that reaches well beyond background applications for other nursing courses to the lifetime careers of nurses. Nearly all medical emergencies involve serious changes in the acid–base status of the blood. The terminology and quantitative values associated with this status are widely encountered both in professional nursing work and in the literature nurses ought to be reading throughout their careers. Where nursing students make up the majority of a class, the acid–base status of the blood must be taught as thoroughly as possible.

People going into inhalation therapy or into any aspect of sports medicine also need the material on acid–base status. If this material is presented in the right way, liberal arts students with any interest in sports will be fascinated by the subject (and so motivated to learn it). I have given January interim courses on the subject where all of the students entered the class simply (and reluctantly) to satisfy a general education requirement. But some were athletes and many had enjoyed strenuous outdoor activities, like mountain skiing and wilderness trekking. Grudgingly, many became "believers," and one senior was a bit rueful over missing the opportunity to "go into chemistry." Imagine that! (If you can get *first-year* humanities majors to take the course for which this book is designed, you may recruit some into chemistry or the life sciences.)

A new Special Topic in this chapter, "Gastric Juice and Ulcer Treatment," describes how the two most common medications used to treat ulcers work. The Special Topic also notes that very recent evidence indicates that these medications should be used in conjunction with antibiotics, because ulcers appear to have a bacterial cause.

- Chapter 19 ("Molecular Basis of Energy for Living") contains an updated but somewhat simpler treatment of the respiratory chain. Exciting details on how ATP forms and is then released from its enzyme are included.

- Chapter 20 ("Metabolism and Molecule Building") has been shortened but contains updated Special Topics on diabetes and on lipoprotein complexes and "good" and "bad" cholesterol.

- Chapter 21 ("Nuclei Acids") updates the terminology of the field, the factors that stabilize duplex DNA, the recent developments in gene therapy and the defective gene in cystic fibrosis, and briefly describes the Human Genome Project. The Special Topic on DNA typing (genetic "fingerprinting") has been updated, but the legal controversy is unresolved.

Other Features of the Text.   There are frequent **margin comments** to restate a point, offer data, or simply remind.

Key terms are highlighted in boldface at those places where they are defined and then discussed. A complete *Glossary* of these terms plus a few others appears at the end of the book. The ***Study Guide*** that accompanies this book also has individual chapter glossaries.

Each section of a chapter begins with a **headline.** This is *not* a one-sentence summary of the section but rather a lead-in to the beginning of the section that tries to state the section's major point.

Each chapter has a **Summary** that uses key terms in a narrative manner. The summaries are not necessarily organized in the same order in which the material occurs in the various chapter sections. The summaries assume that the sections have been studied so that the needed vocabulary is in place. The summaries thus illustrate what was said earlier, namely, that the pedagogy for first-time learning is not necessarily the same as that for reviewing.

The chapters in the first two-thirds of the book have several **worked examples.** In those involving calculations, the **factor-label** method is exploited. New features of these examples, which have always had "Problem" and "Solution" parts, are "Analysis" and "Check" comments. Thus, immediately after the statement of the problem comes the *analysis*. What is the problem really asking? In a multistep solution, what must be done first? Then comes the *solution*. We want to encourage students to see that *solving* a problem (figuring out what to do) occurs *before* the calculations. Following the "Solution" section of an example there is often a *"Check"* section. "Does the *size* of the answer make sense?" This takes the student back over the problem and encourages the use of the mind (as opposed to mechanical use of factor-labels) to see the sense of the analysis and the solution. Among problems in the organic chapters, "Check" sections help students to learn how to double-check their answers.

Nearly all worked examples are followed by **Practice Exercises,** which encourage immediate reinforcements of skills learned in the examples. Answers to all Practice Exercises are in Appendix V. A copious number of **Review Exercises** close each chapter, including some that are "additional," that is, they are not identified by topic. Many additional exercises require the use of material from earlier chapters. Thus you will find stoichiometry problems scattered throughout the book.

I have introduced three **icons** that will draw the attention of a student to places that emphasize various skills that should be mastered or that point out topics of particular interest from either a health or an environmental view.

 This icon, which suggests either a measurement or the application of a skill, draws attention to discussions of chemical calculations, balancing equations, or similar skills.

 This is my "map sign" icon, and it appears almost exclusively in the organic chemistry chapters. I draw an analogy between the representations of functional groups in organic structures (like an alkene group) and the symbols used by map makers. We need only a few map symbols to enable us to read almost any map. Similarly, we can see a functional group symbol as representing a relatively short list of properties conferred on the substance. By knowing structural "map signs," we can "read" structural formulas like a map and predict some properties surprisingly well.

 This icon, suggesting not only planet earth but also all people on it, draws attention to topics whereby applications of chemical knowledge are made to matters of health or earthcare.

SUPPLEMENTARY MATERIALS FOR STUDENTS AND TEACHERS

The complete package of supplements that are available to help students study and teachers teach includes the following.

Laboratory Manual for Elements of General, Organic, and Biological Chemistry, ninth edition.   This thoroughly revised edition was prepared by Dr. Sandra Olmsted, Augsburg College. An *Instructor's Guide* to this laboratory manual is contained in the *Instructor's Manual* described below.

Study Guide for Elements of General, Organic and Biological Chemistry, ninth edition.   This softcover book contains chapter objective, chapter glossaries, additional worked examples and exercises, sample examinations, and answers to all of the Review Exercises.

Instructor's Manual for Elements of General, Organic, and Biological Chemistry, ninth edition.   This softcover supplement by Sandra Olmsted and John R. Holum is available to teachers, and includes answers to all text problems and exercises. It also contains a complete *Instructor's Guide* for the laboratory manual mentioned above, including samples of all pre- and postlaboratory reports.

Test Bank.   Available in both hard copy and software (Macintosh and IBM compatible) versions, this test resource contains roughly 900 questions.

Transparencies.   Instructors who adopt this book may obtain from Wiley, without charge, a set of more than 100 four-color transparencies that duplicate key illustrations from the text.

ACKNOWLEDGMENTS

My wife Mary has been my strongest supporter, and I am deeply grateful to this wonderful woman. My daughters, Liz, Ann, and Kathryn, now grown, also have been strong champions, and I thank them for what they have meant to Mary and me.

At Augsburg College, I have always enjoyed unstinting support from the former Chair of the Chemistry Department, Dr. Earl Alton (now Assistant Dean), from the Academic Dean, Dr. Ryan LaHurd (now the President of Lenore Rhyne College), and from the President, Dr. Charles Anderson. Dr. Arlin Gyberg, Dr. Joan Kunz, and Dr. Sandra Olmsted of the Chemistry Department have been important sources of suggestions and corrections.

Extraordinarily nice people are all over the place at John Wiley & Sons. I think particularly of my Chemistry Editor, Nedah Rose, her Administrative Assistant, Marianne Stepanian, and my Supplements' Editor, Joan Kalkut.

The overall design was the responsibility of Laura Nicholls with whom I have worked with great pleasure on this and other books. Sigmund Malinowski has been skillful, artistic, and faithful in handling the line drawing art work. Mary Ann Price, Photo Editor, produced such a rich supply of outstanding choices for photographs that my choosing became difficult, yet exciting and pleasurable. My copy editor, Josephine Della Peruta, handled her assignment with grace. The production was supervised by Bonnie Cabot, and her job is surely one of the most difficult in textbook publishing. Coordinating copy editing, art work, photos, the setting of galleys, proofreading, the preparation of page dummies and final pages, and printing and binding, without letting anything fall between the slats, requires the patience of Job, the accuracy of a computer, and the discipline of a wagon master.

Dr. Sandra Olmsted, an outstanding proofreader, saved me from innumerable embarrassments. Dr. Melinda Lee (Saint Cloud State University) has checked the answers to the Practice Exercises and Review Exercises, and she also prevented many glitches. It's hard to imagine that any errors remain but, based on experience, no doubt some do. They are now entirely my responsibility. Please use a letter to my Chemistry Editor to let me know about them.

The professional critiques of many teachers are part of the process of preparing a manuscript. I am most pleased to acknowledge and to thank the following people for their work.

Scott Davis
Mansfield College

Sharmaine Cady
East Stroudsburg State College

Ricardo Rodriguez
Texas Wesleyan University

Jack Dalton
Boise State University

Lorraine Brewer
University of Arkansas

Kent Thomas
Kansas–Newman College

William L. Haag
Lake Superior State College

Ronald Offley
New Mexico State University
    at Alamogordo

Don Harriss
University of Minnesota/Duluth

Herman Knoche
University of Nebraska—Lincoln

Larry Jackson
Montana State University

John R. Holum
Minneapolis, MN

CONTENTS

1

GOALS, METHODS, AND MEASUREMENTS

We commonly think that all of nature's wonders are "out there," and one of them is certainly this view from the Toroweap Overlook at Arizona's Grand Canyon. At the molecular level of life deep within us, however, lie vistas and wonders equally grand. We'll explore many of them in this text.

1.1   CHEMISTRY AND THE MOLECULAR BASIS OF LIFE

The *theme* of this book is the molecular basis of life.

Centuries ago, people surely noticed that many *different* animals drank at the same water holes, breathed the same air, ate the same kinds of food, and enjoyed the same salt licks. Ancient farmers knew that animal droppings nourished plants and that many animals prospered by eating plants.

Evidently, at some deep level of existence, living things can exchange parts, not organs and tissues, but smaller things called *molecules* made of even tinier particles called *atoms*. All of life, whether plant or animal, has a *molecular* basis, and chemistry has been the route to its discovery. **Chemistry** is the study of that part of nature dealing with substances, their compositions and structures, and their abilities to be changed into other substances. There are, however, so many different substances, that we have to have a plan of study.

■ Well over 6 million chemical substances are known.

Our Strategy. Life at the molecular level involves molecules and chemical reactions that are often complicated. The symbols we use for them, however, are actually less complex than many symbol systems you have already mastered, like those used to draw maps. You learned how to read and understand dozens of maps by mastering just a few map symbols. Our symbols for molecules are like maps because the same pieces of molecules, like molecular "map signs," occur over and over again. Before we study some of the more complicated molecules in nature (Chapters 14 through 22), it would be a good idea to learn these molecular map signs among simpler substances. Our chapters on organic compounds (Chapters 9 through 13) do this.

As we said earlier, molecules are made of atoms. It really isn't possible to understand molecules without first learning about atoms and how their own (even tinier) parts become reorganized into molecules. This study occurs mainly in the first eight chapters together with the essential background about a variety of substances such as acids, bases, salts, and solutions. All these studies rest on experimental evidence that was obtained by taking measurements of physical quantities. In this chapter we learn about some of the measurements that have been useful in chemistry.

■ The atoms of all of the kinds of matter are made of varying combinations of just three extremely tiny particles: electrons, protons, and neutrons.

1.2   PROPERTIES AND PHYSICAL QUANTITIES

A *physical property* differs from a *chemical property* by being observable without changing a substance into a different substance.

A **property** is any characteristic of something that we can use to identify and recognize it when we see it again. The observations of some properties, however, actually change a substance into something else. We can measure, for example, how much gasoline it takes to drive a car 100 miles, but this measurement uses up the gasoline. As it burns, gasoline changes into water and carbon dioxide (the fizz in soda pop). A property that cannot be observed without substances changing into new substances is called a **chemical property.** What is being observed is a **chemical reaction.** A chemical property of iron, for example, is that it rusts in moist air; it changes slowly into a reddish, powdery substance, iron oxide, quite unlike metallic iron. Chemistry is the study of these kinds of changes in substances, how they occur, and how atoms become reorganized as a result of these changes.

Properties like color, height, and weight, which can be observed without changing the object into something different, are called **physical properties.** We usually rely on physical properties to recognize and name things. Some physical properties of liquid water, for example, are that it is colorless and odorless; it dissolves sugar and table salt but not butter; it makes a thermometer read 100 °C (212 °F) when it boils (at sea level); and if mixed with gasoline, water will sink, not float. If handed a glass containing a liquid with these properties, your initial hypothesis would undoubtedly be that the liquid is water. Think of how often each day you recognize things (and people) by simply observing physical properties.

Notice how much a description of water's properties depends on human senses, our abilities to see, taste, feel, and sense hotness or coldness. Our senses, however, are limited, so inventors have developed instruments that extend the senses and make possible finer and sharper observations. These devices are equipped with scales or readout panels, and the data we obtain by using them are called *physical quantities.*

A **physical quantity** is a property to which we can assign both a numerical value *and a unit.* Your own height is a simple example. Suppose you have a friend who is 5.5 feet tall. The numerical value of height (5.5) and the unit (feet) together tell us at a glance how much greater the friend's height is than an agreed-on *reference* of height, the foot.

The unit in a physical quantity is just as important as the number. If you said that your height is "2," people would ask, "2 what?" If you said "2 *yards,*" they would know what you meant (provided they knew what a yard is). (But they might ask, "*Exactly* 2 yards?") This example shows that we cannot describe a physical property by a physical quantity without giving both a number and a unit.

$$\text{Physical quantity} = \text{number} \times \text{unit}$$

■ 2 = a number
2 yards = a physical quantity

Physical Quantities Are Obtained by Measurements.   A measurement, is an operation by which we compare an unknown physical quantity with one that is known. As you were growing up, someone no doubt measured your height by comparing it with how many sticks, probably 1-foot rulers, it took to equal your height. Usually the number of sticks did not match your height exactly, so fractions of sticks called inches (each with their own fractions) were also used. It's probably quite obvious to you by now that somebody has decided what an inch, a foot, or a yard is and that the rest of us have agreed to use these *definitions.* That's just what they are, of course, definitions. We'll learn those that are the most useful in chemistry in the next section.

1.3   UNITS AND STANDARDS OF MEASUREMENT

The fundamental quantities of measurement are *base quantities,* and each has an official *reference standard* for one unit of base quantity.

Mass, Length, Time, and Temperature Are Base Quantities.   The most fundamental measurements in chemistry are those of mass, volume, temperature, time, and amount of chemical substance.

Mass is the measure of the inertia of an object. Anything said to have a lot of inertia, such as a train engine, a massive boulder, or an ocean liner, is very hard to get

Large inertia goes with large mass.

■ *Quantitative* describes something expressible by a number and a unit.

into motion, or if it is in motion, it is difficult to slow it down or make it change course. It is this inherent resistance to any kind of change in motion that we call **inertia,** and *mass* is our way of describing inertia quantitatively. A large inertia means a large *mass.*

A large mass doesn't always mean a large *weight.* Your mass does not depend on where you are in the universe, but your weight does. Your weight is a measure of the gravitational force of attraction that the earth exerts on your body. This gravitational force is less on the moon, which is a smaller object than the earth—about one-sixth of that on earth. But the mass of an astronaut, the fundamental resistance to any change in motion, is the same on the moon as on the earth. When we use a laboratory balance to *weigh* something, we are actually measuring mass because we are comparing two weights *at the same place on the earth* and, therefore, under the same gravitational influence. One weight is the quantity being measured, and the other is a "weight" (or set of weights) built into the weighing balance (Figure 1.1). Although we commonly call the result of the measurement a "weight," we would more properly call it the *mass* of the object or the sample. We generally speak of masses, not weights, in this book, but we follow the traditional practice of using "weighing" (not "massing") for the operation of obtaining a mass.

The **volume** of an object is the space it occupies, and space is described by means of a more basic physical quantity, *length.* The volume of a cube, for example, is the product of (length) × (length) × (length), or (length)^3. **Length** is a physical quantity that describes how far an object extends in some direction, or it is the distance between two points.

A fundamental quantity like mass or length is a **base quantity.** Any other quantity that is described in terms of one or more base quantities, like volume, is a **derived quantity.**

Another base quantity in science is **time,** our measure of how long events last. We need this quantity to describe how rapidly the heart beats, for example, or how fast some chemical reaction occurs.

Still another important physical quantity is **temperature,** which we use to describe the hotness or coldness of an object.

Volume of a cube = $(l)^3$

(a)

(b)

FIGURE 1.1
(a) A traditional two-pan balance showing a small container on the left pan and some weights on the right pan.
(b) A modern single-pan balance with a digital readout and capable of measurements to the nearest 0.001 g. The balance is fitted with a cover to keep air currents from affecting the measurement.

All of these base quantities are necessary to all sciences, but chemistry has a special base quantity called the *mole* that describes a certain amount of a chemical substance. It consists of a particular (and very large) *number* of tiny particles without respect to their masses or volumes. We will not study this base quantity further until we know more about these particles.

Every Base Unit Has a Reference Standard of Measurement. To measure and report the mass of an object, its temperature, or any of its other base or derived physical quantities, we obviously need some units and some references. By international treaties among the countries of the world, the reference units and standards are decided by a diplomatic organization called the General Conference of Weights and Measures, headquartered in Sèvres, a suburb of Paris, France. The General Conference has defined a unit called a **base unit** for each of seven base quantities, but we need units only for the five that we have already mentioned: mass, length, time, temperature, and mole. We also need some units for the derived quantities, for example, for volume, density, pressure, and heat. The standards and definitions of base and derived quantities and units together make up what is now known as the **International System of Units** or the **SI** (after the French name, *Système Internationale d'Unites*).

Each base unit is defined in terms of a **reference standard,** a physical description or embodiment of the base unit. Long ago, the reference (such as it was) for the *inch* was "three barleycorns, round and dry, laid end to end." Obviously, which three barleycorns were picked had a bearing on values of length under this "system." And if the barley corns got wet, they sprouted. You can see that a reference standard should be entirely free of risks such as corrosion, fire, war, theft, or plain skullduggery, and it should be accessible at any time to scientists in any country.

The SI base unit of length is called the **meter,** abbreviated **m,** and its reference standard is called the *standard meter.* Until 1960, the standard meter was the distance separating two thin scratches on a bar of platinum–iridium alloy stored in an underground vault in Sèvres. This bar, of course, could have been lost or stolen, so the latest reference for the meter is based on a property of light, something available everywhere, in all countries, and that obviously cannot be lost or damaged. This change in reference didn't change the actual length of the meter; it only changed its official reference.[1]

■ The other two SI base quantities are electric current and luminous intensity. Their base units are called the ampere and the candela, respectively.

■ The predecessor to the SI was called the *metric system*. The differences between the two systems are chiefly in the reference standards, not in the sizes of names of the base units.

■ An *alloy* is a mixture of two or more metals made by stirring them together in their molten states.

[1]The SI now defines the standard meter as how far light will travel in 1/299,792,458th of a second. It is thus based on the speed of light as measured by an "atomic clock."

TABLE 1.1   Some Common Measures of Length[a]

SI	U.S. Customary
1 kilometer (km) = **1000** meters (m)	1 mile (mi) = **5280** feet (ft)
1 meter = **100** centimeters (cm)	= **1760** yards (yd)
1 centimeter = **10** millimeters (mm)	1 yard = **3** feet (ft)
	1 foot = **12** inches (in.)
Other Relationships	
1 meter = 39.37 inches	1 inch = **2.54** centimeters

[a]Numbers in boldface are exact.

■ The term *exactly* will become significant when we study Section 1.5 and learn about *significant figures*.

In the United States, older units are now legally defined in terms of the meter. For example, the yard (yd), roughly nine-tenths of a meter, is defined as 0.9144 m (exactly). The foot (ft), roughly three-tenths of a meter, is defined as 0.3048 m (exactly).

In chemistry, the meter is usually too long for convenience, and submultiples are often used, particularly the **centimeter,** or **cm,** and the **millimeter,** or **mm.** Expressed mathematically, these are defined as follows.

$$1 \text{ m} = 100 \text{ cm}$$
$$1 \text{ m} = 1000 \text{ mm}$$
$$1 \text{ cm} = 10 \text{ mm}$$

Notice that the subunits are in fractions based on 10. The millimeter, for example, is one-tenth of a centimeter. As we'll often see, this makes many calculations much easier than they were under older systems (where, for example, the inch was one-twelfth of a foot, and the foot was one-third of a yard).

The inch (in.) is about two and a half centimeters; more exactly,

$$1 \text{ in.} = 2.54 \text{ cm (exactly)}$$

Table 1.1 gives several relationships between various units of length.

The SI base unit of mass is named the **kilogram,** abbreviated **kg.** Its reference is named the *standard kilogram mass,* a cylindrical block of platinum−iridium alloy housed at Sèvres under the most noncorrosive conditions possible (Figure 1.2). This is the only SI reference that could still be lost or stolen, but no alternative has yet been devised. Duplicates made as much like the original as possible are stored in other countries. One kilogram has a mass roughly equal to 2.2 pounds in the U.S. customary system (the avoirdupois system). Table 1.2 gives a number of useful relationships among various units of mass. Unless we state otherwise, however, we will use only the SI and the U.S. customary (avoirdupois) system in this book.

The most often used units of mass in chemistry are fractions of the SI kilogram, namely, the **gram (g),** the **milligram (mg),** and the **microgram (μg),** defined as follows.

$$1 \text{ kg} = 1000 \text{ g}$$
$$1 \text{ g} = 1000 \text{ mg}$$
$$1 \text{ mg} = 1000 \text{ μg}$$

Lab experiments in chemistry usually involve grams or milligrams of chemical substances.

1 kg of butter

■ One cubic meter holds a little more than 250 gallons.

The SI unit of volume, one of the important derived units, is the cubic meter (m^3) but this is much too large for convenient use in chemistry. An older unit, the **liter,** abbreviated **L,** is accepted as a *unit of convenience.* The liter occupies a volume of 0.001 m^3 (exactly), and one liter is almost the same as one liquid quart: 1 quart (qt) = 0.946 L.

TABLE 1.2 Some Common Measures of Mass[a]

SI
1 kilogram (kg) = **1000** grams (g)
1 gram = **1000** milligrams (mg)
1 milligram = **1000** micrograms (μg, γ, or mcg)[b]

U.S. Customary (Avoirdupois)[c]
1 short ton = **2000** pounds (lb avdp)
1 pound = **16** ounces (oz avdp)

Other Relationships
1 kilogram = 2.205 lb
1 lb avdp = 453.6 grams

[a]Numbers in boldface are exact.

[b]The microgram is sometimes called a *gamma* in medicine and biology.

[c]These are the common units in the United States.

Paper clip, 0.4 g

Penny, 3.4 g

FIGURE 1.2
The SI standard kilogram mass of the International Bureau of Weights and Measures in France.

Even the liter is often too large for convenience in chemistry, and two submultiples are used, the **milliliter (mL)** and the **microliter (μL).** These are related as follows.

$$1 \text{ L} = 1000 \text{ mL}$$
$$1 \text{ mL} = 1000 \text{ μL}$$

In routine chemistry work, the milliliter is by far the most common unit you will encounter. Table 1.3 gives several other relationships among units of volume. Figure 1.3 shows apparatus used to measure volumes in the lab.

The SI unit of time is called the **second,** abbreviated **s.** The SI *definition,* however, involves complexities of atomic physics that are entirely beyond our needs. Fortunately, the *duration* of the SI second is the same as before, for essentially all purposes.

TABLE 1.3 Some Common Measures of Liquid Volume[a]

SI
1 cubic meter (m³) = **1000** liters (L)
1 liter = **1000** milliliters (mL)
1 milliliter = **1000** microliters (μL)

U.S. Customary
1 gallon (gal) = **4** liquid quarts (liq qt)
1 liquid quart = **2** liquid pints (liq pt)
1 liquid pint = **16** liquid ounces (liq oz)

Other Relationships
1 cubic meter = 264.2 gallons
1 liter = 1.057 liquid quarts
1 liquid quart = 946.4 milliliters
1 liquid ounce = 29.57 milliliters

[a]Numbers in boldface are exact.

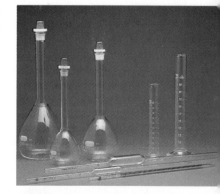

FIGURE 1.3
Some apparatus used to measure liquid volumes. In the back are three volumetric flasks (*left*) and two graduated cylinders (*right*). Lying on the surface are three volumetric pipets.

One drop of water is about 60 uL.

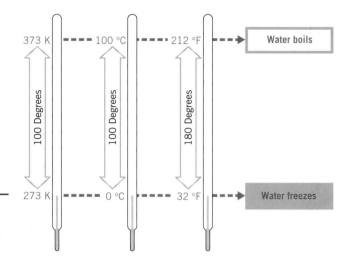

FIGURE 1.4
Relationships between the Kelvin, Celsius, and Fahrenheit scales of temperature.

The second is 1/86,400 of a mean solar day. Decimal-based multiples and submultiples of the second are used in science, but so are old units such as minute, hour, day, week, month, and year.

■ The *kelvin* is named after William Thomson, Baron Kelvin of Largs (1842–1907), a British scientist.

The SI unit for temperature is called the **kelvin, or K.** (Be sure to notice that the abbreviation is K, not °K.) The kelvin is the same size as the **degree Celsius,** which was once called the degree centigrade (also °C). Then it was defined as 1/100th the interval between the freezing point of water (designated 0 °C) and the boiling point of water (called 100 °C). The most extreme coldness possible is −273.15 °C, and this is called *absolute zero* or 0 K on the Kelvin scale. Thus the *numbers* assigned to points on the Celsius and Kelvin scales differ, but always remember that the *sizes* of the two *degrees* are identical (Figure 1.4).

Because 0 K corresponds to −273.15 °C, we have the following simple relationships between kelvins and degrees Celsius (where we follow the common practice of rounding 273.15 to 273).

$$°C = K − 273$$
$$K = °C + 273$$

■ PRACTICE EXERCISE 1   For many people, normal body temperature is 37 °C. What is this in kelvins?

The Kelvin scale is used in chemistry mostly to describe temperatures of gases. The Celsius scale is more popular for most other uses and is supplanting the older Fahrenheit scale in medicine. The **degree Fahrenheit (°F)** is five-ninths the size of the degree Celsius. To convert a Celsius temperature, t_C, to a Fahrenheit temperature, t_F, we can use either of the following equations. (Practice exercises for these equations occur at the end of Section 1.6.)

$$t_C = \frac{5\ °C}{9\ °F}\ (t_F − 32\ °F)$$

$$t_F = \frac{9\ °F}{5\ °C} \times t_C + 32\ °F$$

TABLE 1.4   Some Common Temperature Readings in °C and °F

	°F	°C
Room temperature	68	20
Very cold day	−20	−29
Very hot day	100	38
Normal body temperature	98.6[a]	37
Hottest temperature the hands can stand	120	49

[a]A revision in this value is currently underway. In some healthy people, the normal temperature is as low as 98.2 °F, and in others, as high as 99 °F.

Table 1.4 gives some common temperatures in both degrees Celsius (°C) and degrees Fahrenheit (°F).

1.4   SCIENTIFIC NOTATION

Scientific notation expresses very large or very small numbers in exponential form to make comparisons and calculations easier.

The typical human red blood cell has a diameter of 0.000008 m. Whether we want to write it, say it, or remember it, 0.000008 m is an awkward number. To make life easier scientists have developed a method called **scientific notation** for recording very small or very large numbers. In scientific notation (sometimes called exponential notation), a number is written as the product of two numbers. The first is a decimal number with a value usually between 1 and 10, although some times a wider range is used. Following this number is a times ($\times$) sign and then the number 10 with an exponent or power. For example, we can write 4000 as

■ Appendix I has a review of exponential numbers.

$$4000 = 4 \times 1000 = 4 \times 10 \times 10 \times 10$$
$$= 4 \times 10^3$$

Notice that the exponent 3 is the number of places to the *left* we have to move the decimal point in 4000 to get to 4, which is a number in the desirable range.

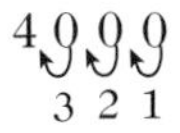

If our large number is 42,195, the number of meters in a marathon distance, we can rewrite it as follows after figuring out that we have to move the decimal point four places to the left to get a decimal number between 1 and 10.

■ When the decimal point is omitted, we assume that it is after the last digit in the number.

$$42,195 \text{ m} = 4.2195 \times 10^4 \text{ m}$$

In rewriting numbers smaller than 1 in scientific notation, we have to move the decimal point to the *right* to get a number in the acceptable range of 1 to 10. This number of moves is the value of the *negative* exponent of 10. For example, we can rewrite 0.000008 as

$$0.000008 = 8 \times 10^{-6}$$

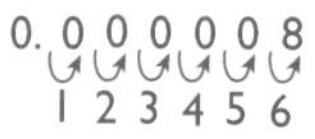

You should not continue until you are satisfied that you can change large or small numbers into scientific notation. For practice, do the following exercises.

■ PRACTICE EXERCISE 2   Express each number in scientific notation.[2] Let the decimal part be a number between 1 and 10.

(a)   545,000,000   (b)   5,670,000,000,000   (c)   6454

(d)   25   (e)   0.0000398   (f)   0.00426

(g)   0.168   (h)   0.00000000000987 (See footnote 2.)

Prefixes to the Names of SI Base Units Are Used to Specify Fractions or Multiples of These Units.   If we rewrite 3000 m as 3×10^3 m and try to pronounce the result, we have to say "three times ten to the third meters." There is nothing wrong with this, but it's clumsy. This is why the SI has names for several exponential expressions, not independent names but prefixes that can be attached to the name of any unit. For example, 10^3 has been assigned a prefix of *kilo-*, abbreviated *k-*. Thus, 1000 or 10^3 meters can be called 1 kilometer. Abbreviated, this becomes 10^3 m = 1 km. Notice that the "k" in "km" is a substitute for 10^3.

With just a few exceptions, the prefixes defined by the SI go with exponentials that involve powers of 3, 6, 9, 12, 15, and 18 or powers of $-3, -6, -9, -12, -15$, and -18. These are all divisible by 3. Table 1.5 lists the SI prefixes and their symbols. Those given in boldface are so often encountered in chemistry that they should be learned now.

Notice that there are four prefixes that do not go with powers divisible by 3. The SI hopes their usage will gradually fade away, but this hasn't happened yet. The two in boldface have to be learned. In chemistry, however, *centi* is used almost entirely with only one physical quantity, the centimeter, and *deci* is seen in the deciliter unit of volume (100 mL or 1/10 L). Clinical chemists often use the deciliter because it saves space on clinical report sheets to abbreviate 100 mL to 1 dL.

To take advantage of the SI prefixes, we sometimes have to modify a rule used in converting a large or small number into scientific notation. The goal in this conversion will now be to get the exponential part of the number to match one with an SI prefix, even if the decimal part of the number isn't between 1 and 10. For example, we know that the number 545,000 can be rewritten as 5.45×10^5, but 5 isn't divisible by 3, and there isn't an SI prefix to go with 10^5. If we counted 6 spaces to the left, however, we could use 10^6 as the exponential part.

$$5\,4\,5\,0\,0\,0 = 0.545 \times 10^6$$
$$\quad 6\ 5\ 4\ 3\ 2\ 1$$

■ 1 dL = 1 × 10^{-1} L
= 1/10th liter
But 1/10th liter = 100 mL
Therefore,
1 dL = 100 mL

■ We usually put a zero in front of a decimal point in numbers that are less than 1, such as in 0.545. The zero helps us to remember that the decimal point is there.

Now we could rewrite 545,000 m as 0.545×10^6 m or 0.545 Mm (megameter), because the prefix *mega*, abbreviated *M*, stands for 10^6. "M" *actually replaces or substitutes for* "$\times 10^6$" in 0.545×10^6 m and is attached directly to the "m" to make "Mm."

---

[2]Some of the numbers in this exercise illustrate a small problem that the SI is trying to get all scientists to handle in a uniform way. In part (h), for example, you might become dizzy trying to count closely spaced zeros. The SI recommends—and most European scientists have accepted the suggestion— that the digits in numbers having four or more digits be grouped in threes separated by thin spaces. For large numbers, simply omit the commas. The number 545,000,000, by this convention, would be written as 545 000 000. The number 0.00000000000987 becomes 0.000 000 000 009 87. You will not soon see this as common usage in the United States, but when you do you'll know what it means. Incidentally, European scientists use a comma instead of a period to locate the decimal point. You might see this yourself soon when you first weigh something in the lab. If the weighing balance was made in Europe, a reading such as 1,045 g means 1.045 g.

TABLE 1.5 SI Prefixes for Multiples and Submultiples of Base Units[a]

Relationship	Prefix	Symbol
1 000 000 000 000 000 000 = 10^{18}	exa	E
1 000 000 000 000 000 = 10^{15}	peta	P
1 000 000 000 000 = 10^{12}	tera	T
1 000 000 000 = 10^9	giga	G
1 000 000 = 10^6	**mega**	**M**
1 000 = 10^3	**kilo**	**k**
100 = 10^2	hecto	h
10 = 10^1	deka	da
0.1 = 10^{-1}	**deci**	**d**
0.01 = 10^{-2}	**centi**	**c**
0.001 = 10^{-3}	**milli**	**m**
0.000 001 = 10^{-6}	**micro**	**μ**
0.000 000 001 = 10^{-9}	nano	n
0.000 000 000 001 = 10^{-12}	pico	p
0.000 000 000 000 001 = 10^{-15}	femto	f
0.000 000 000 000 000 001 = 10^{-18}	atto	a

[a]The most commonly used prefixes and their symbols are in boldface. Thin spaces instead of commas are used to separate groups of three zeros to illustrate the format being urged by the SI (but not yet widely adopted in the United States).

We also could have rewritten 545,000 m as 545×10^3 m. If we do this, 545,000 m could be rewritten as 545 km (kilometers) because we can replace "$\times 10^3$" in "545 $\times 10^3$ m" with "k." The prefix *kilo* and its abbreviation *k* go with 10^3. *Be sure to use this approach when you learn the relationships of prefix symbols and their associated exponential terms.* (Thus, "c goes with $\times 10^{-2}$"; "μ goes with $\times 10^{-6}$"; and so on.)

EXAMPLE 1.1   Rewriting Physical Quantities Using SI Prefixes

Bacteria that cause pneumonia have diameters roughly equal to 0.0000009 m. Rewrite this using the SI prefix that goes with 10^{-6}.

ANALYSIS   In straight exponential notation, 0.0000009 m is 9×10^{-7} m, but -7 is not divisible by 3 and no SI prefix goes with 10^{-7}. If we move the decimal six places instead of seven to the right, however, we get 0.9×10^{-6} m, and -6 is divisible by 3.

SOLUTION   The SI prefix for 10^{-6} is *micro* with the symbol μ, so we simply replace "$\times 10^{-6}$" in "0.9×10^{-6} m" by μ, joining it directly to "m" to make "μm."

$$0.0000009 \text{ m} = 0.9 \times 10^{-6} \text{ m} = 0.9 \text{ μm}$$

The diameter of one of these bacteria is 0.9 μm (0.9 micrometer).

■ **PRACTICE EXERCISE 3**   Complete the following conversions to exponential notation by supplying the exponential parts of the numbers.

(a) $0.0000398 = 39.8 \times$ _____   (b) $0.000000798 = 798 \times$ _____

(c) $0.000000798 = 0.798 \times$ _____   (d) $16500 = 16.5 \times$ _____

■ **PRACTICE EXERCISE 4**   Write the abbreviation of each of the following.

(a)   milliliter   (b)   microliter   (c)   deciliter   (d)   millimeter

(e)   centimeter   (f)   kilogram   (g)   microgram   (h)   milligram

---

■ **PRACTICE EXERCISE 5**   Write the full name that goes with each of the following abbreviations.

(a)   kg   (b)   cm   (c)   dL   (d) μg

(e)   mL   (f)   mg   (g)   mm   (h) μL

■ **PRACTICE EXERCISE 6**   Rewrite the following physical quantities using the SI abbreviated forms to incorporate the exponential parts of the numbers.

(a) 1.5×10^6 g   (b) 3.45×10^{-6} L   (c) 3.6×10^{-3} g

(d) 6.2×10^{-3} L   (e) 1.68×10^3 g   (f) 5.4×10^{-1} m

■ **PRACTICE EXERCISE 7**   Express each of the following physical quantities in a way that uses an SI prefix.

(a) 275,000 g   (b) 0.0000625 L   (c) 0.000000082 m

1.5   ACCURACY, ERROR, UNCERTAINTY, AND PRECISION IN MEASUREMENTS

The way in which the number part of a physical quantity is expressed says something about the *uncertainty* of the measurement but nothing about its *accuracy* or *precision*.

Most people use the terms *accuracy* and *precision* as if they meant the same thing, but they don't. The same is true about the terms *error* and *uncertainty*. **Accuracy** refers to the closeness of a measurement to the actual value. The most accurate measurement is made when the instrument is exceptionally faithful to the related reference standard; its scale, gauge, or needle is very steady; and the experimenter is very experienced in using the instrument. The **error** in a measurement is the difference between an experimental value and the actual value. In an accurate measurement, the error is small.

Gauges, readout panels, and scales can vary among instruments meant for measuring the same physical quantity, like volume or mass. The **uncertainty** in a measurement is expressed by the *range* in values that must be recorded because of the need to estimate the last digit being read. If you have ever tried to test the *accuracy* of an automobile or bicycle odometer (mileage gauge) against mile posts set up by the local highway department, you have sensed the problem—the *uncertainty*—of estimating the reading beyond the first decimal place. You might have to record the odometer reading as 5.1 ± 0.1 mi as you pass the 5-mile post, because you judge that you cannot read the odometer more closely. The symbol "±" stands for "plus or minus," and what follows this symbol indicates how much uncertainty is carried in

■ We assume here that you are able to begin the test with your body exactly opposite the 0-mile post and that your body is exactly opposite the 5-mile post when you take your odometer reading. We also assume that the highway department has accurately positioned the posts. You can see that there are many sources of uncertainty in even an ordinary measurement such as this.

the last digit. By recording 5.1 ± 0.1 mi, the mileage is said to be between 5.0 and 5.2, so the range of uncertainty is 2 in the tenths position. The tenths position has the first uncertain digit. You can see that *uncertainty* does not directly supply information about the *accuracy* of the measurement. *Uncertainty* is only an estimate of how finely the number could be read at the time it was taken. *Accuracy,* we repeat, is the closeness of a measurement to the actual value, and only the care of the highway department in planting its mile posts, your own care in reading, and the steadiness of the odometer bear on the actual value of the mileage.

When a measurement can be repeated, the experimenter takes several measurements as carefully as possible. This strategy produces data needed to calculate the *precision* with which a physical quantity is known from the measurements. **Precision** is a measure of how *reproducible* the measurements are when several measurements are taken. The results of the measurements are averaged, and the *precision* is calculated by some index, such as the average absolute deviation from the mean or a standard deviation. These indices are described in the field of statistics; they will no longer concern us because in routine laboratory work involving measurements of mass or volume, more than one measurement is seldom taken of the same quantity. It is important to realize, however, that *precision* and *accuracy* are not the same concept. One could have an average of several measurements all agreeing very closely with each other (and so of high *precision*), but still have an average value grossly different from the actual value (and so of great *error*) because the instrument happens to have been inaccurately manufactured.

Figure 1.5 illustrates the difference between accuracy and precision in the measurement of someone's height. Each dot represents one measurement. In the first set, the dots are tightly clustered close to or exactly at the actual value, and obviously a skilled person was at work with a carefully manufactured meter stick. This set illustrates both high precision and great accuracy. In the second set, a skilled person, without realizing it, used a faulty meter stick, one mislabeled by a few centimeters. The precision is as great as that shown by the first set, because the successive measurements agree well with each other. But they're all untrue, so the accuracy is poor. In the third set of measurements, someone with a good meter stick did careless work. Only by accident do the values average to the true value, so the accuracy, in terms of the average, turned out to be high, but the precision is terrible and no one would really trust the average. The last set displays no accuracy and no precision.[3]

No matter what physical quantities we use, we want to be able to judge how accurately they were measured, but this presents a problem. When we read the value of some physical quantity in a report or a table, we have no way of telling *from it alone* if it is the result of an accurate measurement. Someone might write, for example, "4.5678 mg of antibiotic," but in spite of all its digits we can't tell from this report alone if the balance was working or if the person using it knew how to handle it and read it correctly. A skilled and careful experimenter frequently checks the instruments against references of known accuracy. Thus the question of *accuracy* is a human problem. We learn to trust the *accuracy* of data by employing trained people, giving them good instruments, requiring that they prove they are doing consistently accurate work, and rewarding consistently good results.

The Number of Significant Figures in a Physical Quantity Is the Number of Digits Known to Be Accurate plus One More.   When we do not wish (or need) to specify the *range* of uncertainty in a measurement, we need an understanding about this range.

---

[3]Suggested additional reading: Charles J. Guare, "Error, Precision, and Uncertainty," *Journal of Chemical Education,* August 1991, p. 649.

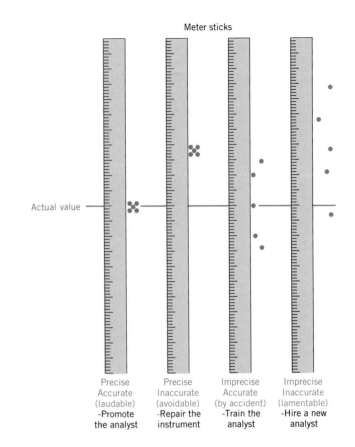

FIGURE 1.5
Accuracy and precision.

Then we can round off the numerical part of a physical quantity to leave it with a particular number of *significant figures*. The number of **significant figures** in a physical quantity is the number of digits known with complete certainty to be accurate plus one more, the digit representing the uncertainty in the measurement. Suppose, for example, you read a report that refers to "4.56 mg of antibiotic." You would be in a quandary because you would wonder about the range of uncertainty in this quantity, about what should be in place of the question mark in "4.56 ± ?. We solve the problem in this text by assuming that the question mark is replaceable by *one unit of the last decimal place* shown; in our example, the reported quantity would, by this rule, mean 4.56 ± 0.01 mg. The quantity "4.56 mg," therefore, has three significant figures. The first two, the 4 and the 5, are known to be accurate but, in the last digit, the analyst is acknowledging a small uncertainty (±0.01). This report, 4.56 ± 0.1 mg, means that the actual measurement of the mass of this one sample gave a value closer to 4.56 mg than to 4.55 mg or 4.57 mg. If the mass is reported as 4.560 mg, then it carries four significant figures. The 4, 5, and 6 are certainly accurate, but there is some uncertainty in the last digit, 0. The measurement, as reported, means a value closer to 4.560 mg than to 4.559 mg or 4.561 mg. Thus "4.560 mg" implies a greater certainty or fineness of measurement than 4.56 mg.

Figuring out how many significant figures there are in a number is easy, provided we have an agreement on how to treat zeros. Are all the zeros counted as *significant* in such quantities as 4,500,000 people, or 0.0004500 L, or 400,005 m? We will use the following rules to decide.

RULES GOVERNING SIGNIFICANT FIGURES

1. **Zeros sandwiched between nonzero digits are always counted as significant.** Thus, both 400,005 and 400.005 have six significant figures. Both 4056 g and 4506 g have four significant figures.

2. **Zeros that do no more than set off the decimal point on their *left* are never counted as significant figures.** Although such zeros are necessary to convey the general *size* of a quantity, they do not say anything about the *certainty* of the measurement. Thus, such quantities as 0.045 mL, 0.0045 mL, and 0.00045 mL all have only two significant figures.

3. **Trailing zeros to the *right* of the decimal point are always significant.** Trailing zeros are any that come to the right of a decimal point at the very end of the number, as in 4.56000. This number has three trailing zeros, and because they are to the right of the decimal point, all are significant. The number 4.56000 has six significant figures and represents considerable certainty or fineness of measurement.

4. **Trailing zeros that are to the *left* of the decimal point are counted as significant only if the author of the book or article has somewhere said or implied so.** *In this text, when we leave trailing zeros before the decimal point, count them as significant figures.* Otherwise, the zeros that disappear when you convert a number into scientific (exponential) notation are *not* significant.

■ The other zeros are needed to locate the decimal points in these numbers, and they definitely are important in this sense. They just have nothing to do with precision.

Rule 4 is really the only tricky rule. The zeros in 4,500,000, for example, are trailing zeros to the left of the decimal point, but are they significant? Suppose this number stands for the population of a city. A city's population changes constantly as people are born and die, and as they move in and out. No one could claim to know a population is *exactly* 4,500,000—not 4,499,999 and not 4,500,001, but 4,500,000. Most scientists handle this problem by restating the number in scientific notation so that any desired trailing zeros can be placed *after* the decimal point. By doing this, as many or as few of such zeros can be given to convey the proper degree of certainty. Suppose that the census bureau feels that the population is known to be closer to 4,500,000 people than to 4,400,000 or to 4,600,000 people. No better certainty than this is possible. If so, then only two significant figures would be allowed in the reported result, for example, 4.5×10^6 people. Giving the population as 4.50×10^6 people indicates a greater certainty—to three significant figures. There are four significant figures in 4.500×10^6.

Not everyone agrees with this way of handling trailing zeros that stand to the *left* of the decimal point, so you have to be careful. Some say that they aren't significant unless the decimal point is actually given, as in 45,000. L. When the decimal point is the last item in a number, however, it is easily forgotten at the time of making the record. This problem is avoided by switching to scientific (exponential) notation so that all of the trailing zeros that should be retained to specify significant figures come *after* the decimal point. This is the practice we usually follow in this book, unless noted otherwise or unless the context makes the intent very clear.

A Few Rules Govern the Rounding Off of Calculated Physical Quantities.   When we mathematically combine the values of two or more measurements, we usually have to round the result so that it expresses the same amount of uncertainty allowed by the data. Normally, such rounding is done at the *end* of a calculation (unless specified otherwise) to minimize the errors that can accumulate and grow when we round at intermediate steps in a multistep calculation. We use four simple rules for rounding calculated quantities.[4]

RULES FOR ROUNDING CALCULATED RESULTS

1. When we multiply or divide quantities, the result is allowed no more significant figures than carried by the least certain quantity (the one with the fewest significant figures).
2. When we add or subtract numbers, the result is allowed no more decimal places than are in the number having the fewest decimal places.
3. When the first of the digits to be removed by rounding is 5 or higher, round the digit to its left *upward* by one unit. Otherwise, drop it and all others after it.
4. Treat *exact numbers* as having an infinite number of significant figures.

An **exact number** is any that we *define* to be so, and we usually encounter exact numbers in statements relating units. For example, all of the numbers in the following expressions are exact and, for purposes of rounding calculated results, have an infinite number of significant figures.

$$1 \text{ in.} = 2.54 \text{ cm (exactly, as } \textit{defined by law})$$
$$1 \text{ L} = 1000 \text{ mL (exactly, by the } \textit{definition} \text{ of mL)}$$

The significance of having an infinite number of significant figures lies in our not letting such numbers affect how we round results. It would be silly to say that the "1" in "1 L" has just one significant figure when we intend, by definition, that it be an exact number.

■ We use the period in the abbreviation of inch (in.) to avoid any confusion with the preposition *in*, which has the same spelling.

EXAMPLE 1.2 Rounding the Result of a Multiplication or a Division

A floor is measured as 11.75 m long and 9.25 m wide. What is its area, correctly rounded by our rules?

SOLUTION

$$\text{Area} = \text{length} \times \text{width}$$
$$= 11.75 \text{ m} \times 9.25 \text{ m}$$
$$= 108.6875 \text{ m}^2 \text{ (not rounded)}$$

But the measured width, 9.25 m, has only three significant figures whereas the length, 11.75 m, has four. By our rules, we have to round the calculated area to three significant figures.

$$\text{Area} = 109 \text{ m}^2 \text{ (correctly rounded)}$$

■ Resist the impulse that some owners of new calculators have of keeping all the digits they paid for.

---

[4]When the actual range in uncertainty is known from experiment for each piece of data, the uncertainty in a calculated result can be expressed with greater care. See the reference in footnote 3.

EXAMPLE 1.3 Rounding the Result of an Addition or a Subtraction

Samples of a medication having masses of 1.12 g, 5.1 g, and 0.1657 g are mixed. How should the total mass of the resulting sample be reported?

SOLUTION The sum of the three values, obtained with a calculator, is 6.3857 g, which shows four places following the decimal point. One mass is precise only to the first decimal place, however, so by our rules we have to round to this place. The final mass should be reported as 6.4 g. Notice that the reported mass of the second sample, 5.1 g, says nothing about the third or fourth decimal places. We don't know whether the mass is 5.101 g or 5.199 g, or what; the sample simply wasn't measured precisely. This is why we can't know anything beyond the first decimal place in the sum.

■ PRACTICE EXERCISE 8 The following numbers are the numerical parts of physical quantities. After the indicated mathematical operations are carried out, how must the results be expressed?

(a) 16.4×5.8

(b) $5.346 + 6.01$

(c) 0.00467×5.6324

(d) $2.3000 - 1.00003$

(e) $16.1 + 0.004$

(f) $(1.2 \times 10^2) \times 3.14$

(g) $9.31 - 0.00009$

(h) $\dfrac{1.0010}{0.0011}$

1.6 FACTOR-LABEL METHOD IN CALCULATIONS

In calculations involving physical quantities, the units are multiplied or canceled as if they were numbers.

Many people have developed a mental block about any subject that requires the use of mathematics. They know perfectly well how to multiply, divide, add, and subtract, but the problem is in knowing *when,* and no pocket calculator tells this. We said earlier that the inch is defined by the relationship 1 in. = 2.54 cm. This fact has to be used when a problem asks for the number of centimeters in some given number of inches, but for some people the problem arises in knowing whether to divide or multiply.

Science teachers have worked out a method called the *factor-label* method for correctly setting up such a calculation and *knowing* that it is correct. The **factor-label method** takes a relationship between units stated as an equation (such as 1 in. = 2.54 cm), expresses the relationship in the form of a fraction, called a **conversion factor,** and then multiplies some given quantity by this conversion factor. In this multiplication, identical units (the "labels") are multiplied or canceled as if they were numbers. If the units that remain for the answer are right, then the calculation was correctly set up. We can learn how this works by doing a simple example, but first let's see how to construct conversion factors.

■ Some call the factor-label method the unit-cancellation or the factor-unit method.

■ When we divide both sides of the equation 2.54 cm = 1 in. by 2.54 cm, we get:

$$\frac{2.54\ cm}{2.54\ cm} = \frac{1\ in.}{2.54\ cm}$$

This only restates the relationship between the centimeter and the inch; it does not change it. The use of a conversion factor just changes *units*, not actual quantities.

The relationship 1 in. = 2.54 cm can be restated in either of the following two ways, and both are examples of conversion factors.

$$\frac{2.54\ cm}{1\ in.} \quad or \quad \frac{1\ in.}{2.54\ cm}$$

If we read the divisor line as "per," then the first conversion factor says "2.54 cm per 1 in.," and the second says "1 in. per 2.54 cm." These are merely alternative ways of saying that "1 in. equals 2.54 cm." *Any relationship between two units can be restated as two conversion factors.* For example,

$$1\ L = 1000\ mL \qquad \frac{1000\ mL}{1\ L} \quad or \quad \frac{1\ L}{1000\ mL}$$

$$1\ lb = 453.6\ g \qquad \frac{453.6\ g}{1\ lb} \quad or \quad \frac{1\ lb}{453.6\ g}$$

■ PRACTICE EXERCISE 9   Restate each of the following relationships in the forms of their two possible conversion factors.
(a) 1 g = 1000 mg   (b) 1 kg = 2.205 lb

Suppose we want to convert 5.65 in. into centimeters. The first step is to write down what is given, 5.65 in. Then multiply the given by the one conversion factor relating inches to centimeters that lets us cancel the unit no longer wanted and leaves the unit we want.

$$5.65\ \cancel{in.}\ \frac{2.54\ cm}{1\ \cancel{in.}} = 14.4\ cm\ (rounded\ correctly\ from\ 14.351\ cm)$$

Notice how the units of "in." cancel. Only "cm" remains, and it is on top in the numerator where it has to be.
Suppose we had used the wrong conversion factor.

■ The arithmetic is correct, but the result is still all wrong.

$$5.65\ in. \times \frac{1\ in.}{2.54\ cm} = 2.22\ \frac{in.^2}{cm}\ (correctly\ rounded)$$

That's right. We *must* do to the units exactly what the times sign and the divisor line tell us, and (in.) times (in.) equals (in.)^2 just as $2 \times 2 = 2^2$. Of course, the units in the answer, (in.)^2/cm, make no sense, so we know with certainty that we can't set up the solution this way. The reliability of the factor-label method lies in this use of the units (the "labels") as a guide to setting up the solution. Now let's work an example.

EXAMPLE 1.4 Using the Factor-Label Method

How many grams are in 0.230 lb?

ANALYSIS   From Table 1.2, we find that 1 lb = 453.6 g, so we have our pick of two conversion factors.

$$\frac{453.6\ g}{1\ lb} \quad or \quad \frac{1\ lb}{453.6\ g}$$

To change 0.230 lb into grams, we want "lb" to cancel and we want "g" in its place in the numerator. Therefore, we pick the first conversion factor; it's the only one that can give this result.

SOLUTION

$$0.230 \, \cancel{lb} \times \frac{453.6 \, g}{1 \, \cancel{lb}} = 104 \, g \text{ (correctly rounded)}$$

There are 104 g in 0.230 lb. (We rounded from 104.328 g to 104 g because the given value, 0.230 lb, has only three significant figures. Remember that the "1" in "1 lb" has to be treated as an exact number because it's in a definition.)

■ PRACTICE EXERCISE 10   The *grain* is an old unit of mass still used by some pharmacists and physicians, and 1 grain = 0.0648 g. How many grams of aspirin are in a tablet containing 5.00 grains of aspirin?

Often there is no single conversion factor that does the job, and two or more have to be used. For example, we might want to find out how many kilometers are in, say, 26.22 miles, but our tables do not list a direct relationship between kilometers and miles. If, however, we can find in a table that 1 mile = 1609.3 m and that 1 km = 1000 m, we can still work the problem. We see in the next example how we can string two (or more) conversion factors together before doing the calculation that gives the final answer.

EXAMPLE 1.5   Using the Factor-Label Method: Stringing Conversion Factors

How many kilometers are there in 26.22 miles, the distance of a marathon race? Use the fact that 1 mile equals 1609.3 meters and any other relationships that are available.

ANALYSIS   The fact that 1 mile equals 1609.3 meters gives us the following conversion factors.

$$\frac{1 \text{ mile}}{1609.3 \text{ m}} \qquad \frac{1609.3 \text{ m}}{1 \text{ mile}}$$

The "given" is 26.22 miles, so we write it down first. Then we see that the second conversion factor makes the following calculation possible to convert miles into meters.

$$26.22 \, \cancel{\text{mile}} \times \frac{1609.3 \text{ m}}{1 \, \cancel{\text{mile}}} = ? \text{ m}$$

If, however, we pause to carry out this calculation now, the answer would not be specified in kilometers (km) but in meters (m) instead. Therefore, *before doing the calculation,* we look for another conversion factor that, if possible, directly relates meters to kilometers. We need a conversion factor that lets us cancel "m" and replace it by "km." Knowing that "kilo-" stands for 1000, we know that the relationship 1 km = 1000 m will supply the correct factor.

SOLUTION   The final setup has two conversion factors strung together.

$$26.22 \text{ mile} \times \frac{1609.3 \text{ m}}{1 \text{ mile}} \times \frac{1 \text{km}}{1000 \text{m}} = 42.20 \text{ km (correctly rounded)}$$

The marathon distance is 42.20 km.

■   PRACTICE EXERCISE 11   Using the relationships between units given in the exercises or in tables in this chapter, carry out the following conversions. Be sure that you express the answers in the correct number of significant figures.
(a) How many milligrams are in 0.324 g (the aspirin in one normal tablet)?
(b) A long-distance run of 10.0×10^3 m is how far in feet? (This is the 10-km distance.)
(c) A prescription calls for 5.00 fluidrams of a liquid. What is this in milliliters? (The *fluidram* is a measure of volume in the old apothecary system: 8 fluidrams = 1 liquid ounce.)
(d) One drug formulation calls for a mass of 10.00 drams. If only an SI balance is available, how many grams have to be weighed out? (The *dram* is a measure of mass in the apothecary system: 16 drams = 1 ounce).
(e) How many microliters are in 0.00478 L?

On page 8, equations were given relating degrees Celsius and degrees Fahrenheit. The use of these equations illustrates further examples of how to cancel units no longer wanted, as you can demonstrate by using those equations to work the following Practice Exercises.

■   PRACTICE EXERCISE 12   A child has a temperature of 104 °F. What is this in degrees Celsius?

■   PRACTICE EXERCISE 13   If the water at a beach is reported as 15 °C, what is this in degrees Fahrenheit? (Would you care to swim in it?)

1.7   HEAT ENERGY

Heat is a factor in most chemical changes, and one common unit of heat is the calorie.

If you place two objects with different temperatures together, their temperatures eventually equalize. The colder object becomes warmer and the warmer becomes colder. Something flows between the two to make their temperatures equal, and what transfers is not matter but one of several forms of energy called **heat.**

■ *Kinetic* is from the Greek *kine-tikos,* meaning "of motion."

Energy of Motion Is Called Kinetic Energy.   We define **energy** as a capacity for causing change. For example, a falling rock has energy because it can make other objects in its way move aside or break apart. The energy associated with motion is called **kinetic energy** (or **KE**), and its amount depends on both the mass and the velocity of what is in motion. The more massive is the rock or the more rapidly it is moving, the more it is able to cause changes, and its kinetic energy is related to its mass *(m)* and its velocity *(v)* by the following equation:

$$KE = \tfrac{1}{2}mv^2 \qquad (1.1)$$

If the rock is a piece of hot, newly solidified volcanic lava, it has the capacity to cause any cooler object nearby to become warmer. It has a temperature-changing capacity, and we say that it has *thermal energy* or *heat.*

If the rock is so hot that it glows, it has an illumination-changing capacity. It emits *light energy,* still another form of energy.

During its fall, the rock might also make a sound, especially if it cracks apart as it cools. Now the rock has a capacity to change the noise level, and it emits *sound energy.*

The rock might land on a narrow ledge partway down the face of a cliff. Although it is motionless, soundless, and soon cool, no one would linger long beneath the ledge because the rock still has energy. It is still able to cause change. As this capacity is inactive, we speak of a quiet form of energy called **potential energy,** which is energy in storage as in the rock on a ledge. Potential energy is an inactive form of energy the quantity of which is related to location or, as we see next, to chemical makeup.

Chemical Energy Is One Kind of Potential Energy. The rock that we have endowed with so many kinds of energy has still another kind. If the rock were an iron meteorite, and it fell into a vat of concentrated acid instead of onto the ledge, we would see all sorts of interesting changes. The acid would become hotter, a gas would bubble away from the iron's surface, and the iron would dissolve. If the rock, however, were a gold nugget, none of these changes would happen. Thus, the iron meteorite has a potential for causing change simply because it is iron. Potential energy that exists because of composition is called **chemical energy.**

The Calorie Is a Common Unit of Heat Energy. Our foods have chemical energy, and our bodies can convert them into other forms such as electrical energy (when the nervous system works), kinetic energy (when muscles flex), sound energy (when we speak), and the chemical energy of other substances, which the body makes from foods.

One way to measure the chemical energy in a food is to burn a weighed sample and see by how much the evolving heat changes the temperature of a weighed sample of water. This works because heat is the one form of energy into which all other forms can be entirely converted. This is why scientists have developed a general unit of energy related specifically to heat, namely, the *calorie.* One **calorie,** or **cal,** of energy is the energy that changes the temperature of one gram of water by one degree Celsius (specifically, the degree between 14.5 and 15.5 °C). This isn't much energy, so a multiple, the kilocalorie, is often used. The "calorie" used in nutrition is actually the kilocalorie (see Special Topic 1.1).

$$1 \text{ kcal} = 10^3 \text{ cal}$$

The SI unit for energy is based on Equation 1.1 for kinetic energy given earlier. A mass of 2 kg moving with a speed of 1 m/s (meter per second) has a kinetic energy of 1 **joule,** symbolized **J.**

$$KE = \tfrac{1}{2}mv^2 = \tfrac{1}{2}(2 \text{ kg})(1 \text{ m/s})^2 = 1 \text{ J}$$

The joule is even smaller than the calorie, about one-fourth as much energy.

$$1 \text{ cal} = 4.184 \text{ J (exactly)}$$

Basal Metabolism Consists of the Most Basic Activities of Metabolism The minimum activities inside the body needed to maintain muscle tone, control body temperature, circulate the blood, breathe, run the brain and the nervous system, make compounds or break them down, and otherwise operate tissues and glands during periods of rest are called the body's *basal activities*. The sum total of all the chemical reactions that supply the energy for basal activities is called the body's *basal metabolism*. The *basal metabolic* rate is the rate at which chemical energy is used for basal activities. In the figures that follow, remember that the chemist's kilocalorie is the same as the food "calorie," which is sometimes written and abbreviated with a capital "C"—Calorie and Cal—to differentiate it from the "small calorie," the chemist's ordinary calorie (cal).

Basal Activities Consume More Than Half of Our Energy Intake A 70-kg (154-lb) adult male has a basal metabolic rate of 1.0 to 1.2 kcal/min. With 24 h/day × 60 min/h or 1440 minutes per day, the basal metabolic rates mean that an adult male expends (ignoring the niceties of rounding) from 1440 to 1728 kcal of energy per day just to carry out the most basic functions, even if he's a total couch potato. The rate for a 58-kg (128 lb) woman is 0.9 to 1.1 kcal/min, which translates to 1296 to 1584 kcal of energy per day. The daily energy expenditures are higher, of course, because the body generally does much more than basal work. The rate soars to as high as 12 kcal/min for pick and shovel work. Normal walking on a level surface requires 2 to 5 kcal/min depending on one's weight. On the average, an adult male needs 2700 kcal per day; the adult female needs roughly 2000 kcal per day. You can see that the basal activities alone use up slightly more than half of these amounts of energy.

The Foods We Eat Provide the Energy for All Activities Carbohydrates (sugars and starch) and proteins (e.g., egg white, milk, and meat products) deliver 4 kcal/g; fats and oils (like salad oil, margarine, and butter) provide 9 kcal/g. Alcohol rates 7 kcal/g. If you eat a pat of butter (about 5 g) you take in 9 Cal/g × 5 g or 45 Cal (45 kcal; 45,000 cal). How do we know this? It's really quite simple in principle and rather easy in a lab with the right equipment. The butter pat is burned in oxygen, a process called *combustion,* which liberates carbon dioxide, water, and heat. The heat is measured. (This could be done, for example, by measuring how many grams of ice the heat could melt. Ice at 0 °C requires 80 cal/g to melt and become liquid water at 0 °C.) It is only because our bodies produce the identical products from metabolizing a pat of butter as are formed during combustion—carbon dioxide, water, and heat—that we can use data from a controlled combustion in the lab as a measure of how much heat energy is available from the pat of butter.

How do we determine the caloric content of a mixed food like a hamburger sandwich, a food that combines some carbohydrate, protein, and an oil or fat? You could burn the sandwich, but there is an easier way. Chemists can separate the water, salts, carbohydrates, proteins, and fats from a particular food, like hamburger meat, and determine the percentages of each (meaning the number of grams of each component in 100 g of the product). A cooked "quarter-pounder" of regular hamburger, for example, has roughly 23 g of fat, so its fat alone carries 23 g × 9 Cal/g or 207 Cal. It's easy to see that this is 10% of the average daily Calorie needs.

You can get information about the composition and caloric content per serving of packaged foods directly from the label. The data on a typical carton of skim milk, for example, will disclose that each serving

For larger quantities of energy, the kilojoule (kJ) is usually employed.

$$1 \text{ kJ} = 10^3 \text{ J}$$

For most purposes in this text, we use the calorie or the kilocalorie.

Each Substance Has a Thermal Property Called Its Specific Heat. You probably know that our bodies have to control their temperatures carefully, yet they must function both in very hot climates, when heat flows into the body, and under very cold conditions, when the body loses heat. Substances differ widely in their abilities to absorb (or lose) heat when the temperature changes, and to compare these abilities

TABLE 1	Recommended Daily Energy Intakes	
	Calories per Day	
Age	**Male**	**Female**
15–22	3000	2100
22–50	2700	2000
51+	2400	1800

(1 cup) has 9 g of protein (for 9 g × 4 Cal/g = 36 Cal), 12 g of carbohydrate (for 12 g × 4 Cal/g = 48 Cal), and 1 g of fat (for 1 g × 9 Cal/g = 9 Cal). These numbers add up to 93 Cal per serving. (The label may say "90 Cal," but, as we said, the niceties of rounding are not observed in this area. The label, incidentally, will also write "cal," not "Cal," but you now know what is meant.)

An Intake of Calories in Excess of Needs Means "You Know What" Table 1 gives the dietary intakes of energy recommended for teenagers and adults by the U.S. Food and Nutrition Board. (Pregnant women add 300 Cal and lactating women add 500 to 1000 Cal.) Suppose that the best average for you is 2400 Cal per day, but you take in 3000 Cal per day. What does the body do with the excess? The excess represented by food fat calories is put into storage as body fat. Most of the excess obtained from carbohydrates is changed to body fat and stored. Not all is converted to fat because the chemical energy in a small fraction (about one-sixth) is needed for the chemical conversion work. Some of the excess obtained as protein is also changed to fat. The rule of thumb is that for every 3500 Cal of unused energy, you gain a pound of body weight. So if you're taking in 600 Cal per day too much, it takes only 3500 Cal × 1 day/600 Cal or 5.8 days to put on an extra pound. Even if you take in only 135 Cal too much per day (which is provided by the fat alone in one serving of roasted peanuts), you'll gain a pound in about 26 days.

FIGURE I
A hamburger obtained at a fast food restaurant is rich in calories.

Just Say "NO" or Get Out and Run? Suppose that you are able to run at a rate of 6 miles per hour. This is equivalent to taking 10 minutes for 1 mile, a relatively slow running pace. The U.S. Food and Nutrition Board rates running as an activity requiring between 7 and 12 Cal per minute. Suppose you "burn" 8 Cal per minute. At a running pace of 10 minutes per mile, you'd be burning up 8 Cal/min × 10 min/mile or 80 Cal per mile. To use up the Calories in the single serving of peanuts (135 Cal), you'd have to run 1 mile/80 Cal × 135 Cal or 1.7 miles. You can eat the peanuts and run 1.7 miles, say "no" to the peanuts, or accept a steadily increasing weight. It's up to you. Nature is generous with options but rigorous about consequences.

we use the concept of specific heat. The **specific heat** of a substance is the amount of heat that changes the temperature of one gram of the material by one degree Celsius. Because, by definition, one calorie changes the temperature of one gram of water by one degree Celsius, the specific heat of water is one calorie per gram per degree.

$$\text{Specific heat of water} = \frac{1 \text{ cal}}{\text{g } °\text{C}}$$

The specific heat of water is higher than for almost any other known material. For example, the heat that changes the temperature of 1 g of water by 1 °C changes that of 1 g of iron by 10 °C. The specific heat of iron is 0.10 cal/(g °C). Thus, on an equal

Specific Heats of Common Substances

	Specific Heat (cal/g °C)
Alcohol	0.58
Gold	0.03
Granite	0.19
Iron	0.10
Olive oil	0.47
Water	1.0

mass basis, iron has one-tenth the ability of water to absorb heat for the same change in temperature.

Water's High Specific Heat Is Important to Us.   The adult body is about 60% water, and because of water's high specific heat, we can absorb heat or lose it with little life-threatening changes in temperature. Water's high specific heat also explains why cold water is such a good coolant, which we know by experience when we reach for a cooling drink or plunge into a pool on a hot day. The management of the body's heat budget is a vital topic at the molecular level of life, and we return to it in a later chapter.

1.8   DENSITY

One of the important physical properties of a liquid is its density, the amount of mass per unit volume.

An Object's Density Is the Ratio of Its Mass to Its Volume.   There are two kinds of physical properties of objects, *intensive* and *extensive*. An extensive property depends on the size of the object; mass and volume, for example, are extensive properties. The mass of a sample of sugar clearly depends on the sample size. An intensive property is unrelated to size; an object's temperature, for example, is an intensive property.

One particularly useful intensive property of a chemical substance, particularly if it is a fluid, is its *density.* **Density** is the ratio of mass to volume; it is the mass per unit volume of a substance.

$$\text{Density} = \frac{\text{mass}}{\text{volume}} \qquad (1.2)$$

The density of mercury, the silvery liquid used in most thermometers, is 13.5 g/mL, which makes mercury one of the most dense substances known. The density of liquid water is 1.0 g/mL.

Don't make the mistake of confusing *heaviness* with *denseness.* A pound of mercury is just as heavy as a pound of water or a pound of feathers, because a pound is a pound. But a pound of water occupies 13.5 times the volume of a pound of mercury.

The density of a substance varies with temperature, because the volume of a sample of almost any substance changes with temperature but the mass does not change. You can see from Equation 1.2 that if the denominator of the fraction (the volume) changes, but the numerator stays the same, the ratio of the two, the density, must change.

Most substances expand in volume when warmed and contract when cooled. The effect isn't great if the substance is a liquid or a solid. The density of mercury, for example, only changes from 13.60 to 13.35 g/mL when its temperature changes from 0 to 100 °C, a density change of only about 2%. Note from the data in the margin how the density of water, to two significant figures, is 1.0 g/mL in the (liquid) range 0 to 30 °C (32 to 86 °F).

Water is unusual in that its density *decreases* when it is cooled from 3.98 °C to 0 °C, where it freezes. This decrease occurs because of a small increase in volume, the term in the denominator of Equation 1.2. When liquid water at 0 °C freezes, it expands

Density of Water at Various Temperatures

Temperature (°C)	Density (g/mL)
0	0.9987
3.98	1.0000
10	0.9973
20	0.99823
25	0.99707
30	0.99567
35	0.99406
45	0.99025
60	0.98324
80	0.97183
100	0.95838

another 10% in volume. Ice, therefore, has a density less than that of liquid water, so ice floats on water. If this did not happen, the entire ecology of northern lakes and the Arctic Ocean would be different. If their waters froze and the ice sank instead of floated, the ice would likely never melt.

One of the uses of density is in calculating what volume of a liquid to take when the problem or experiment specifies a certain mass. Often it is easier (and sometimes safer) to measure a volume than a mass, as we will note in the next example.

EXAMPLE 1.6 Using Density to Calculate Volume from Mass

Concentrated sulfuric acid is a thick, oily, and very corrosive liquid that no one would want to spill on the pan of an expensive balance to say nothing of the skin. It is an example of a liquid that is usually measured by volume instead of by mass, but suppose an experiment called for 25.0 g of sulfuric acid. What volume (in mL) should be taken to obtain this mass? The density of sulfuric acid is 1.84 g/mL.

ANALYSIS The given value of density means that 1.84 g acid = 1.00 mL acid. This gives two conversion factors:

$$\frac{1.84 \text{ g acid}}{1 \text{ mL acid}} \qquad \frac{1 \text{ mL acid}}{1.84 \text{ g acid}}$$

The "given" in our problem, 25.0 g acid, should be multiplied by the second of these conversion factors to get the unit we want, mL.

SOLUTION The final setup starts with the given, 25.0 g acid, and multiplies it by the proper conversion factor. (Show the cancel lines yourself.)

$$25.0 \text{ g acid} \times \frac{1 \text{ mL acid}}{1.84 \text{ g acid}} = 13.6 \text{ mL acid}$$

Thus, if we measure 13.6 mL of acid, we obtain 25.0 g of acid. (The pocket calculator result is 13.58695652 but, by our rules, we must round the result to three significant figures.)

■ The milliliter (mL) is identical to one cubic centimeter (1 cm^3 or 1 cc), so you will sometimes see the symbols cm^3 and cc used for mL.

■ PRACTICE EXERCISE 14   An experiment calls for 16.8 g of methyl alcohol, the fuel for fondue burners, but it is easier to measure this by volume than by mass. The density of methyl alcohol is 0.810 g/mL, so how many milliliters have to be taken to obtain 16.8 g of methyl alcohol?

■ PRACTICE EXERCISE 15   After pouring out 35.0 mL of corn oil for an experiment, a student realized that the mass of the sample also had to be recorded. The density of the corn oil is 0.918 g/mL. How many grams are in the 35.0 mL?

Specific gravity is a property of a fluid that is very similar to density. It is not used very often in chemistry, but in clinical work the specific gravity of liquid specimens (such as urine) helps to reveal the nature of an illness, as described in Special Topic 1.2.

SPECIAL TOPIC 1.2
SPECIFIC GRAVITY

The **specific gravity** of a liquid is the ratio of the mass contained in a given volume to the mass of the identical volume of water at the same temperature. If we arbitrarily say that the "given volume" is 1.0 mL, then the water sample has a mass of 1.0 g (or extremely close to this over a wide temperature range). This means that dividing the mass of some liquid that occupies 1.0 mL by the mass of an equal volume of water is like dividing by 1, but all the units cancel. Specific gravity has no units, and the value of the specific gravity of something is numerically so close to its density that we usually say they are numerically the same. This fact has resulted in rather limited use of the concept of specific gravity, but one use occurs in medicine.

In clinical work, the idea of specific gravity surfaces most commonly in connection with urine specimens. Normal urine has a specific gravity in the range 1.010 to 1.030. It's slightly higher than that of water because the addition of wastes to water usually increases its mass more rapidly than its volume. Thus, the more wastes in 1 mL of urine, the higher is its specific gravity.

Figure 1 illustrates the traditional apparatus used to measure the specific gravity of a urine specimen, a urinometer; however, it has largely been supplanted by the refractometer, which needs only one or two drops of urine for the measurement. (The refractometer is an instrument that measures the ratio of the speed of light through air to its speed through the sample being tested. This ratio can be correlated with the concentration of dissolved substances in the urine. *How* the refractometer does this is beyond the scope of our study.)

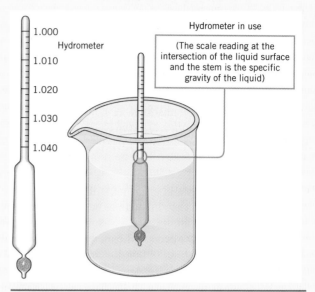

FIGURE 1
A hydrometer designed to serve as a urinometer.

One of the important functions of the kidneys is to remove chemical wastes from the bloodstream and put them into the urine being made. The kidneys' mechanism for doing this does not remove those substances from the blood that ought to remain in the blood. The clinical significance of a change in the concentration of substances dissolved in the urine is that it indicates a change in the activity of the kidneys. This might be the result of a kidney disease that causes substances that should stay in the blood to leak into the urine being made. Or it might mean that wastes are being generated somewhere else in the body more rapidly than the kidneys can remove them.

SUMMARY

Chemistry and the Molecular Basis of Life   Down at the level of nature's tiniest particles, we find the "parts"—molecules—that nature shuffles from organism to organism in the living world. One of the many ways of looking at life is to examine its molecular basis, the way in which well-being depends on chemicals and their properties.

Physical Quantities   Physical properties are those that can be studied without changing the substance into something else. They include mass, volume, time, temperature, color, and density. For our purposes, the important base quantities are (with the names of the SI base units given in parentheses) mass (kilogram), length (meter), time (second), temperature (kelvin), and quantity of chemical substance (mole).

Special prefixes can be attached to the names of the base units to express multiples or submultiples of these units. To select a prefix we have to be able to convert very large or very small numbers into scientific notation.

Precision and Accuracy   Whether we obtain data from direct measurements or by calculations, we have to be careful not to imply too much precision by using the incorrect number of significant figures. When we add or subtract numbers, the decimal places in the result can be no more than the least number of decimal places among the original numbers. When we multiply or divide, we have to round the result to show the same number of significant figures as in the least precise original number.

Factor-Label Method   The units of the physical quantities involved in a calculation are multiplied or canceled as if they were numbers. To convert a physical quantity into its equivalent in other units, we multiply the quantity by a conver-

sion factor that permits the final units to be correct. The conversion factor is obtained from a defined relationship between the units.

Heat Energy   Heat is the energy that transfers from one object to another when there is a difference in temperature between them. When the heat raises the temperature of one gram of water one degree Celsius, one calorie of energy has moved. The SI unit of energy is the joule (J): 1 cal = 4.184 J. The "calorie" used in nutrition is the same as the kilocalorie.

The specific heat of a substance, a thermal property, is the number of calories per gram that transfer when the temperature of the substance changes by one degree Celsius. The high specific heat of water helps living systems maintain steady temperatures.

Density   Density, an intensive physical property, is the ratio of mass to volume. Because volumes of substances change with temperature, density is temperature dependent. The change, however, is usually quite small.

REVIEW EXERCISES

The answers to Review Exercises whose numbers are in color are found in Appendix V. The answers to the other Review Exercises are found in the Study Guide that accompanies this book. The more challenging questions are marked with asterisks.

Molecular Basis of Life

1.1   On which aspects of nature do chemists focus most?

1.2   What kinds of observations of nature led people to suppose that living things must have many common features?

1.3   Concerning living things, at what level of existence are "parts" most freely exchanged between members of different species? What are these parts, in the most general terms?

Physical Quantities, Properties, and Measurements

1.4   When we speak of the *properties* of some substance, what is meant?

1.5   What is the basis for distinguishing between *chemical* and *physical* properties?

1.6   What marks the difference between a *physical property* and a *physical quantity*?

1.7   In a word, what operation must be done to obtain a value for a physical quantity?

1.8   What is meant by the *inertia* of an object, and what is the name of the physical quantity used to describe this property?

1.9   Your *weight* on the moon would be less than your weight on earth, yet your *mass* is the same at both locations. Explain.

1.10   What makes it possible for us to say that we determine the *mass* of an object when the actual operation we use is *weighing* (and the instrument is a two-pan balance)?

1.11   What is the general name we give to those fundamental quantities in terms of which all other physical quantities are defined? Name five of these physical quantities that are defined or mentioned in this chapter.

1.12   For a physical quantity to have any meaning or any usefulness, what must be defined for it?

1.13   What is the *name* of the base unit for the following physical quantities?
(a) length   (b) time   (c) mass
(d) temperature   (e) amount of chemical substance

1.14   In general terms, the *definition* of a base unit involves what kind of a standard? What organization has been responsible for these definitions?

1.15   Why is *volume* not called a *base* quantity?

1.16   What are the *names* of the SI base unit of length and the corresponding reference standard?

1.17   What are the *names* of the SI base unit of mass and its corresponding reference standard?

1.18   Scientists regard the SI reference standard for length as far more satisfactorily specified than the SI reference standard for mass. Give the reasons.

1.19 Examine each pair of quantities and state which is larger.
(a) meter and yard   (b) inch and centimeter
(c) gram and ounce   (d) millimeter and centimeter
(e) pound and kilogram (f) kilogram and ton
(g) liter and quart   (h) microliter and milliliter
(i) ounce and pound   (j) gram and kilogram

1.20 How many milliliters are in 1 liter?

1.21 How many micrograms are in 1 milligram?

1.22 How many grams are in 1 kilogram?

Degrees and Scales of Temperature

1.23 What is the value given to the point on a mercury-filled thermometer where the mercury level eventually comes to rest after the thermometer is immersed in an ice-water slush on each of the following scales of temperature?
(a) Celsius   (b) Fahrenheit   (c) Kelvin

1.24 When a mercury-filled thermometer is immersed into a container of boiling water (at sea level), what is the value given to the level the mercury eventually reaches on each of the following scales?
(a) Celsius   (b) Fahrenheit   (c) Kelvin

1.25 How many degree divisions (arbitrarily) separate the mercury levels for the freezing and the boiling points of water on each of the following scales?
(a) Celsius   (b) Fahrenheit   (c) Kelvin

1.26 Which is the larger degree, the Celsius degree or the Fahrenheit degree? By how much is it larger?

1.27 Which is the larger degree, the kelvin or the Fahrenheit degree? By how much is it larger?

1.28 When expressed in degrees Celsius, the zero point on the Kelvin scale has what value?

1.29 What is true about a temperature of 0 K?

1.30 A Canadian weather report said that the temperature at one reporting station was −40 °C. What is this in °F?

1.31 If you read a Kelvin thermometer in your room as 278 K, would you be comfortable without a coat or sweater? (First convert 278 K into °C and then convert the answer into °F. As part of your answer, give the results of these calculations.)

1.32 In testing an American recipe, a French baker had to decide how to set the French oven for a recipe specification of 320 °F. What oven setting in °C was needed?

1.33 An American visitor to Germany wanted to set a room thermostat for the degree Celsius equivalent of 68 °F. What should the setting be in °C?

*1.34 A clinical thermometer was used to take the temperature of a patient, and it registered 40 °C. Did the patient have a fever? (Do the calculation. Normal body temperature has traditionally been taken to be 98.6 °F, but many healthy individuals have normal temperatures slightly lower or slightly higher than this.)

Scientific Notation and SI Prefixes

1.35 Rewrite the following physical quantities with their units abbreviated.
(a) 2.5 deciliters of solution
(b) 31 milligrams of medication
(c) 46 centimeters in length
(d) 110 kilometers in distance
(e) 35 microliters of solution
(f) 75 micrograms of progestin
(g) 25 millimeters wide

1.36 Rewrite the following physical quantities with their units written out in full.
(a) 110 mL of Ringer's solution
(b) 150 mg of sugar
(c) 16 km to the airport
(d) 50 μg of vitamin K
(e) 1.5 dL of saline solution
(f) 2.5 kg of salt
(g) 75 μL of serum

*1.37 Restate the following physical quantities in scientific notation in which the decimal part of the number is between 1 and 10.
(a) 523 g   (b) 0.0450 L
(c) 1562 m   (d) 0.0000093 g

*1.38 How would the following physical quantities be re-expressed in scientific notation in which the decimal part of the number is between 1 and 10?
(a) 0.130 L   (b) 3568.5 m
(c) 0.0000042 g   (d) 0.0045 g

*1.39 Use a suitable SI prefix to express each of the quantities in Review Exercise 1.37.

*1.40 Use a suitable SI prefix to restate each of the quantities in Review Exercise 1.38.

Significant Figures

1.41 The number of tickets sold for a concert was 25,342. Restate this number in scientific notation but retain only three significant figures.

1.42 The population of the world constantly changes, but at one moment it was 5,154,689 people. Reexpress this in scientific notation, retaining two significant figures.

*1.43 Study the following numbers.
(A) 4.55×10^8 (B) 0.0455   (C) 45,500
(D) 0.00455   (E) 4550   (F) 4.550×10^{-3}
(G) 4.550   (H) 0.45500 (I) 4.5500×10^7
(a) Which of these numbers has three significant figures? (Identify them by their letters.)
(b) Which has four significant figures?
(c) Which has five significant figures?

*1.44 Rewrite the following number according to the number of significant figures specified in each part. Express your answers in scientific notation.

16,560,010.01

(a) one   (b) two   (c) three
(d) four   (e) five   (f) six

*1.45   Rewrite the following number according to the number of significant figures specified in each part. Give your answers in scientific notation.

199,898.9091

(a) three   (b) four   (c) five
(d) six   (e) eight   (f) nine

1.46   The relationship between the milliliter and the microliter is given by

$$1 \text{ mL} = 1000 \text{ μL}$$

How many significant figures are considered to be in each number?

Precision and Accuracy

*1.47   Consider that the following mathematical operations are calculations that involve physical quantities. (The units have been omitted.) Determine the significant figures that can be retained in the answer in each part according to our rules, and express the results of the calculations in the proper way. Use scientific notation in which the decimal part of the number is between 1 and 10.
(a) $4.665 \times 3.2 \times 10^{-5}$ (b) $6.3 \times 5.6000 \times 10^3$
(c) $4.005 \times 6.23 \times 10^{23}$ (d) $4.5 + 62.003$
(e) $6.004 - 3.2$ (f) $45.0023 + 0.023$
(g) $90.00 \div 3.0$ (h) $0.00050 \div 0.005$
(i) $6.40 \div 3.200$

*1.48   The following mathematical operations are calculations that involve physical quantities. (The units have been omitted.) Determine how many significant figures can be retained, and express the results of the calculations in the proper way. Use scientific notation in which the decimal part of the number is between 1 and 10.
(a) $9600.00 \div 320.0000$ (b) $45.0 \div 1.50$
(c) $45.0 + 1.50$ (d) 45.0×1.50
(e) $45.0 - 1.50$ (f) 0.000009×1.1

*1.49   When a scale was used to take six successive measurements of a person's mass, the following data were recorded.

59.85kg,   59.70kg,   59.91kg,   59.73kg,
59.94kg,   59.91kg

The balance had earlier been tested against a set of official reference standard masses and found to be working exceptionally well. The actual value of the mass was verified as 59.86 kg.
(a) Can the measurements be described as *accurate?* Explain.
(b) What, if anything, do the data disclose about their *uncertainty?*

1.50   What is the specific problem when a measurement is known to be in *error?*

Converting between Units

1.51   Conversion factors relate identical physical amounts that are expressed in different units. Write each of the following relationships between units in the forms of two conversion factors.
(a) 39.37 in. = 1 m   (b) 1 L = 1.057 quart
(c) 1 g = 1000 mg   (d) 1 kg = 2.205 lb
(e) 1 grain = 0.0648 g

*1.52   Given the relationships expressed in Review Exercise 1.51, which of the two quantities that follow the symbol, ≈ ("equals approximately") most closely matches the quantity in the first column? You should develop the skill to make these kinds of judgments without doing an actual calculation using conversion factors.
(a) 0.50 m ≈ 20 in. or 80 in.
(b) 0.5 lb ≈ 4.4 kg or 0.23 kg
(c) 6 g ≈ 170 oz or 0.20 oz
(d) 4500 mg ≈ 4.5 g or 4.5×10^6 g

*1.53   Given the relationships that can be found among the tables in this chapter, what quantity most nearly matches what is given in a different unit before the symbol ≈?
(a) 1 cup ≈ 50 mL or 250 mL (1 quart = 4 cup)
(b) 1 mile ≈ 1.6 km or 0.66 km
(c) 10 mm ≈ 2.5 in. or 0.25 in.
(d) 1000 mL ≈ 1 μL or 1 L

1.54   Make the following conversions using relationships found in the tables in this chapter. Do all calculations to three significant figures.
(a) Convert 163 cm into inches (the height of an adult female).
(b) Convert 154 lb into kilograms (the mass of an adult male).

1.55   Make the following conversions using relationships found in the tables in this chapter. Do all calculations to three significant figures.
(a) Convert 111.5 lb into kilograms (the mass of an adult female).
(b) Convert 192 cm into inches (the height of an adult male).

1.56   The normal content of cans of popular soft drinks is 12.0 liquid ounces. How much is this in milliliters (to three significant figures)?

1.57   The popular-size bottles of mineral water hold 296 mL. How much is this in liquid ounces (to the proper number of significant figures)?

1.58   The gasoline tank of a small car holds 12.0 U.S. gallons. How much is this in liters?

*1.59   While driving on a country road in a European country you come to a bridge limited to 1.4×10^3 kg. Your vehicle has a mass of 4.5×10^3 lb. Should you cross? (Do the calculation.)

*1.60   A physician prescribed 0.50 g of valinomyocin. The pharmacy dispenses valinomyocin in 250-mg tablets. How many tablets are needed for one prescribed dose?

*1.61   Valium is available in tablets containing 5 mg of this medication. How many tablets must be administered to give a dose of 0.015 g of Valium?

*1.62   An IV (intravenous) solution contains 25 mg of a drug per 5.0 mL of solution. You are to administer 0.75 g of the drug. What volume of the solution should be used?

*1.63   A vial of a medication carries a label instructing the user to add 7.50 mL of water to the contents of the vial to obtain a solution containing 25.0 mg of the active drug per milliliter of the solution. How many milliliters of this solution must be taken to obtain 0.175 g of the drug?

1.64   The highest mountain in the world, Mount Everest in Nepal, is 8847.7 m. What is this in feet?

1.65   The highest mountain in the United States is Alaska's Mount McKinley, 20,322 ft. How high is it in meters?

1.66   One pound of butter can be made into 128 equal-sized pats of butter. What is the mass of each pat in grams?

1.67   A diamond rated as 2.50 carats has a mass of how many grams? (1 carat = 200 mg)

1.68   What condition is necessary for us to say that heat *flows* from some object to another?

1.69   What is the significance of the unit of *heat energy* when it comes to the measurement of other forms of energy?

1.70   How many kilocalories will increase the temperature of 1.0 kg of water from 14.5 to 15.5 °C?

1.71   What is the specific heat of water, and what is the significance of this value being unusually high when compared with other substances?

*1.72   The nutritionists' "calorie," or Cal, is actually the kilocalorie. How many kilojoules correspond to a daily diet of 2500 Cal?

Density

1.73   Mass is called an extensive property, and density is an intensive property. Explain the difference.

1.74   The density of aluminum is 2.70 g/cm³. A block of aluminum with a volume of 250 mL (about 1 cup) has a mass of how many grams? Pounds?

1.75   Corrosive chemical solutions are usually more safely measured by volume than by mass. To obtain 25.0 g of sulfuric acid (density 1.84 g/mL), how many milliliters should be measured?

*1.76   Corn oil has a density of 7.60 lb/gal.
(a) Calculate its density in g/mL.
(b) How many milliliters of corn oil must be taken to obtain 250 g?

1.77   Liquids and solutions generally expand in volume as they are warmed.
(a) Assuming no loss by evaporation, does the mass of a sample change as it is warmed?
(b) When a liquid sample is warmed, does its density increase, decrease, or stay the same?

Metabolism and Food Calories (Special Topic 1.1)

1.78   What distinguishes *basal metabolism* from metabolism in general?

1.79   The brain consumes the equivalent of 120 g of glucose, a carbohydrate, per day. How many calories does this amount represent? If the basal metabolic rate is 1 Cal/min, what percentage of the daily *basal* energy needs are used to operate the brain?

*1.80   The label on a popular brand of creamy peanut butter says that one serving (32 g, 2 tablespoons) contains 9 g of protein, 6 g of carbohydrate, and 16 g of fat. How many Calories (kcal) are in one serving?

*1.81   A "quarter-pounder with cheese" contains 30.7 Cal of energy in the fat alone. If this represented Calories that you do not need and use, how many days will it take, at a rate of one of these sandwiches per day, to gain 1.00 lb.?

Specific Gravity (Special Topic 1.2)

1.82   Specific gravity is defined such that the numerical value of something's density is virtually the same as its specific gravity. Explain.

1.83   If a urine specimen has an abnormally low value of specific gravity, what is known about the specimen?

1.84   A urinometer float rides higher in what kind of fluid, one with a high density or one with a low density?

1.85   Scarcely any change in volume occurs when 3.00 g of sugar is dissolved in 100 mL of water. Assuming no volume change, what is the specific gravity of this solution?

Additional Exercise

*1.86   Rehydration therapy is a life-saving procedure for victims of cholera, who typically lose large amounts of fluid. An English physician, Thomas Latta, was the first to use this procedure during the cholera epidemic in London in 1832. The solution he had the victims drink contained 3.0 drachmas of sodium chloride and 2.0 scruples of sodium bicarbonate per 6.0 pints of water. [1 drachma = 60 grains; 1 ounce (avdp) = 480 grains; 1 ounce (avdp) = 28.35 g; 1 scruple = 20 grains]
(a) How many milligrams of sodium chloride were in each 6-pint unit of the solution? How many would be in 1 L of solution?
(b) How many milligrams of sodium bicarbonate were in each 6-pint unit of the solution? How many would be in 1 L of solution?

2

THE NATURE OF MATTER: THE ATOMIC THEORY

Matter, Its Kinds
  and States

Atomic Theory

Electron
  Configurations
  of Atoms

Elements

Periodic Law and
  Periodic Table

Is there any limit to the number of times one could subdivide a large ingot of gold *and still obtain smaller pieces of gold?* Ancient philosophers said "No." To this day, we use the word *atom,* after the Greek *atomas* for not-cuttable, to designate a piece of a chemical element that is uncuttable, even in principle. To understand the molecular basis of life, we must first learn about the atoms that make up molecules.

2.1 MATTER, ITS KINDS AND STATES

Elements and compounds have definite compositions; mixtures do not.

Matter is anything that occupies space and has mass, and there are three kinds of matter called *elements, compounds,* and *mixtures.*

Elements Are the Building Blocks of Compounds. **Elements** are substances that cannot be broken down into anything simpler, yet are stable enough to obtain, store (sometimes only under special conditions), and use in experiments. Familiar examples are aluminum, iron, copper, silver, and gold. **Compounds** are invariably made from two or more elements, and compounds obey the **law of definite proportions,** one of the important laws of chemical combination.

■ *Stable* here means only that the element can be stored at room temperature, but many elements further require special equipment and environments.

LAW OF DEFINITE PROPORTIONS
In a given chemical compound, the elements are invariably combined in the same proportion by mass.

Water is a typical *compound.* Regardless of its source, when water is broken down into its (physically quite different) *elements,* hydrogen and oxygen, they are always obtained in a mass ratio of 1.01 g of hydrogen to 8.00 g of oxygen.

Mixtures Do Not Obey the Law of Definite Proportions. If great care is taken to exclude sparks and ultraviolet light (the kind that gives a sunburn), hydrogen and oxygen can be blended into a gaseous mixture without producing water. Like the blending of perfume with air, **mixtures** are invariably made by the blending of two or more compounds or elements in any particular proportion we choose. This blending is only a **physical change,** meaning a change not accompanied by a chemical reaction. The separation of a mixture into its original parts is likewise done by the use of physical changes and operations, not chemical reactions.

If we made a mixture of hydrogen and oxygen in a mass ratio of 1.01 g of hydrogen and 8.00 g of oxygen and then exposed the mixture to a spark, it would detonate. When everything cooled down, only droplets of water would be present. No oxygen or hydrogen would remain. (Had we used different mass ratios of hydrogen to oxygen, we would have still obtained water, but either some hydrogen or some oxygen would have been left over.) The spark set off the change of the two elements into a compound, and this change is an example of a *chemical reaction.* As we learned in Chapter 1, a **chemical reaction** is an event in which substances change into other substances. The substances that change, like hydrogen and oxygen, are called the **reactants,** and the substances that form are called the **products.**

A chemical reaction is the *only* way that a compound can be made from its elements, and it is the only way that a compound can be broken down into its elements. There are many kinds of chemical reactions, each with hundreds of examples, and we begin our study of a selected few in the next chapter.

The reaction of hydrogen with oxygen that produces water also generates heat, so it's described as an **exothermic reaction.** Many reactions, however, do not occur unless the reactants are continuously heated. In other words, such reactions require a continuous input of heat and so are called **endothermic reactions.** Some of the chemical reactions that convert cake batter into a cake are endothermic.

Substance	Freezing Point
Water	0 °C
Oxygen	−218 °C
Hydrogen	−259 °C

One of the major features of all chemical reactions is the **law of conservation of mass.**

LAW OF CONSERVATION OF MASS
Mass is neither gained nor lost in a chemical reaction; mass is conserved.

When 1.01 g of hydrogen combines, for example, with 8.00 g of oxygen, exactly 9.01 g of water forms. There is no loss in mass as the reactants change into the product.

The Three Kinds of Matter Can Occur in Any One of Three Physical States.   Everyone is familiar with ice, liquid water, and steam (Figure 2.1), and these illustrate the three possible physical **states of matter.** Matter in the **gaseous state** depends entirely on its container for both the shape and the volume of the sample. In the **liquid state,** a sample of matter has the *shape* of its container but possesses a definite volume. In the **solid state,** matter has both a definite shape and a definite volume.

The changes that convert water into steam or ice are physical, not chemical, because they do not change water into something that isn't water.

We've introduced a number of fundamental terms in this section that apply to substances in bulk, to samples that we can handle and see. We must now shift levels and study atoms, the smallest particles of matter. Remember that our ultimate goal is the molecular basis of life. But molecules are made from atoms, so to understand molecules, we need to know about atoms first.

■ *Kinds* and *states* are different. The *kinds* of matter are elements, compounds, and mixtures. The *states* are solid, liquid, and gas.

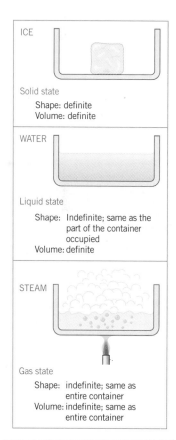

FIGURE 2.1
The three states of matter, as illustrated by water.

2.2   ATOMIC THEORY

Atoms of the same element have the same average mass; those of different elements have different average masses.

Centuries ago some Greek philosophers wondered whether matter, like a piece of gold, could be cut into an infinite number of pieces. Or is there a limit? Is there some small piece that, if broken, would not give simply smaller pieces of the same substance? They decided that there is a limit, that gold, for example, consists of tiny invisible pieces that cannot be cut further, even in the imagination, and still give gold pieces. The Greek word for "not cuttable" is *atomos,* which is the origin of the word *atom.*

The Laws of Chemical Combination Support Our Belief in Atoms.   The Greek atom was only an idea. Scientific evidence came centuries later, emerging slowly as the laws of definite proportions and conservation of mass became established. John Dalton (1766–1844), an English chemist, reasoned that the laws of chemical combination actually compel us to believe in the existence of atoms. The laws make sense, Dalton said, if each kind of atom has an unchangeable mass and if atoms lose no mass as they form compounds. Then it would follow that if each chemical compound is made by combining its atoms in a definite proportion *by numbers,* the compound cannot help but contain its elements in a definite proportion *by mass.* Here is the full list of Dalton's postulates. We'll use them as a starting point for our discussion and then see what modifications were required by later discoveries.

John Dalton

■ The proton and neutron themselves are made of even more fundamental particles called *quarks*.

Like charges repel.

> ### DALTON'S ATOMIC THEORY
> 1. Matter consists of definite particles called atoms.
> 2. Atoms are indestructible. They can become relocated in chemical reactions, but not broken up.
> 3. All atoms of a particular element are identical in mass.
> 4. Atoms of different elements are different in mass.
> 5. By joining together in different ratios as whole particles, atoms form different compounds.

Dalton did not speculate if atoms themselves are made of even smaller things, and in his time there was no way to find out.

Atoms Are Made of Subatomic Particles Called *Electrons*, *Protons*, and *Neutrons*. Unknown to Dalton, atoms are destructible. To break an atom apart, however, requires the enormous energies that only atom smashing machines can provide. When atoms are broken up, the debris consists of **subatomic particles,** which consist of the same small set of particles *regardless of the element.* This remarkable fact makes the study of atoms vastly simpler.

Three subatomic particles, the **electron,** the **proton,** and the **neutron,** are particularly important to the chemical properties of the elements. All three have mass. Two, the electron and the proton, are electrically charged, meaning that they can exert pushes and pulls, forces of repulsion and attraction, on each other without physically touching.

Objects whose electrical charges are opposite in character experience a force of attraction for each other. We say, in short, that *unlike charges attract.* An electron and a proton thus attract each other because they bear opposite charges. A proton, however, exerts a force of repulsion on another proton. An electron also repels another electron. In short, *like charges repel.* The two simple laws about charged particles explain much about how atoms of different elements are able to join to make compounds.

The charge on the proton has the same intensity as that on the electron but is opposite in character. The amount (intensity) of charge on a proton or an electron is defined as one unit, and so to symbolize their equal but opposite electrical charges we use the symbols 1+ and 1−. The proton has a charge of 1+ and the electron, a charge of 1−.

The masses in grams of the three subatomic particles are extremely small (Table 2.1), so there's an advantage to reexpressing their values in a new unit of mass, the **atomic mass unit** or **u,** defined as follows.

$$1 \text{ u} = 1.6605665 \times 10^{-24} \text{ g}$$

TABLE 2.1 Properties of Subatomic Particles

Particle	Mass (g)	Mass (u)	Electrical Charge	Symbol
Electron	$9.1093897 \times 10^{-28}$	0.0005485712	1−	$_{-1}^{0}e$
Proton	$1.6726430 \times 10^{-24}$	1.0072725	1+	$_{1}^{1}\text{H}, _{1}^{1}p$
Neutron	1.674954×10^{-24}	1.008664	0	$_{0}^{1}n$

In the atomic mass unit, despite its seeming awkwardness in grams, the mass of the proton or the neutron, when rounded to two significant figures, is 1.0 u. The mass of the proton, for example, when converted from grams to atomic mass units, becomes

$$1.6726430 \times 10^{-24} \frac{g}{\text{proton}} \times \frac{1 \text{ u}}{1.6605665 \times 10^{-24} \text{ g}} = 1.0072725 \text{ u/proton}$$

This rounds to 1.0 u, for the mass of one proton, a much easier number to use. (Do the conversion of the neutron's mass from g to u as an exercise.) The mass of the electron, a much lighter particle, is about 1/1836th u. The practical result of the much smaller mass of an electron is that we can ignore the contributions that electrons make to the masses of whole atoms.

■ Even the 109 electrons in the largest known atom contribute only 0.02% to the atom's mass.

An Atom's Protons and Neutrons Make Up Its Atomic *Nucleus*.   Early in this century, British scientists led by Ernest Rutherford (1871–1937) discovered that atoms are mostly empty space, that essentially all of the atom's mass is in an extremely dense central core, a particle they named the **nucleus**. It has all of the atom's heavy, subatomic particles, its protons and neutrons. Because the atom's protons are in its nucleus, the nucleus also has all of the atom's positive charge. Because the charge on one proton is 1+, the total nuclear charge equals the number of protons.

■ The 1908 Nobel prize in chemistry went to Ernest Rutherford.

Charge on an atomic nucleus = number of protons

Each Element Has a Unique *Atomic Number*.   We can now identify something special about a given element. All atoms of the same element have identical nuclear charges, meaning identical numbers of protons. Atoms of different elements have different nuclear charges or different numbers of protons. Thus, each element owns a unique number, called its **atomic number,** the number of protons in one of its atoms.

Atomic number of element = positive charge on its atomic nuclei
= number of protons per atom

The Number of an Atom's Electrons Also Equals the Atomic Number.   All atoms are electrically neutral. This means that the positive charge on the nucleus is exactly balanced by the negative charge contributed by the atom's electrons. The number of protons, therefore, must equal the number of electrons, so that each charge of 1+ is neutralized in an electrical sense by each charge of 1−. The atomic number, therefore, also tells us how many electrons an atom has.

Atomic number of an element = number of protons
= number of electrons

Each Kind of Atom Has a Unique Mass.   The sum of an atom's neutrons and protons is its **mass number.**

Mass number = protons + neutrons

■ You are not expected to memorize atomic numbers or mass numbers for the various elements.

Because each neutron and proton has a mass of 1.0 u, the mass number is the same as the mass of the atom in atomic mass units, u. Only when a precision requiring three or more significant figures is needed would we have to modify this statement.

To summarize what we have learned about atoms thus far, we can say that atoms are tiny neutral particles consisting of nuclei and electrons, and atomic nuclei consist of protons and neutrons. Each element has its own atomic number, which equals its number of protons. Each kind of atom has a mass number, which equals the sum of its numbers of protons and neutrons.

■ PRACTICE EXERCISE 1   What are the mass numbers of atoms that have the following nuclear compositions?
(a) 7 protons and 8 neutrons   (b) 12 protons and 12 neutrons
(c) 11 protons and 13 neutrons

■ PRACTICE EXERCISE 2   How many neutrons are in each of these atoms?
(a) Atomic number 4, mass number 9
(b) Atomic number 17, mass number 35
(c) Atomic number 17, mass number 37

2.3   ELECTRON CONFIGURATIONS OF ATOMS

Each element has a unique electron configuration, and this determines its chemical properties.

The space taken by an atomic nucleus is a small fraction of the space occupied by the entire atom. If an atomic nucleus were a tennis ball, the outer edge of the atom would be at least 30 football fields away. Existing in this space are the electrons.

Electrons Are Confined to Particular *Energy Levels*.   The electrons do not have unlimited freedom. They normally do not escape, for example, because they are attracted toward the oppositely charged nucleus. They do not fall into the nucleus either, because they are in rapid motion of a type that tends to hurl them outward. They are also largely confined to particular spaces, to certain regions of space near the nucleus. Niels Bohr (1885–1962), a Danish scientist, was the first to suggest *specific* locations for electrons, and he called them the *allowed energy states* or the **energy levels** of an atom. He further postulated that as long as electrons remain in the allowed levels of *lowest energy,* an atom will neither emit nor absorb energy. The atom enjoys its greatest stability as long as its electrons stay in their lowest energy states. When they do not, chemical reactions can occur, which is why we're studying where an atom's electrons normally are.

To help the public visualize his view of the atom, Bohr likened the allowed energy levels to the orbits followed by planets around the sun. His picture of the atom was therefore dubbed the *solar system model* (Figure 2.2). This model has long since been discarded, but Bohr's postulates remain valid: an atom has only certain allowed energy states for electrons, and an atom is most stable when the electrons remain in their lowest states.

The energy level nearest the nucleus is called the *first level,* or level 1. It corresponds to the lowest energy that an electron in an atom can have. An electron in level

■ Niels Bohr won the 1922 Nobel prize in physics.

TABLE 2.2   The Principal Energy Levels

Principal level number	1	2	3	4	5	6	7
Maximum number of electrons actually observed in nature[a]	2	8	18	32	32	18	8

[a]In theory, levels 5, 6, and 7 could accommodate 50, 72, and 98 electrons, respectively, but such maxima do not occur among any of nature's elements. The maximum *observed* numbers of electrons for levels 1, 2, 3, and 4 shown in this table are also their respective *theoretical* maximum numbers of electrons.

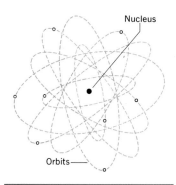

1 is held more firmly and is more stable than an electron anywhere else. Additional levels are numbered in order as they are found farther from the nucleus: level 2, level 3, and so forth. Each level has its own capacity for electrons. Level 1 can hold up to 2 electrons; level 2 can have 8; and level 3 can have 18 (Table 2.2). Let's now look briefly at how the existence of energy levels was deduced. The discovery bears on our understanding of *how* substances can absorb or release energy.

Electrons Can Change Energy Levels When Atoms Absorb or Emit Energy.   Scientists inferred the energy states of atoms by studying the energies that are absorbed or released by elements under controlled conditions. Different energy states explained a striking observation, namely, a given element absorbs not just any quantity of energy but specific quantities unique to the given element. If it is true that different and unique energy states exist in atoms, then the separation (in energy terms) of any two states necessarily corresponds to a specific quantity of energy. When light possessing this quantity strikes an atom, an electron in the lower state is forced to the higher state. By this mechanism, the atom absorbs the light energy. The atom becomes an *excited atom* and is no longer in its most stable state. In time, the electron falls back to the lower level, and the atom then emits light energy in an amount again identical to the energy separation of the two levels.

A common way to make electronically excited atoms is by heat. If you heat an iron rod hot enough, for example, it glows first red, then orange, and eventually "white hot," depending on the temperature. The heat promotes electrons of the iron atoms in huge numbers to higher energy states. As the electrons "relax" to lower energy states, energy is emitted in the form of the light we observe.

A Few Rules Give the Pattern of the Electron Configurations of Elements 1 Through 20.   The distribution of electrons among the available levels is called the **electron configuration** of the atom. We'll study those of only the 20 simplest kinds of atoms. These include nearly all of the elements found in living systems, so we do not have to say too much about atoms of the higher atomic numbers.

Given the atomic number of any of the 20 simplest atoms, we can figure out the electron configuration by using the following rules.

1. Use the atomic number as the number of electrons to arrange.
2. Place electrons one by one into the energy levels, filling the lowest level first and then moving out. Let a maximum of 2 electrons go into level 1 and a maximum of 8 go into level 2.
3. At level 3, stop at 8 electrons. It can hold more, but more do not enter it until we reach atomic numbers above 20, and these are beyond our present interest.
4. Use level 4 for the 19th and 20th electrons.

FIGURE 2.2
This solar system model of the atom, after Niels Bohr, is commonly used in popular print to depict an atom.

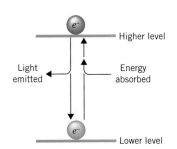

Electrons shift to higher levels when the correct amount of energy is absorbed, and they emit this energy when they drop back again.

■   Level 3 will resume filling at element 21, and it becomes filled at element 29 (copper).

Electrons do not move in simple orbits, like planets about the sun. Their motions are too complicated to know with exactness, so we give attention to the regions where they most probably reside instead, regardless of how they move. These individual regions are called **atomic orbitals.**

The principal energy levels discussed in Section 2.3 have one or more sublevels, and each sublevel is made up of one or more atomic orbitals. The first energy level has only one sublevel. (It is its own sublevel, in other words.) And this sublevel has just one orbital, called the 1s orbital. The 1 refers to main level 1 and the s denotes the shape of the orbital. It looks like a sphere with the atomic nucleus in the center (Figure 1).

The second energy level has two sublevels, designated 2s and 2p. The 2s sublevel, like the 1s, has just one orbital, so we can call it the 2s orbital. It looks like the 1s orbital from the outside—spherical. The 2p sublevel has three orbitals, each with a figure-eight-cross-sectional shape, seen in Figure 1. The axes of these three are mutually perpendicular, like the x, y, and z axes, so the 2p orbitals are individually named the $2p_x$, $2p_y$, and $2p_z$ orbitals. These differ only in orientation, not in energy. An electron can be in any of these orbitals and have the same energy.

Energy level 3 has three sublevels and three kinds of orbitals, one of the s type, three of the p type, and five of still another type, the d orbitals (whose shapes we do not discuss because they would have no bearing

on the properties of the first 20 elements. There is a fourth kind of orbital, the f orbital, at level 4 and higher, which we also leave to more comprehensive books.)

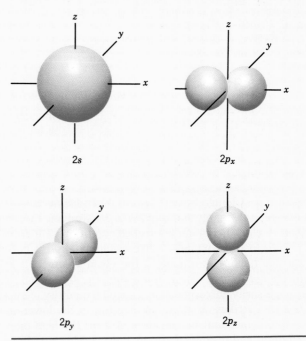

FIGURE 1
Two kinds of orbitals, the s and the p. All main energy levels have one s orbital. Level 2 and higher levels all have three p orbitals.

Figure 2.3 shows two ways we could represent an electron configuration. The dots in part *a* are tedious to count, however, so we'll use the symbolism given in part *b*. What we're really after is the situation in the highest *occupied* level, the **outside level,** or outer level. When we know the number of electrons in an atom's outside level, we can make some sense of the chemistry of the element. Table 2.3 gives the electron configurations for the first 20 elements. For additional details, see Special Topic 2.1.

FIGURE 2.3
Two ways to display electron configurations. (*a*) The electron-dot symbol of an atom of atomic number 12 and mass number 24. (*b*) A concise display of the same electron configuration.

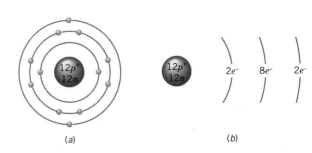

Each atomic orbital can hold up to two electrons, no more. When two are present, they must spin in opposite directions. The spin of an electron is like the spin of the earth about its axis. When two electrons spin in opposite directions they behave like two magnets oriented to attract each other. This attraction helps to compensate for the electrical repulsion that exists between two electrons.

Electron configurations by orbitals are written as follows, using nitrogen (atomic number 7) as an example:

$$1s^2 2s^2 2p_x{}^1 2p_y{}^1 2p_z{}^1$$

The superscripts give the number of electrons in each orbital. You can see that level 2 of the nitrogen atom has a total of five electrons, the number predicted by the use of the rules given in Section 2.3. Level 2 of the neon atom (atomic number 10) has eight electrons and all orbitals of levels 1 and 2 are full. When the three p orbitals at a given main energy level are full (each has two electrons), the condition is abbreviated to p^6.

The accompanying table gives the electron configurations, orbital by orbital, of the first 20 elements. Notice at atomic numbers 6 and 7 or 14 and 15 that two electrons do not go into the same orbital if another of *identical energy* is available.

Electron Configurations by Orbitals of Elements 1 to 20

Atomic Number	Element	Electron Configuration
1	H	$1s^1$
2	He	$1s^2$
3	Li	$1s^2 2s^1$
4	Be	$1s^2 2s^2$
5	B	$1s^2 2s^2 2p_x{}^1$
6	C	$1s^2 2s^2 2p_x{}^1 2p_y{}^1$
7	N	$1s^2 2s^2 2p_x{}^1 2p_y{}^1 2p_z{}^1$
8	O	$1s^2 2s^2 2p_x{}^2 2p_y{}^1 2p_z{}^1$
9	F	$1s^2 2s^2 2p_x{}^2 2p_y{}^2 2p_z{}^1$
10	Ne	$1s^2 2s^2 2p_x{}^2 2p_y{}^2 2p_z{}^2$
11	Na	$1s^2 2s^2 2p^6 3s^1$
12	Mg	$1s^2 2s^2 2p^6 3s^2$
13	Al	$1s^2 2s^2 2p^6 3s^2 3p_x{}^1$
14	Si	$1s^2 2s^2 2p^6 3s^2 3p_x{}^1 3p_y{}^1$
15	P	$1s^2 2s^2 2p^6 3s^2 3p_x{}^1 3p_y{}^1 3p_z{}^1$
16	S	$1s^2 2s^2 2p^6 3s^2 3p_x{}^2 3p_y{}^1 3p_z{}^1$
17	Cl	$1s^2 2s^2 2p^6 3s^2 3p_x{}^2 3p_y{}^2 3p_z{}^1$
18	Ar	$1s^2 2s^2 2p^6 3s^2 3p_x{}^2 3p_y{}^2 3p_z{}^2$
19	K	$1s^2 2s^2 2p^6 3s^2 3p^6 4s^1$
20	Ca	$1s^2 2s^2 2p^6 3s^2 3p^6 4s^2$

EXAMPLE 2.1 Writing an Electron Configuration

PROBLEM An atom has an atomic number of 19 and a mass number of 39. What are the composition of its nucleus and its electron configuration?

SOLUTION The nucleus has 19 protons (the atomic number) and 20 neutrons (mass number − atomic number). There are 19 electrons. Two electrons go into level 1 and 8 go into level 2. This leaves 9 electrons. Only 8 of these go into level 3 (rules 3 and 4). The last electron is in level 4. Using the kind of display of Figure 2.3b, we write the configuration as shown in the margin.

■ PRACTICE EXERCISE 3 What are the composition of the nucleus and the electron configuration of each of the following atoms? (Use this exercise to learn the rules; do not use the table.)

(a) atomic number 7, mass number 14

(b) atomic number 13, mass number 27

(b) atomic number 20, mass number 40

TABLE 2.3   Electron Configurations of Elements 1 to 20, Main Energy Levels

Element	Symbol	Atomic Number	Level Number			
			1	2	3	4
Hydrogen	H	1	1			
Helium	He	2	2			
Lithium	Li	3	2	1		
Beryllium	Be	4	2	2		
Boron	B	5	2	3		
Carbon	C	6	2	4		
Nitrogen	N	7	2	5		
Oxygen	O	8	2	6		
Fluorine	F	9	2	7		
Neon	Ne	10	2	8		
Sodium	Na	11	2	8	1	
Magnesium	Mg	12	2	8	2	
Aluminum	Al	13	2	8	3	
Silicon	Si	14	2	8	4	
Phosphorus	P	15	2	8	5	
Sulfur	S	16	2	8	6	
Chlorine	Cl	17	2	8	7	
Neon	Ne	18	2	8	8	
Potassium	K	19	2	8	8	1
Calcium	Ca	20	2	8	8	2

2.4 ELEMENTS

Atoms of the same element have identical atomic numbers and electron configurations.

As we move more deeply into the nature of atoms and elements, we can make our definitions sharper. For example, we said earlier that an element cannot be broken into anything that is simpler, yet stable. We now add that an **element** is a substance whose atoms all have identical nuclear charges and electron configurations. An **atom,** we can now say, is the smallest representative sample of an element; it has only one nucleus, and it is always an electrically neutral particle.

The atomic numbers go as high as 109, and this many different elements are known. Of these, 90 occur naturally, and the rest have been prepared using high-energy devices.

Most of the Elements Are Metals.   **Metals** are substances that conduct electricity and that can be polished, drawn into wires, and hammered into sheets. About 20 elements are **nonmetals.** They are poor conductors of electricity, and the solid nonmetals shatter when struck. All of the gaseous elements and one liquid element are nonmetals. A few elements are difficult to classify. Carbon, for example, is brittle, but in the form of graphite it conducts electricity.

Each Element Has an Atomic Symbol.   Each element has a shorthand symbol consisting of either one or two letters. The first is always capitalized and the second letter, if any, is always lowercase. Table 2.4 lists several, and these should be learned now.

TABLE 2.4   Names and Symbols of Some Common Elements[a]

C	Carbon	Al	Aluminum	Cl	Chlorine	Ag	Silver (*argentum*)
H	Hydrogen	Ba	Barium	Mg	Magnesium	Cu	Copper (*cuprum*)
O	Oxygen	Br	Bromine	Mn	Manganese	Fe	Iron (*ferrum*)
N	Nitrogen	Ca	Calcium	Pt	Platinum	Pb	Lead (*plumbum*)
S	Sulfur	Li	Lithium	Zn	Zinc	Hg	Mercury (*hydrargyrum*)
P	Phosphorus	Si	Silicon	As	Arsenic	K	Potassium (*kalium*)
I	Iodine	Co	Cobalt	Cs	Cesium	Na	Sodium (*natrium*)
F	Fluorine	Ra	Radium	Cr	Chromium	Au	Gold (*aurum*)

[a]The names in parentheses in the last column are the Latin names from which the atomic symbols were derived.

Nearly Every Element Consists of a Mixture of a Small Number of Its *Isotopes*.   According to Dalton's third postulate (page 34), "all atoms of a particular element are identical in mass." We now know that this is untrue for virtually all of the elements. Nearly any particular element we care to name is actually a mixture of atoms differing in mass. The atoms of varying mass of a given element, however, do not differ in atomic number (number of protons), nor do they have different electron configurations. They differ only in their numbers of neutrons and so have different mass numbers. Substances whose atoms are identical in atomic number and electron configuration *but different in mass number* are called **isotopes.**

Chlorine (atomic number 17), for example, consists chiefly of two isotopes called chlorine-35 and chlorine-37, where the numbers 35 and 37 are *mass numbers*. The atoms of both isotopes possess 17 protons (the shared atomic number), but an atom of chlorine-35 has 18 neutrons ($35 - 17$) and the chlorine-37 atom has 20 neutrons ($37 - 17$). Wherever chlorine is found in nature, regardless of its state of chemical combination, the *ratio* of chlorine-35 atoms to chlorine-37 atoms is very close to 3:1. The ratios of the isotopes for each of the other elements are different from 3:1, and many elements have more than two isotopes, *but for any given element, the ratios of its isotopes are essentially constant throughout nature.*

Because they have the same electron configuration, the chlorine isotopes have identical *chemical* properties. Other than those rare situations involving *how fast* the reactions of isotopes occur and not what happens chemically, *the isotopes of any given element are identical in chemical properties.* This is why the existence of isotopes could not have affected Dalton's theory. Dalton's postulates were based on *chemical* properties and on the mass relationships observed in *chemical* reactions. Isotopes have no bearing on these. We need to know about isotopes for one reason, to understand an important physical property, an element's *atomic mass.*

The *Atomic Mass* of an Element Reflects the Relative Abundances of Its Isotopes.   It's quite a remarkable fact about our world that the isotopes of any given element occur in the same proportion regardless of where the element is found or its state of chemical combination. Some elements conform less well to this "rule" than others but, as we illustrated earlier, the two chlorine isotopes occur *everywhere,* all over the earth, in the same ratio of close to 3:1. For every four chlorine atoms in nature, three are chlorine-35 and one is chlorine-37. To find the *average* mass of these four atoms, the only kind of mass that could be obtained by purely *chemical* experiments, we'll assume that the ratio is exactly 3:1.

$$\text{Mass of 3 atoms of chlorine-35: } 3 \times 35 \text{ u} = 105 \text{ u}$$
$$\text{Mass of 1 atom of chlorine-37: } 1 \times 37 \text{ u} = \underline{37 \text{ u}}$$
$$\text{Mass of these four atoms} = 142 \text{ u}$$

■ Be sure to distinguish between *substances* and *particles*. Elements, compounds, and mixtures are *substances*. Atoms, protons, neutrons, and electrons are the *particles* of which substances are made.

■ Isotopes are often symbolized as follows, where the superscripts are mass numbers and the subscripts are atomic numbers.

$$^{35}_{17}\text{Cl} \qquad ^{37}_{17}\text{Cl}$$

■ A table of atomic masses and numbers is inside the front cover of this book.

■ The element xenon has nine stable isotopes, the most of any element.

Dividing 142 u by 4 gives us 35.5 u as the average mass of the chlorine atoms as this element occurs naturally anywhere on earth. The ratio of atoms is not exactly 3 to 1 so the actual average is 35.457 u. The average mass of the atoms of the various isotopes of any given element, as they occur in their natural proportions, is called the **atomic mass** of the element. Unlike mass numbers, atomic masses are not simple, whole numbers because of the existence of isotopes.

It is important to realize that isotopes are *substances*. Isotopes are not particles; rather, they consist of particles (atoms) having identical protons but different numbers of neutrons. Thus, an element, which is one of the three kinds of *matter* (and not one of a few kinds of particles), usually consists of two or more *substances,* its *isotopes,* physically mixed. About 250 isotopes occur in nature. More than a thousand have been made by nuclear transformations that we will study in Chapter 22. Many are useful in medicine.

■ PRACTICE EXERCISE 4   Suppose that element number 25 consists of two isotopes in a 50:50 ratio. One has 25 neutrons and the other has 30. What is the atomic mass of this element?

2.5   PERIODIC LAW AND PERIODIC TABLE

In a given family of representative elements, the atoms of all members have the same outside-level electron configurations.

If each of the 109 elements were completely unlike the others, the study of chemistry would be extremely difficult. Fortunately, the elements can be sorted into a small number of families whose members have much in common. Dimitri Mendeleev, a Russian scientist, was the first to notice this as he wrote a chemistry textbook published in 1869.

The Properties of the Elements Recur Periodically.   When Mendeleev organized some of the physical and chemical properties of the known elements, he observed that the properties seemed to go through cycles from the elements of lowest atomic mass to those of highest atomic mass. The cycles weren't geometrically perfect, but there were definite rises and falls. One improvement, which came with the discovery of atomic numbers after Mendeleev's time, was the use of atomic numbers instead of atomic masses to rank the elements. As seen in Figure 2.4*a*, the temperatures at which the first 20 elements boil do not increase ever higher as the atomic numbers increase. Instead, the boiling temperatures rise and fall. They vary *periodically,* not linearly, with increasing atomic number.

The ionization energies of the elements display a similar fluctuation (see Figure 2.4*b*). An element's ionization energy is the energy required to make one electron leave each atom in a sample whose mass (in grams) equals the element's atomic mass. Enough other properties of elements also vary periodically with atomic number to establish a general law of nature, the **periodic law.**

■ The mass of an element *in grams* numerically equal to its atomic mass is called one *mole* of the element, and the mole (abbreviated *mol*) is the SI base unit for *amount of substance.*

PERIODIC LAW
The properties of the elements are a periodic function of their atomic numbers.

This natural law makes possible a useful way to organize and study the elements, as we'll see next.

The *Periodic Table* Organizes the Elements to Show Off Their Periodic Properties. Notice that in the plot of boiling points versus atomic numbers, helium, neon, and argon are at the bottoms of cycles in Figure 2.4*a*, and at the tops in Figure 2.4*b*. They seem to form a set of elements that behave alike. Let us now start to line up the elements horizontally by atomic number, but start new rows so as to make a set of similar elements fall into the same *vertical column*. When we do this, the rest of the columns *automatically* are made up of other sets of similar elements. The resulting display of the elements is called the **periodic table,** shown inside the front cover of this book.

Each horizontal row in the periodic table is called a **period,** and each vertical column is called a **group** or a **chemical family.** What makes this organization so helpful is that a study of chemistry can focus first on just a few families, rather than on dozens of individual elements.

Some Families of Elements Are *Representative Elements*. The periods in the periodic table are not all of the same length, and several must be split to create vertical columns that contain only elements with similar properties, particularly *chemical* properties. Thus, period 1, with only hydrogen and helium, is very short and, like the next two periods, is split into two parts.

The groups have both numbers and letters. One set makes up an A-series: IA, IIA, IIIA, and so forth up to VIIA. This set plus Group 0 are called the **representative elements.** The groups in the B-series, clustered near the middle of the periodic table, are called the **transition elements.**

The two rows of elements placed outside the table are the **inner transition elements.** (The table would not fit well on the page if the inner transition elements were

■ If one element in a group or chemical family forms a compound with, say, chlorine, then all the others in the family do, too, and all the compounds have the same atom ratios or formulas.

■ The transition elements represent successive fillings of *inner* energy levels; their outside levels hold one or two electrons.

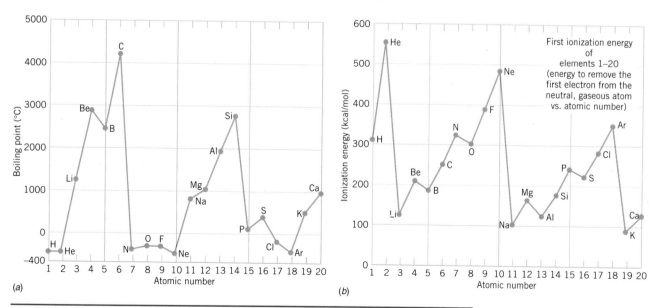

FIGURE 2.4
Periodicity and two properties of elements 1 to 20. (*a*) Boiling points versus atomic numbers.
(*b*) Ionization energies versus atomic numbers.

■ The lanthanides are also called the *rare earth* elements, and some are essential in electronic devices.

not handled this way.) Elements 58 through 71 constitute the *lanthanide series,* named after element 57, which just precedes this series. The series of elements 90 to 103 is the *actinide series.*[1]

Several of the groups of representative elements have family names. Those in Group IA, for example, are the **alkali metals,** because they all react with water to give an alkaline or caustic (skin-burning) solution. The elements in Group IIA are the **alkaline earth metals,** because their compounds are commonly found in "earthy" substances, like limestone.

The elements in Group VIIA are called the **halogens** after a Greek word signifying their salt-forming ability. For example, chlorine in Group VIIA is present in table salt (in a chemically combined form with sodium).

The elements in Group 0, all gases, were discovered after Mendeleev's work. Except for a few compounds that xenon and krypton form with fluorine and oxygen, the Group 0 elements react with nothing. Hence, they are called the **noble gases** (*noble* signifying limited activity).

Other groups of representative elements are named after their first members, for example, the **carbon family** (Group IVA), the **nitrogen family** (Group VA), and the **oxygen family** (Group VIA).

■ PRACTICE EXERCISE 5   Referring to the periodic table, pick out the symbols of the elements as specified.
(a) A member of the carbon family: Sr, Sn, Sm, S
(b) A member of the halogen family: C, Ca, Cl, Co
(c) A member of the alkali metals: Rn, Ra, Ru, Rb
(d) A member of the alkaline earth metals: Mg, Mn, Mo, Md
(e) A member of the noble gas family: Ac, Al, Am, Ar

Electron Configurations Cause the Periodic Law.   The numbers of electrons housed in the principal energy levels of several representative elements are given in Table 2.5. When we look at the outside levels, family by family, we note two very striking facts. First, *the atoms of each family all have identical numbers of electrons in their outside levels.* Second, *the numbers in the outside levels differ from family to family.*

The atoms of Group IA metals, for example, all have one electron in their outside levels. In Group IIA, all atoms have two outside-level electrons. In fact, *among the representative elements, the group number (I, II, and so forth through VII) equals the number of electrons in the outside levels of their atoms.*

■ The periodic table was established before electron configurations were known.

All atoms of the noble gases (except those of helium) have eight outer-level electrons, an *outer octet.* Helium atoms have just two outer-level electrons, but this level (number 1) cannot hold more than two anyway. The noble gas elements, as we said, are the least chemically reactive of all elements, so we might suspect that having eight electrons in an outer electron level somehow relates to chemical stability. As we'll see in the next chapter, the outer octet will be one of the most significant features of atomic structure we will study.

The electron configurations of the B-series elements, such as the various transition elements, have their own fairly regular patterns, but our study will not require knowledge of them.

---

[1]There is an international effort to do away with the A and B organization. Until this movement succeeds in changing the standard references and the standard chemistry examinations in the United States, we feel bound to stay with the A and B system.

TABLE 2.5 Electron Configurations of Four Families of Representative Elements

Family	Element	Atomic Number	Principal Energy Level						
			1	2	3	4	5	6	7
Group IA	Lithium	3	2	1					
Alkali metals	Sodium	11	2	8	1				
	Potassium	19	2	8	8	1			
	Rubidium	37	2	8	18	8	1		
	Cesium	55	2	8	18	18	8	1	
	Francium	87	2	8	18	32	18	8	1
Group IIA	Beryllium	4	2	2					
Alkaline earth	Magnesium	12	2	8	2				
metals	Calcium	20	2	8	8	2			
	Strontium	38	2	8	18	8	2		
	Barium	56	2	8	18	18	8	2	
	Radium	88	2	8	18	32	18	8	2
Group VIIA	Fluorine	9	2	7					
Halogens	Chlorine	17	2	8	7				
	Bromine	35	2	8	18	7			
	Iodine	53	2	8	18	18	7		
	Astatine	85	2	8	18	32	18	7	
Group 0	Helium	2	2						
Noble Gases	Neon	10	2	8					
	Argon	18	2	8	8				
	Krypton	36	2	8	18	8			
	Xenon	54	2	8	18	18	8		
	Radon	86	2	8	18	32	18	8	

EXAMPLE 2.2 Finding Information in the Periodic Table

PROBLEM How many electrons are in the outside level of an atom of iodine?

SOLUTION Until you become more familiar with the locations of certain elements in the periodic table, you'll have to use the Table of Atomic Masses and Numbers (inside the front cover of this book) to find the atomic number of a given element. Doing this, we find that the atomic number of iodine is 53. Now we use the periodic table and find that iodine is in Group VIIA. Being one of the A-type elements, we know that iodine is a *representative* element, which means that its group number is the same as the number of outside-level electrons, namely, 7.

■ PRACTICE EXERCISE 6 How many electrons are in the outside level of an atom of each of the following elements?
(a) potassium (b) oxygen (c) phosphorus (d) chlorine

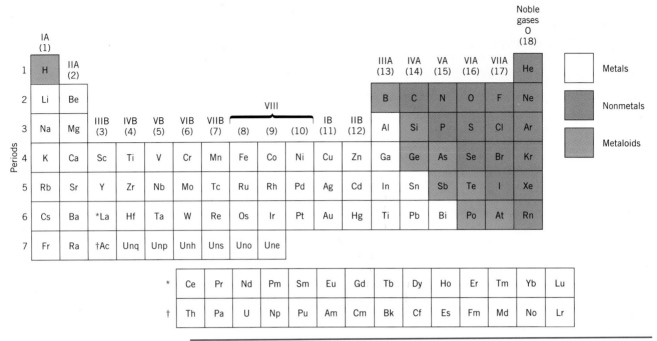

FIGURE 2.5
Locations of metals, nonmetals, and metalloids in the periodic table.

Metals and Nonmetals Are Segregated in the Periodic Table.   Another use of the periodic table is to tell which elements are metals and which are not. The nonmetals, for example, are all in the upper right-hand corner of the table (Figure 2.5). All of the noble gases (Group 0) and the halogens (Group VIIA) are nonmetals. The few elements along the borderline between metals and nonmetals are sometimes called **metalloids,** and they have properties that are partly metallic and partly nonmetallic.

If you compare the locations of the elements in Figure 2.5 with the electron configurations given in Table 2.5, you will see that all of the nonmetals, except hydrogen and helium, have four to eight outer electrons. Except for hydrogen and helium, all atoms with one, two, or three outer electrons belong to metals. Only among some elements with high atomic numbers do we find metals with more than three outside-level electrons. Tin and lead of Group IVA are common examples.

SUMMARY

Matter   Matter, anything with mass that occupies space, can exist in three physical states: solid, liquid, and gas. Broadly, the three kinds of matter are elements, compounds, and mixtures. Elements and compounds obey the law of definite proportions. Mixtures do not. A chemical reaction, an event in which substances change into different substances, occurs without any loss of mass.

Dalton's Atomic Theory   The law of definite proportions and the law of conservation of mass in chemical reactions led John Dalton to the theory that all matter consists of discrete, noncuttable particles called atoms. When atoms of different elements combine to form compounds, they combine as *whole* atoms; they do not break apart.

Atomic Structure   Atoms are tiny electrically neutral particles that are the smallest representatives of an element that can display the element's chemical properties. Each atom has one nucleus, a dense inner core, surrounded by enough electrons to balance the positive nuclear charge. All of the atom's protons and neutrons are in the nucleus, and each of these subatomic particles has a mass of 1.0 u. The mass of the electron is only 1/1836th the mass of a proton. The proton has a charge of 1+, the electron's charge is 1−, and the neutron is electrically neutral.

Atomic Numbers and Atomic Masses   The atomic number of an element is the number of protons in each atom. It also equals the number of electrons in an atom. Nearly all elements occur as a mixture of a small number of isotopes. The isotopes of an element share the same atomic number, but their atoms differ slightly in their numbers of neutrons. To fully characterize an isotope, we have to specify both the atomic number and the mass number, the number of protons plus neutrons. The atomic mass of an element is numerically equal to the average mass (in u) of the atoms of all its isotopes as they occur naturally together.

Electron Configurations   To write an electron configuration of an atom, we place its electrons one by one into the available energy levels. We start with the level of lowest energy, level 1, which can hold up to 2 electrons. Level 2 can hold 8 electrons, and level 3 can handle 18 (but is limited to 8 until atomic number 21 is reached). As long as electrons remain in their levels, an atom neither absorbs nor radiates energy.

Periodic Properties   Because many properties of the elements are periodic functions of atomic numbers, the elements fall naturally into groups or families that we can organize into vertical columns in a periodic table. Atoms of the representative elements that belong to the same family have the same number of outside-level electrons. This number equals the A-group number itself. (Helium does not fit this generalization.)

The horizontal rows of the periodic table are called periods. The nonmetallic elements are in the upper right hand corner of the periodic table. The metals, the great majority of the elements, make up the rest of the table. At the border between metals and nonmetals occur the metalloids, which have both metallic and nonmetallic properties.

REVIEW EXERCISES

The answers to Review Exercises whose numbers are in color are found in Appendix V. The answers to the other Review Exercises are found in the Study Guide that accompanies this text. The more challenging questions are marked with asterisks.

States and Kinds of Matter

2.1   It is important to distinguish between the *states* of matter and the *kinds* of matter.
(a) What are the names of the states of matter?
(b) What are the names of the kinds of matter?

2.2   Among which, the states of matter or the kinds of matter, are the distinctions basically in terms of *physical* properties?

2.3   Which of the following events are chemical changes (as well as being physical changes)?
(a) When heated in a pan, sugar turns brown (caramelizes).
(b) When stirred in water, table salt seems to disappear.
(c) When struck with a hammer, an ice cube shatters.
(d) A bleaching agent causes a colored fabric to lose its color.

2.4   How are compounds different from elements?

2.5   How are compounds different from mixtures?

2.6   Sodium reacts with chlorine to give sodium chloride (table salt).

(a) Physical changes obviously occur as the metal and the gas change into a white solid, so why is this event also called a *chemical* reaction?
(b) Which substances are the *reactants* and which are the products?
(c) What is always true about the *ratio of the masses* of sodium and chlorine that combine to form sodium chloride?

2.7   The fact that 2.0 g of hydrogen combines with 16.0 g of oxygen, no more, no less, to give 18.0 g of water illustrates what important laws of chemical combination?

2.8   What law of chemical combination most surely distinguishes compounds from mixtures?

Dalton's Atomic Theory

2.9   What are the postulates of Dalton's atomic theory?

2.10   What law of chemical combination contributed most to Dalton's postulate that compounds include their elements in definite ratios *by atoms*?

2.11   The fact that the ratio of the elements *by mass* in compounds made of only two elements is almost never 1:1 suggested what postulate of Dalton's?

*2.12   If the ratio *by mass* of two elements, X and Y, in the hypothetical compound XY were exactly 1:1, what would be true in Dalton's theory about the relative masses of the *atoms* of X and Y?

2.13   Dalton said that atoms are indestructible, but we now know that they can be broken up into three (actually more) particles.
(a) What is the *general* name for such particles?
(b) What are the names and electrical conditions of the three particles of greatest interest in understanding the chemical properties of substances?
(c) Two of these particles attract each other. What are their names?
(d) Two simply stated rules govern the behaviors of electrically charged particles toward each other. What are they?

*2.14**   In Chapter 3 we will learn that particles having the following compositions exist.
Particle X: 11 protons, 12 neutrons, and 10 electrons
Particle Y: 17 protons, 18 neutrons, and 18 electrons
(a) What is the *net* electrical charge carried by particle X?
(b) What is the *net* electrical charge carried by particle Y?
(c) Would the particles X and Y attract each other, repel each other, or be indifferent to each other? How can you tell?
(d) What are the mass numbers of X and Y?

2.15   The atomic mass unit has approximately what size, 10^{-230} g, 10^{-23} g, 10^{-3} g, or 10^{23} g?

Atomic Theory

2.16   It costs energy to make an electron leave an atom, but which of the following changes would cost the *least* energy, and why?
(a) Removal of an electron from a particle having 12 protons and 12 electrons
(b) Removal of an electron from a particle having 12 protons and 11 electrons

2.17   Rutherford found that essentially all of the mass of an atom is in its nucleus. The following calculations will give you an idea of what this means.
(a) Calculate the density in g/cm^3 of the nucleus of a hydrogen atom from the following data. Assume that the nucleus is a perfect sphere so that you can calculate its volume by the standard equation, volume of sphere = $\frac{4}{3}\pi r^3$, where r = radius and π = 3.14. The radius of the hydrogen nucleus is 5×10^{-14} cm and the mass of the nucleus is 1.67×10^{-24} g.
(b) Convert the answer to part (a) into units of metric tons per cubic centimeter (1 metric ton = 1000 kg).

2.18   What fact lets us ignore the masses of electrons in an atom when we compute an atomic mass from its subatomic particles?

2.19   To five significant figures, calculate the mass in grams of a sample of 6.0220×10^{23} hydrogen atoms, each one having just one proton. Ignore the mass of the electron. How does the result compare with the atomic mass of hydrogen? (To two significant figures, are the results the same or different?)

2.20   Who discovered the existence of atomic nuclei and what did he conclude about the nature of an atom?

2.21   What causes the positive charge on an atomic nucleus, and why is it always a *whole* number?

2.22   When we know the specific arrangement of the electrons around an atomic nucleus, what do we know about the atom? (What short term is used?)

2.23   Niels Bohr suggested two postulates about the arrangements of electrons in atoms. What are they? Are they still true?

2.24   When light is emitted from, say, a red-hot bar of iron, what happens at the atomic level?

2.25   What analogy did Bohr suggest to illustrate his view of atomic structure? Do scientists still use this analogy?

Electron Configurations

2.26   What is the maximum number of electrons that can be in each?
(a)   principal level 2   (b)   principal level 3

2.27   Write the atomic number, the mass number, and the electron configuration of an atom with six protons and seven neutrons.

2.28   What are the atomic number, mass number, and electron configuration of an atom with 18 protons and 21 neutrons?

2.29   Write the electron configuration of silicon (atomic number 14).

2.30   The iron atom has 2 electrons in the first level, 8 in the second, 14 in the third, and 2 in the fourth. Using only this information and general facts about electron configurations, answer the following questions.
(a) What is the atomic number of iron?
(b) Is principal level 3 completely filled?

2.31   Write the electron configurations of the elements that have the following atomic numbers, using only these numbers.
(a)   11   (b)   8   (c)   17   (d)   20

2.32   Write the electron configurations of the elements that have the following atomic numbers.
(a)   12   (b)   19   (c)   9   (d)   3

Elements and Their Isotopes

2.33   Roughly, how many elements are known: 80, 100, 200, or 2000?

2.34   Roughly, how many elements are not solids under ordinary room conditions, but are either liquids or gases: 2, 13, 44, 75, 180, or 1885?

2.35   Indium is a substance that can be given a shiny finish, and it conducts electricity well. Is indium more likely to be a metal or a nonmetal?

2.36 Phosphorus is a dark-reddish, powdery substance that doesn't conduct electricity. Is it more likely to be a metal or a nonmetal?

2.37 Elements and isotopes are *substances,* not tiny particles, although they consist of such particles. What is the distinction between the terms *element* and *isotope of an element*?

2.38 Which of the following are pairs of isotopes? (Use the hypothetical symbols for your answer.)
M has 12 protons and 13 neutrons
Q has 13 protons and 13 neutrons
X has 12 protons and 12 neutrons
Z has 13 protons and 12 neutrons

2.39 The members of which of the following pairs of elements would be expected to have identical chemical properties and why?
Pair 1. *A* (10 protons and 9 neutrons)
        B (10 protons and 10 neutrons)
Pair 2. *C* (7 protons and 6 neutrons)
        D (6 protons and 6 neutrons)

2.40 Carbon, atomic number 6, has three isotopes that are found in nature. The most abundant (98.89%) has a mass number of 12. The isotope with a mass number of 13 makes up 1.11% of naturally occurring carbon. The third isotope is carbon-14, which is obviously present in the merest trace, because 98.89% + 1.11% = 100.00%. But carbon-14 makes possible the dating of ancient artifacts.
(a) What is the same about these isotopes?
(b) In what specific feature of atomic structure do they differ?

2.41 Later, when we want to describe some *chemical* reaction of sulfur, which consists principally of two isotopes (mass numbers 32 and 34), we will use the symbol S. Why won't we have to specify which isotope?

Atomic Symbols

2.42 Of the symbols NO and No, which stands for an element and which for a compound?

2.43 Write the symbols of the following elements.
(a) phosphorus (b) calcium (c) bromine
(d) platinum   (e) carbon (f) barium

2.44 What are the symbols of the following elements?
(a) lead   (b) mercury (c) fluorine
(c) potassium (d) hydrogen (e) iron

2.45 Write the symbols of the following elements.
(a) sulfur   (b) iodine (c) manganese
(d) sodium (e) copper (e) magnesium

2.46 What are the symbols of the following elements?
(a) nitrogen (b) oxygen (c) silver
(d) zinc   (e) lithium (f) chlorine

2.47 What are the names of the elements with the following symbols?
(a) I (b) Pb (c) Li
(d) N (e) Zn (f) Ba
(f) C (h) Ca (i) Cl
(j) F (k) Cu (l) Fe

2.48 Write the names of the elements with these symbols.
(a) H (b) Mg (c) Al
(d) Hg (e) Mn (f) Na
(g) O (h) Ag (i) Br
(j) P (k) K (l) Pt

Atomic Masses

2.49 Mass numbers are always whole numbers. Atomic masses almost never are. Explain.

2.50 Iodine (atomic number 53), which is needed in the diet to have a healthy thyroid gland, occurs in nature as only one isotope, iodine-127. Using just this information, what is the atomic mass of iodine (to three significant figures)?

2.51 Bromine, in the same chemical family as iodine, occurs in nature as a nearly 1:1 mixture of two isotopes, bromine-79 and bromine-81. Using just this information, what is the atomic mass of bromine (to two significant figures)?

2.52 An element consists of two isotopes, *X* and *Y*, in the ratio 4:1. Atoms of *X* have 30 protons and 34 neutrons. Those of *Y* have 30 protons and 38 neutrons. What is the atomic mass of this element (to three significant figures)?

2.53 What is the atomic mass of an element that consists of three isotopes in the ratio 1:1:1 if the atomic number is 30, and one isotope has 30 neutrons, the second has 36, and the third has 38?

2.54 To two significant figures, the atomic mass of carbon is 12 and the atomic mass of magnesium is 24.
(a) How many times heavier are magnesium atoms than carbon atoms?
(b) If we counted out in separate piles 10^{23} atoms of carbon and the identical number of magnesium atoms, which pile of atoms would have the larger mass, and by what factor?
(c) If we weighed out 2.0 g of carbon and we wanted a sample of magnesium that had the identical number of atoms, how many grams of magnesium should we weigh out?

*\ **2.55** How many times heavier are carbon atoms than hydrogen atoms (on the average)? Calculate the answer to two significant figures. Use data from the Table of Atomic Masses and Numbers inside the front cover.

*\ **2.56** Carbon and oxygen atoms can chemically combine to form particles consisting of these atoms in the ratio 1:1. (The substance that forms is the poisonous gas carbon monoxide.)
(a) How much heavier are oxygen atoms than carbon atoms (to three significant figures)?

(b) If all of the atoms in a sample of 12.0 g of carbon are to combine entirely with oxygen atoms, how many grams of oxygen are needed?

The Periodic Table

2.57 What general fact about the chemical elements makes possible the stacking of these elements in the kind of array seen in the periodic table?

2.58 What do the terms *group* and *period* refer to in the periodic table?

2.59 Without consulting the periodic table, is an element in Group VA a representative or a transition element?

2.60 Without consulting the periodic table, is an element in Period 2 a representative or a transition element?

2.61 Give the group number and the chemical family name of the set of elements to which each of the following belongs.
(a) sodium (b) bromine
(c) sulfur   (d) calcium

2.62 For each of the following elements give the group number and the name of the family to which it belongs.
(a) iodine   (b) phosphorus
(c) magnesium (d) lithium

2.63 The electron configuration of a representative element is

$$2 \quad 8 \quad 7$$

Answer the following questions without looking at the periodic table.
(a) How many electrons are in level 2, and is it filled?
(b) Is the element more likely to be a metal or a nonmetal? Why?
(c) What are the group number and the family name associated with this element?

2.64 The electron configuration of a representative element is:

$$2 \quad 8 \quad 18 \quad 18 \quad 8 \quad 2$$

Answer the following questions without consulting the periodic table.
(a) Is this element more likely a metal or a nonmetal? How can you tell?
(b) What is the group number of this element?

*2.65 The following table is a section from the periodic table where the numbers are atomic numbers. The numbers of one row have been given hypothetical atomic symbols. You should be able to answer the following questions without referring to the actual periodic table.

5	6	7	8	9
13 a	14 b	15 g	16 d	17 e
31	32	33	34	35

(a) Give the numbers of the elements in the same period as g.
(b) What are the numbers of the elements in the same group as b?
(c) Give the numbers of the elements in the same family as e.
(d) Above each box of the top row of elements, write the group numbers of the elements, including the A or B designation.
(e) How many electrons are in the highest occupied principal energy level of the element that would stand immediately to the left of a?
(f) How many electrons would be in the outside shell of the element standing immediately below d in the periodic table?
(g) Which element is more likely to be a nonmetal, element 9 or 31?

Atomic Orbitals (Special Topic 2.1)

2.66 Principal energy level 2 has how many sublevels? How many orbitals are in each sublevel, and how are they named?

2.67 If an orbital is to hold two electrons, what must be true about them?

2.68 The three $2p$ orbitals are alike in what ways? How are they different?

2.69 Write the electron configurations by atomic orbitals for carbon (atomic number 6) and sodium (atomic number 11).

3

THE NATURE OF MATTER: COMPOUNDS AND BONDS

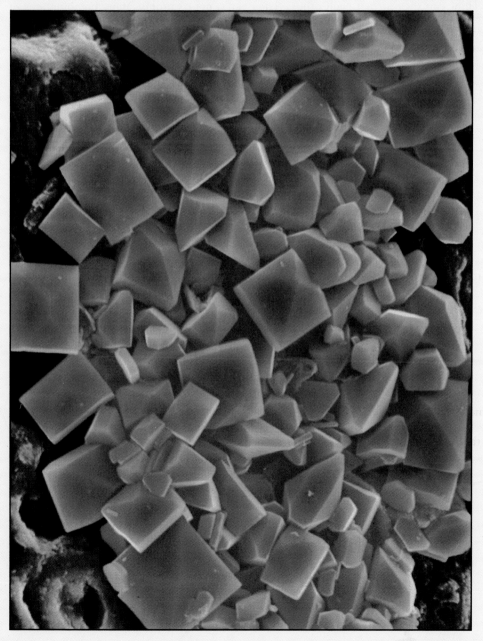

Octet Rule

Ions and Ionic Compounds

Names and Formulas of Ionic Compounds

Molecules and Molecular Compounds

Polar Molecules

At the molecular level of life, we must have salt. Some of the properties that make salt indispensable also give salt crystals very precise shapes. These crystals of sea salt, colored blue by trace impurities, have regular shapes because their constituent ions cannot help but collect in an orderly manner. How atoms form ions and molecules is studied in this chapter.

3.1   OCTET RULE

Atoms and ions whose outside energy levels hold eight electrons are substantially more stable than those that do not.

Whatever holds us together—skin and bones or cell membranes and blood vessels—must handle stresses and strains, both physical and chemical. We now ask: What must be true at the atomic level to explain how some kinds of matter stick together in bulk, and other kinds do not?

Our first clue is the law that unlike charges attract. Electrical forces of attraction are at the heart of what holds matter together. But atoms are electrically *neutral,* so how can they attract each other? They can't, not *as atoms.* Many combinations of atoms, however, can reorganize their electron configurations to form new particles that are able to attract each other electrically. Strong forces of attraction called **chemical bonds** can exist between such reorganized atoms. Our broad goal in this chapter is to learn the nature of these bonds and the laws that govern their formation.

Molecular and Ionic Compounds Involve the Major Kinds of Bonds.   The electrons and nuclei of atoms can become reorganized into compounds in principally two ways. One, normally found between atoms of *nonmetals,* results in a new kind of small particle called a *molecule.* A **molecule** is an electrically neutral particle made up of two or more atomic nuclei surrounded by enough electrons to make the tiny package neutral. The bonds holding molecules together are called *covalent bonds,* which we will study in Section 3.4. Compounds consisting of molecules are called **molecular compounds.** Sugar, vitamin C, cholesterol, and aspirin are examples; all are made from the atoms of nonmetal elements, C, H, and O. Water, made of H and O, is also a molecular compound.

■ *Molecule* is from a Latin term meaning "little mass."

The other way to reorganize the electrons and nuclei of atoms into compounds occurs when metals react with nonmetals. Neutral atoms of metals give up electrons to atoms of nonmetals, forming tiny particles called **ions** that have opposite electrical charges. Thus atoms of sodium metal, Na, can change into *sodium ions,* Na^+. Atoms of chlorine, Cl, can become *chloride ions,* Cl^-. Unlike charges attract, so oppositely charged ions cannot help but attract each other, building aggregations of ions that make up **ionic compounds.** A crystal of table salt or sodium chloride, for example, is a combination of Na^+ and Cl^- ions. They occur in the ratio 1:1 because only this ratio ensures that the collection of ions has a net electrical charge of zero. The force of attraction between oppositely charged ions in such compounds is called the *ionic bond.*

■ *Ion* is from the Greek *ienai,* meaning "to go." In solution ions can move under the influence of an electric current, but molecules can't.

The Outer Octet Is a Condition of Unusual Stability.   We wrote an electrical charge of 1+ for the sodium ion and a charge of 1− for the chloride ion. Why just these charges? Why not Na^{2+}, or Na^-, or Cl^+, or Cl^{2-}? The answer begins with a correlation between the ionic charges permitted by nature for the reactive elements and the electron configurations of the least reactive elements, those in Group 0. These elements, the noble gases, do not form ions; in fact, they *have almost no chemical reactions.* Behind this remarkable but still poorly understood fact about our world stand the electron configurations of the noble gases. The atoms of all but one of them have **outer octets,** meaning eight electrons in the *outside* level. Atoms of the exception are of the simplest noble gas, helium, for which the outside level is level 1, which cannot hold eight electrons. Yet helium is also chemically unreactive. Its atoms have a *filled* level 1 as the *outside* level, namely, two electrons in level 1.

■ Krypton and xenon of the noble gas family form a few compounds with the most reactive of all elements, fluorine and oxygen.

Evidently the two kinds of configurations of the noble gases—the outer octet or the outer filled level 1—confer great chemical stability. We can only point this out; scientists don't really know why this is so. However, *many other chemical species, like ions with the same noble gas configurations, are similarly stable.* When reactive elements interact to form ions or molecules, achieving a noble gas configuration is such a common pattern that we have a powerful "rule of thumb." Traditionally called the **octet rule** but, better, the **noble gas rule**, it works best among the *representative* or *A-series* elements. (There are exceptions to the rule, so it's not a law of nature.)

OCTET RULE OR NOBLE GAS RULE
The atoms of the reactive representative elements tend to undergo those chemical reactions that most directly give them electron configurations of the nearest noble gas.

Let's see how this rule correlates with the existence of certain kinds of ions but not others.

3.2 IONS AND IONIC COMPOUNDS

Electron transfers between atoms generate ions.

Sodium chloride is a typical ionic compound. Its parent elements, sodium and chlorine, cannot be stored in the same container because they react violently. The following changes in electron configurations occur among billions of billions of sodium and chlorine atoms.

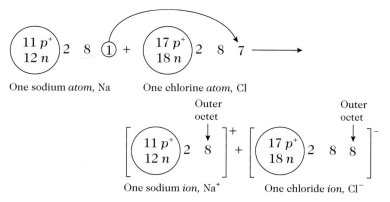

One sodium *atom*, Na   One chlorine *atom*, Cl

One sodium *ion*, Na$^+$ One chloride *ion*, Cl$^-$

■ For purposes of illustration, we have picked the sodium-23 and chlorine-35 isotopes. Sodium is a soft, silvery metal that has to be stored out of contact with air or water. Chlorine is a greenish-yellow poisonous gas.

The new particle with the sodium nucleus is no longer an *atom,* because it is no longer electrically neutral. It is an *ion.* An **ion** is an *electrically charged* particle of atomic or molecular size. What forms from the sodium atom is the *sodium ion,* Na$^+$, with a net charge of 1+. (The 11+ of the 11 nuclear protons is balanced by only 10− from the 10 remaining electrons.) Notice that Na$^+$ *has an outer octet.* In fact, Na$^+$ has the same electron configuration as neon, a noble gas. Whatever makes the outer octet confer stability on the neon *atom* evidently works on the sodium *ion,* because Na$^+$ is also chemically very stable. It's not stable *in isolation,* however. No ion is. If an ion is to be stable at all, it must be surrounded by something of opposite charge. In our example, the other particle that forms provides exactly this environment.

■ The reaction of sodium with chlorine is very violent.

Neon *Atoms* or Sodium *Ions*

Level	Electrons
I	2
2	8

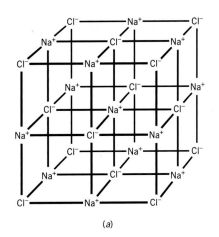

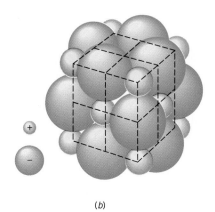

(a) (b) (c)

FIGURE 3.1
Structure of a crystal of sodium chloride. *(a)* This schematic drawing shows the alternating pattern of Na^+ and Cl^- ions. *(b)* Sodium ions have much smaller diameters than chloride ions. As seen here, the sodium ions are surrounded by chloride ions as nearest neighbors, and the like-charged ions are just a little farther apart. *(c)* The regular shape of a sodium chloride crystal comes from the regular arrangment of its ions.

Argon *Atoms* or Chloride *Ions*

Level	Electrons
I	2
2	8
3	8

■ Even a speck-sized sample of sodium has at least 10^{19} atoms.

The other new particle is also no longer an *atom;* it's the chloride *ion,* Cl^-, with a charge of $1-$. (It has 17 protons for $17+$ but 18 electrons for $18-$, so the net charge is $1-$.) The chloride ion, however, has an outer octet; its electron configuration is identical to that of argon, another noble gas. Like argon, the chloride ion is unusually stable, *but not in isolation.* As we said, ions are stable only when very close to something of opposite charge, and Cl^- ions become surrounded by Na^+ as they both form.

The spontaneous reaction of sodium with chlorine leads to greater stability for the atoms of both. Their atoms change from particles that do not have noble gas configurations to those that do. *The atoms sacrifice neutrality for stability.*

The Force of Attraction between Oppositely Charged Ions Is Called an Ionic Bond. The reaction between sodium and chlorine is typical of those between any element in Group IA and any in Group VIIA. Electron transfers occur from the metal to the nonmetal, and ions form. Of course, atoms are so small that any visible sample, even the tiniest speck, has billions of billions of them. When actual samples of reactive atoms are mixed, therefore, a storm of electron transfers occurs, and countless numbers of oppositely charged ions form.

Because like charges repel, the new sodium ions repel each other. The chloride ions repel each other also. Because unlike charges attract, however, sodium ions and chloride ions attract each other. Out of all of these attractions and repulsions, the storm of newly formed ions subsides to produce firm, hard, and regularly shaped crystals of sodium chloride. Spontaneously, the unlike charged ions, Na^+ and Cl^-, nestle together as closest neighbors and the like-charged ions stay just a little farther apart (Figure 3.1). The Na^+ ions have Cl^- ions as closest neighbors, and the Cl^- ions have Na^+ ions as nearest neighbors. As close neighbors, they cannot help but attract each other.

The forces of *repulsion* within a crystal of sodium chloride, Na^+ repelling Na^+ and Cl^- repelling Cl^-, are still present, or course. But because they act over slightly longer distances, they cannot overcome the attractions between Na^+ and Cl^- ions. Thus, a net force of attraction called the **ionic bond** develops between oppositely charged ions.

TABLE 3.1 **Important Monatomic Ions**

Group	Element	Symbol for Neutral Atom	Symbol for its Common Ion	Name of Ion
IA	Lithium	Li	Li^+	Lithium ion
	Sodium	Na	Na^+	Sodium ion
	Potassium	K	K^+	Potassium ion
IIA	Magnesium	Mg	Mg^{2+}	Magnesium ion
	Calcium	Ca	Ca^{2+}	Calcium ion
	Barium	Ba	Ba^{2+}	Barium ion
IIIA	Aluminum	Al	Al^{3+}	Aluminum ion
VIA	Oxygen	O	O^{2-}	Oxide ion
	Sulfur	S	S^{2-}	Sulfide ion
VIIA	Fluorine	F	F^-	Fluoride ion
	Chlorine	Cl	Cl^-	Chloride ion
	Bromine	Br	Br^-	Bromide ion
	Iodine	I	I^-	Iodide ion
Transition elements	Silver	Ag	Ag^+	Silver ion
	Zinc	Zn	Zn^{2+}	Zinc ion
	Copper	Cu	Cu^+	Copper(I) ion (cuprous ion)[a]
			Cu^{2+}	Copper(II) ion (cupric ion)
	Iron	Fe	Fe^{2+}	Iron(II) ion (ferrous ion)
			Fe^{3+}	Iron(III) ion (ferric ion)

[a]The names in parentheses are older names, but are still in use.

Crystals of Ionic Compounds Have Regular Shapes. As you can see in Figure 3.1, *the ions end up evenly spaced in the crystal.* This maximizes attractive forces and minimizes repulsive forces and so produces the most stable system. Given the necessary freedom, nature always tends to maximize the stabilities of its systems. This is why the crystal has a *regular* shape, as seen in Figure 3.1c. Such crystals can be shattered, of course. A hammer blow makes one layer of ions shift so that like-charged ions momentarily become closest neighbors. Along this layer the net force is now one of repulsion, not attraction, and the crystal splits.

Many familiar substances besides sodium chloride are ionic. Sodium bicarbonate in baking soda, barium sulfate in "barium X-ray cocktails," sodium hydroxide in lye and drain cleaners, and calcium sulfate in plaster of paris are other examples of ionic compounds.

The Symbols and Names of Monatomic Ions Are Like the Names of Their Elements.

When an ion has just one nucleus, it's called a *monatomic ion,* and it's named after its parent element, either the same name or a name close to it. All ions derived from metals have the same name as the element (plus the word *ion*). Monatomic ions from nonmetals have names that end in *-ide,* as in *chloride ion,* whose parent element is chlorine. The names, symbols, and electrical charges of several common ions are listed in Table 3.1, and they must be learned now.

TABLE 3.2   Electron Configurations among Four Families of Representative Elements

Family	Common Element		Atomic Number	Atoms—Electron Configurations						Ion	Ions—Electron Configurations					Nearest Noble Gas
				1	2	3	4	5	6		1	2	3	4	5	
IA	Lithium	Li	3	2	1					Li^+	2					Helium
Alkali	Sodium	Na	11	2	8	1				Na^+	2	8				Neon
metals	Potassium	K	19	2	8	8	1			K^+	2	8	8			Argon
IIA	Magnesium	Mg	12	2	8	2				Mg^{2+}	2	8				Neon
Alkaline	Calcium	Ca	20	2	8	8	2			Ca^{2+}	2	8	8			Argon
earth	Strontium	Sr	38	2	8	18	8	2		Sr^{2+}	2	8	18	8		Krypton
metals	Barium	Ba	56	2	8	18	18	8	2	Ba^{2+}	2	8	18	18	8	Xenon
VIA	Oxygen	O	8	2	6					O^{2-}	2	8				Neon
Oxygen	Sulfur	S	16	2	8	6				S^{2-}	2	8	8			Argon
family																
VIIA	Fluorine	F	9	2	7					F^-	2	8				Neon
Halogens	Chlorine	Cl	17	2	8	7				Cl^-	2	8	8			Argon
	Bromine	Br	35	2	8	18	7			Br^-	2	8	18	8		Krypton
	Iodine	I	53	2	8	18	18	7		I^-	2	8	18	18	8	Xenon
0	Helium	He	2	2						—						
Noble	Neon	Ne	10	2	8					—						
gases	Argon	Ar	18	2	8	8				—						
	Krypton	Kr	36	2	8	18	8			—						
	Xenon	Xe	54	2	8	18	18	8		—						
	Radon	Rn	86	2	8	18	32	18	8	—						

Notice in Table 3.1 that the ions derived from elements within the same family of the periodic table have the same electrical charges. All Group IA metals, for example, are like sodium; their ions bear charges of 1+. Those in Group IIA form ions with charges of 2+. The atoms of Group IIA metals must lose *two* electrons to achieve noble gas configurations, not *one* and not *three* electrons, but exactly two. When we continue this kind of comparison between the electron configurations of atoms and ions, family by family, we can see why the reactive elements form the monatomic ions with the charges they have (Table 3.2). The table shows how many electrons must be lost or gained by the elements of a particular family to give ions having noble gas configurations. Groups IVA and VA are not included in Table 3.2 because the non-metals of these groups do not readily form ions (as we'll study shortly). Let's continue our study of the octet rule by working some additional examples that apply it.

EXAMPLE 3.1   Using the Octet Rule to Predict the Charge on an Ion

When nutritionists speak of the calcium requirement of the body, they always mean the calcium *ion* requirement. Calcium has atomic number 20. What charge does the calcium ion have, and what is the symbol of this ion?

ANALYSIS 1 Two methods are available to solve this kind of problem, and you should learn both. The first is to work with the electron configuration of the atom, and the second is to use the periodic table. Using the rules learned in Chapter 2, we write the electron configuration of element 20, calcium, as follows.

number of electrons	2	8	8	2
level	1	2	3	4

This gives us what we need to know; the calcium atom has two electrons in the outside level (level 4) and eight in the next inner level (level 3). Only by losing *both* of the outer electrons can this atom get a new outside level with an octet. Losing one electron won't do. Neither will losing three or more. The only stable ion that calcium can form, therefore, is one with the following electron configuration.

Number of electrons	2	8	8
Level	1	2	3

SOLUTION By losing two electrons, the net charge on the particle becomes 2+, so the symbol for the calcium ion is Ca^{2+}

ANALYSIS 2 The second way to arrive at the answer is to find calcium in the periodic table. It's in Group IIA, so it's a representative element and the relationship between the group number and the ionic charge can be used. All of the Group IIA elements have two outside-level electrons, and all can acquire configurations of the nearest noble gases by losing these two. Thus, the Group IIA ions all bear charges of 2+: Be^{2+}, Mg^{2+}, Ca^{2+}, Sr^{2+}, Ba^{2+}, and Ra^{2+}.

EXAMPLE 3.2 Using the Octet Rule to Predict the Charge on an Ion

Oxygen can exist as the oxide ion in such substances as calcium oxide, an ingredient in cement. What is the symbol for the oxide ion, including its electrical charge?

ANALYSIS 1 As in Example 3.1, we can solve this problem in two ways. The first is by using the electron configuration of an oxygen atom (atomic number 8), which is

Number of electrons	2	6
Level	1	2

The outside level of the oxygen atom has six electrons, just two short of an octet and a neon configuration. We could also say, of course, that the oxygen atom has six too many electrons to have the helium configuration of two electrons in level 1 as the outside level. However, gaining two electrons to become like neon is much easier than losing six, so oxygen achieves a noble gas configuration *most directly* by accepting two electrons from some metal atom donor. The configuration of a stable *ion* of oxygen, therefore, is

Number of electrons	2	8
Level	1	2

SOLUTION The two extra electrons make the ionic charge 2−, so the symbol for the oxide ion is O^{2-}.

■ Remember that our procedures for figuring out likely charges on ions work only for the representative elements (the A series elements).

ANALYSIS 2 Using the Periodic Table to figure out a reasonable charge on the oxide ion, we see that oxygen is in Group VIA and is a representative element. Hence we know that oxygen's outside level has six electrons. It must pick up two electrons—not just one and not more than two—to have a noble gas configuration. These two extra electrons give the particle a charge of 2−, so we can write O^{2-} directly.

Nonmetal Elements in Groups IVA and VA Almost Never Occur as Ions. Hardly any element is involved in more compounds than carbon (Group IVA). More than six million carbon compounds are known. Yet carbon atoms very rarely occur in compounds as carbon *ions*. Why not?

Carbon is in Group IVA so it is a representative element whose atoms have four outside-level electrons. To achieve a noble gas configuration, a carbon atom must either lose these four (and become like helium) or gain four (and become like neon). Neither option is exercised because *both are too costly in energy terms*. To *gain* four electrons, the fourth must come to a particle already with three extra electrons and so with a charge of 3−, a strongly repelling charge for the fourth electron. To *lose* four electrons, the fourth must pull away from a particle already short of three electrons and so with a strongly attracting positive charge of 3+. As we said, either of these options costs too much energy and so they are avoided. Carbon only rarely is found as a monatomic ion (and nowhere in our future study).

■ In rare situations, carbon exists as the C^{4-} ion, called the methide ion.

It is equally rare for ions of Group VA elements to form because their atoms either would have to acquire three electrons or lose five to achieve outer octets, changes that are also too costly in energy terms. We can therefore state a corollary to the octet rule: *nonmetal atoms in Groups IVA and VA do not form ions.* The *metals* in these groups are able to form ions, but we need not be concerned about them.

■ Tin and lead, metals in Group IVA, form ions with charges of 2+ and 4+.

■ **PRACTICE EXERCISE 1** Write the electron configuration of an atom of each of the following elements, and from it deduce the charge on the corresponding ion. If the atom isn't expected to have a corresponding ion, state so. The numbers in parentheses are atomic numbers. Do not use the periodic table for this exercise.

(a) potassium (19)   (b) sulfur (16)   (c) silicon (14)

■ **PRACTICE EXERCISE 2** Write the electron configurations of the *ions* of the elements in Practice Exercise 1 that can form ions.

■ **PRACTICE EXERCISE 3** Relying on their locations in the periodic table, write the symbols of the ions of each of the following elements. Always remember that no symbol of an ion is complete without its electrical charge. (The numbers in parentheses are atomic numbers.)

(a) cesium (55)   (b) fluorine (9)   (c) phosphorus (15)   (d) strontium (38)

Electron Transfer Reactions Are Examples of Redox Reactions. Because electron transfers are such important chemical events, they have a special vocabulary. When an atom loses an electron, we call the change an **oxidation.** When an atom gains an electron, we call it a **reduction.** An electron lost from one atom cannot simply disappear. It has to have a place to go, an electron-acceptor atom. Electrons always transfer from something to something. It takes two substances for an electron transfer reaction.

Reactions that involve electron transfers are **redox reactions,** a contraction of reduction–oxidation reaction. In every such reaction, one substance is the **oxidizing agent,** because it causes the oxidation of the other substance by accepting electrons from it. Chlorine is the oxidizing agent in the reaction of sodium and chlorine. By accepting electrons from sodium atoms, it oxidizes them to sodium ions, Na^+.

Another species in a redox reaction is called the **reducing agent,** because it causes the reduction of something by giving it electrons. Sodium is the reducing agent in our reaction. It reduces chlorine by giving electrons to chlorine atoms.

In the reaction of sodium with chlorine we can assign these new terms as follows.

■ Chlorine bleaches such as Clorox work by oxidizing dyes to uncolored products.

Oxidizing agent: chlorine   Substance oxidized: sodium
Reducing agent: sodium   Substance reduced: chlorine

The oxidizing agent is always itself reduced; the reducing agent is always itself oxidized.

Because oxidation and reduction events involve the transfer of electrons and the formation of ions, the charges borne by monatomic ions are often called their **oxidation numbers.** Somewhat perversely, however, the *charge* on an ion is written with the digit first, like 1+ or 2−, but an oxidation number is written in the reverse order. Thus, the oxidation number of Na in NaCl is +1 and that of Cl is −1. Except for this, the charge on a monatomic ion and the oxidation number are the same. The most common oxidation numbers for the represented elements are shown in the table in the margin.

Group Number	Oxidation Number
IA	+1
IIA	+2
IIIA	+3
VIA	−2
VIIA	−1
0	0

3.3   NAMES AND FORMULAS OF IONIC COMPOUNDS

The ions in an ionic compound must assemble in whatever ratio lets the compound be electrically neutral.

Shorthand symbols for compounds are called **chemical formulas,** and they are made from the symbols of their elements in a pattern that shows the ratios *by atoms* present in the compound.

Because compounds are electrically neutral, the formula of an ionic compound must have its ions in a ratio that balances plus and minus charges. The formula of sodium chloride, as we noted earlier, must show its Na^+ and Cl^- ions in a 1:1 ratio to ensure neutrality, because then each 1+ charge is balanced by a 1− charge.

The formula of an ionic compound is made up of the symbols of its ions, but the electrical charges are omitted. They are "understood." The formula of sodium chloride is thus written as NaCl, not as Na^+Cl^-. By convention, the positive ion is placed first. The formula Na_2Cl_2 also has a 1:1 ion ratio but, by convention, the ratio is nearly always expressed by the *smallest* whole numbers.

The formula of a compound formed by calcium ions, Ca^{2+}, and oxide ions, O^{2-}, must also have the ions in the ratio 1:1. Only this ratio balances the 2+ of each calcium ion with the 2− of each oxide ion. The formula is CaO.

■ The *net* charge on any compound is zero.

■ A formula that uses the *smallest* whole numbers to give the ratio of atoms is called an **empirical formula.**

Subscripts Are Used in Formulas When Atom Ratios Are Not 1:1. A compound made of calcium ions and chloride ions must have *two* Cl^- ions to one Ca^{2+} ion, because two minus charges are needed for every particle with a 2+ charge. To show this ratio, the formula includes numbers called **subscripts.** These follow and are placed half a line below their associated atomic symbols. The subscript 1 is never used because just writing a symbol means that you are taking at least one of it. Remember that the positive ion goes first, so the formula of the compound of the calcium ion and the chloride ion is $CaCl_2$ (not Cl_2Ca nor $Ca^{2+}Cl^-_2$ nor Ca_2Cl_4).

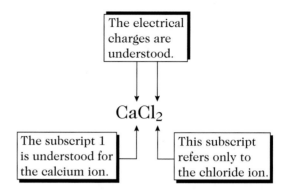

The electrical charges are understood.

$$CaCl_2$$

The subscript 1 is understood for the calcium ion.

This subscript refers only to the chloride ion.

Ionic Compounds Are Named after Their Ions.   To name an ionic compound, we assemble the names of the ions in the order in which they appear in the formula (except the word *ion* is omitted). For example,

■ Notice again that the charges on the ions are omitted in the formula.

NaCl	sodium chloride (sodium *ion* plus chloride *ion*)
CaO	calcium oxide (calcium *ion* plus oxide *ion*)
$CaCl_2$	calcium chloride (calcium *ion* plus two chloride *ions*)

Four metal ions at the bottom of Table 3.1 are given two names, for example, copper(I) ion and cuprous ion. The first name is the modern name, but the second is still seen. Both should be learned. The Roman numeral in "copper(I)" and "iron(III)" stands for the positive charge on the ion, Cu^+ and Fe^{3+}. [Notice there is no space between "copper" and "(I)" in "copper(I)."] In the common names of the ions of copper and iron, the *-ous* ending goes with the ion of lower charge and the *-ic* ending is for the ion of higher charge.

There are two kinds of operations involving names and formulas that you must learn: how to write formulas from names and how to write names from formulas. We limit the next examples and exercises to those involving the ions in Table 3.1.

EXAMPLE 3.3   Writing Formulas from Names of Ionic Compounds

Write the formula of aluminum oxide.

ANALYSIS   The name of the compound suggests two facts. First, the compound is made of a metal and a nonmetal, so it is almost certainly *ionic*. The rules for ionic compounds therefore apply. Second, the name of the compound also gives us the names of the ions. They must be the aluminum ion and the oxide ion. Knowing this, we must next be able to recall their electrical charges, Al^{3+} and O^{2-}. These charges pose a small (but easy) problem; we can't simply write AlO directly as the formula because 3+ isn't canceled by 2−. (The net charge on AlO would be 1+, and compounds *must* be neutral.) We can, however, use the lowest common multiple of 2 and 3, namely, 6 ($2 \times 3 = 6$), to figure out the correct ratio of ions. To balance the charges, we pick the smallest number of aluminum ions that give a total charge of 6+ and the smallest number of oxide ions that give a total charge of 6−. We need 2 Al^{3+}, because 2 (×)(3+) = 6+, and we must use 3 O^{2-}, because 3 (×) (2−) = 6−. The ratio, must therefore, be 2 Al^{3+} to 3 O^{2-}.

SOLUTION Writing Al first, with a subscript of 2, and O next, with a subscript of 3, the formula is Al_2O_3 (the answer).

The strategy in Example 3.3 to find and use the lowest common multiple of the ionic charges always works. Try Practice Exercise 4 to develop experience.

■ PRACTICE EXERCISE 4   Write the formulas of the following compounds.

(a) silver bromide (a light-sensitive chemical used in photographic film)

(b) sodium oxide (a very caustic substance that changes to lye in water)

(c) ferric oxide (the chief component in iron rust)

(d) copper(II) chloride (an ingredient in some laundry-marking inks)

EXAMPLE 3.4   Writing Names from Formulas of Ionic Compounds

Write the name of $FeCl_3$, using both the modern and the older forms.

ANALYSIS   Because Fe is the symbol of iron and Cl is that of chlorine, the formula tells us that the substance is made of a metal and a nonmetal. We therefore use the rules for naming ionic compounds. The symbol Fe stands for one of the two ions of iron, however, so we must figure out which one, Fe^{2+} or Fe^{3+}. Here we can use our knowledge of the charge on one ion, Cl^-, to figure out the charge on the other. The Cl_3 part of $FeCl_3$ tells us there must be a total negative charge of 3−, because 3 × (1−) = 3−. The Fe part of $FeCl_3$ has to neutralize this, so we must be dealing with the Fe^{3+} ion, named either the iron(III) ion or the ferric ion.

SOLUTION   The modern name of $FeCl_3$ is iron(III) chloride; the older name is ferric chloride.

■ Notice that there is no space between "iron" and "(III)" in iron(III).

■ PRACTICE EXERCISE 5   Write the names of each of the following compounds. When a name can be given in either a modern form or an older form, write both.

(a) CuS   (b) NaF   (c) FeI_2 (d) $ZnBr_2$ (e) Cu_2O

The technique used in Example 3.4 to find the electrical charge of an unfamiliar ion from the known charge of another in an ionic compound greatly reduces the number of ions and their charges that have to be memorized. Practice Exercise 6 provides further examples. It involves unfamiliar ions not listed in Table 3.1.

■   PRACTICE EXERCISE 6   What are the charges on the *metal* ions in each of these substances?

(a) Cr_2O_3 (a green pigment in stained glass)

(b) HgS ("Chinese red," a bright, scarlet-red pigment)

(c) $CoCl_2$ (an ingredient in invisible ink)

A Small Particle Made from the Atoms Shown in a Formula Is Called a Formula Unit.   It is not possible to have a sample of NaCl that we could manipulate in the lab that consists of just one sodium ion and one chloride ion. Yet it is still useful to have a name for such a particle. We call it a *formula unit,* and we define this term so that it can apply to any kind of substance—an element, an ionic compound, or a molecular compound. A **formula unit** is the very small particle made only of the kinds and numbers of atoms specified by the formula, regardless of how the atoms are held together. One formula unit of NaCl thus consists of one sodium ion and one chloride ion. One formula unit of sodium metal, Na, consists of one atom of sodium. One formula unit of water, H_2O, consists of one molecule of water.

■   Clinical chemists can measure the concentrations of several ions in the blood, and changes in concentrations can signify particular health problems.

Ionic Compounds Are Electrolytes.   People in professional health care fields speak of the "electrolyte balance" of this or that body fluid, where **electrolyte** refers to any substance that can furnish ions in a solution in water. Every fluid in every living thing contains dissolved ions. Even among the large molecules in living organisms, molecules of proteins, for example, there generally are several electrically charged sites. Thus, at the molecular level of life, electrolytes and ions are everywhere. As we will see later, the term *electrolyte* derives from the ability of solutions of electrolytes to conduct electricity.

3.4   MOLECULES AND MOLECULAR COMPOUNDS

In molecules, pairs of atoms that could not become ions by electron transfer share pairs of electrons in *covalent bonds.*

The widespread occurrence of carbon in nature's compounds means that nature is able to form other kinds of bonds than the ionic bond. Such other bonds usually occur among compounds made entirely of nonmetals, compounds that consist of *molecules,* not ions. Molecules are the *formula units* of molecular compounds. Some elements are also molecular, as we will see next.

■   The definitions of *molecule* and *molecular compound* are on page 52.

■   The three fundamental *chemical* particles are the atom, the ion, and the molecule.

Electron Density Becomes Concentrated between Two Nuclei in the Covalent Bond.   Unlike the noble gas elements (Group 0), the halogens (Group VIIA) consist of diatomic molecules with the formulas F_2, Cl_2, Br_2, and I_2. The other diatomic elements are oxygen, O_2, hydrogen, H_2, and nitrogen, N_2. The diatomic particles of these elements cannot be explained in terms of ions, because we would need *positively* charged as well as negatively charged ions. No nonmetal atom forms a positively charged ion. If, for example, a fluorine atom (Group VIIA) with seven outer electrons became F^+, it would have six outside-level electrons, not an outer octet. Fluorine, therefore, cannot form F^+ that could attract F^- and so form an ionic F_2 formula unit. Yet we know that to fashion any kind of chemical bond we must have an attraction between opposite charges. So if we can't use F^+ and F^- to make F_2, there must be another way to induce an attraction between two F atoms.

The source of such an attraction was first proposed by an American chemist, G. N. Lewis (1875–1946). He suggested that a pair of outer-level electrons, one from each atom, can be *shared* between two nuclei. Being "shared " means that *the pair spends most of its time between the nuclei.* The pair thus gives to this region a relatively high density of negative charge, a high **electron density.** The nuclei are naturally attracted to this region, because unlike charges attract. The electrical force of attraction created by sharing a pair of electrons is called a **covalent bond.** As each participating atom gives up an electron for the shared pair, the two initially *separated* atoms lose their individual identities, their nuclei remain permanently near each other, and a *molecule* results. The covalent bond made by sharing *one* pair of electrons is also called a **single bond.**

■ *Co-* from *cooperative; -valent,* from the Latin *valere,* "to be strong," signifying strong binding.

There are two ways to depict the sharing of a pair of electrons between chlorine atoms in a molecule of Cl_2. Structure **1** is an *electron-dot* structure in which all inner-level electrons are stripped away and left "understood." Only the outside-level electrons are shown, and they are grouped in pairs.

$$:\ddot{\underset{..}{Cl}}:\ddot{\underset{..}{Cl}}:\qquad Cl-Cl$$
$$12$$

Structure **2** shows only the shared pair, now represented simply by a single line. The remaining outside-level electrons are "understood." The single line is a particularly useful symbol for a covalent bond because it so easily lets us describe the connections within molecules having many bonds.

The Octet Rule Applies to Covalent Bonds.   If we count both electrons of a shared pair for each atom, the octet rule can be used for molecules. We can see this in the chlorine molecule, for example.

The circles enclose octets
when the shared pair is
counted for each atom.

The hydrogen molecule, **3** or **4,** illustrates how a shared pair or electrons gives each hydrogen atom the helium configuration.

$$H \mathbin{:} H\qquad H-H$$
$$34$$

Although there is no outer octet in H_2, each H atom has one of the two noble gas configurations, namely, that of helium. For a discussion in greater depth of the covalent bond, based on atomic orbitals that merge to form molecular orbitals, see Special Topic 3.1.

Structural Formulas Show the Patterns of Connections in Molecules.   Because molecules made of several atoms would require us to write a large number of dots, chemists seldom use this symbolism. Instead, all shared pairs are represented as straight lines. The resulting symbol is called a **constitutional formula,** or a **structural formula,** or, simply, a **structure.** A structure depicts an atom-to-atom sequence in a molecule (Figure 3.2).

SPECIAL TOPIC 3.1
MOLECULAR ORBITALS

In Special Topic 2.1 we learned about atomic orbitals, the homes of individual electrons in atoms. Here we'll learn about the homes of the shared electron pairs of covalent bonds in molecules, *molecular orbitals*.

Suppose we have two atoms approaching each other on a collision course. Each atom has an atomic orbital with just one electron in it. In this circumstance as the two orbitals hit, they partly merge to form a new kind of orbital, a **molecular orbital**, which encloses the nuclei of both atoms. The two electrons, one from each individual atom's merging orbital, now move within this space. The two electrons now experience an attraction toward two nuclei. This attraction toward two nuclei is what we mean by the *sharing* of a pair of electrons between the covalently bonded atoms.

The shared electrons are naturally between the two nuclei most of the time, so this is where the electron density is the greatest. This electron density is what attracts and holds the nuclei and creates the bond.

Figure 1 illustrates how the molecular orbital of a hydrogen molecule forms. Two $1s$ atomic orbitals, each with an electron, partially merge to give a space within which the two hydrogen nuclei now are found. This partial merging is sometimes called the *overlapping* of atomic orbitals.

Figure 2 shows how two p orbitals can overlap to create the molecular orbital for the covalent bond in the fluorine molecule.

In other situations (not illustrated), an s orbital can overlap with one lobe of a p orbital.

In double bonds, two pairs of atomic orbitals overlap. And in triple bonds, three pairs of orbitals overlap.

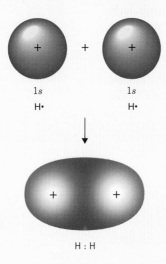

The molecular orbital in the hydrogen molecule is created by the partial overlapping of the $1s$ orbitals of two hydrogen atoms. The shared pair of electrons spends most of the time between two nuclei, so these nuclei are attracted toward each other.

FIGURE 1

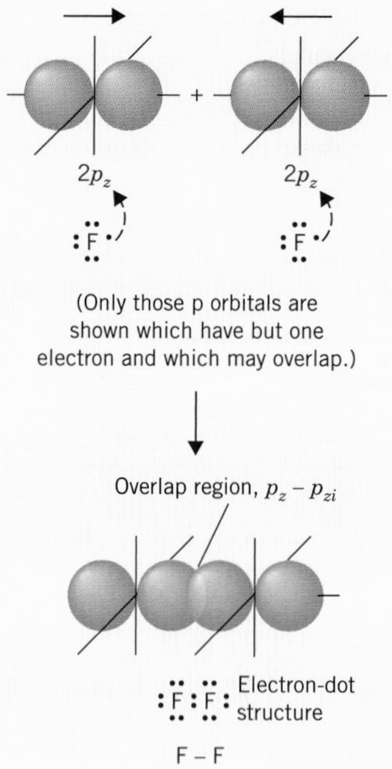

FIGURE 2

Every structural formula has a **molecular formula,** but a molecular formula gives only the *composition* of one molecule and no information about the internal sequences of atoms. When a molecule has three or more atoms, more than one structure can usually be written, so a structural formula gives more information than its molecular formula. The molecular formula of water, for example, is H_2O, and with no additional instructions we might imagine that the structure is either H—H—O or

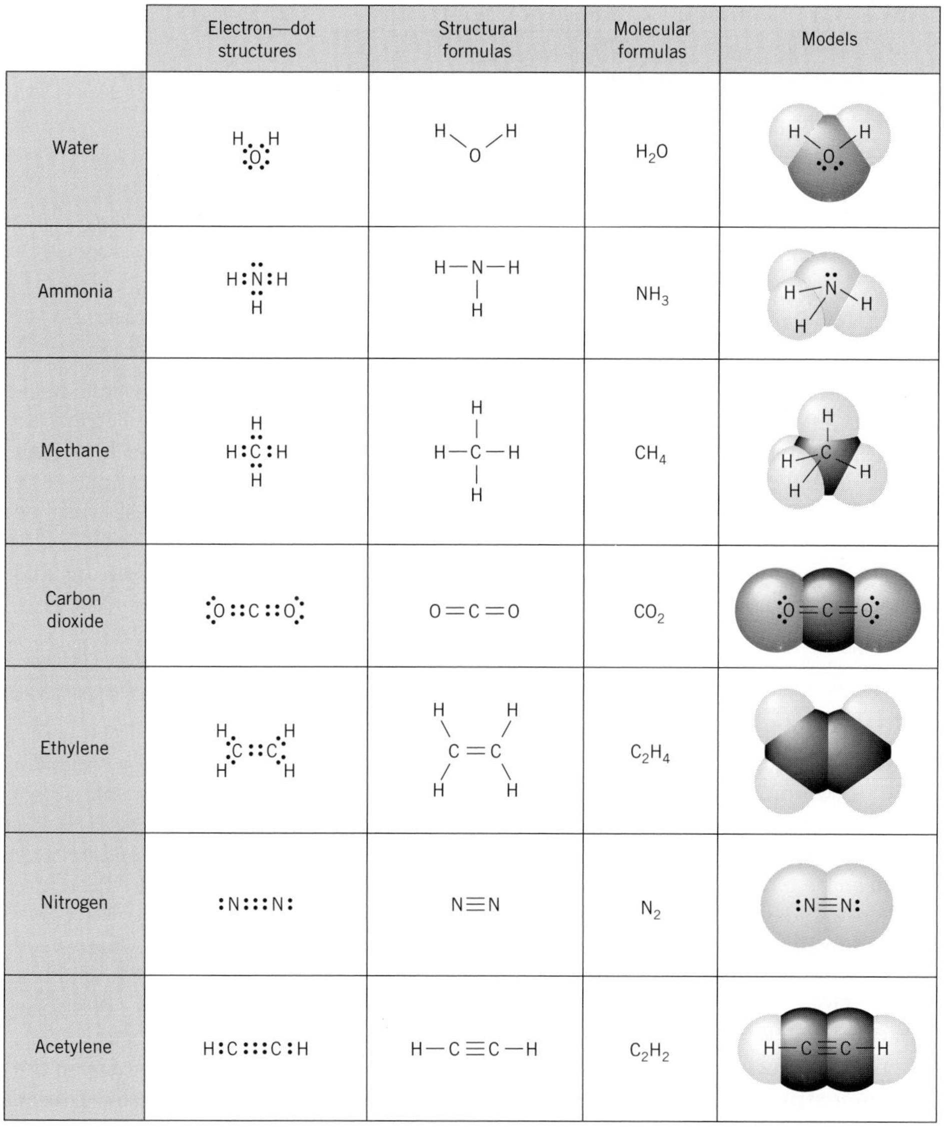

	Electron—dot structures	Structural formulas	Molecular formulas	Models
Water	H $\cdot\cdot$ H $\overset{\cdot\cdot}{\underset{\cdot\cdot}{O}}$	H $\diagdown$ O $\diagup$ H	H_2O	
Ammonia	H $\colon$ N $\colon$ H H	H — N — H $\vert$ H	NH_3	
Methane	H H $\colon$ C $\colon$ H H	H $\vert$ H — C — H $\vert$ H	CH_4	
Carbon dioxide	$\colon\colon$ O $\colon\colon$ C $\colon\colon$ O $\colon\colon$	O $=$ C $=$ O	CO_2	
Ethylene	H H C $\colon\colon$ C H H	H H $\diagdown$ $\diagup$ C $=$ C $\diagup$ $\diagdown$ H H	C_2H_4	
Nitrogen	$\colon$ N $\colon\colon\colon$ N $\colon$	N $\equiv$ N	N_2	$\colon$ N $\equiv$ N $\colon$
Acetylene	H $\colon$ C $\colon\colon\colon$ C $\colon$ H	H — C $\equiv$ C — H	C_2H_2	H — C $\equiv$ C — H

FIGURE 3.2
Common molecular compounds in which single, double, and triple bonds occur. The last column displays space-filling molecular models in which the relative volumes of the atoms of the molecules are shown.

H—O—H. The correct structure is H—O—H and, as we'll learn soon, the octet rule helps us reject H—H—O.

Two or Three Pairs of Electrons Can Be Shared. Many molecules have adjacent atoms sharing two or three pairs of electrons. When two pairs are shared, we have a **double bond,** and with three shared pairs, we have a **triple bond.** As seen in Figure 3.2, carbon dioxide has two double bonds, ethylene has one double bond, and nitrogen and acetylene each have one triple bond. In all cases, as the electron-dot structures show, the octet rule is obeyed.

The Number of Covalent Bonds That an Atom Can Have in a Molecule Is Called Its *Covalence.* As Figure 3.2 shows, atoms of nonmetals differ in their electron-sharing abilities. Oxygen can form two bonds; nitrogen can form three; and carbon, four.

■ Double bonds between carbon atoms occur often among the compounds at the molecular level of life.

TABLE 3.3   **Common Covalences of Nonmetals**

Periodic Table Group Number									
IA		**IVA**		**VA**		**VIA**		**VIIA**	
H	1	C	4	N	3	O	2	F	1
		Si	4	P	3	S	2	Cl	1
								Br	1
								I	1

■ To match the term *covalence*, the term *electrovalence* is sometimes used for the charge on a monatomic ion.

■ Atoms of nonmetals in Periods (rows) 3 and higher in the periodic table have outer levels capable of holding more than eight electrons. Hence, in some of their compounds they have more covalent bonds than predicted by the octet rule.

■ Each C in acetylene has one triple bond and one single bond.

H—C≡C—H

Acetylene

Halogen atoms and hydrogen atoms can form one covalent bond. The covalent combining ability of an element, called its **covalence**, equals the number of electrons its atoms must obtain by sharing to have a noble gas configuration. Thus, a hydrogen atom must have a share of one more electron to achieve a helium configuration, so its covalence is 1. Several common covalences are given in Table 3.3. These must be learned.

The Group VIIA elements (the halogens) all have seven outer-level electrons and all need a share in one more to have outer octets. Thus, their covalences are 1 also. Notice how the following equation gives the covalence of a nonmetal.

$$\text{Covalence number} = 8 \, (-) \, \text{group number}$$

Thus with the halogens, the covalence is 8 − 7, or 1.

Group VIA elements (oxygen family) have six outer-level electrons. They need a share in two more for octets, so they have covalences of 2. This also fits the above equation because 8 − 6 is 2, 6 being the group number.

Group VA elements (nitrogen family) have five outer-level electrons and so they need a share in three more electrons (8 − 5) for octets. Group VA elements therefore have covalences of 3, as in the ammonia molecule, NH_3 (see Figure 3.2).

Group IVA elements (carbon family) have four outer-level electrons. They need four more electrons (by sharing) to achieve octets, so their covalences are 4, which is 8 − 4. Thus, the methane molecule (CH_4) has four bonds from carbon, all single. Carbon also has four bonds (two double bonds) in carbon dioxide (see Figure 3.2).

Think of the covalence as the number of lines (bonds) that must radiate from the atom in a structure. Thus, a nitrogen atom, with a covalence of 3, must have three lines. They can be three single bonds, a single plus a double bond, or one triple bond, as long as there are three. Examine Figure 3.2 for additional examples. We can now see why we cannot accept H—H—O as the structure for water, H_2O. One H atom has two lines (bonds) from it (one too many), and the O atom has only one line (one bond too few). But the structure H—O—H satisfies the covalences.

■ PRACTICE EXERCISE 7   Convert the following molecular formulas into structures using the covalences listed Table 3.3. (*Hint:* Draw in the bonds on the atom of highest covalence first and use those of covalence 1 last.)

(a) CCl_4 (b) H_2S (c) NCl_3 (d) $CHBr_3$

The Atoms in Polyatomic Ions Are Held Together by Covalent Bonds.   We didn't mention earlier that several ions, the **polyatomic ions**, consist of many atoms because we needed to know about the covalent bond to explain them. These ions are clusters of atomic nuclei and electrons held together by covalent bonds, but the total plus charge on all nuclei is not perfectly balanced by the total number of electrons. There are either one, two, or three too many electrons, or (rarely) one or two too few. A simple example is the hydroxide ion OH^-:

$$\left[:\overset{..}{\underset{..}{O}} : H \right]^-$$

5

In **5**, oxygen has an octet, and hydrogen has the helium configuration, so the conditions of stability are met. The OH^- ion, however, has 10 electrons (2 in oxygen's inner level plus 2 shared electrons plus 6 more shown as dots). To balance a total charge of $10-$, OH^- has only 9 protons (8 in the oxygen nucleus and 1 in the hydrogen nucleus). The net charge is therefore $1-$.

Other polyatomic ions can be analyzed this way, and important examples are shown in Table 3.4. Their names, formulas, and charges must now be learned. Many are involved as electrolytes in body fluids.

Parentheses Sometimes Enclose Polyatomic Ions in Formulas. In ammonium sulfide the ratio of ions is two NH_4^+ ions to one S^{2-} ion. To show this ratio without getting subscripts confused, we place parentheses as follows.

$$(NH_4)_2S$$
Ammonium sulfide

TABLE 3.4 Important Polyatomic Ions

Name	Formula
Ammonium ion	NH_4^+
Hydronium ion[a]	H_3O^+
Hydroxide ion	OH^-
Acetate ion[b]	$CH_3CO_2^-$
Carbonate ion	CO_3^{2-}
Bicarbonate ion[c]	HCO_3^-
Sulfate ion	SO_4^{2-}
Hydrogen sulfate ion[d]	HSO_4^-
Phosphate ion	PO_4^{3-}
Monohydrogen phosphate ion	HPO_4^{2-}
Dihydrogen phosphate ion	$H_2PO_4^-$
Nitrate ion	NO_3^-
Nitrite ion	NO_2^-
Hydrogen sulfite ion[e]	HSO_3^-
Sulfite ion	SO_3^{2-}
Cyanide ion	CN^-
Permanganate ion	MnO_4^-
Chromate ion	CrO_4^{2-}
Dichromate ion	$Cr_2O_7^{2-}$

[a]This ion is known only in a water solution.

[b]The formula of the acetate ion is commonly also written as $C_2H_3O_2^-$.

[c]Formal name: hydrogen carbonate ion.

[d]Common name: bisulfate ion.

[e]Common name: bisulfite ion.

■ Such polyatomic ions as HCO_3^-, SO_4^{2-}, $H_2PO_4^-$, and HPO_4^{2-} occur in body fluids.

The 4 goes with H to show a ratio of 4 H to 1 N *within* the ammonium ion, but the 2 is outside the parenthesis to show that it goes with the entire ammonium ion. Thus in $(NH_4)_2S$ there are two NH_4^+ ions giving a total of 2 N and 8 H.

■ **PRACTICE EXERCISE 8**   Using Tables 3.1 and 3.4, write formulas of the following compounds. (Eventually, you must be able to do this without referring to the tables.)

(a) sodium nitrate   (b) potassium hydroxide   (c) calcium hydroxide

(d) magnesium carbonate   (e) sodium sulfate   (f) ammonium phosphate

---

■ **PRACTICE EXERCISE 9**   Write the names of the following compounds.

(a) Li_2CO_3 (b) $NaHCO_3$ (c) $KMnO_4$ (d) NaH_2PO_4 (e) $(NH_4)_2HPO_4$

Rock candy consists of large crystals of sugar.

3.5 POLAR MOLECULES

Even electrically neutral molecules can attract each other if they are polar.

If molecules are neutral, how can they adhere to each other? Sugar molecules, for example, stick to each other in an orderly manner to make beautiful crystals. What holds the molecules together? The answer is deeply relevant to explaining the strengths of muscle fibers and nylon threads, the resilience of skin, and even the molecular basis for the action of hormones.

Shared Pairs between Unlike Atoms Are Usually Not Equally Shared.   The space between two nuclei occupied by shared electrons is a region with a high density of negative charge. The space is sometimes said to have an **electron cloud,** because we can imagine the electrons swarming like a cloud of bugs. Some clouds, as you know, are thicker than others, and clouds are mobile. Electron clouds are special, however. Unlike charges attract, and electron clouds are always in the vicinity of oppositely charged atomic nuclei. These influence the shape, thickness, and location of any nearby electron cloud.

In a *symmetrical* molecule, like H—H, the electron cloud is not distorted because it is equally attracted to the two identical nuclei at the ends of the bond. In an *unsymmetrical* molecule, like H—F, however, the electron cloud of the bonding pair is distorted; the F end of the molecule has a positive charge (9+) that is nine times that of the H end of the molecule. To be sure, the F atom has many nonbonding electrons, but they are not positioned in the H—F molecule to shield entirely the bonding electrons from the 9+ charge of the F nucleus. Thus in H—F, the electron cloud of the bond is pulled somewhat away from the H end toward the F end. This leaves the electron cloud by H thinner than it is at either end of the bond in H—H. With a thinner electron cloud, the H end of H—F has a fraction of the plus charge of the H nucleus not neutralized. A fractional charge is called a *partial charge,* symbolized as $\delta+$.

A partial negative charge, $\delta-$, occurs at the F end of H—F, where excess electron density has been drawn. The two partial charges in H—F are opposite in sign but equal in size, so their algebraic sum is zero; the molecule is neutral. Because the H—F molecule has opposite partial charges, it is said to have an **electrical dipole.**

A covalent bond with $\delta+$ at one end and $\delta-$ at the other is called a **polar bond.**

■ The atomic number of F is 9; that of H is 1.

■ The Greek lowercase delta, δ, is used for "partial"; thus, $\delta+$ is pronounced "delta plus."

■ Remember, molecules taken as a whole are neutral.

We can call attention to a bond's polarity in either of two ways, illustrated by structures **6** and **7** for hydrogen fluoride.

$$\begin{array}{cc} \delta+ \;\; \delta- & \longmapsto \\ \text{H}-\text{F} & \text{H}-\text{F} \\ \mathbf{6} & \mathbf{7} \end{array}$$

In **7**, the arrow points in the direction toward which the electron cloud is drawn and so toward the end of the bond that is richer in electron density.

When Bonded Atoms Have Different Electronegativities, the Bond Is Polar. The extent to which a bonding pair's electron cloud is distorted by nearby nuclei depends on how protected the nuclei are by their nonbonding electrons, particularly those of inner energy levels. When the two atoms held by a bond are from the *same* period, however, the distortion depends only on the nuclear charge (the atomic number). The nucleus with the *larger* positive charge pulls the electron cloud somewhat away from the other nucleus. The ability of an atom of an element to draw electron density toward itself in a covalent bond is called the **electronegativity** of the element. Each element has been assigned a numerical value of *relative electronegativity,* and those of a commonly used scale are shown in the margin for the elements of greatest importance to our study.

■ The two atoms being compared must be of the same period so that they have the same number of *inner-level* electrons shielding the nuclei from those in the outer level.

Fluorine has the highest electronegativity of all of the elements. Its large nuclear positive charge is shielded by the electrons of only levels 1 and 2. Oxygen, just to the left of fluorine in the periodic table, has the next highest electronegativity. Its nuclei have one less positive charge than fluorine while being shielded in the same way. Nitrogen, just to the left of oxygen, has the third highest electronegativity. The lowest values of relative electronegativity belong to the elements of Groups IA (0.9–1.0) and IIA (0.9–1.5). Such relatively low values "make sense" because these metals certainly have little tendency to *accept* electrons (or electron density). They tend rather to *donate* electrons and form positive ions. Thus, *ionic* compounds of metals and nonmetals, like NaCl or CaF_2, generally involve elements having the largest differences in their relative electronegativities.

■ Relative electronegativities
F 4.0
O 3.5
N 3.0
C 2.5
H 2.1

We won't need to know the numerical values of relative electronegativities, but you should at least learn the following *order;* it involves the nonmetals that we'll most often encounter.

$$O > N > C > H$$

The difference between C and H is very small, as the table in the margin shows.

■ This order of electronegativities should be learned.

It's easy to tell if a covalent *bond* is polar; it always is if the atoms it holds have different electronegativities. Symmetrical diatomic molecules, like H—H and F—F, cannot be polar, of course, but H—F, H—Cl, and H—I molecules are.

■ **PRACTICE EXERCISE 10** Examine the bonds in each structure. Write $\delta+$ and $\delta-$ signs at the correct ends of each bond.

    (a)   H   **(b)** H—F **(c)** H—O—O—H (hydrogen peroxide)
            |
     H—N—H

Whether a Polyatomic Molecule Has a Net Polarity Depends on Both Its Atoms and Its Geometry. Only with *diatomic* molecules can we be certain that when the bond is polar, so is the molecule. For others, like NH_3, H_2O, and CO_2, having polar bonds,

The octet theory doesn't explain why the water molecule has a bent and not a linear geometry.

$$--H—O—H--$$
180°

Linear molecule
(incorrect)

H—O—H
104.5°

Bent molecule
(correct)

The easiest way to understand the correct geometry is in terms of a simple theory with a very long name—**valence-shell electron-pair repulsion theory,** or **VSEPR** for short. *Valence shell* refers to the outer-level electrons of an atom. *Electron-pair* refers to the strong tendency for electrons in valence shells in molecules to occur in pairs, each pair making up one electron cloud. *Repulsion* describes the effect of each cloud on other clouds.

VSEPR theory says that the directions taken by bonds at a central atom (such as the O atom in H_2O, the N atom in NH_3, and the C atom in CH_4) are determined by the need for valence-shell electron-pair clouds to avoid each other as much as possible. Whether they are occupied by shared pairs or unshared pairs, the electron-pair clouds try to keep apart. A simple analogy involving balloons will help. If you have four balloons of the same shape and tie them together at a common point, the balloons spontaneously take up the positions shown in Figure 1. Now imagine that lines—they're called *axes*—project from the common point to the farthest points of the balloons' surfaces. These axes intersect at angles of

109.5°, and their ends are at the corners of a regular tetrahedron (Figure 2). Now imagine that each balloon represents an electron-pair cloud, and the common point is the nucleus of some central atom, like O, N, or C. The clouds, like the balloons, try to stay out of each other's way. In other words, the natural orientation for the axes of four electron-pair clouds is tetrahedral. They make angles of 109.5°.

When all four electron-pair clouds are involved in molecular orbitals—covalent bonds—to the central atom, as in CH_4, the bond angles are exactly tetrahedral, 109.5° (Figure 3a).

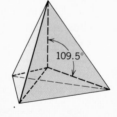

FIGURE 1
When four balloons are tied together, their axes do not lie in a plane but, instead, point to the corners of a regular tetrahedron (see Figure 2).

109.5°

FIGURE 2
A regular tetrahedron, a four sided figure bounded by identical equilateral triangles. Any two lines from corners to the center make an angle of 109.5°.

as all three do, does not guarantee a net polarity for the whole *molecule.* Whether larger molecules are polar in an overall sense depends not just on polar bonds but also on the geometry of the molecule. Given the appropriate geometry, *the polarities of individual bonds can cancel each other.* When all bond polarities cancel, the molecule as a whole is nonpolar. A **polar molecule** is one in which bond polarities do not cancel. Consider, for example, the carbon dioxide molecule, **8.**

$$O=C=O \qquad H—O$$
8 · · · · · · 9
· · · · · · H

Each carbon–oxygen double bond in carbon dioxide is polar, but the two dipoles point in exactly opposite directions. So they cancel each other's effects, leaving the molecule, as a whole, nonpolar. The water molecule, **9,** on the other hand, is angular. Therefore, its two individual O—H bond polarities, do not cancel, and the water molecule is polar.

When one electron pair of the central atom is unshared and three are in covalent bonds, as in NH_3, the cloud of the unshared pair, not being involved with another atomic nucleus, slightly pushes the other clouds away and squeezes the bond angles from 109.5° to 107.3° (Figure 3b).

When two electron pairs of the central atom are unshared and two are in bonds, as in H_2O, there is a slightly greater squeezing effect on the shared electron pairs in the bonds. Thus, the bond angle in water is forced from 109.5° to 104.5° (Figure 3c).

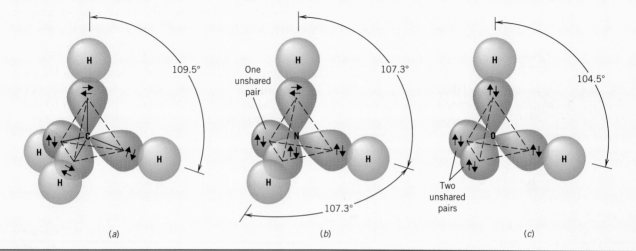

(a)

(b)

(c)

FIGURE 3
How VSEPR theory explains the geometries of (a) methane, CH_4, (b) ammonia, NH_3, and (c) water, H_2O.

When Molecules Are Polar, They Adhere to Each Other. We are now close to seeing how otherwise neutral molecules are able to adhere, sometimes strongly enough to make the substance a rigid solid, like a sugar crystal. A simple molecule with an electrical dipole, like H—F, is analogous to a magnet with a magnetic dipole, two poles labeled north and south. If you have ever played with toy magnets, you know that two magnets tend to stick to each other *if they are lined up properly,* meaning that unlike poles are nearest neighbors and like poles are farther apart. Electrically polar molecules adhere to each other just like magnets, because unlike charges attract (Figure 3.3).

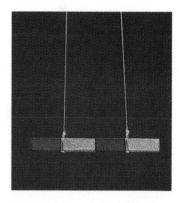

Magnets stick together when the north pole of one is touched by the south pole of the other.

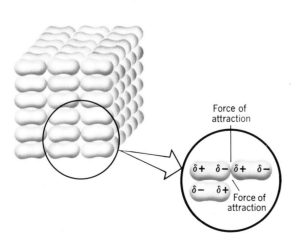

FIGURE 3.3
Polar molecules attract each other in a crystal of a molecular compound.

■ A center of electrical charge is like a center of gravity, a single point from which all charge (or mass, in the case of gravity) can be thought of as acting.

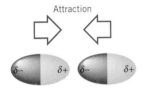

Attraction

Two polar molecules can attract each other.

Another useful way to think about the polarity of a molecule involves the idea of a *center of density of charge,* something like a "balance point" for electrical charge. The symmetry of the carbon dioxide molecule, **8**, tells us that all the positive charges on the three nuclei balance around the center of the carbon nucleus. Similarly, all of the negative charges contributed by all the electrons must also balance *around the identical point.* These two centers are thus in the same place, so the molecule is nonpolar. The angularity of the water molecule, **8**, tells us that the two centers of electrical charge cannot be at the same point, so the water molecule is polar.

Molecular Shapes Affect How Near Two Polar Molecules Can Approach Each Other.
How strongly polar molecules attract each other depends not only on the sizes of the partial charges but also on the shapes of the molecules. Some molecules that we'll study later have projecting nonpolar parts that tend to bury the sites of $\delta+$ and $\delta-$ charges. Even when these charges are large, the molecules might not be able to get close enough for forces of attraction to cause the molecules to adhere strongly. How molecules acquire and hold unique shapes is explained in Special Topic 3.2.

SUMMARY

Ionic Bonds and Ionic Compounds   A reaction between a metal and a nonmetal usually occurs by transfer of an electron from the metal atom to the nonmetal atom. The metal atom changes into a positively charged ion and the nonmetal atom becomes a negatively charged ion. The oppositely charged ions aggregate in whatever whole-number ratio ensures that the product is electrically neutral. The electrical force of attraction between the ions is called the ionic bond, and compounds made of ions are called ionic compounds. They are also electrolytes.

Octet Rule   Among the representative elements, a useful guide in predicting how many electrons transfer is that the resulting ions must have electron configurations of the nearest noble gases. Metal atoms lose one, two, or three outer-level electrons to achieve this, and nonmetal atoms gain enough new outer-level electrons to acquire such configurations. However, the nonmetal elements of Groups IVA and VA seldom become ions.

Formulas   The formulas of ionic compounds begin with the symbol of the metal ion (without the sign). Subscripts are used to give the ratios of the ions. When two or more polyatomic ions are in the formula, parentheses must enclose their symbols. The names of ionic compounds are based on the names of the ions (except that the word *ion* is omitted). The name of the positively charged ion precedes that of the other ion.

Redox Reactions   Electron transfers are called redox reactions. An atom that loses electrons is oxidized, and one that gains electrons is reduced. Anything that causes an oxidation is an oxidizing agent; a reducing agent is anything that causes a reduction. Metals tend to be reducing agents and nonmetals are oxidizing agents. Simple, monatomic ions have oxidation numbers that equal their electrical charges. Elements have oxidation numbers of zero.

Molecular Compounds   Atoms of nonmetals can form molecules by sharing pairs of outer-level electrons. Each shared pair constitutes one covalent bond, and each pair is counted as the joint property of both atoms when octets are checked. The shared pair creates a region of relatively high electron density between the two atoms toward which the nuclei are electrically attracted, and this attraction is called a covalent bond.

Covalence Numbers   The minimum number of covalent bonds a nonmetal atom can have equals the number of electrons it must get by sharing to achieve an octet. Sometimes two or three pairs of electrons are shared, which means that double or triple bonds occur. In polyatomic ions, the atoms are joined by covalent bonds, but the overall numbers of electrons and protons do not balance.

Polar Molecules   If individual bond polarities, caused by electronegativity differences between the joined atoms, do not cancel, the otherwise neutral molecule is polar and it can stick to adjacent molecules much as magnets can stick together.

REVIEW EXERCISES

The answers to Review Exercises whose numbers are in color are found in Appendix V. The answers to the other Review Exercises are found in the Study Guide that accompanies this text. The more challenging questions are marked with asterisks.

Octet Rule

3.1   What natural law concerning electrically charged particles is particularly important for understanding the nature of the chemical bond?

3.2   In general terms, what parts of two atoms become reorganized relative to each other when a chemical bond forms between the two atoms? Once the bond forms, can we say that the two *atoms* are still present? Explain.

3.3   What two fundamental kinds of compounds are recognized, based on the way that electrons and nuclei become reorganized?

3.4   What is the name of the fundamental particle that carries all of the chemical properties of a molecular compound?

3.5   The chemical properties of which family of elements is most closely associated with the octet rule? Explain.

3.6   State the octet rule.

3.7   Do the atoms of the elements in Groups VIA and VIIA tend to gain or lose electrons to form ions? Explain.

3.8   If M is the symbol of some representative metal, and the symbol of its ion is M^{3+}, in what group in the periodic table is M?

3.9   An atom of the representative nonmetal X can accept two electrons and become an ion. In what group in the periodic table is X?

3.10   An element is in Group IIA. What charges can its ions have?

3.11   Using only what can be deduced from electron configurations built from the atomic numbers, write electron configurations of each of the following *ions*. The atomic numbers are given in parentheses.
(a) calcium ion (20)   (b) aluminum ion (13)

3.12   In certain compounds, the hydrogen atom can exist as a negatively charged ion called the *hydride ion*. What is likely to be the electron configuration of this ion? Write a symbol for the ion.

Ions and Ionic Compounds

3.13   Assuming that sodium atoms interact with fluorine atoms in the same way they react with chlorine atoms, use changes in electron configurations to illustrate how Na and F react. Use electron configurations (including the compositions of nuclei) as we did in Section 3.2 for the reaction of sodium with chlorine to show how a sodium atom and a

fluorine atom can change to particles that can attract each other. What are the names of the particles that form?

3.14   In terms of their fundamental structures, what is the difference between a sodium atom and a sodium ion?

*3.15   Show how magnesium atoms and fluorine atoms can cooperate to form ions that will aggregate in the correct ratio. (Follow the directions given in Review Exercise 3.13.)

*3.16   Write diagrams to show how oxygen atoms and sodium atoms can cooperate to form ions that will aggregate in the correct ratio. (Follow the directions given in Review Exercise 3.13.)

Oxidation, Reduction, and Oxidation Numbers

3.17   If the number of electrons of a particle increases, has it been oxidized or reduced?

3.18   In a reaction in which an oxidation occurs, what other event must also happen?

3.19   What are the oxidation numbers of the following ions?
(a) potassium ion   (b) aluminum ion   (c) oxide ion
(d) copper(II) ion   (e) chloride ion   (f) mercury(II) ion

3.20   Determine the oxidation number of the metal component in each of the following compounds.
(a) $AuCl_3$ (b) $GaCl_3$ (c) Cr_2O_3
(d) PbF_4 (e) $ZrOCl_2$ (f) V_2O_5

3.21   What would be the formula of an oxide of manganese, Mn, if the oxidation number of Mn is +7? (The oxidation number of oxygen is always −2.)

3.22   Tungsten, W, has an oxidation number of +5 in one of its oxides. Oxygen always has an oxidation number of −2 in metal oxides. What is the formula of this tungsten oxide?

3.23   Some substances are described as *electrolytes*. What kinds of compounds are most likely to be electrolytes?

Names and Formulas Involving Monatomic Ions

3.24   It is common in studies of heart conditions to hear scientists speak of the *sodium level* of the blood. *Level* refers to the concentration, the ratio of substance to volume. To what specifically does *sodium* refer?

3.25   Write the symbols with the correct charges for the following ions.
(a) iodide ion   (b) sodium ion
(c) silver ion   (d) calcium ion
(e) zinc ion   (f) aluminum ion
(g) sulfide ion   (h) lithium ion
(i) bromide ion   (j) chloride ion

(k) potassium ion   (l) barium ion
(m) oxide ion   (n) cupric ion
(o) fluoride ion   (p) copper(I) ion
(q) ferric ion   (r) magnesium ion
(s) iron(II) ion

3.26 What are the names of the following ions? (Give both the older and the modern name for ions with both.)
(a) K^+ (b) S^{2-} (c) Al^{3+} (d) Fe^{2+} (e) Br^-
(f) Fe^{3+} (g) Ba^{2+} (h) Na^+ (i) Cl^- (j) Mg^{2+}
(k) Li^+ (l) Cu^+ (m) Zn^{2+} (n) Ca^{2+} (o) Ag^+
(p) O^{2-} (q) I^- (r) F^- (s) Cu^{2+}

3.27 Two ions of tin are commonly known, Sn^{4+} and Sn^{2+}. Which is the stannous ion and which is the stannic ion? (These common names stem from *stannum*, Latin for "tin.")

3.28 What would be the modern names for the following ions for which common names are given in parentheses?
(a) Pb^{2+} (plumbous ion)   (b) Au^{3+} (auric ion)
(c) Hg_2^{2+} (mercurous ion)   (d) Pb^{4+} (plumbic ion)

3.29 What are the formulas of the following compounds?
(a) cuprous sulfide   (b) barium oxide
(c) aluminum oxide   (d) calcium chloride
(e) ferrous bromide   (f) sodium iodide

3.30 Write the formulas of each of the following compounds.
(a) magnesium fluoride   (b) lithium oxide
(c) cupric sulfide   (d) ferric chloride
(e) sodium bromide   (f) calcium oxide

3.31 Write the names of the following compounds. Where two names are possible, one based on modern terminology and the other a common name, write both names.
(a) AgI   (b) $FeCl_2$ (c) Al_2O_3
(d) $BaCl_2$ (e) CaS   (f) KI

3.32 What are the names of the following compounds? If both a modern and a common name are possible, write both.
(a) NaF   (b) Li_2O (c) $CuBr_2$
(d) $MgCl_2$ (e) ZnO   (f) $FeBr_3$

Molecules and Molecular Compounds

3.33 In what ways are a molecule and an atom alike and in what ways are they different?

3.34 In what ways are a molecule and an ion alike and how do they differ?

3.35 How do molecular elements and molecular compounds differ?

3.36 How do molecular compounds and ionic compounds differ?

3.37 What kind of force of attraction holds the two nuclei in one hydrogen molecule, H_2, quite near each other? What special name do we give to this force of attraction when it operates within molecules?

3.38 In your own words, describe how the force of attraction arises as two hydrogen atoms combine to form a hydrogen molecule.

3.39 How does the octet rule work with molecular compounds?

3.40 Draw figures to illustrate a brief discussion of how two fluorine atoms develop a covalent bond in F_2.

3.41 Discuss how the hydrogen atom and a chlorine atom interact to make the H—Cl molecule. Draw figures to illustrate your discussion.

3.42 Hydrazine, N_2H_4, has been used as a rocket fuel. Which of the following structures is correct for hydrazine, **A**, **B**, or **C**?

3.43 The raw material for polypropylene, widely used in making indoor–outdoor carpeting, is propylene, C_3H_6. Which of the following structures for propylene is correct, **A** or **B**?

3.44 Without referring to the periodic table or any other table, write the electron configuration of an atom of atomic number 14, and predict its covalence.

3.45 A molecular compound between germanium and hydrogen has the following structure.

(a) What is the covalence of Ge in this compound?
(b) Germanium is one of the representative elements. Without referring to the periodic table, using only the answer to part (a), to what group in the periodic table does germanium belong?

3.46 The molecular formula of phosphine is PH_3. Write a structural formula for this compound.

3.47 The molecular formula of carbon disulfide is CS_2. Write a structural formula that is consistent with the covalences of carbon and sulfur.

*3.48 Antimony, atomic number 51, is a representative element. It forms a compound with hydrogen called stibine. Using the location of antimony in the periodic table as a clue, write the molecular and structural formulas of stibine.

*3.49 One of the components of a certain brand of chlorine bleach has the molecular formula HOCl, hypochlorous acid. Write a structural formula for this compound that is consistent with the covalences of H, Cl, and O, which were given in this chapter

Polyatomic Ions and Formulas of Compounds Involving Them

3.50 Write the names of the following ions.
(a) OH^- (b) NH_4^+ (c) CN^- (d) MnO_4^-
(e) HSO_4^- (f) HSO_3^- (g) $H_2PO_4^-$ (h) H_3O^+
(i) $C_2H_3O_2^-$ (j) HCO_3^- (k) SO_4^{2-} (l) NO_3^-
(m) HPO_4^{2-} (n) CrO_4^{2-} (o) CO_3^{2-} (p) NO_2^-
(q) PO_4^{3-} (r) $Cr_2O_7^{2-}$

3.51 Write the formulas of the following ions.
(a) sulfite ion
(b) acetate ion
(c) nitrite ion
(d) bicarbonate ion
(e) hydroxide ion
(f) ammonium ion
(g) carbonate ion
(h) nitrate ion
(i) phosphate ion
(j) cyanide ion
(k) hydronium ion
(l) monohydrogen phosphate ion
(m) hydrogen sulfate ion
(n) dihydrogen phosphate ion
(o) dichromate ion
(p) hydrogen sulfite ion
(q) chromate ion
(r) sulfate ion

3.52 Write the formulas of the following compounds.
(a) potassium phosphate
(b) sodium carbonate
(c) calcium sulfate
(d) ammonium cyanide
(e) lithium nitrite
(f) sodium hydrogen sulfite
(g) calcium dichromate
(h) magnesium acetate

3.53 What are the formulas of the following compounds?
(a) sodium monohydrogen phosphate
(b) ammonium carbonate
(c) sodium hydrogen sulfate
(d) ammonium dihydrogen phosphate
(e) sodium permanganate
(f) aluminum hydroxide
(g) lithium bicarbonate
(h) calcium nitrate

3.54 Write the names of the following compounds.
(a) Na_2CO_3 (b) $(NH_4)_2NO_3$ (c) $Mg(OH)_2$
(d) $BaSO_4$ (e) $KHCO_3$ (f) $Ca(CH_3CO_2)_2$
(g) $NaNO_2$ (h) $(NH_4)_3PO_4$

3.55 What are the names of the following compounds?
(a) $KHSO_4$ (b) Li_2HPO_4 (c) $Ca(CN)_2$
(d) $Na_2Cr_2O_7$ (e) Na_2SO_3 (f) $BaCrO_4$
(g) $Al_2(SO_4)_3$ (h) $KMnO_4$

3.56 What is the total number of atoms of all kinds in one formula unit of each of the following compounds?
(a) $(NH_4)_2CO_3$ (b) $Al(CH_3CO_2)_3$ (c) $Ba(H_2PO_4)_2$

3.57 One formula unit of each of the following compounds has how many atoms of all kinds?
(a) $Al_2(CO_3)_3$ (b) $Ca(NO_3)_2$ (c) $(NH_4)_3PO_4$

*3.58 The *Merck Index,* an encyclopedia of chemicals, drugs, and biologicals, gives the formula of ferrous gluconate, a hematinic agent (promotes the formation of red blood cells), as $Fe[HOCH_2(CHOH)_4CO_2]_2$. How many atoms of all kinds are in one formula unit of this compound?

Polar Molecules

3.59 What is the underlying cause of the polarity of a covalent bond?

3.60 Atoms X and Y are atoms of elements in the same period of the periodic table. Atom X has the larger atomic number. Which has the higher electronegativity? Briefly explain.

3.61 Suppose that X and Y form a diatomic molecule $X—Y$, and that X is less electronegative than Y.
(a) Is the $X—Y$ molecule polar?
(b) If so, where are the $\delta+$ and the $\delta-$ charges located?

3.62 Arrange the atomic symbols C, F, H, N, and O in the order of the relative electronegativities of the corresponding elements by placing the symbols in the correct order, left to right, with the most electronegative being on the left.

*3.63 The relative electronegativity of chlorine, atomic number 17, is 3.0, not quite as high as that of oxygen, atomic number 8. Yet chlorine has a much larger positive charge on its nucleus than does oxygen. What feature about a chlorine atom explains why its nucleus is not as effective in making chlorine more electronegative than oxygen?

3.64 Suppose that the difference in electronegativity between X and Y in compound XY is so great that the electron cloud is pulled toward one end effectively to an extent of 100%. Can we call the bond between X and Y a covalent bond anymore? If not, then what is a better name for the bond?

Molecular Orbitals (Special Topic 3.1)

3.65 What event brings a molecular orbital into existence? Give an example.

3.66 In molecular orbital theory, what does *sharing* of a bonding pair of electrons mean? How does this create the covalent *bond?*

3.67 What atomic orbitals overlap to make the molecular orbital for the bond in the following molecules? (a) H—H (b) F—F (c) H—F

VSEPR Theory (Special Topic 3.2)

3.68 What do the letters VSEPR stand for, and what feature of molecular structure is VSEPR theory meant to explain?

3.69 Briefly explain how VSEPR theory explains the bond angle in CH_4.

3.70 The water molecule does not have a linear geometry. Why?

4

CHEMICAL REACTIONS: EQUATIONS AND MASS RELATIONSHIPS

Chemical Equations

Avogadro's Number

Formula Masses and
  Molecular Masses

The Mole

Reactions in Solution

Molar Concentration

Chemists often work with substances in dissolved states, as
seen here in an analytical laboratory. We'll learn about reacting quantities of substances
and their solutions in this
chapter.

4.1 CHEMICAL EQUATIONS

The *coefficients* of a balanced equation give the proportions of the formula units in the reaction.

Mass relationships in chemical reactions are at the heart of almost all practical applications of chemistry to matters of health and medicine. Our own study of mass relationships began in Chapter 2 when we learned that atoms combine only in whole-number ratios to form compounds. The formula units of compounds behave in the same way; they react only in whole-number units. Because the ratios of these units are given by *chemical equations,* we need to learn how to read and write equations first.

■ The technical term for the study of mass relationships in chemistry is *stoichiometry* after the Greek *stoicheion* for "element" and *metron* for "measure".

All Reactant Atoms Are Somewhere among the Products in a Balanced Equation. A **chemical equation** is a condensed description of a reaction that uses the chemical formulas of reactants and products. The formulas of the reactants, separated by plus signs, are on one side of an arrow, and the formulas of the products, also separated by plus signs, are on the arrowhead side of the arrow. For example, the formation of iron(II) sulfide from iron and sulfur is described by the following equation.

$$\text{Fe} + \text{S} \xrightarrow{\text{to give}} \text{FeS}$$

Translation: Iron reacts with sulfur in the ratio of one atom of iron to one of sulfur ⟶ one formula unit of iron(II) sulfide.

A more complicated example is the equation for the formation of aluminum sulfide, Al_2S_3, from its elements. The reactants are not now in a simple 1:1 ratio but in a 2:3 ratio. Numbers called **coefficients** are therefore placed in front of the appropriate formulas to show the proportions by *formula units* that are involved.

$$2\text{Al} + 3\text{S} \xrightarrow{\text{to give}} \text{Al}_2\text{S}_3$$

Translation: Aluminum reacts with sulfur in the ratio of two atoms of aluminum to three atoms of sulfur ⟶ one formula unit of aluminum sulfide.

■ As with subscripts, whenever a coefficient is 1, the 1 isn't written; it is understood.

A coefficient is a multiplier for the entire associated formula. The symbol $2C_3H_8$, for example, occurs in the equation for the combustion (burning) of propane to water and carbon dioxide.

$$2C_3H_8 + 10O_2 \rightarrow 8H_2O + 6CO_2$$
Propane

In "$2C_3H_8$," the numbers 3 and 8 are subscripts and the 2 is a coefficient, a multiplier for the whole formula. In $2C_3H_8$, therefore, there are 6C and 16H atoms.

We are never allowed to change subscripts, once we have the correct formulas, to get an equation to balance. If we change H_2O to H_2O_2, for example, we change the formula for water to the formula for hydrogen peroxide, an entirely different substance. To write a balanced equation we must use correct formulas and then adjust only the coefficients, never the subscripts.

EXAMPLE 4.1 Balancing a Chemical Equation

Iron, Fe, can be made to react with oxygen, O_2, to form an oxide with the formula Fe_2O_3. Write the balanced equation for this reaction.

■ Oxygen, remember, exists as molecules, O_2, not as atoms.

■ O_3 is the formula of ozone, an extremely dangerous gas that occurs in trace amounts in smog.

ANALYSIS We first write down the correct formulas in the format of an equation.

$$Fe + O_2 \longrightarrow Fe_2O_3 \quad \text{(unbalanced)}$$

We cannot change O_2 on the left side to O_3 because the correct formula of oxygen is given as O_2. We do note, however, that oxygen has a subscript of 2 in O_2 and a subscript of 3 in Fe_2O_3. To get a balance, let's switch these numbers. Let's use the 3 as a coefficient for O_2 and the 2 as a coefficient for Fe_2O_3.

$$Fe + 3O_2 \longrightarrow 2Fe_2O_3 \quad \text{(unbalanced)}$$

Now there are six oxygen atoms on the left (in $3O_2$) and six on the right (in $2Fe_2O_3$). Of course, the coefficient of 2 in the formula on the right also means that there are 4 Fe atoms on the right. To fix this, we simply use a coefficient of 4 for Fe on the left.

SOLUTION The balanced equation is

$$4Fe + 3O_2 \longrightarrow 2Fe_2O_3$$

CHECK 4Fe on the left equals (2×2)Fe or 4Fe on the right; (3×2)O or 6O on the left matches (2×3)O or 6O on the right.

■ PRACTICE EXERCISE 1 In the presence of an electrical discharge like lightning, oxygen, O_2, can be changed into ozone, O_3. Write the balanced equation for this reaction.

■ PRACTICE EXERCISE 2 Aluminum, Al, reacts with oxygen to give aluminum oxide, Al_2O_3. Write the balanced equation for this change.

Sometimes, when we adjust coefficients to balance an equation, we get an equation with coefficients all divisible by the same whole number. Suppose, for example, that we had obtained the following as we tried to balance the equation for the reaction of iron with oxygen in Example 4.1.

$$8Fe + 6O_2 \longrightarrow 4Fe_2O_3$$

The equation is surely balanced and all of its formulas are correct, so there is nothing basically wrong with it. Chemists, however, generally (but not always) write balanced equations using the set of *smallest* whole numbers as coefficients. We will normally follow this rule. In the preceding equation, the coefficients are all divisible by 2.

When the formulas in an equation include groups of atoms inside parentheses, and when it is obvious that the groups do not themselves change, treat the groups as whole units in balancing equations.

■ The symbol *(aq)* stands for aqueous solution, meaning a solution in which water is the solvent. The symbol *(s)* means a solid, one that forms directly and is not in solution.

EXAMPLE 4.2 Balancing Equations Involving Polyatomic Ions

When water solutions of $(NH_4)_2SO_4(aq)$, and $Pb(NO_3)_2(aq)$, are mixed, a white solid that separates has the formula $PbSO_4(s)$. The other product is $NH_4NO_3(aq)$, but it remains dissolved as indicated by the *(aq)*. Represent this reaction by a balanced equation, showing the physical states.

ANALYSIS   As usual, we start by simply writing the correct formulas in the format of an equation.

$$(NH_4)_2SO_4(aq) + Pb(NO_3)_2(aq) \rightarrow PbSO_4(s) + NH_4NO_3(aq)$$

Because polyatomic ions are involved, as the parentheses indicate, we next examine the formulas to see whether any such ions change or if they all react as whole units. We can see here that they do remain as intact units, so the subscript 2 in $(NH_4)_2SO_4$ suggests that we use 2 as the coefficient in the formula on the right where NH_4 occurs.

SOLUTION   We place 2 as a coefficient for $NH_4NO_3(aq)$.

$$(NH_4)_2SO_4(aq) + Pb(NO_3)_2(aq) \rightarrow PbSO_4(s) + 2NH_4NO_3(aq)$$

This automatically brought into balance the units of NO_3 on each side of the arrow. The equation is now balanced.

CHECK   The 2 NH_4^+ ions on the left match the 2 NH_4^+ ions on the right; the 2 NO_3^- ions on the left match the 2 NO_3^- ions on the right. The other ions also match.

■ One formula unit of ammonium sulfate, $(NH_4)_2SO_4$, consists of 2 NH_4^+ ions and 1 SO_4^{2-} ion.

■ PRACTICE EXERCISE 3   Balance each of the following equations.
(a) $Ca + O_2 \rightarrow CaO$
(b) $KOH + H_2SO_4 \rightarrow H_2O + K_2SO_4$
(c) $Cu(NO_3)_2 + Na_2S \rightarrow CuS + NaNO_3$
(d) $AgNO_3 + CaCl_2 \rightarrow AgCl + Ca(NO_3)_2$
(e) $Al + H_2SO_4 \rightarrow Al_2(SO_4)_3 + H_2$
(f) $CH_4 + O_2 \rightarrow H_2O + CO_2$

4.2   AVOGADRO'S NUMBER

Avogadro's number, 6.02×10^{23}, **is the number of formula units in a sample of a pure substance with a mass numerically equal to the formula mass in grams.**

We have learned that the coefficients of a balanced equation disclose the proportions of the formula units involved. Thus, the equation for the reaction of sodium with chlorine,

$$2Na + Cl_2 \rightarrow 2NaCl$$

can be interpreted as follows.

$$2 \text{ atoms of } Na + 1 \text{ molecule of } Cl_2 \rightarrow 2 \text{ formula units of } NaCl$$

■ The formula unit of chlorine is a molecule, Cl_2, not an atom.

We cannot, however, go into the lab and carry out a reaction on a scale as small as individual atoms or molecules. We must use much larger numbers of particles to obtain samples that we can manipulate. We could pick almost any number we pleased, provided it was large enough to give us workable samples of formula units. The question is, What number would be the most useful as a standard number of formula units?

■ Amadeo Avogadro (1776–1856) was an Italian physicist at the University of Turin.

The number to which chemists have long agreed is called *Avogadro's number*. Its value (to three significant figures) is 6.02×10^{23}.

$$\text{Avogadro's number} = 6.02 \times 10^{23}$$

Avogadro's number is a pure number with a special name, like the name *dozen* for 12. In fact, Avogadro's number is sometimes called the chemist's "dozen." Just as *dozen* can be used to signify 12 of anything, so *Avogadro's number* can be used to signify 6.02×10^{23} of anything—electrons, protons, atoms, virus particles, anything.

The usefulness of a number this complicated lies in its relationship to atomic masses. *Avogadro's number of atoms delivers a mass of atoms numerically equal to something familiar about the element, its atomic mass in grams.* We can demonstrate this for any element, but we use sodium, atomic mass 22.99 (from the table inside the front cover). Therefore, in atomic mass units, one atom of sodium has a mass of 22.99 u, and we learned earlier that 1 u = 1.6606×10^{-24} g. So the mass in grams of one atom of sodium is calculated as follows.

■ Draw in the cancel lines for the units yourself.

$$\frac{22.99 \text{ u}}{1 \text{ atom Na}} \times \frac{1.6606 \times 10^{-24} \text{ g}}{1 \text{ u}} = \frac{3.818 \times 10^{-23} \text{ g}}{1 \text{ atom Na}}$$

The mass of Avogadro's number of sodium atoms is then

$$\frac{3.818 \times 10^{-23} \text{ g Na}}{1 \text{ atom Na}} \times 6.02 \times 10^{23} \text{ atoms Na} = 22.98 \text{ g Na}$$

■ The difference between the calculated result, 22.98, and the value in the table, 22.99, is caused by rounding.

Thus, Avogadro's number of sodium atoms weighs 22.98 g, which is *numerically equal* to the atomic mass of sodium, a number that naturally belongs to this element. Similar calculations could be carried out for any other element.

The footnotes to the table of elements inside the front cover explain why values of atomic masses differ so much in their numbers of significant figures. We thus need a policy for rounding atomic masses when we use them in calculations.

POLICY ON ROUNDING ATOMIC MASSES
Round atomic masses to the first decimal place *before* using them in any calculation. (Round the atomic mass of H, however, to 1.01.)

■ Our rules for rounding are on page 16.

EXAMPLE 4.3   Relating Masses to Avogadro's Number

How many carbon atoms are in 6.00 g of naturally occurring carbon?

ANALYSIS   This is our first encounter with a *chemistry* calculation. Like almost all such problems, it basically calls for a transformation of units. The units in this problem, "atoms" and "grams," suggest what we need, a conversion factor that connects numbers of atoms to numbers of grams. You have just learned that this connection is Avogadro's number. *Virtually all problems in chemistry can be analyzed by returning to basic definitions and connections. Whenever you are stuck and don't know what to do, go back to definitions and look for connections among units. These connections are sources of conversion factors.*

We solve this problem by working with the basic meaning of Avogadro's number. Applied to elements, Avogadro's number gives us the number of atoms present in whatever number of grams of the element equal its atomic mass. So we first look up the atomic mass of carbon (12.011) and round it by our rule to the first decimal place, 12.0. Now we can write the connecting relationship between the mass of a carbon sample and the number of atoms in it.

$$12.0 \text{ g of C} = 6.02 \times 10^{23} \text{ atoms of C}$$

As we said, whenever we have a specific connection between two quantities like this, we can devise a choice of two conversion factors:

$$\frac{6.02 \times 10^{23} \text{ atoms C}}{12.0 \text{ g C}} \quad \text{or} \quad \frac{12.0 \text{ g C}}{6.02 \times 10^{23} \text{ atoms C}}$$

SOLUTION If we multiply what is given, 6.00 g C, by the first conversion factor, the unit "g C" cancels and our answer is in the unit "atoms of C." (Draw the cancel lines yourself.)

$$6.00 \text{ g C} \times \frac{6.02 \times 10^{23} \text{ atoms C}}{12.0 \text{ g C}} = 3.01 \times 10^{23} \text{ atoms C}$$

Thus, 6.00 g of carbon contains 3.01×10^{23} atoms of carbon.

CHECK Does the answer make sense? Yes, 6.00 g of C is *less* than a whole mole of C atoms (exactly half as much, in fact). So the answer should be *less* than Avogadro's number (exactly half, in fact), and 3.01×10^{23} is half of 6.02×10^{23}. We will sometimes make "sense checks" like this; it's a habit that you will want to develop not just to avoid embarrassing mistakes but also to further your understanding.

■ PRACTICE EXERCISE 4 How many atoms of gold are in 1.00 oz of gold? (1.00 oz = 28.4 g.)

4.3 FORMULA MASSES AND MOLECULAR MASSES

The sum of the atomic masses of all of the atoms in a chemical formula is the *formula mass* of the substance.

Because atoms lose no mass when they form compounds, we can extend the idea of atomic mass to any formula unit and calculate a formula mass for any substance. The **formula mass** of a compound is simply the sum of the atomic masses of all of the atoms in one formula unit. The formula mass of NaCl, for example, is calculated as follows.

1 atom of Na in NaCl gives	23.0
1 atom of Cl in NaCl gives	35.5
Total	58.5

■ Remember our rule about rounding atomic masses (page 80) before using them in calculations

The formula mass of NaCl is 58.5, so *one formula unit* of NaCl has a mass of 58.5 u. Alternatively, we can say that *Avogadro's number* of NaCl formula units has a mass of 58.5 g. Either meaning can be taken, depending on the situation. The idea of a formula mass is general; it applies to anything with a definite formula, even elements. A synonym for formula mass, namely, **molecular mass,** is used by many scientists.

■ We use the term formula mass because it is a more general term than molecular mass.

EXAMPLE 4.4   Calculating a Formula Mass

Some baking powders contain ammonium carbonate, $(NH_4)_2CO_3$. Calculate its formula mass.

ANALYSIS   A formula mass is the sum of the atomic masses of all the atoms expressed in the formula. So we must look up and write down the atomic masses of all of the elements present, rounding each by our rule (page 80).

N, 14.0   H, 1.01   C, 12.0   O, 16.0

Each atomic mass must be multiplied by the number of times the atom appears in the formula, and then the resulting numbers are added.

SOLUTION   In one formula unit of $(NH_4)_2CO_3$, N occurs 2 times, H occurs 8 times, C occurs 1 time, and O occurs 3 times. Therefore,

$$2\,N \quad + 8\,H \quad + 1\,C \quad + 3\,O \quad = (NH_4)_2CO_3$$
$$2 \times 14.0 + 8 \times 1.01 + 1 \times 12.0 + 3 \times 16.0 = 96.1 \text{ (correctly rounded)}$$

The formula mass of ammonium carbonate is 96.1, and one formula unit has a mass of 96.1 u. Avogadro's number of $(NH_4)_2CO_3$ formula units has a total mass of 96.1 g.

CHECK   To check a formula mass calculation, make sure that you have counted all of the atoms and have used the correct atomic masses before you double-check the arithmetic.

■ PRACTICE EXERCISE 5   Calculate the formula masses of the following compounds.
(a) $C_9H_8O_4$ (aspirin)
(b) $Mg(OH)_2$ (milk of magnesia)
(c) $Fe_4[Fe(CN)_6]_3$ (ferric ferrocyanide or Prussian blue, an ink pigment)

4.4   THE MOLE

The *mole* is the name of the SI base unit for *amount of chemical substance* and equals the formula mass of a substance in grams.

When we weigh out the formula mass in grams of any chemical, whether element or compound, we obtain one **mole** (abbreviated **mol**) of it. The *mole* is the SI base unit for *amount of substance*. Because each substance has its own formula mass, *the actual mass that equals one mole varies from substance to substance*. What does not vary is the number of formula units. Regardless of the chemical, if you have 1 mol of it, you have Avogadro's number of its formula units. Thus 1 mol of Na has a mass of 23.0 g and consists of 6.02×10^{23} Na atoms; 1 mol of Cl_2 has a mass of 71.0 g, but it also consists of 6.02×10^{23} molecules, but this time Cl_2 molecules; and 1 mol of NaCl has a mass of 58.5 g, but it likewise consists of 6.02×10^{23} formula units, those of NaCl.

In practical work, chemists think of the *mole* as the lab-sized unit of a chemical

■ *mol* stands for both the plural and the singular.

■ What is *constant* about one mole of any substance is not the mass but the number of formula units.

expressed in grams, a mass of the substance that can be manipulated experimentally, either by itself or as by any of its fractions or multiples. For example, the formula mass of H_2O is 18.0, so 1 mol of H_2O has a mass of 18.0 g. A smaller sample, say 1.80 g, would contain 0.100 mol of water, because 1.80 is 1/10th of 18.0. If we take 36.0 g of H_2O, then we have 2.00 mol, because 36.0 is 2 times 18.0. The *millimole (mmol)* is 10^{-3} mol. (Remember, the SI prefix "m" goes with 10^{-3}, so we join "m" to "mol" to create "mmol.")

An Equation's Coefficients Tell Us the Mole Proportions.   The mole concept lets us think about the coefficients in an equation in two ways at the same time. In the following equation for the reaction of Fe with S, each formula has a coefficient of 1. We can now interpret these coefficients in any of the ways given beneath the formulas in the equation.

Fe	+ S	→ FeS
1 atom of Fe	+ 1 atom of S	→ 1 formula unit of FeS
1 dozen atoms of Fe	+ 1 dozen atoms of S	→ 1 dozen formula units of FeS
6.02×10^{23} atoms Fe	+ 6.02×10^{23} atoms S	→ 6.02×10^{23} formula units of FeS
1 mol of Fe	+ 1 mol of S	→ 1 mol of FeS

Notice that the *proportions* of the formula units all remain the same. All that changes is the *scale* of the reaction, the actual numbers of formula units used. The last line in the series expresses what is most important to our future work: *the coefficients in a balanced equation give us the proportions of substances in moles.* We interpret the following equation, for example, to mean that for every *2 mol* of Na that reacts, *1 mol* of Cl_2 also reacts and *2 mol* of NaCl forms.

■ The *mole* interpretation of coefficients is what we'll use almost entirely from now on.

$$2Na + Cl_2 \rightarrow 2NaCl$$

Three kinds of calculations involving moles need to be learned. One uses an equation's coefficients to calculate how many moles of one substance must be involved if a certain number of moles of another are used. *The equation's coefficients will give us any conversion factors we need.*

EXAMPLE 4.5   Using the Mole Concept with Equations

How many moles of oxygen are required to combine with 0.500 mol of hydrogen in the reaction that produces water by the following equation?

$$2H_2 + O_2 \rightarrow 2H_2O$$

ANALYSIS   We need the connection between moles of H_2 (the given) and moles of O_2 (sought). The coefficients tell us that 2 mol of H_2 combines with 1 mol of O_2 in this particular reaction. They tell us that 2 mol of H_2 is *chemically equivalent* to 1 mol of O_2. So we can symbolize the connection as follows.

$$2 \text{ mol } H_2 \Leftrightarrow 1 \text{ mol } O_2$$

It would not be proper to say "2 mol of H_2 *is equal to* 1 mol of O_2" and so use an equals sign to write 2 mol H_2 = 1 mol O_2. Hydrogen and oxygen are not "equal." Their ability to react in a 2 to 1 ratio, however, tells us that 2 mol of H_2 *requires* exactly 1 mol of O_2, so *for this reaction only,* we know that 1 mol of O_2 is the chemical equivalent of 2 mol of H_2. A connection, of course, is a connection, whether it is symbolized by "=" or "⇔," and we can use the relationship

■ Read the symbol ⇔ as "chemically equivalent to."

to construct conversion factors. From: 2 mol H_2 ⇔ 1 mol O_2, we can construct and select from the following conversion factors.

$$\frac{2 \text{ mol } H_2}{1 \text{ mol } O_2} \qquad \frac{1 \text{ mol } O_2}{2 \text{ mol } H_2}$$

We choose one of them to multiply by the given, 0.500 mol of H_2, to calculate how many moles of O_2 are needed.

SOLUTION   The correct conversion factor is the second.

$$0.500 \text{ mol } H_2 \times \frac{1 \text{ mol } O_2}{2 \text{ mol } H_2} = 0.250 \text{ mol } O_2$$

In other words, 0.500 mol of H_2 requires 0.250 mol of O_2 for this reaction.

CHECK   Is the size of the answer sensible? Yes. We can see by the balanced equation that half as many moles of O_2 is needed for the given moles of H_2, and half of 0.500 mol is 0.250 mol.

■ PRACTICE EXERCISE 6   How many moles of H_2O are made from the 0.250 mol of O_2 in Example 4.5?

■ PRACTICE EXERCISE 7   Nitrogen and oxygen combine at high temperature in an automobile engine to produce nitrogen monoxide, NO, an air pollutant. The equation is $N_2 + O_2 \longrightarrow 2NO$. To make 8.40 mol of NO, how many moles of N_2 are needed? How many moles of O_2 are also needed?

■ PRACTICE EXERCISE 8   Ammonia, an important nitrogen fertilizer, is made by the following reaction: $3H_2 + N_2 \longrightarrow 2NH_3$. To make 300 mol of NH_3, how many moles of H_2 and how many moles of N_2 are needed?

■ If balances were calibrated in *moles,* we'd need a separate balance, marked for moles, for each and every possible formula mass!

The Molar Mass of a Substance Is the Number of Grams per Mole.   Laboratory balances do not read in moles but in grams. Therefore, to carry out any experiment involving specific numbers of moles, we must calculate how many grams to weigh out to get the number of moles we want. In other words, we must do a moles-to-grams calculation.

In a moles-to-grams calculation, we use a formula mass in a new way. We write the units of grams per mole (g/mol) after a formula mass, and when we do we have the **molar mass** of the substance, the number of grams per mole. The molar mass of sodium, for example, is 23.0 g Na/mol Na because its atomic mass is 23.0. In other words,

■ The = sign rather than the ⇔ symbol is appropriate here because a specific quantity of the *same* chemical is on both sides of the sign.

$$1 \text{ mol Na} = 23.0 \text{ g Na}$$

We now can devise two conversion factors, as we will see in the next worked example.

EXAMPLE 4.6   Converting Moles to Grams

About 21% of the air we breathe is oxygen, O_2. How many grams of O_2 are in 0.250 mol of O_2?

ANALYSIS   The problem comes down to the following question:

$$0.250 \text{ mol } O_2 = ? \text{ g } O_2$$

For any given substance, what *always* connects grams to moles is the formula mass. So we need the formula mass of O_2; it's two times the atomic mass of O, 16.0, or $2 \times 16.0 = 32.0$. This tells us that

$$1 \text{ mol } O_2 = 32.0 \text{ g } O_2$$

This equation gives us the following conversion factors.

$$\frac{1 \text{ mol } O_2}{32.0 \text{ g } O_2} \qquad \frac{32.0 \text{ g } O_2}{1 \text{ mol } O_2}$$

If we now multiply what is given, 0.250 mol of O_2, by whichever conversion factor lets us cancel "mol O_2" we'll have the answer to the question in the desired final unit, "g O_2." The second conversion factor is what we need.

SOLUTION

$$0.250 \text{ mol } O_2 \times \frac{32.0 \text{ g } O_2}{1 \text{ mol } O_2} = 8.00 \text{ g } O_2$$

Thus, 0.250 mol of O_2 has a mass of 8.00 g of O_2.

CHECK   Does the size of the answer, 8.00 g O_2, make sense? Of course. A whole mole of O_2 has a mass of 32.0 g, so a quarter of a mole of O_2 is 1/4th of 32.0 g, or 8.00 g.

■   PRACTICE EXERCISE 9   An experiment calls for 24.0 mol of NH_3. How many grams is this?

The next worked example shows how to use a formula mass to convert grams to moles.

EXAMPLE 4.7   Converting Grams to Moles

A student was asked to prepare 12.5 g of NaCl. How many moles is this?

ANALYSIS   The question can be restated as follows.

$$12.5 \text{ g NaCl} = ? \text{ mol NaCl}$$

The connection between grams and moles of NaCl is given by the formula mass of NaCl, which we have already calculated to be 58.5. This tells us that

$$58.5 \text{ g NaCl} = 1 \text{ mol NaCl}$$

Therefore, we have these two conversion factors.

$$\frac{58.5 \text{ g NaCl}}{1 \text{ mol NaCl}} \qquad \frac{1 \text{ mol NaCl}}{58.5 \text{ g NaCl}}$$

If we multiply the given, 12.5 g NaCl, by the second ratio, the units cancel properly and the result is the number of moles of NaCl in 12.5 g NaCl.

SOLUTION

$$12.5 \text{ g NaCl} \times \frac{1 \text{ mol NaCl}}{58.5 \text{ g NaCl}} = 0.214 \text{ mol of NaCl (from 0.2136752137)}$$

Thus, 12.5 g of NaCl consists of 0.214 mol of NaCl.

CHECK   Is the answer reasonable? Yes. The sample is about 2/10th of one mole (58.5 g NaCl), so the answer must certainly be *less,* not more, than 58.5 g NaCl.

■   PRACTICE EXERCISE 10   A student was asked to prepare 6.84 g of aspirin, $C_9H_8O_4$. How many moles is this?

We can now put these kinds of calculations together in an example of a very common laboratory situation: How many grams of one substance are needed to make a given mass of another according to some equation? We'll see how in the next example.

EXAMPLE 4.8   Mole Calculations Using Balanced Equations

How many grams of aluminum are needed to make 24.4 g of Al_2O_3 by the following equation?

$$4Al + 3O_2 \longrightarrow 2Al_2O_3$$

ANALYSIS   *All mole problems involving equations must first be worked at the mole level, because the equation's coefficients refer to moles, not masses.* Thus, we must first find out how many *moles* are in 24.4 g of Al_2O_3, which we calculate by a grams-to-moles conversion (as in Example 4.7). Then we use the connection given by the coefficients,

$$2 \text{ mol } Al_2O_3 \Leftrightarrow 4 \text{ mol Al}$$

to do a moles-to-moles conversion and find out how many *moles* of Al are chemically equivalent to the number of *moles* of Al_2O_3 that we found are in 24.4 g of Al_2O_3. When we know the number of *moles* of Al, we convert this to *grams* of Al, using the connection provided by aluminum's atomic mass.

$$1 \text{ mol Al} = 27.0 \text{ g Al}$$

Before you continue, carefully study Figure 4.1, which outlines the "flow." In short, our calculation "flow trip" is

$$24.4\text{g of }Al_2O_3 \xrightarrow{\text{g to mol}} ? \text{ mol of } Al_2O_3 \xrightarrow{\text{mol to mol}} ? \text{ mol of Al} \xrightarrow{\text{mol to g}} ?\text{g of Al}$$

Knowing that we'll be needing formula masses, it's usually a good idea at the start of such a problem to compute any needed formula masses. The atomic mass of Al is 27.0; the formula mass of Al_2O_3 is 102.0.

SOLUTION To determine the number of moles of Al_2O_3 in 24.4 g Al_2O_3, we devise a conversion factor based on the connection between moles and mass: 1 mol Al_2O_3 = 102.0 g Al_2O_3. Then we carry out the following calculation.

$$24.4 \text{ g } Al_2O_3 \times \frac{1 \text{ mol } Al_2O_3}{102.0 \text{ g } Al_2O_3} = 0.239 \text{ mol of } Al_2O_3$$

Now we can use the connection, given by the equation's coefficients, between the number of moles of Al_2O_3 and the number of moles of Al, 2 mol Al_2O_3 ⇔ 4 mol Al, to calculate how many moles of Al are needed. The connection gives us the following conversion factors.

$$\frac{4 \text{ mol Al}}{2 \text{ mol } Al_2O_3} \qquad \frac{2 \text{ mol } Al_2O_3}{4 \text{ mol Al}}$$

So we multiply 0.239 mol of Al_2O_3 by the first factor.

$$0.239 \text{ mol } Al_2O_3 \times \frac{4 \text{ mol Al}}{2 \text{ mol } Al_2O_3} = 0.478 \text{ mol Al}$$

The problem called for the answer in grams of Al, not moles of Al, so we next have to convert 0.478 mol of Al into grams of Al. We use the molar mass of Al, 1 mol Al = 27.0 g Al, to devise the correct conversion factor.

$$0.478 \text{ mol Al} \times \frac{27.0 \text{ g Al}}{1 \text{ mol Al}} = 12.9 \text{ g Al}$$

This is the answer; it takes 12.9 g of Al to prepare 24.4 g of Al_2O_3 according to the given equation.

CHECK It's harder to do a "head check" of the sense of this answer, but we can ask if the *size* of the answer makes sense. The answer we found is close to

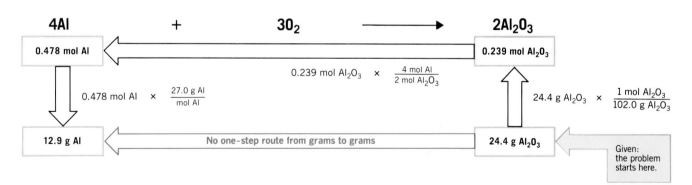

FIGURE 4.1
All calculations involving masses of reactants and products that participate in a chemical reaction must be worked out at the mole level. There is no direct route from grams of one substance to grams of another.

half a mole of Al. The equation's coefficients tell us that half again of this amount of Al is chemically equivalent to Al_2O_3, which has a formula mass of 102.0. In other words, half a mole of Al is chemically equivalent to a quarter of a mole of Al_2O_3. Is this what we began with? Yes, a quarter of a mole of Al_2O_3 is about 25 g, close to the 24.4 g of Al_2O_3 actually given.

■ PRACTICE EXERCISE 11   How many grams of oxygen are needed for the experiment described in Example 4.8? Use a diagram of the solution in the style of Figure 4.1 as you work out the answer.

■ PRACTICE EXERCISE 12   If 28.4 g of Cl_2 is used up in the following reaction, how many grams of Na are also used up, and how many grams of NaCl form?

$$2Na + Cl_2 \longrightarrow 2NaCl$$

4.5   REACTIONS IN SOLUTION

Virtually all of the chemical reactions studied in the lab and those that occur in living systems take place in an *aqueous solution*.

If the atoms, ions, or molecules of one substance are to react with those of another, they must have enough freedom to move about and find and hit each other. Such freedom exists in gases and liquids, but not in solids. Normally, to get solids to react, we put them into solution first. Then their formula units can move around. To study mass relationships when the reactants are in solution (next section), we will first learn some common terms used to describe solutions.

■ You may already have encountered these terms in the laboratory.

Solutions Consist of *Solutes* Dissolved in *Solvents*.   A **solution** is a uniform mixture of particles that are of atomic, ionic, or molecular size. A minimum of two substances are needed. One is called the *solvent*, and the other(s) is called the *solute*. The **solvent** of a solution is the medium into which the other substances are mixed or dissolved. Normally, the solvent is a liquid. When it is water, we have an **aqueous solution.**

■ Unless we state otherwise, we'll always mean *aqueous* solution when we talk about solutions.

A **solute** is anything dissolved by the solvent. In an aqueous solution of sugar, the solute is sugar and the solvent is water. The solute can be a gas. Club soda is a solution of carbon dioxide in water. The solute can be a liquid. Some brands of antifreeze, for example, are mostly solutions of liquid propylene glycol in water.

Solutions Can Be *Dilute, Concentrated, Unsaturated,* and *Saturated*.   Several terms are used to describe a solution. A **dilute solution,** for example, is one in which the ratio of solute to solvent is very small, like a few crystals of sugar dissolved in a glass of water. Most of the aqueous solutions in living systems have more than two solutes, and they are dilute in each of them. In a **concentrated solution,** the ratio of solute to solvent is large. Syrup, for example, is a concentrated solution of sugar in water.

Some solutions are **saturated solutions,** which means that it isn't possible to dissolve more of the solute at the given temperature. If we add more solute to a saturated solution, what we add will only remain separate. If it's a solid, it will sink to the bottom and stay there. The only way to dissolve more solute into a saturated solution is to change the solution's temperature.

An **unsaturated solution** is one in which more solute could be dissolved without a change in temperature. The ratio of solute to solvent is thus lower in an unsaturated solution than in the corresponding saturated solution.

■ The maximum ratio of solute to solvent is different at different temperatures.

A *Solubility* Can Be Expressed as the Maximum Grams of Solute in 100 Grams of Solvent.   The amount of solute needed to give a saturated solution in a given quantity of solvent at a specific temperature is called the **solubility** of the solute. The units commonly are grams solute/100 grams solvent. Table 4.1 gives some examples that show how widely solubilities vary. Notice particularly that a saturated solution can still be very dilute. Less than a milligram of barium sulfate, for example, can dissolve in 100 g of water. Notice also that the solubilities of solids generally increase with temperature. Gases, on the other hand, become less soluble in water with increasing temperature.

Supersaturated Solutions Are Unstable.   Most solids, as we said, become less soluble as the temperature of a solution is reduced. When we cool *saturated* solutions of such solids, therefore, we normally expect the now excess solute to leave the solution. Such a separation of a solid from a solution is called **precipitation,** and the solid is called the **precipitate.**

If we are very careful when we cool a saturated solution, one completely free of dirt, lint, or specks of solute, sometimes the excess solute does not precipitate. The cooled solution now has more solute than presumably it is able to hold. A solution whose concentration exceeds that of saturation is called a **supersaturated solution,** but it's an unstable system. If we scratch the inner wall of the container with a glass rod, or if we add a "seed" crystal of the pure solute, the excess solute will usually separate immediately. It's both dramatic and pretty (Figure 4.2). The seed crystal provides a starting surface on which excess solute particles can form more precipitate.

TABLE 4.1   Solubilities of Some Substances in Water

Solute	Solubility (g/100 g water)			
	0 °C	20 °C	50 °C	100 °C
Solids				
Sodium chloride, $NaCl$	35.7	36.0	37.0	39.8
Sodium hydroxide, $NaOH$	42	109	145	347
Barium sulfate, $BaSO_4$	0.000115	0.00024	0.00034	0.00041
Calcium hydroxide, $Ca(OH)_2$	0.185	0.165	0.128	0.077
Gases				
Oxygen, O_2	0.0069	0.0043	0.0027	0
Carbon dioxide, CO_2	0.335	0.169	0.076	0
Nitrogen, N_2	0.0029	0.0019	0.0012	0
Sulfur dioxide, SO_2	89.9	51.8	4.3	1.8[a]
Ammonia, NH_3			28.4	7.4[b]

[a]At 90 °C.

[b]At 96°C.

FIGURE 4.2

Supersaturation. A "seed" crystal has been added to a supersaturated solution in the glass dish (left photo). Whatever solute present in excess of that normally allowed in a saturated solution quickly separates as beautiful crystals. Any solution remaining in contact with the crystals in the last photo is now a saturated solution.

4.6   MOLAR CONCENTRATION

The ratio of *moles per liter* is the most useful way to describe a solution's concentration.

How are mass relationships handled when a chemical reaction occurs in a solution? The answer requires information about the solution's concentration, which we will study next.

The *Molarity* of a Solution Is the Moles of Solute per Liter of Solution.   The **concentration** of a solution is the ratio of the quantity of solute to some given unit of the solution. The units in this ratio can be anything we wish, but the most useful are those of *moles of solute per liter of solution,* a ratio called the solution's **molar concentration,** or **molarity,** abbreviated **M.**

■ The standard abbreviation for "solution" is *soln.*

$$M = \frac{\text{mol solute}}{\text{L soln}} = \frac{\text{mol solute}}{1000 \text{ mL soln}}$$

■ M = moles per liter
mol = moles

A bottle might, for example, have the label "0.10 *M* NaCl," meaning a concentration of 0.10 mol of NaCl per liter of solution (or per 1000 mL of solution). The actual *volume* of solution in the bottle might be small or large, but in either case the *ratio* is the same: 0.10 mol NaCl per liter of solution. A value of molarity always gives two conversion factors.

$$\frac{0.10 \text{ mol NaCl}}{1000 \text{ mL NaCl soln}} \qquad \frac{1000 \text{ mL NaCl soln}}{0.10 \text{ mol NaCl}}$$

A Volumetric Flask Is Used to Make a Solution of Known Molarity.   Figure 4.3 shows how to make a solution having a known molarity. The mass of solute corresponding to the desired moles of solute is weighed out. The sample is then placed in a *volumetric flask,* a special piece of glassware pictured in Figure 4.3, the size being determined by the desired final volume. Enough solvent is added to the flask to

■ Volumetric flasks are manufactured in a large number of fixed capacities.

(a)

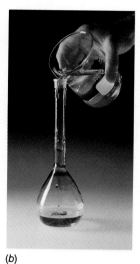
(b)

(c)

(d)

(e)

FIGURE 4.3

Preparation of a solution of known molarity. The volumetric flask has an etched line on its neck that marks the liquid level at which the flask will hold the specified volume. (*a*) The solute, accurately weighed, has been placed in the flask. (*b*) Some water (distilled or deionized) is added. (*c*) The flask is agitated so that the solution dissolves. (*d*) Enough water is added to bring the level to the etched line. (*e*) After the flask is stoppered, it is shaken so that the solution will be uniform.

dissolve all of the solute, and then more solvent is used to bring the liquid level exactly to the mark etched on the flask.

The concept of molarity will become clearer by studying how to do some of the calculations associated with it. In the next worked example we'll see what kinds of calculations are done in order to go into the lab and prepare a certain volume of a solution that has a given molar concentration.

EXAMPLE 4.9 Preparing a Solution of Known Molar Concentration

How many grams of sodium bicarbonate, $NaHCO_3$, must be taken to prepare 500 mL of 0.125 *M* $NaHCO_3$?

ANALYSIS The 0.125 *M* on the label indirectly refers to *moles*, but the question asks for the answer in grams. Before we can calculate the grams needed, we have to find out how many *moles* of $NaHCO_3$ are required. Here is where the given concentration provides what we now need, a conversion factor to calculate the moles of $NaHCO_3$ in the given volume, 500 mL, of 0.125 *M* $NaHCO_3$. The molarity gives us the following conversion factors.

$$\frac{0.125 \text{ mol NaHCO}_3}{1000 \text{ mL NaHCO}_3 \text{ soln}} \qquad \frac{1000 \text{ mL NaHCO}_3 \text{ soln}}{0.125 \text{ mol NaHCO}_3}$$

If we multiply the given volume, 500 mL of $NaHCO_3$ solution, by the first conversion factor, the volume units cancel, and we learn how many moles of $NaHCO_3$ are needed.

SOLUTION   We carry out the calculation. (Draw in the cancel lines yourself.)

$$500\,\text{mL NaHCO}_3 \text{ soln} \times \frac{0.125 \text{ mol NaHCO}_3}{1000 \text{ mL NaHCO}_3 \text{ soln}} = 0.0625 \text{ mol NaHCO}_3$$

In other words, the 500 mL of solution must contain 0.0625 mol of $NaHCO_3$. To weigh out this much $NaHCO_3$, we first convert 0.0625 mol into grams of $NaHCO_3$. The formula mass of $NaHCO_3$, 84.0, gives us the following choice.

$$\frac{84.0 \text{ g NaHCO}_3}{1 \text{ mol NaHCO}_3} \quad \text{or} \quad \frac{1 \text{ mol NaHCO}_3}{84.0 \text{ g NaHCO}_3}$$

When we multiply 0.0625 mol of $NaHCO_3$ by the first factor, the unit "mol $NaHCO_3$" cancels and our answer is in the unit we want, "g $NaHCO_3$."

$$0.0625 \text{ mol NaHCO}_3 \times \frac{84.0 \text{ g NaHCO}_3}{1 \text{ mol NaHCO}_3} = 5.25 \text{ g NaHCO}_3$$

Thus, to prepare 500 mL of 0.125 M $NaHCO_3$, we must weigh out 5.25 g of $NaHCO_3$ and dissolve it in some water in a 500-mL volumetric flask. Then we carefully add water until the liquid level reaches the mark, making sure that the contents become well mixed.

CHECK   Is the *size* of the answer, 5.25 g $NaHCO_3$, reasonable? Yes. The molarity is 0.125 M $NaHCO_3$, and the formula mass of $NaHCO_3$ is 84 g. If the molarity were 0.100 M instead of 0.125 M, then we'd need 8.4 g of solute for a whole liter, and 4.2 g for the specified half a liter. But the molarity is larger than 0.100 M, so we need a mass of $NaHCO_3$ somewhat larger than 4.2 g, which is what the answer is. (By checking the *size* of the answer, you get an idea if you multiplied when you should have divided, or vice versa.)

■   PRACTICE EXERCISE 13   How many grams of each solute are needed to prepare the following solutions?

(a) 250 mL of 0.100 M H_2SO_4 (b) 100 mL of 0.500 M glucose ($C_6H_{12}O_6$)

Another calculation is to find the volume of a solution of known molar concentration that will deliver a certain quantity of its solute. The next worked example shows how this is done.

EXAMPLE 4.10   Using Solutions of Known Molar Concentration

In an experiment to see whether mouth bacteria can live on mannitol ($C_6H_{14}O_6$), a student needed 0.100 mol of mannitol. It was available as a 0.750 M solution. How many milliliters of this solution must be used to obtain 0.100 mol of mannitol?

ANALYSIS   The two conversion factors provided by the given concentration are

$$\frac{0.750 \text{ mol mannitol}}{1000 \text{ mL mannitol soln}} \qquad \frac{1000 \text{ mL mannitol soln}}{0.750 \text{ mol mannitol}}$$

We must multiply the given, 0.100 mol of mannitol, by the second conversion factor to calculate the answer.

SOLUTION

$$0.100 \text{ mol mannitol} \times \frac{1000 \text{ mL mannitol soln}}{0.750 \text{ mol mannitol}} = 133 \text{ mL mannitol solution}$$

Thus, 133 mL of 0.750 M mannitol holds 0.100 mol of mannitol.

CHECK If 1000 mL of solution holds 0.750 mol of solute, we need roughly one-seventh as much (0.100/0.750 is about one-seventh) to hold 0.100 mol, and one-seventh of 1000 is about 130 mL.

■ PRACTICE EXERCISE 14 To test sodium carbonate, Na_2CO_3, as an antacid, a scientist needed 0.125 mol of Na_2CO_3. It was available as 0.800 M Na_2CO_3. How many milliliters of this solution are needed for 0.125 mol of Na_2CO_3?

Once solutions of known molar concentration are prepared, then one common calculation involves a reaction when at least one reactant is in a solution of known molar concentration.

EXAMPLE 4.11 Calculations Involving Molar Concentrations

Potassium hydroxide, KOH, reacts with hydrochloric acid, HCl, as follows:

$$HCl(aq) + KOH(aq) \rightarrow KCl(aq) + H_2O$$

How many milliliters of 0.100 M KOH are needed to react with the acid in 25.0 mL of 0.0800 M HCl?

ANALYSIS The question can be restated as follows.

$$25.0 \text{ mL of } 0.0800 \ M \text{ HCl} \Leftrightarrow ? \text{ mL of } 0.100 \ M \text{ KOH}$$

To work at the mole level, *as we must,* we have to "translate" the given, 25.0 mL of 0.0800 M HCl solution, into "moles of HCl." This many moles of HCl will be identical to "moles of KOH," because of the 1:1 mole ratio of HCl to KOH in the equation. Finally, we translate "moles of KOH" into "mL of 0.100 M KOH solution." The calculation flow is as follows, beginning with the "given" for the HCl solution.

$$\text{vol of HCl soln} \xrightarrow[\substack{M \text{ and} \\ \text{vol to} \\ \text{mol HCl}}]{} \text{mol of HCl} \xrightarrow{\text{mol to mol}} \text{mol of KOH} \xrightarrow[\substack{\text{mol solute} \\ \text{and } M \text{ to} \\ \text{vol KOH soln}}]{} \text{vol KOH soln}$$

SOLUTION The number of moles of HCl in 25.0 mL of 0.0800 M HCl is found by using the first of the following conversion factors.

■ These ratios flow from the definition of molarity and the given value, 0.0800 M HCl.

$$\frac{0.0800 \text{ mol HCl}}{1000 \text{ mL HCl soln}} \qquad \frac{1000 \text{ mL HCl soln}}{0.800 \text{ mol HCl}}$$

So we multiply the given, 25.0 mL of HCl solution, by the first factor:

$$25.0 \text{ mL HCl soln} \times \frac{0.800 \text{ mol HCl}}{1000 \text{ mL HCl soln}} = 0.00200 \text{ mol HCl}$$

As noted earlier, 0.00200 mol of HCl requires 0.00200 mol of KOH because the coefficients of the equation tell us that 1 mol HCl ⇔ 1 mol KOH.

To calculate the number of milliliters of KOH solution that contain 0.00200 mol of KOH, we use a conversion factor obtained from the molarity of the KOH solution.

■ The concentration of 0.100 M KOH supplies these conversion factors.

$$\frac{0.100 \text{ mol KOH}}{1000 \text{ mL KOH soln}} \qquad \text{or} \qquad \frac{1000 \text{ mL KOH soln}}{0.100 \text{ mol KOH}}$$

So we multiply 0.00200 mol of KOH by the second factor.

$$0.00200 \text{ mol KOH} \times \frac{1000 \text{ mL KOH soln}}{0.100 \text{ mol KOH}} = 20.0 \text{ mL KOH soln}$$

Thus, 20.0 mL of 0.100 M KOH solution provides exactly the right amount of KOH to react with the acid in 25.0 mL of 0.0800 M HCl. Check this out yourself. Is the *size* of the answer roughly correct?

The next worked example shows how to solve a problem in which the relevant mole ratio in the equation is not 1:1.

EXAMPLE 4.12   Calculations That Involve Molar Concentrations

Sodium hydroxide, NaOH, reacts with sulfuric acid, H_2SO_4, by the following equation:

$$H_2SO_4(aq) + 2NaOH(aq) \longrightarrow Na_2SO_4(aq) + 2H_2O$$

How many milliliters of 0.125 M NaOH provides enough NaOH to react completely with the sulfuric acid in 16.8 mL of 0.118 M H_2SO_4 by the given equation?

ANALYSIS   This is just like Example 4.11. The question is really the following:

$$16.8 \text{ mL of } 0.118 \text{ M } H_2SO_4 \Leftrightarrow \text{? mL of } 0.125 \text{ M NaOH}$$

We start by asking how many moles of sulfuric acid are in 16.8 mL of 0.118 M H_2SO_4. Then we can relate the number of moles of H_2SO_4 to the number of moles of NaOH that match it in the equation, noting that the coefficients tell us that 1 mol H_2SO_4 ⇔ 2 mol NaOH. Finally, we will find out how many milliliters

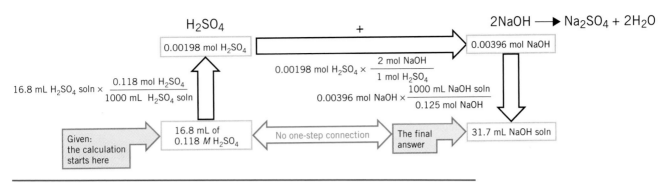

FIGURE 4.4
Calculation flow diagram for Example 4.12.

of the NaOH solution hold these calculated number of moles of NaOH. Figure 4.4 provides a pictorial summary—a calculation flowchart—of the steps for solving this problem.

SOLUTION   First, the number of moles of H_2SO_4 that react is

$$16.8 \text{ mL } H_2SO_4 \text{ soln} \times \frac{0.118 \text{ mol } H_2SO_4}{1000 \text{ mL } H_2SO_4 \text{ soln}} = 0.00198 \text{ mol } H_2SO_4$$

Next, the number moles of NaOH that chemically matches 0.00198 mol of H_2SO_4 based on the fact that 1 mol $H_2SO_4 \Leftrightarrow$ 2 mol NaOH, is

$$0.00198 \text{ mol } H_2SO_4 \times \frac{2 \text{ mol NaOH}}{1 \text{ mol } H_2SO_4} = 0.00396 \text{ mol NaOH}$$

Finally, the volume of 0.125 M NaOH solution that holds 0.00396 mol NaOH is

$$0.00396 \text{ mol NaOH} \times \frac{1000 \text{ mL NaOH soln}}{0.125 \text{ mol NaOH}} = 31.7 \text{ mL NaOH soln}$$

Thus, 31.7 mL of 0.125 M NaOH solution is needed to react with all of the sulfuric acid in 16.8 mL of 0.118 M H_2SO_4. As a check, look again at the conversion factors used and check if the units have properly canceled.

■ PRACTICE EXERCISE 15   Blood isn't supposed to be acidic, but in some medical emergencies it tends to become so. To stop and reverse this trend, the emergency care specialist might administer a dilute solution of sodium bicarbonate intravenously. Sodium bicarbonate destroys (neutralizes) acids. For example, it reacts with sulfuric acid (which is *not* present in blood) as follows:

$$2NaHCO_3(aq) + H_2SO_4(aq) \rightarrow Na_2SO_4(aq) + 2CO_2(g) + 2H_2O$$

How many milliliters of 0.112 M H_2SO_4 will react with 21.6 mL of 0.102 M NaHCO$_3$ *according to this equation?*

SUMMARY

Equations   Chemical equations use the formulas of reactants (given before an arrow) and the products (after the arrow) to describe a chemical reaction. Coefficients in front of formulas ensure that all atoms among the reactants occur among the products, and they give the mole proportions of the substances involved. Subscripts within formulas must never be changed just to balance an equation.

Mass Relationships   A quantity of a substance equal to its formula mass taken in grams is one mole of the substance, so to calculate a molar mass just find the formula mass and attach the units grams per mole (g/mol). One mole of any pure substance, element, or compound, consists of 6.02×10^{23} of its formula units. This number is named Avogadro's number. In working problems involving balanced equations and quantities of substances, be sure to solve them at the mole level where the coefficients can be used. Then, as needed, convert moles to grams.

Solutions   A solution is made up of a solvent and one or more solutes, and the ratio of quantity of solute to some unit quantity of solvent or of solution is called the concentration of the solution. A solution can be described as dilute or concentrated depending on whether its ratio of solute to solvent is small or large.

A solution can also be described as unsaturated, saturated, or supersaturated depending on whether it can be made to dissolve any more solute (at the same temperature). Each substance has a particular solubility in a given solvent at a specified temperature, and this is often expressed as the grams of solute that can be dissolved in 100 g of the solvent.

Molar Concentration   The most useful quantitative description of the concentration of a solution is the ratio of the moles of solute per liter (or 1000 mL) of solution. This is the molar concentration or the molarity of the solution.

REVIEW EXERCISES

The answers to Review Exercises whose numbers are in color are found in Appendix V. The answers to the other Review Exercises are found in the Study Guide that accompanies this text. The more challenging questions are marked with asterisks.

Balanced Equations

4.1   State the information given in the following equation in words.

$$S + O_2 \longrightarrow SO_2 \text{ (sulfur dioxide)}$$

4.2   Write in your own words what the following equation says.

$$2NO \quad + O_2 \longrightarrow \quad 2NO_2$$
$$\text{(nitrogen} \qquad\qquad\quad \text{(nitrogen}$$
$$\text{monoxide)} \qquad\qquad \text{dioxide)}$$

4.3   The following equation is balanced, but what would be a more acceptable way to write it?

$$4H_2SO_4 + 8NaOH \longrightarrow 4Na_2SO_4 + 8H_2O$$

4.4   Balance the following equations.
(a) $N_2 + O_2 \longrightarrow NO$
(b) $MgO + HNO_3 \longrightarrow Mg(NO_3)_2 + H_2O$
(c) $CaBr_2 + AgNO_3 \longrightarrow Ca(NO_3)_2 + AgBr$
(d) $HI + Mg(OH)_2 \longrightarrow MgI_2 + H_2O$
(e) $CaCO_3 + HBr \longrightarrow CaBr_2 + CO_2 + H_2O$

4.5   Balance each of the following equations.
(a) $P + O_2 \longrightarrow P_4O_{10}$
(b) $Fe_3O_4 + H_2 \longrightarrow Fe + H_2O$
(c) $Al_2S_3 + H_2SO_4 \longrightarrow Al_2(SO_4)_3 + H_2S$
(d) $HNO_3 \longrightarrow N_2O_5 + H_2O$
(e) $KHCO_3 + H_2SO_4 \longrightarrow K_2SO_4 + CO_2 + H_2O$

Avogadro's Number

4.6   What is it about chemical substances and chemical reactions that makes the study of the *numbers* of formula units in a given mass of a substance important?

4.7   When rounded to one significant figure, what is Avogadro's number?

Formula Masses[1]

4.8   What law of chemical combination permits us simply to add atomic masses to calculate formula masses?

4.9   Calculate the formula masses of the following substances.
(a) HCl (b) KOH (c) $MgBr_2$
(d) HNO_3 (e) $NaHCO_3$ (f) $Ba(NO_3)_2$
(g) $(NH_4)_2HPO_4$ (h) $Ca(CH_3CO_2)_2$ (i) $C_6H_{12}O_2$

---

[1]Remember that the policy is to round values of atomic masses to their first decimal point before starting any calculations (except that we round the atomic mass of H to 1.01).

4.10 Calculate the formula masses of the following compounds.
(a) Na_2CO_3 (b) H_2SO_4 (c) $(NH_4)_3PO_4$
(d) $Mg_3(PO_4)_2$ (e) $Al(CH_3CO_2)_3$ (f) $Ca(ClO_4)_2$
(g) $(NH_4)_2SO_3$ (h) $K_2Cr_2O_7$ (i) $Fe_4(OH)_2(SO_4)_5$

Moles of Chemical Substances

*4.11 The *mole* is the SI base unit for amount of chemical substance. All other SI base units, like the meter and the kilogram mass, have constant values, but the mass of one mole varies from substance to substance. Explain. Is there any feature about a mole that is *constant* from substance to substance?

*4.12 What is the fundamental reason why we must calculate the amount of *mass* in a mole in connection with running chemical reactions in the laboratory?

4.13 How is the quantity of mass in one mole of some substance determined?

4.14 After calculating the formula mass of sodium hydroxide, NaOH, what two conversion factors can we prepare for chemical calculations?

4.15 Calculate the number of grams in 0.125 mol of each of the substances in Review Exercise 4.9.

4.16 How many grams are in 0.750 mol of each of the compounds in Review Exercise 4.10?

4.17 Calculate the number of moles in 50.0 g of each of the compounds in Review Exercise 4.9.

4.18 Calculate the number of moles in 1.50 g of each of the compounds in Review Exercise 4.10.

4.19 How many *molecules* of N_2 are there in 1.00 g of N_2, roughly the amount of nitrogen in one liter of air?

4.20 How many *molecules* of water are in one drop, which we can assume has a volume of 0.0625 mL and a density of 1.00 g/mL?

*4.21 At a level of only 0.5 μg of ozone, O_3, in one cubic meter of air, the air is considered dangerous for active children to breathe. How many molecules of ozone are in 0.5 μg?

4.22 A "5-grain" aspirin tablet holds about 180 mg of aspirin ($C_9H_8O_4$). How many moles and how many molecules are in 180 mg of aspirin?

Mole and Mass Relationships of Chemical Reactions

4.23 The natural gas piped to homes for heating and cooking purposes is generally methane, CH_4. When it burns in a plentiful supply of oxygen, the products are CO_2 and H_2O.
(a) Write the balanced equation for this reaction.

(b) What pairs of conversion factors express the mole relationships between the following?
   CH_4 and O_2
   CH_4 and CO_2
   CH_4 and H_2O

4.24 In one brand of stomach antacid the active ingredient is calcium hydroxide, $Ca(OH)_2$. The stomach acid is hydrochloric acid, HCl, which is neutralized (destroyed) by the following reaction:

$$Ca(OH)_2 + 2HCl \longrightarrow CaCl_2 + 2H_2O$$

What conversion factors describe the *mole* relationship between $Ca(OH)_2$ and HCl?

4.25 Gasohol is a fuel consisting of various hydrocarbons and ethyl alcohol, C_2H_6O. The ethyl alcohol burns in oxygen to give only carbon dioxide and water.
(a) Write the balanced equation for this reaction.
(b) The burning of 5.00 mol of ethyl alcohol uses up how many moles of oxygen?
(c) How many moles of carbon dioxide are produced by the burning of 5.00 mol of ethyl alcohol?

*4.26 The rusting of iron involves the reaction of oxygen with iron. Although the process is complicated, the following equation can be used to represent the overall results.

$$4Fe + 3O_2 \longrightarrow 2Fe_2O_3$$

(a) If 0.556 mol of iron is changed in this way, how many moles of oxygen are consumed?
(b) How many moles of Fe_2O_3 are produced from 0.556 mol of iron?

4.27 Butane, C_4H_{10}, the fuel in lighters, burns according to the following equation.

$$2C_4H_{10} + 13O_2 \longrightarrow 8CO_2 + 10H_2O$$

(a) If 3.00 mol of O_2 is to be consumed by this reaction, how many moles of butane will be used up?
(b) To produce 1.15 mol of CO_2 by this reaction requires how many moles of butane and how many moles of oxygen?

*4.28 Aluminum metal is made industrially by passing a current of electricity through a solution of aluminum oxide, Al_2O_3, in a special solvent. The other product is molecular oxygen.
(a) Complete and balance the following equation for this reaction.

$$Al_2O_3 \xrightarrow{\text{electric current}}$$

(b) How many grams of aluminum can be made from 100 g of aluminum oxide?
(b) How many grams of oxygen are produced from 100 g of aluminum oxide?
(c) What is the total mass of aluminum plus oxygen produced from 100 g of aluminum oxide? Compare the answer

with the amount of aluminum oxide used. What law of chemical combination is illustrated by this?

4.29   One chemical reaction that is used industrially to make iron from iron oxide is the reduction of iron(III) oxide by carbon monoxide according to the following equation.

$$Fe_2O_3 + 3CO \longrightarrow 2Fe + 3CO_2$$

(a) How many grams of iron can be made from 750 g of Fe_2O_3?
(b) How many grams of carbon monoxide are needed to reduce 750 g of Fe_2O_3 by this reaction?
(c) How many grams of carbon dioxide are produced by this reaction from 750 g of iron(III) oxide?

*4.30   The chemical reaction that causes silver to tarnish is between silver metal, oxygen in the air, and traces of hydrogen sulfide, also in the air. The black tarnish consists of silver sulfide.

$$4Ag + 2H_2S + O_2 \longrightarrow 2Ag_2S + 2H_2O$$

(a) If 4.68 mg of Ag tarnishes by this reaction, how many milligrams of hydrogen sulfide are needed?
(b) How many milligrams of silver sulfide form from 4.68 mg of silver?

*4.31   When a small amount of an acid is accidentally spilled onto a laboratory bench, it should be promptly destroyed (neutralized) before further cleanup is tried. One common way to do this that poses little danger is to sprinkle the acid spill with sodium carbonate until the fizzing caused by escaping carbon dioxide stops. Sulfuric acid, for example, reacts as follows with sodium carbonate.

$$H_2SO_4 + Na_2CO_3 \longrightarrow Na_2SO_4 + CO_2 + H_2O$$

The sodium sulfate and water produced by this reaction and any leftover sodium carbonate can be safely wiped up with paper toweling and discarded into a suitable waste receptacle. Suppose that a spill of 30.0 g of sulfuric acid occurs. What is the minimum number of grams of sodium carbonate needed to destroy the acid?

Solutions

4.32   When water is the dissolving medium for something like sugar, the water itself is designated in what way? How is the sugar designated? What general term can be used to describe any solution for which water is the dissolving medium?

4.33   Sugar (sucrose) is very soluble in water; 100 g of water dissolves 200 g of sugar.
(a) A solution made up at this concentration would be described as *supersaturated, saturated,* or *unsaturated?*
(b) A solution made up at this concentration would be described as *dilute* or *concentrated?*

4.34   Using the information in Table 4.1, name a compound (not a gas) that can form the most dilute solution that could still be called saturated.

4.35   What laboratory operation *not* involving the use of any added solute or solvent could be used to convert a saturated solution of sodium hydroxide into a supersaturated solution? Into an unsaturated solution?

4.36   Suppose that for a series of experiments you needed to have on hand a saturated solution of sodium chloride in water. How could such a solution be prepared in the certain knowledge that it is saturated without actually weighing out the solute?

Molar Concentrations

4.37   What is another term for *molar concentration?*

4.38   Distinguish between the terms *molar concentration, molarity, mole,* and *molecule.*

4.39   Do the units of molar concentration refer to moles per liter of solvent or moles per liter of solution?

*4.40   Calculate the number of moles and the number of grams of solute needed to prepare the given volumes of the following solutions.
(a) 500 mL of 0.125 M NaCl
(b) 250 mL of 0.100 M $C_6H_{12}O_6$
(c) 100 mL of 0.250 M H_2SO_4
(d) 125 mL of 0.500 M Na_2CO_3

*4.41   How many moles and how many grams of solute are needed to prepare the stated volume of each of the following solutions?
(a) 500 mL of 0.200 M $NaCH_3CO_2$
(b) 250 mL of 0.125 M HNO_3
(c) 100 mL of 0.100 M NaOH
(d) 50.0 mL of 0.250 M $NaHCO_3$

4.42   How many milliliters of 0.150 M HNO_3 contain 0.0100 mol of HNO_3?

4.43   To obtain 0.125 mol of H_2SO_4, how many milliliters of 0.440 M H_2SO_4 would have to be taken?

4.44   An experiment called for 0.100 mol of Na_2CO_3, which was available in a solution with a concentration of 0.250 M. How many milliliters of this solution are required?

4.45   The stock solution of hydrochloric acid is 0.500 M HCl. If 100 mL of this solution is taken, how many moles of HCl are taken?

4.46   A student obtained 50.0 mL of 6.00 M Na_2CO_3 solution. How many moles of Na_2CO_3 are in this quantity?

4.47   How many moles of glucose, $C_6H_{12}O_6$, are in 100 mL of 0.100 M glucose solution?

*4.48   If a stock solution of nitric acid, HNO_3, has a concentration of 1.00 M, how many milliliters of this solution are needed to obtain 5.00 g of HNO_3?

*4.49   The stock supply of sulfuric acid, H_2SO_4, has a concentration of $0.500 M$. To obtain 1.00 g of sulfuric acid in the form of this solution, how many milliliters have to be taken?

4.50   To obtain 10.0 g of HCl, how many milliliters of 12.0 M HCl have to be taken?

Stoichiometry of Reactions in Solution

4.51   The label on a reagent bottle reads "$0.250 M$ HCl." What two conversion factors are available from this information? (Base these factors on the milliliter unit for the volume.)

*4.52   Barium sulfate, the ingredient in a "barium cocktail" given to patients about to undergo an X-ray of the intestinal tract, can be made by the following reaction.

$$Ba(NO_3)_2(aq) + Na_2SO_4(aq) \rightarrow BaSO_4(s) + 2NaNO_3(aq)$$

The desired product, as you can see, is a water-insoluble compound that can be separated from the other substances by filtration (letting the mixture flow through filter paper). To prepare 1.00 g of $BaSO_4$, how many milliliters of $0.100 M$ $Ba(NO_3)_2$ and how many milliliters of $0.150 M$ Na_2SO_4 must be mixed together?

*4.53   Gold is attacked by very few chemicals. A mixture of concentrated nitric acid and hydrochloric acid, called *aqua regia* ("royal water"), however, dissolves gold by the following equation.

$$Au(s) + 3HNO_3(aq) + 4HCl(aq) \rightarrow$$
$$HAuCl_4(aq) + 3NO_2(g) + 3H_2O$$

To dissolve 28.4 g of Au (1.00 oz) by this reaction, what is the minimum number of milliliters of 12.0 M HCl and of 16.0 M HNO_3 needed?

*4.54   The active ingredient in milk of magnesia, an over-the-counter antacid, is finely divided magnesium hydroxide slurried in water. The acid it destroys in the stomach by the following equation is 0.1 M HCl. (A minimum recommended dose of milk of magnesia is 2 tablespoons or 30 mL.)

$$Mg(OH)_2(s) + 2HCl(aq) \rightarrow MgCl_2(aq) + 2H_2O$$

How many milliliters of $0.100 M$ HCl can be destroyed by 30.0 mL of milk of magnesia when this medication contains 1.20 g of solid $Mg(OH)_2$ per 15.0 mL of milk of magnesia slurry? (Normally, nearly 2 L of 0.1 M HCl is secreted per day into the stomach.)

*4.55   The concentration of sodium bicarbonate in pancreatic juice, one of the digestive juices, can reach up to 0.120 M $NaHCO_4$. It reacts with the hydrochloric acid delivered in the stomach contents as they move into the upper intestinal tract by the following reaction.

$$NaHCO_3(aq) + HCl(aq) \rightarrow NaCl(aq) + CO_2(g) + H_2O$$

How many liters of 0.120 M $NaHCO_3$ solution provide enough solute to react with the solute in 1.25 L of 0.100 M HCl?

5

KINETIC THEORY AND CHEMICAL REACTIONS

When confined gases are heated strongly, explosive pressures build up that hurl molten rock high into the sky. Knowing how the pressure, volume, and temperature of a gas interact, a subject of this chapter, is also important to anyone involved in respiratory care.

5.1 GASEOUS STATE AND PRESSURE

The earth's atmosphere exerts a pressure on the earth's surface that gives us a unit for describing the pressure of any gas.

Because the air we breathe is a mixture of gases, our understanding of *respiration* depends on a knowledge of the gaseous state. Unlike liquids and solids, all gases obey a number of the same natural laws, regardless of their chemical identities. The laws are stated in terms of four physical quantities: pressure *(P)*, volume *(V)*, temperature *(T)*, and moles *(n)*. We've already learned about the last three, so before we can continue with our study of the laws of gases we must study the concept of pressure.

Pressure Is Defined as Force per Unit Area. Air is matter, so it has mass, and like anything with mass it is subject to the earth's gravitational attraction. This causes air to exert a force on each unit of the area of the earth's surface, which we call the *atmospheric pressure.* The force per unit area exerted by any mass is called **pressure.**

A column of air 1 in.² in cross section that extends from sea level to the edge of outer space exerts a force that averages 14.7 pounds, so the air pressure at sea level is 14.7 lb/in.². At higher elevations, less air is in the column, so the air pressure is less. On the top of Mt. Everest, for example, the earth's highest mountain, the air pressure is about one-third that at sea level.

One way to observe atmospheric pressure is with a Torricelli barometer (Figure 5.1). At 0 °C and sea level, the pressure of the atmosphere supports a column of mercury in the barometer averaging 760 mm high, a value now defined as one **standard atmosphere,** abbreviated **atm.**

$$1 \text{ atm} = 760 \text{ mm Hg} \qquad (\text{at } 0 \text{ °C})$$

In scientific work and medicine, gas pressures are often expressed in a smaller unit, the **millimeter of mercury,** abbreviated **mm Hg,** defined by the following equation:

$$1 \text{ mm Hg} = \tfrac{1}{760} \text{ atm}$$

The air pressure in the lab might, for example, be 740 mm Hg (0.974 atm) one day and 747 mm Hg (0.983 atm) on another, depending on the weather. At the top of Mt. Everest, the pressure is close to 250 mm Hg.

■ Pressure, volume, temperature, and moles are called the *variables* of the gaseous state because they can be varied according to the experiment.

■ By *outer space* we mean space beyond the earth's atmosphere, where the atmosphere is so thin as to be almost nonexistent.

■ The temperature is specified because mercury, like all metals, expands and contracts with temperature (and so is used in thermometers).

■ Outside of health-related fields, **torr** is used as a synonym for *mm Hg.* This avoids the use of a unit of length to describe something (pressure) that isn't a length.

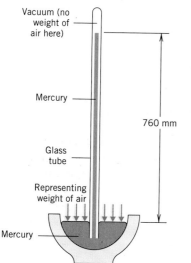

FIGURE 5.1
Torricelli barometer (named after Evangelista Torricelli, 1608–1647, an Italian physicist). The long tube, sealed at one end, is filled with mercury and inverted into a dish of mercury. Some mercury runs out of the tube, but the remainder is held up by air pressure. No air is above the mercury inside the tube, so no air pressure acts inside to oppose the air pressure outside. The column of mercury in the tube stands 760 mm high at sea level and 0 °C.

■ Just before popping, the pressure inside a popcorn kernel is about 9 atm (at a temperature of 250 °C).

The SI unit of pressure is the pascal (Pa).

$$1 \text{ mm Hg} = 133.3224 \text{ Pa}$$
$$1 \text{ atm} = 101{,}325.024 \text{ Pa}$$

This makes the standard atmosphere about 101 kilopascals (kPa).[1]

In a Mixture of Gases, Each Gas Exerts Its *Partial Pressure* Independently of the Others.   Air is a mixture principally nitrogen and oxygen but with small traces of a few other gases. It's possible to separate 100 liters of dry air into 21 liters of oxygen and 79 liters of nitrogen, so on a volume basis air is 21% oxygen and 79% nitrogen. *These percentages are true at all altitudes of the earth's atmosphere.* What changes with altitude is not the *volume* ratio of oxygen to nitrogen but their individual pressures, called their *partial pressures.*

The **partial pressure** of one particular gas in a mixture like air is the contribution that this gas makes to the total pressure. It's the pressure that the gas would exert if all other gases were removed from the container to leave the one gas all alone in the same container (at the same temperature). When dry air is at a total pressure of 760 mm Hg, the partial pressure of oxygen is 160 mm Hg and that of nitrogen is 600 mm Hg. Notice that the sum, 160 mm Hg + 600 mm Hg, is 760 mm Hg.

■ Note that 160 mm Hg is 21% of 760 mm Hg and 600 mm Hg is 79% of 760 mm Hg. Thus, the partial pressure is directly proportional to the volume percentages, 21% O_2 and 79% N_2.

John Dalton (of atomic theory fame) was the first to notice that partial pressures add up to the total pressure of a gas mixture.

DALTON'S LAW OF PARTIAL PRESSURES   The total pressure exerted by a mixture of gases is the sum of their individual partial pressures.

$$P_{\text{total}} = P_a + P_b + P_c + \text{etc.}$$

P_a, P_b, and so on refer to the partial pressures of the individual gases, a, b, etc. When the gases are known, then their formulas can be substituted for the small letters. Thus, PO_2 refers to the partial pressure of oxygen. The value of PO_2 in the atmosphere decreases with altitude until oxygen-enriched air eventually becomes essential for breathing (see Special Topic 5.1).

■ Other references will express PO_2 as P_{O_2} and even pO_2.

The Air We Exhale Includes Carbon Dioxide.   Table 5.1 gives partial pressure data for inhaled and exhaled air as well as for air in the lung's air sacs (alveoli). The air we inhale is generally not dry air, and its water vapor makes a small contribution to the total pressure.[2]

Notice in Table 5.1 the decrease in PO_2 between inhaled and exhaled air and the relatively large increase in PCO_2. The PO_2 decreases, of course, because oxygen is removed from the air in the lungs and is absorbed by the blood. The value of PCO_2 increases, however, because carbon dioxide, a gaseous waste product of metabolism, is removed from the blood at the lungs and transferred to the air about to be exhaled. The warm, moist environment in the alveoli loads the exhaled air with water vapor, as the data in Table 5.1 also show. But notice that the total pressure, regardless of the location, is still the sum of the partial pressures.

■ The removal of CO_2 by the lungs is an important factor in controlling the acidity of the blood.

[1]Although most professional organizations urge health care specialists to switch from *mm Hg* to *kPa* to report pressures of respiratory gases, relatively few have made the change. The *mm Hg* still appears to be the most widely used unit in medicine, which is why we use it in this book.

SPECIAL TOPIC 5.1
BREATHING AT HIGH ALTITUDE

When the total pressure of dry air is 760 mm Hg, the value of PO_2 is 160 mm Hg. At an altitude of 5.5 km (3.4 miles; 18,000 ft), the total pressure is one-half as much, 380 mm Hg, and so the partial pressure of oxygen is also one-half as much, one-half of 160 mm Hg or 80 mm Hg. At 18,000 ft, air must be enriched to 42% oxygen to give its PO_2 the same value as it has at sea level.

For the lung systems of most people, a partial pressure of 80 mm Hg for oxygen in air is not high enough to force atmospheric oxygen out of the lungs and into the bloodstream at a rate fast enough to support life. Nearly all people have to use pressure tanks that deliver oxygen-enriched air to the lungs when they operate even above 14,000 ft (Figure 1), although with conditioning at altitude, experienced mountain climbers are able, unaided by such air, to handle peaks this high and higher. Even Mt. Everest has been climbed by several people without supplemental oxygen.

FIGURE I

To survive and operate at high altitude, most people need to breathe oxygen-enriched air.

5.2 GAS LAWS

The pressure, volume, and temperature of one mole of a gas interact with each other nearly identically for all gases.

Gas Pressure Varies Inversely with Volume at Constant Temperature. Robert Boyle discovered the following relationship between gas pressure and volume, now called the **pressure–volume law** or **Boyle's law**.

■ Robert Boyle (1627–1691) was an English scientist.

TABLE 5.1 **The Composition of Air during Breathing**

Gas	Partial Pressure (mm Hg)		
	Inhaled Air	Exhaled Air	Alveolar Air[a]
Nitrogen	594.70	569	570
Oxygen	160.00	116	103
Carbon dioxide	0.30	28	40
Water vapor	5.00	47	47
Total	760.00	760	760

[a]Alveolar air is air within the alveoli, thin-walled air sacs enmeshed in beds of fine blood capillaries. These sacs, little more than bubbles of tissue, are the terminals of the successively branching tubes that make up the lungs. We have about 300 million alveoli in our lungs.

[2]The partial pressure of water (5.00 mm Hg) in inhaled air given in Table 5.1 corresponds to air at room temperature with a 20% relative humidity, a condition in which air is holding 20% of the maximum amount of water vapor it is able to hold at the given temperature.

■ "Fixed amount" means a fixed number of moles (or grams) of the gas.

PRESSURE–VOLUME LAW (BOYLE'S LAW)   The volume of a fixed amount of gas is inversely proportional to its pressure when its temperature is kept constant.

Put in mathematical form, Boyle's law states that

$$V \propto \frac{1}{P} \tag{5.1}$$

■ The symbol $\propto$ stands for *is proportional to*.

where P is pressure and V is volume. Or, after rearranging terms and introducing a constant of proportionality,

$$PV = \text{a constant}$$

This can be expressed in the following form, which is easier to use in calculations. The subscripts 1 and 2 represent before and after (initial and final) states.

$$P_1V_1 = P_2V_2 \tag{5.2}$$

■ At the heart of Boyle's law is the *experimental* fact that the two sides in Equation 5.2 equal the same constant.

Equation 5.2 is true because each side equals the same constant.

EXAMPLE 5.1   Doing Pressure–Volume Law Calculations

A given mass of oxygen occupies 500 mL at 760 mm Hg at 20 °C. At what pressure will it occupy 450 mL at the same temperature?

ANALYSIS   The conditions for using Boyle's law are satisfied; the mass (or moles) and temperature are fixed and only P and V vary. So we can use equation 5.2. We have the following data:

$$P_1 = 760 \text{ mm Hg} \qquad P_2 = ?$$
$$V_1 = 500 \text{ mL} \qquad V_2 = 450 \text{ mL}$$

Because we want to calculate P_2, we rearrange Equation 5.2 to

$$P_2 = \frac{P_1V_1}{V_2}$$

SOLUTION   We simply insert the data into this equation and do the arithmetic.

$$P_2 = \frac{(760 \text{mmHg})(500 \text{mL})}{450 \text{mL}}$$

$$P_2 = 844 \text{ mm Hg} \qquad \text{(correctly rounded)}$$

Thus, the new pressure is 844 mm Hg.

■ PRACTICE EXERCISE 1   If 660 mL of helium at 20 °C is under a pressure of 745 mm Hg, what volume will the sample occupy (at the same temperature) if the pressure is changed to 375 mm Hg?

Gas Volume Varies Directly with the Kelvin Temperature at Constant Pressure. Jacques Charles discovered how a gas changes its volume when its temperature is changed while both the pressure and number of moles are kept constant. His discovery is now called the **temperature–volume law** or **Charles' law.**

TEMPERATURE–VOLUME LAW (CHARLES' LAW)   The volume of a fixed amount of any gas is directly proportional to its Kelvin temperature, if the pressure is kept constant.

> ■ The interest of Jacques Charles (1746–1823), a French scientist, in the relationship between volume and temperature stemmed from a curiosity about hot air balloons, which were just being developed in his day.

Charles' law can be expressed in mathematical form as follows.

$$V \propto T$$

By rearranging terms and introducing a constant of proportionality we have:

$$\frac{V}{T} = \text{a constant}$$

> ■ T *must* be in kelvins, K. K = °C + 273.

A useful way to express this, where we use the subscripts 1 and 2 as we did in Equation 5.2, is

$$\frac{V_1}{T_1} = \frac{V_2}{T_2} \quad \text{(constant } P \text{ and } n \text{)} \tag{5.3}$$

> ■ n = number of moles of gas.

Charles' law calculations are handled much as those of Boyle's law. Remember that temperatures in degrees Celsius must be changed to kelvins.

■   PRACTICE EXERCISE 2   A sample of cyclopropane, an anesthetic, with a volume of 575 mL at a temperature of 30 °C, was cooled to 15 °C at the same pressure. What was the new volume?

Gas Pressure Varies Directly with Kelvin Temperature.   Joseph Gay-Lussac discovered the way in which an increase in the temperature of a confined gas changes its pressure, a relationship now called the **temperature–pressure law** or **Gay-Lussac's law.**

> ■ Joseph Gay-Lussac (1778–1850) was another French scientist.

TEMPERATURE–PRESSURE LAW (GAY-LUSSAC'S LAW)
The pressure of a fixed amount of gas is directly proportional to its Kelvin temperature if the volume is kept constant.

Gay-Lussac's law can be written in mathematical form as follows.

$$P \propto T \quad (V \text{ and } n \text{ are constant)}$$

The direct proportionality of pressure to temperature is the property of a gas that underlies the warning not to put an aerosol can into an incinerator. The can is sealed, and it still holds residual gas. As its temperature increases so does its internal pressure. Eventually, the can explodes, which might damage the incinerator or possibly hurt bystanders.

> ■ When a chemical explosive (e.g., TNT) is detonated, a solid of small volume suddenly changes into a large-volume mixture of hot gases at high temperature and pressure.

Under Identical Pressure and Temperature, Equal Volumes of Gases Have the Same Number of Moles.   The gas laws so far studied assume a constant amount of gas in moles, n. If we want to increase the number of moles, something else must change. The volume must increase, for example, if we want to add more gas but keep the pressure and temperature constant. In fact, at constant pressure and temperature, the volume of a gas is directly proportional to the number of moles.

$$V \propto n \quad \text{(at constant } P \text{ and } T)$$

This relationship becomes an equation when we insert a proportionality constant, C.

$$V = Cn \quad \text{(at constant } P \text{ and } T)$$

Amedeo Avogadro discovered an important fact about this equation: *C is the same for all gases.* A 45-L sample of oxygen, for example, has the same number of moles as a 45-L sample of nitrogen or any other gas (under the same temperature and pressure). Thus we have another important gas law, now called **Avogadro's principle.**

AVOGADRO'S PRINCIPLE (VOLUME–MOLE RELATION-SHIP)   Equal volumes of gases have equal numbers of moles when compared at the same pressure and temperature.

At 273 K and 1 Atm, One Mole of a Gas Occupies 22.4 L.   The actual volume occupied by one mole of any gas varies, of course, with pressure and temperature. To make comparisons, scientists have therefore agreed to reference conditions called the **standard conditions of temperature and pressure** or **STP.** Standard pressure is 1 atm and standard temperature is 273 K (0 °C). At STP, one mole of any gas occupies 22.4 L, a value called the **standard molar volume** of a gas. We now have a basis for learning about a much more general equation that combines the four gas laws that we have studied.

■ The standard molar volume of 22.4 at STP is an *experimental* value, the average of the molar volumes of several gases at STP.

The Ratio, PV/nT Is the Same for All Gases.   As scientists worked more and more with the laws of Boyle, Charles, Gay-Lussac, and Avogadro, they eventually discovered a more general relationship. For any gas, the result of multiplying P and V and dividing by T is proportional to the moles of the gas, n.

$$\frac{PV}{T} \propto n$$

Using a proportionality constant, R, we have:

$$\frac{PV}{T} = nR \tag{5.4}$$

This equation, usually rewritten as $PV = nRT$, is the **universal gas law,** and the constant, R, is the **universal gas constant.** Its value (together with its units) is

$$R = \frac{PV}{nT} = 6.23 \times 10^4 \ \frac{\text{mm Hg mL}}{\text{mol K}}$$

■ For this value of R to be true, P must be in mm Hg, V in mL, T in kelvins, and n in moles.

In nearly all experiments with gases, however, the experimenter holds the value of n constant. (No gas is added and none is permitted to escape.) Under this condition, the right side of Equation 5.4 becomes the product of two constants and so is itself another constant.

$$\frac{PV}{T} = nR = \text{another constant} \qquad (5.5)$$

Using subscripts 1 and 2 in the usual way, the form of Equation 5.5 particularly convenient for calculations is an equation called the **general gas law:**

$$\frac{P_1V_1}{T_1} = \frac{P_2V_2}{T_2} \qquad \text{(at constant } n) \qquad (5.6)$$

■ Both sides of Equation 5.6 equal $n \times R$, so they must equal each other.

Equation 5.6 includes the laws of Boyle, Charles, and Gay-Lussac as special cases. Notice, for example, when both n and T are constant (Boyle's law conditions), the temperatures can be canceled in Equation 5.6, leaving $P_1V_1 = P_2V_2$, an expression of Boyle's law. (You should show how Equation 5.6 similarly incorporates the laws of Charles and Gay-Lussac.)

EXAMPLE 5.2   Using the General Gas Law

PROBLEM   A sample of an anesthetic gas with a volume of 925 mL at 20.0 °C and 750 mm Hg pressure is warmed to 37.0 °C at a pressure of 745 mm Hg. What is the final volume of the gas?

ANALYSIS   It's best to collect all of the data first to isolate what is to be calculated.

■ When temperatures are given in degrees Celsius, they must always be converted to kelvins before working any gas law problem.

$$V_1 = 925 \text{ mL} \qquad\qquad V_2 = ?$$
$$P_1 = 750 \text{ mm Hg} \qquad P_2 = 745 \text{ mm Hg}$$
$$T_1 = 293 \text{ K } (20.0 °C) \qquad T_2 = 310 \text{ K } (37.0 °C)$$

Because n is a constant but P, V, and T all change, the appropriate equation is the general gas law equation, 5.6.

SOLUTION   Using Equation 5.6,

$$\frac{750 \text{ mm Hg} \times 925 \text{ mL}}{293 \text{ K}} = \frac{745 \text{ mm Hg} \times V_2}{310 \text{ K}}$$

Solving for V_2, we find:

$$V_2 = \frac{750 \text{ mm Hg} \times 925 \text{ mL} \times 310 \text{ K}}{745 \text{ mm Hg} \times 293 \text{ K}} = 985 \text{ mL}$$

CHECK   Notice that there is an *increase* in volume from 925 to 985 mL. Is this result sensible? Yes, the pressure itself *decreases*, and Boyle's law tells us that the volume should therefore increase. The temperature *increases*, and Charles's law says that this also should work to increase the volume. (A "head check" like this is trickier when the individual effects tend to work oppositely.)

■ PRACTICE EXERCISE 3   A fire occurred in a storage room where a steel cylinder of oxygen-enriched air was kept. The pressure of the gas in the cylinder was 300 atm at 20.0 °C. To what value does the pressure change if the temperature increases to 200 °C but the volume of the cylinder does not change?

5.3   KINETIC THEORY OF GASES

All of the gas laws can be explained in terms of a model of a gas described by the *kinetic theory of gases.*

As the gas laws unfolded over the decades, more and more scientists asked: What must gases be like to make these laws true? As we have said, what is remarkable about the gas laws is that *they hold for all gases,* particularly when a gas is not close to condensing to a liquid (not under high pressure or very cold). There are no general laws like the gas laws for liquids and solids.

An Ideal Gas Would Obey the Gas Laws Exactly.   When measurements of accuracy and high precision are made, no gas obeys the gas laws *exactly* over all ranges of pressure and temperature. Many come so close, however, that it was natural for scientists to theorize about a hypothetical gas that would obey the gas laws exactly under all conditions. Such a gas is called an **ideal gas,** and scientists used the behavior of real gases to postulate the following characteristics of the ideal gas.

■ The noble gases (the Group 0 elements, like helium and neon) behave most nearly as ideal gases. Polar gases, like water vapor and ammonia are least ideal.

Gas
—Mostly empty space
—Random motions

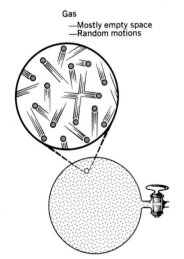

MODEL OF AN IDEAL GAS
1.   The ideal gas consists of a large number of extremely tiny particles in a state of chaotic, utterly random motion.
2.   The particles are perfectly hard, and when they collide they lose no energy because of friction.
3.   The particles neither attract nor repel each other.
4.   The particles move in accordance with the known laws of motion.

Because the kinetic model said that gases consist of tiny particles *in motion,* it came to be called the **kinetic theory of gases.** Let's see how it explains the gas laws.

Collisions with the Container of Moving Gas Molecules Create Gas Pressure.   The pressure of a gas arises from the innumerable collisions its atoms or molecules make with the walls of the container (Figure 5.2). If we imagine one wall forced in to create a smaller gas volume, the number of collisions *per unit wall area* must increase. So if we move the wall in to decrease the gas volume, the gas pressure cannot help but increase. But this is just what Boyle's law says, a larger pressure (at constant T and n) must be associated with a correspondingly smaller volume.

  If the atoms or molecules of a gas generally do not attract or repel each other—the third postulate of the kinetic theory—they *must* act independently. Hence, they must make *independent* contributions to the total pressure, just as Dalton's law of partial pressure describes.

■ Dalton's law, in fact, was the experimental evidence for the third postulate.

Heating a Gas Makes Its Molecules Move Faster.   One result of the calculations based on the model of an ideal gas is a mechanism for how a gas absorbs energy when heated. *Heat causes an increase in the average kinetic energy of the molecules of the gas.* As we've learned, the kinetic energy, KE, of a moving object is

$$KE = \tfrac{1}{2}mv^2$$

where m = mass and v = velocity. The *mass* of each gas molecule cannot change, so a change in temperature can alter the average kinetic energy only by changing the

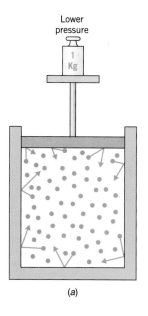

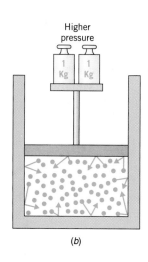

FIGURE 5.2
Kinetic theory and the pressure–volume relationship (Boyle's law). The pressure of the gas is proportional to the frequency of collisions per unit area. When the gas volume is made smaller in going from *(a)* to *(b)*, the frequency of collisions per unit area of the container's walls increases. This is how the pressure increase occurs.

average *velocity* of the molecules. Thus when a gas is heated, its now more energetic molecules move with a greater average velocity. If a gas is cooled, the molecules move with a lower average velocity. At $-273\ ^\circ\text{C}$, all molecular motions in a gas stop, which is the reason for calling this temperature *absolute zero* (and making it equal to 0 K on the Kelvin scale).

■ The relationship between temperature and average molecular velocity bears strongly on the effect of temperature on the rates of chemical reactions.

With a Higher Collision Energy, Molecules of a Hotter Gas Cause a Higher Pressure. When a gas is made hotter, its molecules have a greater average energy, as we just saw. The molecules in the hotter gas, therefore, must cause an increase in the force they exert when they strike a unit area of the container's walls. If the walls cannot move, the result is an increase in gas pressure with temperature, exactly as observed by Gay-Lussac.

With a Higher Collision Energy, Molecules of a Hotter Gas Can Move Walls. The harder-hitting molecules of a hotter gas ordinarily would increase the pressure, as we just learned, but when a wall can give way, the hotter gas instead takes up more room. The gas pressure thus stays constant. At constant pressure, in other words, a gas volume must increase with temperature, as Charles' law says.

5.4 LIQUID AND SOLID STATES AND KINETIC THEORY

As the molecules in liquids move about randomly, some escape and thus create the *vapor pressure* of a liquid.

At the molecular level, the chief physical difference between liquids and gases is that the molecules of a liquid are nearly always in contact with neighbor molecules (Figure 5.3). Otherwise, molecules in a liquid move around randomly, like those of a gas. Gas

Densely packed; random motions

FIGURE 5.3
The liquid state as viewed by the kinetic model.

molecules, in contrast, touch only when they collide. In fact, the reason why all gases obey the same laws is that there is so much empty space in a gas.

A Liquid's Evaporation Creates a Vapor Pressure.   A fraction of the molecules at the surface of a liquid happen to be moving upward. With enough velocity they can escape into the space above. If they do not return at an equal rate, the liquid evaporates. **Evaporation** is the gradual change of a liquid to a gas, which is called a **vapor** whenever the substance is normally a liquid (or solid) at room temperature. A liquid that readily evaporates, like ether or alcohol, is called a *volatile* liquid. One that does not evaporate at any noticeable rate, like salad oil, is said to be *nonvolatile*.

Evaporation ensures that the space above a liquid's surface contains molecules of vapor, which behave as a gas and so add their own partial pressure to the air. Thus, in accordance with Dalton's law of partial pressure, the liquid's vapor contributes a partial pressure, called the liquid's **vapor pressure,** to the total pressure above the liquid.

The higher a liquid's temperature is, the higher is its vapor pressure (Figure 5.4). At body temperature, the vapor pressure of water is 47 mm Hg. This is the partial pressure of water vapor in both alveolar and exhaled air, as we saw in Table 5.1, so alveolar and exhaled air are saturated in water.

Over a Boiling Liquid, the Vapor Pressure Is at a Maximum.   When a liquid is heated to a temperature high enough to make its vapor pressure equal to the atmospheric pressure, the liquid boils. Vapor pockets and bubbles now form *beneath* the liquid's surface. As they rise they create the familiar turbulence we see in a boiling liquid. The temperature at which a liquid's vapor pressure equals 760 mm Hg is called the **normal boiling point.**

If the atmospheric pressure is lower, as it is at higher altitudes, the liquid's vapor pressure becomes equal to the outside atmospheric pressure at a lower temperature. Thus, liquids have lower than normal boiling points at lower pressures. At the top of Mount Everest, water boils at only 69 °C (156 °F) because the pressure is so low there.

Water Has a High *Heat of Vaporization.*   When a liquid boils, its temperature no longer increases as it absorbs more heat. All the heat now being absorbed is used only to cause the change in state. The heat needed to convert a liquid to its vapor is called the liquid's **heat of vaporization.** Water, for example, has a heat of vaporization at its normal boiling point of 540 cal/g, one of the highest of all known values for any liquid.

■ It's helpful to think of a liquid's vapor pressure as its *escaping tendency.*

Vapor Pressure of Water

Temperature (°C)	Vapor Pressure (mm Hg)
20	17.5
30	31.8
37	47.1
40	55.3

■ In the mile-high city of Denver, Colorado, water boils at about 95 °C.

■ The heat of vaporization of ethyl alcohol is 204 cal/g at its normal boiling point.

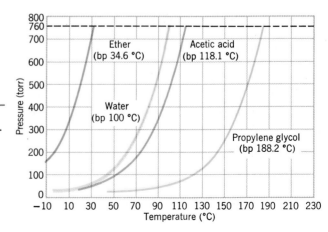

FIGURE 5.4

Vapor pressure versus temperature. (Ether was once widely used as an anesthetic. Acetic acid is the sour component in vinegar. Propylene glycol is present in several brands of antifreeze mixtures.)

Water, of course, can change from its liquid to vapor form (evaporate) at temperatures below the boiling point, and a certain heat of vaporization is needed for evaporation at any temperature. At body temperature, for example, about 500 cal/g is needed to vaporize water.

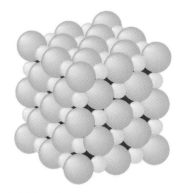

Solid (ionic)
—Densely and orderly packed
—Vibrations about fixed points

Ions and Molecules in Solids Vibrate about Fixed Points.   The ions and molecules in a solid do not move around. They are fixed relative to each other, but they still vibrate. When we heat a solid, the intensities of the vibrations increase. Eventually, at a particular temperature unique for each solid, the vibrations are vigorous enough to overcome the forces of attraction that keep the crystal rigid. The particles now slip and slide, abandon old neighbors, and get new ones, and the solid melts. The minimum temperature that causes this to happen is called the **melting point** of the solid.

Some solids, like "dry ice" (solid carbon dioxide), do not melt but change directly to the vapor state, a change called *sublimation*. (The verb is "to sublime.")

■ When a cooler object is put into contact with one that is warmer, we say that heat "flows" into the cooler object.

The Heat of Fusion of Water Is Unusually High.   A quantity of heat called the **heat of fusion** is required to make a solid melt. If we want to refreeze the substance, we have to remove the identical amount of energy by cooling. The heat of fusion of water is 80 cal/g, one of the highest of all known heats of fusion. This makes ice a particularly efficient coolant. To melt only one gram of ice requires 80 cal, so when a little ice is put into an ice pack, it can remove as it melts a great quantity of heat from an inflamed area.

Heat Generated by Metabolism Is Removed by Evaporation, Radiation, Conduction, and Convection.   The thermal properties of water are as vital to life as any other properties because they help the body manage the heat produced by metabolism. **Metabolism** is the continuing chemical activity of all kinds that occurs in the body. Not only does it provide the energy for operating nerves, muscles, vital organs, and the synthesis of biochemicals, metabolism also tends to increase body temperature. If this increase is not prevented, the body experiences *hyperthermia*. In the opposite condition, *hypothermia*, the body temperature is too low. See Special Topic 5.2 for more details. Both hyperthermia and hypothermia are life threatening.

■ *hyper,* over or above
therm, heat
hypo, lower

The body loses heat by four mechanisms: evaporation of water, radiation, conduction, and convection. Evaporation occurs by perspiration at the skin or from the inner surfaces of the lungs, the latter giving moistness to exhaled air. Perspiration can be *sensible* or *insensible;* that is, it can be obvious and noticeable in beads of sweat, or it can go unnoticed. Either form is effective, because each gram of water that evaporates and leaves the body carries with it about 500 cal (0.5 kcal) of heat, water's heat of vaporization at body temperature. We ingest about 2.5 L of water every day. We lose about 1.0 L by perspiration, so this removes about 20 to 25% of the heat produced daily by metabolism.

■ The vaporization of water from the skin and the lungs is a major mechanism for the removal of heat from the body.

Radiation refers to energy like light energy. *Thermal* radiation ("heat rays") is called infrared radiation because it is just a little less energetic than red light. It is invisible to the eye but special camera film can detect it. Any warm object, including the body, radiates this form of energy, and its loss tends to cool the body.

Heat loss by *conduction* occurs whenever a warmer object is in contact with one that is colder as, for example, when cold machinery is handled or when the body is in contact with cold air.

Convection is the loss by wind or draft of the warm, thin layer of air next to the skin. When this loss is prevented, the air layer is an excellent insulation. Waffle weave underclothing is designed, for example, to trap this layer in place. Woolen fabrics also work well.

■ Nearly half of the heat loss from the body occurs through the uncovered head. Outdoors in winter, when your feet are cold, put on a hat.

The body responds to a fall in its temperature by trying to increase its rate of metabolism so that more heat is generated internally. Uncontrollable shivering is the outward sign of this response, and it begins after the body temperature has decreased 2 to 3 °F (measured rectally). If the temperature continues to drop, the shivering will be violent for a period. Loss of memory—amnesia—sets in at about 95 to 91 °F. The muscles become more rigid as the core temperature drops to the range 90 to 86 °F. The individual must now have outside help immediately, because mental ability to take life-saving steps is gone. The heart beat becomes erratic, the person becomes unconscious, (87 to 78 °F), and below 78 °F death occurs by heart failure or pulmonary edema.

Death by hypothermia, often called death by *exposure,* can happen even when the air temperature is above freezing. If you become soaked by perspiration or rain and the wind comes up, an outside temperature of 40 °F is dangerous. Those who fall overboard in cold water (32 to 35 °F) seldom live longer than 15 to 30 min.

The legendary St. Bernard dogs who brought little casks of brandy to blizzard victims in the Swiss Alps were more agents of death than life to anyone who drank the brandy. A shot of brandy in a hypothermic individual worsens the situation. Alcohol *dilates* (enlarges) blood capillaries. When the capillaries near the skin's surface, which are loaded with the most chilled blood in the hypothermic body, suddenly enlarge, the chilled blood moves quickly to the body's core. This rapid drop in *core* temperature is particularly life threatening.

A victim of hypothermia who is conscious and able to swallow food or drink should be given warm, nonalcoholic fluids and sweet foods. As quickly as possible, get the victim dry and out of the wind. Get into a dry sleeping bag with the victim, so that your own body warmth can be used. It is a genuine medical emergency, and prompt aid is vital.

5.5 KINETIC THEORY AND RATES OF CHEMICAL REACTIONS

The rate of a chemical reaction increases with temperature and with reactant concentrations.

In chemical reactions, electrons and nuclei become reorganized. Old bonds change and different bonds form. The electron configurations of the reactants switch over to those of the products. If a gentle touching of the reactant particles were all it took for this to happen, no reactant would be stable in the presence of anything else. Yet many substances are stable and can be stored in the presence of air, moisture, glass, people, and other potential reactants.

The kinetic theory helps us understand why some combinations of reactants do nothing to each other, why others can stand each other until the temperature increases too much, and why still other combinations can't be stored in any circumstances. The field of chemistry that deals with the rates of chemical reactions is called *kinetics.*

The Conversion of Kinetic to Potential Energy Makes Colliding Molecules React. For the particles of two reactants to change each other chemically, they must collide. Only in this close encounter, however brief, can the kinetic energy provided by the collision make electrons of the reactant particles relocate relative to their nuclei. Generally, very light collisions do not work.

We know that the law of conservation of energy operates in nature, so what happens to the kinetic energy of colliding particles when they slow down or stop? Is it lost? If so, what of the law of conservation of energy?

The energy that existed as *kinetic* energy is not lost; it is transformed into potential energy. In the realm of colliding molecules, the kinetic energy existing at the moment of collision usually changes otherwise stable electron–nuclei arrangements to less stable arrangements, at least temporarily. Now chemical things can happen.

The particles, of course, might simply revert to their original configurations. It happens often, and the particles bounce away from each other without permanent change. The potential energy in the temporary and unstable arrangement at the instant of collision reconverts to kinetic energy of motion. This is how a ball can hit something hard, momentarily stop, be temporarily deformed, and then bounce away, still as a ball and not something else (Figure 5.5). In like manner, many collisions of molecules lead to nothing, chemically.

Particles deformed by collision might also undergo a reorganization of electrons and nuclei to give different chemical species. Thus, the conversion of kinetic energy into potential energy during a collision can make a chemical reaction possible. The collisions that give products are called *successful* collisions or *effective* collisions. The *rate of a reaction* is the number of effective collisions that occur each second in each unit of volume.

A Certain Minimum Collision Energy Is Needed to Make a Reaction Occur. Almost no reaction is instantaneously rapid (explosive). Each reaction has its own minimum potential energy that must develop out of collisions before electrons can relocate and form new bonds. This minimum value of collisional kinetic energy that must convert to potential energy is called the reaction's **energy of activation**, symbolized by E_{act}. Figure 5.6a shows what it means, and you can see why the energy of activation is sometimes called the "energy hill" or the energy barrier of a reaction.

The vertical axis in Figure 5.6a represents changes in the *fraction* of all collisions that occur as the collision energy changes, changes given by the horizontal axis. The collision energies vary from zero values (barely taps) on the left to large values on the right, values approaching infinity. Collision energies vary because different particles of the reactants happen to have different speeds as they whiz around, much like hockey or soccer players at any instant are moving with different speeds (and colliding). The speeds vary from low values (sometimes even zero) to very high values. Thus, some collisions are such slight taps that virtually no kinetic energy changes into potential energy. However, the *fraction* of such collisions is very small, essentially zero, so the curve starts at the zero point where the two axes intersect. As the value of the collision energy increases (moving to the right in Figure 5.6a), the fractions of collisions with particular energies also increase until a maximum fraction is reached. Then, as we continue to move to the right in Figure 5.6a, actual collisions having increasingly higher energies become less and less likely. The fractions of collisions with ever higher energies thus decline, and the curve moves back down. At some point on the curve we are at a value of collision energy that provides the exact minimum potential energy needed for the electron–nuclei rearrangement (the chemical reaction) to occur. This minimum value of collision energy is what we mean by the reaction's *energy of activation*.

■ The *total* energy—kinetic plus potential—remains constant throughout the change, but it becomes apportioned differently.

FIGURE 5.5
Although severely deformed by its collision with the racquet, the tennis ball leaves the collision as a tennis ball, not as something else. Similarly, most collisions between reacting particles cause no significant changes in the particles.

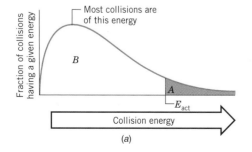

(a)

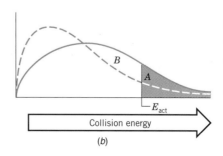

(b)

FIGURE 5.6
Energy of activation. (a) Only a small fraction of all of the collisions, represented by the ratio of areas, A/(A + B), has enough energy for reaction. (b) This fraction greatly increases when the temperature of the reacting mixture is increased.

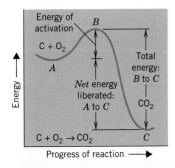

FIGURE 5.7
Progress of reaction diagram for the exothermic reaction of carbon with oxygen that produces carbon dioxide.

When a collision provides the energy of activation or any higher amount of energy, enough energy has been made available for the reaction to take place. Colliding particles that do not achieve the energy of activation simply bounce away, chemically unchanged.

Of course, it isn't enough to have sufficient energy. The colliding particles must hit each other just right, much as the runners in a relay race must pass the baton correctly regardless of how rapidly or slowly they are moving at this critical moment.

Only a portion of all collisions have enough energy to be effective (result in a reaction). We can represent this portion as $A/(A + B)$, the ratio of the shaded area, A, in Figure 5.6a to the total area, $(A + B)$. If the energy of activation were much higher, meaning a high energy barrier, the shaded area, A, would be even smaller. Then the fraction $A/(A + B)$ would also be much smaller and the reaction would be slower. On the other hand, if the energy of activation were very low, the fraction $A/(A + B)$ would be large, and the reaction would be more rapid. In the extreme, where E_{act} of a reaction equaled zero, A would equal $(A + B)$, the ratio would equal 1, and every well-oriented collision would be successful. In practical terms, such a reaction would be very rapid—an explosion, in fact—and it would occur at the instant the reactants were mixed.

At the other extreme, E_{act} would be so high that $A/(A + B)$ would be virtually zero. Then no reaction would occur, and the "reactants" would be eternally stable in each other's presence.

An Increase in the Temperature of the Reactants Makes the Fraction of Successful Collisions Higher. It is well known that increasing the temperature of the reactants increases the rate of their reaction. As the reactant molecules acquire a higher average kinetic energy, the average collision energy also increases (Figure 5.5b. The curve thus flattens out, and its maximum point shifts to the right. *The energy of activation, however, remains essentially unchanged.* Therefore, as higher temperatures shift the curve to the right, a larger fraction of the area under the curve moves to the *right* of E_{act}. The fraction of collisions given by $A/(A + B)$ thus increases, and so the reaction rate increases. The effect of temperature is large. As a rule of thumb, an increase of only 10 °C doubles or triples most rates.

■ The higher the temperature, the flatter the curve becomes and the greater is the percentage of collisions that are effective.

Metabolism Accelerates When Body Temperature Increases. A large rate acceleration for a small temperature change has serious implications for health. An increase in body temperature of only 1 °F raises the rate of metabolism so much that the oxygen requirement of tissue increases by 7%. This places an extra strain on the heart, because the heart beat must increase to speed up the delivery of oxygen from the lungs. To counteract this, a patient with high fever is sometimes even immersed in very cold water to decrease the body temperature rapidly.

■ The heart rate increases between four or five beats per minute for each 1 °C increase in both temperature.

All reactions in a healthy human body occur at a nearly constant temperature, the body temperature. Chemicals that do not react outside the body at this temperature, such as sugar and oxygen, do react inside. Why? We'll answer this question in Section 5.6, but we need more background.

■ Traditionally, normal body temperature has been taken to be 37 °C (98.6 °F). Actually among healthy adults (ages 18–40), the body temperature varies as much as 2.7 °C and even varies with the time of day.

Each Reaction Has a Characteristic *Heat of Reaction*. Figure 5.7 helps us make an important distinction between a reaction's energy of activation and the heat of reaction. The figure shows a kind of plot new to our study, a *progress of reaction diagram.* Its vertical axis gives *relative* values of the potential energies of the substances, either the reactants or the products, depending on where in the plot we are. The horizontal axis indicates the direction of the chemical change. The symbols of the species are included to keep the nature of the reaction in view.

To follow the progress of the reaction diagrammed in Figure 5.7, begin on the left at site *A* with the reactants, carbon and oxygen in our example. We know that they are quite stable in each other's presence at or near room temperature. Coal (mostly carbon), after all, can be stored with no trouble in air (with its 21% oxygen). As you know, to make carbon and oxygen react, we have to heat them; we ignite them, in other words. This gives their particles higher kinetic energies, and as the temperature increases, more and more collisions come closer to being successful. We are now moving in the diagram both upward toward higher potential energies of the reactants and rightward in the direction of the products. The colliding reactant particles are climbing the energy hill as the collision progresses.

Eventually, enough kinetic energy is changed to potential energy to provide the energy of activation, and we are at the top of the energy barrier at *B*. The reactants' electrons and nuclei can now rearrange to give product molecules, carbon dioxide. A great deal of the potential energy in the complex of electrons and nuclei at the top of the barrier now transforms into the kinetic energy of the newly forming molecules of carbon dioxide. There is quite a drop in potential energy now as the reaction progresses to products at site *C*. Some of this potential energy goes to repay the cost of climbing the energy hill, but a net excess that is liberated from the mixture. This net energy difference (*A* to *C*) between the reactants and the products is called the **heat of reaction.**

As we know, once the reaction of real samples of carbon and oxygen starts, it continues spontaneously. The reaction is *exothermic,* and some of the energy represented in Figure 5.7 by the distance from *B* to *C* activates still unchanged particles of the reactants. The heat of reaction in this case represents the conversion of some of the chemical energy in carbon and oxygen into the kinetic energy of the molecules of CO_2.

Not all reactions liberate energy. Many chemical changes won't occur unless there is a continuous input of energy. One example is the conversion of potassium chlorate into potassium chloride and oxygen, as shown in the progress of reaction diagram of Figure 5.8. In this example, a good share of the energy of activation (*A* to *B* in this figure) is permanently retained by the product molecules as their own internal or potential energy. This net retained energy is represented in Figure 5.8 by the distance between levels *A* and *C*. The reaction is endothermic, because there is a conversion of energy supplied by the heat into the potential (chemical) energy of the products. Thus, you can see that both exothermic and endothermic reactions have energies of activation, but in the exothermic reaction there is still a net release of energy, whereas in the endothermic reaction there is a net absorption of energy.

Reactions Usually Go Faster When the Reactants Are Concentrated.   Besides adding heat, another way of increasing the frequency of effective collisions (the reaction rate) is simply to increase the frequency of *all* collisions. Even if we do not increase their average violence, by making collisions of *all* kinds occur more often, we will make successful collisions happen more frequently. We can do this by increasing the concentrations of the reactants. If the molar concentration of one reactant is doubled, the frequency of all collisions must double because twice as many of its particles are present *in the same volume.* It's like going from a stroll down a lonely country lane to an aisle of a very crowded store. An increase in the concentration of moving people increases the "excuse-me" kind of bumps and collisions.

One of the spectacular results of increasing the concentration of a reactant can be seen in the contrast between something burning in pure oxygen and something burning in air. Red-hot steel wool that only glows and gives off sparks when held in a bunsen burner flame bursts into flame when thrust into pure oxygen (Figure 5.9). Someone has estimated that if the atmosphere were 30% oxygen instead of 21%, no forest fire could ever be put out, and eventually all of the world's forests would

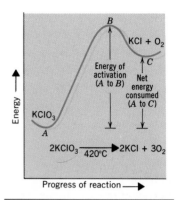

FIGURE 5.8
Progress of reaction diagram for the endothermic conversion of potassium chlorate into potassium chloride and oxygen.

FIGURE 5.9
Steel wool that has been heated in the flame of a bunsen burner burns brightly when dipped into an atmosphere rich in oxygen.

■ People shouldn't even wear shoes with cleats or carry metal cigarette lighters (which might be accidentally dropped and made to light) in rooms where oxygen-enriched air is in use.

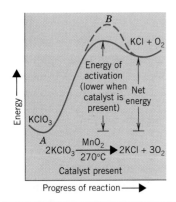

FIGURE 5.10
Progress of reaction diagram for the endothermic, catalyzed conversion of potassium chlorate into potassium chloride and oxygen. The dashed line curve shows where the energy barrier was in the uncatalyzed reaction sketched in Figure 5.8. Notice that the net energy consumed, the heat of reaction, is identical to that of the uncatalyzed reaction, but the energy of activation is lower.

■ Dilute solutions of hydrogen peroxide are sold in drugstores as a bleach and disinfectant.

■ The names of nearly all enzymes end in -ase, as in catalase.

■ Sometimes the special conditions for a reaction are written above or below the arrow in the equation.

disappear. Obviously, if you are ever where oxygen-enriched air is in use, you will want to exclude all flames, as well as all possible sources of sparks.

Not all reactions require energy from *collisions* to occur. Many reactions in nature, like photosynthesis (see Special Topic 14.1), are initiated when molecular species absorb light. As we noted on page 37, electrons can be promoted to higher energy states by absorbing light energy. The resulting species are in *excited states* and sometimes have remarkably different chemical properties than species in their fully "relaxed" states. (Not unlike people, one may note.)

5.6   CATALYSTS AND REACTION RATES

Catalysts **let reactions go faster by permitting lower energies of activation.**

Catalysts Accelerate Reaction Rates.   Now that we have the background of the previous section, we can deal with the question of why chemicals that do not react at body temperature outside the body react readily inside. One of the interesting and most important phenomena in all of nature is the acceleration of a reaction rate by a trace amount of some chemical that does not permanently change as the reaction proceeds. This phenomenon is called **catalysis**, the chemical responsible for it is called a **catalyst**, and the verb is *to catalyze*.

Enzymes Function as Catalysts.   The catalysts in living systems are called *enzymes*, and a special enzyme is involved in virtually every reaction in living things. Nearly all enzymes are members of a large family of biochemicals called proteins. You can easily see the catalytic effect of an enzyme when a dilute solution of hydrogen peroxide (H_2O_2) comes in contact with any blood product. Hydrogen peroxide decomposes as follows.

$$2H_2O_2 \rightarrow 2H_2O + O_2$$

The reaction is very slow at room temperature or body temperature, so if you look at a sample of hydrogen peroxide, you won't notice any bubbling action. But if you use hydrogen peroxide to disinfect a wound, an enzyme in blood called *catalase* causes frothing to occur as the decomposition accelerates and oxygen rapidly evolves (and kills bacteria). Hydrogen peroxide actually is a product in certain reactions of metabolism. Because it is toxic, however, it has to be broken down quickly. Catalase, which is present in liver, catalyzes this detoxification reaction.

Catalysts Also Let Reactions Occur under Milder Conditions.   A catalyst like catalase makes a reaction occur much faster at the same temperature than it does in the absence of a catalyst. A catalyst can also make a reaction take place at a much lower temperature than otherwise. A classic example is the decomposition of potassium chlorate ($KClO_3$) into potassium chloride (KCl) and oxygen that we mentioned in the previous section. Notice in the following equations how the temperature at which the reaction occurs varies with the presence of manganese dioxide (MnO_2).

Without MnO_2, the temperature must be 420 °C or more:
With MnO_2, the temperature need be only 270 °C:

$$2KClO_3 + heat \xrightarrow{420°C} 2KCl + 3O_2$$

$$2KClO_3 + heat \xrightarrow[MnO_2]{270°C} 2KCl + 3O_2$$

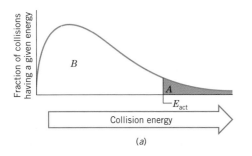
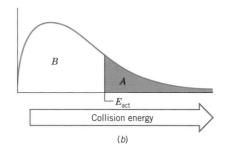

(a)   (b)

FIGURE 5.11

Effect of lowering the energy of activation on the frequency of successful collisions—the rate of the reaction. (a) When the energy of activation is high, the ratio of areas in the drawing that represents the fraction of successful collisions, A/(A + B), is small and the rate of reaction is slow. (b) The ratio is much larger—the rate is much faster—when the energy of activation is less. Catalysts lower energies of activation and so increase reaction rates.

The rates of the evolution of oxygen are approximately equal under the two sets of conditions given here. Thus, the rate at 420 °C in the absence of MnO_2 is observed at a substantially lower temperature in the presence of MnO_2.

With or without the catalyst, the decomposition of potassium chlorate is endothermic, as depicted in Figure 5.8. Figure 5.10 shows the progress of reaction diagram when the MnO_2 catalyst is present. The catalyst does not change the heat of reaction but lowers the energy of activation, instead. This is why the reaction proceeds faster. *This is how all catalysts work:* they lower the energy barrier so the fraction of all collisions with enough energy is larger than without the catalyst. See also Figure 5.11 and compare it with Figure 5.6 to see how a reduction in E_{act} increases the fraction of collisions with energies equal to or higher than E_{act}.

In summary, a catalyst either makes a reaction go faster at the same temperature or permits the reaction to go at the same rate at a lower temperature. It does this by reducing the energy of activation of a reaction, but a catalyst does not affect the heat of reaction. So we add the effect of a *catalyst* to the effects of *temperature* and *concentration* as the important factors that govern reaction rates.

None of the preceding discussion explained *how* a catalyst works. Each works by a different mechanism? All we have discussed thus far is what catalysts do. When you study the subject of enzymes later, you will learn something about their mechanism of action.

SUMMARY

Gas Properties The four important variables for describing the physical properties of gases are moles (n), temperature $(T$, in kelvins), volume (V), and pressure (P). We express pressure—force per unit area—in atmospheres (atm) or millimeters of mercury (mm Hg). Other important physical quantities in the study of gases are partial pressures, standard pressure and temperature (STP = 273 K and 1 atm), and the molar volume at STP (22.4 L).

Gas Laws All real gases obey, more or less, some important laws. Gas pressure is inversely proportional to volume (when n and T are fixed)—the pressure–volume law

(Boyle's law). Gas volume is directly proportional to the Kelvin temperature (when n and P are fixed)—the temperature–volume law (Charles' law). Gas pressure is directly proportional to the Kelvin temperature (when n and V are fixed)—the temperature–pressure law (Gay-Lussac's law). Gas volume is also directly proportional to the number of moles (when T and P are fixed)—Avogadro's principle. In fact, 1 mol of any gas under the same conditions of T and P has as many particles as 1 mol of any other gas. According to another gas law, the total pressure of a mixture of gases equals the sum of their partial pressures (Dalton's law).

Boyle's and Charles' laws and Avogadro's principle combine in the universal gas law, $PV = nRT$, where R, the universal gas constant, holds for all gases. A variation of this, the general gas law, is represented by the equation $P_1V_1/T_1 = P_2V_2/T_2$.

Kinetic Theory   An ideal gas consists of a huge number of very tiny, very hard particles in random, chaotic motion that do not attract or repel each other. The Kelvin temperature of a gas is directly proportional to the average kinetic energy of the gas particles.

Liquid State   Liquids do not follow general laws like gases, because essentially no space separates liquid particles from each other. Liquids can evaporate, and this "escaping tendency" causes vapor pressure, which varies with temperature. When its vapor pressure equals the external pressure, a liquid boils. When this external pressure is 1 atm, the boiling temperature is called the normal boiling point. A specific amount of heat, the heat of vaporization, is needed to convert a gram of liquid into its vapor state.

Solid State   The particles in a solid vibrate about fixed points, but if the solid is heated, these vibrations eventually become so violent that the particles enter the liquid state. A specific amount of heat, the heat of fusion, is needed to convert a gram of solid into its liquid state.

Managing the Heat of Metabolism   The body keeps from becoming hyperthermic by using radiation, conduction, convection, and perspiration to lose some of the heat generated from metabolism. Evaporation of water at the skin and via moist exhaled air uses up some of the body's heat as heat of vaporization. Both hyperthermia and hypothermia are serious conditions.

Kinetic Theory and Chemical Reactions   Virtually all chemical reactions have an energy of activation. The more concentrated the reactant particles are, and the more energetically they collide, the greater is the frequency with which they surmount the barrier and the faster is their reaction. Raising the temperature increases the frequency of collisions energetic enough to be successful.

Catalysts   A catalyst (e.g., an enzyme) accelerates a reaction by lowering the energy of activation, but it does not affect the heat of reaction.

REVIEW EXERCISES

The answers to Review Exercises whose numbers are in color are found in Appendix V. The answers to the other Review Exercises are found in the Study Guide that accompanies this text. The more challenging questions are marked with asterisks.

Pressure, Partial Pressure, and Other Gas Variables

5.1   When the word *gas* is used in chemistry, does it refer to a kind of matter or a state of matter? What are the three kinds of matter? What are the three states of matter?

5.2   How is *pressure* defined?

5.3   What causes the atmosphere to have a pressure on the earth?

5.4   Why is the air pressure on top of a high mountain less than at sea level?

5.5   What is one standard atmosphere of pressure?

5.6   How many mm Hg are in 1 atm?

5.7   Why doesn't all of the mercury run out of a Torricelli barometer?

5.8   The atmospheric pressure on the summit of Mount Everest, the earth's highest mountain (8848 m, 29,029 ft), is roughly 250 mm Hg. What is this in atm?

5.9   In an unpressurized aircraft, a pilot's upper limit on altitude (assuming the availability of oxygen-enriched air) is about 40,000 ft (13 km, 8 mi). The air pressure up there is roughly 0.20 atm. What is this in mm Hg?

5.10   To what kind of a gas sample is the law of partial pressures relevant, and why is this law important for an understanding of the molecular basis of life?

5.11   What was it that John Dalton discovered about a mixture of gases?

5.12   The partial pressure of nitrogen in clean, dry air at 0 °C is 601 mm Hg when the total pressure is 760 mm Hg. The only other gas present to any appreciable extent is oxygen. If a sample of this air were sealed into a container at a pressure of 760 mm Hg and then all of the oxygen were removed, what would be the pressure exerted by the residual gas?

5.13   A gas mixture at 745 mm Hg consists of at least two gases, helium and argon. Their partial pressures are 350 mm Hg for helium and 375 mm Hg for argon. Is it likely that any other gas is also present? How can you tell?

5.14   What four gases are in exhaled air?

*5.15   The value of P_{N_2} at the summit of Mt. McKinley is 277 mm Hg on a day when the atmospheric pressure there is 350 mm Hg. Assuming that the air is made up only of nitrogen and oxygen, what is the partial pressure of oxygen in mm Hg up there?

5.16   At an elevation of 40,000 ft, the partial pressure of nitrogen is 119 mm Hg on a day when the air pressure at this elevation is 150 mm Hg. What is the value of P_{O_2}?

The Individual Gas Laws

5.17 State the following laws both in words and in the forms of equations most useful for gas law calculations. By what other name is each law known?
(a) pressure–volume law
(b) temperature–pressure law
(c) temperature–volume law
(d) law of partial pressures

5.18 In each of the individual gas laws, one or more of the four variables involving gases are assumed to be held constant. Which are they for the following gas laws?
(a) Dalton's law   (b) Gay-Lussac's law
(c) Boyle's law   (d) Charles' law
(e) Avogadro's principle   (f) combined gas law

5.19 If, at constant temperature, the pressure on a sample of gas is changed from 745 to 730 mm Hg, by what factor will the volume of the gas change, by 745/730 or by 730/745?

5.20 The pressure on a 2.56-L sample of nitrogen is changed from 750 to 740 mm Hg while the temperature is left constant. Calculate the new volume of the gas in liters.

5.21 A gas sample under a pressure of 745 mm Hg has a volume of 16.4 L and a temperature of 25 °C. If the pressure on this sample is changed at 25 °C to 760 mm Hg, what is its new volume?

5.22 If 2.45 L of a gas at 25.0 °C is heated under constant pressure to 50 °C, what is the new volume of the gas?

5.23 A heavy-walled steel cylinder with a volume of 25.0 L and containing oxygen at 25.0 °C and a pressure of 125 atm was in a building that caught fire. The temperature of the cylinder rose to 375 °C. If the cylinder did not burst, what was the final pressure of the oxygen?

Combined and Ideal Gas Laws

5.24 A sample of argon at a pressure of 740 mm Hg in a volume of 2.75 L was heated from 25.0 to 78.0 °C. The volume of the container expanded to 3.24 L. What was the final pressure in mm Hg of the gas?

5.25 What must be the final volume of a sample of neon (in L) if 2.55 L at 744 mm Hg and 24.0 °C is heated to 365 °C under conditions that let the pressure change to 760 mm Hg?

5.26 When 250 mL of nitrogen at 740 mm Hg and 19.0 °C is heated to 35.0 °C, the pressure changes to 760 mm Hg. What is the final volume in mL?

5.27 A sample of helium with a volume of 5.28 L, a pressure of 745 mm Hg, and a temperature of 25.0 °C expanded to a volume of 7.75 L with a pressure of 645 mm Hg. What final temperature (in °C) was needed to achieve this?

Universal Gas Law

5.28 What are the standard conditions of temperature and pressure?

5.29 What is meant by *a molar volume*?

5.30 In what circumstances is the molar volume equal to 22.4 L?

5.31 What is the equation for the universal gas law?

5.32 To use the value of R that we calculated in this chapter, in what specific units must P, V, n, and T be?

5.33 At STP how many molecules of hydrogen are in 22.4 L?

5.34 A sample of 2.50 mol of O_2 at 20.0 °C occupies a volume of 12.5 L. Under what pressure (in atm) is this sample?

5.35 A steel cylinder with a volume of 12.5 L contains O_2 at a pressure of 127 atm at 25 °C. The tank contains how many moles and how many kilograms of O_2?

5.36 When the pressure in a certain gas cylinder with a volume of 5.00 L reaches 450 atm, the cylinder is likely to explode. Is it on the verge of exploding if it contains 44.5 mol of nitrogen at 25.0 °C? (Calculate the pressure in atm.)

*5.37 A steel cylinder of a oxygen with a volume of 15.0 L was available for medical purposes. The cylinder pressure decreased from 40.6 to 38.5 atm during which time the temperature remained at 24.0 °C. How many moles of oxygen had been removed?

*5.38 Under suitable conditions, water can be broken down by an electric current into hydrogen and oxygen according to the following equation.

$$2H_2O \rightarrow 2H_2 + O_2$$

In one experiment, 875 mL of a dry sample of one of the product gases was collected at 748 mm Hg and 23.0 °C.
(a) How many moles of the gas were in this sample?
(b) The sample was found to have a mass of 1.133 g. What is the formula mass of the gas in the sample? (Recall that formula mass is the ratio of grams to moles.)
(c) Which gas was it, hydrogen or oxygen?
(d) Based on stoichiometry and Avogadro's principle, how many moles of the other gas were also collected?
(e) How many grams of the other gas were collected?

*5.39 The following equation shows how ammonia can be made from nitrogen and hydrogen.

$$N_2(g) + 3H_2(g) \xrightarrow[\text{heat}]{\text{high pressure}} 2NH_3(g)$$

(a) If 6.00 mol of H_2 is consumed, how many moles of NH_3 are produced?
(b) If 250 L of nitrogen at 745 mm Hg and 25.0 °C is consumed, how many liters of ammonia can be made when the measurements of V and T are made at 745 mm Hg and 25.0 °C?
(c) If 75.0 g of N_2 is consumed, how many grams of H_2 are required and how many grams of NH_3 can be made?

*5.40 Calcium carbonate decomposes as follows when it is strongly heated.

$$CaCO_3(s) \xrightarrow{\text{heat}} CaO(s) + CO_2(g)$$

In one experiment, 246 mL of CO_2 was collected at 740 mm Hg and 24.0 °C.
(a) How many moles of CO_2 formed?
(b) How many moles of $CaCO_3$ decomposed?
(c) How many grams of $CaCO_3$ decomposed?
(d) How many grams of CaO formed?

Kinetic Theory of Gases

5.41 Scientists asked, "What must gases be like for the gas laws to be true?" What was their answer?

5.42 What is true about an ideal gas that is not strictly true about any real gas?

5.43 Dalton's law of partial pressures implies that gas molecules from different gases actually leave each other alone in the mixture, both physically and chemically (except at moments of collisions, when they push each other around). Which one of the three postulates in the model of an ideal gas is based on Dalton's law?

5.44 How does the kinetic theory of gases explain the phenomenon of gas pressure?

5.45 How does the kinetic theory of gases account for Boyle's law (in general terms)?

5.46 Those working out the kinetic theory found that for 1 mol of an ideal gas, the product of pressure and volume is proportional to the average molecular kinetic energy of the ideal gas particles.
(a) To which of the four physical quantities used to describe a gas is the product of pressure and volume for 1 mol of a gas also proportional, according to the universal gas law (which makes no mention of kinetic energy)?
(b) If the product of P and V is proportional both to the average kinetic energy of the ideal gas particles and to the Kelvin temperature, what does this say about the relationship between the average kinetic energy and this temperature?

5.47 What happens to the motions of gaseous molecules at 0 K?

5.48 How does the kinetic theory explain (in general terms) the temperature–volume law?

5.49 The temperature–pressure law (Gay-Lussac's law) can be explained in terms of the kinetic theory in what way (in general terms)?

The Liquid State and Vapor Pressure

5.50 Why aren't there universal laws for the physical behavior of liquids (or solids) as there are for gases?

5.51 How does the kinetic theory explain
(a) How vapor pressure arises?
(b) Why vapor pressure rises with increasing liquid temperature?

Kinetic Theory and the Liquid and Solid States

5.52 The following are some common observations. Using the kinetic theory, explain how each occurs in terms of what molecules are doing.
(a) Moisture evaporates faster in a breeze than in still air.
(b) Ice melts much faster if it is crushed than if it is left in one large block.
(c) Even if hung out to dry in below-freezing weather, wet clothes will become completely dry even though they freeze first.

5.53 At room temperature, nitrogen is a gas, water is a liquid, and sodium chloride is a solid. What do these facts tell us about the relative strengths of electrical forces of attraction in these substances?

Metabolism and the Body's Heat Budget

5.54 What constitutes the body's *metabolism,* in general terms?

5.55 What name is given to the loss of body water by evaporation that does not involve the sweat glands?

5.56 Name the body's three mechanisms for losing heat that do not involve evaporation directly.

5.57 What is the difference between radiation and conduction as means for losing heat from the body?

5.58 Wearing woolen clothing minimizes heat loss from the body by chiefly what mechanism?

5.59 What is hypothermia, and why is it dangerous to life?

5.60 What is hyperthermia, and how can it be life threatening?

Factors Affecting Reaction Rates

5.61 In terms of what we visualize as happening when two molecules interact to form products, how do we explain the existence of an energy barrier to the reaction—an energy of activation?

5.62 Study the accompanying progress of reaction diagram for the conversion of carbon monoxide and oxygen to carbon dioxide, and then answer the questions. The equation for the reaction is

$$2CO(g) + O_2(g) \rightarrow 2CO_2(g)$$

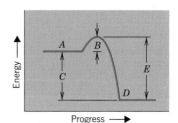

(a) What substance or substances occur at position A?
(b) What substance or substances occur at position D?

(c) Which letter labels the arrow that represents the heat of reaction?

(d) Which letter labels the arrow that stands for the energy of activation?

(e) Is this reaction endothermic or exothermic? How can you tell?

(f) Which letter labels the arrow that would correspond to the energy of activation if the reaction could go in reverse?

*5.63 Suppose that the following hypothetical reaction occurs.

$$A + B \longrightarrow C + D$$

Suppose further that this reaction is endothermic and that the energy of activation is numerically twice as large as the heat of reaction. Draw a progress of reaction diagram for this reaction, and draw and label arrows that correspond to the energy of activation and the heat of reaction.

5.64 The reaction of X and Y to form Z is exothermic. For every mole of Z produced, 10 kcal of heat is generated. The energy of activation is 3 kcal. Sketch the energy relationships on a progress of reaction diagram.

5.65 How do we explain the rate-increasing effect of a rise in temperature?

5.66 As a rule of thumb, how much of a temperature increase doubles or triples the rates of most reactions?

5.67 Explain how a rise in body temperature can lead to a strain on the heart.

5.68 How can we increase the frequency of all collisions in a reacting mixture without raising the temperature?

5.69 When an increase in the concentration of one or more reactants causes an increase in the rate of a reaction, how do we explain this?

5.70 When an increase in the rate of a reaction has been caused by an increase in the concentration of one of the reactants, which of the following factors has been changed? (Identify them by letter.)

A Energy of activation
B Heat of reaction
C Frequency of collisions
D Frequency of successful collisions

5.71 In what way, if any, does a catalyst affect the following factors of a chemical reaction?
(a) heat of reaction
(b) energy of activation
(c) frequency of collisions
(d) frequency of successful collisions

5.72 What is the general name for the catalysts found in living systems?

5.73 Once we have selected a particular reaction, we have to accept whatever energy of activation and heat of reaction go with it. There are, however, three things that we might try to help speed up the reaction. What are they?

Breathing at High Altitude (Special Topic 5.1)

5.74 How many liters of oxygen can be obtained from 100 L of air at an altitude of 18,000 ft, both volumes being measured under the same conditions of temperature and pressure?

5.75 Why must the air breathed at high altitude be enriched in oxygen? Be as specific as possible.

Hypothermia (Special Topic 5.2)

5.76 What is hypothermia?

5.77 With a decrease of about how many degrees in the body's core temperature does hypothermia set in? What are some early signs?

5.78 Why should one not give, say, brandy to a victim of hypothermia?

Additional Exercises

5.79 A sample of a gas was collected in a 220-mL gas bulb under a pressure of 575 mm Hg at a temperature of 25.0 °C. Its mass was found to be 0.299 g. What is the formula mass of the gas? (*Hint:* Use one of the gas laws to calculate the moles of gas, and then compute the ratio of grams to moles.)

5.80 The label on a cylinder of one of the noble gases became illegible, so a student allowed some of the gas to flow into a 300-mL gas bulb until the pressure was 685 mm Hg. The sample now weighed 1.45 g. What is the formula mass of this gas? Which of the Group 0 gases was it?

5.81 A student needs to prepare some CO_2 and intends to use the following reaction in which $CaCO_3$ is heated strongly.

$$CaCO_3(s) \longrightarrow CO_2(g) + CaO(s)$$

The question concerns the size of the flask needed to accept the gas. How large should the container be in milliliters to hold the CO_2 if 1.25 g of $CaCO_3$ is to be used? The final pressure is to be 740 mm Hg. The final temperature is to be 25.0 °C. (*Hint:* How many *moles* of CO_2 can be obtained from the given mass of $CaCO_3$? Is there a connection between *moles of gas* and gas volume, given pressure and temperature?)

5.82 In one lab, the gas-collecting apparatus used a gas bulb with a volume of 250 mL. How many grams of $Na_2CO_3(s)$ would be needed to prepare enough $CO_2(g)$ to fill this bulb when the pressure is 738 torr and the temperature is 23 °C? The equation is

$$Na_2CO_3(s) + 2HCl(aq) \longrightarrow 2NaCl(aq) + CO_2(g) + H_2O$$

6

WATER, SOLUTIONS, AND COLLOIDS

A beautiful sunset reminds a chemist that the solution we call air, when it holds micro-droplets of water and dust particles, is also a colloidal dispersion. At the molecular level of life, solutions and colloidal dispersions hold a similar beauty for those who understand them.

6.1 WATER

Many physical properties of water relate to its polarity and to *hydrogen bonds* between its molecules

We take in more water than all other materials combined. We use it as the fluid in all cells, as a heat-exchange agent, and as the carrier in the bloodstream for distributing oxygen and all molecules from food, all hormones, minerals, and vitamins, and all disease-fighting agents.

Water is a superb solvent. It can dissolve at least trace amounts of almost anything, including rock. It is particularly good at dissolving ionic substances and the more polar molecular compounds.

In this chapter we focus on water and some of the physical properties of aqueous solutions. To understand many aspects of life at the molecular level, we need to know why water dissolves some things well but not others, and to achieve this goal we must study the polarity of water in greater detail.

Water's Boiling Point Is Unusually High. Evidence for the polarity of the water molecule is given by boiling point data. As a rule, boiling points increase as formula masses increase as long as we're comparing otherwise similar substances. You can see this trend, for example, in the boiling points (BPs) of the Group 0 elements (noble gases) given in the margin. Three hydrides of the Group IVA elements—CH_4, SiH_4, and GeH_4—also follow the rule, as shown by the lowest plot of data in Figure 6.1. (A *hydride* is a compound between hydrogen and another element.) Significant departures from the rule occur, however, among the hydrides of other nonmetal groups, most notably the hydride of oxygen in Group VIA, namely, water. The group VIA hydrides are H_2O, H_2S, H_2Se, and H_2Te, and all but H_2O have boiling points that fall roughly on a straight line in accordance with the rule (see Figure 6.1). If we extend the line (by the dashed line) to the formula mass of water, the boiling point of water "should" be about $-100\ °C$. But water boils 200 degrees higher than this. Similarly, those hydrides of the Group VA and VIIA elements of lowest atomic mass, N and F respectively, also break the rule. Their hydrides are NH_3 and HF, and they also boil much higher than they "should," if we expected their boiling points to follow the same trends displayed by the other hydrides of their particular groups (see the other plots in Figure 6.1).

■ Of the 2500 mL of water we take in each day, 1200 mL is in the liquid we drink, 1000 mL is in the food we eat, and 300 mL is made by the reactions of metabolism.

Noble Gas Boiling Points

Element	Formula Mass	BP (K)
Helium	4	4
Neon	20	27
Argon	40	87
Krypton	84	166
Xenon	131	211

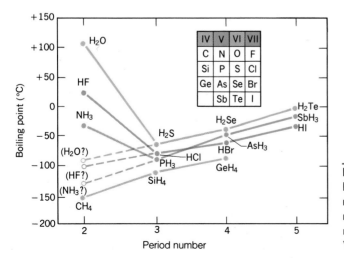

FIGURE 6.1
Boiling points versus formula masses for binary, non-metal hydrides of the elements in Groups IVA, VA, VIA, and VIIA.

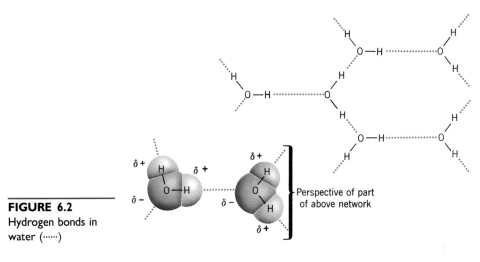

FIGURE 6.2
Hydrogen bonds in water (······)

Perspective of part of above network

Hydrogen Bonds Exist between Water Molecules. The three hydrides, H_2O, NH_3, and HF, are rogues toward the rule because their molecules are particularly polar, relative to those of the other hydrides in the respective groups. The polarities of H_2O, NH_3, and HF molecules stem from the presence of the three most strongly electronegative elements, in order: F > O > N.

Oxygen is much more electronegative than hydrogen, and each of the two O—H bonds in H_2O is very polar, as we learned on page 69. Because the water molecule is *angular,* its individual bond polarities do not cancel, so the water molecule as a whole is polar. Two polar water molecules, in fact, attract each other with a force strong enough to be called a bond. It's not a covalent bond or an ionic bond but, instead, the attraction of *partial* charges, $\delta+$ and $\delta-$. The attraction between a $\delta+$ site on H (when held by F, O, or N) and a $\delta-$ site on F, O, or N is given its own name, **hydrogen bond,** after the location of the $\delta+$ charge. It's important to remember that the $\delta+$ on H is large enough to make a hydrogen bond possible only when H is bonded to the most electronegative elements, F, O, or N.

The hydrogen bond is a bridging bond between molecules (Figure 6.2). Although, like all bonds, it is an electrical force of attraction, the hydrogen bond is not nearly as strong a force as a covalent bond. It's only about 5% as strong, but this is strong enough to make a major difference not only to the properties of water but also to the properties of such vital substances as muscle proteins, cotton fibers, and the chemicals of genes (DNA). Largely because of the hydrogen bond, water has the unusually high heats of fusion and vaporization described in Section 5.4. An extra large input of heat per gram is necessary to melt or boil water because energy is needed to overcome its innumerable hydrogen bonds.

A Water Surface Acts Like a Skin Because of Hydrogen Bonding. All liquids possess a *surface tension,* but that of water is unusually high. **Surface tension** is what gives a liquid's surface the properties of a thin, invisible, elastic membrane. Water's polar molecules tend to jam together where they meet the air because the polar forces that pull surface molecules inward aren't counterbalanced by forces from nonpolar air molecules to pull them outward (Figure 6.3). The resulting net inward pull causes the jam-up of water molecules responsible for water's surface tension.

The same inward pull makes water form beads on a greasy or waxed surface. Wax and grease molecules, like those of air, are nonpolar and cannot attract water molecules and make the water spread out. So the water draws inward on itself and forms a bead. A clean *glass* surface, on the other hand, has innumerable polar sites that strongly attract water molecule. Water thus spreads out on *clean* glass. When water forms beads on glass, you can be sure that the surface is unclean, that it has an invisible greasy coating.

■ VSEPR theory (Special Topic 3.2) explains why there is an angle in the water molecule.

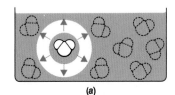

(a)

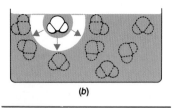

(b)

FIGURE 6.3
Surface tension. *(a)* In the interior of a sample of water, individual water molecules are attracted equally in all directions. *(b)* At the surface, nothing in the air counterbalances the downward pull on the surface molecules in water.

Surface-Active Agents Reduce the Surface Tension of Water. The molecules that make up the membranes of the air sacs (alveoli) in the lungs are, like glass, dotted with innumerable polar sites. Unlike glass, however, the membrane is flexible. The attraction that water molecules have for the membrane might otherwise be expected to pull on the entire membrane, causing the lung to collapse The membrane surface, however, also is coated with a substance that reduces water's surface tension. In some diseases, however, this substance is defective and lungs do collapse.

Anything that reduces the surface tension of water is called a **surface-active agent,** or a **surfactant** for short. All soaps and detergents are surfactants, for example. Sections of their molecules are able to attract water molecules and other sections can act to "dissolve" in oils and greases, the materials that glue soil to fabrics and skin. This double action makes surfactants excellent cleansing agents. Bile, one of the digestive juices, contains extremely powerful surfactants called *bile salts.* Without them, our digestive systems would be much less able to digest the fats and oils in our diets or to wash them from the particles of other kinds of food. (Special Topic 15.2 covers more about how detergents work.)

■ Bile is secreted into the upper intestinal tract from the gallbladder.

6.2 WATER AS A SOLVENT

Water dissolves best those substances whose ions or molecules attract water molecules and form solvent cages.

When crystalline table salt, NaCl, is added to water, the water molecules bombard the crystal surfaces and begin to dislodge Na^+ and Cl^- ions (Figure 6.4). In the crystal environment, however, the ions have oppositely charged ions as nearest neighbors. Unless something else substitutes for this environment, Na^+ and Cl^- ions will not leave the crystal. Water molecules are polar enough to provide a new environment for the ions.

Water Molecules Can Hydrate Ions. Because water molecules are very polar, they can surround Na^+ ions, letting the $\delta-$ sites on the O atoms of H_2O molecules point toward the positively charged ion (see Figure 6.4). Similarly, Cl^- ions also attract water molecules (see Figure 6.4); the $\delta+$ sites on H in H_2O molecules point toward the Cl^- ions.

Once in solution, Cl^- ions no longer have Na^+ ions as nearest neighbors, but they have several water molecules performing the same service, giving each Cl^- an environment of opposite charge. Similarly, Na^+ ions no longer have Cl^- ions as nearest neighbors in solution but, instead, the electron-rich oxygen atoms in water molecules. The phenomenon of the attraction of water molecules to ions is called **hydration,** and it leads to the formation of a cage of water molecules about each ion in solution.

Of all of the common solvents, only water has molecules both polar enough and small enough to form effective solvent cages around ions. Of course, in doing this, water molecules give up some of their attractions for each other. Only ionic substances or compounds made of very polar molecules (Figure 6.5) can break up hydrogen bonds between water molecules. Thus, the formation of an aqueous solution isn't just the separation of the solute particles from each other. It is also, to some extent, the separation of water molecules from each other as well as the attraction between solute particles and solvent molecules.

Hydrates Are Compounds with Water Molecules in Their Crystals. If we let water evaporate from an aqueous solution of any one of several substances, the crystalline

■ The attraction of an ion to a partial opposite charge is called an *ion–dipole attraction.*

■ *Hydr-* is from the Greek *hydor* for "water."

■ Usually, when both ions of an ionic compound carry charges of two or three units, the compound isn't very soluble in water. The ions find more stability by remaining in the crystal than they can replace by accepting solvent cages.

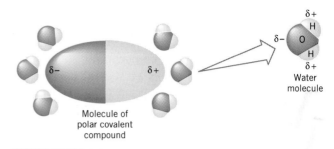

FIGURE 6.5
Hydration of a polar molecule helps polar molecular substances to dissolve in water.

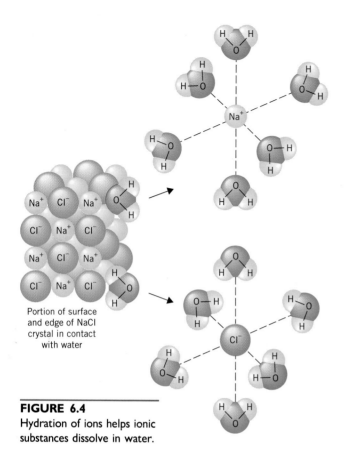

FIGURE 6.4
Hydration of ions helps ionic substances dissolve in water.

residue contains intact water molecules. They are held within the crystals in *definite* proportions. Such water-containing solids are called **hydrates,** and they are true compounds, not mixtures, because they obey the law of definite proportions.

The formulas of hydrates are written to show that intact water molecules are present. The formula of the pentahydrate of copper(II) sulfate, for example, is written as $CuSO_4 \cdot 5H_2O$, where a raised dot separates the two parts of the formula. Table 6.1 lists a number of other common hydrates.

The water present in a hydrate is called the **water of hydration.** It usually can be expelled by heat to leave a residue called the **anhydrous form** of the compound. For example:

$$CuSO_4 \cdot 5H_2O(s) \longrightarrow CuSO_4(s) \quad + \quad 5H_2O(g)$$

Copper(II) sulfate
pentahydrate
(deep blue crystals)

Copper(II) sulfate (as steam)
(anhydrous form
is white)

Many anhydrous forms of hydrates readily take up water and re-form the hydrates. Plaster of paris, for example, although not completely anhydrous, contains relatively less water than gypsum. When we mix plaster of paris with water, it soon sets into a hard, crystalline mass according to the following reaction:

$$(CaSO_4)_2 \cdot H_2O + 3H_2O \longrightarrow 2CaSO_4 \cdot 2H_2O$$

Plaster of paris   Gypsum

■ The reaction of plaster of paris with water is exothermic, as anyone who has had a plaster of paris cast know.

Some compounds in their anhydrous forms are used as drying agents or **desiccants,** substances that remove moisture from air by forming a hydrate. Any substance that can "draw" water in this manner is described as **hygroscopic.** Anhydrous calcium chloride, $CaCl_2$, sometimes used to dehumidify damp basements, is a common desiccant. In humid air it draws enough water to form a liquid solution. Any substance this active as a desiccant is also said to be *deliquescent.*

TABLE 6.1 Some Common Hydrates

Formula	Name	Decomposition Mode (Temperature, °C[a]	Uses
$(CaSO_4)_2 \cdot H_2O$	Calcium sulfate hemihydrate (plaster of paris)	$-H_2O$ (163)	Casts, molds
$CaSO_4 \cdot 2H_2O$	Calcium sulfate dihydrate (gypsum)	$-2H_2O$ (163)	Casts, molds, wallboard
$CuSO_4 \cdot 5H_2O$	Copper(II) sulfate pentahydrate (blue vitriol)	$-5H_2O$ (150)	Insecticide
$MgSO_4 \cdot 7H_2O$	Magnesium sulfate heptahydrate (epsom salt)	$-6H_2O$ (150) $-7H_2O$ (200)	Cathartic in medicine Used in tanning and dyeing
$Na_2CO_3 \cdot 10H_2O$	Sodium carbonate decahydrate (washing soda)	$-H_2O$ (33.5)	Water softener
$Na_2S_2O_3 \cdot 5H_2O$	Sodium thiosulfate pentahydrate (photographer's hypo)	$-5H_2O$ (100)	Photographic developing

[a]Loss of water is indicated by the minus sign before the symbol, and the loss occurs at the temperature (in °C) given in parentheses.

6.3 DYNAMIC EQUILIBRIA IN SOLUTIONS

Equilibrium exists between the dissolved and undissolved states in a saturated solution.

A solution is *saturated* when it holds its maximum of solute. When excess undissolved solute is also present, the system illustrates something new to our study, a *dynamic equilibrium*. A dynamic equilibrium occurs among enough situations at the molecular level of life to make its special vocabulary one goal of this section.

Dynamic Equilibrium Is Much Activity Without Any Net Change.   Imagine that we put more solid sodium chloride into a water sample than is needed to make a saturated solution. Some salt dissolves, and its ions enter the solution and move randomly about as the crystals become smaller. Some of the dissolved ions, however, find their way back to the crystals. They reattach to a crystal surface and the crystal grows in size again. The more concentrated the dissolved ions become as the salt dissolves, the more frequently they bump into crystals and become reattached. Eventually, the rates at which ions leave crystals and return become equal. Now the crystals regrow as fast as they dissolve, and such a situation exists in any saturated solution *in contact with undissolved solute* (Figure 6.6).

In spite of considerable coming and going, the activities in a saturated sodium chloride solution oppose or cancel each other. No longer is there any net change either in the solution's concentration or in the mass of undissolved solute. Any situation with opposing activities but no net change is called a **dynamic equilibrium.** In a **saturated solution** in contact with undissolved solute, dynamic equilibrium exists between undissolved and dissolved solute.

Dynamic Equilibria Have Both a *Forward* Reaction and a *Reverse* Reaction.   We represent an equilibrium by an *equilibrium expression* or *equilibrium equation*, one

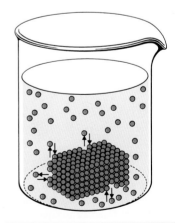

FIGURE 6.6

In a saturated solution, dynamic equilibrium exists between the ions or molecules of the solute in solution and those in the undissolved state.

■ At 20 °C, 36 g of NaCl in 100 g of water makes a saturated solution.

with a pair of half-barbed arrows. We describe, for example, the physical equilibrium in a saturated solution as follows.

$$solute_{undissolved} \rightleftharpoons solute_{dissolved}$$

The oppositely pointing arrows signify that equilibrium exists between the materials on one side and those on the other. The change from left to right is called the *forward reaction,* and the opposing change is the *reverse reaction.*

Equilibria Respond to Stresses.   Once an equilibrium becomes established, no net change occurs *spontaneously.* We can still *force* a change, however, by altering the conditions. When we do, we say that we have *upset* the equilibrium. Any change in conditions that upsets an equilibrium is called a *stress* on the equilibrium. Adding heat, for example, is a common stress and will upset the equilibrium in a saturated solution. Let's consider the case of a solid in contact with its saturated solution in which the forward reaction (dissolving) is endothermic. When the forward change is endothermic, we may even write "heat" as a "reactant." For example,

■ Although heat often *evolves* when solids are added to *pure water,* nearly saturated solutions require the addition of heat to make more solute dissolve.

$$solute_{undissolved} + heat \rightleftharpoons solute_{dissolved}$$

The addition of heat to this system causes some solid to dissolve because dissolving uses up heat. The stress of more heat causes the rate of the forward change (dissolving) to speed up over the rate of the reverse change (coming out of solution). We say that the equilibrium *shifts* in favor of the forward change. However, at least temporarily, the system is actually no longer in equilibrium because a net change (dissolving) is occurring.

If we maintain the temperature of the now more concentrated solution at a constant higher value, the rate of the return of solute to the crystals eventually catches up to the rate of dissolving. The opposing rates become equal again. Although *both* the forward and reverse changes are faster at the higher temperature, no further net change occurs. Once again there is dynamic equilibrium. It doesn't matter how rapidly the opposing changes occur; equilibrium exists as long as they take place at *equal* rates.

■ *Dynamic* signifies much activity, and *equilibrium* means no net change.

The removal of heat by cooling can also be a stress on the equilibrium. If we cool the solution that we're studying, additional solute will leave its dissolved state and join what is already undissolved. Now the rate of the reverse change is temporarily faster than the rate of the forward. In time, the rates of the forward and reverse changes again become equal (and slower). We have equilibrium once again, but now the solution is less concentrated.

Le Châtelier's Principle Lets Us Predict How a Stressed Equilibrium Will Shift.   In which direction a stress will shift an equilibrium can be predicted by an insight credited to Henri Louis Le Châtelier (1850–1936), a French chemist.

■ Many chemical reactions in the body are actually shifts in chemical equilibria in response to chemical stresses.

LE CHÂTELIER'S PRINCIPLE   An equilibrium always responds to a stress by shifting in whatever direction *absorbs* the stress.

The stress of *adding* heat to our saturated solution is relieved by a heat-*consuming* reaction, the forward change, whereby more solute dissolves. The stress of *removing* heat or cooling the same system is absorbed by a heat-*supplying* change, the reverse reaction, by which some solute comes out of solution. Notice that whenever a system at equilibrium shifts in response to a stress, materials on one side of the double arrows increase at the expense of those on the other side. Heat is by no means the only stress; some equilibria are affected by changes in pressure.

Gas Solubility Increases with the Partial Pressure of the Gas. Important applications of Le Châtelier's principle involve the solubility of a gas in water and the exchange of gas between the liquid and gaseous regions. The solubilities of gases in water vary widely. In 100 g of water at 20 °C, for example, only 4.3 mg of oxygen dissolves, but 169 mg of carbon dioxide and 10.6 g of ammonia dissolve. Gas solubility, however, depends not only on the nature of the gas but also on the gas pressure. According to the **gas solubility–pressure law** or **Henry's law**, *the mass of a gas that dissolves in water is directly proportional to the partial pressure of the gas.*

■ The solubility data are for solutions under a total pressure of 1 atm and in contact with the pure gas.

The increase in gas solubility with pressure occurs because *pressure is a volume-reducing stress.* The volume occupied by the gas above its solution can be reduced in response to an increase in its pressure if more of the gas dissolves. Dissolving absorbs at least some of the stress of the higher pressure. Reducing the pressure, on the other hand, allows dissolved gas to leave a solution. For example, in bottled "club water," a carbonated beverage, the CO_2 is under pressure, so more has dissolved. When the bottle is uncapped, the pressure is released and CO_2 fizzes out of solution.

■ William Henry (1774–1832) was an English chemist who wrote a textbook in chemistry that went through eleven editions.

There is a corollary to Henry's law that deals with the partial pressure of a gas generated by some means within the solution and able to escape from it. *The more concentrated a solution is in some gas, the higher will the gas's partial pressure be above the solution at equilibrium.* The partial pressure of a gas above a solution is called the **gas tension** of the *solution.* In the fluid of the average human cell during rest, for example, the oxygen tension is about 30 mm Hg. In the lungs, the partial pressure of oxygen in alveolar air is slightly more than 100 mm Hg. Because gases diffuse from a higher to a lower pressure, oxygen for this reason alone tends to move from the lungs to the cells. (There are other forces at work, too.) At high altitudes, however, the partial pressure of oxygen in air is too low to force oxygen into the blood at a rate rapid enough for life (as we discussed in Special Topic 5.1 in the previous chapter).

■ *Diffusion* is the spontaneous intermingling of one substance with another.

Carbon dioxide, on the other hand, has a gas tension in the fluids of cells at rest of about 50 mm Hg. In alveolar air in the lungs, the partial pressure of CO_2 is about 40 mm Hg. The natural tendency of carbon dioxide, therefore, is to move from cells to lungs, from a region of higher to one of lower gas tension. But this is the direction in which waste CO_2 should move in order to be removed from the body.

One application of Henry's law affects those who work in high-pressure environments, like deep sea divers. They run serious risks when exposed to high air pressure because more air (both its nitrogen and oxygen) dissolves in their blood. Special Topic 6.1 discusses decompression sickness (the "bends"), which can disable or kill its victims.

Sometimes patients with crushing injuries or victims of carbon monoxide poisoning are placed in *hyperbaric chambers* where they receive 100% oxygen at pressures up to 2 atm (as opposed to oxygen's partial pressure of 0.2 atm in air as we ordinarily breathe it). Under the higher oxygen pressure, the patient's blood takes up oxygen more efficiently for delivery to tissues.

■ *Hyperbaric* denotes a condition of higher than normal ("hyper") barometric pressure.

Gas Solubility Decreases with Temperature. Heating an aqueous solution of a gas drives the gas out of solution. All gases dissolve exothermically, so we can represent the equilibrium as follows in which heat is shown as a "product."

$$gas_{undissolved} \rightleftharpoons gas_{dissolved} + heat$$

People who work where the air pressure is high must return to normal atmospheric pressure slowly and carefully. Otherwise, they could experience the bends, or decompression sickness—severe pains in muscles and joints, fainting, and even deafness, paralysis, or death. At risk are deep-sea divers and those who work in deep tunnels where air pressures are increased to help keep out water.

Figure 1 describes the changes in solubilities of nitrogen and oxygen as the pressure increases. The straight-line plots, of course, are exactly in accordance with Henry's law. As you can see in the figure, under high pressure, the blood dissolves more nitrogen and oxygen than at normal pressure. If blood thus enriched in nitrogen and oxygen is too quickly exposed to lower pressures, these gases suddenly come out of solution in blood. Their microbubbles block the tiny blood capillaries, close off the flow of blood, and lead to the symptoms we described.

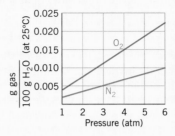

FIGURE 1
Solubilities of oxygen and nitrogen in water versus pressure.

If the return to normal pressure is made slowly, the gases leave the blood more slowly and can be removed as they emerge. The excess oxygen can be used by normal metabolism, and the excess nitrogen has a chance to be gathered by the lungs and removed by normal breathing. For each atmosphere of pressure above normal that the person was exposed to, about 20 minutes of careful decompression is usually recommended.

■ Because hot water holds less dissolved oxygen than cold water, game fish tend to avoid it when the summer is hottest, and they move to cooler depths.

When heat is *added,* the equilibrium must shift to the *left* to absorb the heat. As the temperature of a solution of gas in water approaches 100 °C, the gas loses essentially all of its solubility.

Some Gases, Like CO_2 and NH_3, Dissolve Partly by Reacting with Water. As we noted earlier, carbon dioxide and ammonia are much more soluble in water than oxygen. Carbon dioxide and ammonia molecules, however, do not simply intermingle with water molecules. A small percentage of the CO_2 molecules, for example, change to carbonic acid as follows, and this reaction helps to pull CO_2 into solution.

$$CO_2(aq) + H_2O \rightleftharpoons H_2CO_3(aq)$$
Carbonic acid

We'll consider the situation in aqueous ammonia in Chapter 7.

6.4 PERCENTAGE CONCENTRATIONS

The number of grams of solute in 100 g of solution is the weight/weight percentage concentration of the solution.

■ A positive test might be a change in color or odor, the appearance of a precipitate, the generation of heat, or the evolution of a gas.

Many times in the lab, a *reagent* is used in a test tube test. A **reagent** is simply a chemical system, usually a solution, prepared in advance for testing purposes, such as detecting the presence of a compound in a mixture. A small amount of the reagent, a few drops, might be added to the mixture. A particular observation, a "positive" test, would mean that the compound is present. We'll mention some reagents later, but no doubt you've used some in the lab already. We're interested here only in ways by which reagent concentrations are sometimes expressed.

Percentage Concentration Expressions Are Often Used for Reagents. A variety of percentage concentration expressions have been developed for reagents to supplement molar concentrations when the latter information is unnecessary. The most common is the **weight/weight percent (w/w %) concentration** of a solution, defined as the number of grams of solute in 100 g of the solution. A 10.0% (w/w) glucose solution, for example, has a concentration of 10.0 g of glucose in 100 g of solution. To make 100 g of this solution, you would mix 10.0 g of glucose with 90.0 g of the solvent for a total mass of 100 g.

■ We now use *percent* instead of *percentage* to conform to common usage.

EXAMPLE 6.1 Using Weight/Weight Percents

A special kind of salt solution, 0.900% (w/w) NaCl, is sometimes used in medicine. How many grams of 0.900% (w/w) NaCl solution contain 0.250 g of NaCl?

■ A 0.900% NaCl solution is called *isotonic saline solution.*

ANALYSIS We first identify the *goal,* a certain number of grams of the *solution;* we want this mass of solution to contain 0.250 g of NaCl, the *solute.* We need a connection between *grams of solute* and *concentration.* Conversion factors provide connections, and a concentration *in any units* can always be expressed as two conversion factors, two ratios. Working from the basic definition of weight/weight percent, we have the following conversion factors implicit in "0.900% (w/w)".

$$\frac{0.900 \text{ g NaCl}}{100 \text{ g NaCl soln}} \qquad \frac{100 \text{ g NaCl soln}}{0.900 \text{ g NaCl}}$$

Whenever you're stuck in trying to solve this kind of problem, try starting with this approach; use a basic definition to give you options of conversion factors. Simply looking at them might suggest your next step.

SOLUTION We multiply the "given," 0.250 g NaCl, by the second factor so that the final units will be what we seek, "g NaCl soln."

$$0.250 \text{ g NaCl} \times \frac{100 \text{ g NaCl soln}}{0.900 \text{ g NaCl}} = 27.8 \text{ g NaCl soln}$$

Thus, 27.8 g of 0.900% (w/w) NaCl contains 0.250 g of NaCl.

CHECK Does the answer make sense? Be sure to ask this kind of question. If 100 g of solution holds 0.900 g of solute, we'd need *less* than 100 g of solution to hold only 0.250 g—very roughly one-third of 100 g, as we calculated.

EXAMPLE 6.2 Preparing Weight/Weight Percent Solutions

How would you prepare 750 g of 0.900% (w/w) NaCl solution?

ANALYSIS Once again, we have to work from the basic definition of weight/weight percent and translate the given concentration, 0.900% (w/w) NaCl, into conversion factors.

$$\frac{0.900 \text{ g NaCl}}{100 \text{ g NaCl soln}} \qquad \frac{100 \text{ g NaCl soln}}{0.900 \text{ g NaCl}}$$

We are to calculate the number of grams of NaCl we must weigh out and dissolve in water to make the final mass of the solution equal 750 g, a "given" in this problem.

SOLUTION   We multiply the desired mass of the NaCl *solution* by the first conversion factor.

$$750 \text{ g NaCl soln} \times \frac{0.900 \text{ g NaCl}}{100 \text{ g NaCl soln}} = 6.75 \text{ g NaCl}$$

Thus, if we dissolve 6.75 g of NaCl in water and add enough water to make the final mass of the solution equal to 750 g, the label on the bottle can read 0.900% (w/w) NaCl.

CHECK   If we need only 0.9 g of NaCl to make 100 g of NaCl solution (from the *meaning* of 0.900% (w/w) NaCl solution), we'll surely need more than 0.9 g to make 750 g of solution, 7.5 times as much, in fact.

■   PRACTICE EXERCISE 1   Sulfuric acid (H_2SO_4) can be purchased from a chemical supply house as a solution that is 96.0% (w/w) H_2SO_4. How many grams of this solution contain 9.80 g of H_2SO_4 (or 0.100 mol)?

■   PRACTICE EXERCISE 2   How many grams of glucose and how many grams of water are needed to prepare 500 g of 0.250% (w/w) glucose?

A **volume/volume percent (v/v)%** gives us the number of volumes of one substance present in 100 volumes of the mixture. For example, when we say that air is 21% (v/v) in oxygen, we mean that there are 21 volumes of oxygen in 100 volumes of air.

A Number of "Hybrid" Percentage Concentrations Are in Use.   A weight/volume percent (w/v %) describes a concentration as the number of grams of solute in 100 mL of the solution (not 100 *grams* of solution but 100 *milliliters*). Thus a 0.90% (w/v) NaCl solution has 0.90 g of NaCl in every 100 mL of the solution. A weight/volume percent isn't a true percent because the units do not cancel. Volume/volume and weight/volume percent problems are handled through conversion factors just as we did for weight/weight percent problems.

■   The units *g/100 mL* are identical to *g/dL* (grams per deciliter) because 100 mL = 1 dL.

■   PRACTICE EXERCISE 3   How many grams of solute are needed to prepare each of the following solutions?
(a) 100 mL of 2.00% (w/v) $KMnO_4$ (b) 25 mL of 1.0% (w/v) NaOH

You may occasionally encounter special concentration expressions that are used when the solutions are very dilute. Mercifully, they are being phased out; but you may still see them. *Milligram percent,* for example, means the number of milligrams of solute in 100 mL of the solution. For an extremely dilute solution, the concentration might be given in *parts per million (ppm),* the number of parts (in any unit) in a million parts (the same unit) of the solution. *Parts per billion (ppb)* similarly means parts per billion parts, such as grams per billion grams.

■   One part per billion is like one penny in $10 million (1 billion pennies).

Because of the many ways the term *percent* can be taken, there is a trend away from using it. It's better simply to state the units in full. Thus, instead of describing a concentration of, say, 10.0% (w/v) KCl, the label or the report should read 10.0 g KCl/100 mL or 10.0 g KCl/dL. *When no units are given, just a percent, you have to assume that weight/weight percent is meant.*

6.5 COLLOIDAL DISPERSIONS

The sizes of the particles in a mixture, as distinct from their chemical identities, determine several properties.

There are two chief kinds of **homogeneous mixtures,** mixtures in which any small sample has the same composition and properties as any other sample of the same size taken anywhere else in the mixture. A **heterogeneous mixture** is a mixture of uneven composition throughout. The solution is one kind of homogeneous mixture and the other is the *colloidal dispersion.* Another kind of mixture, the *suspension,* is heterogeneous (unless vigorously stirred) because its components separate fairly quickly. All three types of mixtures—solution, colloidal dispersion, and suspension—are found in the body. They differ fundamentally in the sizes of the particles involved, differences that alone cause interesting and important changes in properties. Our chief interest in this section is with colloidal dispersions, but to understand them we need to contrast them with solutions.

In Solutions, the Dispersed Particles Are Smallest. The ions and molecules that make up solutions have formula masses of no more than a few hundred and diameters of 0.1 to 1 nm. We usually think of solutions as being liquids but, in principle, the solvent can be in any state—solid, liquid, or gas—as can the solute. Air, for example, is a solution of gases in each other.

■ 1 nanometer (nm) = 10^{-9}

Solutions are generally transparent—you can see through them—but are often colored. Solutes do not separate from the solvent under the influence of gravity, and the solutes can't be separated by passing the solution through ordinary filter paper. The blood carries many substances in solution, including the sodium ion, the chloride ion, and molecules of glucose, the chief sugar in blood.

In Colloidal Dispersions, the Particle Sizes Are Larger. A **colloidal dispersion** is a homogeneous mixture in which the dispersed particles are very large clusters of ions or molecules or are *macromolecules* (molecules with formula masses in the thousands or hundreds of thousands). The dispersed particles have diameters in the range 1 to 1000 nm.

Table 6.2 gives several examples of colloidal dispersions. They include many familiar substances, like whipped cream, milk, dusty air, jellies, and pearls. The blood also carries many substances in colloidal dispersions.

When colloidal dispersions are in a fluid state—liquid or gas—the dispersed particles, although large, are still not large enough to be trapped by ordinary filter paper during filtration. They are large enough, however, to reflect and scatter light (Figure 6.7). Light scattering by a colloidal dispersion is called the **Tyndall effect,** after British scientist John Tyndall (1820–1893). The Tyndall effect is responsible for the milky, partly obscuring character of smog and or the way sunlight sometimes seems to stream through a forest canopy.

■ Solutions do not exhibit the Tyndall effect because the solute particles are too small.

The large, dispersed particles in a fluid colloidal dispersion eventually separate

TABLE 6.2 Colloidal Systems

Type	Dispersed Phase[a]	Dispersing Medium[b]	Common Examples
Foam	Gas	Liquid	Suds, whipped cream
Solid foam	Gas	Solid	Pumice, marshmallow
Liquid aerosol	Liquid	Gas	Mist, fog, clouds, some air pollutants
Emulsion	Liquid	Liquid	Cream, mayonnaise, milk
Solid emulsion	Liquid	Solid	Butter, cheese
Smoke	Solid	Gas	Dust in smog
Sol	Solid	Liquid	Starch in water, jellies,[c] paints
Solid sol	Solid	Solid	Black diamonds, pearls, opals, alloys

[a]The colloidal particles constitute the dispersed phase.

[b]The continuous matter into which the colloidal particles are scattered is called the dispersing medium.

[c]Sols that adopt a semisolid, semirigid form (e.g., gelatin desserts, fruit jellies) are called **gels.**

■ Robert Brown (1773–1858), an English botanist, first observed this phenomenon when he saw the trembling of particles inside grains of pollen that he viewed with a microscope.

under the influence of gravity, but this can take time, anywhere from many hours to many decades! One factor that keeps the particles dispersed is that they are constantly buffeted and pushed around by collisions with solvent molecules. This motion of colloidal particles is called **Brownian movement.** Evidence for it can be seen with a good microscope. (What you see are the light scintillations caused as the colloidal particles move erratically about.)

In the most stable colloidal systems, all of the particles bear *like* electrical charges. They repel each other and so cannot coagulate into particles large enough to come under any rapid influence of gravity. Electrically charged, colloidally dispersed particles are common among proteins in living systems. (To balance the charges, other dissolved species of opposite charge, such as small ions, are also present.)

■ The stabilities of other emulsified sauces such as hollandaise sauce, béarnaise sauce, and salad dressing also depend on the lecithin of egg yolk.

Some colloidal dispersions are stabilized by *emulsifying agents.* **Emulsions** are colloidal dispersions of two liquids in each other, liquids that do not dissolve each other, like oil and vinegar. Mayonnaise, for example, is a dispersion of olive oil (and spices) in vinegar stabilized by a component of egg yolk called lecithin. Vinegar is

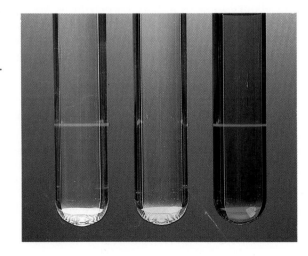

FIGURE 6.7
Tyndall effect. The tube on the left contains a colloidal starch dispersion, and the tube on the right has a colloidal dispersion of Fe_2O_3 in water. The middle tube has a solution of Na_2CrO_4, a colored solute. The thin red laser light is partly scattered in the two colloidal dispersions so it can be seen, but it passes through the middle solution unchanged.

TABLE 6.3   Characteristics of Three Mixtures: Solutions, Colloidal Dispersions, and Suspensions

Particle Sizes Become Larger →		
Solutions	**Colloidal Dispersions**	**Suspensions**
All particles are on the order of atoms, ions, or small molecules (0.1–1 nm)	Particles of at least one component are large clusters of atoms, ions, or small molecules or are very large ions or molecules (1–1000 nm)	Particles of at least one component may be individually seen with a low-power microscope (larger than 1000 nm)
Most stable to gravity	Less stable to gravity	Unstable to gravity
Most homogeneous	Also homogeneous but borderline	Homogeneous only if well stirred
Transparent (but often colored)	Often translucent or opaque, but may be transparent	Often opaque, but may appear translucent
No Tyndall effect	Tyndall effect	Not applicable (suspension cannot be transparent)
No Brownian movement	Brownian movement	Particles separate unless system is stirred
Cannot be separated by filtration	Cannot be separated by filtration	Can be separated by filtration
Homogeneous to Heterogeneous →		

about 5% acetic acid and so is mostly water, which has polar molecules. The molecules of olive oil are nonpolar, but lecithin molecules coat the oil droplets in such a way that their surfaces no longer present a nonpolar but a polar "face" to the water. Under the careful whipping of the slowly mixed ingredients as the mayonnaise is made, the oil droplets are made extremely tiny, all become coated with lecithin, and all now accumulate cages of water molecules. In this form, the droplets do not readily coalesce. If they did, the mayonnaise would separate.

Unstirred Suspensions Readily Separate. In **suspensions**, the dispersed or suspended particles are more than 1000 nm in average diameter, and they separate under the influence of gravity. They are also large enough to be trapped by filter paper. A suspension such as clay in water has to be stirred constantly to keep it from separating. The blood, while it is moving, is a suspension, besides being a solution and a colloidal dispersion. Suspended in circulating blood are its red and white cells and its platelets.

See Table 6.3 for a summary of the features of solutions, colloidal dispersions, and suspensions and how their properties change with particle sizes.

6.6   OSMOSIS AND DIALYSIS

The migration of ions and molecules through membranes is an important mechanism for getting nutrients inside cells and waste products out.

■ From the Greek *kolligativ,* "depending on number" and not on identity or nature.

Colligative properties are those properties of solutions and colloidal dispersions that depend not on the chemical identities of the solutes but on nothing more than their concentrations. Two examples of colligative properties are the depression of the freezing point and the elevation of the boiling point, meaning that solutions have lower melting points and higher boiling points than their pure solvents. Aqueous solutions, for example, freeze not at 0 °C but at slightly lower temperatures. Their normal boiling points are also not at 100 °C but at slightly higher temperatures.

■ Both NaCl and KBr break up into two ions as they dissolve.

The magnitude of a colligative property, as we said, depends on concentration, not chemical identities. For example, if we prepare two solutions, one of 1.0 mol of NaCl in 1000 g water and the other of 1.0 mol of KBr in 1000 g water, each freezes at −3.4 °C, and each boils at 101 °C (at 760 mm Hg). Both the freezing and the boiling points are the same despite the chemical difference in the solutes, *because the ratios of the moles of ions to the moles of water in both solutions are identical.* Otherwise, the identities of the solutes are immaterial.

■ The temperature at equilibrium of a mixture made from 33 g of NaCl and 100 g of ice is about −22 °C (−6 °F).

When concentrations are very high, the colligative effects can be large and important. For example, antifreeze in a radiator works because the presence of the solute makes the freezing point of the coolant much less than that of water. A 50% (v/v) aqueous solution of antifreeze doesn't freeze until about −40 °F (−40 °C).

Osmosis Is the Diffusion of Water Molecules through Membranes.   The colligative properties of the highest importance at the molecular level of life are osmosis and a closely related phenomenon, *dialysis.* Cells in the body are enclosed by membranes. On both sides of such membranes are aqueous systems with substances in both solution and colloidal dispersion. Materials and water have to be able to move through membranes in either direction so that nutrients can enter cells and wastes can leave.

One way that ions and molecules get through cell membranes is by **active transport.** *Active* signifies the participation in the membrane itself, of components, usually protein molecules, that form channels. They accept and pass ions and molecules through the membrane using energy obtained from metabolism. We do no more in this section than mention active transport, but it is very important and we'll return to it in Chapter 16.

■ From the Latin *permeare,* "to go through."

The other means of passing things through membranes is spontaneous dialysis, and the simplest form of dialysis is called *osmosis.* Membranes of cells are **semipermeable;** they can let some but not all kinds of molecules and ions through. Cellophane is an example of a synthetic, semipermeable membrane. In contact with an aqueous solution, cellophane allows only water molecules and other small molecules and ions through. It stops molecules of colloidal size. Evidently, cellophane has ultrafine pores just large enough for small particles but too small for colloidal particles.

Some semipermeable membranes have pores so small that only water molecules can pass through. Ions, being hydrated, have effective sizes that are apparently too large. No other molecules can get through either. A semipermeable membrane so selective that only water molecules pass through is called an **osmotic membrane.**

When two aqueous solutions with different concentrations of solute particles are separated by an osmotic membrane, osmosis occurs. **Osmosis** is the net migration of water from the solution of lower concentration of solute into the solution of higher concentration. If osmosis could continue long enough, the concentrated solution

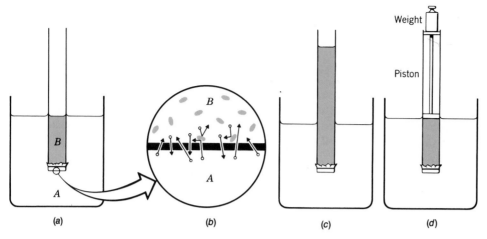

(a)   (b)   (c)   (d)

FIGURE 6.8
Osmosis and osmotic pressure. *(a)* In the beaker, *A*, there is pure water and in the tube, *B*, there is a solution. An osmotic membrane closes the bottom of the tube. *(b)* A microscopic view of an osmotic membrane shows how solute particles interfere with the movement of water molecules from *B* to *A*, but not from *A* to *B*. *(c)* The level in *A* has fallen and that in *B* has risen because of osmosis. *(d)* To prevent osmosis, a back-pressure is needed, and the exact amount of pressure is the osmotic pressure of the solution in *B*.

would become dilute enough and the other solution (by losing *water*) would become concentrated enough to make the two concentrations equal.

Figure 6.8 helps to explain why the net flow of water in osmosis is into the more concentrated solution. The figure shows the special case in which pure water is on one side of the membrane. Water molecules can move in *both* directions, but the solute particles on the solution side interfere with the movement of water molecules into and through the membrane. Water molecules, therefore, cannot leave as frequently from this side as they can come in from the side holding pure water. Thus, more water molecules enter the concentrated solution than leave it, and the concentrated solution becomes increasingly diluted. Eventually, the rising column of water shown in Figure 6.8 exerts a high enough back-pressure to prevent any further rise, and osmosis stops.

The back-pressure necessary to prevent osmosis is called the **osmotic pressure** of the solution. Even dilute solutions can have high osmotic pressures. For example, a $0.100\,M$ solution of sugar in water has an osmotic pressure of nearly 2000 mm Hg (2.6 atm). This dilute solution could, in the right apparatus, support a column of water 25.3 m (83.0 ft) high!

■ This way of thinking about the *direction* of the net flow of water in osmosis, incidentally, is perhaps the best way to remember how it goes. *The net flow is such as to make the concentrated solution more dilute.*

■ Osmotic pressure is one factor that causes sap to rise in trees.

Separated Ions as Well as Molecules Contribute Individually to Osmotic Pressure.
The osmotic pressure of a solution has to be understood not as something that the solution is actually exerting, like some hand pushing on a surface. Instead, osmotic pressure is a *potential* directly related to the concentration of the solute particles. The potential actually is fully activated only when an osmotic membrane separates the solution from pure water.

The particles responsible for osmosis can be ions, molecules, or macromolecules. When the solute is an ionic compound, like NaCl, the particles in solution are not molecules but separated Na^+ and Cl^- ions. Two solute particles are thus in solution for each formula unit of NaCl that dissolves. The molar concentration of solute particles, therefore, is twice the molar concentration of the salt. In $0.10\,M$ NaCl, for example, the concentration of all ions is 2×0.10 mol/L $= 0.20$ mol/L. The osmotic pressure of $0.10\,M$ NaCl is therefore twice as large as that of $0.10\,M$ glucose, which does not break up into ions. An ionic compound like Na_2SO_4 breaks up into three ions when one formula unit dissolves, two Na^+ ions and one SO_4^{2-} ion. The concentration of all ions in $0.10\,M$ Na_2SO_4 is therefore 3×0.10 mol/L $= 0.30$ mol/L.

Osmolarity Relates Better to the Osmotic Pressure of a Solution Than Molarity.
Because *molarity* does not reveal enough about a solution when we think about its

■ Sometimes the *milliosmole, mOsmol* is the unit used: 1 Osmol = 10^3 mOsmol.

osmotic pressure, the **osmolarity** of a solution, abbreviated **Osm**, is used to express the molar concentration of all osmotically active particles. A solution containing 0.10 mol of NaCl/L, for example, has a molarity of 0.10 M but an osmolarity numerically twice as large, 0.20 Osm. The osmolarity of 0.10 M Na_2SO_4 is 0.30 Osm.

■ PRACTICE EXERCISE 4   Assuming that any *ionic* solutes in this exercise break up completely into their constituent ions when they dissolve in water, what is the osmolarity of each solution?

(a) 0.010 M NH_4Cl (which ionizes as NH_4^+ and Cl^-)

(b) 0.005 M Na_2CO_3 (which ionizes as $2Na^+$ and CO_3^{2-})

(c) 0.100 M fructose (a sugar and a molecular substance)

(d) A solution that contains both fructose and NaCl with concentrations of 0.050 M fructose and 0.050 M NaCl

---

Small Solute Ions and Molecules Pass through Dialyzing Membranes.   **Dialysis** is like osmosis except that ordinary sized ions and molecules can move through a dialyzing membrane. A dialyzing membrane can be thought of as having larger pores than an osmotic membrane. Cell membranes are largely but not entirely dialyzing membranes. They always incorporate components that selectively block the migrations of some small ions and molecules but allow others to pass through. Sometimes the special blocking components, usually proteins, are controlled by hormones.

■ Most natural membranes are dialyzing membranes. Osmotic membranes are difficult to prepare.

Dialysis produces a net migration of water only if the fluid on one side of the dialyzing membrane has a higher concentration in *colloidal* substances than the other. Colloidal size particles are blocked by dialyzing membranes, so they get in the way of the movements of smaller particles through the membrane. The *net* flows of water and small solute particles are in opposite directions. The solutes move toward the side of the membrane having the more dilute solution; water (as a *net* flow) moves oppositely.

The contribution made by colloidal particles to an imbalance in concentration gives the system a **colloidal osmotic pressure,** which is similar in meaning to osmotic pressure. In the next section we present some situations involving life at the molecular level where osmotic pressure relationships are very critical and depend on the colloidal osmotic pressure of blood.

6.7   DIALYSIS AND THE BLOOD

When the osmotic pressure of blood varies too much, the result can be shock or harmful damage to red blood cells.

The body tries to maintain the concentrations of all of the substances in the blood within fairly narrow limits. To do this, the body has a thirst mechanism to help bring in water, and it has the machinery of diuresis and perspiration to let water leave. A number of hormones are involved in maintaining the integrity of blood, but in this section we look briefly at two problems that occur when the osmotic pressure of the blood changes too much.

■ *Diuresis* is the formation of urine.

A Loss of the Blood's Macromolecules Triggers the Shock Syndrome.   When a person goes into shock, blood capillaries have become unable to prevent the loss of macromolecules from the blood. They are mostly molecules of albumin, a protein, and

■ A *syndrome* is the whole collection of symptoms that characterize a disease.

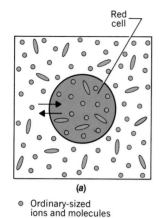
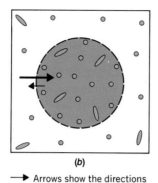
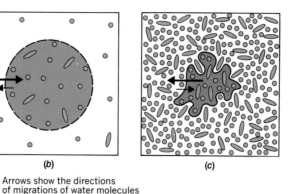

Red cell

(a)   (b)   (c)

○ Ordinary-sized ions and molecules

⬭ Macromolecules

→ Arrows show the directions
← of migrations of water molecules

FIGURE 6.9
Dialysis. *(a)* The red cell is in an isotonic environment. *(b)* Hemolysis is about to occur because the red cell is swollen by extra fluids brought into it from its hypotonic environment. *(c)* The red cell experiences crenation when it is in a hypertonic environment.

their loss quickly reduces the colloidal osmotic pressure of blood. When blood has a lower colloidal osmotic pressure, water is less able to diffuse back into it from the surrounding spaces. Yet water is still able to *leave* the blood. The net result is a loss in blood volume, and this upsets the mechanisms that bring nutrients to the brain and carry wastes away. The result to the nervous system is called *shock*.

Red Blood Cells Hemolyze in a Medium of Low Osmolarity. In some clinical situations, body fluids need replacement or nutrients have to be given by intravenous drip. It is important that the osmolarity of the solution being added to the blood matches that of the fluid inside the red cells. Otherwise, *hemolysis* or *crenation* of red blood cells occurs. Let's see what these terms mean.

Millions of red blood cells circulate in the bloodstream, and their membranes behave as dialyzing membranes. Within each red cell is an aqueous fluid containing dissolved and colloidally dispersed substances (Figure 6.9a). Although the dispersed particles are too large to dialyze, they contribute to the colloidal osmotic pressure. They help, therefore, to determine the *direction* in which water flows as dialysis occurs through the membrane of a red cell. When red cells are placed in pure water, for example, enough water will *enter* the red cells to make them burst open (Figure 6.9b). The rupturing of red cells is called **hemolysis,** and we say that the cells *hemolyze*.

Dialysis also occurs if we put red cells in a medium with an osmolarity *higher* than their own fluid. Water *leaves* the red cells and moves into the solution. Now the cells, losing fluid volume, shrivel and shrink, a process called **crenation** (Figure 6.9c).

Two solutions of equal osmolarity are called **isotonic solutions.** If one has a lower osmotic pressure than the other, the first is said to be **hypotonic** with respect to the second. A hypotonic solution has a lower osmolarity than the one with which it is compared. Red cells hemolyze if placed in a hypotonic environment, including pure water.

If one of the two solutions has a higher osmotic pressure than the other, it is described as **hypertonic** compared with the second. Thus, 0.14 *M* NaCl is hypertonic with respect to 0.10 *M* NaCl. Red cells undergo crenation when they are in a hypertonic environment.

A 0.9% (w/w) NaCl solution, called **physiological saline solution,** is isotonic with respect to the fluid inside a red cell. Any solution that is administered into the bloodstream of a patient normally should be similarly isotonic.

All of the topics we have studied in this and the preceding section are important factors in the operation of artificial kidney machines, which are discussed in Special Topic 6.2.

The kidneys cleanse the bloodstream of nitrogen waste products such as urea and other wastes. If the kidneys stop working efficiently or are removed, these wastes build up in the blood and threaten the life of the patient. The artificial kidney is one remedy for this.

The overall procedure is called hemodialysis—the dialysis of blood—and Figure 1 shows how it works. The bloodstream is diverted from the body and pumped through a long, coiled cellophane tube that serves as the dialyzing membrane. (The blood is kept from clotting by an anticlotting agent such as heparin.) A solution called the dialysate circulates outside of the cellophane tube. This dialysate is very carefully prepared not only to be isotonic with blood but also to have the same concentrations of all the essential substances that should be left in solution in the blood. When these concentrations match, the rate at which such solutes migrate out of the blood equals the rate at which they return. In this way several key equilibria are maintained, and there is no net removal of essential components. Figure 2 shows how this works. The dialysate, however, is kept very low in the concentrations of the wastes, so the rate at which they leave the blood is greater than the rate at which they can get back in. In this manner, hemodialysis slowly removes the wastes from the blood.

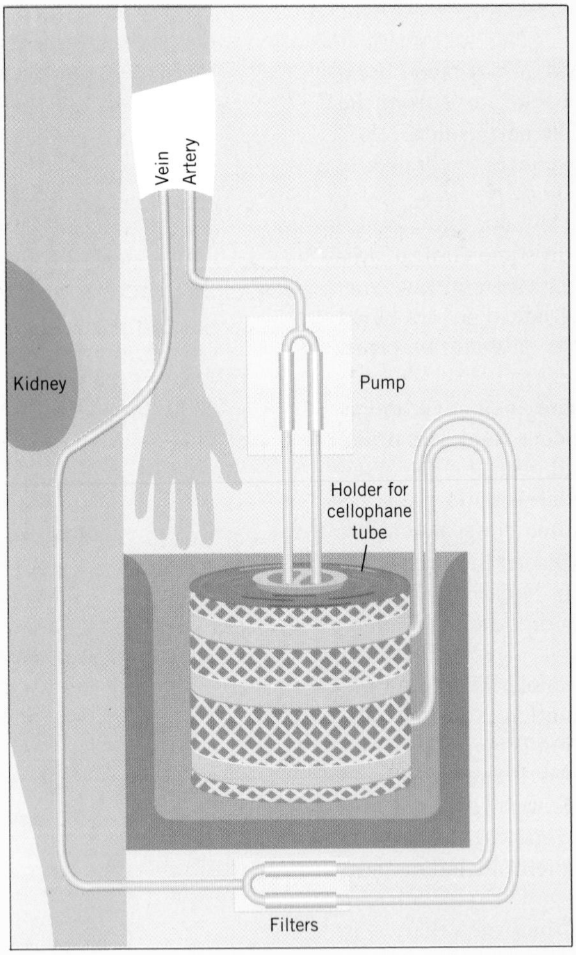

FIGURE 1
Schematic of an artificial kidney. (Courtesy Artificial Organs Division, Travenol Laboratories, Inc.)

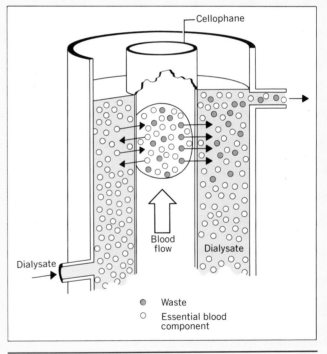

FIGURE 2
Waste molecules move out of the blood faster than they can return, but essential substances leave and return at equal rates.

SUMMARY

Water The higher electronegativity of oxygen over hydrogen and the angularity of the water molecule make water a very polar compound, so polar that hydrogen bonds exist between the molecules. The high polarity of water explains many of its unusual thermal properties, such as its high heats of fusion and vaporization, its high surface tension, and its ability to dissolve ionic and polar molecular compounds.

Hydrogen Bonds When hydrogen is covalently bonded to atoms of any of the three most electronegative elements (O, N, or F), its partial positive charge is large enough to be attracted rather strongly to the partial negative charge on an atom of O, N, or F on a nearby molecule. This force of attraction is the hydrogen bond. The most common errors are to consider this bond as the bond within a molecule of hydrogen or to think of it as a covalent bond within some molecule. It's neither. The hydrogen bond is a force of attraction between the $\delta+$ on H in H—O, H—N, or H—F to the $\delta-$ of another O, N, or F.

Hydration The attraction of water molecules to ions or to polar molecules leads to a loose solvent cage that shields the ions or molecules from each other. This phenomenon is called hydration. Sometimes water of hydration is present in a crystalline material in a definite proportion to the rest of the formula unit, and such a substance is a hydrate. Heat converts most hydrates to their anhydrous forms. And some anhydrous forms serve as drying agents or desiccants.

Solutions Ions and molecules of ordinary size, if soluble in water at all, form solutions, homogeneous mixtures that neither gravity nor filtration can separate. The solubilities of most solids increase with temperature, because the process of their dissolving is usually endothermic. (More energy is needed to break up the crystal than is recovered as the solvent cages form about the ions or molecules.)

Le Châtelier's Principle A stress on a dynamic equilibrium causes the equilibrium to shift in whatever direction absorbs the stress.

Gas Solubilities According to Henry's law, the solubility of a gas is directly proportional to its partial pressure in the space above the solution. Some gases do more than mechanically dissolve in water; part of what dissolves reacts with water to form soluble species. The availability of a gas from an aqueous solution can be expressed as the gas tension of the gas in the solution.

Percent Concentration A variety of concentration expressions have been developed to provide ways of describing a concentration without going into molar concentrations. These include weight/weight percents, volume/volume percents, and hybrid descriptions that aren't true percentages: weight/volume percent, milligram percent, parts per million, and parts per billion.

Colloidal Dispersions Large clusters of ions or molecules or macromolecules do not form true solutions but colloidal dispersions. These can reflect and scatter light (Tyndall effect), experience Brownian movement, and (in time) succumb to the force of gravity (if the medium is fluid). Protective colloids sometimes stabilize these systems. If the dispersed particles grow to an average diameter of about 1000 nm, they slip over into the category of suspended matter, and such systems must be stirred to maintain the suspension.

Osmosis and Dialysis When a semipermeable membrane separates two solutions or dispersions of unequal osmolarities, a net flow occurs in the direction that, if continued, would produce solutions of identical osmolarities. When the membrane is osmotic, only the solvent can migrate, and the phenomenon is osmosis. The back pressure needed to prevent osmosis is called the osmotic pressure, and it's directly proportional to the concentration of all particles of solute that are osmotically active—ions, molecules, and macromolecules.

REVIEW EXERCISES

The answers to Review Exercises whose numbers are in color are found in Appendix V. The answers to the other Review Exercises are found in the Study Guide that accompanies this text. The more challenging questions are marked with asterisks.

Hydrogen Bond

6.1 The hydrogen molecule, H—H, does not become involved in hydrogen bonding.

(a) What kind of bond occurs in a hydrogen molecule?

(b) Why can't this molecule become involved in hydrogen bonding?

6.2 The methane molecule, CH_4, does not become involved in hydrogen bonding. Why not?

6.3 Draw the structures of two water molecules. Write in $\delta+$ and $\delta-$ symbols where they belong. Then draw a correctly positioned dotted line between the two molecules to symbolize a hydrogen bond.

6.4 Hydrogen bonds exist between two molecules of ammonia, NH_3. Draw the structures of two ammonia molecules. (You don't have to try to duplicate their tetrahedral geometry.) Put $\delta+$ and $\delta-$ signs where they should be located. Then draw a dotted line that correctly connects two points to represent a hydrogen bond.

*6.5** The hydrogen bond between two molecules of ammonia must be much weaker than the hydrogen bond between two molecules of water.
(a) How do boiling point data suggest this?
(b) What does this suggest about the relative sizes of the $\delta+$ and $\delta-$ sites in molecules of water and ammonia?
(c) Why are the $\delta+$ and $\delta-$ sites different in their relative amounts of fractional electric charge when we compare molecules of ammonia and water?

6.6 If it takes roughly 100 kcal/mol to break the covalent bond between O and H in H_2O, about how many kilocalories per mole are needed to break the hydrogen bonds in a sample of liquid water?

6.7 Explain in your own words how hydrogen bonding helps us understand each of the following.
(a) The high heats of fusion and vaporization of water
(b) The high surface tension of water

6.8 Explain in your own words why water forms tight beads on a waxy surface but spreads out on a clean glass surface.

6.9 What does a surfactant do to water's surface tension?

6.10 What common household materials are surfactants?

6.11 What surfactant is involved in digestion? What juice supplies it? How does it aid digestion?

6.12 Bile comes from the gallbladder, and when a patient has the gallbladder removed, he or she is put on a diet that is relatively low in fats and oils. Why?

Aqueous Solutions

6.13 In a crystal of sodium chloride the chloride ions are surrounded by oppositely charged ions (Na^+) as nearest neighbors. What replaces this kind of electrical environment for chloride ions when sodium chloride dissolves in water?

6.14 When we say that a sodium ion in water is *hydrated*, what does this mean? (Make a drawing as part of your answer.)

6.15 Carbon tetrafluoride, CF_4, has unusually polar C—F bonds, and its molecules are tetrahedral, like those of meth-

ane, CH_4. This substance does not dissolve in water. Why won't water let CF_4 molecules in?

6.16 Suppose that you do not know and do not have access to a reference in which to look up the solubility of sodium nitrate, $NaNO_3$, in water at room temperature. Yet you need a solution that you know beyond doubt is saturated. How can you make such a saturated solution and know that it is saturated?

Hydrates

6.17 Epsom salt is magnesium sulfate heptahydrate. Write the equation for the decomposition of this hydrate to its anhydrous form. (*Hepta-* denotes seven.)

6.18 Why are hydrates classified as compounds and not as wet mixtures?

6.19 When water is added to anhydrous sodium sulfate, the decahydrate of this compound forms. Write the equation. (*Deca-* denotes ten.)

6.20 Anhydrous calcium chloride is hygroscopic. What does this mean? Does this property make it useful as a desiccant?

6.21 Sodium hydroxide is sold in the form of small pellets about the size and shape of split peas. It is a very deliquescent substance. What can happen if you leave the cover off of a bottle of sodium hydroxide pellets?

*6.22** When 6.47 g of the hydrate of compound X was strongly heated to drive off all of the water of hydration, the residue, the anhydrous form of X, had a mass of 3.90 g. What number should y be in the formula of the hydrate, $X \cdot yH_2O$? The formula mass of X is 82.0.

*6.23** When all of the water of hydration was driven off of 4.32 g of a hydrate of compound Z, the residue, the anhydrous form of Z, had a mass of 3.09 g. What is the formula of the hydrate (using the symbol Z as part of it)? The formula mass of Z is 90.0.

Dynamic Equilibria and Le Châtelier's Principle

6.24 State Le Châtelier's principle.

6.25 Consider the following dynamic equilibrium that describes a saturated solution of sodium chloride in water.

$$NaCl(s) + heat \rightleftharpoons Na^+(aq) + Cl^-(aq)$$

(a) Write the equation for the forward reaction.
(b) Write the equation for the reverse reaction.
(c) Which reaction, forward or reverse, is endothermic? What does this mean?
(d) If a beaker containing this saturated system at 40 °C were cooled to 5 °C, which reaction, forward or reverse, would become more rapid than the other?
(e) After the system rested at 5 °C for a long period, what would become of the rates of the forward and reverse reactions? (Would they become equal or remain unequal?)

(f) Would the concentration of the NaCl solution at equilibrium at 5 °C be higher or lower than it is at 40 °C?

(g) If more solid sodium chloride were added to the solution at equilibrium at 5 °C, would the solution become more concentrated?

(h) If the solution at equilibrium at 5 °C were warmed, what specific stress would be placed on the equilibrium?

(i) The stress in part (h) causes the equilibrium to do what? (What verb is used to describe this?) This change would be in what direction?

6.26 The following expression describes the equilibrium that exists when ice and water, both at 0 °C, are mixed.

$$H_2O(s) + heat \rightleftharpoons H_2O(l)$$

(a) What could be done, experimentally, to make the rate of the exothermic process greater than the rate of the endothermic change?

(b) According to Le Châtelier's principle, which change, forward or reverse, becomes favored by the removal of heat from this system? Does the ratio of the mass of ice to liquid water increase or decrease as a result?

(c) The addition of NaCl (which dissolves) puts Na^+ and Cl^- ions into the liquid water but not into the ice crystals. These ions slow down the rate of the reverse process but not the forward change. (The ions interfere with the return of water molecules to the crystal surface, but not with their escape from these surfaces.) To reestablish the equilibrium what has to be done to the rate of the forward change? Would this be done by adding heat or removing heat? In other words, would you have to raise or lower the temperature of the system to reestablish equilibrium? (Another way of asking this is "Does salty water freeze at a temperature above or below 0 °C?")

6.27 If the solubility of a compound *decreases* with increasing temperature, how would you write the equilibrium expression, as A or as B?

(A) solid + heat $\rightleftharpoons$ solution
(B) solid $\rightleftharpoons$ solution + heat

6.28 Ammonium chloride dissolves in water endothermically. Suppose that you have a saturated solution of this compound, that its temperature is 30 °C, and that undissolved solute is present. Write the equilibrium expression for this saturated solution, and use Le Châtelier's principle to predict what will happen if you cool the system to 20 °C.

6.29 We have to distinguish between *how fast* something dissolves in water and *how much* can dissolve to make a saturated solution. The speed with which we can dissolve a solid in water increases if we (a) crush the solid to a powder, (b) stir the mixture, or (c) heat the mixture. Use the kinetic theory as well as the concept of forward and reverse processes to explain these facts.

Gas Solubilities

6.30 What is Henry's law? How is advantage taken of it in the technology of hyperbaric chambers?

6.31 Explain why carbon dioxide is more soluble in water than is nitrogen.

6.32 Using Le Châtelier's principle, explain why the solubility of a gas in water should decrease with decreasing partial pressure of the gas.

6.33 If the gas tension of O_2 in blood is described as 80 mm Hg, what specifically does this mean?

6.34 If in one region of the body the gas tension of oxygen over blood is 79 mm Hg and in a second region it is 60 mm Hg, which region (the first or the second) has a higher concentration of oxygen in the blood itself?

6.35 A patient is placed in a *hyperbaric chamber*. What does this mean and why is it done?

Percent Concentrations

6.36 If a solution has a concentration of 0.915% (w/w) NaOH, what two conversion factors can we write based on this value?

6.37 A solution bears the label 1.42% (w/v) KCl. What two conversion factors can be written for this value?

6.38 A solution of rubbing alcohol in water is described as 30% (v/v). What two conversion factors are possible from this value?

6.39 How many grams of solute are needed to prepare each of the following solutions?
(a) 250 g of 0.900% (w/w) NaCl
(b) 500 g of 3.22% (w/w) $NaCH_3CO_2$
(c) 125 g of 6.75% (w/w) NH_4Cl
(d) 250 g of 1.25% (w/w) Na_2CO_3

6.40 Calculate the number of grams of solute needed to make each of the following solutions.
(a) 250 g of 0.625% (w/w) NaI
(b) 125 g of 0.375% (w/w) NaBr
(c) 100 g of 1.00% (w/w) $C_6H_{12}O_6$ (glucose)
(d) 50 g of 8.50% (w/w) H_2SO_4

6.41 How many grams of solute have to be weighed out to make each of the following solutions?
(a) 250 mL of 12.5% (w/v) NaCl
(b) 500 mL of 1.50% (w/v) KBr
(c) 100 mL of 1.12% (w/v) $CaCl_2$
(d) 500 mL of 0.900% (w/v) NaCl

6.42 To prepare the following solutions, how many grams of solute are required?
(a) 125 mL of 6.00% (w/v) $Mg(NO_3)_2$
(b) 250 mL of 2.25% (w/v) NaBr
(c) 100 mL of 5.00% (w/v) KI
(d) 75.0 mL of 1.25% (w/v) $Ca(NO_3)_2$

6.43 How many milliliters of ethyl alcohol have to be used to make 500 mL of 20.0% (v/v) aqueous ethyl alcohol solution?

6.44 A sample of 750 mL of 6.25% (v/v) aqueous methyl alcohol contains how many milliliters of pure methyl alcohol?

*6.45 A chemical supply room has supplies of the following solutions: 4.00% (w/w) NaOH, 10.00% (w/w) Na_2CO_3, and 4.00% (w/v) glucose. If the densities of these solutions can be taken to be 1.00 g/mL, how many milliliters of the appropriate solution would you have to measure out to obtain the following quantities?
(a) 3.00 g of NaOH
(b) 0.325 g of Na_2CO_3
(c) 0.255 g of glucose
(d) 0.115 mol of NaOH
(e) 0.200 mol of glucose ($C_6H_{12}O_6$)

*6.46 The stockroom has the following solutions: 3.00% (w/w) KOH, 0.600% (w/w) HCl, and 1.00% (w/v) NaCl. Assuming that the densities of these solutions are all 1.00 g/mL, how many milliliters of the appropriate solution have to be measured out to obtain the following quantities of solutes?
(a) 0.100 g of KOH
(b) 0.200 g of HCl
(c) 0.115 mol of NaCl
(d) 0.125 mol of KOH

*6.47 A student needed 75.0 mL of 12.5% (w/w) aqueous sodium acetate, $NaCH_3CO_2$. (The density of this solution is 1.05 g/mL.) Only the trihydrate of this compound, $NaCH_3CO_2 \cdot 3H_2O$, was available, and the student knew that the water of hydration would just become part of the solvent once the solution was made. How many grams of the trihydrate would have to be weighed out to prepare the needed solution?

*6.48 How many grams of $Na_2SO_4 \cdot 10H_2O$ have to be weighed out to prepare 250 mL of 5.00% (w/w) Na_2SO_4 in water? (The density of this solution is 1.09 g/mL.)

Solutions and Colloidal Dispersions

6.49 What is always true about solutions that is not necessarily true about mixtures in general?

6.50 Particle *size* is one basis for distinguishing among solutions, colloidal dispersions, and (stirred) suspensions. Explain why size works for this purpose.

6.51 Which of kind of mixture, solution, colloidal dispersion, or suspension
(a) can be separated into its components by filtration?
(b) exhibits the Tyndall effect?
(c) shows observable Brownian movement?
(d) has the smallest particles of all kinds?
(e) is likeliest to be the least stable at rest over time?

6.52 What kinds of particles make the most stable colloidal dispersions? Explain.

6.53 The blood is simultaneously a solution, a colloidal dispersion, and a suspension. Explain.

6.54 A colloidal dispersion, but not a solution, exhibits the Tyndall effect. Explain.

6.55 What causes Brownian movement?

6.56 What simple test could be used to tell if a clear, colorless solution contains substances in colloidal dispersion?

6.57 What is an emulsion? Give some examples

6.58 What is a sol? Give some examples.

6.59 What is a gel? Give an example.

6.60 If a solution that contains 1.00 mol of sucrose in 1000 g of water freezes at $-1.86\ °C$, what is the freezing point of a solution that contains 1.00 mol of glucose in 1000 g of water? (Both are compounds that do not break up into ions when they dissolve.)

6.61 A solution that contains 1.00 mol of glucose in 1000 g of water has a normal boiling point of 100.5 °C. Another solution that contains 1.00 mol of an unknown compound in 1000 g of water has a normal boiling point of 101.0 °C. What is the likeliest explanation for the higher boiling point of the second solution?

Osmosis and Dialysis

6.62 Explain in your own words and drawings how osmosis gives a net flow of water from pure water into a solution on the other side of an osmotic membrane.

6.63 In general terms, how does an osmotic membrane differ from a dialyzing membrane?

6.64 Explain in your own words why the osmotic pressure of a solution should depend only on the concentration of its solute particles and not on its chemical properties.

6.65 Why is the osmolarity of 1.0 M NaCl not the same as its molarity?

*6.66 Which has the higher osmolarity, 0.10 M NaCl or 0.080 M Na_2SO_4? Explain.

*6.67 Which solution has the higher osmotic pressure, 5.0% (w/w) NaCl or 5.0% (w/w) KI? Both NaCl and KI break up in water in the same way, two ions per formula unit.

*6.68 Solution *A* consists of 0.60 mol of NaCl, 0.12 mol of $C_6H_{12}O_6$ (glucose, a molecular substance), and 0.055 mol of starch (a colloidal, macromolecular substance), all in 1000 g of water. Solution *B* is made of 0.60 mol of NaBr, 0.12 mol of $C_6H_{12}O_6$ (fructose, a molecular substance related to glucose), and 0.005 mol of starch, all in 1000 g of water. Which solution, if either, has the higher osmotic pressure? Explain.

6.69 What happens to red blood cells in crenation?

6.70 Physiological saline solution has a concentration of 0.90% (w/w) NaCl.
(a) Is a solution that is 1.1% (w/w) NaCl described as hypertonic or hypotonic with respect to physiological saline solution?
(b) What would happen, crenation or hemolysis, if a red blood cell were placed (1) in 0.5% (w/w) NaCl? (2) In 1.5% (w/w) NaCl?

6.71 Explain how the loss of macromolecules from the blood can lead to the increased loss of water from blood and a reduction in blood volume.

*6.72 For rehydration therapy for cholera patients, the World Health Organization (WHO) uses an aqueous solution with the following concentrations: 3.5 g NaCl/L, 2.5 g

$NaHCO_3$/L, 1.5 g KCl/L, and 20 g of glucose/L. Assuming that the ionic compounds break up fully into their ions (Na^+, Cl^-, HCO_3^-, and K^+), calculate the osmolarity of the solution.

Decompression Sickness (Special Topic 6.1)

6.73 The solubilities of which gases increase in blood to cause decompression sickness? Why do they increase?

6.74 How does the increased solubility of a gas in blood cause a problem when the individual comes back to normal pressure?

6.75 How does slow decompression reduce the possibility of decompression sickness?

6.76 What is the "rule of thumb" about the rate of decompression needed to avoid decompression sickness?

Hemodialysis (Special Topic 6.2)

6.77 What does *hemodialysis* mean?

6.78 During hemodialysis, what is the *dialysate*?

6.79 With respect to the following solutes in blood, what should be the concentration of the dialysate for effective hemodialysis, more or less concentrated or the same concentration?
(a) Na^+ (b) Cl^- (c) urea

Additional Exercises

*6.80 What is the osmolarity of a solution prepared by dissolving 10.0 g of KNO3 in a final volume of 500 mL of solution? (KNO_3 entirely breaks up as K^+ and NO_3^- ions.)

*6.81 Calculate the molarity and the osmolarity of 9.00% (w/w) NaCl. (The density of the solution is 1.0633 g/mL.)

7

ACIDS, BASES, AND SALTS

Sources of Ions and Electrolytes

Common Aqueous Acids and Bases

Chemical Properties of Aqueous Acids and Bases

Relative Strengths of Acids and Bases

Salts

Carbonate salts come out of solution as ground water slowly seeps and drips within limestone caverns. The solubilities of salts and how they can be made from acids and bases are among the topics of this chapter.

7.1 SOURCES OF IONS AND ELECTROLYTES

The principal ion producers in water are acids, bases, and salts.

Almost all water contains dissolved ions, whether the water is in lakes and rivers or in the fluids of living systems. Most experiments done in the lab and nearly all clinical analyses involve ions in solution. Sometimes the tiniest imbalances in the concentrations of certain ions in cells or in the blood cause the gravest medical emergencies.

We begin here a rather extensive study of ions, a study that spreads over more than one chapter. In this chapter we learn about the most general reactions of the principal families of ionic compounds—*acids, bases,* and *salts* and how they behave in water toward each other.

Even the Purest Water Contains Traces of H_3O^+ and OH^- Ions. Before we study the families of compounds that supply ions in water, we have to learn about the ions that form just from water molecules. When two water molecules collide powerfully enough (Figure 7.1), a proton, H^+, transfers from one molecule to another. Two ions form in a 1:1 ratio, the **hydronium ion,** H_3O^+, and the **hydroxide ion,** OH^-. Any reaction in which ions form from neutral molecules is called **ionization,** and this self-ionization of water is the forward reaction in a chemical equilibrium.

$$2H_2O \rightleftharpoons H_3O^+ \ (aq) + OH^- (aq)$$

The forward reaction is definitely not favored, because the concentration of each ion is only 1.0×10^{-7} mol/L (at 25 °C). This may seem too small to mention, but life hinges on holding the concentrations of H_3O^+ and OH^- in body fluids at about this value. In pure water, their concentrations are exactly equal, but acids, bases, and *even many salts* upset the equality.

The Acidity of a Solution Depends on the Ratio of H_3O^+ to OH^-. **Acids** make the molar concentration of H_3O^+ in aqueous systems higher than that of OH^-. **Bases or alkalies** do the opposite; they make the molar concentration of OH^- greater than that of H_3O^+. A **neutral solution** is one in which the molar concentrations of H_3O^+ and OH^- are equal. These definitions can be summarized as follows, where we use brackets, [ ], about a formula to mean its *molar concentration.*

Acidic solutions	$[H_3O^+] > [OH^-]$
Neutral solutions	$[H_3O^+] = [OH^-]$
Basic solutions	$[H_3O^+] < [OH^-]$

■ H· = hydrogen atom.
H^+ = hydrogen ion, a bare proton.

■ > means "greater than."
< means "less than."

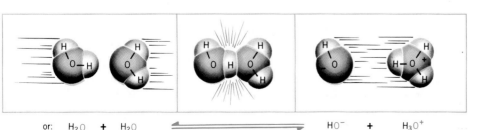

or: H_2O + H_2O ⟶ HO^- + H_3O^+

FIGURE 7.1
In the self-ionization of water, H^+ transfers from one water molecule to another. Unusually violent collisions are needed for successful transfers, and because these are relatively infrequent, the concentrations of H_3O^+ and OH^- are extremely small at equilibrium.

Acidic Solutions Have Some Common Properties. When hydronium ions are in sufficient excess over hydroxide ions in water, the solutions have some common properties, all related to this excess. One is the ability of acidic solutions to behave identically toward a dye called *litmus,* which can exist in either a red or a blue form. Acidic solutions turn litmus in its blue form to red, one of the oldest tests for acids and why litmus is called an **acid–base indicator.** It "indicates" that the solution is acidic.

Acidic solutions also have tart or sour tastes, not that anyone should perform this test on an unknown system. The "unknown" might be a strong poison! But citric acid causes the tart tastes of citrus fruits. Sour milk has lactic acid. Vinegar contains acetic acid. Rhubarb is sour because of oxalic acid. Sour apples have malic acid. Gastric juice contains hydrochloric acid. You can see that acids are common substances.

Basic Solutions Also Have Common Properties. Basic solutions all turn red litmus blue, they generally have bitter tastes, and they are slippery or soapy between the fingers.

Perhaps the most common chemical property of acids and bases is their reaction with each other to produce a *salt*. When taken in the right molar proportions, aqueous acids and bases combine to produce *neutral* solutions, so the reaction is called **acid–base neutralization.** A typical example is the reaction of hydrochloric acid with sodium hydroxide. The base, sodium hydroxide, neutralizes the acid, and sodium chloride forms.

$$\underset{\text{Hydrochloric acid}}{\text{HCl}(aq)} \quad + \quad \underset{\text{Sodium hydroxide}}{\text{NaOH}(aq)} \quad \longrightarrow \quad \underset{\text{Sodium chloride}}{\text{NaCl}(aq)} \quad + \quad \text{H}_2\text{O}$$

The resulting solution, consisting only of sodium chloride in water, has no effect on litmus.

Sodium chloride is but one example of a very broad family of ionic compounds, the salts. **Salts** consist of any positive ion, except H_3O^+, and any negative ion, except OH^- or O^{2-}. The most familiar salt, "table salt" or sodium chloride, consists of Na^+ and Cl^- ions. Among thousands of other examples of salts are calcium sulfate (in plaster board) and sodium bicarbonate (baking soda). All salts are crystalline solids at room temperature, because the attractions of oppositely charged ions for each other are very strong within salt crystals.

Aqueous Solutions of Acids, Bases, and Salts Conduct Electricity. One of the important general properties of aqueous solutions of ion producing compounds is their ability to conduct electricity. The electricity in lightning travels well through wet ground because soil water contains ions. Electrocardiograms can be obtained by attaching wires to the *outside* of the body only because the fluids in the skin and inside the body contain ions that can conduct very small electric currents safely.

Unlike an electric current in a metal, a current moving through a solution of ions is not directly carried by electrons. Instead it is carried by dissolved ions, which move and which actually carry the electrons. Let's see how ions do this. Figure 7.2 shows a typical setup. The plates or wires that dip into the solution are called **electrodes.** A source of electricity, such as a battery, forces electrons to one of these electrodes, called the **cathode,** which becomes electron rich. Positive ions, therefore, are attracted to the cathode (because opposite charges attract), explaining why positive ions are called **cations** (pronounced CAT-ions).

The electrons that make the cathode electron rich are pumped from the other electrode, the **anode,** which becomes electron poor as a result. Negative ions are naturally attracted to the anode and are called **anions** (AN-ions).

When cations arrive at the cathode, they take electrons from its electron-rich surface. When anions reach the anode, they deposit electrons at the electron-poor

■ Litmus paper is made by dipping porous paper into litmus solution and letting the paper dry.

■ In the lab, *never ever* taste a chemical.

■ With equal validity we could say that the acid neutralizes the base.

■ The positive ion in a salt is nearly always a *metal ion.*

■ The electricity from lightning that strikes trees often follows the root systems in the ground.

■ The word *ion* is from the Greek *ienai,* "to move."

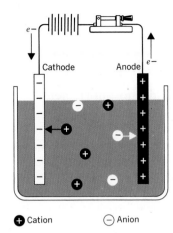

FIGURE 7.2
Electrolysis. Cations, positive ions, migrate to the cathode and remove electrons. Anions, negative ions, migrate to the anode and deposit electrons. The effect is as if electrons flowed through the solution.

surface. Notice carefully that if something takes electrons from one electrode and something (else) puts them on the other, the effect is the same as if the electrons themselves were actually moving through the solution. This is how a solution of ions carries an electric current.

■ Electricity in metal wires is a flow of electrons.

The passage of an electric current through a fluid is called **electrolysis,** and substances that permit electrolysis are called **electrolytes.** For electrolysis to happen, ions must be mobile. When ions are immobilized in crystals, no electrolysis can occur. Ions are made mobile either by melting ionic crystals or by dissolving ion-producing compounds in water. Thus, *electrolyte* refers either to a solution of ions or to the solid ionic compound. (The term does not apply to metals. Metals are called *conductors.*)

■ Ions are chemically changed by electrolysis, and many useful substances are made this way.

Electrolytes Can Be Strong or Weak According to Their Abilities to Supply Ions in Water. Even in solutions of equal molar concentrations, electrolytes are not equally good at carrying electricity. A "good" or **strong electrolyte** is one that readily supplies ions, so it assists the flow of a strong current. An electrolyte is *strong* when essentially 100% of its formula units break up into ions as they dissolve.

■ Sodium hydroxide, hydrochloric acid, and sodium chloride are all strong electrolytes, for example.

When a substance is a **weak electrolyte,** even though it might be very soluble in water, only a small percentage of its formula units break up into ions. Acetic acid, the acid that makes vinegar sour, is a typical weak electrolyte that happens to be very soluble in water. Acetic acid solutions, however, are poor conductors of electricity compared with solutions of hydrochloric acid of equal molar concentration. Aqueous solutions of ammonia are also poor conductors, so aqueous ammonia is a weak electrolyte.

Many substances are **nonelectrolytes.** Their aqueous solutions do not conduct ordinary currents of electricity at all (i.e., household currents). Alcohol and sugar are examples. Pure water itself is a nonelectrolyte. Although the purest of water does contain traces of ions from water's self-ionization, their concentrations are too low to make water a conductor.

We can summarize what we have learned about strong and weak electrolytes as follows.

Strong electrolyte	One that is strongly ionized in water—a high percentage ionization
Weak electrolyte	One that is weakly ionized in water—a low percentage ionization
Nonelectrolyte	One that does not ionize in water—essentially zero percentage ionization

■ Be sure to notice the emphasis on *percentage* ionization as the feature that dominates these definitions.

7.2   COMMON AQUEOUS ACIDS AND BASES

Hydrochloric acid, nitric acid, sulfuric acid, and phosphoric acid are common acids, and sodium hydroxide is a common base.

■ Svante Arrhenius won the third Nobel prize in chemistry (1903) for proposing the idea of ions. To this day, the phrase *Arrhenius theory of acids and bases* is assigned to his identification of acids as suppliers of hydrogen ions and bases as suppliers of hydroxide ions.

An *Acid* Is Any Species That Can Donate H⁺.   Svante Arrhenius (1859–1927), a Swedish chemist, was the first to propose the existence of ions. He explained that all acids have properties *in common* because they all supply the same ion, the hydrogen ion, H^+. But we now know that H^+ is nothing more than a naked proton, a subatomic particle. Its positive charge is concentrated into an extremely small volume, much smaller than that of any other positively charged ion. (Other cations have at least one electron energy level to puff up their ionic volumes.) Because of its high concentration of charge, H^+ is powerfully attracted to electrons. It *never* exists as a *bare* proton when anything with electrons, like a water molecule, is around. Although protons can be *transferred* from one particle to another, they have no existence as separate entities in solution any more than do electrons. H^+ is *always* held by an electron-pair bond to something, such as a water molecule in the form of H_3O^+ or to units of other acids.

It is the ability of an acid to *transfer* or *to donate* H^+ that makes it an acid, not some ability to send H^+ spinning free. H^+, however, is so easily available by transfer from H_3O^+ that chemists today commonly use the terms *proton, hydrogen ion,* and *hydronium ion* interchangeably. We often use *hydrogen ion* as a convenient nickname for *hydronium ion* ourselves and use the symbol $H^+(aq)$ as a simpler way of writing $H_3O^+(aq)$. But remember, $H^+(aq)$ always stands for $H_3O^+(aq)$.

■ Thomas Lowry, an English chemist, and Niels Bjerrum from Denmark independently proposed the same concept of acids and bases.

A *Base* Is Any Species That Can Accept H⁺.   If the ability of something to transfer or donate H^+ makes it an acid, then the ability of something to *accept* and bind H^+ makes it a base. The hydroxide ion is one such proton acceptor or base. These relationships give us simple and extremely useful definitions of acids and bases, definitions for which Johannes Brønsted (1879–1947), a Danish chemist, usually receives the largest credit.

> **Acids** are proton donors.
> **Bases** are proton acceptors.

When hydrochloric acid neutralizes sodium hydroxide, a proton transfer to the base occurs, a water molecule forms, and the other ions (Cl^- and Na^+) remain in solution. The definitions, which are often referred to as the *Brønsted concept of acids and bases,* apply regardless of the solvent and even in the absence of a solvent. Gaseous hydrogen chloride, for example, reacts with gaseous ammonia in the following proton transfer reaction.

■ (g) = gas, (l) = liquid, (s) = solid, (aq) = aqueous solution

$$\underset{\substack{\text{(proton} \\ \text{donor)}}}{HCl(g)} + \underset{\substack{\text{(proton} \\ \text{acceptor)}}}{NH_3(g)} \longrightarrow \underset{\text{(a salt)}}{NH_4Cl(s)}$$

The product is a crystalline solid that forms in a cloud of microcrystals when fumes of ammonia and hydrogen chloride intermingle (Figure 7.3). Molecules of $HCl(g)$ are proton donors; $NH_3(g)$ molecules are proton acceptors. When H^+ transfers from the acid, $HCl(g)$, to the base, $NH_3(g)$, NH_4^+ ions and Cl^- ions form. No solvent is involved, and crystals of $NH_4Cl(s)$ form.

■ NH_4Cl illustrates a salt in which the positive ion, NH_4^+, is not a metal ion.

FIGURE 7.3
The reaction of $NH_3(g)$, from the bottle on the left, with $HCl(g)$, from the bottle on the right, produces a cloud of microcrystals of $NH_4Cl(s)$ by a neutralization reaction in which air, not water, is the "solvent." Gaseous NH_3 and gaseous HCl are always present in the air spaces above their concentrated aqueous solutions, aqueous ammonia and hydrochloric acid.

Acids and Bases Vary Widely in Strength. Acids and bases are quite different in their abilities to function as proton donors or acceptors. Water molecules, for example, are extremely weak as either; the self-ionization of water (page 147) barely produces any ions. Molecules of gaseous hydrogen chloride, $HCl(g)$, on the other hand, so readily donate H^+ that even such a weak acceptor as a water molecule is able to take H^+ from $HCl(g)$. Essentially 100% of all hydrogen chloride molecules that dissolve in water react with water as follows.

$$HCl(g) + H_2O \longrightarrow H_3O^+(aq) + Cl^-(aq)$$

It is the *solution* of HCl in water that is called *hydrochloric acid* and symbolized by $HCl(aq)$. Otherwise, HCl refers to the gas, symbolized as $HCl(g)$.

Because water is the solvent nearly always used in acid–base chemistry, *we classify acids as strong or weak according to the extent of their ionization reactions with water.* If the general symbol HA is used for any acid, whether it is a gas, liquid, or solid, then *all strong acids react as follows essentially 100%.*

$$HA + H_2O \longrightarrow H_3O^+(aq) + A^-(aq)$$

We define a **strong acid** as one that is 100% ionized in water, or nearly so.

When sufficiently concentrated, *all strong acids can cause chemical burns to the skin and gravely harm the eyes,* a reminder to use protective eye wear in the lab. If you ever handle strong acids in their concentrated, commercially available forms, wear protective gloves, too.

A **weak acid** is one for which only a small percentage of its molecules ionize by reacting with water. You may have already noticed that the terms *strong* and *weak* are used alike with both acids and electrolytes. They refer to *percentage* ionization. All strong acids are strong electrolytes. All weak acids are weak electrolytes. Let's now look further at a few of the most common acids and bases.

Hydrochloric Acid and Nitric Acid Are Strong Monoprotic Acids. Hydrochloric acid is a **monoprotic acid** because the ratio of H_3O^+ ions to Cl^- ions in an aqueous solution of hydrogen chloride is $1:1$. Other strong monoprotic acids are those that can be made by dissolving $HBr(g)$ and $HI(g)$ in water to give hydrobromic acid, $HBr(aq)$, and hydriodic acid, $HI(aq)$, respectively.

■ A lot of the time, however, you'll see "HCl" used alone, without *(aq)*, to stand for the aqueous solution, so let the context be your guide.

■ The concentrated $HCl(aq)$ of commerce is 12 M and is a saturated solution of $HCl(g)$ in water. Handle it very carefully. It can cause severe chemical burns.

■ PRACTICE EXERCISE 1   Depict the ionization reactions of $HBr(g)$ and $HI(g)$ in water as we did earlier for $HCl(g)$.

■ The concentrated $HNO_3(aq)$ of commerce is 16 M, which is 71% (w/w) in HNO_3. Handle it carefully.

Nitric acid, $HNO_3(aq)$, is also a strong, monoprotic acid. We can represent its ionization as follows.

$$H_2O + HNO_3(aq) \longrightarrow H_3O^+(aq) + NO_3^-(aq)$$

Nitric acid   Hydronium ion   Nitrate ion

Nitric acid in water

■ **Organic compounds** are compounds of carbon and other nonmetals, excluding the oxides. Other compounds are **inorganic compounds.**

Acetic Acid Is a Weak Monoprotic Acid. Acetic acid is a typical organic acid. Although its molecule has four hydrogen atoms, only one is attached to an oxygen atom. Only this H can transfer to a water molecule, leaving the acetate ion behind.

$$
\underset{\substack{CH_3CO_2H \\ \text{Acetic acid}}}{H-\overset{\overset{\displaystyle H}{|}}{\underset{\underset{\displaystyle H}{|}}{C}}-\overset{\overset{\displaystyle :O:}{\|}}{C}-\ddot{O}-H}
\qquad
\underset{\substack{CH_3CO_2^- \\ \text{Acetate ion}}}{H-\overset{\overset{\displaystyle H}{|}}{\underset{\underset{\displaystyle H}{|}}{C}}-\overset{\overset{\displaystyle :O:}{\|}}{C}-\ddot{O}:^-}
$$

(the H in color is transferable)

Like virtually all organic acids, acetic acid is a weak acid. The H—O bond in acetic acid holds the proton much more strongly than the proton is held in $HCl(g)$. Acetic acid molecules, therefore, do not as readily transfer H^+ to H_2O as do molecules of HCl. The expression "not as readily" means that more powerful collisions must occur for the proton transfer to work in aqueous acetic acid than when $HCl(g)$ molecules enter water. In the random bumpings around, such extra energetic collisions are less frequent in aqueous acetic acid. We actually have a chemical equilibrium, represented as follows.

$$H_2O + \underset{\text{Acetic acid}[1]}{CH_3CO_2H} \rightleftharpoons H_3O^+(aq) + \underset{\text{Acetate ion}}{CH_3CO_2^-(aq)}$$

■ Vinegar is about 5% (w/w) acetic acid in water.

In 0.1 M acetic acid, only 0.5% of all acetic acid molecules are ionized, but there are the coming and going characteristic of all dynamic equilibria.

Both water and acetic acid have H—O bonds, but acetic acid is vastly more acidic. The question is, "Why?" That acetic acid is acidic at all can be credited to the $C{=}O$ group, a group made electronegative by its oxygen atom. The CO group therefore pulls electron density away from the oxygen of the H—O group in acetic acid. The H—O bond in acetic acid is thus weakened, so the transfer of a proton from CH_3CO_2H to H_2O occurs more readily than the transfer of H from one H_2O molecule to another (water's self-ionization).

$$
\underset{\text{Sulfuric acid}}{H-\overset{..}{\underset{..}{O}}-\overset{\overset{\displaystyle :O:}{\|}}{\underset{\underset{\displaystyle :O:}{\|}}{S}}-\overset{..}{\underset{..}{O}}-H}
$$

Sulfuric acid

$$
\underset{\text{Hydrogen sulfate ion}}{H-\overset{..}{\underset{..}{O}}-\overset{\overset{\displaystyle :O:}{\|}}{\underset{\underset{\displaystyle :O:}{\|}}{S}}-\ddot{O}:^-}
$$

Hydrogen sulfate ion

Sulfuric Acid Molecules Have Two Protons to Donate. Sulfuric acid, H_2SO_4, is the only common yet stable **diprotic acid,** one able to release two protons per formula unit. The ionization of the first hydrogen ion is so easy that we don't even write equilibrium arrows for the reaction; sulfuric acid is a strong acid.

$$H_2O + \underset{\text{Sulfuric acid}}{H_2SO_4} \longrightarrow H_3O^+(aq) + \underset{\text{Hydrogen sulfate ion}}{HSO_4^-(aq)}$$

The four oxygens on sulfur in the sulfuric acid molecule help make it a much stronger acid than acetic acid, which has only one electronegative oxygen.

[1]Some references use the formula $HC_2H_3O_2$ for acetic acid and $C_2H_3O_2^-$ for the acetate ion instead of the full structures. The formulas CH_3CO_2H and $CH_3CO_2^-$, however, give more structural information. In any case, remember that acetic acid is *monoprotic* and that only one hydrogen is active in acid–base reactions.

The Hydrogen Sulfate Ion Is a Proton Donor.   The hydrogen sulfate ion is itself a monoprotic acid and our first example of an acid that starts out as an *ion* instead of a molecule. The transfer of H^+ from HSO_4^-, however, requires a positively charged particle (H^+) to leave one already oppositely charged (HSO_4^-). Therefore this transfer is harder than that of the first H^+ from the uncharged H_2SO_4 molecule. Yet it still happens to a greater percentage than the ionization of acetic acid. In a solution of sodium hydrogen sulfate, $NaHSO_4$, the following equilibrium exists. (We won't show the sodium ion because nothing happens to it.)

$$H_2O + HSO_4^-(aq) \rightleftharpoons H_3O^+(aq) + SO_4^{2-}(aq)$$

Hydrogen sulfate ion   Sulfate ion

The same ionization equilibrium of HSO_4^- exists in $H_2SO_4(aq)$ also. The *reactants* are favored, but not by much. When the concentration of HSO_4^- is $0.10\,M$ (as it is in $0.10\,M$ $NaHSO_4$), the percentage ionization of the HSO_4^- ion is slightly less than 40%.

Phosphoric Acid Is a Moderately Strong, Triprotic Acid.   A **triprotic acid** is one that is able to give up three H^+ ions to acceptors, but each H^+ separates with greater difficulty than the previous one. Phosphoric acid, H_3PO_4, is the only common example of an inorganic triprotic acid. Like the ionization of sulfuric acid, that of phosphoric acid occurs in steps, each one more difficult than the previous. Even the first step does not happen to all of the phosphoric acid molecules, so we have to write equilibrium expressions for all steps. In each of the following equilibria, the *reactants* are favored.

$$H_2O + H_3PO_4(aq) \rightleftharpoons H_3O^+(aq) + H_2PO_4^-(aq)$$

Phosphoric   Dihydrogen
acid   phosphate ion

$$H_2O + H_2PO_4^-(aq) \rightleftharpoons H_3O^+(aq) + HPO_4^{2-}(aq)$$

Monohydrogen
phosphate ion

$$H_2O + HPO_4^{2-}(aq) \rightleftharpoons H_3O^+(aq) + PO_4^{3-}(aq)$$

Phosphate ion

The percentage ionization of H_3PO_4 in a $0.10\,M$ solution is roughly 24%, too low to let us call phosphoric acid a strong acid. It's classified as a *moderate acid*.

   It is important that you learn the names and formulas of the three ions available from phosphoric acid, because the phosphate ion system occurs widely in the body.

Carbonic Acid Occurs in Carbonated Beverages.   Carbonic acid, H_2CO_3, is a weak but unstable diprotic acid. It exists only in a solution that holds dissolved carbon dioxide. A small percentage of the CO_2 molecules react with water to form carbonic acid, and the following equilibrium exists in "carbonated water," like club soda or Perrier water.

$$CO_2(aq) + H_2O \rightleftharpoons H_2CO_3(aq)$$

Carbonic acid

A small fraction of the carbonic acid molecules ionizes, a fraction small enough to make carbonic acid a *weak* acid but yet large enough to give carbonated water a slightly tart taste. The ionization of carbonic acid establishes the following equilibrium.

$$H_2CO_3(aq) + H_2O \rightleftharpoons H_3O^+ + HCO_3^-(aq)$$

Bicarbonate ion

The bicarbonate ion is the chief form in which waste carbon dioxide is carried from

Sulfate ion

Phosphoric acid

Dihydrogen phosphate ion

Monohydrogen phosphate ion

Phosphate ion

■ Its instability, however, is an important property when the body has to manage one of the respiratory gases, carbon dioxide.

Carbonic acid

Bicarbonate ion

TABLE 7.1   Common Acids[a]

Acid	Formula	Percentage Ionization
Strong Acids		
Hydrochloric acid	HCl	Very high
Hydrobromic acid	HBr	Very high
Hydriodic acid	HI	Very high
Nitric acid	HNO_3	Very high
Sulfuric acid[b]	H_2SO_4	Very high
Moderate Acids		
Phosphoric acid	H_3PO_4	27
Sulfurous acid[c]	H_2SO_3	20
Weak Acids		
Nitrous acid[c]	HNO_2	1.5
Acetic acid	CH_3CO_2H	1.3
Carbonic acid[c]	H_2CO_3	0.2

[a]Data are for 0.1 M solutions of the acids in water at room temperature.

[b]*Concentrated* sulfuric acid (99%) is particularly dangerous not only because it is a strong acid but also because it is a powerful dehydrating agent. This action generates considerable heat at the reaction site, and at higher temperatures sulfuric acid becomes even more dangerous. Moreover, concentrated sulfuric acid is a thick, viscous liquid that does not wash away from skin or fabric very quickly.

[c]An unstable acid.

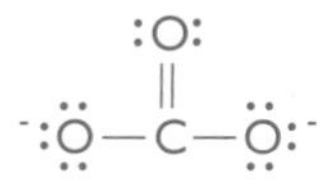

Carbonate ion

body tissues to the lungs. The ion is itself a (very) weak proton donor, so weak that only a strong acceptor, like OH^-, can remove its proton.

$$HCO_3^-(aq) + OH^- \rightleftharpoons CO_3^{2-}(aq) + H_2O$$
$$\text{Carbonate ion}$$

H_2O, in sharp contrast to OH^-, is much too weak to take a proton from HCO_3^- to any but the most minute extent.

Table 7.1 lists the common aqueous acids. You should memorize the names and formulas of all on the list because we'll use this knowledge as we go along. There are very few strong or moderate acids, so once they are learned, you can be fairly certain that any unfamiliar acid will be a weak acid. It's easier to learn a few strong acids than several hundred weak acids.

■ Household lye is sodium hydroxide. It's imperative to keep it out of reach of children.

The Hydroxide Ion Is a Commonly Used Proton Acceptor. Sodium hydroxide and potassium hydroxide are the most frequently used suppliers of OH^- ion in the lab. They and other bases are listed in Table 7.2; their names and formulas should be memorized. The **strong bases** listed there are metal hydroxides that break up nearly 100% in water into metal ions and hydroxide ions. It's their OH^- ion that is the true base, the true proton acceptor. We may represent what happens when solid NaOH and KOH dissolve in water by equations that do not use equilibrium arrows because the breakup is essentially total, as we said.

$$NaOH(s) \xrightarrow{\text{water}} Na^+(aq) + OH^-(aq)$$

$$KOH(s) \xrightarrow{\text{water}} K^+(aq) + OH^-(aq)$$

In sufficient concentration, the hydroxide ion causes a severe chemical burn, severe enough to be extremely hazardous to soft tissues and the eyes, still another reminder to the rule: *always use protective eye wear in the lab.* Both sodium and potassium hydroxides are soluble enough to be able to give concentrated solutions that are

TABLE 7.2   Common Bases

Base	Formula	Solubility[a]	Percentage Ionization
Strong Bases			
Sodium hydroxide	NaOH	109	>90(0.1 M solution)
Potassium hydroxide	KOH	112	>90(0.1 M solution)
Calcium hydroxide	$Ca(OH)_2$	0.165	100 (saturated solution)
Magnesium hydroxide	$Mg(OH)_2$	0.0009	100 (saturated solution)
Weak Base			
Ammonia, aqueous	NH_3	89.9	1.3 $(18\ °C)^b$

[a]Solubilities are in grams of solute per 100 g of water at 20 °C.
[b]The ionization referred to is the equilibrium: $NH_3(aq) + H_2O \rightleftharpoons NH_4^+(aq) + OH^-(aq)$

dangerous chemicals. Protective gloves should also be worn when using concentrated solutions of NaOH or KOH, particularly when they are heated and might spatter.

Two other strong bases are hydroxides of Group IIA metals, magnesium hydroxide, $Mg(OH)_2$, and calcium hydroxide, $Ca(OH)_2$. Although their breakup into ions in water is essentially 100%, they are so insoluble (see Table 7.2) that even their *saturated* solutions provide only traces of hydroxide ions. We may represent what happens when they dissolve as follows. (Only if excess *solid* solute were present would we be justified in writing equilibrium arrows.)

$$Ca(OH)_2(s) \rightarrow Ca^{2+}(aq) + 2OH^-(aq)$$
$$Mg(OH)_2(s) \rightarrow Mg^{2+}(aq) + 2OH^-(aq)$$

Calcium and magnesium hydroxide are so insoluble that not only do they pose no grave danger to the skin, but they are used internally in home remedies. Calcium hydroxide is a component of one commercial antacid tablet. A slurry of magnesium hydroxide in water, called *milk of magnesia,* is used as an antacid and a laxative.

A weak base is defined by its ability, very poor, to accept a proton *from* H_2O. A **weak base** is a poor proton acceptor. Ammonia is the most common example. A solution of ammonia in water, called *aqueous ammonia,* does have some excess hydroxide ion, but only a small percentage of ammonia molecules react to produce it. The following equilibrium exists, and the *reactants* are strongly favored; roughly 99% of all solute molecules exist as NH_3 in 1 M $NH_3(aq)$.

$$\underset{\text{Ammonia}}{NH_3(aq)} + H_2O \rightleftharpoons \underset{\text{Ammonium ion}}{NH_4^+(aq)} + OH^-(aq)$$

A dilute solution (about 5%) of ammonia in water is sold as household ammonia. It's a good cleaning agent, but watch out for its fumes.

Milk of magnesia, a slurry of $Mg(OH)_2$ in water, is a common remedy for acid indigestion. Too frequent doses, however, can upset the levels of magnesium ion in various body fluids.

■ You'll sometimes see aqueous ammonia called "ammonium hydroxide," but it's better not to use this term. The compound NH_4OH is unknown as a pure compound.

7.3   CHEMICAL PROPERTIES OF AQUEOUS ACIDS AND BASES

Neutralizers of acids include hydroxides, bicarbonates, carbonates, and ammonia.

In this section we will principally study the reactions of the hydronium ion, H_3O^+, using the symbol $H^+(aq)$ as shorthand for H_3O^+ in most of the equations. We will see how a variety of substances besides OH^- are able to accept H^+ from H_3O^+ in reactions that neutralize acids. Acid neutralizers include not only the hydroxide ion but also the bicarbonate ion, the carbonate ion, and the ammonia molecule. All are bases, because all accept and bind H^+. We'll also learn in this section that many metals are able to reduce two H^+ ions to H_2 and themselves be oxidized to metal ions. Some metals are so active that they can even abstract protons directly from molecules of water.

We must first master the use of a special kind of equation, the *net ionic equation,* because it has no equal in its usefulness for the study of the chemical properties of ions. We have already used such equations, but now we must learn how to write them.

A Net Ionic Equation Omits "Spectator" Species. The conventional equation for a reaction is called a **molecular equation** because it shows all of the substances in the molecular or empirical formulas that we would need to plan an actual experiment. The molecular equation for the reaction of sodium carbonate decahydrate with hydrochloric acid, for example, is

$$Na_2CO_3 \cdot 10H_2O(s) + 2HCl(aq) \longrightarrow 2NaCl(aq) + CO_2(g) + 11H_2O$$

Similarly, the molecular equation for the reaction of hydrochloric acid with aqueous sodium hydroxide is

$$HCl(aq) + NaOH(aq) \longrightarrow NaCl(aq) + H_2O$$

- We say *molecular equation* even though some of the chemicals in the equation might be ionic.

- Unless otherwise designated, we always omit *(l)* after H_2O in equations, because the liquid state is assumed.

As we now know, however, $HCl(aq)$ is really $H^+(aq)$ and $Cl^-(aq)$, and $NaOH(aq)$ is actually $Na^+(aq)$ and $OH^-(aq)$. We know this because we know that the acid $HCl(aq)$ is a *strong* acid and the base $NaOH(aq)$ is a *strong* base, so both must be essentially fully ionized in solution. We don't have un-ionized molecules of HCl or NaOH in the solution. We also know that $NaCl(aq)$ is fully ionized because all salts are strong electrolytes, and the *(aq)* next to its formula tells us that the NaCl is in solution. The fourth formula in the above equation is that of water, a nonelectrolyte. Its molecules aren't separated into ions. (We ignore, of course, the self-ionization of water, because it occurs to an exceedingly low percentage.)

- The formula of an *insoluble* salt in an equation would have *(s)* next to it.

Using the known facts about reactants and products, we can expand the molecular equation into what is called the **ionic equation,** one that shows all of the dissolved species, whether ionic or molecular. We do this by replacing anything that we know is present as ions by the actual formulas of these ions. Thus the ionic equation for our example is

$$H^+(aq) + Cl^-(aq) + Na^+(aq) + OH^-(aq) \longrightarrow Na^+(aq) + Cl^-(aq) + H_2O$$

| These came from HCl(aq) | These came from NaOH(aq) | These came from NaCl(aq) | Not ionized |

The preparation of an ionic equation is actually only a scratch paper operation, because we next cancel all of the formulas that appear *identically* on opposites sides of the arrow. The foregoing ionic equation shows, for example, that nothing actually happens either to $Na^+(aq)$ or to $Cl^-(aq)$. Therefore, there is no reason to let these ions remain in the equation, if we want to give full attention only to the species that react and form. $Na^+(aq)$ and $Cl^-(aq)$, of course, do serve one function; they give electrical neutrality to their respective compounds. Otherwise, they are nothing more than *spectator particles.* Formulas must be of the same physical state before they can be canceled. We could not, for example, cancel $HCl(g)$ by $HCl(aq)$, because their states are different.

- Sometimes we don't cancel, but only reduce in number. If, for example, an equation has

$$\ldots + 4H_2O \longrightarrow \ldots + 2H_2O$$

we can simplify it to

$$\ldots + 2H_2O \longrightarrow \ldots$$

When we cancel the spectator species from the ionic equation above, we are left with the **net ionic equation,** one that shows only the reacting species and the products they form.

$$H^+(aq) + OH^-(aq) \rightarrow H_2O$$

This equation is a simple description of what happens when hydrochloric acid, *or any strong acid,* neutralizes sodium hydroxide, *or any hydroxide base,* in solution.

Net Ionic Equations Must Be in Balance Both Electrically and Materially. For a net ionic equation to be balanced, two conditions must be met: a *material balance* and an *electrical balance.* We have **material balance** when the numbers of atoms of each element, regardless of how they are chemically present, are the same on both sides of the arrow. We have **electrical balance** when the algebraic sum of the charges to the left of the arrow equals the sum of the charges to the right.

EXAMPLE 7.1   Writing a Net Ionic Equation

Sulfuric acid is the most important acid in industrial use, and sometimes it has to be neutralized by sodium hydroxide. The reaction can be carried out to produce sodium sulfate, $Na_2SO_4(aq)$, and water. Write the molecular, ionic, and net ionic equations. The question calls for three answers, so we'll proceed in steps.

ANALYSIS I   First we figure out the molecular equation from the given facts. The complete formulas of the reactants and products must be assembled in the pattern of an equation, which is then balanced.

SOLUTION I: THE MOLECULAR EQUATION

$$H_2SO_4(aq) + 2NaOH(aq) \rightarrow Na_2SO_4(aq) + 2H_2O$$

ANALYSIS 2   Using our knowledge about which acids and bases are strong and which salts are fully ionized, we disassemble the molecular equation to display all of the ions present or provided. We show H_2SO_4 as breaking up entirely into $2H^+$ and SO_4^{2-} because the given reaction is with a strong base, not with water, and SO_4^{2-} forms, not HSO_4^-.

$H_2SO_4(aq)$ means $2H^+(aq) + SO_4^{2-}(aq)$	H_2SO_4 gives up *two* H^+ ions to the other reactant.
$2NaOH(aq)$ means $2Na^+(aq) + 2OH^-(aq)$	NaOH is a strong base, a fully ionized metal hydroxide.
$Na_2SO_4(aq)$ means $2Na^+(aq) + SO_4^{2-}(aq)$	Na_2SO_4 is a salt, and *aq* tells us that it is in solution; hence, it is fully ionized.
$2H_2O$ means $2H_2O$	No break up of H_2O occurs with this nonelectrolyte.

SOLUTION 2: THE IONIC EQUATION   The analysis gives us the ionic equation.

$$2H^+(aq) + SO_4^{2-}(aq) + 2Na^+(aq) + 2OH^-(aq) \rightarrow$$
$$2Na^+(aq) + SO_4^{2-}(aq) + 2H_2O$$

ANALYSIS 3   Finally, identical species in identical states on opposite sides of the arrow in the ionic equation are canceled.

$$2H^+(aq) + \cancel{SO_4^{2-}(aq)} + \cancel{2Na^+(aq)} + 2OH^-(aq) \rightarrow$$
$$\cancel{2Na^+(aq)} + \cancel{SO_4^{2-}(aq)} + 2H_2O$$

■ More than 400 billion moles (44 million tons) of sulfuric acid was manufactured annually in the United States in the early 1990s.

SOLUTION 3: THE NET IONIC EQUATION   After removing the canceled terms, we have the net ionic equation.

$$2H^+(aq) + 2OH^-(aq) \rightarrow 2H_2O$$

We can divide all of the coefficients by 2 and convert them to smaller whole numbers, so the final net ionic equation is (very simply)

$$H^+(aq) + OH^-(aq) \rightarrow H_2O$$

CHECK   We have both a material and an electrical balance for each equation.

The pattern of Example 7.1 is the same for all of the reactions of ionic compounds.

STEPS TO A NET IONIC EQUATION
1. Use the formulas to prepare a conventional, *molecular equation* and balance it.
2. Analyze the facts about the participants' ionizing properties to see how to disassemble the molecular formulas.
3. Expand the molecular equation to the *ionic equation*.
4. Cancel spectator ions and write the *net ionic equation*.

■ Remember, formulas must be of the same physical state before they can be canceled.

■ PRACTICE EXERCISE 2   Write the molecular, ionic, and net ionic equations for the neutralization of nitric acid by potassium hydroxide. A water-soluble salt, $KNO_3(aq)$, and water form.

The Hydroxide Ion Reacts with Acids to Give Water.   Example 7.1 and Practice Exercise 2 illustrate reactions of strong acids with metal hydroxides. We saw by the net ionic equations that the reactions are those of the acid species, H^+, with a base, OH^-.

$$H^+(aq) + OH^-(aq) \rightarrow H_2O$$

Most common weak acids, like acetic acid, also donate H^+ to OH^-. Thus acetic acid reacts with hydroxide ion as follows. The reaction is an acid–base neutralization, only now the formula of the un-ionized acid (not H^+ nor H_3O^+) must be used in the net ionic equation.

$$CH_3CO_2H(aq) + OH^-(aq) \rightarrow CH_3CO_2^-(aq) + H_2O$$

Net ionic equations for the reactions of the water-*insoluble* metal hydroxides with acid are slightly different from those for the reactions with water-*soluble* bases. The insolubility of such hydroxides prevents us from using $OH^-(aq)$ as a separated species in the equations. For example, solid (and mostly undissolved) calcium hydroxide reacts with hydrochloric acid as follows.

$$Ca(OH)_2(s) + 2HCl(aq) \rightarrow CaCl_2(aq) + 2H_2O$$

The net ionic equation retains $Ca(OH)_2(s)$ and H_2O, but shows separate ions for the rest. (The chloride ions cancel.)

$$Ca(OH)_2(s) + 2H^+(aq) \rightarrow Ca^{2+}(aq) + 2H_2O$$

■ PRACTICE EXERCISE 3   When milk of magnesia is used to neutralize hydro-chloric acid (stomach acid), solid magnesium hydroxide in the suspension reacts with the acid. Write the molecular and net ionic equations for this reaction.

The Bicarbonate Ion Neutralizes Acids, Giving CO_2 and H_2O.   All metal bicarbonates react the same way with strong, aqueous acids. The products are carbon dioxide, water, and the salt formed from the cation of the bicarbonate and the anion of the acid. For example, sodium bicarbonate and hydrochloric acid react as follows.

$$HCl(aq) + NaHCO_3(aq) \longrightarrow CO_2(g) + H_2O + NaCl(aq)$$

Potassium bicarbonate and hydrobromic acid give a similar reaction.

$$HBr(aq) + KHCO_3(aq) \longrightarrow CO_2(g) + H_2O + KBr(aq)$$

What is believed to form initially, besides a salt, is not CO_2 and H_2O but H_2CO_3, carbonic acid. However, almost all of the H_2CO_3 promptly decomposes to CO_2 and H_2O, so the solution fizzes strongly as CO_2 evolves.

In writing net ionic equations for reactions between strong acids and metal bicarbonates, we'll treat all metal bicarbonates as ionized in water to give the metal ion and the bicarbonate ion. The metal ion will always be a "spectator." The bicarbonate ion is the base, the proton acceptor. *Because of its involvement in the chemistry of respiration and associated medical emergencies, this reaction of the bicarbonate ion with acid is one of the most important acid–base reactions we study.*

■ The fizzing when an Alka-Seltzer tablet is dropped into water is caused by CO_2 gas formed by the reaction of two ingredients, citric acid (a solid) and crystalline $NaHCO_3$, whose ions are set free as they dissolve.

EXAMPLE 7.2   Writing Equations for the Reactions of Bicarbonates with Strong Acids

What are the molecular and the net ionic equations for the reaction of potassium bicarbonate with hydriodic acid?

ANALYSIS I   The products of all such reactions are CO_2, H_2O, and a salt. The formula of the salt must be KI, a combination of the bicarbonate's metal ion, K^+, and the acid's anion, I^-.

SOLUTION: THE MOLECULAR EQUATION   The balanced molecular equation is

$$KHCO_3(aq) + HI(aq) \longrightarrow CO_2(g) + H_2O + KI(aq)$$

ANALYSIS 2   To prepare the ionic equation from the molecular equation, we analyze each formula to see which can be disassembled into ions.

$KHCO_3(aq)$ means $K^+(aq)$ and $HCO_3^-(aq)$	As we were told.
$HI(aq)$ means $H^+(aq) + I^-(aq)$	Because HI is a strong, fully ionized acid.
$KI(aq)$ means $K^+(aq) + I^-(aq)$	The *aq* tells us that KI is a water-soluble salt and so is fully ionized.
$CO_2(g)$ and H_2O stay the same	Neither is ionized.

Using these facts, we expand the molecular equation into the ionic equation.

$$[K^+(aq) + HCO_3^-(aq)] + [H^+(aq) + I^-(aq)] \longrightarrow$$
$$CO_2(g) + H_2O + [K^+(aq) + I^-(aq)]$$

The $K^+(aq)$ and the $I^-(aq)$ can now be canceled from each side of the arrow.

SOLUTION: THE NET IONIC EQUATION   Canceling leaves

$$H^+(aq) + HCO_3^-(aq) \longrightarrow CO_2(g) + H_2O$$

CHECK   The equation is balanced both materially and electrically.

The equation produced by Example 7.2 is *the same net ionic equation for the reaction of all bicarbonates with all strong aqueous acids.*

$$H^+(aq) + HCO_3^-(aq) \longrightarrow CO_2(g) + H_2O$$

Because the reaction destroys hydrogen ions, it clearly is a neutralization reaction. In fact, the familiar "bicarb" used as a home remedy for acid stomach is nothing more than sodium bicarbonate. Stomach acid is roughly 0.1 *M* HCl, and bicarbonate ion neutralizes this acid by the reaction we have just studied. An overdose of "bicarb" must be avoided because it can cause a medical emergency involving the respiratory gases.

■ Solid sodium carbonate has been pumped onto serious acid spills from tank car shipments to neutralize the acid before it further harms the environment.

■ PRACTICE EXERCISE 4   Write the molecular, ionic, and net ionic equations for the reaction of sodium bicarbonate with sulfuric acid in which sodium sulfate, $Na_2SO_4(aq)$, is one of the products.

The Carbonate Ion Reacts with Acids to Give CO_2 and H_2O   Carbonates react with hydrogen ions to give the same products as bicarbonates, namely, CO_2, H_2O, and the salt that forms from the carbonate's metal ion and the acid's anion. Only the mole proportions change. The CO_3^{2-} ion is the base or proton acceptor and mole for mole, it can neutralize twice as much H^+ as can the HCO_3^- ion. Thus, the reaction of sodium carbonate with nitric acid is

$$2HNO_3(aq) + Na_2CO_3(aq) \longrightarrow CO_2(g) + H_2O + 2NaNO_3(aq)$$

The net ionic equation—do the analysis as an exercise—is

$$2H^+(aq) + CO_3^{2-}(aq) \longrightarrow CO_2(g) + H_2O$$

The identical net ionic equation describes the reactions of all strong, aqueous acids with all of the carbonates of group IA metals.

■ PRACTICE EXERCISE 5   Write the molecular, ionic, and net ionic equations for the reaction of aqueous potassium carbonate, $K_2CO_3(aq)$, with sulfuric acid to give potassium sulfate, $K_2SO_4(aq)$, a water-soluble salt, and the other usual products.

■ Sometimes powdered lime-stone is strewn from a helicopter onto a lake polluted by acid rain to neutralize excess acid.

Only the carbonates of Group IA metal ions (as well as ammonium carbonate) are very soluble in water. Most others carbonates are water insoluble. Calcium carbonate, $CaCO_3$, for example, is the chief substance in limestone and marble. Despite its insolubility in water, calcium carbonate reacts readily with strong, aqueous acids (with their hydrogen ions, of course). The products are soluble in water, so the insoluble carbonates dissolve by this reaction.

$$CaCO_3(s) + 2HCl(aq) \longrightarrow CO_2(g) + H_2O + CaCl_2(aq)$$

We have to write the entire formulas of water-insoluble carbonates in net ionic equations, so the net ionic equation for the above reaction is

$$CaCO_3(s) + 2H^+(aq) \rightarrow CO_2(g) + H_2O + Ca^{2+}(aq)$$

■ **PRACTICE EXERCISE 6**   Dolomite, a limestone-like rock, contains both calcium and magnesium carbonates. Magnesium carbonate is attacked by nitric acid. The salt that forms is water soluble. Write the molecular, ionic, and net ionic equations for this reaction.

■ Adding a few drops of strong acid to a rock sample is a field test for carbonate rocks. A positive test is an odorless fizzing reaction.

A weak acid, like acetic acid, also is neutralized by either bicarbonates or carbonates, but the net ionic equations must use the un-ionized species, CH_3CO_2H, not H^+ nor H_3O^+. For example,

$$CH_3CO_2H(aq) + HCO_3^-(aq) \rightarrow CH_3CO_2^-(aq) + CO_2(g) + H_2O$$
$$2CH_3CO_2H(aq) + CO_3^{2-}(aq) \rightarrow 2CH_3CO_2^-(aq) + CO_2(g) + H_2O$$

Ammonia Neutralizes Aqueous Acids.   An aqueous solution of ammonia is an excellent reagent for neutralizing acids. For example, ammonia reacts with hydrochloric acid as follows.

$$NH_3(aq) + HCl(aq) \rightarrow NH_4Cl(aq)$$

Because all ammonium salts are soluble in water, $NH_4Cl(aq)$ is actually broken up in solution into $NH_4^+(aq)$ and $Cl^-(aq)$ ions. The latter, therefore, are only spectators, so the net ionic equation is

$$NH_3(aq) + H^+(aq) \rightarrow NH_4^+(aq)$$

We need to learn this reaction because many biochemicals have ammonia-like molecules that neutralize hydrogen ions in the same way.

■ **PRACTICE EXERCISE 7**   Write the molecular and net ionic equations for the reaction of aqueous ammonia with (a) $HBr(aq)$ and (b) $H_2SO_4(aq)$.

Active Metals React with Acids to Give Hydrogen and a Salt.   Many metals are attacked more or less readily by the hydrogen ion in solution. The products are generally hydrogen gas and a salt made of the cation from the metal and the anion from the acid. Zinc, for example, reacts with hydrochloric acid as follows:

$$Zn(s) + 2HCl(aq) \rightarrow H_2(g) + ZnCl_2(aq)$$

The net ionic equation is

$$Zn(s) + 2H^+(aq) \rightarrow H_2(g) + Zn^{2+}(aq)$$

Aluminum is also attacked by acids. Its reaction with nitric acid, for example, can be written as

$$2Al(s) + 6HNO_3(aq) \rightarrow 2Al(NO_3)_3(aq) + 3H_2(g)$$

The net ionic equation is

$$2Al(s) + 6H^+(aq) \rightarrow 2Al^{3+}(aq) + 3H_2(g)$$

■ PRACTICE EXERCISE 8   Write the molecular and the net ionic equations for the reaction of magnesium with hydrochloric acid.

Metals Form an Activity Series in Their Reactions with Acids.   Metals differ greatly in their tendencies to react with aqueous hydrogen ions. When they do react, atoms of the metal are oxidized because they lose electrons and become metal ions. The oxidizing agent is H^+. The electrons are transferred to H^+, taken from H_3O^+ ions (sometimes from H_2O), and these protons are reduced and made electrically neutral. Two H atoms combine and emerge as a molecule of hydrogen gas, H_2. Thus the metal is oxidized by H^+ to a cation and H^+ is reduced by the metal to H_2.

■ Remember, a loss of electrons is oxidation, and a gain of electrons is reduction.

Group IA metals such as sodium and potassium include the most reactive metals of all. They not only reduce H^+ taken from hydronium ions, but they also are able to pry H^+ ions directly from water molecules. No acid need be present in the solution. Sodium, for example, reacts with water as follows.

$$2Na(s) + 2H_2O \longrightarrow 2NaOH(aq) + H_2(g)$$

The reaction is extremely violent (Figure 7.4), and it should never be attempted except by an experienced chemist working with safety equipment, including a fire extinguisher.

Gold, silver, and platinum, in contrast, are stable not only toward water but also toward hydronium ions. Lead and tin react very slowly with acids. Figure 7.5 shows how the reactivities of iron, zinc, and magnesium differ toward the same acid, 1 M HCl.

The vast differences in the reactivities of metals toward acids make it possible to arrange the metals in order of reactivity. The result is the **activity series** of the metals (Table 7.3.) Atoms of any metal above hydrogen in the series can transfer electrons to H^+, either from H_2O or from H_3O^+, to form hydrogen gas, and the metal atoms change to ions. The farther a metal is above hydrogen, the more reactive it is. Metals below hydrogen in the activity series do not transfer electrons to H^+.

FIGURE 7.4
When bits of sodium metal are dropped into water, enough heat is evolved by the reaction with water to melt the unreacted sodium skittering on the surface and ignite the evolving hydrogen gas. The result is a shower of light.

FIGURE 7.5
Metals vary widely in their ease of oxidation. Iron is in the first tube, zinc in the second, and magnesium in the third, and all are exposed to HCl (aq) at the same molarity. All these metals can be oxidized to their metal ion states by hydrogen ion, which is reduced to hydrogen gas. However, iron, the least readily oxidized of these metals, reacts so slowly as to produce hardly any visible fizzing of hydrogen gas. Zinc, the next most easily oxidized of the three, reacts rather well. Magnesium, the most easily oxidized of the three, reacts very vigorously.

TABLE 7.3 The Activity Series of the Common Metals

Greatest tendency to become ionic	Potassium	React violently with water
	Sodium	
	Calcium	Reacts slowly with water
React with hydrogen ions to liberate H_2	Magnesium	React very slowly with water
	Aluminum	
	Zinc	
	Chromium	
	Iron	
	Nickel	
	Tin	
	Lead	
	HYDROGEN	
	Copper	
Do not react with hydrogens ions	Mercury	
	Silver	
Least tendency to become ionic	Platinum	
	Gold	

(Increasing Tendency to Become Ionic ↑)

7.4   RELATIVE STRENGTHS OF ACIDS AND BASES

The anion of a strong acid is a weak base; the anion of a weak acid is a relatively strong base.

The Relative Strengths of Unfamiliar Acids and Bases Can Be Predicted.   A strong acid leaves behind a weak base when a proton transfers away. The particle left behind, for example, when the strong acid, HCl(g), gives up its proton to water is the chloride ion, a *very* weak base. In fact, $Cl^-(aq)$ is not a base at all; it cannot even take back a proton from H_3O^+, the best proton donor we can have in an aqueous solution.

Reciprocally, the particle released when a weak acid gives up a proton is a strong base. The water molecule, for example, although an extremely weak acid, does leave behind the OH^- ion, a very strong base, when a proton transfers from one H_2O molecule to another during self-ionization (page 147). These examples illustrate the existence of pairs of important reciprocal relationships between the protonated forms of species, acids, and their nonprotonated forms, bases.

RECIPROCAL ACID–BASE RELATIONSHIPS
1. The stronger an acid is, the weaker is its anion as a base.
2. The weaker an acid is, the stronger is its anion as a base.
3. The stronger a base is, the weaker is its protonated form as an acid.
4. The weaker a base is, the stronger is its protonated form as an acid.

Let's illustrate these generalizations by working some examples and see how they give us the power to predict important chemical properties of unfamiliar species.

First, let's review how easy it is to tell if an *acid* is a weak or a strong proton donor. For example, lactic acid is the acid responsible for the tart taste of sour milk. Is it a strong or a weak acid? Remembering from memory the names of the strong acids (see Table 7.1), we don't find lactic acid on the list. Hence, we can be almost positive that it is a weak acid and a poor proton donor. It's as simple as that (and we'd very seldom err).

When the question deals with a potential proton acceptor or *base,* we find the answer in a roundabout fashion. We accept this approach because the alternative would be to memorize a rather long list of the stronger bases. Here's how to do it.

EXAMPLE 7.3   Deducing if an Ion or Molecule Is a Strong or a Weak Proton Acceptor

Is the bromide ion a strong or a weak proton acceptor or base?

ANALYSIS   The key tactic *is to pretend that the potential base actually functions as one.* Write the formula of what forms if Br^- accepts H^+. If Br^- actually were a base, and it accepted H^+, it would change to un-ionized HBr. Now comes the crucial question. Is HBr a strong acid? We apply what we know, namely, the short list of strong acids, and HBr is on it. Being a strong acid, HBr *easily gives up a proton.* And this is how we figure out that Br^-, the species that remains when H^+ (readily) leaves HBr, is a poor proton binder.

SOLUTION Br^- is a weak base.

EXAMPLE 7.4   Deducing if an Ion or a Molecule Is a Strong or a Weak Proton Acceptor

Is the phosphate ion, PO_4^{3-}, a relatively strong or weak base?

ANALYSIS   Using the tactic described in Example 7.3, we pretend that PO_4^{3-} actually is a base, a proton acceptor. So we give it a proton, and write the result, HPO_4^{2-}. (Notice that combining H^+ with PO_4^{3-} not only adds H to the formula but changes the net charge.) The HPO_4^{2-} ion, the protonated form of PO_4^{3-}, is not on the list of strong proton donors or acids. So we conclude that HPO_4^{2-} is a *weak* acid. This means, reciprocally, that PO_4^{3-} is a good proton binder (holding the proton as HPO_4^{2-}).

SOLUTION   A good proton binder is a *strong* base, so our answer to the question is that PO_4^{3-} is a relatively strong base.

■ Aqueous trisodium phosphate (Na_3PO_4) is a powerful cleaner, but very unkind to hands. Wear rubber gloves if you use it.

■ PRACTICE EXERCISE 9   Classify the following particles as strong or weak acids.
(a) HSO_3^- (b) HCO_3^- (c) $H_2PO_4^-$

■ PRACTICE EXERCISE 10   Classify the following ions as strong or weak bases.
(a) I^- (b) NO_3^- (c) CN^- (d) NH_2^-

TABLE 7.4 Relative Strengths of Some Acids and Bases

Acid (Proton Donor)		Base (Proton Acceptor)	
Name	Formula	Name	Formula
Hydrogen iodide	HI	Iodide ion	I^-
Hydrogen bromide	HBr	Bromide ion	Br^-
Sulfuric acid	H_2SO_4	Hydrogen sulfate ion	HSO_4^-
Hydrogen chloride	HCl	Chloride ion	Cl^-
Nitric acid	HNO_3	Nitrate ion	NO_3^-
HYDRONIUM ION	H_3O^+	WATER	H_2O
Hydrogen sulfate ion	HSO_4^-	Sulfate ion	SO_4^{2-}
Phosphoric acid	H_3PO_4	Dihydrogen phosphate ion	$H_2PO_4^-$
Acetic acid	CH_3CO_2H	Acetate ion	$CH_3CO_2^-$
Carbonic acid	H_2CO_3	Bicarbonate ion	HCO_3^-
Dihydrogen phosphate ion	$H_2PO_4^-$	Monohydrogen phosphate	HPO_4^{2-}
Ammonium ion	NH_4^+	Ammonia	NH_3
Bicarbonate ion	HCO_3^-	Carbonate ion	CO_3^{2-}
Monohydrogen phosphate ion	HPO_4^{2-}	Phosphate ion	PO_4^{3-}
WATER	H_2O	HYDROXIDE ION	OH^-
Methyl alcohol	CH_3OH	Methoxide ion	CH_3O^-
Ammonia	NH_3	Amide ion	NH_2^-

Increasing Acid Strength (arrow pointing up, left side)

Increasing Base Strength (arrow pointing down, right side)

The relative strengths of several acids and bases are given in Table 7.4. The acids above H_3O^+ in the table are all about equally strong in water. In water, any acid that is a better proton donor than H_3O^+ will spontaneously and rapidly transfer H^+ to H_2O to make H_3O^+. This is why H_3O^+ was described earlier as the strongest acid that can exist as a proton donor in water.

All bases above H_2O are roughly equally (very) weak bases in water. In fact, they're not really bases at all. And the acids listed below H_2O are not really acids in water. They're neutral. The bases below the OH^- are such powerful proton acceptors that they cannot even exist in water. They react with water, take protons from it, and leave hydroxide ions in their place. This is why the strongest base we can have in water is the hydroxide ion.

The Ammonium Ion Is a Proton Donor. The ammonium ion occupies a special place in our study because many biochemicals, like proteins, have a molecular part that is very much like this ion. Although the ammonium ion is a weak acid (see Table 7.4), it still is a proton donor, and it can neutralize the hydroxide ion. When we mix aqueous solutions of sodium hydroxide and ammonium chloride, the following reaction occurs.

$$NaOH(aq) + NH_4Cl(aq) \longrightarrow NH_3(aq) + H_2O + NaCl(aq)$$

The net ionic equation is

$$OH^-(aq) + NH_4^+(aq) \longrightarrow NH_3(aq) + H_2O$$

This reaction neutralizes the hydroxide ion, and it leaves a solution of the weaker base, NH_3.

In some medical emergencies, when the blood has become too alkaline or too basic, an isotonic solution of ammonium chloride is administered by intravenous drip. Its ammonium ions are used to neutralize excess base in the blood caused by the emergency and to bring the acid−base balance back to normal.

■ If the initial solution is concentrated enough, the final solution has a strong odor of ammonia.

Electrolyte Balance The ions that are present in blood in the highest concentrations and that contribute most to the aggregate electrical charges borne by ions are Na^+, Cl^-, and HCO_3^-. By comparison, K^+, Ca^{2+}, and Mg^{2+} have small (yet extremely important) concentrations. Clinical chemists and medical technologists can determine the concentrations of these six ions relatively easily. Anions of organic acids, like the acetate, citrate, and lactate ions, are also present, but they are much harder to analyze. The term *electrolyte balance* broadly refers to the *relative* levels of the ions.

Equivalents and Milliequivalents To describe the concentrations of ions, a commonly used expression is based not on the number of *moles* of an ion per liter but on the number of *equivalents* or the number of *milliequivalents* per liter. One **equivalent** of an ion, abbreviated **eq**, is that amount carrying Avogadro's number or one mole of electrical charge. For example, when the charge is 1, either 1+ or 1−, Avogadro's number of ions or one mole of the ions bears an aggregate of Avogadro's number of charges. The mass of 1 equivalent of singly charged ions such as Na^+, K^+, Cl^-, and Br^- is the same as the mass of one mole of each ion. Thus, the mass of one mole and the mass of one equivalent of Na^+ are identical, 23.0 g.

When an ion has a double charge, either 2+ or 2−, one mole of the ion carries an aggregate of two moles of electrical charges, so the mass of one equivalent of ions such as Ca^{2+} or CO_3^{2-} equals the mass of one-half mole. For example, 1 mol of CO_3^{2-} ion has a mass of 60.0 g, so the mass of 1 eq of CO_3^{2-} ion is 30.0 g. This mass of carbonate ion carries Avogadro's number of negative charges. The extension to ions of higher charges is obvious, and the accompanying table gives the mass of one equivalent for a number of ions.

Because the concentrations of ions in body fluids are small, the **milliequivalent**, abbreviated **meq**, is usually used.

$$1 \text{ eq} = 1000 \text{ meq}$$

Equivalents of Ions

Ion	g/mol	g/eq
Na^+	23.0	23.0
K^+	39.1	39.1
Ca^{2+}	40.1	20.1
Mg^{2+}	24.3	12.2
Al^{3+}	27.0	9.00
Cl^-	35.5	35.5
HCO_3^-	61.0	61.0
CO_3^{2-}	60.0	30.0
SO_4^{2-}	96.1	48.1

Glance at the data inside the back cover giving the "normals," or normal concentrations of several species in blood, and you'll see "meq" used for some. The normal concentration Na^+, for example, can be expressed as in the range 135 to 145 meq/L; that of Cl^-, 100 to 108 meq/L; and that of HCO_3^-, 21 to 29 meq/L. In contrast, the levels of K^+, Ca^{2+}, Mg^{2+}, and "phosphate" ions (total of all anions of H_3PO_4) are in the range of only 1 to 5 meq/L each, with the sum of the levels of the three cations exceeding that of the total of the various anions of H_3PO_4.

The advantage of the concept of the equivalent is the simplicity of a 1-to-1 ratio. Solutions are electrically neutral, so the aggregate of charges on all positively charged ions *must* equal the aggregate of all negative charges. In other words, an absolute requirement of any solution or fluid is

number equivalents of (+) charge = number of equivalents of (−) charge

Alternative equations that say the same thing about an invariable condition of a solution are

eq of cations = eq of anions
meq of cations = meq of anions

Anion Gap The levels of the harder-to-measure organic anions in blood tend to increase in several met-

---

The acid–base balance or acid–base status of the blood is only one aspect of a larger concern, the *electrolyte balance* of body fluids. To deal with this, primary health care specialists—doctors and nurses—take advantage of a special way of describing concentrations or "levels" of ions. Special Topic 7.1 discusses how ion concentrations given in milliequivalents per liter (meq/L) enable them to estimate difficult-to-measure concentrations from those readily done in clinical laboratories and to use such data as an aid in making diagnoses.

abolic disturbances, such as diabetes and kidney disease. The combined levels of the organic anions, however, can be estimated without a direct measurement by calculating a quantity known as the **anion gap**, defined by the following equation:

$$\text{anion gap} = \frac{\text{meq of Na}^+}{L} - \left(\frac{\text{meq of Cl}^-}{L} + \frac{\text{meq of HCO}_3^-}{L} \right)$$

To determine the anion gap in a patient's blood, a sample is analyzed for the concentrations in meq/L of Na^+, Cl^-, and HCO_3^-, which, as we said, are the most abundant ions but relatively easy to analyze. (The concentrations of K^+, Mg^{2+}, Ca^{2+}, SO_4^{2-}, and the anions of the phosphoric acid system have little net effect on the anion gap.) Then the concentration data are fed into the anion gap equation. Suppose, for example, that the analyses of a blood sample found the following data: $Na^+ = 137$ meq/L; $Cl^- = 100$ meq/L; $HCO_3^- = 28$ meq/L. Then the anion gap is found by

$$\text{anion gap} = \frac{137\ \text{meq}}{L} - \left(\frac{100\ \text{meq}}{L} + \frac{28\ \text{meq}}{L} \right)$$
$$= 9\ \text{meq/L}$$

The 9 meq/L anion gap is an estimate of the concentration of anions that must be present if the solution is to be electrically neutral. The normal range for the anion gap is 5 to 14 meq/L, so an anion gap of 9 meq/L falls within the normal range.

When certain metabolic disturbances cause the levels of organic anions to increase, the body must retain cations in the blood and excrete some of the more common anions, like Cl^- and HCO_3^-. This is necessary to meet the absolute requirement that the blood be electrically neutral. In other words, the more difficult-to-measure negative organic ions tend to expel other negative ions but retain whatever positive ions are available. This is how the anion gap widens in metabolic disturbances that generate organic anions. The anion gap routinely rises above 14 meq/L in untreated diabetes. In severe exercise, the anion gap rises above normal, too, but it goes back down again in time. Thus an above normal anion gap has to be interpreted in the light of other facts.

You can see that by using rather easily measured data on the meq/L concentrations of Na^+, Cl^-, and HCO_3^-, and calculating the anion gap from these data, the clinical chemist can inform the health care professionals of any unusual buildups in anions which indicate possible disease.

7.5 SALTS

A very large number of ionic reactions can be predicted from a knowledge of the solubility rules of salts.

Salts, as we have said, are ionic compounds whose cations are any except H^+ and whose anions are any except OH^- or O_2^-. A **simple salt** is one made of only *two* kinds of oppositely charged ions, for example, NaCl, $MgBr_2$, and $CuSO_4$. *Mixed salts* are those that have three or more different ions. Alum, used in water purification, is an example: $K_2SO_4 \cdot Al_2(SO_4)_3 \cdot 24H_2O$. As the formula of alum illustrates, the salt family includes hydrates. Some salts of practical value are given in Table 7.5.

■ The oxide ion, O^{2-}, reacts completely with water to give the hydroxide ion: $O^{2-} + H_2O \rightarrow 2OH^-$

Salts Can Be Made by Combining Acids with Hydroxides, Carbonates, Bicarbonates, or Active Metals.   In the laboratory, salts are obtained whenever an acid is used in any of the following ways. We summarize and review these methods here and show their similarities.

$$\text{Acid} + \text{metal hydroxide} \rightarrow \text{a salt} + H_2O$$
$$\text{Acid} + \text{metal bicarbonate} \rightarrow \text{a salt} + H_2O + CO_2$$
$$\text{Acid} + \text{metal carbonate} \rightarrow \text{a salt} + H_2O + CO_2$$
$$\text{Acid} + \text{metal} \rightarrow \text{a salt} + H_2$$

If the salt is soluble in water, we have to evaporate the solution to dryness to isolate it. Sometimes, however, the salt precipitates. To predict when to expect the precip-

TABLE 7.5   Some Salts and Their Uses

Formula	Name	Uses
$BaSO_4$	Barium sulfate	Used in the "barium cocktail" given prior to X-ray of the gastrointestinal tract
$MgSO_4 \cdot 7H_2O$	Magnesium sulfate heptahydrate (epsom salt)	Purgative
$(CaSO_4)_2 \cdot H_2O$	Calcium sulfate hemihydrate (plaster of paris)	Plaster casts; wall stucco; wall plaster
$AgNO_3$	Silver nitrate	Antiseptic and germicide; used in eyes of infants to prevent gonorrheal conjunctivitis; photographic film
$NaHCO_3$	Sodium bicarbonate (baking soda)	Baking powders; effervescent salts; stomach antacid; fire extinguishers
$Na_2CO_3 \cdot 10H_2O$	Sodium carbonate decahydrate (soda ash, sal soda, washing soda)	Water softener; soap and glass manufacture
$NaCl$	Sodium chloride	Manufacture of chlorine and sodium hydroxide; preparation of food
$NaNO_2$	Sodium nitrite	Meat preservative

itation of a salt and to introduce another way to make salts, we turn to the solubility rules for salts. When a *counter ion* is referred to in the following rules, it means the *unnamed* ion of the ionic compound. For example, in the *lithium* salt LiCl, the counter ion is an anion, Cl^-. In the *hydroxide* $Ca(OH)_2$, the counter ion is a cation, Ca^{2+}.

■ By *soluble*, remember, we mean the ability to form a solution with a concentration of at least 3 to 5% (w/w).

SOLUBILITY RULES FOR IONIC COMPOUNDS IN WATER
1. All lithium, sodium, potassium, and ammonium salts are soluble regardless of the counterion.
2. All nitrates and acetates are soluble, regardless of the counterion.
3. All chlorides, bromides, and iodides are soluble, *except* when the counterion is lead, silver, or mercury(I).
4. All sulfates are soluble, *except* those of lead, calcium, strontium, mercury(I), and barium.
5. All hydroxides and metal oxides are insoluble, *except* those of the Group IA cations and those of calcium, strontium, and barium.
6. All phosphates, carbonates, sulfites, and sulfides are insoluble, *except* those of the Group IA cations and NH_4^+.

■ By developing the skill of *predicting* reactions from a few facts, we sharply reduce the quantity of facts that should be memorized.

Although there are exceptions, we will seldom be wrong in applying the solubility rules. One of their many applications is in predicting possible reactions involving ionic compounds.

Salts Can Form by *Double Replacement* ("Exchange of Partners") Reactions. In addition to the acid–base neutralization reactions already studied that give salts, salts also form by a "change of partners" reaction called **double replacement**. For example, if we mix aqueous solutions of sodium carbonate and calcium chloride, both soluble in water, the following reaction occurs. It takes place because the CO_3^{2-} ion and the Ca^{2+} ion cannot remain in solution in each other's presence beyond extremely trace concentrations. Their combination, $CaCO_3$, is insoluble, which we learn from solubility rule 6.

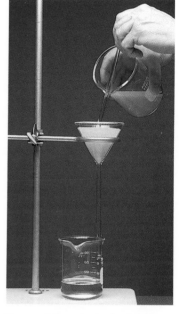

The precipitate in the beaker is being separated from the liquid by *filtration*. The clear liquid passing through the filter paper cone is the *filtrate*.

Combining CO_3^{2-} with Ca^{2+}

$$Na_2CO_3(aq) + CaCl_2(aq) \longrightarrow CaCO_3(s) + 2NaCl(aq)$$

Combining Na^+ with Cl^-

An ion from each salt combines with an ion from the other salt, which gives the informal name "exchange of partners" to the reaction or, more formally, *double replacement*. The ionic equation more clearly shows the exchange. (Cancel the spectator ions yourself.)

$$[2Na^+(aq) + CO_3^{2-}(aq)] + [Ca^{2+}(aq) + 2Cl^-(aq)] \rightarrow$$
$$CaCO_3(s) + [2Na^+(aq) + 2Cl^-(aq)]$$

The net ionic equation is

$$Ca^{2+}(aq) + CO_3^{2-}(aq) \rightarrow CaCO_3(s)$$

The other product, NaCl, stays in solution because its ions easily tolerate each other without forming a precipitate (solubility rule 1 or 3). We could isolate the precipitate of calcium carbonate by filtration and then evaporate the clear filtrate to dryness to obtain crystalline NaCl.

Not all combinations of solutes give double replacement reactions. If you mixed solutions of NaI and KCl, no combination of oppositely charged ions is insoluble (rule 1 or 3), so the solution would contain only separated (and hydrated) ions of Na^+, K^+, I^-, and Cl^-.

EXAMPLE 7.5 Predicting Double Replacement Reactions of Salts

What happens if we mix aqueous solutions of sodium sulfate and barium nitrate?

ANALYSIS By solubility rules 1 and 2, we know that both sodium sulfate and barium nitrate are soluble in water, so their solutions contain their separated ions. When we pour the two solutions together, four ions experience attractions and repulsions. Hence, we must examine each possible combination of oppositely charged ions to see which might make a water-insoluble salt. If we find one, a reaction is predicted. So we then write an equation. Here are the possible combinations when Ba^{2+}, NO_3^-, Na^+, and SO_4^{2-} ions intermingle in water.

$$Ba^{2+} + 2NO_3^- \xrightarrow{?} Ba(NO_3)_2(s)$$ This possibility is obviously out, because barium ions and nitrate ions do not precipitate together from water. ("All nitrates are soluble," rule 2.)

■ A slurry of barium sulfate in flavored water is the barium "cocktail" a patient drinks before having an X ray taken of the gastrointestinal tract.

$$2Na^+ + SO_4^{2-} \xrightarrow{?} Na_2SO_4(s)$$

This possibility is also out. ("All sodium salts are soluble," rule 1.)

$$Na^+ + NO_3^- \xrightarrow{?} NaNO_3(s)$$

No. ("All sodium salts are soluble," rule 1.)

$$Ba^{2+} + SO_4^{2-} \xrightarrow{?} BaSO_4(s)$$

Yes. Barium sulfate, $BaSO_4$, is not in any of the categories of water-soluble salts, so we conclude that it is an insoluble salt (rule 4.)

Because we predicted that $BaSO_4$ can form a precipitate, we first write a molecular equation. We'll use some connector lines to show how partners exchange—how *double* replacement occurs.

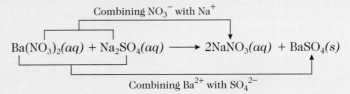

SOLUTION   The net ionic equation is a better way to describe what happens.

$$Ba^{2+}(aq) + SO_4^{2-}(aq) \rightarrow BaSO_4(s)$$

The sodium and nitrate ions are only spectators.

■ PRACTICE EXERCISE 11   If solutions of sodium sulfide, Na_2S, and copper(II) nitrate, $Cu(NO_3)_2$, are mixed, what if anything will happen chemically? Write a molecular equation and a net ionic equation for any reaction.

One of the many uses of the solubility rules is to understand what it means for water to be called *hard water* and what it means to *soften* such water. These are discussed in Special Topic 7.2.

To Summarize, the Chief Reactions of Ions Are Those That Form Gases, Molecules in Solution, or Precipitates.   Our study of the reaction of sodium sulfate and barium nitrate in Example 7.5 illustrates the power of knowing just a few facts for the sake of predicting an enormous number of others with a high probability of success. The following facts summarize those that should now be well learned.

1.   The *solubility rules* of ionic compounds (because then we can assume that all the other ionic compounds are insoluble)

2.   The *five strong acids* in Table 7.1 (because then we can assume that all the other acids, including organic acids, are weak)

3.   The *first two strong bases* in Table 7.2 (because then we can assume that all the other bases are either weak or are too insoluble in water to matter much)

To further summarize, we expect ions to react with each other if any one of the following possibilities is predicted.

1.   A *gas* forms that (mostly) leaves the solution. It could be
    (a) Carbon dioxide—from acids reacting with carbonates or bicarbonates—or
    (b) Hydrogen—from the action of acids on metals

Ground water that contains magnesium, calcium, or iron ions at a high enough level to interact with ordinary soap to form scum is called **hard water**. In **soft water**, these "hardness ions" — Ca^{2+}, Mg^{2+}, Fe^{2+}, and Fe^{3+} — are either absent or are present in extremely low concentrations. (The anions that most frequently accompany the hardness ions are SO_4^{2-}, Cl^-, and HCO_3^-.)

Hard water in which the principal anion is the bicarbonate ion is called **temporary hard water**. Hard water in which the chief negative ions are anything else is called **permanent hard water**. When temporary hard water is heated near its boiling point, as in hot boilers, steam pipes, and instrument sterilizers, the bicarbonate ion breaks down to the carbonate ion. And this ion forms insoluble precipitates with the hardness ions. Their carbonate salts form, come out of solution, and deposit as scaly material that does not conduct heat well and can even clog the equipment (Figure 1). The equations for these changes are as follows.

The breakdown of the bicarbonate ion is represented by

$$2HCO_3^-(aq) \rightarrow CO_3^{2-}(aq) + CO_2(g) + H_2O$$

The formation of the scaly precipitate (using the calcium ion to illustrate) is represented by

$$CO_3^{2-}(aq) + Ca^{2+}(aq) \rightarrow CaCO_3(s)$$

Water Softening Removes the Hardness Ions Chemically Hard water can be softened in various ways. Most commonly, excess soap is used. Some scum does form, but then the extra soap does the cleansing work. To avoid the scum altogether, softening agents are added before the soap is used. One common water-softening chemical is sodium carbonate decahydrate, known as washing soda. Its carbonate ions take out the hardness ions as insoluble carbonates by the kind of reaction for which we wrote the previous net ionic equation.

Another home water-softening agent is household ammonia, 5% (w/w) NH_3. We've already learned about the following equilibrium in such a solution.

$$NH_3(aq) + H_2O \rightleftharpoons NH_4^+(aq) + OH^-(aq)$$

In other words, aqueous ammonia has some OH^- ions, and the hydroxides of the hardness ions are not solu-

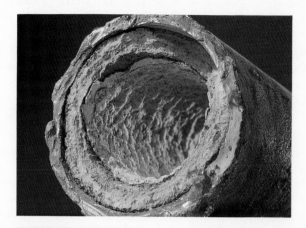

FIGURE I
Boiler scale building up on the inside of a water pipe.

ble in water. Therefore, when aqueous ammonia is added to hard water, the following kind of reaction occurs (illustrated using the magnesium ion this time):

$$Mg^{2+}(aq) + 2OH^-(aq) \rightarrow Mg(OH)_2(s)$$

As hydroxide ions are removed by this reaction, more are made available from the ammonia–water equilibrium. (A loss of OH^- ion from this equilibrium is a stress, and the equilibrium shifts to the right in response, as we'd predict on the basis of Le Châtelier's principle.)

Still another water-softening technique is to let the hard water trickle through zeolite, a naturally occurring porous substance that is rich in sodium ions. When the hard water is in contact with the zeolite, sodium ions go into the water and the hardness ions leave solution and attach themselves to the zeolite. Later, the hardness ions are flushed out by letting water that is very concentrated in sodium chloride trickle through the spent zeolite, and this restores the zeolite for reuse. Synthetic ion-exchange materials are also used to soften water by roughly the same principle.

Perhaps the most common strategy in areas where the water is quite hard is to use synthetic detergents instead of soap. Synthetic detergents do not form scums and precipitates with the hardness ions.

2. An un-ionized, *molecular compound* forms that remains in solution. It could be
   (a) Water—from acid–base neutralizations or
   (b) A weak acid—by the action of H^+ on a strong base, the conjugate base of any weak acid; or
   (c) Ammonia—by the reaction of OH^- with NH_4^+.
3. A *precipitate* forms. It is some water-insoluble salt or one of the water-insoluble hydroxides.

■ PRACTICE EXERCISE 12   When a solution of hydrochloric acid is mixed in the correct molar proportions with a solution of sodium acetate, $NaCH_3CO_2$, essentially all of the hydronium ion concentration vanishes. What happens and why? Write the net ionic equation.

■ PRACTICE EXERCISE 13   What, if anything, happens chemically when each pair of solutions is mixed? Write net ionic equations for any reactions that occur.

(a) NaCl and $AgNO_3$
(b) $CaCO_3$ and HNO_3
(c) KBr and NaCl

SUMMARY

Self-Ionization of Water   Trace concentrations of hydronium ions, H_3O^+, and hydroxide ions, OH^-, are always present in water. In neutral water, their molar concentrations are equal (and very low). In writing equations, we usually write H_3O^+ as H^+, calling the latter either the hydrogen ion or the proton.

Acidic solutions are those in which $[H_3O^+] > [OH^-]$. In basic solutions, $[OH^-] > [H_3O^+]$. Neutral solutions have these two ions in equal (and extremely low) concentrations.

Electrolytes   Salts in their molten states and the aqueous solutions of all soluble salts and of all strong acids and strong, soluble bases conduct electricity, and are called strong electrolytes. Solutions of weaker acids or of slightly soluble strong bases are weak electrolytes. All molecular substances, unless they change into ions by reacting with water, are nonelectrolytes. Pure water is a nonelectrolyte.

Acids and Bases as Proton Donors and Acceptors   An acid is any species, molecule or ion, that can donate a proton, and a base is any that can accept a proton. A strong base is able to accept and bind a proton strongly. In water, the strongest base is the hydroxide ion. Water is a weak base and the chloride ion in water is an even weaker base. A strong acid has a strong ability to give up or donate a proton. The hydronium ion is an example of a strong proton donor, and

gaseous hydrogen chloride is even stronger. A weak acid has a weak or poor ability to donate a proton. In solutions of weak acids, only a tiny percentage of solute molecules have given up protons to water molecules. Acetic acid is a weak acid. The water molecule is so weak that we don't even call it an acid in the usual sense; it has no effect on litmus.

Bases include the hydroxide ion, the bicarbonate ion, the carbonate ion, and ammonia plus any of the anions of weak acids. What is left behind when a weak acid donates a proton is a relatively strong base, and what forms when a strong acid gives up a proton is a weak base.

Common Aqueous Acids and Bases   The five most common strong acids are hydrochloric, hydrobromic, hydriodic, sulfuric, and nitric acids. All are monoprotic except sulfuric acid, which is diprotic.

The strongest bases are those that supply the hydroxide ion, the hydroxides of Group IA and IIA metals, particularly NaOH, KOH, $Ca(OH)_2$, and $Mg(OH)_2$.

Reactions of Common Aqueous Acids   The aqueous acids react with

metal hydroxides, to give a salt and water

metal bicarbonates, to give a salt, carbon dioxide, and water metal carbonates, to give a salt, carbon dioxide, and water

metals, to give the salt of the metal and hydrogen

A solution of an acid is neutralized when any sufficiently strong proton binding species is added in the correct mole proportion to make the concentrations of hydrogen ion and hydroxide ion equal (and very small). The ammonium ion also neutralizes OH^-.

Carbonic Acid and Carbonates   Carbonic acid, H_2CO_3, is both a weak and an unstable acid. When it is generated in water by the reaction of any stronger acid with a bicarbonate or a carbonate salt, virtually all of the carbonic acid decom-poses to carbon dioxide and water, and most of the carbon dioxide fizzes out. The carbonate ion and the bicarbonate ion are both proton acceptors or bases.

Salts   Salts are ionic compounds consisting of any positive ion except H^+ and any negative ion except OH^- or O^{2-}. The chemical properties of salts in water are the properties of their individual ions. Salts can be produced by any of the reactions of strong acids summarized above as well as by double replacement reactions. The solubility rules for salts are guides for the prediction of their reactions. If a combination of oppositely charged ions can lead to an insoluble salt, an un-ionized species that stays in solution, or a gas, then the ions react.

REVIEW EXERCISES

The answers to Review Exercises whose numbers are in color are found in Appendix V. The answers to the other Review Exercises are found in the Study Guide that accompanies this text. The more challenging questions are marked with asterisks.

Sources of Ions

7.1   What families of compounds are the principal sources of ions in aqueous solutions?

7.2   Review the differences between atoms and ions by answering the following questions.
(a) Are there any atoms that have more than one nucleus? If so, give an example.
(b) Are there any ions with more than one nucleus? If so, give an example.
(c) Are there any ions that are electrically neutral? If so, give an example.
(d) Are there any atoms that are electrically charged? If so, give an example.

7.3   Write the equilibrium equation for the self-ionization of water, and label the ions that are present.

7.4   Tell whether each of the following solutions is acidic, basic, or neutral.
(a) $[H^+] = 7.9 \times 10^{-6}$ mol/L and $[OH^-] = 1.3 \times 10^{-9}$ mol/L
(b) $[H^+] = 1.0 \times 10^{-7}$ mol/L and $[OH^-] = 1.0 \times 10^{-7}$ mol/L
(c) $[H^+] = 7.6 \times 10^{-8}$ mol/L and $[OH^-] = 1.2 \times 10^{-7}$ mol/L

7.5   What features do the common aqueous acids share?

7.6   Salts are all crystalline solids at room temperature. Why?

7.7   In general terms, what are the products of an acid–base neutralization?

Electrolytes

7.8   The word *electrolyte* can be understood in two ways. What are they?

7.9   To which electrode do anions migrate?

7.10   The cathode has what electrical charge, positive or negative?

7.11   An electrode that is positively charged attracts what kinds of ions, cations or anions?

7.12   Explain in your own words how the presence of cations and anions in water enables the system to conduct electricity.

7.13   When KOH(*s*) is dissolved in water, the solution is an excellent conductor of electricity, but when ethyl alcohol is dissolved in water, the solution won't conduct electricity at all. What does this behavior suggest about the structural nature of KOH and ethyl alcohol, whose structure is given below? (Notice that both appear to have OH groups in their formulas.)

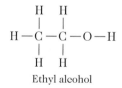
Ethyl alcohol

7.14   When lithium nitrate dissolves in water, its crystals break up entirely into Li^+ and NO_3^- ions. Do we call this compound a *weak* or a *strong* electrolyte?

7.15   In the liquid state, tin(IV) chloride, $SnCl_4$, is a nonconductor. What does this suggest about the structural nature of this compound?

Acids and Bases

7.16   In the context of acid–base discussion, what are two other names that we can use for "hydrogen ion?"

7.17   Acids have a set of common reactions, and so do bases, but not salts. Explain.

7.18 How does litmus paper work to tell whether a solution is acidic, basic, or neutral?

7.19 How did Brønsted define an acid? A base?

7.20 What is the essential difference between an aqueous solution of a strong acid and that of a weak acid?

7.21 Explain why a weak acid must also be a weak electrolyte.

7.22 What is the difference between hydrochloric acid and hydrogen chloride?

7.23 Write the equation for the ionization of nitric acid in water.

7.24 Why is sulfuric acid with only 2 H in its formula called a diprotic acid but acetic acid, which has 4 H in its formula, is a monoprotic acid?

7.25 If we represent all diprotic acids by the symbol H_2A, write the equilibrium expressions for the two separate ionization steps.

7.26 Would the ionization of the second proton from a diprotic acid occur with greater ease or with greater difficulty than the ionization of the first proton? Explain.

7.27 Write the equations for the progressive ionizations of sulfuric acid. Include the names of the ions.

7.28 Write the equations for the progressive ionizations of phosphoric acid, including the names of the ions.

7.29 Write the equilibrium expression for the solution of carbon dioxide in water that produces some carbonic acid.

7.30 Write the equilibrium expressions for the successive steps in the ionization of carbonic acid.

7.31 KOH is a strong base and a strong electrolyte. What do these terms mean in connection with this compound?

7.32 Calcium hydroxide is only slightly soluble in water, and yet it is classified as a *strong* base. Explain.

7.33 What are the names and formulas of two bases that are both strong and capable of forming relatively concentrated solutions in water?

7.34 What is meant by aqueous ammonia? Why don't we call it "ammonium hydroxide?"

7.35 Write the names and formulas of the five strong acids that we have studied.

7.36 What are the four strong bases—both the names and formulas? Which are quite soluble in water?

Net Ionic Equations

7.37 Consider the following net ionic equation.

$$6H^+(aq) + Cu(s) + 3NO_3^-(aq) \rightarrow$$
$$Cu^{2+}(aq) + 3NO_2(g) + 3H_2O$$

(a) Does it have material balance?
(b) Does it have electrical balance?
(c) Is it a balanced equation?

*7.38 Complete and balance the following molecular equations, and then write the net ionic equations.
(a) $HNO_3(aq) + KOH(aq) \rightarrow$
(b) $HCl(aq) + NaHCO_3(aq) \rightarrow$
(c) $HBr(aq) + MgCO_3(s) \rightarrow$
(d) $HNO_3(aq) + KHCO_3(aq) \rightarrow$
(e) $HBr(aq) + NH_3(aq) \rightarrow$
(f) $HNO_3(aq) + Ca(OH)_2(s) \rightarrow$
(g) $HCl(aq) + Mg(s) \rightarrow$

*7.39 Complete and balance the following molecular equations, and then write the net ionic equation.
(a) $NaOH(aq) + H_2SO_4(aq) \rightarrow$
(b) $K_2CO_3(aq) + HNO_3(aq) \rightarrow$
(c) $NaHCO_3(aq) + HBr(aq) \rightarrow$
(d) $CaCO_3(s) + HI(aq) \rightarrow$
(e) $NH_3(aq) + HI(aq) \rightarrow$
(f) $Mg(OH)_2(s) + HBr(aq) \rightarrow$
(g) $Al(s) + HCl(aq) \rightarrow$

7.40 What are the net ionic equations for the following reactions of strong, aqueous acids? (Assume that all reactants and products are soluble in water.)
(a) With metal hydroxides
(b) With metal bicarbonates
(c) With metal carbonates
(d) With aqueous ammonia

7.41 Write net ionic equations for the reactions of all the water-insoluble Group IIA carbonates, where you use $MCO_3(s)$ as their general formula, with nitric acid (chosen so that all the products are soluble in water).

7.42 If we let $M(OH)_2(s)$ represent the water-insoluble Group IIA metal hydroxides, what is the general net ionic equation for all of their reactions with hydrochloric acid (chosen so that all the products are soluble in water)?

7.43 If we let $M(s)$ represent either calcium or magnesium metal, what net ionic equation represents the reaction of either with hydrochloric acid?

*7.44 Sodium and potassium in Group IA are higher in the activity series than calcium and magnesium in Group IIA.
(a) What does it mean to be higher in the activity series?
(b) If you check back to Figure 2.4b on page 43, you will see that sodium and potassium have lower ionization energies than calcium and magnesium. In what way does this fact correlate with their *higher* position in the activity series of the metals?

7.45 Zinc metal reacts more rapidly with which acid, 1 *M* hydrochloric acid or 1 *M* phosphoric acid? Explain.

7.46 How many moles of potassium bicarbonate can react quantitatively with 0.355 mol of HCl?

7.47 How many moles of sodium hydroxide can react quantitatively with 0.256 mol of H_2SO_4 (assuming that both H^+ in H_2SO_4 are neutralized)?

7.48 How many grams of sodium carbonate will neutralize 5.24 g of HCl?

7.49 How many grams of calcium carbonate react quantitatively with 7.98 g of HNO_3?

7.50 How many grams of sodium bicarbonate does it take to neutralize all the acid in 28.9 mL of 1.05 M H_2SO_4?

7.51 How many grams of potassium carbonate will neutralize all of the acid in 32.9 mL of 0.435 M HCl?

*7.52 How many milliliters of 0.165 M NaOH are needed to neutralize the acid in 28.6 mL of 0.212 M HNO_3?

*7.53 How many milliliters of 0.115 M KOH are needed to neutralize the acid in 14.6 mL of 0.161 M H_2SO_4?

*7.54 For an experiment that required 13.5 L of dry CO_2 gas (as measured at 745 mm Hg and 24 °C), a student let 4.62 M HCl react with marble chips, $CaCO_3$.
(a) Write the molecular and net ionic equations for this reaction.
(b) How many grams of $CaCO_3$ and how many milliliters of the acid are needed?

*7.55 How many liters of dry CO_2 gas are generated (at 740 mm Hg and 25 °C) by the reaction of $Na_2CO_3(s)$ with 325 mL of 5.85 M HCl? Write the molecular and net ionic equations for the reaction, and calculate how many grams of Na_2CO_3 are needed.

Relative Strengths of Acids and Bases

7.56 What are the formulas of the protonated forms of the following?
(a) HSO_3^- (b) Br^- (c) H_2O (d) $CH_3CO_2^-$

7.57 Write the formulas of the protonated forms of the following.
(a) HSO_4^- (b) HCO_3^- (c) I^- (d) NO_2^-

7.58 Write the formulas of the deprotonated forms of the following.
(a) NH_3 (b) HNO_2 (c) HSO_3^- (d) H_2SO_3

7.59 What are the deprotonated forms of the following? Write their formulas.
(a) H_2CO_3 (b) $H_2PO_4^-$ (c) NH_4^+ (d) OH^-

7.60 Which member of each pair is the stronger base?
(a) NH_3 or NH_2^- (b) OH^- or H_2O (c) HS^- or S^{2-}

7.61 Which member of each pair is the stronger base?
(a) Br^- or HCO_3^- (b) $H_2PO_4^-$ or HSO_4^-
(c) NO_2^- or NO_3^-

7.62 Which member of each pair is the stronger acid?
(a) H_2CO_3 or HCl   (b) H_2O or OH^-
(c) HSO_4^- or HSO_3^-

7.63 Study each pair and decide which is the stronger acid.
(a) $H_2PO_4^-$ or HPO_4^{2-} (b) H_2SO_3 or HSO_3^-
(c) NH_4^+ or NH_3

Salts

*7.64 Write the names and formulas of three compounds that, by reacting with hydrochloric acid, give a solution of

potassium chloride. Write the molecular equations for these reactions.

*7.65 Write the names and formulas of three compounds that will give a solution of sodium bromide when they react with hydrobromic acid. Write the molecular equations for these reactions.

7.66 Which of the following compounds are insoluble in water (as we have defined solubility)?
(a) KOH   (b) NH_4Cl (c) Hg_2Cl_2
(d) $Mg_3(PO_4)_2$ (e) NaBr   (f) Li_2SO_4

7.67 Which of the following compounds are insoluble in water?
(a) $(NH_4)_2SO_4$ (b) $NaNO_2$ (c) LiBr
(d) AgBr   (e) $Ca_3(PO_4)_2$ (f) KNO_3

7.68 Identify the compounds that do not dissolve in water.
(a) $BaCO_3$ (b) NH_4NO_3 (c) K_2CO_3
(d) $PbCl_2$ (e) Na_2SO_4 (f) $LiCH_3CO_2$

7.69 Which of the following compounds do not dissolve in water?
(a) K_2CrO_4 (b) AgBr   (c) $FeCO_3$
(d) $Na_2Cr_2O_7$ (e) Li_2CO_3 (f) NH_4I

*7.70 Assume you have separate solutions of each compound in the pairs below. Predict what happens chemically when the two solutions of a pair are poured together. If no reaction occurs, state so. If there is a reaction, write its net ionic equation.
(a) KCl and $AgNO_3$ (b) KNO_3 and $MgCl_2$
(c) NaOH and H_2SO_4 (d) $Pb(NO_3)_2$ and NaCl
(e) NH_4Cl and K_2SO_4 (f) Na_2S and $CuSO_4$
(g) Na_2SO_4 and $Ba(NO_3)_2$ (h) NaOH and HI
(i) Na_2S and $NiCl_2$ (j) $AgNO_3$ and NaBr
(k) $LiHCO_3$ and HI   (l) $MgCl_2$ and KOH

*7.71 For each pair, you have separate solutions of the compounds and then mix the two together, what (if anything) happens chemically? If no reaction occurs, state so, but if there is a reaction write its net ionic equation.
(a) H_2S and $Cu(NO_3)_2$ (b) LiOH and HBr
(c) Na_2SO_4 and $Ba(NO_3)_2$ (d) $Pb(CH_3CO_2)_2$ and Na_2SO_4
(e) $Ba(NO_3)_2$ and NaCl   (f) $KHCO_3$ and H_2SO_4
(g) Na_2S and $Cd(NO_3)_2$ (h) NaOH and HBr
(i) $Hg(NO_3)_2$ and KCl   (j) $NaHCO_3$ and HI
(k) KBr and NaCl   (l) $Pb(NO_3)_2$ and Na_2CrO_4

Milliequivalents of Ions and the Anion Gap (Special Topic 7.1)

7.72 The concentration of potassium ion in blood serum is normally in the range 0.0035 to 0.0050 mol K^+/L. Express this range in units of milliequivalents of K^+ per liter.

7.73 The concentration of calcium ion in blood serum is normally in the range 0.0042 to 0.0052 eq Ca^{2+}/L. Express this range in units of milliequivalents of Ca^{2+} per liter.

7.74 The level of chloride ion in blood serum is normally quoted as 100 to 106 meq/L. How many grams and how many milligrams constitute 106 meq of Cl^-?

7.75 The sodium ion level in the blood is normally 135 to 145 meq/L. How many grams and how many milligrams of sodium ion constitute 135 meq of Na^+?

*7.76 The potassium ion level of blood serum normally does not exceed 0.196 g of K^+ per liter. How many milliequivalents of K^+ ion are in 0.196 g of K^+?

*7.77 The magnesium ion level in plasma normally does not exceed 0.0243 g of Mg^{2+}/L. How many milliequivalents of Mg^{2+} are in 0.0243 g of Mg^{2+}?

7.78 The analysis of the blood from a young man recovering from polio found 137 meq of Na^+/L, 34 meq of HCO_3^-/L, and 93 meq of Cl^-/L. Calculate the anion gap. Does it suggest a serious disturbance in his metabolism?

7.79 A patient on a self-prescribed diet consisting essentially only of protein was found to have the following blood analyses after 2 weeks of the diet: Na^+, 174 meq/L; Cl^-, 135 meq/L; HCO_3^-, 20 meq/L. Calculate the anion gap. Does it suggest a disturbance in metabolism?

Hard Water (Special Topic 7.2)

7.80 What is hard water?

7.81 What are the formulas of the "hardness ions?"

7.82 What chemical property of these ions and of ordinary soap makes it difficult to use such soap in hard water?

7.83 What is temporary hard water? Why is it designated temporary?

7.84 What is permanent hard water?

7.85 What is meant by water softening?

7.86 Concerning aqueous ammonia as a water-softening agent,
(a) What is the composition of aqueous ammonia?
(b) What is the specific species in aqueous ammonia that is the active softening agent?
(c) How does this species arise in aqueous ammonia? (Write an equilibrium expression.)
(d) What is the net ionic equation for its work in water where the hardness is caused by Ca^{2+}?

7.87 In general terms, what is a zeolite and how does it work in water softening?

Additional Exercises

*7.88 Compounds A and B are both white solids that dissolve in water. One is an ionic compound and the other is molecular. Discuss an experiment that could be conducted to find out which is molecular and mention any possible drawbacks to the kind of experiment you select. (How might the experiment give ambiguous results?)

*7.89 A white solid is either KNO_3 or K_2O. Describe an experiment that you could carry out using only test tubes and aqueous solutions that would tell which compound is present. Assume that the lab has whatever other chemicals you need.

*7.90 A white solid is either Na_2CO_3 or $NaHCO_3$. A sample of the solid with a mass of 0.144 g requires 15.0 mL of 0.114 M HCl to react with it fully until the exact point is reached when no more CO_2 forms. Which compound is it?

*7.91 A white solid was a mixture of K_2CO_3 and KNO_3. A 0.624-g sample of the mixture consumed 21.5 mL of 0.156 M HCl before CO_2 stopped evolving. How many grams of K_2CO_3 were in the mixture?

8

ACIDITY: DETECTION, CONTROL, MEASUREMENT

pH Concept

Effects of Ions on pH

Buffers: Preventing
Large Changes in pH

Acid–Base Titrations

The pH concept, studied in this chapter, gives us a method for describing the acidities of solutions, including those that rain down on forests. These dead fir trees of Most, Czech Republic, stand as mute witnesses to the ravages of air pollution and acid rain caused by unconstrained industrialization.

8.1 pH CONCEPT

Very low levels of H⁺ are more easily described and compared in terms of pH values than as molar concentrations.

At the molecular level of life we must have almost perfect control over the acid–base balance of body fluids. Blood, for example, must have a molar concentration of hydrogen ion, $[H^+]$, of 4.47×10^{-8} mol/L or extremely close to this. If the value of $[H^+]$ goes up to 10×10^{-8} mol/L or drops to 1×10^{-8} mol/L, death is very near. Yet we can change the acidity of water far more than these figures represent by adding only one drop of concentrated hydrochloric acid or one drop of concentrated sodium hydroxide to a liter of pure water.

In Acidosis, the Blood Becomes More Acidic. So important is the acid–base balance of the blood that a special vocabulary exists to describe small shifts away from normal. When the molar concentration of H^+ in the blood is higher than normal, the condition is called **acidosis,** a characteristic of untreated diabetes, emphysema, and many other conditions. (Remember that H^+ is our "nickname" symbol for H_3O^+.)

When the molar concentration of H^+ in the blood is lower than normal, the condition is called **alkalosis.** An overdose of bicarbonate, exposure to the low partial pressure of oxygen at high altitudes, or prolonged hysteria can cause alkalosis.

In advanced stages of acidosis and alkalosis, respiration is impaired enough to constitute a medical emergency. **Respiration** is what brings oxygen into the body, chemically uses it, and removes waste carbon dioxide. All of these activities are intimately and beautifully linked at the molecular level of life. We need to develop more background in biochemistry in later chapters before we can study the chemistry of this important work of the body. Here, we continue to develop the concepts and vocabulary used to describe acid–base balance.

The Product of the Molar Concentrations of H⁺ and OH⁻ in Water Is a Constant. The word *balance* suggests a teeter-totter or seesaw weighted at each end sufficiently evenly to cause a stalemate. If some impulse occurs to cause motion, we know that when one side goes up, the other goes down. The *balance* in the term *acid–base balance* refers to an analogous *and desired* stalemate in the molar concentrations of hydrogen and hydroxide ion. Normally, the value of $[H^+]$ is slightly greater than that of $[OH^-]$, but if one concentration goes up the other inevitably and invariably goes down. Such "seesawing" is simply a fact about the equilibrium present in pure water. As we learned on page 147, the following equilibrium exists.

$$H_2O \rightleftharpoons H^+(aq) + OH^-(aq)$$

What is remarkable about this equilibrium is that in an aqueous solution, *regardless of any solute,* the result of multiplying the molar concentrations of hydrogen and hydroxide ions is a constant. Its called the **ion product constant of water,** symbolized by K_w.

$$K_w = [H^+][OH^-] \tag{8.1}$$

The values of $[H^+]$ and $[OH^-]$ in any aqueous solution "seesaw." We can alter the $[H^+]$ as we please by adding either acid or base to a solution, yet the value of $[OH^-]$ automatically and rapidly adjusts in the opposite direction. The *product* of the two molar concentrations, $[H^+][OH^-]$, remains constant, namely, K_w. The only way to

■ The brackets in $[H^+]$ signify that the concentration is specifically moles per liter.

■ One drop of concentrated hydrochloric acid changes the pH of a liter of water from 7 to 4.

■ Starvation, alcohol overdose, aspirin overdose, and severe diarrhea also cause acidosis.

■ The mismanagement of a nasogastric suction device can also cause alkalosis.

change K_w is to change the temperature, as data in the margin show. At 25 °C, both $[H^+]$ and $[OH^-]$ equal 1.0×10^{-7} mol/L. Therefore,

$$K_w = (1.0 \times 10^{-7})(1.0 \times 10^{-7})$$
$$= 1.0 \times 10^{-14} \text{ (at 25 °C)}$$

In most of our work, we assume a temperature of 25 °C. Knowing the value of K_w, we can calculate either of the two concentrations, $[H^+]$ or $[OH^-]$, if we know the other.

■ **K_w AT VARIOUS TEMPERATURES**

Temperature (°C)	K_w
0	1.5×10^{-15}
10	3.0×10^{-15}
20	6.8×10^{-15}
25	1.0×10^{-14}
30	1.5×10^{-14}
40	3.0×10^{-14}

EXAMPLE 8.1 Using the Ion Product Constant of Water

The value of $[H^+]$ of blood (when measured at 25 °C, not body temperature) is 4.5×10^{-8} mol/L. What is the value of $[OH^-]$, and is the blood acidic, basic, or neutral?

ANALYSIS We simply use the value of $[H^+]$ in the equation for K_w.

SOLUTION

$$K_w = 1.0 \times 10^{-14} = (4.5 \times 10^{-8}) \times [OH^-]$$
$$[OH^-] = \frac{1.0 \times 10^{-14}}{4.5 \times 10^{-8}} \text{ mol/L}$$
$$= 2.2 \times 10^{-7} \text{ mol/L}$$

Because the value of $[H^+]$, 4.5×10^{-8} mol/L, is less than that of $[OH^-]$, 2.2×10^{-7} mol/L, the blood is (very slightly) basic.

■ **PRACTICE EXERCISE 1** For each of the following values of $[H^+]$, calculate the value of $[OH^-]$ and state whether the solution is acidic, basic, or neutral.

(a) $[H^+] = 4.0 \times 10^{-9}$ mol/L

(b) $[H^+] = 1.1 \times 10^{-7}$ mol/L

(c) $[H^+] = 9.4 \times 10^{8}$ mol/L

The pH of an Acidic Solution Is Less than 7.00. When we compare quantities like 4.5×10^{-8} and 2.2×10^{-7} mol $[H^+]$/L to see which is larger (Example 8.1), we need to look in two places for each number. We have to compare the *exponents* of 10 and then the *coefficients,* here the numbers that appear before each 10. Yet quantities this small must often be compared in dealing with the relative acidities of body fluids. To make such comparisons easier, S. P. L. Sorenson invented the concept of pH. Two equivalent equations define pH, but we need only one, Equation 8.2.

■ S. P. L. Sorenson (1868–1909) was a Danish biochemist.

$$[H^+] = 1 \times 10^{-pH} \tag{8.2}$$

In words, the **pH** of a solution is the negative power (the *p* in pH) to which the number 10 must be raised to express the molar concentration of a solution's hydrogen ions (hence, the *H* in pH).[1] When $[H^+] = 1.0 \times 10^{-7}$ mol/L, the pH is 7.00. A pH of 7.00

---

[1]If the logs of both sides of Equation 8.2 are taken, after a change in sign, we get the second (equivalent) equation for pH: $pH = -\log[H^+]$

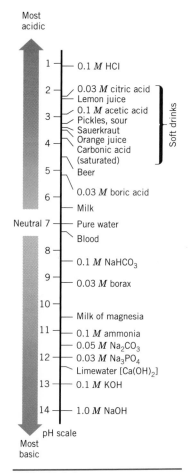

FIGURE 8.1
The pH scale and the pH values of several common substances.

■ A review of exponents can be found in Appendix I.

therefore corresponds to a neutral solution (or pure water) at 25 °C. The number of significant figures in any value of pH is the number of figures that *follow* the decimal point.[2]

Because pH occurs as a *negative* exponent in Equation 8.2, a pH *less* than 7.00 corresponds to an acidic solution. A pH greater than 7.00 corresponds to a basic solution. In pH terms, then, we have the following definitions of acidic, basic, and neutral solutions (at 25 °C).

Acidic solution	pH < 7.00
Neutral solution	pH = 7.00
Basic solution	pH < 7.00

When the value of $[H^+]$ is 1 mol/L or higher, the pH concept isn't used. The exponents would no longer be negative, so there would be none of the confusion that Sorenson addressed when he invented pH. The pH values of several common substances are shown in Figure 8.1.

The definition of pH in Equation 8.2 has the number 1 as the coefficient: $[H^+] = 1 \times 10^{-pH}$. Only when "1" occurs before the times sign can we simply take the exponent itself as identical to the pH. But what if the "1" isn't there—in the real world, it almost never is—but some other number is? For example, what can we say about the pH if $[H^+]$ equals, say, 2.0×10^{-8} mol/L? We certainly cannot state that the pH is 8; however, we can use Equation 8.2 to show that when $[H^+]$ equals 2.0×10^{-8} mol/L, $[OH^-]$ equals 5.0×10^{-7} mol/L. By comparing these two values, we see that $[OH^-]$ exceeds $[H^+]$, so the solution is basic and its pH must be greater than 7. That's all we can say so far about the pH when $[H^+]$ is 2.0×10^{-8} mol/L. In fact, the pH is 7.70, but to calculate this value from 2.0×10^{-8} mol H^+/L we have to use the second definition of pH given in footnote 1. This is easy with a small scientific calculator,[3] but in none of the uses to which we will put the pH concept will we have to carry out such a calculation. Our needs will be met simply by being able to tell if a pH value corresponds to an acidic, neutral, or basic solution.

We'd face a similar need for a calculator if we had to calculate $[H^+]$ from a pH when the latter is not a simple whole number. What if the pH is, say, 4.56? If we substitute 4.56 for pH into Equation 8.2, we obtain

$$[H^+] = 1 \times 10^{-4.56} \text{ mol/L}$$

Although this is a perfectly good number, we normally reexpress exponentials so that the exponents are *whole* numbers. We can tell from the quantity as it stands, $1 \times 10^{-4.56}$ mol/L, that the value of $[H^+]$ is between 1.0×10^{-4} and 1.0×10^{-5} mol/L, but we can't tell exactly where between these limits without using the second definition

[2]The 7 in a pH value of 7.00 comes from the *exponent* in 1.0×10^{-7}, so it actually does nothing more than set off a decimal point when we rewrite the number as 0.00000010. Hence, the 7 in the pH value of 7.00 can't be counted as a significant figure (rule 2, page 15). A pH value of 7.00 therefore has only two significant figures, those that *follow* the decimal point, just as there are only two significant figures in the value of a molar concentration of H^+ equaling 1.0×10^{-7} mol/L. To repeat, the number of significant figures in any value of pH is the number of figures that *follow* the decimal point.

[3]If your calculator has keys marked [exp], [±], and [log], first enter 2.0, use [exp] and then [±], and enter 8. This completes the entering of 2.0×10^{-8}. (The [exp] key really translates as "times 10 to the," and [±] merely ensures that the next number you enter will go in as a negative exponent.) Now use [log]; the screen should read −7.698970004. *Change the sign* and trim the digits to two figures following the decimal point; the pH is 7.70. Some calculators go about this slightly differently. If you have one, your instruction book will show you how to do it.

of pH (footnote 2). We can only say that the solution is acidic, because the pH is less than 7, but that is all the skill we'll need.[4]

■ PRACTICE EXERCISE 2 A patient suspected of having taken an overdose of barbiturates was found to have blood with a pH of 7.10.

(a) Between what limits of values for the molar concentration of H^+ does this pH correspond to?

(b) Is the blood slightly basic, acidic, or neutral?

(c) If the pH of blood normally is 7.35, is the patient experiencing acidosis or alkalosis?

■ PRACTICE EXERCISE 3 The patient in Practice Exercise 2 had been admitted to a teaching hospital, and the attending physician asked a group of students which of two isotonic reagents, sodium bicarbonate or ammonium chloride, might be administered by intravenous drip to restore the pH of the patient's blood to a normal value of 7.35. Which solution could be used? Explain how it would work using a net ionic equation.

Small Changes in pH Correspond to Large Changes in [H^+]. One of the very deceptive features of the pH concept is that the actual hydrogen ion concentration changes greatly—by a factor of 10—for each change of only one unit of pH. For example, if the pH of a solution is zero (meaning that [H^+] = 1×10^0 mol/L), only 1 L of water is needed to contain 1 mol of H^+. When the pH is 1, however, then 10 L of water (about the size of an average wastebasket) is needed to hold 1 mol of H^+. At a pH of 5 when [H^+] = 1×10^{-5} mol/L, it takes a large railroad tanker full of water to include just 1 mol of H^+. If the pH of the water flowing over Niagara Falls, New York, were 10 (which, of course, it isn't), an entire 1-hour flowage would be needed for 1 mol of H^+ to pass by. And at a pH of 14, the volume that would hold 1 mol of H^+ is about one-quarter of the volume of Lake Erie, one of the Great Lakes. You can see that seemingly small changes in pH values signify enormous changes in real concentrations of hydrogen ions.

■ Large railroad tankers hold more than 30,000 gallons.

■ Lake Erie isn't alkaline. The pH of 14 is what 1 M NaOH (lye) is.

pH Is a Measure of the *Acidity* of a Solution, Its H_3O^+ Concentration. A simple correlation exists between solute molarity and pH when the solute is a *strong* acid. Strong acids are 100% ionized. Suppose we have 0.010 M HCl, for example, which can be restated as 1.0×10^{-2} mol HCl(aq)/L. The molar concentration of H^+ ions must be numerically the same, 1.0×10^{-2} mol H^+/L, because each molecule of HCl that goes into solution breaks up into one H^+ ion and one Cl^- ion. When [H^+] = 1.0×10^{-2} mol/L, the pH is simply the negative of the exponent, or 2.00 (using two digits *after* the decimal point to specify two significant figures). Similarly, a solution that is 0.00010 M HNO_3, also a *strong* monoprotic acid, has a pH of 4.00, because 0.00010 can be reexpressed as 1.0×10^{-4}.

■ $0.010 = 1.0 \times 10^{-2}$

No simple correlation exists between the molar concentration of a *weak* acid and the pH of the solution. When the solute is a weak acid, only a small percentage of its molecules are ionized at equilibrium. In 0.10 M acetic acid, for example, the pH is about 2.9, not 1.0. Acetic acid would have to be a strong acid and 100% ionized for a solution of 0.10 M acetic acid to have a pH of 1.0 (because 0.10 mol H^+/L = 1.0 ×

■ Remember, acetic acid (CH_3CO_2H) is a *monoprotic* acid.

---

[4]Enter 4.56 into your calculator, *change the sign* by using the $\boxed{\pm}$ key (because pH is a *negative* exponent), and then use the $\boxed{10^x}$ key. The screen should read 2.754228703 −05, meaning (after rounding) 2.8×10^{-5} (mol/L, of course). Thus, $1.0 \times 10^{-4.56}$ mol H^+/L = 2.8×10^{-5} mol H^+/L.

FIGURE 8.2
Some common acid–base indicators.

pH 8.2   pH 10.0
Phenolphthalein

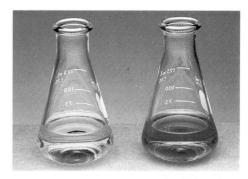

pH 6.0   pH 7.6
Bromothymol blue

FIGURE 8.3
A pH meter.

pH 3.2   pH 4.4
Methyl orange

pH 9.4   pH 10.6
Thymolphthalein

■ If we add the proper amount of NaOH to aqueous acetic acid, the OH^- ions can take protons not only from H_3O^+ ions but also from every molecule of CH_3CO_2H.

■ Phenolphthalein (fee-noll-THAY-lean)

■ Some commercial pH test papers contain several indicator dyes in the same test strip, which makes possible a whole spectrum of colors according to the pH color code on the container.

10^{-1} mol H^+/L). But acetic acid is a *weak* acid, so the pH of 0.10 M acetic acid, namely 2.9, is *not* a measure of the molarity of the *solute* but of the molarity of the *hydrogen* ions that the solute liberates. When we talk about the **acidity** of a solution, we thus refer to the molar concentration of H_3O^+ of which the pH is a measure.

Several Dyes Indicate a Solution's pH.   In the previous chapter we introduced litmus as an indicator, but it is only one of many **acid–base indicators** used to find out whether a solution is acidic, basic, or neutral. The dye phenolphthalein, for example, has a bright pink color at a pH above 10.0 and is colorless below a pH of 8.2. In the pH range 8.2 to 10.0, phenolphthalein undergoes a gradual change from colorless to pale pink to deep pink. Each of the many indicators has its own pH range and set of colors (Figure 8.2). Litmus is blue above a pH of about 8.5 and red below a pH of 4.5. The range of the color change of litmus thus straddles a pH of 7.0, so litmus is a good indicator for a *neutral* solution.

When the solutions to be tested for pH are themselves highly colored, we cannot use indicators. Moreover, we often need more than a rough idea of pH, which is all that indicators can actually give. For such situations there are instruments called pH meters equipped with special electrodes that can be dipped into the solution to be tested (Figure 8.3). With a good pH meter, pH values can be read to the second decimal place.

8.2 EFFECTS OF IONS ON pH

The balance between $[H_3O^+]$ and $[OH^-]$ can be upset by a number of species other than OH^- or H^+.

In many laboratory situations, an aqueous solution of some salt is prepared only to have the solution turn out to be acidic or basic toward litmus. If we weren't aware that this could happen, we might unknowingly prepare a salt solution that could be corrosive to metals, or could harm living things, or could make food, drink, or medications unfit. In the most general terms, salts are able to upset the 1:1 mole balance between $[H^+]$ and $[OH^-]$ whenever *one* of the salt's ions, the anion or the cation, can react even to a small percentage with water. The reaction of ions with water is called the **hydrolysis of ions,** and we say that this or that ion *hydrolyzes.* The hydrolysis of an ion affects pH. Because of our interest in acid–base balance, we must be concerned about anything that can alter the pH of a solution. Our next tasks, therefore, are to learn a simple way to *identify* any ion that can hydrolyze and to predict the direction in which its hydrolysis changes the pH.

■ "Hydrolysis" is from the Greek *hydro,* "water," and *lysis,* "loosening or breaking"—breaking or loosening by water.

The Anions of Weak Acids Are Strong Enough Proton Acceptors to Hydrolyze. Aqueous solutions of such sodium salts as Na_3PO_4, Na_2CO_3, Na_2HPO_4, and $NaHCO_3$ all test basic to litmus. Yet none shows the presence of the hydroxide ion in its formula. The *anions* of these sodium salts, however, hydrolyze; they react with water to create an excess of $[OH^-]$ over $[H^+]$. The bicarbonate ion, for example, hydrolyzes to establish the following equilibrium. (Assume that Na^+ is the spectator ion.)

$$HCO_3^-(aq) + H_2O \rightleftharpoons H_2CO_3(aq) + OH^-(aq)$$

The forward reaction, to be sure, is not favored. After all, HCO_3^- is not a strong proton acceptor, and H_2O is a *very poor* proton donor. The extent, however, to which the forward reaction occurs *at all* generates some OH^- ions and so causes an excess of OH^- ions over H^+ ions. A solution of $NaHCO_3$, therefore, tests basic to litmus.

■ The hydrolysis of HCO_3^- increases the pH of the solution above 7 because some OH^- is generated.

Notice that all of the anions in the list of sodium salts given above are anions derived from *weak* acids—PO_4^{3-} from HPO_4^{2-}, CO_3^{2-} from HCO_3^-, HPO_4^{2-} from $H_2PO_4^-$, and HCO_3^- from H_2CO_3. We learned in Chapter 7 that the anions left behind when weak acids donate protons to something are relatively strong proton acceptors. We now learn that such anions are strong enough proton acceptors to hydrolyze to some percentage and so generate trace excesses of OH^- ion. Thus, *the anions of weak acids are able to make a solution basic to litmus.* It is not necessary, in other words, to add a metal *hydroxide* to water to increase the pH; salts of weak acids do this also.

Anions such as Cl^-, Br^-, I^-, NO_3^-, and SO_4^{2-}, which are anions produced when *strong* acids ionize, are all extremely weak bases, too weak to react with water. Unable to hydrolyze, the anions of strong acids cannot upset the 1:1 mole ratio of H^+ to OH^- in water. To summarize,

---

Anions derived from all weak acids hydrolyze in water and tend to make a solution basic.

Anions of strong acids do not hydrolyze and so do *not* affect a solution's pH.

Most Transition Metal Cations and the Ammonium Ion Hydrolyze to Generate H_3O^+ Ions and Lower the pH of the Solution.   A solution of ammonium chloride, NH_4Cl, in water turns blue litmus red, so in this solution, $[H^+]$ is greater than $[OH^-]$. The excess hydrogen ions come from the forward reaction of the following equilibrium involving the cation of NH_4Cl, namely NH_4^+. (Cl^- ion is the spectator ion.)

$$NH_4^+ + H_2O \rightleftharpoons NH_3(aq) + H_3O^+(aq)$$

■ A proton transfer from NH_4^+ to H_2O occurs in the forward reaction. The transfer occurs in the opposite direction in the reverse reaction.

To be sure, the ammonium ion is a *weak* proton donor, but not so weak that it cannot produce enough hydronium ions by hydrolysis to turn blue litmus red. The pH of 0.10 M NH_4Cl is about 5.1, not so acidic as to be dangerous to the skin, but certainly more acidic than water (roughly 100 times more acidic). (If the blood had a pH of 5.1, you'd be dead!)

Many other cations also hydrolyze. Recall that all ions are hydrated in water; they gather water molecules about themselves to form solvent cages (page 125). The *hydrated cations* are what hydrolyze and make the solutions of most metals test acidic to litmus. The aluminum ion, for example, exists in water largely as the hydrate $[Al(H_2O)_6]^{3+}$. Its central metal ion is Al^{3+} whose dense, high positive charge attracts some electron density from the H—O bonds of the H_2O molecules of the hydrated ion. These H—O bonds are thus weakened enough so that a water molecule in $[Al(H_2O)_6]^{3+}$ can donate one H^+ to a solvent water molecule as follows.

$$[Al(H_2O)_6]^{3+}(aq) + H_2O \rightarrow [Al(H_2O)_5(OH)]^{2+} + H_3O^+(aq)$$

In this way H_3O^+ forms by the hydrolysis of $[Al(H_2O)_6]^{3+}(aq)$, and the solution is made acidic to litmus. A 0.1 M solution of $AlCl_3$, for example, has a pH of about 3, the same as 0.1 M acetic acid, but there is nothing about the formula $AlCl_3$ to indicate that aqueous $AlCl_3$ can affect pH. The metal ions of the Group IA elements, like Na^+ and K^+, although hydrated in water, do not affect the pH. Let's pause to see why.

■ The relatively high positive charge on the cation draws in the surrounding electrons, causing shrinkage of the volume of the ion compared with the volume of the neutral atom.

All hydrated metal ions with charges of $3+$ and most with charges of $2+$ generate hydronium ions by hydrolysis in water. The reason why these and not cations of only $1+$ charge tend to make solutions acidic lies in their higher positive charge densities. Metal ions with a $3+$ or a $2+$ charge are generally smaller than those with a $1+$ charge, so the *density* of positive charge, the charge *per unit volume,* is higher in the multicharged cations. *A site of high positive charge density is strongly electronegative.* It draws electron density from H—O bonds in surrounding water molecules, and so weakens these bonds. When weakened enough, an H—O bond becomes a proton donor sufficiently strong to generate H_3O^+ ions in water and so lower the pH.

■ The Be^{2+} ion is particularly small and so has a higher positive charge density than the other cations of the Group IIA metals.

The only common metal ions that do *not* hydrolyze to give acidic solutions are those of Groups IA and IIA (except Be^{2+} of Group IIA); for example, Li^+, Na^+, and K^+, Mg^{2+}, Ca^{2+}, and Ba^{2+} do not hydrolyze. Evidently, except for Be^{2+}, the cations of Groups IA and IIA do not have sufficiently high positive charge densities. We can summarize what we have learned about the hydrolysis of cations as follows.

> Metal ions from Group IA or IIA (except Be^{2+}) do *not* hydrolyze.
> Expect other metal ions as well as NH_4^+ to hydrolyze, generate H^+, and so tend to make a solution acidic.

With these rules we can generally predict correctly whether a given salt will affect the pH of an aqueous solution. The only difficult cases involve salts where *both* cation and anion can hydrolyze, like ammonium acetate. Such salts must be analyzed on a case-by-case basis with the result hinging on the relative strengths of the cation as a proton producer and the base as a proton neutralizer. We will not work with such salts. But let's see how we can predict the hydrolysis of salts that respond to a simpler analysis.

EXAMPLE 8.2 Predicting How a Salt Affects the pH of Its Solution

Sodium phosphate, Na_3PO_4 ("trisodium phosphate"), is a strong cleaning agent for walls and floors. Is its aqueous solution acidic or basic to litmus?

ANALYSIS Na_3PO_4 involves Na^+ and PO_4^{3-} ions. The Na^+ ion does not hydrolyze, but PO_4^{3-} does. We predict this because HPO_4^{2-}, the protonated form of PO_4^{3-}, is not on our list of strong acids and so PO_4^{3-} is the anion of a weak acid. We therefore expect PO_4^{3-} to be a relatively strong base and we expect it to hydrolyze as follows:

$$PO_4^{3-}(aq) + H_2O \rightarrow HPO_4^{2-}(aq) + OH^-(aq)$$

SOLUTION This reaction generates some OH^- ions, so the solution is basic to litmus.

Notice in Example 8.2 that we did not need a table of bases to predict that PO_4^{3-} is a strong enough proton acceptor to hydrolyze and make the solution test basic to litmus. We used our knowledge of just a few facts—which acids are strong acids in water and which cations do not hydrolyze—to figure out what we needed to know. We will work another example to practice using the list of strong aqueous acids to decide whether a given salt can affect the pH of its solution.

EXAMPLE 8.3 Predicting Whether a Salt Affects the Acidity of Its Solution

Chromium(III) nitrate, $Cr(NO_3)_3$, is soluble in water. Does this salt make its aqueous solution acidic or basic to litmus?

ANALYSIS $Cr(NO_3)_3$ dissociates into $Cr^{3+}(aq)$ and three $NO_3^-(aq)$ ions. Because the nitrate ion is the anion of a *strong* acid, HNO_3, we know that NO_3^- is *unable* to react with water to generate hydroxide ions. The nitrate ion, in other words, does not hydrolyze. The Cr^{3+} ion, however, is not from Group IA or IIA. Moreover, Cr^{3+} has a high positive charge density, like Al^{3+}, so we expect Cr^{3+} to hydrolyze and generate some excess hydronium ion.

SOLUTION An aqueous solution of $Cr(NO_3)_3$ is predicted to test acidic to litmus.

■ PRACTICE EXERCISE 4 Determine without the use of tables whether each ion can hydrolyze. If so, state whether it tends to make the solution acidic or basic. (a) CO_3^{2-} (b) S^{2-} (c) HPO_4^{2-} (d) Fe^{3+} (e) NO_2^- (f) F^-

■ PRACTICE EXERCISE 5 Is a solution of potassium acetate, $KC_2H_3O_2$, acidic, basic, or neutral to litmus?

■ PRACTICE EXERCISE 6 Is a solution of copper(II) nitrate, $Cu(NO_3)_2$, acidic, basic, or neutral to litmus?

■ PRACTICE EXERCISE 7   Ammonium sulfate, $(NH_4)_2SO_4$, is a nitrogen fertilizer. Could the application of an aqueous solution of this fertilizer affect the pH of the soil? If so, will it increase or decrease the pH?

8.3 BUFFERS: PREVENTING LARGE CHANGES IN pH

The pH of a solution is held relatively constant when it contains a *buffer.*

There are several life-threatening situations in which the blood is forced to accept or retain either more acid or more base than normal. The pH of blood, however, cannot be allowed to seesaw by more than 0.2 to 0.3 pH units. Wider variations, fortunately, are prevented by the activities of *buffers.* Let's see what they are and how they work.

Buffers Have Components That Neutralize Either H^+ or OH^-.   **Buffers** are combinations of solutes that prevent large changes in pH when strong acids or bases are added to an aqueous solution. One component of the buffer system neutralizes H^+ and another neutralizes OH^-.

The Phosphate Buffer Works Inside Cells.   The **phosphate buffer,** the principal buffer inside cells, consists of the pair of ions, HPO_4^{2-} and $H_2PO_4^-$, the monohydrogen and the dihydrogen phosphate ions. Notice that *the buffer's two component ions differ by only one proton; this is always true of buffer pairs.* Of the two, $H_2PO_4^-$ has more protons and so is the better proton-*donating* member of the pair and the component that neutralizes hydroxide ion. Thus, if the cell receives more OH^- than normal because of some change in metabolism, the following reaction neutralizes OH^- and prevents the pH from increasing.

$$H_2PO_4^-(aq) + OH^-(aq) \rightarrow HPO_4^{2-}(aq) + H_2O$$

Actually, this is the *forward reaction* of an equilibrium in which the second member of the buffer pair is one product.

$$H_2PO_4^-(aq) + OH^-(aq) \rightleftharpoons HPO_4^{2-}(aq) + H_2O$$

With this understanding, we can view the influx of more OH^- ions as putting a *stress* on the equilibrium. In accordance with Le Châtelier's principle (page 128), the equilibrium must shift to the right; only such a change can neutralize OH^- and so absorb the stress.

Because of its higher negative charge, the HPO_4^{2-} ion is the better proton acceptor of the phosphate buffer. If the cell receives more H^+ than normal, the following reaction neutralizes it.

$$HPO_4^{2-}(aq) + H^+(aq) \rightarrow H_2PO_4^-(aq)$$

Once again, this reaction is actually the forward reaction of an equilibrium in which the other member of the buffer pair is a product.

$$HPO_4^{2-}(aq) + H^+(aq) \rightleftharpoons H_2PO_4^-(aq)$$

Responding to the stress posed by an influx of H^+ ions, the equilibrium shifts to the

■ Fluids that contain buffers are said to be *buffered* against changes in pH.

■ The chief counter ion for the phosphate buffer inside a cell is K^+.

right. This shift consumes H^+ and so prevents the cell's contents from becoming more acidic.

The Carbonate Buffer Works in the Blood.   Buffer duties in blood are handled largely by the **carbonate buffer,** which consists of dissolved carbon dioxide, $CO_2(aq)$, and bicarbonate ion, HCO_3^-. Although carbon dioxide has no proton to donate, it is still a neutralizer of OH^- because it reacts with hydroxide ion to form the bicarbonate ion, the other member of the carbonate buffer pair.

$$CO_2(aq) + OH^-(aq) \rightarrow HCO_3^-(aq)$$

By now you've probably guessed that this reaction is also the *forward reaction* of an equilibrium.

$$CO_2(aq) + OH^-(aq) \rightleftharpoons HCO_3^-(aq) \tag{8.3}$$

If for any reason excess OH^- appears in the blood, Equilibrium 8.3 cannot help but shift to the right in accordance with Le Châtelier's principle. This shift, of course, means that OH^- is neutralized, thus preventing alkalosis.[5]

The proton acceptor or base in the carbonate buffer is HCO_3^-. If excess acid comes into the blood, the following equilibrium is stressed and so shifts to the right, which neutralizes H^+ and prevents acidosis.

$$HCO_3^-(aq) + H^+(aq) \rightleftharpoons CO_2(aq) + H_2O \tag{8.4}$$

The Expulsion of CO_2 by Exhaling Must Also Occur to Prevent Acidosis.   When acid is neutralized by a shift of Equilibrium 8.4 to the right, an increase occurs in the level of dissolved carbon dioxide, represented by $CO_2(aq)$. When the blood is moving through the lungs, however, dissolved $CO_2(aq)$ becomes quickly involved in the following equilibrium with gaseous CO_2.

$$CO_2(aq) \rightleftharpoons CO_2(g) \tag{8.5}$$

Thus when Equilibrium 8.4 shifts to the *right,* the extra $CO_2(aq)$ made by it puts a stress on Equilibrium 8.5. To relieve this stress, Equilibrium 8.5 must shift to the right, putting more excess CO_2 into its gaseous state. The normal course of exhaling a breath then carries excess $CO_2(g)$ into the outside air. This final act, exhaling, does not normally involve an equilibrium. The irreversible loss of CO_2 from the body thus draws Equilibrium 8.5 to the right and, in a cascade of shifting equilibria, *helps to draw Equilibria 8.3 and 8.4 to the right also.* The blood thus uses *two* mechanisms to handle excess acid. It neutralizes it by the work of the carbonate buffer, and it uses *ventilation* to make the neutralization permanent. Because CO_2 is able to neutralize OH^- ion and so acts as an acid, the loss of CO_2 by exhaling is sometimes called the loss of acid. Reciprocally, the retention of CO_2 by the blood would be called the retention of acid.

The Brain Changes the Rate of Ventilation in Response to Acidosis or Alkalosis. **Ventilation** is the circulation of air into and out of the lungs. The rate of ventilation normally is controlled by a site in the brain called the *respiratory center.* It works first by monitoring the level of $CO_2(aq)$ in the blood. When this level increases, the brain instructs the breathing apparatus to breathe more rapidly and deeply, a response

■ The principal counter ion for the carbonate buffer in blood is Na^+.

■ If you put a bag over your head and rebreathe your own waste $CO_2(g)$, Equilibrium 8.5 will tend to shift to the *left*, which means that your blood will retain $CO_2(aq)$.

---

[5]A very small fraction, less than 0.5%, of the $CO_2(aq)$ molecules in blood exist as carbonic acid or H_2CO_3 [from $CO_2(aq) + H_2O \rightleftharpoons H_2CO_3(aq)$]. We ignore this; the principal base-neutralizing species of the carbonate buffer is overwhelmingly $CO_2(aq)$.

Enormous quantities of sulfur dioxide are generated worldwide from the combustion of coal and oil, which contain relatively small quantities of sulfur compounds. This sulfur becomes oxidized to gaseous SO_2 as the fuels burn. Although the sulfur content of a fuel seldom exceeds 3%, often much less, the vast tonnages of fuels consumed worldwide annually release hundreds of millions of tons of SO_2 per year into the atmosphere. It is a major contributor to "acid rain."

Sulfur dioxide dissolves in water by forming hydrates, $SO_2 \cdot n H_2O$, where n varies with concentration, temperature, and pH. The hydrates are in equilibrium with some hydronium ion and hydrogen sulfite ion, HSO_3^-, whose presence has long been explained simply in terms of $H_2SO_3(aq)$, sulfurous acid. Actual molecules of this species—H_2SO_3—have never been detected in or out of water, however. Nevertheless, for convenience in writing chemical equations, the formula H_2SO_3 is widely used for the solute in aqueous sulfur dioxide. It is the first ionization of sulfurous acid that generates virtually all of the hydrogen ion that this acid can produce in water.

$$H_2SO_3(aq) \rightleftharpoons H^+(aq) + HSO_3^-(aq)$$

Sulfurous acid is a *moderate* acid, not a weak acid. Thus when rain washes gaseous SO_2 from the atmosphere, the rainwater is acidic. Moreover, both oxygen and ozone (O_3) in smog convert some SO_2 to SO_3, particularly in sunlight when fine dust is present. When SO_3 reacts with water, sulfuric acid forms. This strong acid, of course, also contributes to the acidity of rain where air pollution occurs.

Nitrogen Dioxide Is Another Major Air Pollutant That Contributes to Acid Rain. As you know, nitrogen and oxygen—the chief components of the air we breathe—are very stable toward each other *at ordinary temperatures and pressures*. When fuels are burned in vehicles, however, high temperatures and higher pressures in the engines cause some reaction to occur between nitrogen and oxygen to give nitrogen monoxide, NO.

$$N_2(g) + O_2(g) \rightarrow 2NO(g)$$

When the vehicle exhaust leaves the engine and enters the outside air, nitrogen monoxide reacts with oxygen to give nitrogen dioxide.

$$NO_2(g) + O_2(g) \rightarrow 2NO_2(g)$$

Nitrogen dioxide is responsible for the reddishness of smog. In water, NO_2 reacts to give two acids—HNO_3, a strong acid, and HNO_2, a weak acid.

$$2NO_2(g) + H_2O \rightarrow HNO_3(aq) + HNO_2(aq)$$

Thus, oxides of both nitrogen and sulfur are chiefly responsible for **acid rain.** Rain as acidic as lemon juice (pH 2.1) was observed in 1964 in the northeastern section of the United States, and rain as acidic as vinegar (pH 2.4) fell at Pitlochry, Scotland, in 1974.

A Better Term for Acid Rain Is *Acid Deposition.* Dry dust particles settling on buildings, metals, and soil also carry acidic materials adhering to their surfaces. Thus, acid deposition refers to all means whereby acids enter the earth's system. Acid deposition is an acute problem in regions downwind from major users of sulfur-containing fuels. Southern parts of the Scandinavian peninsula receive acid deposition from Germany's Ruhr valley, the English Midlands, and countries of eastern Europe. Parts of southern

---

■ Vigorous exercise generates waste CO_2 faster than normal, so hyperventilation is invoked by the respiratory system to get rid of it.

called **hyperventilation**. This increases the rate at which CO_2 is exhaled from the lungs, and this response pulls all of the carbonate equilibria in their acid-neutralizing directions.

The successive shifts of Equilibria 8.3 and 8.4, set in motion by the rightward shift of Equilibrium 8.5, have a major consequence: one H^+ ion is permanently neutralized for each CO_2 molecule that leaves the body. All of these steps are summarized in Figure 8.4. Notice that the neutralized H^+ ends up in a molecule of water. The ability of this water molecule to form and so ensure the neutralization of acid thus depends on the loss of the CO_2 molecule from the body.

One of many lessons we can learn from these facts is that anything that might interfere with the loss of $CO_2(g)$ would also inhibit the neutralization of the H^+ ion and cause acidosis. One such interference occurs in people with emphysema. Their lungs work poorly and their breathing becomes slow and shallow, a condition called **hypoventilation**. When their lungs are not expelling $CO_2(g)$ at a fast enough rate, the

■ A pattern of slow, deep breathing, sometimes seen in an individual with untreated diabetes, is called *Kussmaul breathing*.

Canada and the northern United States receive acid deposition from the great industrial belt curving from Boston to Chicago. Industrial areas of Poland, the Czech Republic, Slovakia, and Russia have released enormous loads of sulfur and nitrogen oxides into their atmospheres.

Acid deposition affects lakes, soil, vegetation, and building stone. It makes bodies of water too acidic for much aquatic life. Because CO_2 from the air is naturally present in water, the pH of fresh water in equilibrium with air is about 5.7. Below a pH of 5.5, the hatchlings of most game fish are unable to live, so such fish have disappeared from heavily affected lakes. Acids in ground water also leach calcium and magnesium ions from soil and thus adversely affect vegetation. Severe damage has occurred to many mature forests in or near heavily industrialized nations.

Acid deposition is corrosive to exposed metals such as railroad rails, vehicles, and machinery, as well as to stone building materials (Figure 1). Limestone and marble are particularly sensitive, because they are chiefly calcium carbonate, and carbonates are dissolved by acids.

$$CaCO_3(s) + 2H^+(aq) \rightarrow Ca^{2+}(aq) + CO_2(g) + H_2O$$

Several major cathedrals in Europe need constant repair because of the attack of deposited acids on their limestone and marble.

No Easy Alternatives Exist to Sulfur-Containing Fuels. Less reliance on sulfur-containing coal and oil might be thought to be a solution to the acid deposition problem. No doubt it would help considerably. The alternatives to coal and oil are a drastic cutback in energy consumption, made possible by a simpler, less consuming lifestyle; a heavier reliance on nuclear power; or a switch to a new technology not yet tested on a huge scale. Nuclear power bears its own pollution

FIGURE I
Acid deposition increased the rate of decay of this gargoyle on Old City Hall in Munich, Germany.

ills and it presently is costlier in every way than power obtained from the burning of coal or oil. Meanwhile, as coal and oil are used, the removal of most of the SO_2 from smokestack gases is possible. SO_2, for example, is absorbed by wet calcium hydroxide by the following reaction.

$$SO_2(g) + Ca(OH)_2(s) \rightarrow CaSO_3(s) + H_2O$$

Not all SO_2 is removed, however, and given the enormous quantities of coal and oil burned worldwide, emissions of SO_2 still occur. In a *technological* sense, the problem is controllable. The remaining issues concerning emissions of the sulfur oxides are mostly personal, political, economic, and diplomatic.

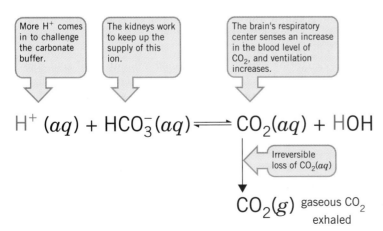

FIGURE 8.4
The irreversible neutralization of H^+ through the loss of CO_2 is one way the carbonate buffer system handles acidosis. It is the last step, the change of dissolved CO_2 into gaseous CO_2, which is exhaled, that draws all of the equilibria to the right and makes H^+ "disappear" into H_2O.

level of $CO_2(aq)$ increases in their blood. Remember that CO_2 neutralizes hydroxide ion and so, functionally, *the retention of carbon dioxide amounts to the retention of acid.* People unable to breathe deeply enough, therefore, experience acidosis.

Any cause of shallow breathing, either because the tubes to the lungs are blocked (asthma), the lungs are filled with fluid (pneumonia), or the brain's respiratory centers have been knocked out (by narcotics or barbiturates), means acidosis.

A too rapid and sustained expulsion of carbon dioxide from the blood can also constitute a medical emergency. Remember, the excessive loss of carbon dioxide functionally means too much loss of acid, so the problem now is alkalosis, not acidosis. Any time hyperventilation becomes uncontrolled and involuntary, as in hysterical fits or overbreathing at high altitudes, alkalosis results. Just how acidosis and alkalosis interfere with respiration are topics whose study we must postpone until we know more about the chemistry of the blood. We return to this topic in Chapter 18.

Not only human life is threatened by changes in pH. All forms of life can be affected, as the discussion of *acid rain* in Special Topic 8.1 describes.

8.4 ACID–BASE TITRATIONS

At the *end point* of an acid–base *titration*, the number of moles of an acid's protons should match the number of moles of base.

One very common kind of chemical analysis is the determination of the concentration of an acid or a base. The purpose is to find the molarity of some whole solute, like the number of moles of acetic acid per liter of solution, not just to measure the pH of the solution.

A Solution's Capacity to Neutralize OH^- *Is Its Neutralizing Capacity.* An acid is classified as *weak* or *strong* according to its ability to transfer a proton to one particular and very weak base, H_2O. When sodium hydroxide is mixed with an acid, however, we are adding a very strong base, OH^-. It can take H^+ not only from H_3O^+ but also from the un-ionized molecules of a weak acid, like acetic acid. Thus, the **neutralizing capacity** of a solution of a weak acid, its capacity to neutralize OH^-, is considerably greater than its actual concentration of hydronium ion.

■ One liter of 1 *M* acetic acid has only 0.004 mol of H_3O^+, but it can still neutralize 1 mol of NaOH because of the proton-donating ability of un-ionized molecules of acetic acid.

The Titration of an Acid with a Base Gives Data from Which Concentrations Can Be Calculated. The procedure used to measure the total acid (or base) neutralizing capacity of a solution is called **titration**. It measures the volume of a *standard solution* of one reactant required to neutralize a known volume of the solution of unknown molarity. A **standard solution** is simply one whose concentration is accurately known.

The apparatus for titration is shown in Figure 8.5. When a titration is used for an acid–base analysis, a carefully measured volume of the solution of unknown acidity (or basicity) is placed in a beaker or a flask. A very small amount of an acid–base indicator, like phenolphthalein, is added. Then a standard solution of the neutralizing reagent is added through a stopcock, portion by portion, from a special tube called a *buret*. The addition from the buret is continued until a change in color occurs in the receiving flask. The change, caused by the acid–base indicator, signals that the unknown has been exactly neutralized. This is the stopping point or the **end point** of the titration.

■ Burets are typically etched at 1-mL and 0.01-mL intervals.

End Points Ideally Occur at Equivalence Points. With a carefully selected acid–base indicator, the color change in an acid–base titration occurs at the **equivalence**

point, when the theoretically available hydrogen ions have been neutralized by the base

A well-chosen indicator is one whose color at the equivalence point is the same as it would impart to a solution containing only the *salt* that forms in the titration (and at the same concentration). If this salt has an ion that hydrolyzes, the equivalence point cannot be at pH 7.00. One mole of sodium acetate, for example, has been made when one mole of acetic acid has been neutralized by one mole of sodium hydroxide. Because the acetate ion hydrolyzes (but not the sodium ion), a sodium acetate solution is slightly basic to litmus and has a pH greater than 7.00. So it would be unwise to use an indicator that changes color at a pH less than 7.

In Section 4.6 we learned how to do calculations involving volumes and concentrations of solutions with an emphasis on calculating *volumes*. Acid–base titrations, however, are usually done to determine *concentrations*, so we'll work through an example to see how it's done.

■ Phenolphthalein's color change occurs over a *basic* range of pH, 8.2 to 10, so it is a good indicator in the titration of a weak acid by a strong base.

EXAMPLE 8.4   Calculating Molarities from Concentration Data

A student titrated 25.0 mL of sodium hydroxide solution with standard sulfuric acid. It required 13.4 mL of 0.0555 M H_2SO_4 to neutralize the sodium hydroxide in the solution. What was the molarity of the sodium hydroxide solution? The equation for the reaction is

$$2NaOH(aq) + H_2SO_4(aq) \rightarrow Na_2SO_4(aq) + 2H_2O$$

ANALYSIS   Be sure to understand the goal first. We are to calculate the *molarity* of the NaOH solution. This means, returning to the definition of "molarity," the *ratio* of the moles of NaOH to liters of NaOH solution. We were given (indirectly) the number of *liters* of the NaOH solution because 25.0 mL is the same as 0.025 L. We must use the volume and molarity of the acid consumed, plus the coefficients of the equation, to calculate the number of *moles* of NaOH used up. Once we have "moles of NaOH" and "liters of NaOH solution," we simply take the ratio of the two—moles per liter—to find the molarity of the NaOH solution.

SOLUTION   The molarity of the H_2SO_4 solution, 0.0555 M, gives us the following conversion factors.

$$\frac{0.0555 \text{ mol } H_2SO_4}{1000 \text{ mL } H_2SO_4 \text{ soln}} \qquad \frac{1000 \text{ mL } H_2SO_4 \text{ soln}}{0.0555 \text{ mol } H_2SO_4}$$

If we multiply the given volume of H_2SO_4 solution, 13.4 mL, by the first factor we have the number of moles of H_2SO_4 used up. (Insert the cancel lines yourself.)

$$13.4 \text{ mL } H_2SO_4 \text{ soln} \times \frac{0.0555 \text{ mol } H_2SO_4}{1000 \text{ mL } H_2SO_4 \text{ soln}} = 7.44 \times 10^{-4} \text{ mol } H_2SO_4$$

The balanced equation tells us that 1 mol H_2SO_4 ⇔ 2 mol NaOH, which gives us the following conversion factors.

$$\frac{1 \text{ mol } H_2SO_4}{2 \text{ mol NaOH}} \qquad \frac{2 \text{ mol NaOH}}{1 \text{ mol } H_2SO_4}$$

FIGURE 8.5
The apparatus for titration. By manipulating the stopcock, the analyst controls the rate at which the solution in the buret is added to the flask below.

If we multiply the number of moles of H_2SO_4 just calculated by the second factor, we know how many moles of NaOH are present in the volume of NaOH solution taken. (Again notice how the units properly cancel.)

$$7.44 \times 10^{-4} \text{ mol } H_2SO_4 \times \frac{2 \text{ mol NaOH}}{1 \text{ mol } H_2SO_4} = 14.9 \times 10^{-4} \text{ mol NaOH}$$

Thus, 14.9×10^{-4} mol of NaOH is dissolved in 25.0 mL or 0.025 L of NaOH solution. To find the molarity of this solution, we take the following ratio of moles to liters:

$$\frac{14.9 \times 10^{-4} \text{ mol NaOH}}{0.025 \text{ L NaOH soln}} = 0.0596 \; M \text{ NaOH}$$

The molarity of the NaOH solution is 0.0596 M.

■ PRACTICE EXERCISE 8   If it takes 24.3 mL of 0.110 M HCl to neutralize 25.5 mL of freshly prepared sodium hydroxide solution, what is the molarity of the NaOH solution?

■ PRACTICE EXERCISE 9   If 20.0 mL of 0.125 M solution of NaOH exactly neutralizes the sulfuric acid in 10.0 mL of H_2SO_4 solution, what is the molarity of the H_2SO_4 solution?

SUMMARY

pH   The self-ionization of water produces an equilibrium in which the ion product constant, K_w, the product of the molar concentrations of hydrogen and hydroxide ions, is 1.0×10^{-14} (at 25 °C). If acids or bases are added, the value of K_w stays the same, but individual values of $[H^+]$ and $[OH^-]$ adjust.

A simple way to express very low values for these molar concentrations is by pH, where $[H^+] = 1 \times 10^{-pH}$. When, at 25 °C, pH is less than 7, the solution is acidic. When pH is greater than 7, the solution is basic. To measure the pH of a solution we use indicators, dyes whose colors change over a narrow range of pH, or we use a pH meter.

Hydrolysis of Ions   A dissolved salt affects the pH of the solution if one of its ions reacts more than the other with water to generate extra H^+ or OH^- ions. Cations of Group IA and IIA metals (except Be^{2+}) do not hydrolyze. Nearly all others do. Anions derived from weak acids are relatively good proton acceptors and therefore hydrolyze to give basic solutions.

Buffers   Solutions that contain something that can neutralize OH^- ion and something else that can neutralize H^+ ion

are buffered against changes in pH when either additional base or acid is added. The phosphate buffer, present in the fluids inside cells, consists of HPO_4^{2-} (to neutralize H^+) and $H_2PO_4^-$ (to neutralize OH^-). The carbonate buffer, the chief buffer in blood, consists of HCO_3^- (to neutralize H^+) and CO_2 (to neutralize OH^-).

The blood's $CO_2(aq)$ can be expelled in the lungs where the $CO_2(g)$ is carried out by exhaling. When metabolism or some deficiency in respiration produces or retains H^+ at a rate faster than the blood buffer can neutralize it, acidosis exists. The lungs then try to remove CO_2 (remove "acid") at a faster rate by hyperventilating. Overall, for each molecule of CO_2 exhaled, one proton is neutralized. If the body expels too much CO_2, however, the blood tends to become more alkaline. If the body cannot expel CO_2 rapidly enough, acidosis is the consequence.

Acid–Base Titration   The concentration of an acid or a base in water can be determined by titrating the unknown solution with a standard solution of whatever can neutralize it. The indicator is selected to have its color change occur at whatever pH the final solution would have if it were made from the salt that forms by the neutralization.

REVIEW EXERCISES

Ion Product Constant of Water

8.1 Write the equation that defines K_w.

8.2 What is the value of K_w at 25 °C?

8.3 The higher the temperature, the higher the value of K_w. Why should there be this trend?

8.4 At the temperature of the human body, 37 °C, the concentration of hydrogen ion in pure water is 1.56×10^{-7} mol/L. What is the value of K_w at 37 °C? Is this pure water acidic, basic, or neutral?

*8.5 "Heavy water" or deuterium oxide, D_2O, is used in nuclear power plants. It self-ionizes like water, and at 20 °C the concentration of D^+ ion is 3.0×10^{-8} mol/L. What is the value of K_w for heavy water at 20 °C?

pH and pOH

8.6 What equation defines pH in exponential terms?

8.7 The average pH of saliva is about 6.8. Is saliva acidic, neutral, or basic?

8.8 The pH of pancreatic juice, a digestive juice of the intestinal tract, is in the range 7 to 8. Is pancreatic juice acidic, basic, or neutral?

8.9 What is the pH of 0.001 M HCl(aq), assuming 100% ionization?

8.10 What is the pH of 0.01 M NaOH(aq), assuming 100% ionization?

8.11 Explain why a pH of 7.00 corresponds to a neutral solution at 25 °C.

8.12 Following surgery, a patient experienced persistent vomiting and the pH of his blood became 7.49. (Normally it is 7.35.) Has the blood become more alkaline or more acidic? Is the patient experiencing acidosis or alkalosis?

8.13 A patient brought to the emergency room following an overdose of aspirin was found to have a pH of 7.18 for the blood. (Normally the pH of blood is 7.35.) Has the blood become more acidic or more basic? Is the condition acidosis or alkalosis?

8.14 A certain brand of beer has a pH of 5.0. What is the concentration of hydrogen ion in moles per liter? Is the beer slightly acidic or basic?

8.15 The pH of a soft drink was found to be 4.5. Between what limits of molar concentration does the value of $[H^+]$ occur?

*8.16 A solution of a monoprotic acid was prepared with a molar concentration of 0.010 M. Its pH was found to be 2.00. Is the acid a strong or a weak acid? Explain.

*8.17 The pH of a solution of a monoprotic acid was found to be 5.72, whereas its molar concentration was 0.0010 M. Is this acid a strong or a weak acid? Explain.

8.18 When a soil sample was stirred with pure water, the pH of the water changed to 7.76. Did the soil produce an acidic or a basic reaction with the water?

8.19 A patient entered the emergency room of a hospital after 3 weeks on a self-prescribed low-carbohydrate, high-fat diet and the regular use of the diuretic, acetazolamide. (A diuretic promotes the formation of urine and thus causes the loss of body fluid.) The pH of the patient's blood was 7.18. Was the condition acidosis or alkalosis?

8.20 Explain in your own words why a solution of sodium acetate is slightly basic, not neutral.

*8.21 Predict if each of the following solutions is acidic, basic, or neutral.
(a) KNO_3 (b) NH_4Br (c) $NaHCO_3$
(d) $FeCl_3$ (e) Li_2CO_3

*8.22 Predict if each of the following solutions is acidic, neutral, or basic.
(a) Na_2SO_4 (b) K_2HPO_4 (c) K_3PO_4
(d) $Cr(NO_3)_3$ (e) KCH_3CO_2

Buffers

8.23 In the study of the molecular basis of life, why is the study of buffers important?

8.24 What is acidosis? What is alkalosis?

8.25 In very general terms, why are both acidosis and alkalosis serious?

8.26 What does it mean when we say that a solution of pH 7.45 is *buffered* at this pH?

8.27 What chemical species constitute the chief buffer inside cells?

8.28 Write the net ionic equation that shows how the phosphate buffer neutralizes OH^-.

8.29 How does the phosphate buffer neutralize acid? Write the net ionic equation.

8.30 What two chemical species make up the chief buffer in blood?

8.31 Write the net ionic equations that show how the chief buffer system in the blood works to neutralize hydroxide ion and hydrogen ion.

8.32 Explain in your own words, using equations as needed, how the loss of a molecule of CO_2 at the lungs permanently neutralizes a hydrogen ion.

8.33 What is meant by *ventilation* in connection with respiration? What is hyperventilation? Hypoventilation?

8.34 What does the respiratory center in the brain instruct the lungs to do when the level of CO_2 in blood increases? Why?

8.35 Why does the hypoventilation of someone with emphysema lead to acidosis?

*8.36 In high-altitude sickness, the patient *involuntarily* overbreathes and expels CO_2 from the body at a faster than normal rate. This results in an *increase* in the pH of the blood.
(a) Is this condition alkalosis or acidosis?
(b) Why should excessive loss of CO_2 result in an increase in the pH of the blood? (*Note:* Such a patient should be returned to lower elevations as soon as possible. It helps to rebreathe one's own air, as by breathing into a paper sack, because this helps the system retain CO_2.)

*8.37 A patient with untreated diabetes tends to hyperventilate. This is a natural response to what kind of change in the pH of the blood? Explain.

Acid-Base Titrations[6]

8.38 What does it mean to have a *standard* solution of, say, HCl(*aq*)?

8.39 When doing a titration, how does one know when the endpoint is reached?

8.40 What steps does an analyst take to ensure that the end point and the equivalence point in an acid–base titration occur together?

8.41 Give an example of a titration in which the equivalence point has a pH that is equal to 7.00. (Devise a specific example of an acid and a base that, when titrated together, produce such a solution.)

8.42 Give a specific example of an acid and a base that, when titrated together, produce a solution with a pH greater than 7.00.

8.43 At the equivalence point in the titration of aqueous ammonia and hydrochloric acid, the pH is less than 7.00. Explain.

8.44 Individual aqueous solutions were prepared that contained the following substances. Calculate the molarity of each solution.
(a) 6.892 g of HCl in 500.0 mL of solution
(b) 8.09 g of HBr in 250.0 mL of solution

8.45 What is the molarity of each of the following solutions?
(a) 32.68 g of HI in 750.0 mL of solution
(b) 4.9048 g of H_2SO_4 in 250.0 mL of solution

[6]For all the calculations in the Review Exercises that follow, round atomic masses to their *second* decimal places before adding them to find formula masses.

*8.46 If 20.00 g of a monoprotic acid in 100.0 mL of solution gives a concentration of 0.5000 *M*, what is the formula mass of the acid?

*8.47 A solution with a concentration of 0.2500 *M* could be made by dissolving 4.000 g of a base in 250.0 mL of solution. What is the formula mass of this base?

*8.48 How many grams of each solute are needed to prepare the following solutions?
(a) 1000 mL of 0.2000 *M* HCl
(b) 750.0 mL of 0.1025 *M* HNO_3
(c) 500.0 mL of 0.01125 *M* H_2SO_4

*8.49 To prepare each of the following solutions would require how many grams of the solute in each case?
(a) 100.0 mL of 0.1000 *M* HI
(b) 750.0 mL of 0.2000 *M* H_2SO_4
(c) 500.0 mL of 1.125 *M* Na_2CO_3

*8.50 To determine the molarity of a solution of sodium carbonate, 22.48 mL of the solution was titrated to the endpoint with 19.82 mL of 0.1181 *M* HCl.
(a) Write the molecular equation for the reaction.
(b) Calculate the molarity of the sodium carbonate solution.
(c) How many grams of Na_2CO_3 does it contain per liter?

*8.51 A freshly prepared solution of sodium hydroxide was standardized with 0.1024 *M* H_2SO_4.
(a) If 19.46 mL of the base was neutralized by 21.28 mL of the acid, what was the molarity of the base?
(b) How many grams of NaOH were in each liter of this solution?

Acid Rain (Special Topic 8.1)

8.52 What is the pH of unpolluted water in nature? Why is it less than 7?

8.53 Rain is called acid rain when its pH is less than what value?

8.54 What are the names and formulas of the sulfur oxides in acid rain?

8.55 What are sources of sulfur for making the sulfur oxides in acid rain?

8.56 What is a way to remove much of the sulfur dioxide from exhaust gases at power plants? (Give an equation.)

8.57 What are the names and formulas of two nitrogen oxides in polluted air? Which causes the reddish-brown color of smog?

8.58 How do these nitrogen oxides form in the environment. (Give equations.)

8.59 Which nitrogen oxide decreases the pH of water? How does it do this? (Write the equation.)

9

INTRODUCTION TO ORGANIC CHEMISTRY

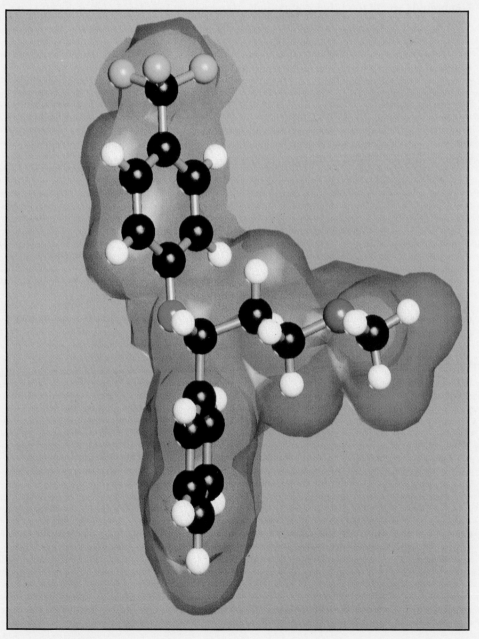

Organic and Inorganic Compounds

Structural Features of Organic Compounds

Isomerism

Few of the several million compounds of carbon, the organic compounds, are as well known for their effects on the mind as Prozac. Because the Prozac molecule is put together in a certain way, shown here in a molecular model, it interacts with nerve cells as it does. This chapter gets us started with the study of interactions at the molecular level of life by introducing the structures of organic molecules.

9.1   ORGANIC AND INORGANIC COMPOUNDS

The major differences between organic and inorganic compounds stem from variations in composition, bond types, and molecular polarities.

Organic compounds are compounds of carbon, and there are more of these than of all the other elements combined, except hydrogen. The name itself, implying *organism,* arose prior to 1828 when scientists believed that organic compounds could be made only by living organisms. They thought that a catalyst-like agency, a *vital force,* is essential to this synthesis, and that only living things possess the vital force. **Inorganic compounds** are all the other compounds, those whose syntheses do not depend on a vital force.

■ *Vita-* is from a Latin root meaning "life."

Wöhler's Experiment Contradicted the Vital Force Theory.   In 1828, Friedrich Wöhler (1800–1828) succeeded in making urea, a white solid that can be isolated from urine and that everyone regarded as an organic compound. Urea was the unexpected result of his attempt to make crystalline ammonium cyanate, NH_4NCO, an inorganic compound. Wöhler prepared an aqueous solution of this salt, which contains the separated ions NH_4^+ and NCO^-. He then boiled off the water, reasonably thinking that the residue would be the desired ammonium cyanate. The product, instead, was not a salt but an entirely different molecular compound, urea. Heat, we now know, had caused the following reaction.

■ Urea is the chief nitrogen waste of the body. It is also manufactured from ammonia and used as a commercial fertilizer.

$$NH_4NCO \xrightarrow{\text{heat}} \begin{array}{ccc} H & O & H \\ | & || & | \\ H-N-C-N-H \end{array}$$

Ammonium cyanate   Urea

After Wöhler's discovery, other organic compounds were made from inorganic substances, and the vital force theory was dead. Today, well over 6 million organic compounds are known, and all have been or could in principle be made from inorganic substances.

Covalent Bonds, Not Ionic Bonds, Dominate Organic Molecules.   The overwhelming prevalence of nonmetal atoms in organic compounds means that their molecular structures are dominated by *covalent* bonds. In contrast, most inorganic compounds are *ionic.* As we will see, carbon–carbon and carbon–hydrogen bonds are the most prevalent in organic molecules. These bonds are essentially nonpolar, so organic compounds tend to be relatively nonpolar, except when atoms of such electronegative elements as oxygen and nitrogen are present. These structural facts are behind several major properties of organic compounds, like melting and boiling points and solubility in water.

■ The covalent bond was studied in Section 3.4.

Most organic compounds have melting points and boiling points well below 400 °C, whereas most ionic compounds melt or boil far above this temperature. The reason for the difference is that the relatively nonpolar molecules in organic substances, unlike oppositely charged ions, are unable to attract each other very strongly.

■ Typically, ionic compounds melt and boil well above 350 °C.

Weakly polar molecules are poorly hydrated, in contrast to ions, so most organic compounds are relatively insoluble in water, whereas many ionic compounds are soluble. This fact is particularly relevant at the molecular level of life where the central fluid is water. We must therefore be alert during our study of organic chemistry to any molecular features that increase solubility in water.

■ Organic *ions* exist and tend to be very soluble in water.

9.2 STRUCTURAL FEATURES OF ORGANIC COMPOUNDS

Organic molecules can have straight or branched chains; they can be open chained or cyclic, saturated or unsaturated; and ring systems can be carbocyclic or heterocyclic.

The uniqueness of carbon among the elements is that its atoms can bond to each other successively many times and still form equally strong bonds to atoms of other nonmetals. A typical molecule of the familiar plastic polyethylene has hundreds of carbon atoms covalently joined in succession. Notice that each carbon binds enough hydrogen atoms to fill out its full complement of four bonds. (Remember that each line represents a shared pair of electrons.

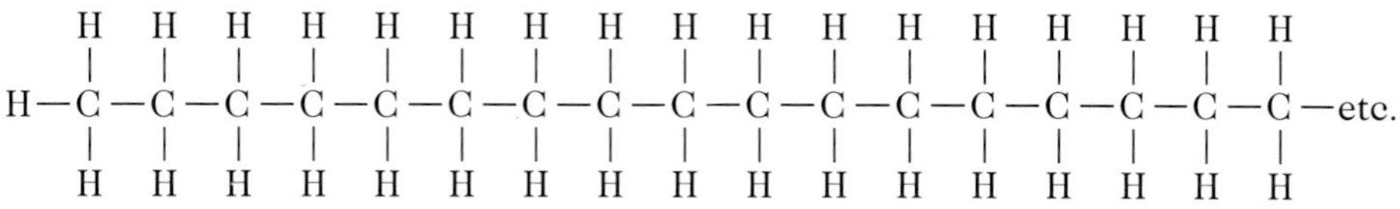

Polyethylene (small segment of one molecule)

■ Only a short segment of a typical molecule of polyethylene is shown here.

Straight chain

The sequence of the heavier atoms, here the carbon atoms, is called the *skeleton* of the molecule, and it holds the hydrogen atoms. Many variations of heavier-atom skeletons occur, and we look at them next.

Carbon Skeletons Can Be in Straight Chains or Branched Chains. The carbon skeleton in the polyethylene molecule is described as **straight chain.** *Straight* has a very limited and technical meaning here: the absence of carbon branches. This means that one carbon follows another, like the pearls in a single-strand necklace, with no additional carbons joined to the skeleton at intermediate points. Pentane illustrates a straight chain of five carbons. The 2-methylpentane molecule has a **branched chain,** a chain with at least one carbon atom joined to the skeleton between the ends of the main chain, like a charm hung on a bracelet.

Branched chain

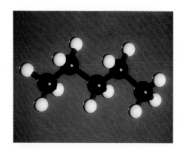
Pentane

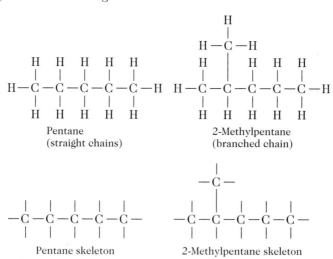

Pentane
(straight chains)

2-Methylpentane
(branched chain)

Pentane skeleton

2-Methylpentane skeleton

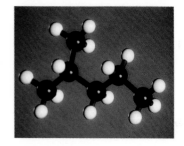
2-Methylpentane

Pentane and 2-methylpentane are sometimes called *open-chain compounds* to contrast them with compounds in which the carbon atoms form a cyclic system called a **ring.** When all ring atoms are carbon, the ring is sometimes called *isocyclic.* I

heterocyclic ring, one or more ring members is a multivalent atom, like O, N, or S. Cyclopentane is an example of a ring compound; tetrahydropyran illustrates a heterocyclic compound. (The six-membered ring with one O atom is common among carbohydrates.).

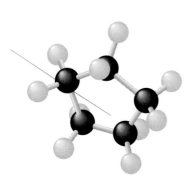

Cyclopentane

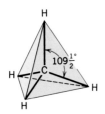

Tetrahedral carbon. When carbon forms four *single* bonds, they project to the corners of a regular tetrahedron and so have angles of 109.5°.

Cyclopentane   Tetrahydropyran

Printed Structural Formulas Usually Ignore Bond Angles. The ball-and-stick models of pentane, 2-methylpentane, and cyclopentane show the correct angle of 109.5° between any two bonds at any carbon with four single bonds (see Special Topic 3.2). To simplify matters, the printed structures nearly always let this detail about bond angles be "understood." We mentally read the correct bond angles into a printed structure.

Free Rotation Occurs about Single Bonds in Open-Chain Compounds. Another fact about molecules left to the imagination when we write their structures is their flexibility. Pieces of open-chain molecules connected by *single* bonds have a property called **free rotation.** Such pieces can rotate with respect to each other around the single bond that joins them. Each of the almost infinitely possible shapes resulting from the kinking and twisting allowed by free rotation is called a *conformation.* Photographs of the models of just three conformations of pentane are shown in Figure 9.1. The internal rotations are generally caused by random collisions between molecules. Because collisions occur constantly, not all of the molecules can be in a fully extended conformation in a sample of pentane. Any specific property of a pentane sample is thus the net effect of all of its variously twisted molecules on whatever physical agent or chemical reactant has been used to observe the property.

Molecular Structures Are Usually Displayed by Condensed Structural Formulas. So far we have shown every bond in a structural formula as a straight line, but we have also seen how it is useful to leave some molecular features to the imagination. We can also leave most of the bonds in a structural formula to the imagination.

With this in mind, we can group the hydrogen atoms held by a particular carbon all together to one side or the other of its symbol. Whenever a carbon holds three hydrogens, for example, we can simplify the system by writing CH_3 (or H_3C, but you don't see this as often). Just remember that the three H atoms are *individually* joined to the carbon. The simplest example of doing this occurs with the structural formula of ethane.

■ Always remember that every carbon must have four bonds.

$$H-\underset{\underset{H}{|}}{\overset{\overset{H}{|}}{C}}-\underset{\underset{H}{|}}{\overset{\overset{H}{|}}{C}}-H \qquad CH_3-CH_3 \ \text{ or } \ H_3C-CH_3$$

Ethane   Ethane
(expanded structure)   (condensed structures)

A carbon holding two hydrogen atoms can be represented as CH_2 or (seen less often) as H_2C. When a carbon holds just one hydrogen, we can write it as CH (or sometimes HC).

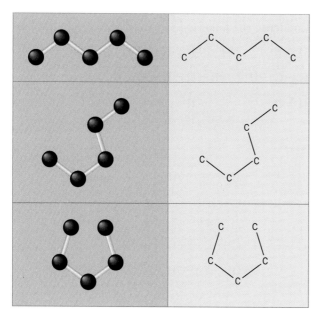

FIGURE 9.1

Free rotation at single bonds. Three of the innumerable conformations of the pentane molecule are shown here. (The hydrogens have been omitted.) Free rotation about single bonds easily converts one conformation into another. When we write the structure of pentane as $CH_3CH_2CH_2CH_2CH_3$, it stands for any of these conformations because some of each (and many others) exist in a sample of pentane.

The result of these simplifications is called a **condensed structure,** or simply the **structure.** These kinds of structures will be used almost exclusively in our continuing study.

EXAMPLE 9.1   Condensing a Full Structural Formula

Condense the structural formula for 2-methylpentane.

$$
\begin{array}{c}
\quad\quad\quad\ \text{H} \\
\quad\quad\quad\ | \\
\quad\quad\ \text{H}-\text{C}-\text{H} \\
\ \text{H}\ \quad | \quad\ \text{H}\ \ \text{H}\ \ \text{H} \\
\ |\ \quad\ |\ \quad\ |\ \ \ |\ \ \ | \\
\text{H}-\text{C}-\text{C}-\text{C}-\text{C}-\text{C}-\text{H} \\
\ |\ \quad\ |\ \quad\ |\ \ \ |\ \ \ | \\
\ \text{H}\ \quad \text{H}\ \quad \text{H}\ \ \text{H}\ \ \text{H}
\end{array}
$$

2-Methylpentane

ANALYSIS Each unit of 3 H attached to the same carbon becomes CH_3. Where 2 H are joined to the same carbon, write CH_2. A carbon holding only one H becomes CH.

SOLUTION

$$
\begin{array}{c}
\quad\quad\ \text{CH}_3 \\
\quad\quad\ | \\
\text{CH}_3-\text{CH}-\text{CH}_2-\text{CH}_2-\text{CH}_3
\end{array}
$$

CHECK We formulate here a general check rule for all molecular structures. *Scan all bond connections to verify that the rules of covalence have been obeyed.* In the answer, each C has four bonds and each H one bond. When you find a violation, the answer cannot possibly be correct, so fix it.

■ Because of free rotation, we have to be able to interpret zigzags. For example,

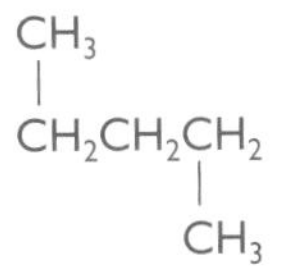

$$CH_3$$
$$|$$
$$CH_2CH_2CH_2$$
$$|$$
$$CH_3$$

is the same molecule as

$$CH_3CH_2CH_2CH_2CH_3$$

■ PRACTICE EXERCISE 1   Condense the following expanded structural formulas.

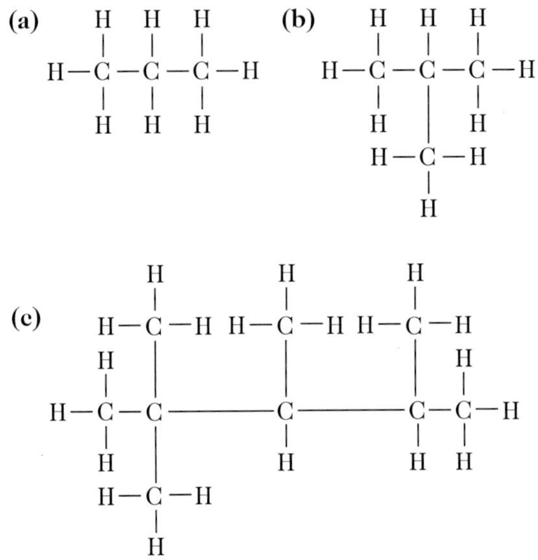

(a)

$$
\begin{array}{ccccc}
 & H & H & H \\
 & | & | & | \\
H- & C- & C- & C- & H \\
 & | & | & | \\
 & H & H & H
\end{array}
$$

(b)

$$
\begin{array}{ccccc}
 & H & H & H \\
 & | & | & | \\
H- & C- & C- & C- & H \\
 & | & | & | \\
 & H & | & H \\
 & & H- & C- & H \\
 & & & | \\
 & & & H
\end{array}
$$

(c)

$$
\begin{array}{c}
\quad H \qquad\qquad H \qquad\qquad H \\
\quad | \qquad\qquad | \qquad\qquad | \\
H-C-H \; H-C-H \; H-C-H \\
\quad H \qquad\qquad | \qquad\qquad H \\
\quad | \qquad\qquad | \qquad\qquad | \\
H-C-C-\!\!-\!\!-\!\!-C-\!\!-\!\!-\!\!-C-C-H \\
\quad | \qquad | \qquad\quad | \qquad\quad | \; | \\
\quad H \quad | \qquad\quad H \qquad H \; H \\
\quad\quad H-C-H \\
\quad\quad\quad | \\
\quad\quad\quad H
\end{array}
$$

Even Most Single Bonds Can Be "Understood."   When a *single* bond appears on a *horizontal* line, we need not write a straight line to represent it; we can leave such single bonds to the imagination. We do not do this, however, for bonds that are not on a horizontal line. Thus we can write the structure of 2-methylpentane (Example 9.1), as follows. Notice that the vertically oriented bond is shown by a line but that all other single bonds are understood.

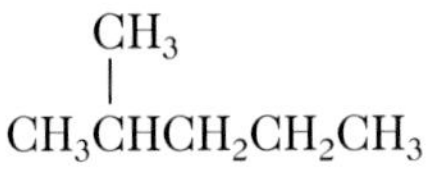

$$CH_3$$
$$|$$
$$CH_3CHCH_2CH_2CH_3 \quad \text{2-Methylpentane}$$

■ PRACTICE EXERCISE 2   Rewrite the condensed structures that you drew for the answers to Practice Exercise 1 and let the appropriate carbon–carbon single bonds be left to the imagination.

■ PRACTICE EXERCISE 3   To be certain that you are comfortable with condensed structures, expand each of the following to make them full, expanded structures with no bonds left to the imagination.

(a) CH_3CH_3 (b) $CH_3CHCHCH_3$ (c) $CH_3CH_2CCH_2CH_2CH_3$

$$
\begin{array}{lll}
 & CH_3 & CH_3 \\
 & | & | \\
\text{(b) } CH_3CHCHCH_3 & \quad \text{(c) } CH_3CH_2CCH_2CH_2CH_3 \\
 & | & | \\
 & CH_3 & CH_3
\end{array}
$$

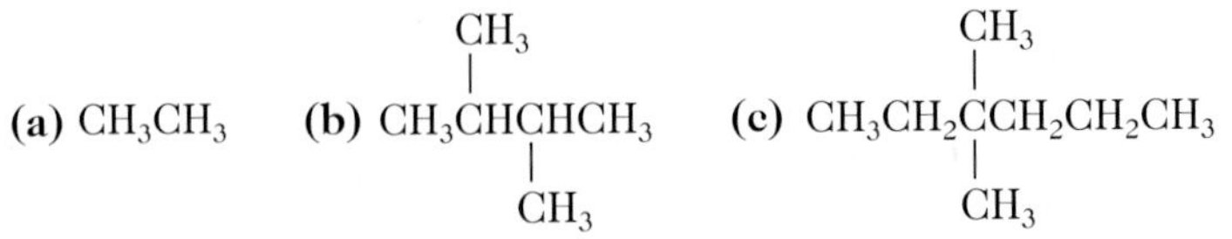

■ The only violations to the rule of four bonds from each carbon occur in rare and unstable species in which the carbon atom is involved with a (+) or a (−) charge or carries an unpaired electron.

As indicated in the *check* part of Example 9.1, an important skill in using condensed structures is the ability to recognize errors. The most common is a violation of the rule that every carbon atom in a structure must have exactly four bonds, no more and no fewer. Do the next Practice Exercise to test your skill in recognizing an error in structure.

■ **PRACTICE EXERCISE 4** Which of the following structures cannot represent real compounds?

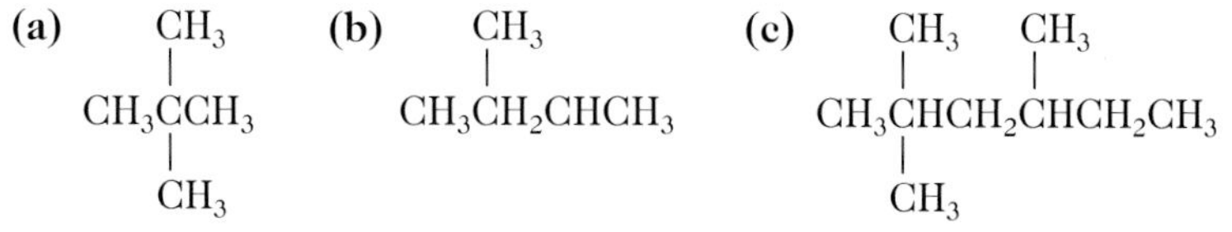

(a)
$$CH_3$$
$$|$$
$$CH_3CCH_3$$
$$|$$
$$CH_3$$

(b)
$$CH_3$$
$$|$$
$$CH_3CH_2CHCH_3$$

(c)
$$CH_3 \quad CH_3$$
$$| \quad |$$
$$CH_3CHCH_2CHCH_2CH_3$$
$$|$$
$$CH_3$$

When atoms other than carbon and hydrogen are present in a molecule, no major new problem arises in writing condensed structures. Remember that every oxygen or sulfur atom carrying no electrical charge must have two bonds, every nitrogen must have three, and every halogen atom must have just one.

■ O and N can have more bonds in certain *ions,* for example, H_3O^+ and NH_4^+.

Double and Triple Bonds Are Seldom Condensed. Another rule about condensed structures is that carbon–carbon double and triple bonds are never left to the imagination.

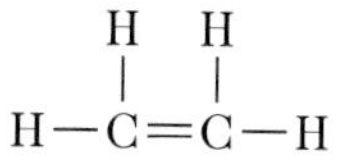

$$\begin{array}{cc} H & H \\ | & | \\ H-C=C-H \end{array}$$ condenses to $CH_2{=}CH_2$ or to $H_2C{=}CH_2$ but not CH_2CH_2

Ethylene (raw material for making polyethylene)

Parentheses Are Sometimes Used to Condense Structures Further. Sometimes two or three identical groups attached to the same carbon are grouped inside a set of parentheses. For example,

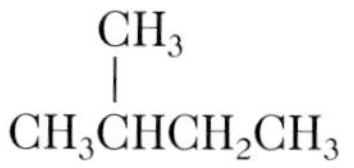

$$CH_3$$
$$|$$
$$CH_3CHCH_2CH_3$$ can be written as $(CH_3)_2CHCH_2CH_3$

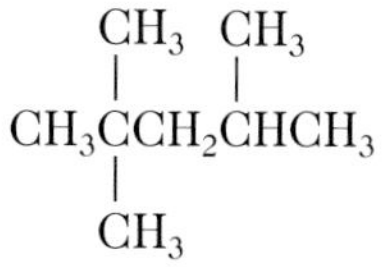

$$CH_3 \quad CH_3$$
$$| \quad |$$
$$CH_3CCH_2CHCH_3$$ can be written as $(CH_3)_3CCH_2CH(CH_3)_2$
$$|$$
$$CH_3$$

Long chains can also be condensed by using parentheses to enclose repeating groups, particularly CH_2 groups that are strung out one after the other. For example, $CH_3CH_2CH_2CH_2CH_2CH_2CH_3$ can be written as $CH_3(CH_2)_5CH_3$.

Functional Groups Define Families of Organic Compounds. Chemicals that react with organic compounds generally attack only certain molecular parts and leave the rest alone. The reacting parts, called **functional groups,** can be double or triple bonds or locations having atoms other than C or H. Sections of organic molecules consisting only of C and H and only single bonds undergo so few reactions that they are called **nonfunctional groups.**

Although more than 6 million organic compounds are known, there are only a handful of functional groups, and each one serves to define a *family* of organic compounds. Our study of organic chemistry will be organized around just a few of these families (Table 9.1). An important example is the *alcohol family.* One member is ethyl alcohol, CH_3CH_2OH, whose molecules have an OH group attached to a chain of two carbons. Chain *length,* however, determines only the name of the specific family member, not the family itself. The chain can be any length you can imagine, and the substance is in the alcohol family if, somewhere on the chain, an OH group is attached

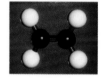

Ethylene

TABLE 9.1 Some Important Families of Organic Compounds

Family	Characteristic Structural Feature[a]	Example
Hydrocarbons	Only C and H Present	
	Families of Hydrocarbons	
	Alkanes: only single bonds	CH_3CH_3
	Alkenes: $C{=}C$	$CH_2{=}CH_2$
	Alkynes: $C{\equiv}C$	$HC{\equiv}CH$
	Aromatic: benzene ring	 ⬡
Alcohols	ROH	CH_3CH_2OH
Ethers	ROR′	CH_3OCH_3
Thioalcohols	RSH	CH_3SH
Disulfides	RS-SR	$CH_3S{-}SCH_3$
Aldehydes	$\overset{\displaystyle O}{\overset{\|}{R}}CH$	$\overset{\displaystyle O}{\overset{\|}{CH_3}}CH$
Ketones	$\overset{\displaystyle O}{\overset{\|}{R}}CR'$	$\overset{\displaystyle O}{\overset{\|}{CH_3}}CCH_3$
Carboxylic acids	$\overset{\displaystyle O}{\overset{\|}{R}}COH$	$\overset{\displaystyle O}{\overset{\|}{CH_3}}COH$
Esters of carboxylic acids	$\overset{\displaystyle O}{\overset{\|}{R}}COR'$	$\overset{\displaystyle O}{\overset{\|}{CH_3}}COCH_3$
Esters of phosphoric acid	$\overset{O}{\overset{\|}{R}}OPOH$, OH	$\overset{O}{\overset{\|}{CH_3}}OPOH$, OH
Esters of diphosphoric acid	$\overset{O\ O}{\overset{\|\ \|}{R}}OPOPOH$, HO OH	$\overset{O\ O}{\overset{\|\ \|}{CH_3}}OPOPOH$, HO OH
Esters of triphosphoric acid	$\overset{O\ O\ O}{\overset{\|\ \|\ \|}{R}}OPOPOP(OH)_2$, HO OH	$\overset{O\ O\ O}{\overset{\|\ \|\ \|}{CH_3}}OPOPOP(OH)_2$, HO OH
Amines	$RNH_2, RNHR', RNR'R''$	CH_3NH_2 ; CH_3NHCH_3 ; $\overset{CH_3}{\overset{\|}{CH_3}}NCH_3$
Amides	$\overset{O\ \ R''(H)}{\overset{\|\ \ \|}{R}}C{-}NR'(H)$	$\overset{O}{\overset{\|}{CH_3}}CNH_2$

[a]R′, and R″ represent hydrocarbon groups—*alkyl groups*—defined in the text. R′(H) or R″ (H) signifies that the substituent can be either a hydrocarbon group or hydrogen.

to a carbon from which only single bonds extend. Four short-chain examples of alcohols are

$$CH_3OH \qquad CH_3CH_2OH \qquad CH_3CH_2CH_2OH \qquad CH_3CH_2CH_2CH_2OH$$

Methyl alcohol Ethyl alcohol Propyl alcohol Butyl alcohol

Organic Families Have Family Symbols. Because all alcohols have the same functional group, they exhibit the same kinds of chemical reactions. When only one of these kinds is learned, it applies to all members of the family, literally to thousands of compounds. In fact, we often summarize a particular reaction for an organic family by using a general family symbol. All alcohols, for example, can be represented by the symbol ROH, where R stands for a carbon chain or ring, any chain of any length or any ring of any size whatsoever. All alcohols, for instance, react with sodium metal as follows.

$$2ROH + 2Na \longrightarrow 2RONa + H_2$$

(RONa is an ionic compound, a combination of RO^- and Na^+). If we wanted to write the specific example of this reaction that involves, say, ethyl alcohol, all we have to do is replace R with CH_3CH_2.

$$2CH_3CH_2OH + 2Na \longrightarrow 2CH_3CH_2ONa + H_2$$

Butyl alcohol, another member of the alcohol family, reacts as follows.

$$2CH_3CH_2CH_2CH_2OH + 2Na \longrightarrow 2CH_3CH_2CH_2CH_2ONa + H_2$$

You can see that the reaction occurs only at the OH group, the functional group of all alcohols. Chain length is not a factor.

■ R is from the German word *Radikal,* which we translate here to mean "group," as in a group of atoms.

9.3 ISOMERISM

Compounds can have identical molecular formulas but different structures.

Ammonium cyanate and urea, the chemicals of Wöhler's important experiment, both have the molecular formula CH_4N_2O, but the atoms are organized differently:

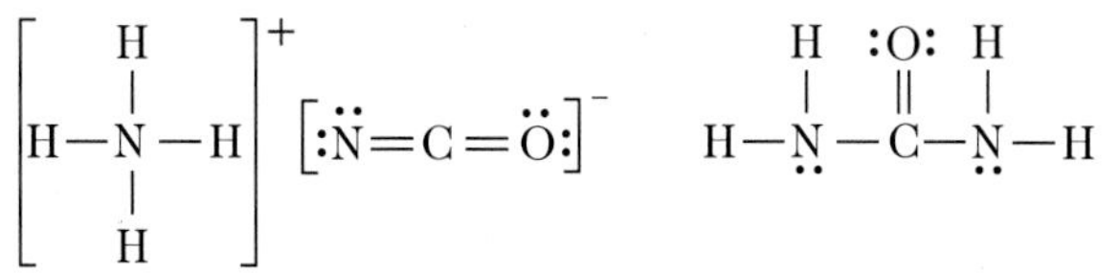

Ammonium cyanate, CH_4N_2O Urea, CH_4N_2O

Compounds that have identical molecular formulas but different structures are called **isomers** of each other, and the existence of isomers is a phenomenon called **isomerism.** Isomerism is one reason why there are so many organic compounds. There are several kinds of isomers, and we consider just one type here, **constitutional isomers.** Constitutional isomers, once called *structural isomers,*[1] differ in the basic atom-to-

■ "Isomer" has Greek roots: *isos,* "the same," and *meros,* "parts"; in other words, "equal parts" (but put together differently).

[1]The term *structural isomer* has fallen into disfavor because it is regarded as too broad in that it implies not only atom-to-atom connectivities but also geometrical differences. Another kind of isomerism, geometrical isomerism, deals with the latter (see Section 10.3).

TABLE 9.2   Properties of Two Isomers: Ethyl Alcohol and Dimethyl Ether

Property	Ethyl Alcohol	Dimethyl Ether
Structure	CH_3CH_2OH	CH_3OCH_3
Boiling point	78.5 °C	−24 °C
Melting point	−117 °C	−138.5 °C
Density (25 °C)	0.79 g/mL (a liquid)	2.0 g/L (a gas)
Solubility in water	Soluble in all proportions	Slightly soluble

atom connectivities. There are three constitutional isomers of C_5H_{12}: pentane, 2-methylbutane, and 2,2-dimethylpropane.

■ The names in parentheses are the common names of these compounds. The letter *n* stands for *normal,* meaning the straight-chain isomer. *Neo* signifies *new,* as in a new isomer.

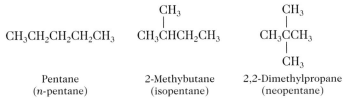

$CH_3CH_2CH_2CH_2CH_3$	$CH_3CHCH_2CH_3$	CH_3CCH_3
Pentane (*n*-pentane)	2-Methybutane (isopentane)	2,2-Dimethylpropane (neopentane)

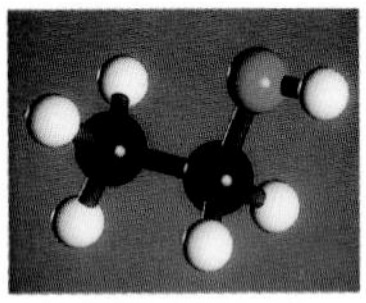

Ethyl alcohol

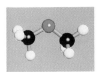

Dimethyl ether

The larger the number of carbon atoms per molecule, the larger the number of constitutional isomers. For example, C_8H_{18} has 18 constitutional isomers; $C_{10}H_{22}$ has 75, and $C_{20}H_{42}$ has 366,319. Someone has figured out that roughly 6.25×10^{13} isomers are possible for $C_{40}H_{82}$ under the rules of covalence. Very few have actually been prepared. It would take nearly 200 billion years to make each one at the rate of one per day. Actually, many could not even exist because, in isomers with a great number of chain branches, the groups would crowd each other too much.

The isomers of pentane or of $C_{40}H_{82}$ have quite similar chemical properties because their molecules all have only C—C and C—H single bonds. Usually, however, isomers have very different properties, because they belong to different families. The atoms in C_2H_6O, for example, can be organized into two isomers, ethyl alcohol and dimethyl ether, that chemically and physically are very different (Table 9.2). At room temperature, ethyl alcohol is a liquid and soluble in water, and dimethyl ether is a gas and relatively insoluble in water. Ethyl alcohol reacts with sodium; dimethyl ether does not. Because it is quite common for isomers to have properties so different, we nearly always use *structural* rather than molecular formulas for organic compounds. Only structures let us see at a glance how the atoms in the molecules are organized. The ability to recognize two structures as identical molecules, as isomers, or as something else is a skill that must be acquired.

EXAMPLE 9.2   Recognizing Isomers

Which pair of structures represents a pair of isomers?

1. CH_3—O—CH_2CH_3 and CH_3CH_2—O—CH_3

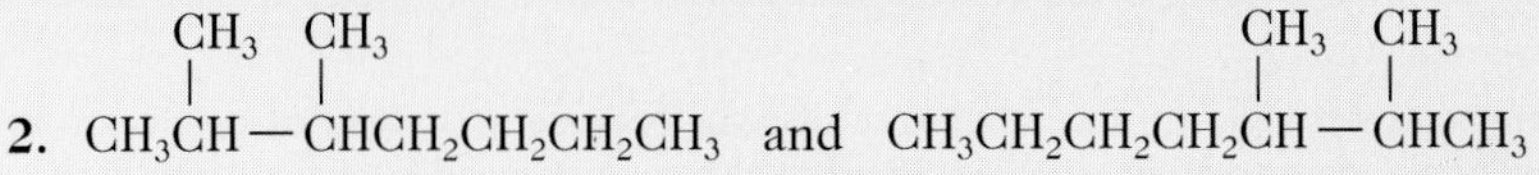

2. CH_3CH—$CHCH_2CH_2CH_2CH_3$ and $CH_3CH_2CH_2CH_2CH$—$CHCH_3$
   with CH_3 and CH_3 branches

$$\overset{\displaystyle CH_3}{\underset{\displaystyle |}{}}$$

$$\overset{\displaystyle CH_3}{\underset{\displaystyle |}{}}$$

3. $CH_3CHCH_2CH_2CH_3$ and $CH_3CH_2CHCH_2CH_3$

$$\overset{\displaystyle CH_3}{\underset{\displaystyle |}{}}$$

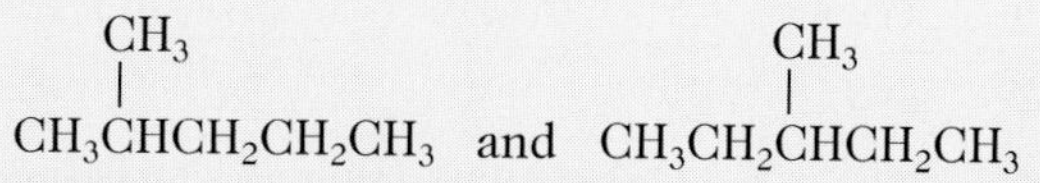

4.  CH_2CH_3 and $CH_3CH_2CH_3$

ANALYSIS Unless you spot a difference that rules out isomerism immediately, the *first step* is to see whether the molecular formulas are the same. If they aren't, the two structures are *not* isomers. If the molecular formulas of the two structures are identical, they might be identical or they might be isomers.

In this problem, the members of each pair share the same molecular formula.

Pair 1: C_3H_8O Pair 2: C_9H_{20} Pair 3: C_6H_{14} Pair 4: C_3H_8

Next, to see whether a particular pair represents isomers, we try to find at least one structural difference. If we can't, the two structures are identical, but they might be oriented differently on the page or their chains might be shown twisted into different conformations. Don't be fooled by an "east-to-west" versus a "west-to-east" type of difference. The difference must be *internal* within the structure.

■ Whether you face east or west, you're the same person.

SOLUTION Pair 1 are identical molecules; they're only oriented differently. (Imagine using a pancake turner to flip the one on the left, left to right; it would then be the structure on the right.)

Pair 2 is also an example of an east versus west difference in orientation. These two structures are identical. Their internal sequences, their atom-to-atom connectivities, are the same.

Pair 3 are isomers. In the first, a CH_3 group joins a five-carbon chain at the chain's second carbon, and in the second, this group is attached at the third carbon.

Pair 4 are identical. The two structures differ only in the conformations of their chains. Recall that free rotation about bonds allows us to imagine the straightening out of a continuous, open chain.

■ PRACTICE EXERCISE 5   Examine each pair to see whether the members are identical, are isomers, or are different in some other way.

(a) $H-O-CH_3$ and CH_3-O-H

(b) $CH_3-NH-CH_3$ and $CH_3-CH_2-NH_2$

$$\overset{\displaystyle CH_2CH_3}{\underset{\displaystyle |}{}}$$

$$\overset{\displaystyle CH_3}{\underset{\displaystyle |}{}}$$

(c) 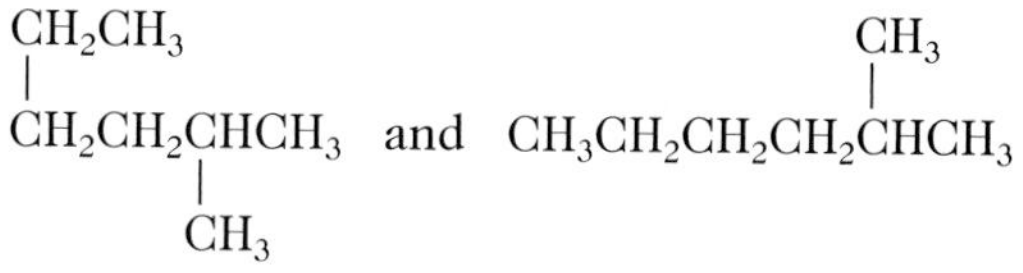 $CH_2CH_2CHCH_3$ and $CH_3CH_2CH_2CH_2CHCH_3$

$$\underset{\displaystyle |}{\displaystyle CH_3}$$

(d) $CH_2=CHCH_2CH_3$ and $CH_3CH=CHCH_3$

$$\overset{\displaystyle O}{\underset{\displaystyle ||}{}}$$

$$\overset{\displaystyle O}{\underset{\displaystyle ||}{}}$$

(e) 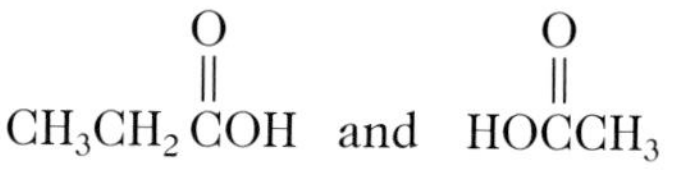 CH_3CH_2COH and $HOCCH_3$

SUMMARY

Organic and Inorganic Compounds   Most organic compounds are molecular and the majority of inorganic compounds are ionic. Molecular and ionic compounds differ in composition, in types of bonds, and in several physical properties.

Structural Features of Organic Molecules   The ability of carbon atoms to join to each other many times in succession in straight chains and branched chains as well as into cyclic rings accounts in a large measure for the existence of several million organic compounds.

When groups within a molecule are joined by a single bond, they can rotate relative to each other about this bond. Full structural formulas of organic compounds are usually condensed by grouping the hydrogens attached to a carbon immediately by it; by letting single bonds on a horizontal line be understood; and by leaving bond angles and conformational possibilities to the informed imagination.

The families of organic compounds are organized around functional groups, which are parts of molecules at which most of the chemical reactions occur. Nonfunctional parts of molecules can sometimes be given the general symbol R, as in ROH, the general symbol for all alcohols.

Isomerism   Differences in the *conformations* of carbon chains do not create new compounds, but differences in *constitution* do. Isomers are compounds with identical molecular formulas but different structures.

REVIEW EXERCISES

The answers to Review Exercises whose numbers are in color are found in Appendix V. The answers to the other Review Exercises are found in the Study Guide that accompanies this text. The more challenging questions are marked with asterisks.

Organic and Inorganic Compounds

9.1   With respect to the *synthesis* of organic compounds, what specifically was the problem that organic chemists faced prior to 1828? What scientific theory had been devised to meet this problem?

9.2   What was Wöhler's goal when he evaporated an aqueous solution of ammonium cyanate to dryness? What happened instead? With respect to *scientific theory* at the time, what specifically did Wöhler accomplish?

9.3   What kind of bond between atoms predominates among organic compounds?

9.4   Which of the following compounds are inorganic?
(a) CH_3CH_2OH (b) CO_2 (c) $CHCl_3$
(d) $KHCO_3$ (e) Na_2CO_3

9.5   Are the majority of all compounds that dissolve in water ionic or molecular? Inorganic or organic?

9.6   Explain why very few organic compounds can conduct electricity either in an aqueous solution or as molten materials.

*9.7   Each compound described below is either ionic or molecular. State which it most likely is, and give one reason.
(a) A compound that melts at 281 °C, and burns in air.
(b) A compound that dissolves in water. When hydrochloric acid is added, the solution fizzes and an odorless, colorless gas is released, which can extinguish a burning flame.
(c) A compound that is a colorless gas at room temperature.
(d) A compound that melts at 824 °C and becomes white when heated.
(e) A compound that is a liquid and does not dissolve in water but does burn.
(f) A compound that is a liquid and does dissolve in water as well as burns.

Structural Formulas

9.8   One can write the structure of propane, a common heating gas, as follows.

$$CH_3-CH_2 \overset{\overset{\textstyle CH_3}{|}}{} \quad \text{Propane}$$

Are propane molecules properly described as straight chain or as branched chain, in the sense in which we use these terms? Explain.

9.9   Which of the following structures are possible, given the numbers of bonds that various atoms can form?
(a) $CH_3CH_2CH_2OCH_3$
(b) $CH_2CH_2CH_2CH_3$
(c) $CH_3{=}CHCH_2CH_3$
(d) $CH_3CH{=}CHCH_2CH_3$
(e) $NH_2CH_2CH_2CH_3$

*9.10   Write full (expanded) structures for each of the following *molecular* formulas. Remember how many covalent bonds nonmetals have in molecules: C, 4; N, 3; O, 2; H, Cl Br, 1 each. In some structures you will have to use double or triple bonds. (*Hint:* A trial-and-error approach will have to be used.)

(a) CH_4O (b) CH_2Cl_2 (c) N_2H_4 (d) C_2H_6
(e) CH_2O (f) CH_2O_2 (g) NH_3O (h) C_2H_2
(i) $CHCl_3$ (j) HCN (k) C_2H_3N (l) CH_5N

9.11 Write, neat condensed structures of the following.

(a)

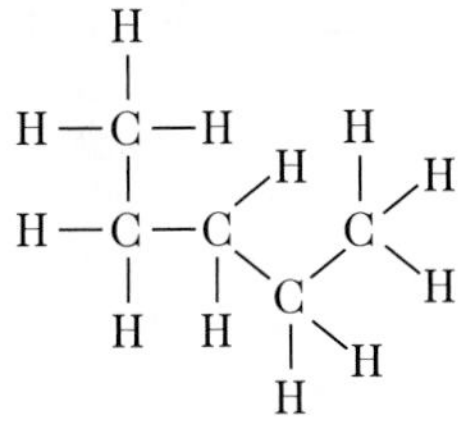

(b)

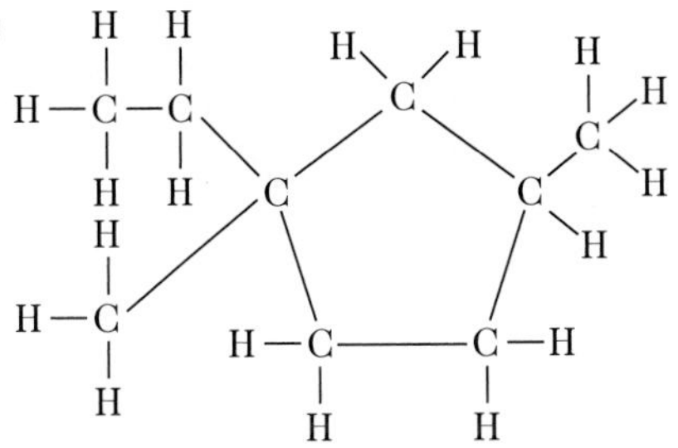

(c)

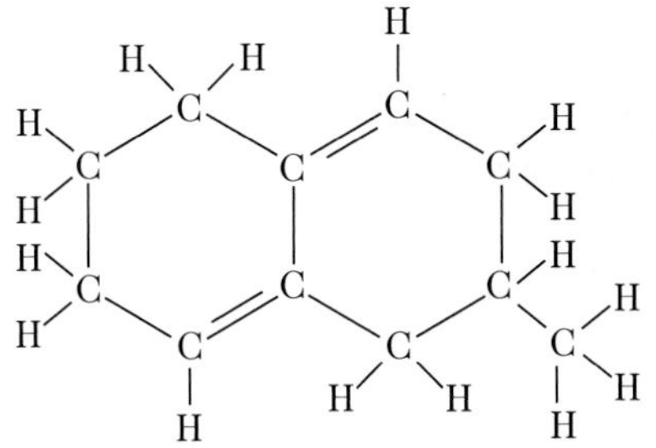

Isomers

*9.12 Decide whether the members of each pair are identical, are isomers, or are neither.

(a) CH_3 and CH_3-OH
       |
       OH

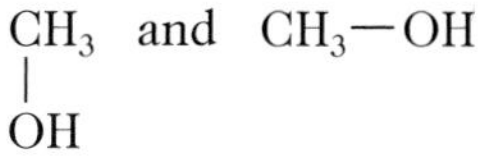

(b)

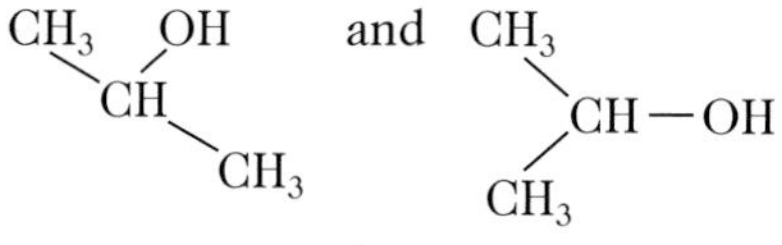

(c) CH_3CH_2SH and $CH_3CH_2CH_2SH$

(d) $CH_3CH=CH_2$ and $CH_2\!-\!CH_2$
                           $\diagdown\;\diagup$
                            CH_2

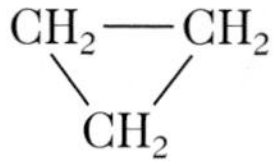

(e)

$CH_3CCH_2CH_3$ and $CH_3CH_2CCH_3$

(f)

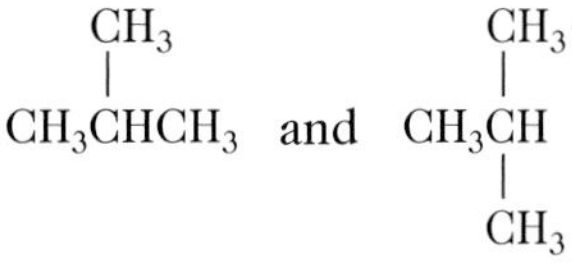

(g) $CH_3CH_2CH_2NH_2$ and $CH_3NHCH_2CH_3$

(h)

CH_3CH_2COH and $HOCH_2CCH_3$

(i)

$HCOCH_2CH_3$ and CH_3CH_2COH

(j)

$HCOCH_2CH_2OH$ and $HOCH_2CH_2COH$

(k)

$CH_3OCH_2CCH_3$ and $CH_3CH_2COCH_3$

(l) $CH_3-CH-CH_2-CH_3$

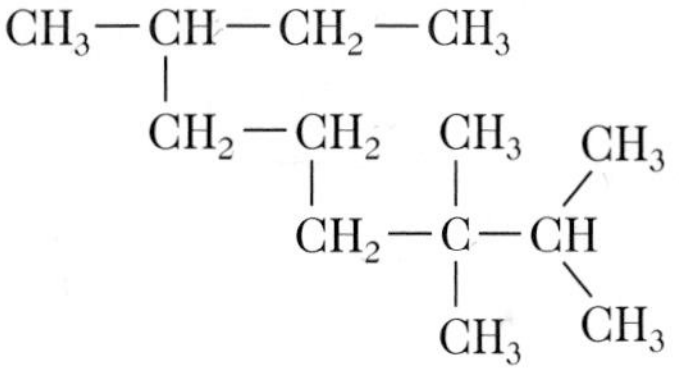

and

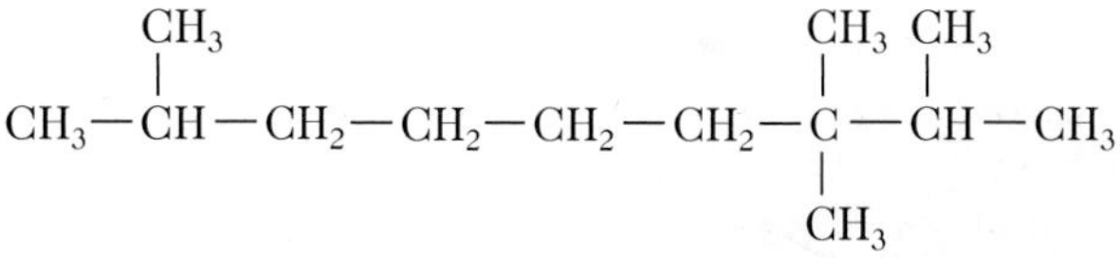

(m) $HO-O-CH_3$ and $H-O-CH_2-OCH_3$

Families of Organic Compounds

9.13 Name the family to which each compound belongs.

(a) $CH_3CH_2CH_3$ **(b)** $HOCH_2CH_2CH_3$

(c) $HCCH_2CH_3$ **(d)** $CH_3OCCH_2CH_3$

(e) $CH_3CH_2CH_2NH_2$ **(f)** $CH_3OCH_2CH_3$

*9.14 Name the families to which the compounds in parts (a)–(l) of Review Exercise 9.12 belong. (A few belong to more than one family.)

10

HYDROCARBONS

Families of Hydrocarbons

Alkanes

Alkenes

Chemical Reactions of the Carbon–Carbon Double Bond

Polymerization of Alkenes

Aromatic Compounds

People today routinely trust their lives to synthetics. Organic polymers have transformed both the gear and the clothing styles of all forms of recreation. The study of synthetic polymers, begun in this chapter, will enable us to understand the natural polymers that make up proteins, carbohydrates, and nucleic acids.

10.1 FAMILIES OF HYDROCARBONS

Hydrocarbons can be saturated or unsaturated, open chain or cyclic, and all are insoluble in water.

Hydrocarbons are organic compounds made only from carbon and hydrogen and found in the *fossil fuels* (see Special Topic 10.1). Molecules of hydrocarbons can have single, double, or triple bonds (Figure 10.1). Hydrocarbons with only single bonds make up the family of **alkanes,** and Table 10.1 gives the structures of the 10 simplest members. **Alkenes** are hydrocarbons with double bonds, and **alkynes** are those with triple bonds.

The ring system in benzene confers such unique properties on the molecule that compounds with the benzene ring, regardless of other functional groups, are in a class by themselves, the **aromatic compounds.** Any compound without the benzene ring (or certain rings similar to it), regardless of its functional groups, is called an **aliphatic compound.** Thus, diethyl ether, once an important anesthetic, isopropyl alcohol (rubbing alcohol), and acetic acid (in vinegar) are all aliphatic.

Compounds Can Be Saturated or Unsaturated. Compounds with only single bonds, like the alkanes, are called **saturated compounds,** regardless of the presence of any other functional groups or rings provided they also have only single bonds. Thus, diethyl ether and isopropyl alcohol are saturated compounds. (You can see how classifications often overlap.)

When one or more multiple bonds are present, the substance is an **unsaturated compound.** All alkenes and alkynes as well as all aromatic compounds are thus *unsaturated hydrocarbons.* The carbon–oxygen double bond in acetic acid makes it an unsaturated compound, too.

Cyclic Hydrocarbons Occur in Various Ring Sizes and Degrees of Unsaturation. Cyclic compounds can have double bonds as in cyclohexene. Rings, of course, can carry attached atoms or groups of atoms called *substituents,* such as the methyl group in methylcyclohexane.

■ The triple bond occurs rarely in nature, and we'll not study it further.

$CH_3CH_2OCH_2CH_3$
Diethyl ether

$$CH_3\overset{\overset{\displaystyle OH}{|}}{C}HCH_3$$
Isopropyl alcohol

$$CH_3\overset{\overset{\displaystyle O}{\|}}{C}OH$$
Acetic acid

■ Remember that alkene double bonds, whether in open chains or rings, are always shown by two lines. They are not "understood."

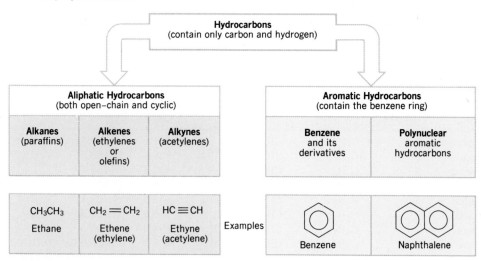

FIGURE 10.1
The several kinds of hydrocarbons. (The circles in the structures for benzene and naphthalene are explained in the chapter.)

$$\text{Cyclohexene} \qquad \text{Methylcyclohexane}$$

In structures of cyclic compounds, the ring system itself is usually represented simply by a polygon, a many-sided figure. A ring carbon atom is understood to be at each corner together with as many H atoms at the corner as are needed to fill out four bonds to the ring carbon. A pentagon, for example, can represent cyclopentane. This symbolism is illustrated in Table 10.2, where you can see how the numbers of atoms making up a ring can vary. (Rings of dozens of atoms are known.) The photograph of the model of methylcyclopentane and its progressively more condensed structures further illustrates the use of a geometric figure to represent a ring.

Three ways to represent the structure of methylcyclopentane

One of the most interesting cyclic molecules is C_{60}, a member of a recently discovered family called the *fullerenes.* The fullerenes, as well as diamond and graphite (Figure 10.2), lack H atoms, so they are not *compounds.* Instead, being variations of ways of organizing the atoms of the *same* element, they are called *allotropes* of carbon.

Hydrocarbons Do Not Dissolve in Water. Carbon–hydrogen bonds and carbon–carbon bonds of all types, single, double, or triple, are almost entirely nonpolar. Hydrocarbons, as a result, are almost completely nonpolar compounds.

TABLE 10.1   Straight-Chain Alkanes

IUPAC Name	Carbons	Molecular Formula	Structure	Boiling Point (°C)	Melting Point (°C)	Density (g/mL, 20 °C)
Methane	1	CH_4	CH_4	−161.5	−182.5	
Ethane	2	C_2H_6	CH_3CH_3	−88.6	−183.3	
Propane	3	C_3H_8	$CH_3CH_2CH_3$	−42.1	−189.7	
Butane	4	C_4H_{10}	$CH_3(CH_2)_2CH_3$	−0.5	−138.4	
Pentane	5	C_5H_{12}	$CH_3(CH_2)_3CH_3$	36.1	−129.7	0.626
Hexane	6	C_6H_{14}	$CH_3(CH_2)_4CH_3$	68.7	−95.3	0.659
Heptane	7	C_7H_{16}	$CH_3(CH_2)_5CH_3$	98.4	−90.6	0.684
Octane	8	C_8H_{18}	$CH_3(CH_2)_6CH_3$	125.7	−56.8	0.703
Nonane	9	C_9H_{20}	$CH_3(CH_2)_7CH_3$	150.8	−53.5	0.718
Decane	10	$C_{10}H_{22}$	$CH_3(CH_2)_8CH_3$	174.1	−29.7	0.730

[a]The molecular formulas of the open-chain alkanes fit the general formula C_nH_{2n+2}, where n = the number of carbon atoms per molecule.

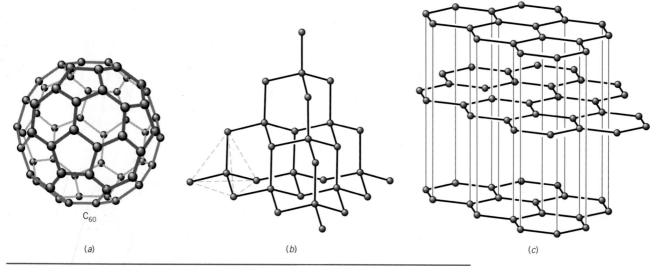

FIGURE 10.2

Three allotropes of carbon. *(a)* C_{60} is a member of the fullerene family, named after Buckminster Fuller, the inventor of the geodesic dome. *(b)* Diamond structure—interlocking and extended networks of six-membered rings of carbon atoms. *(c)* Graphite—sheets of extended benzene-like rings of carbon atoms.

One consequence is that hydrocarbons are insoluble in water. Their molecules have no polar sites that can attract water molecules. Water molecules, strongly attracted to each other by hydrogen bonds, are simply unable to let hydrocarbon molecules in. But in nonpolar solvents, like ether, gasoline, or benzene, hydrocarbon molecules relatively easily slip in and among the solvent molecules, fully intermingle, and so dissolve. Gasoline, itself a mixture of hydrocarbons (mostly alkanes) and therefore nonpolar, is a good solvent for tar and grease. (If you ever use gasoline or lighter fluid to clean tar from surfaces, be sure to keep all flames away.)

■ The vapors of hydrocarbon solvents catch fire very easily, so be careful if you use these solvents for any purpose. In the right proportion in air, hydrocarbon vapors explode when ignited.

TABLE 10.2 Cycloalkanes

IUPAC Name	Structure	Boiling Point (°C)	Melting Point (°C)	Density (20 °C)
Cyclopropane	△	−33	−127	1.809 g/L (0 °C)
Cyclobutane	□	13.1	−80	0.7038 g/mL
Cyclopentane	⬠	49.3	−94.4	0.7460 g/mL
Cyclohexane	⬡	80.7	6.47	0.7781 g/mL
Cycloheptane	⬡	118.5	−12	0.8098 g/mL

SPECIAL TOPIC 10.1
ORGANIC FUELS

The Fossil Fuels *Photosynthesis* is the natural process by which plants use the green pigment chlorophyll to trap energy from the sun. The plants then use the energy to make compounds that they need from carbon dioxide, water, and minerals in soil and water. A good portion of the entrapped solar energy remains as chemical energy within the molecules of the new compounds. Long ago, nature used photosynthesis to transform solar energy into the chemical energy of ancient plants and then locked this energy into the fossilized remains of the plants, the *fossil fuels*—principally petroleum, coal, and natural gas. These fuels are a legacy from the past now being consumed so rapidly that for the first time in history, we are concerned about running out of them and about the impact on civilization if their disappearance occurs too suddenly for us to adapt.

Petroleum and Crude Oil The fossil fuels formed over a span of hundreds of thousands of years during the Carboniferous Period in geologic history, roughly 280 to 345 million years ago. Vast areas of the continents, little more than monotonous plains, then basked in sunlight near sea level. In the oceans, countless tiny, photosynthesizing plants like the diatoms—tons of them per acre of ocean surface during the early spring—soaked up solar energy to power their fugitive lives. Then they died. According to one theory, the death of each such plant released a tiny droplet of oily material that eventually settled into the bottom muds. The muds grew in thickness and compacted, sometimes into a rock called shale, sometimes into limestone and sandstone deposits. Under the pressure and heat of compacting, the oily matter changed into petroleum, a mixture of crude oil, water, and natural gas. (*Petroleum* is from *petra,* "rock," and *oleum,* "oil.") In some parts of the world, the petroleum managed to move slowly through porous rock and collect into vast underground pools to form the great petroleum reserves of our planet. In other regions, this movement could not occur, and the oily substances remain to this day locked in enormous deposits of oil shale and oil sand.

Coal On the marshy lands bordering the ancient oceans, lush vegetation flourished and died in a moist and sunny setting. The rate of decay of the remains of these plants, covered by stagnant, oxygen-poor, often acidic water, was slower than the rate at which the plants died. The slowly rotting mass accumulated to huge depths, became fibrous, and turned to *peat,* a woody material used as fuel in some regions of the world.

Where peat layers became thick enough or were compressed by later deposits of sedimentary rock like limestone and sandstone, the peat changed into lignite ("brown coal"). Although lignite is more than 40% water, it is still an important fuel. Many lignite deposits became thick enough for further compaction to occur, and most of the water was squeezed out. Thus bituminous coal ("soft coal") formed, which has less than 5% water but contains considerable quantities of volatile matter. (*Volatile* means "easily evaporated.") Still further compaction took place, forcing out nearly all of the volatile matter and leading to the formation of anthracite ("hard coal") which is more than 95% carbon.

The energy content in the coal reserves of the world exceeds that of the known petroleum reserves by a wide margin. In estimates that have the available supplies of petroleum lasting less than a century, the coal reserves are considered to have a lifetime of up to three centuries. Coal contains very small quantities of compounds of mercury, sulfur, radioactive elements,

■ "Like" refers to likeness in molecular polarity.

Like Dissolves Like. Grease and tar are relatively nonpolar materials, so their solubility in nonpolar gasoline illustrates the **like dissolves like** rule: *Polar solvents tend to dissolve polar or ionic substances and nonpolar solvents tend to dissolve nonpolar solutes.* Water molecules, for example, are polar, and so water dissolves salt (ionic) and sugar (polar molecular) but not hydrocarbons. Hydrocarbon solvents, with nonpolar molecules, dissolve nonpolar, hydrocarbon-like compounds, but not ionic or polar molecular substances. Neither salt (ionic) nor sugar (molecular) dissolves in gasoline, for example.

Structures Carry Molecular "Map Signs." Our brief survey of the physical properties of hydrocarbons has introduced the first molecular "map sign" of our study, the

Principal Fractions from Petroleum

Boiling Point Range (°C)	Molecular Size	Principal Uses
Below 20	C_1–C_4	Natural gas; heating and cooking fuel; raw materials for other chemicals
20–60	C_5–C_6	Petroleum "ether," a nonpolar solvent and cleaning fluid
60–100	C_6–C_7	Ligroin or light naphtha; nonpolar solvent and cleaning fluid
40–200	C_5–C_{10}	Gasoline
175–325	C_{12}–C_{18}	Kerosene; jet fuel; tractor fuel
250–400	C_{12} and higher	Gas oil; fuel oil; diesel oil
Nonvolatile liquids	C_{20} and up	Refined mineral oil; lubricating oil; grease (a blend of soap in oil)
Nonvolatile solids	C_{20} and up	Paraffin wax; asphalt; road tar; roofing tar

FIGURE 1
Catalytic cracking unit at an Exxon refinery in New Jersey.

and other potential pollutants. Their concentrations are very small, as we said, but because of the enormous quantities of coal annually burned, mostly at electric power plants, air pollution problems arise, which often lead to water pollution as well. The mercury pollution in many lakes north of the industrial regions of the United States stems in part from the mercury released by burning coal.

Natural Gas In both soft coal and petroleum, the most volatile hydrocarbons, methane and ethane, also accumulated. Natural gas is largely methane.

Useful Substances from Crude Oil by Refining Crude oil is a complex liquid mixture of organic compounds, but nearly all are hydrocarbons. Small but vexing amounts of sulfur-containing compounds are also present. The object of refining crude oil is to separate this mixture into products of varying uses. Refinery operations yield mixtures of compounds called *fractions* that boil over certain ranges of temperatures (see the accompanying table). Roughly 500 compounds occur among the fractions boiling up to 200 °C; about a third are alkanes, a third are cycloalkanes, and a third are aromatic hydrocarbons. You can see in the table where the chief fuels for transportation—gasoline, diesel oil, and jet fuel—originate.

The gasoline fraction of crude oil does not provide nearly enough of the world's needs for gasoline. One of the strategies at petroleum refineries to make more gasoline is to subject high boiling petroleum fractions to operations called *catalytic cracking* and *reforming* (Figure 1). In the presence of catalysts and heat, large alkane molecules break up into smaller ones corresponding to lower boiling points that are in the range useful for gasoline engines.

hydrocarbon-like feature. Substances whose molecules are entirely *or even mostly* hydrocarbon-like are insoluble in water but soluble in nonpolar solvents. When a glance at a structure tells us that it is mostly hydrocarbon-like, we can safely predict that the compound is insoluble in water. Consider lauryl alcohol, for example.

$$CH_3CH_2CH_2CH_2CH_2CH_2CH_2CH_2CH_2CH_2CH_2CH_2OH$$
Lauryl alcohol

The long alkane-like chain dominates this molecule. The polar OH group (in this setting, that of a long hydrocarbon chain) cannot contribute enough polarity to the whole molecule to make lauryl alcohol soluble in water. Thus, by learning one very general fact, one molecular "map sign," we do not have to memorize a long list of

■ The abundance of hydrocarbon-like groups in the molecules of cell membranes contributes to both the stability and the flexibility of such systems.

separate (but similar) facts about an equally long list of separate compounds found at the molecular level of life. With the molecular "map sign" in hand–*hydrocarbon-like compounds tend to be insoluble in water*–we can look at the structures of hundreds of complicated compounds and confidently predict particular properties such as the likelihood of their being soluble in some solvent.

■ PRACTICE EXERCISE 1 Which of the following is more soluble in gasoline?

$$\underset{\text{Glycerol}}{HOCH_2\underset{\displaystyle OH}{\underset{|}{C}}HCH_2OH} \qquad \underset{\text{2-Methyl-1-butanol}}{CH_3CH_2\underset{\displaystyle CH_3}{\underset{|}{C}}HCH_2OH}$$

Within a Family, Boiling Points Increase with Formula Mass. Because of their small size and low polarity, hydrocarbons with only one to four carbons per molecule are gases at or near room temperature. Notice in Table 10.1 how the boiling points of the straight-chain alkanes increase with chain length.

The direct correlation between size and boiling point is true of all hydrocarbon families. Hydrocarbons with 5 to about 16 carbon atoms per molecule are normally liquids. When alkane molecules have about 18 or more carbon atoms, the substances are waxy solids at room temperature. Paraffin wax, for example, is a mixture of alkanes with molecules of 20 or more carbon atoms.

■ Most candles are made from paraffin.

Nearly All Hydrocarbons Float on Water. Still another physical property of aliphatic hydrocarbons, as illustrated by the data in Table 10.1, is that all are less dense than water. Such compounds thus float on water. When oil spills occur, some of the material floats away to damage shore habitat.

10.2 ALKANES

The IUPAC name of a compound shows the number of carbon atoms in the parent chain, the kinds and locations of substituents, and the family to which the compound belongs.

Isomerism occurs among alkanes with four and more carbons. Some properties of two isomers of butane and the three of pentane are shown in Table 10.3. The numbers of isomers increase immensely with increasing carbon content, as we learned in Section 9.3. It is, of course, necessary, that each compound have a unique name, one that unmistakably belongs to one and only one structure. Predecessor organizations to the current *International Union of Pure and Applied Chemistry* or *IUPAC*, a multinational organization of the world's chemical societies, recognized this need over a century ago. Today, the IUPAC Commission on Nomenclature of Organic Chemistry has the responsibility of devising rules for naming organic compounds. The rules ensure that only one name, called the IUPAC name, can be devised for a given structure and that only one structure can be written for a given name. To achieve these goals, IUPAC names are made structurally very informative.

■ "Nomenclature" is from the Latin *nomen*, "name," and *calare*, "to call." Wealthy ancient Romans had slaves called *nomenclators* who were to remind their owners of the names of important people who approached them on the street.

IUPAC Rules for Naming the Alkanes.

1. The IUPAC name ending for all alkanes (and cycloalkanes) is *-ane*.

TABLE 10.3   Isomeric Butanes and Pentanes

Structure	IUPAC Name (Common Name)	Boiling Point (°C)
C₄H₁₀ Isomers		
$CH_3CH_2CH_2CH_3$	Butane (*n*-butane)	0
CH_3CHCH_3 with CH_3 branch	2-Methylpropane (isobutane)	−12
C₅H₁₂ Isomers		
$CH_3CH_2CH_2CH_2CH_3$	Pentane (*n*-pentane)	36
$CH_3CHCH_2CH_3$ with CH_3 branch	2-Methylbutane (isopentane)	28
CH_3CCH_3 with two CH_3 branches	2,2-Dimethylpropane (neopentane)	10

2. The *parent chain* is the longest continuous chain of carbons in the structure. For example, the branched-chain alkane

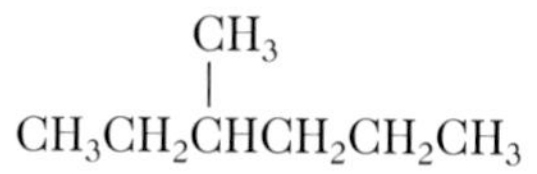

$$CH_3CH_2CHCH_2CH_2CH_3 \text{ (with } CH_3 \text{ on third carbon)}$$

is regarded as being "made" from the *parent* molecule

$$CH_3CH_2CH_2CH_2CH_2CH_3$$

by replacing a hydrogen atom on the third carbon from the left with CH_3.

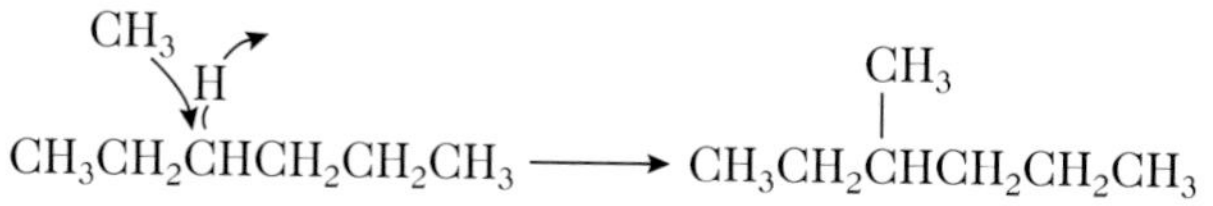

$$CH_3CH_2CHCH_2CH_2CH_3 \longrightarrow CH_3CH_2CHCH_2CH_2CH_3$$

3. Attached to the name ending, *-ane,* is a prefix that specifies the number of carbon atoms in the parent chain. The prefixes through parent chain lengths of 10 carbons are as follows and should be learned. The names in Table 10.1 show their use.

meth–	1 C	hex–	6 C
eth–	2 C	hept–	7 C
prop–	3 C	oct–	8 C
but–	4 C	non–	9 C
pent–	5 C	dec–	10 C

■ We won't need to know the prefixes for the higher alkanes.

Because the parent chain of our example has six carbons, the parent chain is named hexane–*hex* for six carbons and *ane* for being in the alkane family. Thus, the alkane whose name we are devising is regarded as a derivative of this parent, *hexane.*

4. The carbon atoms of the parent chain are numbered starting from whichever end of the chain gives the location of the first branch the lower of two possible numbers. Thus, the correct direction for numbering our example is from left to right.

$$
\begin{array}{c}
CH_3 \\
| \\
CH_3CH_2CHCH_2CH_2CH_3 \\
1 \quad 2 \quad 3 \quad 4 \quad 5 \quad 6
\end{array}
$$

(correct direction of numbering)

Had we numbered from right to left, the carbon holding the branch would have had a higher number, which is not allowed by the IUPAC rules for alkanes.

$$
\begin{array}{c}
CH_3 \\
| \\
CH_3CH_2CHCH_2CH_2CH_3 \\
6 \quad 5 \quad 4 \quad 3 \quad 2 \quad 1
\end{array}
$$

(incorrect direction of numbering)

5. Each branch attached to the parent chain is named. We must now pause and learn the names of some of the *alkyl groups,* groups with alkane-like branches.

The Alkyl Groups.   Any branch that consists only of carbon and hydrogen and has only single bonds is called an **alkyl group,** and the names of all alkyl groups end in *-yl.* Think of an alkyl group as an alkane minus one H.

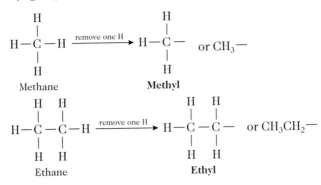

Two alkyl groups can be obtained from propane because the middle position in its chain of three is not equivalent to either of the end positions.

Two alkyl groups can similarly be obtained from butane.

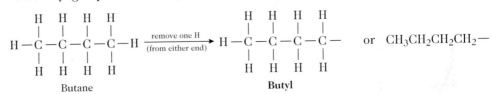

The structural diagrams showing Butane converting to sec-Butyl:

$$H-\underset{\underset{H}{|}}{\overset{\overset{H}{|}}{C}}-\underset{\underset{H}{|}}{\overset{\overset{H}{|}}{C}}-\underset{\underset{H}{|}}{\overset{\overset{H}{|}}{C}}-\underset{\underset{H}{|}}{\overset{\overset{H}{|}}{C}}-H \xrightarrow[\substack{\text{(from either of}\\ \text{the two interior}\\ \text{C atoms)}}]{\text{remove one H}} H-\underset{\underset{H}{|}}{\overset{\overset{H}{|}}{C}}-\underset{\underset{H}{|}}{\overset{\overset{H}{|}}{C}}-\underset{\underset{H}{|}}{\overset{\overset{H}{|}}{C}}-\underset{\underset{H}{}}{\overset{\overset{H}{|}}{C}}-H \quad \text{or} \quad CH_3CH_2CHCH_3$$

Butane   *sec*-Butyl

The last alkyl group is called the *secondary* butyl group (abbreviated *sec*-butyl) because the open bonding site is at a **secondary carbon,** a carbon that is directly attached to just two other carbons. A **primary carbon** is one to which just one other carbon is directly attached. The open bonding site in the butyl group, for example, is at a primary carbon atom. A **tertiary carbon** is one that holds directly three other carbons. We will encounter a tertiary carbon in a group that we will soon study.

Butane is the smallest alkane to have an isomer. The common name of the isomer is isobutane, and we can derive two more alkyl groups from it.

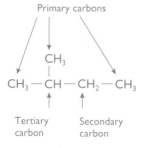

Primary carbons

$$CH_3$$
$$CH_3-CH-CH_2-CH_3$$

Tertiary   Secondary
carbon   carbon

Isobutane → Isobutyl:

$$H-\underset{\underset{H}{|}}{\overset{\overset{\overset{\overset{H}{|}}{H-C-H}}{|}}{C}}... \quad \xrightarrow[\substack{\text{(from any of}\\ \text{the three }CH_3\\ \text{groups)}}]{\text{remove one H}} \quad ... \quad \text{or} \quad CH_3CHCH_2-$$

Isobutane   **Isobutyl**

Isobutane → t-butyl:

$$H-\overset{\overset{H-C-H}{|}}{\underset{\underset{H}{|}}{C}}... \quad \xrightarrow[\substack{\text{(from the}\\ \text{tertiary C}\\ \text{atom)}}]{\text{remove one H}} \quad ... \quad \text{or} \quad CH_3CCH_3$$

Isobutane   *t*-butyl

Notice that the open bonding site in the *tertiary*-butyl group (abbreviated *t*-butyl) occurs at a tertiary carbon.

The names and structures of these alkyl groups must now be learned. If you have access to ball-and-stick models, make models of each of the parent alkanes and then remove hydrogen atoms to generate the open bonding sites and the alkyl groups. Let's now continue with the IUPAC rules for naming alkanes.

6. Attach the name of the alkyl group to the name of the parent as a prefix. Place the location number of the group in front of the resulting name and separate the number from the name by a hyphen. Returning to our original example, its name is 3-methylhexane.

■ Location numbers are called *locants.*

$$CH_3$$
$$CH_3CH_2CHCH_2CH_2CH_3$$
3-Methylhexane

7. When two or more groups are attached to the parent, name each and locate each with a number. The names of alkyl substituents are assembled in their alphabetical order. Always use *hyphens* to separate numbers from words. Here is an application.

$$\text{CH}_3\text{CH}_2 \quad \text{CH}_3$$
$$\underset{7 \quad\;\; 6 \quad\;\; 5 \quad\;\; 4 \quad 3 \quad 2 \quad 1}{\text{CH}_3\text{CH}_2\text{CH}_2\text{CHCH}_2\text{CHCH}_3}$$

4-Ethyl-2-methylheptane

8. When two or more substituents are identical, use such multiplier prefixes as di- (for 2), tri- (for 3), tetra- (for 4), and so forth, and specify the location number of every group. Always separate a number from another number in a name by a *comma*. For example,

$$\text{CH}_3 \quad\;\; \text{CH}_3$$
$$\text{CH}_3\text{CHCH}_2\text{CHCH}_2\text{CH}_3$$

Correct name:	2,4-dimethylhexane
Incorrect names:	2,4-methylhexane
	3,5-dimethylhexane
	2-methyl-4-methylhexane

Groups are alphabetized only by their group names, not by their multiplier prefixes. Thus, "dimethyl" is alphabetized under "m," not under "d".

9. When identical groups are on the same carbon, repeat the number locating this carbon in the name. For example,

$$\text{CH}_3$$
$$\text{CH}_3\text{CCH}_2\text{CH}_2\text{CH}_3$$
$$\text{CH}_3$$

Correct name:	2,2-dimethylpentane
Incorrect names:	2-dimethylpentane
	3,2-methylpentane
	4,4-dimethylpentane

10. To name a cycloalkane, place the prefix *cyclo* before the name of the straight-chain alkane that has the same number of carbon atoms as the ring. This rule was illustrated in Table 10.2.

11. When necessary, give numbers to the ring atoms by giving location 1 to a ring position that holds a substituent. Number around the ring in whichever direction reaches the nearest substituent first. For example,

■ No number is needed when the ring has only one group.

$$\text{CH}_3\text{—}\bigcirc$$

Methylcyclohexane
(*not* 1-methylcyclohexane)

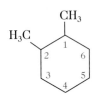

1,2-Dimethylcyclohexane   1,2,4-Trimethylcyclohexane

These are not all of the IUPAC rules for alkanes, but they will handle all of our needs. Study the following examples of correctly named compounds.

$$\text{CH}_3$$
$$\text{CH}_3\text{—}\overset{\displaystyle |}{\underset{\displaystyle |}{\text{C}}}\text{—CH}_3$$
$$\text{CH}_2$$
$$\text{CH}_3$$

2,2-Dimethylbutane
(*not* 2-ethyl-2-methylpropane)

$$\text{CH}_3 \quad \text{CH}_3$$
$$\text{CH}_3\text{—CH}_2 \quad \text{CH}$$
$$\text{CH}_2\text{—CH—CH}_3$$

2,3-Dimethylhexane
(*not* 2-isopropylpentane)

■ Be sure to notice that in choosing the parent chain we sometimes have to go around a corner as the chain zigzags on the page.

$$\text{CH}_3$$
$$\text{CH}_3\text{—}\overset{\displaystyle |}{\underset{\displaystyle |}{\text{C}}}\text{—H}$$
$$\text{CH}_3$$

2-Methylpropane
(*not* 1,1-dimethylethane)

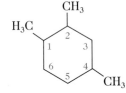

$$\text{CH}_3$$
$$\text{CH}_3\text{CH}_2\text{CH}_2\text{CHCH}_2\text{CHCH}_3$$
$$\text{CH}_3\text{—CH—CH}_3$$

4-Isopropyl-2-methylheptane
(*not* 4-isopropyl-6-methylheptane)

EXAMPLE 10.1   Using the IUPAC Rules to Name an Alkane

What is the IUPAC name for the following compound?

$$\begin{array}{ccc} CH_3 & CH_2CH_2CH_2CH_3 \\ | & | \\ CH_3CHCHCHCHCH_2CH_2CH_3 \\ | & | \\ CH_3 & CH \\ & \diagup \diagdown \\ & CH_3 \ \ CH_3 \end{array}$$

ANALYSIS   The compound is an alkane because it is a hydrocarbon with only single bonds. We must therefore use the IUPAC rules for alkanes.

SOLUTION   The ending to the name must be *-ane*. The next step is to find the longest chain, even if we have to go around corners. This chain is nine carbons long, so the name of the parent alkane is *nonane*. We have to number the chain from left to right, as follows, to reach the first branch with the lower number.

$$\begin{array}{ccc} & \overset{6}{C}H_3 & \overset{7}{C}H_2\overset{8}{C}H_2\overset{9}{C}H_2CH_3 \\ \overset{1}{C}H_3\overset{2}{C}HCHCHCHCH_2CH_2CH_3 \\ & {}^3| \ {}^{4\ 5} \\ CH_3 \ \ CH \\ \diagup \diagdown \\ CH_3 \ \ CH \end{array}$$

At carbons 2 and 3 there are one-carbon methyl groups. At carbon 4, there is a three-carbon isopropyl group (not the propyl group, because the bonding site is at the *middle* carbon of the three-carbon chain). At carbon 5, there is a three-carbon propyl group. (It has to be this particular propyl group because the bonding site is the *end* of the three-carbon chain in the group.) Alphabetically, *isopropyl* comes before *methyl*, which comes before *propyl*, so we must assemble these names as follows to make the final name. (Names of alkyl groups are alphabetized *before* any prefixes such as di- and tri- are affixed.)

4-Isopropyl-2,3-dimethyl-5-propylnonane

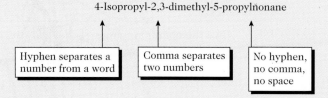

| Hyphen separates a number from a word | Comma separates two numbers | No hyphen, no comma, no space |

CHECK   The most common mistake students make is in not identifying the longest chain. Check your answer to make sure that you have not erred in this. Then be sure you have numbered from the correct end.

■   PRACTICE EXERCISE 2   Write the IUPAC names of the following compounds.

(a)
$$\begin{array}{l} CH_3CH_2 \\ \diagdown \\ CHCH_3 \\ \diagup \\ CH_2CH_2 \\ | \\ CH_3 \end{array}$$

(b)
$$\begin{array}{l} CH_3 \ \ CH_2CH_2CH_3 \\ | \\ \diagdown CHCHCH_2CH_3 \\ \diagup \\ CH_3CH \\ | \\ CH_3 \end{array}$$

(c)
$$\begin{array}{l} CH_3 \ \ CH_3 \ \ CH_3 \\ | \ \ \ \ | \ \ \ \ | \\ CH_3CH_2CHCHCHCH_2CHCH_3 \\ | \\ CH_3CH_2 \end{array}$$

Substituents Other than Alkyl Groups Have IUPAC Names.   When halogen atoms or nitro or amino groups are joined to a carbon of a chain or ring, the following names are used for them in IUPAC nomenclature.

—F	fluoro	—I	iodo
—Cl	chloro	—NO$_2$	nitro
—Br	bromo	—NH$_2$	amino

For example, the IUPAC name of $CH_3CH_2CH_2CHCl_2$ is 1,1-dichlorobutane.

Common Names of Alcohols, Amines, and Haloalkanes Employ the Names of the Alkyl Groups.   For several relatively simple compounds as well as for compounds of great complexity, chemists often use informal names. The so-called "common" names are just that; they are commonly used and yet carry structural information. The following examples of some halogen derivatives of the alkanes, called *haloalkanes*, illustrate how common names are easily constructed by using the names of alkyl groups.

■ Sometimes the name of a straight-chain alkane is preceded by *n-*, as in *n*-butane; the *n* strands for *normal* and is a way of designating the *straight*-chain isomer.

Structure	Common Name	IUPAC Name
CH_3Cl	methyl chloride	chloromethane
CH_3CH_2Br	ethyl bromide	bromoethane
$CH_3CH_2CH_2Br$	propyl bromide	1-bromopropane
CH_3CHCH_3 | Cl	isopropyl chloride	2-chloropropane
$CH_3CH_2CH_2CH_2Cl$	butyl chloride	1-chlorobutane
$CH_3CHCH_2CH_3$ | Br	*sec*-butyl bromide	2-bromobutane
CH_3 | CH_3C—Br | CH_3	*t*-butyl bromide	2-bromo-2-methylpropane

■ **PRACTICE EXERCISE 3**   Give the common names of the following compounds.

(a) $ClCH_2CH_3$ **(b)** $BrCH_2CH_2CH_2CH_3$

(c) CH_3
         |
   CH_3CHCH_2Cl

(d) CH_3
         |
   CH_3CCH_3
         |
         Br

Alkanes and Cycloalkanes Have Very Few Chemical Properties.   The saturated hydrocarbons are the most chemically inert of all organic compounds in our study. Alkanes do burn in air, of course. When sufficient oxygen is available, the reaction, called *combustion,* produces carbon dioxide and water.

■ In insufficient oxygen, combustion also produces carbon monoxide.

No bond in an alkane is polar enough to invite attack by any of the common, ionic (or polar) inorganic acids, bases, oxidizing agents, and reducing agents (at least at room temperature). Alkanes are utterly indifferent to concentrated sulfuric acid, sodium metal, strong alkalies, the permanganate ion, and water. This inertness, in fact, is why the alkanes are called the *paraffins,* from the Latin *parum,* "little," and *affinis,* "affinity." Household paraffin wax, a mixture of alkanes with about 20 carbons per molecule, was once commonly used to seal homemade jellies because of its unreactivity. The chemical stability of alkanes extends to alkyl groups, so we have our second important "map sign." *Alkyl groups are nonfunctional groups.*

Gasoline and diesel fuel are mostly alkanes. When vapors of such fuels mix with strong sunlight, oxygen, and certain oxides of nitrogen, which are present in vehicle exhaust, the ozone in smog is generated (see Special Topic 10.2).

10.3 ALKENES

The IUPAC names of alkenes end in -ene, and the double bond takes precedence over side chains in numbering the parent chain.

The alkene group occurs widely in nature, but usually in molecules that have other functional groups as well. The first four alkenes, like the first four alkanes, are gases at room temperature, and alkenes are generally less dense than water (Table 10.4). Like all hydrocarbons, alkenes are insoluble in water but soluble in nonpolar solvents.

Consistent with the IUPAC rules for any family, the rules for naming the alkenes specify the *name ending,* how to pick out the *parent chain* or *parent ring,* how to *number the chain or ring* and how to designate substituent groups. The IUPAC rules for the open-chain alkenes are as follows.

1. Use the ending -ene for all alkenes.
2. Identify the parent chain as the longest sequence of carbons that *includes the double bond.* Name this chain as if it were that of an alkane and then change the *ane* ending to ene. This gives the basic *name* of the *parent* chain, except that the location of the double bond is yet to be specified. For example, the longest chain *that includes the double bond* in the following structure has six carbons. There is a longer chain of seven carbons, but it does not include the double bond.

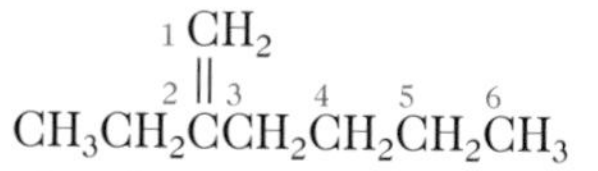

$$\overset{1}{C}H_2$$
$$\underset{2}{\|}\overset{}{}{}_3$$
$$CH_3CH_2\overset{}{C}CH_2CH_2CH_3$$

The parent chain has six carbons, not seven.
The (incomplete) name of the parent chain is hexene.

3. Number the parent chain from whichever end gives the lower number to the first carbon of the double bond. This rule gives precedence to the location of the double bond over the location of the first substituent on the parent chain. For example,

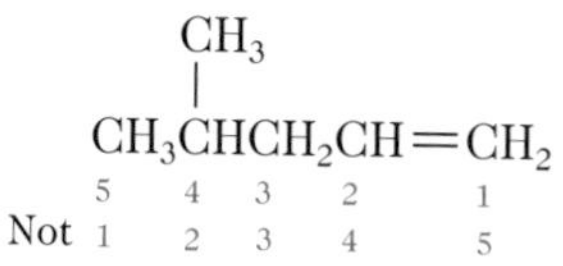

$$\overset{CH_3}{\underset{|}{}}$$
$$CH_3CHCH_2CH{=}CH_2 \qquad \text{The double bond is at position 1, not 4.}$$
$$\underset{5\quad4\quad3\quad2\quad1}{}$$
Not $\underset{1\quad2\quad3\quad4\quad5}{}$

4-Methyl-1-pentene(complete name)
(*not* 2-methyl-4-pentene)

4. To the name begun with rules 1 to 3, place the number that locates the first carbon of the double bond as a prefix, and separate this number from the name by a hyphen.

5. If substituents are on the parent chain, complete the name obtained by rule 4 by placing the names and location numbers of the substituents as prefixes. Remember to separate numbers from numbers by commas, but use hyphens to connect a number to a word.

SPECIAL TOPIC 10.2
OZONE IN SMOG

We have noted how dangerous ozone is to materials, including plants and lung tissue. The U.S. National Ambient[1] Air Quality Standard for ozone is a daily maximum 1-hour average ozone concentration of only 0.12 ppm (120 ppb). Dozens of U.S. urban areas exceed this at least once per year. How does ozone originate in the air where we live? Let's start with the *last* reaction of a multistep process.

The *direct* and only source of ozone in the lower atmosphere is the combination of oxygen atoms with oxygen molecules.

$$O + O_2 + M \longrightarrow O_3 + M \qquad (1)$$

M is any molecule, like N_2 or O_2, that can absorb some of the kinetic energy involved in the collision of O with O_2. The collision that leads to ozone occurs on the surface of M. (Collisions not occurring on a surface like M give ozone molecules that break up instantly because they carry too much energy.) Thus the question, "How does ozone originate?" leads one step back, "How are oxygen atoms generated for Reaction 1?" (Once oxygen atoms are generated, there will be ozone.)

The chief source of oxygen atoms is the breakup of molecules of nitrogen dioxide, an air pollutant, a breakup made possible by solar energy.

$$NO_2 + \text{solar energy} \longrightarrow NO + O \qquad (2)$$

So if we have NO_2 and sunlight, we'll have some oxygen atoms. But where does NO_2 come from (to move back another step)?

Oxides of Nitrogen The NO_2 for equation 2 is made from NO, nitrogen monoxide, which is produced inside vehicle engine cylinders where the direct combination of nitrogen and oxygen is possible because of the high temperature and pressure.

$$N_2 + O_2 \xrightarrow{\text{high temperature}} 2NO \qquad (3)$$

As soon as newly made NO, now in the exhaust gas, hits the cooler outside air, it reacts further with oxygen to give nitrogen dioxide, NO_2. NO_2 gives smog its reddish-brown color.

$$2NO + O_2 \longrightarrow 2NO_2 \qquad (4)$$

There is thus an ample supply of NO_2 in air made smoggy by heavy vehicle traffic.

Ultraviolet Energy Solar energy consists not only of the visible light that we can see but also similar forms of energy, including infrared light ("heat waves") and ultraviolet (UV) light. UV light carries higher energy

---

EXAMPLE 10.2 Naming an Alkene

Write the name of the following alkene.

$$\begin{array}{c} CH_3CCH_3 \\ \| \\ CH_3CHCCH_2CH_2CHCH_3 \\ | \qquad\qquad | \\ CH_3 \qquad\quad CH_3 \end{array}$$

ANALYSIS The longest chain that includes the double bond must be identified and numbered from whichever end gives the first carbon of the double bond the lower of two possible numbers. The parent is an *alkene* with a chain of seven carbons, a heptene. The following numbering of the parent chain gives the double bond position 2. (The alternative numbering, right to left, would have given the double bond position 5.)

$$\begin{array}{c} \overset{1}{C}H_3 \overset{2}{C}CH_3 \\ \| \\ CH_3\overset{}{C}HC\overset{6}{C}H_2\overset{7}{C}H_2CHCH_3 \\ | \;\; \overset{3}{}\;\overset{4}{} \;\;\overset{5}{}\;\; | \\ CH_3 \qquad\quad CH_3 \end{array}$$

■ The location of the double bond takes precedence over any alkyl groups when numbering the chain.

than visible light and can cause grave harm, like severe sunburn, eye burn, and even skin cancer. It takes UV energy to break NO_2 molecules apart by Equation 2.

Virtually all of the dangerous UV radiation coming from the sun is absorbed high in the stratosphere at elevations between 6 and 31 miles (see Special Topic 10.4); however, some UV energy gets to where we live, where it is able to generate ozone in smog-filled air. Here is how.

An interesting feature of Reaction 2 is that it makes not only oxygen atoms but also nitrogen monoxide, NO. One chemical property of nitrogen monoxide might lead us to ask how ozone can develop in smog at all, because nitrogen monoxide is able to *destroy* ozone by Equation 5.

$$NO + O_3 \rightarrow NO_2 + O_2 \qquad (5)$$

If you were to add reactions 1, 2 and 5, canceling identical species on opposite sides of the arrows (try it), *there is no net chemical effect.* How then does ozone form in our atmosphere at all? And it does form, even in clean air.

The answer is that Reactions 1 to 5 proceed at their own *unequal* rates. So in the real world, Reactions 1 to 5 actually do not exactly cancel each other. The net effect is that a small, steady-state concentration of ozone develops even in clean air. The range of ozone concentration in clean air, however, is very low,

between 20 and 50 ppb. In the more polluted urban areas, levels as high as 400 ppb commonly occur for brief periods each year.

Unburned Hydrocarbons and the Ozone in Smog Reaction 5 is the reaction that can make most ozone disappear, *but other substances are able to remove NO from Reaction 5 before it is used to destroy ozone.* Such removal of NO other than by Reaction 5 enables a buildup in the ozone level of the lower atmosphere.

Net destroyers of nitrogen monoxide are themselves present in vehicle exhaust—the unburned or partially oxidized hydrocarbons remaining from the incomplete combustion of the fuel. In sunlight and oxygen, some unburned hydrocarbon molecules are changed into organic derivatives of hydrogen peroxide (which is H_2O_2 or HOOH). These highly reactive organic peroxides (symbolized as ROOH or ROOR) are able to destroy NO molecules (changing them back to NO_2) and so reduce the supply of ozone-destroying NO.

The development of ozone in smog thus depends on hydrocarbon emissions. Specialists generally agree that emissions of hydrocarbons must be significantly reduced before the ozone problem will be appreciably solved. Even the loss of hydrocarbons in the vapors from fuel tanks or at gas pumps has to be reduced.

[1]*Ambient* means "all surrounding, all encompassing."

The parent alkene is thus 2-heptene. It holds two methyl groups (positions 2 and 6) and one isopropyl group (position 3). The names and location numbers are next assembled into the name.

SOLUTION   The correct name is

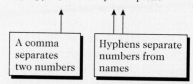

3-Isopropyl-2,6-dimethyl-2-heptene

| A comma separates two numbers | Hyphens separate numbers from names |

CHECK   The most common error that students make is to *fail to find the longest chain that includes the double bond.* The first check step, then, is to go back over the answer to see if there is a chain holding the alkene group that is longer than seven carbons. Use a colored pen to draw an enclosure for this chain so that all substituents are outside. Then move in from either end of the chain, counting carbons, to see which starting point yields the lower number for the beginning of the double bond. All the other numbers must then fall into place. Another common error is failure to identify alkyl groups correctly, so double-check these.

TABLE 10.4   Properties of 1-Alkenes

Name (IUPAC)	Structure	BP (°C)	MP (°C)	Density (g/mL 10 °C)
Ethene	$CH_2\!=\!CH_2$	−104	−169	—
Propene	$CH_2\!=\!CHCH_3$	−48	−185	—
1-Butene	$CH_2\!=\!CHCH_2CH_3$	−6	−185	—
1-Pentene	$CH_2\!=\!CHCH_2CH_2CH_3$	30	−165	0.641
1-Hexene	$CH_2\!=\!CHCH_2CH_2CH_2CH_3$	64	−140	0.673
1-Heptene	$CH_2\!=\!CHCH_2CH_2CH_2CH_3$	94	−119	0.697
1-Octene	$CH_2\!=\!CHCH_2CH_2CH_2CH_2CH_3$	121	−102	0.715
1-Nonene	$CH_2\!=\!CHCH_2CH_2CH_2CH_2CH_2CH_3$	147	−81	0.729
1-Decene	$CH_2\!=\!CHCH_2CH_2CH_2CH_2CH_2CH_2CH_3$	171	−66	0.741
Cyclopentene		44	−135	0.722
Cyclohexene		83	−104	0.811

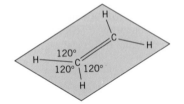

FIGURE 10.3
Geometry at a carbon–carbon double bond.

■ The *side* of a double bond is not the same as the *end* of a double bond.

Same side

One end

■ PRACTICE EXERCISE 4   Write the IUPAC names for the following compounds.

(a) H_3C CH_3
$$C$$
$$\|$$
$$CH_2$$

(b) CH_3 CH_3
$$CH_3CHCH_2CCH_2CHCH_3$$
$$\|$$
$$CH_3CCH_2CH_3$$

(c) $CH_3CH\!=\!CHCl$

(d) $BrCH_2CH\!=\!CH_2$

The Lack of Free Rotation at a Double Bond Causes *Geometric Isomerism*.   The six atoms at a double bond, the two carbons and the four atoms attached to them, all lie in the same plane and all of the bond angles are 120° (Figure 10.3). Unlike the freedom of groups to rotate about a single bond, there is no free rotation about a double bond. When the carbons of the double bond are suitably substituted, restricted rotation makes possible a kind of isomerism new to our study. If we replace a hydrogen at each end of the double bond in ethene by a methyl group, for example, the result is 2-butene, but we can attach the methyl groups in two ways. In one, the methyl groups are on the same side of the double bond, and in the other they are on opposite sides, *and we cannot twist one to make it equivalent to the other.* Thus, there are two geometric forms of 2-butene, and they are truly different compounds.

CH_3 CH_3
$$C\!=\!C$$
H H
cis-2-Butene
(BP 3.7 °C)

H CH_3
$$C\!=\!C$$
CH_3 H
trans-2-Butene
(BP 0.9 °C)

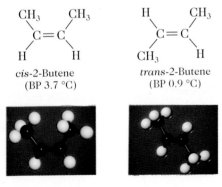

Structurally, *cis*-2-butene and *trans*-2-butene differ only in the *directions* taken by their end-of-chain methyl groups. Isomers that differ in geometry but have identical constitutions, are called **geometric isomers,** and the phenomenon is called **geometric isomerism.**

When two designated substituents are on the same side of the double bond, they are described as *cis* to each other. When they are on opposite sides, they are *trans* to each other. (Sometimes geometric isomerism is called *cis–trans isomerism.*) The designations *cis* and *trans* are made parts of the names of geometric isomers, as in the names *cis*-2-butene and *trans*-2-butene.

When Two *Identical* Groups Are at the Same End of a Double Bond, Geometric Isomers Are Not Possible.   If one end of a double bond has two *identical* groups, like two hydrogens or two methyls, then there is nothing for a group at the other end to be *uniquely* cis or trans to. 1-Butene, for example, has two H atoms at one end of its double bond, so the ethyl group at the other end cannot be positioned to give geometric isomers. We can *write* structures that might appear to be isomers:

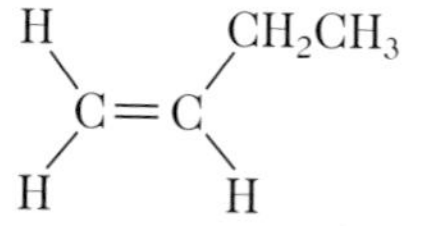

1-Butene	is the same as	1-Butene

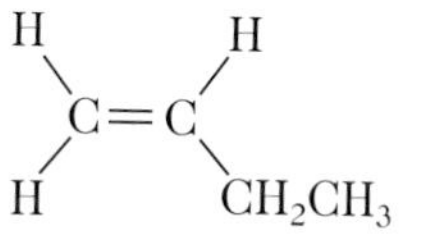

However, they are actually identical. If you mentally flop the *whole* first structure over, top to bottom, you get the second. Whole-molecule flopping, of course, does not reorganize bonds into any new structure or geometry. Thus, there are no geometric isomers of 1-butene.

EXAMPLE 10.3   Writing the Structures of Cis and Trans Isomers

---

Write the structures of the cis and trans isomers, if any, of 2-pentene.

$$CH_3CH\!=\!CHCH_2CH_3$$
2-Pentene

ANALYSIS   Notice first that geometric isomerism is possible in 2-pentene. At *neither* end of the double bond are the two groups identical. At one end there are H and CH_3; at the other end, H and CH_2CH_3. Therefore, we must draw structures of the geometric isomers.

To show the geometry of each isomer correctly, we start by writing a carbon–carbon double bond without any attached groups, *spreading the single bonds at the carbon atoms at angles of about 120°.*

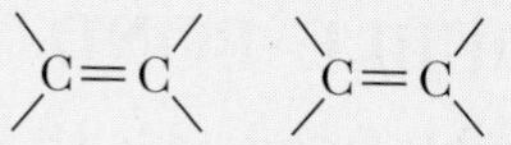

$$\diagup\!\!\!\diagdown C\!=\!C\diagup\!\!\!\diagdown \qquad \diagup\!\!\!\diagdown C\!=\!C\diagup\!\!\!\diagdown$$

Then we attach the two groups that are at one of the ends of the double bond. We attach them *identically* to make identical partial structures.

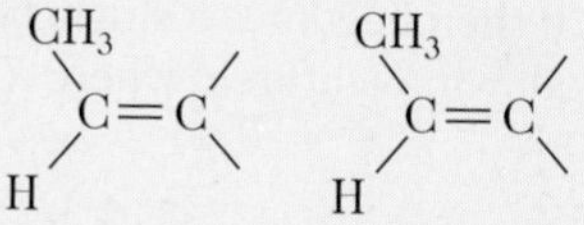

$$\overset{CH_3}{\underset{H}{\diagup}}C\!=\!C\diagup \qquad \overset{CH_3}{\underset{H}{\diagup}}C\!=\!C\diagup$$

Finally, at the other end of the double bond, we draw the other two groups, only this time making sure to switch them in their relative positions.

SOLUTION The geometric isomers of 2-pentene are

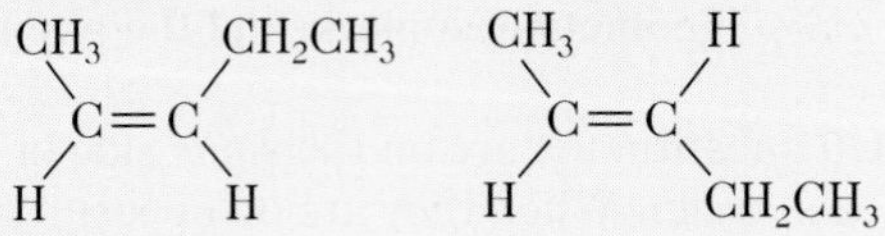

$$CH_3 \quad CH_2CH_3 \qquad CH_3 \quad H$$
$$\underset{H}{\overset{}{C}}=\underset{H}{\overset{}{C}} \qquad \underset{H}{\overset{}{C}}=\underset{CH_2CH_3}{\overset{}{C}}$$

cis-2-Pentene   trans-2-Pentene

CHECK Be sure to check whether the two structures are geometric *isomers* and not two identical structures that are merely flip-flopped on the page.

■   PRACTICE EXERCISE 5   Write the structures of the cis and trans isomers, if any, of the following compounds.

(a) $CH_3CH_2\underset{\underset{CH_3}{|}}{C}=CHCH_3$ (b) $ClCH=CHCl$ (c) $CH_3\underset{\underset{CH_3}{|}}{C}=CH_2$ (d) $Cl\underset{\underset{Cl}{|}}{C}=CHBr$

Cyclic Compounds Can Also Have Geometric Isomers.   The double bond is not the only source of restricted rotation; the ring is another. For example, two geometric isomers of 1,2-dimethylcyclopropane are known, and neither can be twisted into the other without breaking the ring open. This costs too much energy to occur spontaneously even at quite high temperatures.

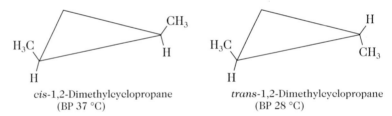

cis-1,2-Dimethylcyclopropane   trans-1,2-Dimethylcyclopropane
(BP 37 °C)   (BP 28 °C)

■   As we have said before, molecular geometry is as important in living processes as functional groups.

We'll see this kind of cis–trans isomerism in the many cyclic structures of carbohydrates; their *geometric* differences alone make most carbohydrates unusable in human nutrition.

10.4   CHEMICAL REACTIONS OF THE CARBON– CARBON DOUBLE BOND

The carbon–carbon double bond adds H₂ and H₂O.

In an **addition reaction,** pieces of an adding molecule become attached to opposite ends of the double bond which then becomes a single bond. All additions to an alkene double bond thus have the following features, where X—Y is the adding molecule:

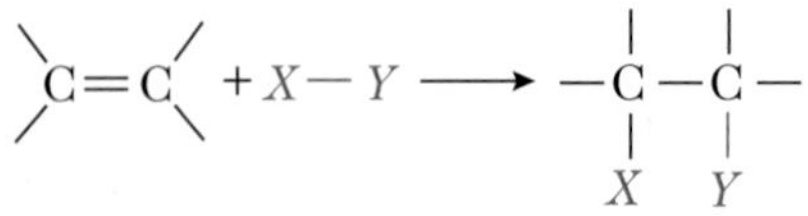

$$\overset{}{\underset{}{C}}=\overset{}{\underset{}{C} + X-Y \longrightarrow -\overset{|}{\underset{|}{C}}-\overset{|}{\underset{|}{C}}-}$$
$$\qquad\qquad\qquad\qquad X \quad Y$$

■   Chlorine and bromine readily add to an alkene group. For example:

$CH_2=CH_2 + Br_2 \rightarrow BrCH_2CH_2Br$ We'll study the additions of hydrogen and water.

Hydrogen Adds to a Double Bond and Saturates It. In the presence of a powdered metal catalyst, like powdered nickel or platinum metal, hydrogen adds to double bonds. The reaction, sometimes called *hydrogenation,* converts an alkene to an alkane as follows.

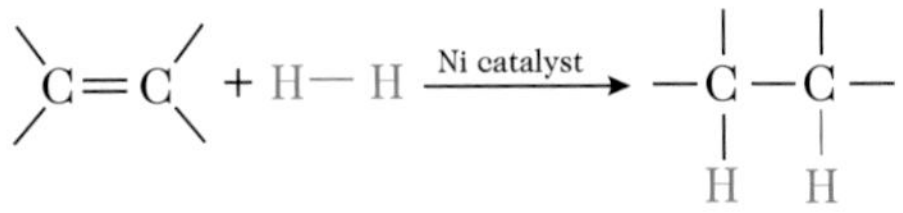

$$\text{\textbackslash}C=C\text{/} + H—H \xrightarrow{\text{Ni catalyst}} —\underset{\underset{H}{|}}{\overset{|}{C}}—\underset{\underset{H}{|}}{\overset{|}{C}}—$$

In the following specific examples be sure to notice how the carbon skeletons of the alkene and the product alkane are identical. The *only* change is the increase of one H in the number of H atoms held by *each* carbon of the original double bond.

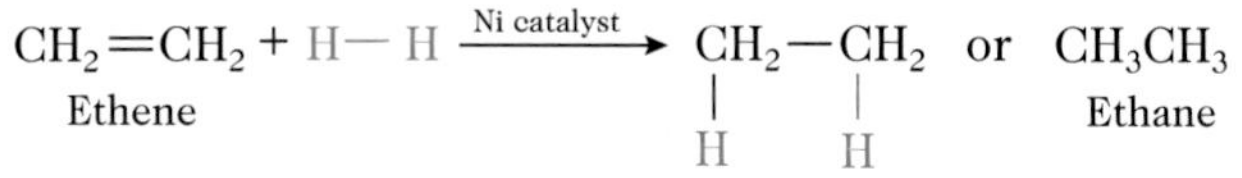

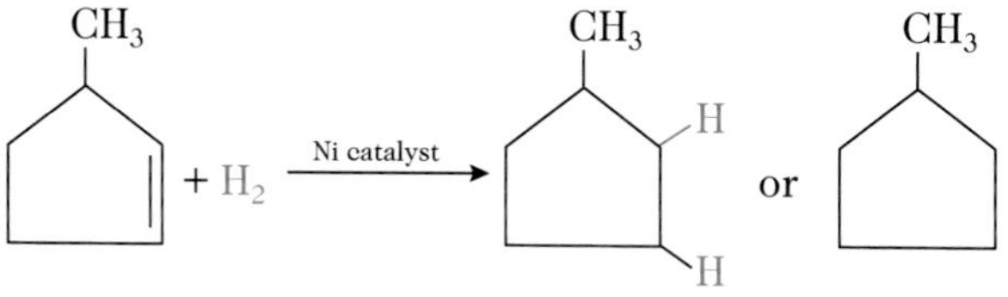

$$CH_2{=}CH_2 + H—H \xrightarrow{\text{Ni catalyst}} \underset{\underset{H}{|}}{CH_2}—\underset{\underset{H}{|}}{CH_2} \quad \text{or} \quad CH_3CH_3$$
Ethene   Ethane

3-Methylcyclopentene   Methylcyclopentane

We have just studied a chemical "map sign" for the carbon–carbon double bond that must be learned: *the alkene group can be made to add hydrogen and change to a single bond as each of its carbon atoms picks up one hydrogen atom.* The best way to learn this *general* property is to apply it to *specific* examples.

■ **PRACTICE EXERCISE 6** Write the structures of the products, if any, of the following.

(a) $CH_3CH{=}CH_2 + H_2 \xrightarrow{\text{Ni catalyst}}$

(b) $CH_3CH_2CH_3 + H_2 \xrightarrow{\text{Ni catalyst}}$

(c)

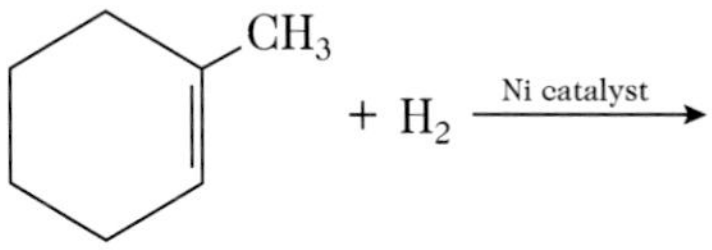

(structure of methylcyclohexene) $+ H_2 \xrightarrow{\text{Ni catalyst}}$

(d) $CH_3(CH_2)_7CH{=}CH(CH_2)_7CO_2H + H_2 \xrightarrow{\text{Ni catalyst}}$

The Net Effect of Hydrogenation Occurs at the Molecular Level of Life. Molecules of H_2 and powdered metal catalysts, of course, are unavailable in the body. Cells, however, have molecular carrier systems that deliver the pieces of a molecule of H—H to carbon–carbon double bonds. One piece is $H:^-$, donated by a carrier enzyme to one end of the double bond. The other piece is H^+, either donated by the carrier or plucked from a proton donor of the surrounding buffer and given to the carbon that was at the other end of the double bond. The *net* effect is the addition of H_2 because $H:^-$ and H^+ together add up to one $H:H$ molecule.

Water Adds to Double Bonds to Give Alcohols. Water adds to the carbon–carbon double bond provided that an acid catalyst (or the appropriate enzyme) is present. The product is an alcohol. Water alone or aqueous bases have no effect on alkenes whatsoever.

SPECIAL TOPIC 10.3
HOW AN ACID CATALYZES THE ADDITION OF WATER TO A DOUBLE BOND

Water does not react with an alkene at all in the absence of an acid catalyst. Remember that the water molecule has unshared electrons on its oxygen atom and so is an electron-rich species. The double bond with *two* shared pairs of electrons is also electron-rich. Thus, the oxygen part of the water molecule is actually repelled by the double bond. Two electron-rich sites, like two like charges, repel. Yet, somehow, a new C—O bond (in the final alcohol product) manages to form. An acid catalyst makes this possible.

The first step in the mechanism of the reaction is a proton transfer *from the acid catalyst.*

$$CH_2{=}CH_2 + \ :\overset{H}{\underset{H}{O}}{}^{+}{-}H \xrightarrow[\text{transfer}]{\text{Proton}} \underset{+}{CH_2}{-}\underset{|}{\underset{H}{CH_2}} + H_2O$$

Ethyl carbocation

The proton is an electron-poor species and so is actually attracted to the double bond where it can form a strong bond to carbon, stronger than its bond to oxygen in the hydronium ion. The product of the proton transfer is a species new to our study, a *carbocation.* In it, a carbon atom is surrounded by only three pairs of electrons, not an octet, and it bears a positive charge. Now a water molecule would be attracted to the electron-poor carbon.

$$\underset{+}{CH_2CH_3} + \ :\overset{H}{\underset{H}{O}}{} \longrightarrow \ H{-}\overset{CH_2CH_3}{\underset{H}{\overset{|}{\underset{+}{O}}}}$$

Now the positive charge is on oxygen, but this is acceptable in this instance because oxygen still has an outer octet. In fact, every atom has a noble gas configuration.

In the last step, the catalyst, H_3O^+, is recovered by another proton transfer, this time from the ethyl-substituted hydronium ion.

$$CH_2CH_3 + \ :\overset{H}{O}{} \longrightarrow CH_2CH_3 + H{-}\overset{H}{\underset{+}{O}}{}$$

An ethyl-substituted hydronium ion　　Ethyl alcohol　　Recovered catalyst

You can see that the function of the acid catalyst is to change a functional group unattractive to water, the alkene group, into one that attracts water, the carbocation. Once the new C—OH bond forms, the catalyst's work is over.

■ Here is another important "map sign": *an alkene group can add water to give an alcohol group.*

$$\underset{\text{Alkene}}{\overset{\diagdown}{\diagup}C{=}C\overset{\diagup}{\diagdown}} + H{-}OH \xrightarrow[\text{heat}]{H^+} \underset{\text{Alcohol}}{-\overset{|}{\underset{|}{C}}{-}\overset{|}{\underset{|}{C}}{-}}$$

The following are specific examples of the reaction.

■ In the body, enzymes catalyze the addition of water to alkene double bonds.

$$\underset{\text{Ethene}}{CH_2{=}CH_2} + H{-}OH \xrightarrow[\substack{240\ °C \\ \text{(closed vessel)}}]{10\%\ H_2SO_4} \underset{\text{Ethyl alcohol}}{CH_3CH_2OH}$$

■ See Special Topic 10.3 for a description of *how* this reaction happens.

$$\underset{\substack{| \\ CH_3 \\ \text{2-Methylpropene}}}{CH_3C{=}CH_2} + H{-}OH \xrightarrow[25\ °C]{10\%\ H_2SO_4} \underset{\substack{| \\ CH_3 \\ t\text{-Butyl alcohol}}}{CH_3\overset{OH}{\overset{|}{C}}CH_3} \left(Not\ \underset{\substack{| \\ CH_3 \\ \text{(isobutyl alcohol)}}}{CH_3CHCH_2OH}\right)$$

■ Vladimer Markovnikov (1838–1904) was a Russian chemist.

Markovnikov's Rule Tells Us Which Isomer Predominates. Although water can add in two possible ways to 2-methylpropene, essentially only one route is followed, and *t*-butyl alcohol is the chief product, not its isomer isobutyl alcohol. This behavior is general and is summarized by **Markovnikov's rule.**

MARKOVNIKOV'S RULE

When an unsymmetrical reactant of the type H—G adds to an unsymmetrical alkene, the carbon with the greater number of hydrogens gets one more H.

The following examples illustrate Markovnikov's rule.

$$CH_3CH{=}CH_2 + H{-}OH \xrightarrow{\;H^+\;} CH_3CHCH_3 \quad (not\ CH_3CH_2CH_2OH)$$
$$\qquad\qquad\qquad\qquad\qquad\qquad\quad |$$
$$\qquad\qquad\qquad\qquad\qquad\qquad\;OH$$

Propene   Isopropyl alcohol   (propyl alcohol)

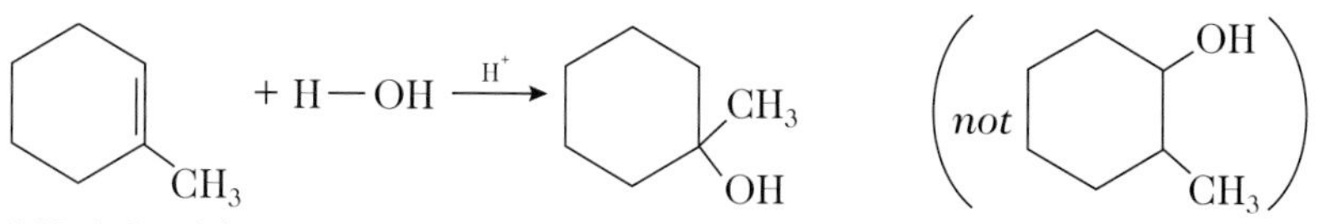

1-Methylcyclohexene   1-Methylcyclohexanol   (2-Methylcyclohexanol)

■ *Concentrated* sulfuric acid adds easily to double bonds in accordance with Markovnikov's rule. For example (where we let $HOSO_3H$ represent sulfuric acid, H_2SO_4):

$$CH_3CH{=}CH_2 + HOSO_3H$$
$$\xrightarrow{\quad} CH_3CHCH_3$$
$$\qquad\qquad\;\; |$$
$$\qquad\qquad OSO_3H$$

EXAMPLE 10.4   Using Markovnikov's Rule

What product forms in the following situation?

$$CH_3CH_2CH{=}CH_2 + H{-}OH \xrightarrow[\text{heat}]{\;H^+\;}$$

ANALYSIS   As in all of the addition reactions that we are studying, the carbon skeleton of the alkene can be copied over intact, except that a single bond is shown where the double bond was.

$$CH_3CH_2CH{-}CH_2 \quad \text{(incomplete)}$$

To decide which carbon of the original double bond gets the H atom from the water molecule, we use Markovnikov's rule. The H atom has to go to the CH_2 end because it has the greater number of H's than the carbon at the other end of the double bond. The latter receives the OH unit from H—OH.

SOLUTION   The product is

$$CH_3CH_2CH{-}CH_2 \quad \text{or} \quad CH_3CH_2CHCH_3$$
$$\qquad\quad |\qquad\; |\qquad\qquad\qquad\qquad |$$
$$\qquad\;\;OH\quad H\qquad\qquad\qquad\;OH$$

sec-Butyl alcohol
(2-butanol)

CHECK   At this stage be sure to see whether each carbon in the product has four bonds. If not, you can be certain that some mistake has been made. This is always a useful way to avoid at least some of the common mistakes made in solving a problem such as this. Then double-check that the carbon of the double bond initially having the greater number of H's has been given one more.

■ PRACTICE EXERCISE 7   Write structures for the product(s), if any, that would form under the conditions shown. If no reaction occurs, write "no reaction."

(a) $CH_2{=}CHCH_2CH_3 + H_2O \xrightarrow{\;H^+\;}$

(b) $CH_3C{=}CH_2 + H_2O \xrightarrow{\;H^+\;}$
$\quad\quad\quad |$
$\quad\quad\; CH_2CH_3$

(c)

$$CH_3CH{=}C{-}\bigcirc + H_2O \xrightarrow{\;H^+\;}$$
$$\quad\quad\quad\; |$$
$$\quad\quad\; CH_3$$

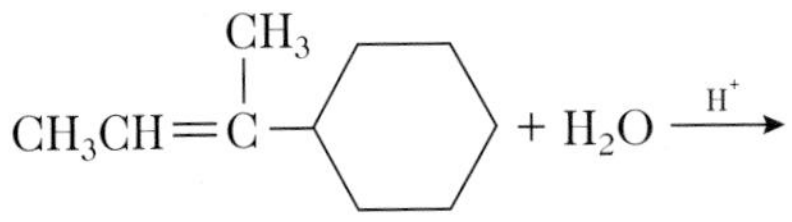

We have looked at only a few of the reactions of the alkene group. Among those omitted are the reactions of this group with strong oxidizing agents, like ozone (O_3) and potassium permanganate ($KMnO_4$). Molecules with alkene groups can be torn apart at their double bonds by these reactants. This makes ozone particularly dangerous to all forms of life because the double bond is so prevalent in cell membranes. How ozone in the stratosphere is *desirable* is explained in Special Topic 10.4, which also discusses the "ozone hole."

10.5   POLYMERIZATION OF ALKENES

Hundreds to thousands of alkene molecules can join together to make one large molecule of a polymer.

Macromolecules Abound in Nature.   A **macromolecule** is simply a distinct molecule with a formula mass in the thousands. Out of all of the many known macromolecular substances there are some, the *polymers,* with a unique structural feature. A **polymer** is a substance consisting of macromolecules *all of which have repeating structural units,* up to many thousands. One of the carbohydrates that we will study, starch, is a *polysaccharide* whose molecules have the following system, where Gl stands for a molecular unit made from a glucose molecule (the details of which we'll leave to Chapter 14).

■ "Polymer" has Greek roots: *poly,* "many," and *meros,* "parts."

etc.—Gl—O—Gl—O—Gl—O—Gl—O—Gl—O—Gl—O—Gl—O—Gl—O—etc.
Section of a polysaccharide

This example illustrates one feature of all polymers, namely, they have *repeating* structural units. Thus, the study of polymers focuses on the origins of these units and how they are joined together. Let's learn the rudiments of this chemistry by studying an easier system.

Polyethylene Is a Simple but Commercially Important Polymer.   Under a variety of conditions, many hundreds of ethylene molecules can reorganize their bonds, join together, and change into one large molecule.

$$n CH_2{=}CH_2 \xrightarrow[\text{trace of } O_2]{\text{heat, pressure}} {-}(CH_2{-}CH_2)_n \quad\quad (n = \text{a large number})$$

Ethylene   Polyethylene
                              (repeating unit)

The *repeating unit* in polyethylene is $CH_2{-}CH_2$, and one after another of these units are joined together into an extremely long chain. Following common practices, we have represented the structure of polyethylene simply by enclosing its repeating unit in parentheses.

SPECIAL TOPIC 10.4
THE OZONE SHIELD

One of the ironies of life on earth is that ozone is an enemy down here but a friend up there, in the stratosphere. In this Special Topic we'll seek to understand why, as well as why chemical events occurring where virtually no one lives—the Antarctic—might pose a threat where people do live.

Stratosphere The *stratosphere* is an envelope of space surrounding the earth between altitudes of 10 and 50 km (6 and 31 miles) and immediately above the region of the atmosphere where we live. The stratosphere contains ultra-small concentrations of several gases, including oxygen, nitrogen, methane, nitrogen dioxide, ozone, and some pollutants. Ozone is a *natural* component of the stratosphere.

Ozone Ozone is O_3, triatomic oxygen. It is an extremely reactive substance, attacking, oxidizing, and splitting apart any molecules with carbon–carbon multiple bonds. A level in air of only 1 μg/L is very dangerous to health. Smog contains some ozone (usually much less than 1 μg/L), and this is but one reason why breathing smoggy air is unhealthy. Stratospheric ozone destroys ultraviolet radiation coming from the sun, radiation to which all living things are vulnerable because it can break covalent bonds in anything.

Ultraviolet Radiation Light, technically called *electromagnetic radiation,* is considered to be a stream of tiny particles having no mass and called *photons,* each photon being a packet of energy. Traditionally, to describe a particular kind of photon, an associated wavelength, symbolized as λ (Greek lambda), is given. We need the term *wavelength* because it's used to define various kinds of light, but we won't develop the concept in detail. What we need to know is that *the energy of a photon is inversely proportional to wavelength.* The *smaller* the value of λ, the *more* energy the photon has.

Ultraviolet photons have wavelengths ranging from 200 nm to nearly 400 nm (where nm means nanometer, 10^{-9} meter). UV photons, as we said, have enough energy to break chemical bonds. The photons of visible light have higher values of wavelength, ranging from about 380 nm to nearly 800 nm. With longer wavelengths, visible light carries less energy per photon than UV light and generally does not break bonds.

Roughly 7% of the solar energy *entering* the *outer* stratosphere, is carried by UV photons. This percentage may seem small, but plant and animal life, including human life, would be almost certainly impossible without a natural mechanism in the stratosphere that removes all incoming solar UV radiation below 290 nm and nearly all below 320 nm. Because it was once

FIGURE 1
Prolonged exposure to sunlight contributes to deep wrinkles.

thought that the UV light nearest the visible region in wavelength was harmless, the UV region above 290 nm was further separated into two adjacent ranges labeled UVA (320–400 nm)—once thought to be harmless—and UVB (290–320 nm). Both kinds are absorbed by the skin, and neither is blocked by glass. Although the ozone in the stratosphere destroys most UV radiation, sunlight reaching us at (or not far from) sea level, includes some of both UVA and UVB, principally that in the region 280 to 315 nm (peaking at 300 nm).

Sunburn, Skin Cancer, and UV Radiation We now know that both UVA light and UVB light carry harm, although UVA poses less harm. In short exposures, UVB causes sunburn and, in longer exposures, the severe, long-term consequences of skin cancer. Photons in the UVA region, as noted, are more abundant than those of UVB and, when excessively absorbed by the skin, also cause sunburn and premature skin hardening and wrinkling (Figure 1). These effects are significantly accelerated in people who smoke. Long-term exposure to UVA also contributes to the onset of various forms of skin cancer. You can no doubt imagine how dangerous it would be if the stratosphere did not screen out as much UV radiation as it does. *What we must try to prevent is anything that would weaken the ability of the stratosphere to perform its task.*

Skin creams are available with ingredients that absorb UVA. Prescription sunglasses and photochromic glasses (they darken in the sun and lighten again in less light) absorb 100% of UVB and more than 90% of UVA. Cosmetic sunglasses—you use them "around town"—take out at least 70% of UVB and 20% of UVA.

The Ozone Cycle and the Ozone Shield A series of chemical reactions called the *stratospheric ozone cycle* converts nearly all incoming UV radiation into heat. The cycle depends for its *initiation* on the formation of oxygen *atoms* from ordinary oxygen (O_2), which is naturally present in the stratosphere. UV photons have sufficient energy to break oxygen molecules into oxygen atoms.

$$O_2 \xrightarrow{\text{UV radiation}} 2O^* \qquad (1)$$

The asterisk is used because the oxygen atoms form in electronically excited states, meaning that they have at least one electron in an energy state higher than normal.

Ozone is made when O^* combines with an O_2 molecule during a collision at the surface of a small particle, M. M can be a molecule of O_2 or N_2.

$$O^* + O_2 + M \rightarrow O_3 + M + \text{heat} \qquad (2)$$

The function of M is to absorb some of the energy of the collision between O and O_2 so that the new O_3 molecule has insufficient energy within itself to break apart as soon as it forms. We want something else to split O_3.

The newly formed ozone is split by the absorption of UV radiation. (Wavelengths from 240 to 320 nm work best.)

$$O_3 + \text{UV energy} \xrightarrow{(\lambda = 240-320 \text{ nm})} O_2 + O^* \quad (3)$$

Notice that the O^* *product* of reaction 3 is a necessary *reactant* for the previous reaction, 2. Reaction 3, therefore, not only absorbs UV photons; *it also supplies what is needed for another occurrence of reaction 2.* So 2 can occur again, making another Reaction 3 possible. Thus Reactions 2 and 3 keep going, together constituting a *chemical chain reaction,* meaning that a reactant for the first step is made by the second step.

The repetitions of Reactions 2 and 3 constitute the *stratospheric ozone cycle,* and a given cycle repeats itself hundreds of times before other events stop it. Notice that if we add Equations 2 and 3, canceling identical species on opposite sides of the arrows, the net result is simply

$$\text{UV energy} \rightarrow \text{heat} \qquad (4)$$

Thus, the net effect of the ozone cycle in the stratosphere is the conversion of UV radiation into heat. Countless individual cycles are occurring all of the time in what we call the *stratospheric ozone shield.* Individual cycles come and go, so *there is a somewhat steady-state concentration or "level" of ozone in the stratosphere.* We take decreases in the ozone level as something of a "litmus test," warning that the stratosphere is becoming less efficient at removing UV radiation on our behalf. Such loss of efficiency, scientists agree, will not only increase the incidence of skin cancer in people; it will also have adverse effects on agriculture.

The stratospheric ozone level varies rather widely with latitude, altitude, and the month of the year, so no single value can be given to it. At latitudes reaching 60° north or south from the equator, the ozone level in 1974 varied roughly between 260 and 360 Dobson units, reaching as high as 500 Dobson units. The *Dobson unit* equals 2.7×10^{16} molecules of O_3 when in a column of air 1 cm^2 in area at its base (the earth's surface) and extending through the stratosphere.

We now ask: "What has occurred in our time, with all good intentions, to pose an avoidable threat to the ozone layer?" What happened is that we began using compounds called chlorofluorocarbons CFCs) as heat-exchange agents in refrigerators and air conditioners, as foaming agents for foamed plastics (e.g., the famous polystyrene coffee cup), and as propellants in aerosol cans. Before we study why the CFCs are a problem, we must say again that natural events also cause the stratospheric ozone level to vary. Volcanic eruptions, for example, often inject enormous tonnages of substances into the stratosphere that enter into stratospheric chemistry. We can't control volcanoes; however, we can do something about the CFCs.

Chlorofluorocarbons and the Ozone Cycle The CFCs are a family of volatile, nonflammable, chemically stable, and essentially odorless and tasteless compounds, properties that made them ideal for their uses. One is called CFC-11 and is $CFCl_3$, boiling at 24 °C. Another is CFC-12, which is CCl_2F_2, boiling at -30 °C. At one time, $CFCl_3$ was used in 50 to 60% of all aerosol cans sold. When refrigerators, air conditioning units, and foamed plastics were junked, to say nothing of the direct releases of aerosols, the CFCs eventually escaped.

The CFCs turned out to be too stable for our own good. Nothing in our lower atmosphere degrades them, and eventually they migrate throughout the entire atmosphere, working their way into the stratosphere. In 1974, chemists M. J. Molina and F. S. Rowland (California) warned that the CFCs could reduce our protection from UV radiation by interfering with the stratospheric ozone cycle. According to Molina and Rowland, the CFCs in the stratosphere *also absorb UV radiation,* which breaks their C—Cl bonds and generates chlorine *atoms.*

$$CCl_3F \xrightarrow{\text{UV radiation}} CCl_2F + Cl$$

$$CCl_2F_2 \xrightarrow{\text{UV radiation}} CClF_2 + Cl$$

Atomic chlorine is able to destroy ozone and so disrupt the ozone cycle by the following series of reactions.

$$Cl + O_3 \rightarrow ClO + O_2 \qquad (5)$$

$$ClO + O \rightarrow Cl + O_2 \qquad (6)$$

$$\text{Net: } O_3 + O \rightarrow 2O_2$$

Notice that this is also a chain reaction, *so the breakup of only one CFC molecule can initiate the destruction of thousands of ozone molecules and ozone-generating chains.*

Because of their inertness, CFCs endure for several decades. Their influence on the stratospheric ozone level will continue throughout the next century, therefore, even if no more are released. The threat is serious, so in 1987, 36 nations signed a treaty called the Montreal Protocol that called for cutting the worldwide CFC production in half by 1998. (This goal was reached in 1993.) In 1989, more deeply alarmed by additional research, 80 nations agreed to ban CFC production by 2000.

The Antarctic Ozone "Hole" The loss of stratospheric ozone is particularly pronounced in a huge column of air over the South Pole during the Antarctic spring, reaching its lowest levels in October each year. (Spring in the Southern Hemisphere occurs during the fall of the Northern Hemisphere.) This ozone-poor column of air is called the *Antarctic ozone hole.* In October of 1993, the ozone level in the Antarctic ozone hole dropped to an all-time low, 90 Dobson units. (For that time of the year, the normal value is 225 Dobson units.) As spring turns to summer in the Southern Hemisphere between November and March, a significant recovery in the ozone level occurs, only to go into a another decline in the following spring.

The ozone hole affects the stratospheric ozone levels elsewhere because of global, atmospheric mixing processes. These bring air richer in ozone into the Antarctic stratosphere so that, overall, the stratospheric ozone level all over the globe has been steadily declining at a rate of more than 2% per decade. This is why what is happening in the Antarctic can affect public health where people live.

Why Over the Antarctic? A circulating wind pattern called the *Antarctic vortex* develops over the South Pole and the Antarctic continent during the Antarctic winter (Figure 2). It is generated by the rotation of the earth and the sharply uneven heating of the atmosphere. Wind velocities as high as 80 m/s (180 mph) occur in the jet stream encircling the South Pole, but very little wind is in the center of the vortex. These winds help to confine the chemical reactions that destroy ozone largely to this vortex.

Between November and March, the polar vortex breaks up and air richer in ozone migrates into the Antarctic stratosphere from midlatitudes. The result, as we indicated, is an overall decrease in the stratospheric ozone level over the Southern Hemisphere. Similar events are occurring over the North Pole. The stratospheric ozone level is believed to have declined

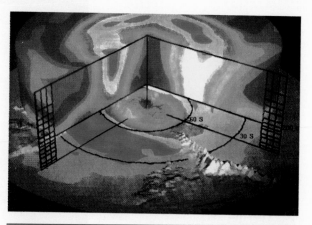

FIGURE 2

The Antarctic polar vortex on October 7, 1989. The cylinder with the large missing wedge is color-coded for wind speeds from darkest blue (no wind) to brightest red (80 m/s). The earth surface beneath the cylinder is color-coded for total ozone in the column above it, ranging from red (highest ozone level) to dark blue (below 200 Dobson units) immediately over the South Pole. Extending to the lower right is the Andes chain of mountains of South America. The tip of Africa is outlined at the lower left. Reprinted by permission from M. R. Schoeberl and D. L. Hartmann, *Science,* Jan. 4, 1991, page 47.)

by about 3% during the last three decades at the latitude of New York City and by about 5% at the latitude of Buenos Aires, Argentina. The U.S. Environmental Protection Agency issued a prediction in 1991 that the increased human exposure to UV radiation would cause an *additional* 200,000 deaths from skin cancer in the United States over the next 50 years.

We've left one question unanswered: "Why during the Antarctic *spring*? The answer is complex, and we'll largely have to leave it to other references. It involves the accumulation of chlorine compounds on the surfaces of microcrystals of nitric acid trihydrate (of all things). These crystals largely make up the thin stratospheric clouds that form when temperatures drop during the sunless Antarctic winter. Then, when the sun reappears in the Antarctic spring, UV photons reinvade the space just as the frozen chlorine compounds are set free. Thus, the generation of chlorine atoms rather suddenly resumes, and the stratospheric ozone level takes another nose dive.

Two References

1. Pamela Zurer, "Ozone Depletion's Recurring Surprises Challenge Atmospheric Scientists," *Chemical & Engineering News,* May 24, 1993, page 8.
3. Jeannette Anders and Eileen Leach, "Sun versus Skin," *American Journal of Nursing,* July 1983, page 1015.

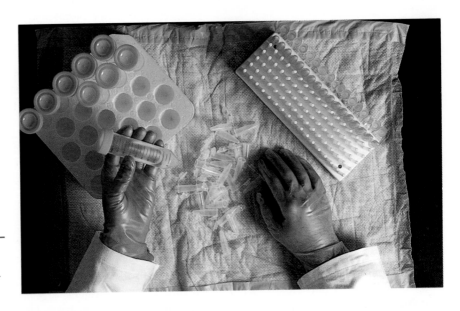

FIGURE 10.4
Polypropylene is used to make many items used in clinics and hospitals.

TABLE 10.5 Polymers of Substituted Alkenes[a]

Polymer[b]	Monomer	Uses
Polyvinyl chloride (PVC) $-(CH_2CH)_n-$ $\|$ Cl	$CH_2{=}CHCl$ Vinyl chloride	Insulation, credit cards, bottles, plastic pipe
Saran $Cl \quad Cl$ $\| \quad \|$ $-(CH_2CCH_2CH)_n-$ $\|$ Cl	$CH_2{=}CCl_2$ Vinylidene chloride and $CH_2{=}CHCl$ Vinyl chloride	Packaging film, fibers, tubing
Teflon $-(CF_2CF_2)_n-$	$F_2C{=}CF_2$ Tetrafluoroethylene	Nonstick surfaces, valves
Orlon $C{\equiv}N$ $\|$ $-(CH_2CH)_n-$	$CH_2{=}CH-C{\equiv}N$ Acrylonitrile	Fabrics
Polystyrene $-(CHCH_2)_n-$ $\|$ C_6H_5	Styrene	Foamed items, insulation
Lucite CH_3 $\|$ $-(CH_2C)_n-$ $\|$ CO_2CH_3	CH_3 $\|$ $CH_2{=}C-CO_2CH_3$ Methyl methacrylate	Windows, coatings, molded items
Natural Polymer		
Rubber CH_3 $\|$ $-(CH_2C{=}CHCH_2)_n-$	CH_3 $\|$ $CH_2{=}C-CH{=}CH_2$ Isoprene	Tires, hoses, boots

[a]The common names rather than the IUPAC names of the monomers are given.

[b]The condensed structures are idealized; the actual polymer consists of molecules similar to them, but with many variations in chain lengths and branchings.

The starting material for making a polymer is a compound called a **monomer,** and the reaction in which a monomer changes to a polymer is called **polymerization.** Because alkenes are nicknamed *olefins,* the polymers of alkenes are usually called *polyolefins* (pol-y-**ol**-uh-fins).

Acid promoters also cause alkenes to polymerize, and we might more easily understand polymerization with reference to them. The reorganization of bonds is initiated when one molecule of ethylene (or other alkene) accepts H^+ from the acid promoter. This opens up one end of ethylene to accepting an electron pair from another molecule of ethylene, a bond-relocating event that now occurs over and over again to build up the long chain in polyethylene.

■ Because the acid "catalyst" is consumed, it isn't really a true catalyst and so is called a *promoter.*

$$G-H + CH_2=CH_2 + CH_2=CH_2 + CH_2=CH_2 + \text{etc.} \longrightarrow H-(CH_2-CH_2)_{\overline{n}}\,G$$

Acid
promoter   Ethylene   Polyethylene

Chain-branching reactions also occur during the formation of polyethylene, so the final product includes both straight- and branched-chain molecules. The molecules in a sample of a commercial polymer like polyethylene are never exact copies of each other. Their chain lengths vary, and the extent of branching varies, but it is still convenient to represent the polymer by showing its most characteristic repeating unit.

Like ethylene, propylene (propene) can also be polymerized, and polypropylene has taken over many uses of polyethylene in both household and hospital applications (Figure 10.4). Some indoor–outdoor carpeting is woven of polypropylene fibers.

Because polypropylene and polyethylene are fundamentally alkanes, they have all of the chemical inertness of this family. These polymers are therefore popular for making containers for food juices and medical fluids; refrigerator boxes and bottles; containers for chemicals; sutures; catheters; various drains; and wrappings for aneurysms.

Table 10.5 gives a few examples of polymers, nicknamed *vinyl polymers,* for which the monomers are substituted alkenes. Dienes also polymerize, and natural rubber is a polymer of a diene called isoprene.

■ The condensed structure of polypropylene is

$$\underset{\displaystyle -(CH_2-\overset{\displaystyle CH_3}{\overset{\displaystyle |}{CH}})_{\overline{n}}}{}$$

■ Proteins, DNA, cellulose, and starch are all polymers.

10.6 AROMATIC COMPOUNDS

The benzene ring undergoes substitution reactions instead of addition reactions despite a high degree of unsaturation.

Historically, the structure of benzene was a problem. Its formula, C_6H_6, indicates that it is quite unsaturated, because the ratio of hydrogens to carbons is much lower than in hexane, C_6H_{14}, or in cyclohexane, C_6H_{12}. But benzene gives essentially none of the reactions of alkenes.

The Benzene Molecule Is Planar and All Its Hydrogens Are Equivalent. The six carbons of benzene are in a ring, and each carbon holds one hydrogen. All six hydrogens are chemically equivalent, meaning that it is possible, for example, to replace any one H by Cl to make the *same* chlorobenzene, C_6H_5Cl. Structure **1,** the basic organization of the atoms in the benzene molecule, is consistent with this fact.

■ The older structure, **2,** is still widely used to represent benzene, although it is usually abbreviated further:

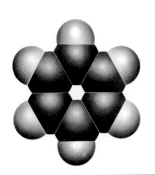

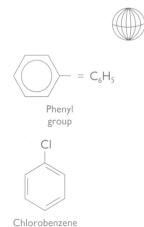

FIGURE 10.5
Scale model of a molecule of benzene.

= C_6H_5

Phenyl group

Chlorobenzene

■ The addition of Cl_2 to $CH_2{=}CH_2$ gives $ClCH_2CH_2Cl$.

1	2	3
(incomplete)	(older structure for benzene)	(new structure for benzene)

The trouble with structure **1** is that each carbon has only three bonds, not four. For decades, scientists simply made each carbon have four bonds by writing in three double bonds, as seen in structure **2**. But **2** says that benzene is a *triene,* a substance with three alkene groups per molecule. Trienes are known compounds, and they have the same kinds of reactions as mono-enes (e.g., ethene) and dienes—addition reactions. But benzene doesn't give addition reactions. Benzene, for example, doesn't add water like an alkene.

Because the three double bonds indicated in structure **2** are misleading in a chemical sense, scientists today often represent benzene simply by a hexagon with a circle inside, structure **3**.

The ring, often referred to as the *benzene ring,* is planar. All six carbons and all six hydrogen atoms lie in the same plane, and all of the bond angles are 120°, as seen in the scale model of Figure 10.5.

The Benzene Ring Resists Addition Reactions.   The typical reactions of benzene are **substitution** reactions, not addition reactions. In a substitution, one hydrogen atom is replaced by another atom or group. Benzene reacts with chlorine or bromine, for example, but only in the presence of iron or an iron(III) chloride catalyst, to give chlorobenzene and HCl. In this equation (and others to come), C_6H_6 stands for benzene, and C_6H_5 represents the **phenyl** group, which is the benzene molecule less one H.

$$C_6H_6 + Cl_2 \xrightarrow[\text{FeCl}_3]{\text{Fe or}} C_6H_5Cl + HCl$$
$$\text{Benzene} \qquad\qquad \text{Chlorobenzene}$$

$$C_6H_6 + Br_2 \xrightarrow[\text{FeBr}_3]{\text{Fe or}} C_6H_5Br + HBr$$
$$\qquad\qquad\qquad\qquad \text{Bromobenzene}$$

Although it was mentioned only in a margin comment (page 226), bromine and chlorine both react readily with alkenes at room temperature, without extra catalysts, by *addition* not substitution. Clearly the benzene ring is in a class by itself.

Aromatic Compounds Are Any with Benzene-like Rings.   Any compounds with planar, highly unsaturated rings, like benzene, that *give substitution reactions at the ring instead of addition reactions* are now classified as **aromatic compounds**. This term is a holdover from the days when most of the known compounds of benzene actually had aromatic fragrances, but the odor test no longer applies. A few particularly important aromatic compounds related to benzene are listed in Table 10.6

Many heterocyclic compounds also have unsaturated, benzene-like rings and give substitution instead of addition reactions. They, too, are classified as aromatic. All other compounds, as we noted earlier, are **aliphatic compounds**.

TABLE 10.6   Important Aromatic Compounds

Name	Structure	Uses
Toluene	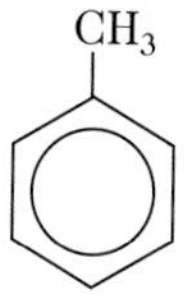 CH$_3$ (benzene ring)	Solvent; raw material for making other aromatic compounds
Phenol	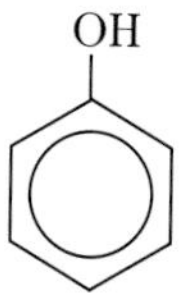 OH (benzene ring)	Bactericide (Lister's original antiseptic): raw material for making aspirin
Aniline	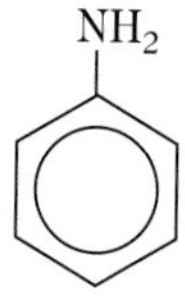 NH$_2$ (benzene ring)	Manufacture of aniline dyes and many pharmaceuticals
Benzoic acid	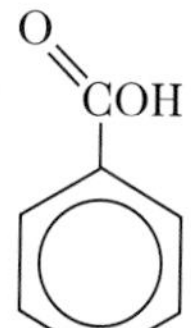 COOH (benzene ring)	In some ointments to soften the skin; raw material for manufacture of other aromatic compounds

Why Benzene Resists Addition Reactions.   Although structure **2** for benzene isn't right in a chemical sense, it does correctly indicate that we must account for three shared pairs of electrons, those of the second bonds of the three double bonds in **2**. None of these pairs, however, is restricted to the space between two ring atoms, as structure **2** suggests. All three pairs move in a space that encompasses the entire ring. Thus, they actually exist in more space in the benzene ring than most shared electrons. The circle in structure **3** is meant to indicate this feature.

Giving electrons more space means that their mutual repulsions are not as great, and this is what stabilizes the system. If something added just to one of the "double bonds" of structure **2**, this closed circuit, ring-encompassing space would be destroyed, because then one "double bond" would become a single bond. The powerful resistance of the benzene ring to addition reactions stems from the extra stability that three pairs of electrons find in being able to exist in a larger space than usual.

SUMMARY

Hydrocarbons   The carbon frameworks of hydrocarbon molecules can be straight chains, branched chains, or rings. When only single bonds occur, the substance is said to be saturated; otherwise it is unsaturated. Regardless of how much we condense a structure, each carbon atom must always have four bonds. Most single bonds need not be shown, but carbon–carbon double and triple bonds are always indicated by two or three lines.

Being nonpolar compounds, the hydrocarbons are all insoluble in water, and many mixtures of alkanes are common, nonpolar solvents. The rule *like dissolves like* lets us predict solubilities.

Nomenclature of Alkanes   In the IUPAC system, a compound's family is always indicated in the name of a compound by the ending; the number of carbons in the parent chain is shown by a prefix; and the locations of side chains or groups are specified by numbers assigned to the carbons of the parent chain. Alkane-like substituents are called alkyl groups, and the names and formulas of those having from one to four carbon atoms must be learned. Common names are still popular, particularly when the IUPAC names are long and cumbersome to use in conversation.

Chemical Properties of Alkanes Alkanes and cycloalkanes are generally unreactive at room temperature toward concentrated acids and bases, toward oxidizing and reducing agents, toward even the most reactive metals, and toward water. They burn, giving off carbon dioxide and water.

Alkenes The lack of free rotation at a double bond makes geometric (cis–trans) isomers possible, but they exist only when the two groups at *either* end of the double bond are not identical. Cyclic compounds also exhibit cis–trans isomerism.

Alkenes and cycloalkenes are given IUPAC names by a set of rules very similar to those used to name their corresponding saturated forms. However, the double bond takes precedence both in selecting and in numbering the main chain (or ring). The first unsaturated carbon encountered in moving down the chain or around the ring through the double bond must have the lower number.

Addition Reactions Hydrogen and water can be made to add to carbon–carbon double bonds. The addition of hydrogen gives alkanes. The addition of water, which follows Markovnikov's rule, gives alcohols.

Polymerization of Alkenes The polymerization of an alkene is like an addition reaction. The alkene serves as the monomer, and one alkene molecule adds to another and so on until a long chain with a repeating unit forms—the polymer molecule.

Aromatic Properties When aromatic compounds undergo reactions at the benzene ring, substitutions rather than additions occur. In this way, the closed-circuit electron network of the ring remains unbroken.

Reactions of Unsaturated Hydrocarbons

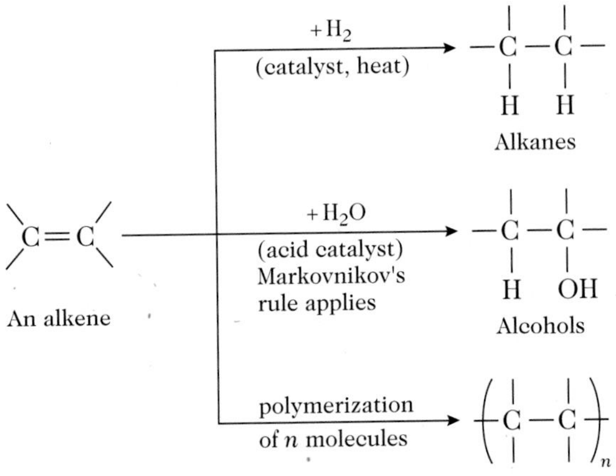

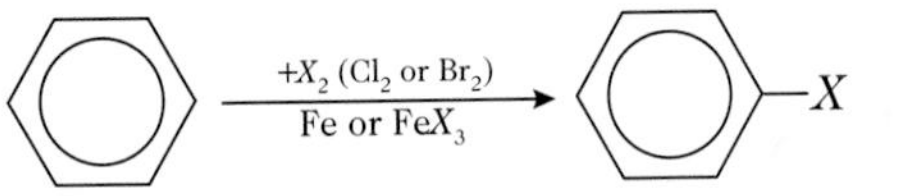

REVIEW EXERCISES

The answers to Review Exercises whose numbers are in color are found in Appendix V. The answers to the other Review Exercises are found in the Study Guide that accompanies this text. The more challenging questions are marked with asterisks.

Types of Compounds

10.1 Name the family of hydrocarbons to which each of the following compounds belongs.

(a)

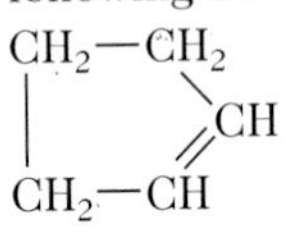

$$
\begin{array}{c}
CH_2{-}CH_2 \\
| \quad\quad \searrow CH \\
CH_2{-}CH \quad \nearrow
\end{array}
$$

(b)

$$
\begin{array}{c}
CH_2{-}CH_2 \\
| \quad\quad\quad \searrow CH{-}CH_3 \\
CH_2{-}CH_2 \quad \nearrow
\end{array}
$$

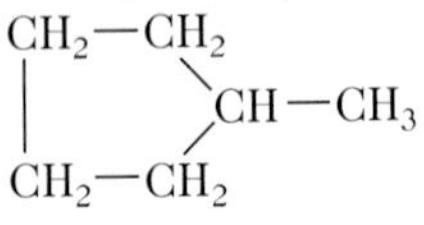

(c) $CH_3CH_2CH_2CH_3$ (d) $HC{\equiv}CCH_2CH_3$

10.2 In what family of hydrocarbons are the following compounds?

(a)

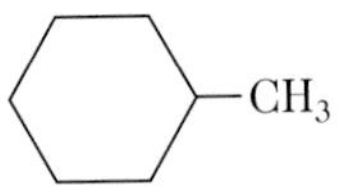

(b)

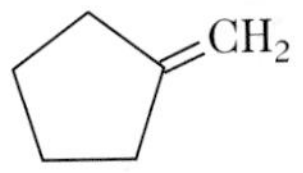

(c) $CH_2{=}CCH_2CH_3$ (d) $CH_3C{\equiv}CCHCH_3$
$\quad\quad\quad\;\; | \qquad\qquad\qquad\qquad\quad |$
$\quad\quad\quad CH_3 \qquad\qquad\qquad\qquad\; CH_3$

10.3 Which of the following compounds are saturated compounds?

(a)
$$
\begin{array}{c}
O \\
\parallel \\
CH_3CCH_3
\end{array}
$$
Acetone

(b)
$$
\begin{array}{c}
OH \\
| \\
CH_3CHCH_3
\end{array}
$$
Isopropyl alcohol

(c) $CH_3CH_2OCH_2CH_3$ (d) $CH_3CH_2CH_2NH_2$
Diethyl ether   Propylamine

10.4 Which of the following compounds are unsaturated?

(a)

Cyclopentadiene

(b)

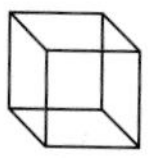

Cubane

(c)

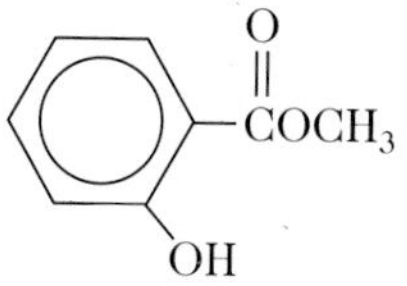

Oil of wintergreen

(d)

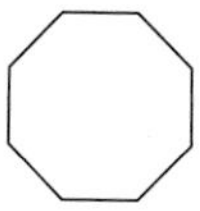

Cyclooctane

Structures for Cyclic Compounds

10.5 Expand the following structure of nicotinamide, one of the B-vitamins.

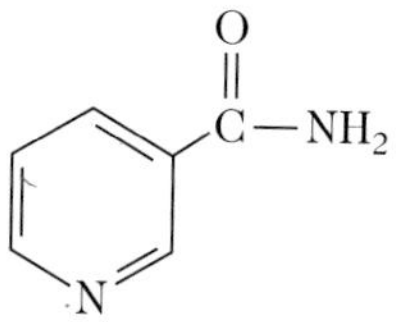

Nicotinamide

10.6 Expand the structure of thiamine, vitamin B_1. Notice that one nitrogen has four bonds so it has a positive charge.

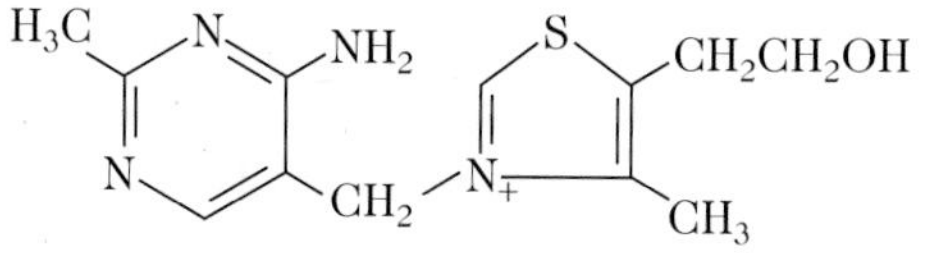

Thiamine

Physical Properties and Structure

10.7 Which compound must have the higher boiling point? Explain.

$$CH_3CH_2CH_3 \qquad CH_3CH_2CH_2CH_2CH_3$$
$$\text{A} \qquad\qquad\qquad \text{B}$$

10.8 Which compound must be less soluble in gasoline? Explain.

$$ClCH_2CH_2CH_2CH_2CH_2Cl \qquad HOCH_2CH_2CH_2CH_2CH_2OH$$
$$\text{A} \qquad\qquad\qquad\qquad \text{B}$$

10.9 Suppose that you are handed two test tubes containing colorless liquids, and you are told that one contains hexane and the other holds dilute sulfuric acid. How can you use just water to tell which tube contains which compound without carrying out any chemical reaction?

10.10 Suppose that you are given two test tubes and are told that one holds ethyl alcohol, CH_3CH_2OH, and the other decane. How can water be used to tell these substances apart without carrying out any chemical reaction?

Nomenclature of Alkanes

10.11 There are five isomers of C_6H_{14}. Write their condensed structures and their IUPAC names.

*10.12 There are nine isomers of C_7H_{16}. Write their condensed structures and their IUPAC names.

*10.13 Write the IUPAC names of the following compounds.

(a)

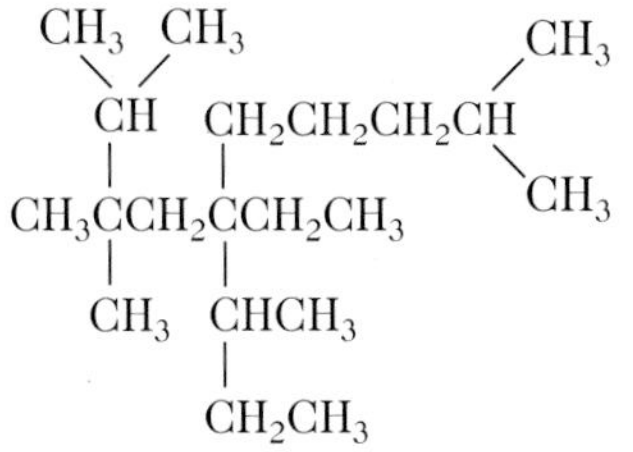

(b)

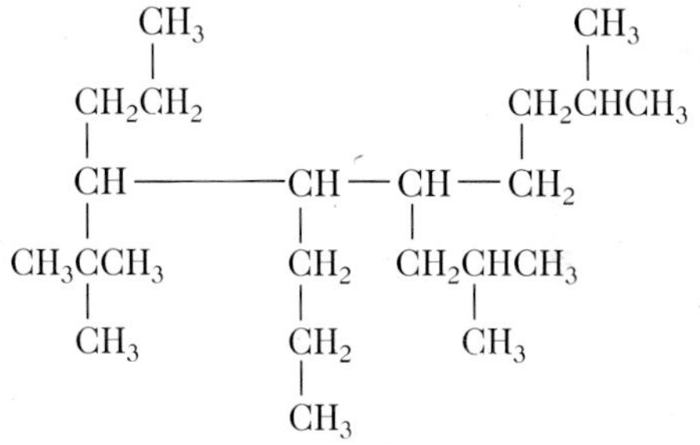

10.14 Write condensed structures for the following compounds.
(a) butyl bromide   (b) methyl iodide
(c) sec-butyl chloride   (d) isopropylcyclohexane

10.15 Write condensed structures for the following compounds.
(a) propyl chloride   (b) isobutyl iodide
(c) t-butyl bromide   (d) ethyl bromide

*10.16 The following are incorrect efforts at naming certain compounds. What are the most likely condensed structures and correct IUPAC names?
(a) propyl chloride   (b) isobutyl iodide
(c) t-butyl bromide   (d) ethyl bromide

10.17 The following names cannot be the correct names, but it is still possible to write structures from them. What are the correct IUPAC names and the condensed structures?
(a) 1-chloroisobutane
(b) 2,4-dichlorocyclopentane
(c) 2-ethylbutane
(d) 1,3-6-trimethylcyclohexane

Reactions of Alkanes

10.18 Write the balanced equation for the complete combustion of heptane, a component of gasoline.

10.19 Gasohol is a mixture of ethyl alcohol, CH_3CH_2OH, in gasoline. Write the equation for the complete combustion of ethyl alcohol.

10.20 What reaction, if any, does heptane give with the following reactants?
(a) H_2O (b) concd H_2SO_4 (c) $NaOH(aq)$ (d) $HCl(aq)$

Cis–Trans Isomerism

10.21 Which of the following pairs of structures represent identical compounds or isomers?

(a) 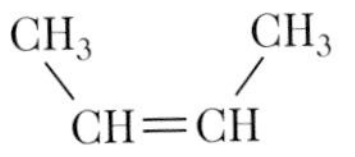
$$CH_3 \quad CH_3$$
$$\diagdown \qquad \diagup$$
$$CH=CH$$
and
$$CH=CH$$
$$\diagup \qquad \diagdown$$
$$CH_3 \qquad CH_3$$

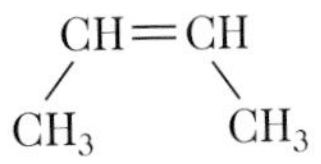

(b) 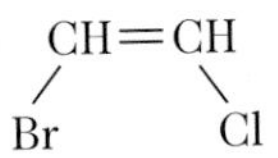
$$CH=CH$$
$$\diagup \qquad \diagdown$$
$$Br \qquad Cl$$
and
$$CH=CH$$
$$\diagup \qquad \diagdown$$
$$Cl \qquad Br$$

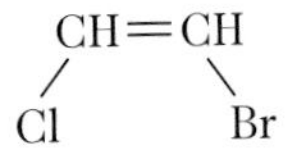

(c) 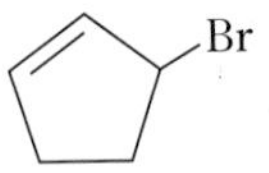 (cyclopentene with Br) and (cyclopentene with Br)

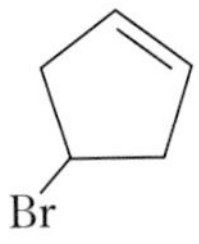

(d) 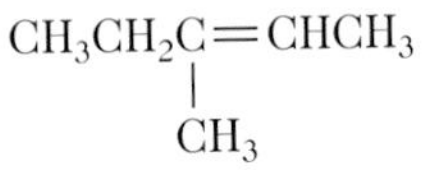 $CH_3CH_2C=CHCH_3$ with CH_3 and $CH_3CH=CCH_2CH_3$ with CH_3

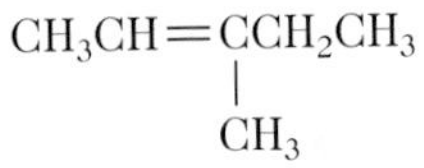

(e) 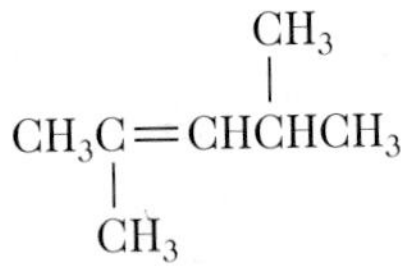
$$CH_3$$
$$|$$
$$CH_3C=CHCHCH_3$$
$$|$$
$$CH_3$$
and
$$CH_3$$
$$|$$
$$CHCHCH_3$$
$$||$$
$$CH_3CCH_3$$

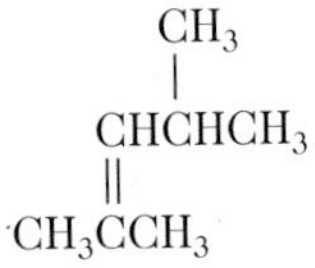

10.22 Study the following structures to discover which are able to exhibit cis–trans isomerism. For those that do, write the structures of the cis and trans isomers.

(a) $CH_3CH=CHCH_2CH_3$

(b) 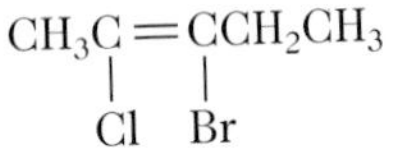
$$CH_3C=CCH_2CH_3$$
$$| \quad |$$
$$Cl \quad Br$$

(c) 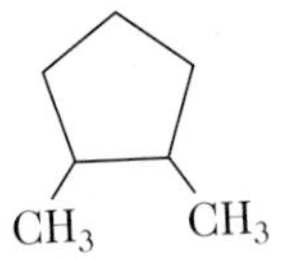 (cyclopentane with two CH₃ groups)

(d) 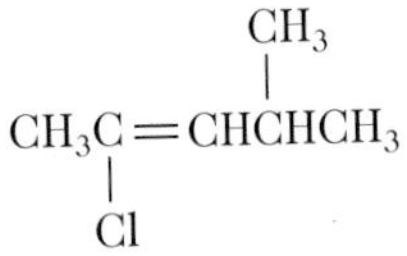
$$CH_3$$
$$|$$
$$CH_3C=CHCHCH_3$$
$$|$$
$$Cl$$

10.23 Identify which of the following compounds can exist as cis and trans isomers and write the structures of these isomers.

(a) $FBrC=CHCl$

(b) 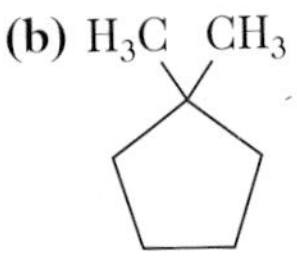 $H_3C \quad CH_3$ (cyclopentane)

(c) $CH_3CH=CHCH=CH_2$

Names of Alkenes

10.24 Write the condensed structures of the following compounds.
(a) 3-methylcyclohexene (b) 1,6-dimethylcyclohexene
(c) *trans*-2-hexene (d) 3-bromo-2-pentene

10.25 Write the IUPAC names of the following compounds.
(a) $CH_2=CH(CH_2)_5CH_3$

(b) 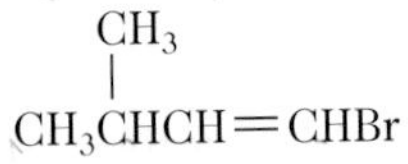
$$CH_3$$
$$|$$
$$CH_3CHCH=CHBr$$

(c) 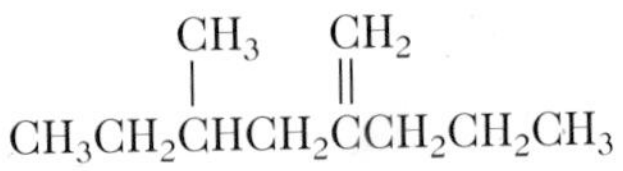
$$CH_3 \quad CH_2$$
$$| \qquad ||$$
$$CH_3CH_2CHCH_2CCH_2CH_2CH_3$$

(d) 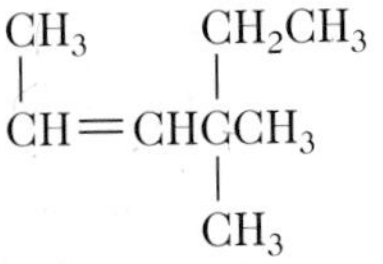
$$CH_3 \quad CH_2CH_3$$
$$| \qquad |$$
$$CH=CHCCH_3$$
$$|$$
$$CH_3$$

*10.26 Write the condensed structures and the IUPAC names for all the isomeric pentenes, C_5H_{10}. Include cis and trans isomers.

Reactions of the Carbon–Carbon Double Bond

10.27 Write equations for the reactions of 2-methylpropene with the following reactants.
(a) Hydrogen, in the presence of a nickel catalyst
(b) Water, in the presence of an acid catalyst

10.28 Write equations for the reactions of 1-methylcyclopentene with the reactants listed in Review Exercise 10.27.

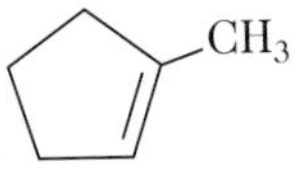

(cyclopentene ring with CH₃) 1-Methylcyclopentene

10.29 Write equations for the reactions of 2-methyl-2-butene with the compounds given in Review Exercise 10.27.

10.30 Write equations for the reactions of 1-methylcyclohexene with the reactants listed in Review Exercise 10.27. (Do not attempt to predict whether cis or trans isomers form.)

*10.31 One of the raw materials for the synthesis of nylon, adipic acid, can be made from cyclohexene by oxidation using potassium permanganate. The balanced equation for the first step in which $K_2C_6H_8O_4$, the potassium salt of adipic acid, forms is

$$3C_6H_{10} + 8KMnO_4 \rightarrow 3K_2C_6H_8O_4 + 8MnO_2 + 2KOH + 2H_2O$$

How many grams of potassium permanganate are needed for the oxidation of 11.2 g of cyclohexene, assuming that the reaction occurs exactly and entirely as written?

*10.32 Referring to Review Exercise 10.31, how many grams of the potassium salt of adipic acid can be made if 21.0 g of $KMnO_4$ is used in accordance with the equation given?

Polymerization

10.33 Rubber cement can be made by mixing some polymerized 2-methylpropene with a solvent such as toluene. When the solvent evaporates, a very tacky and sticky residue of the polymer (called polyisobutylene) remains, which soon hardens and becomes the glue. The structure is quite regular, like polypropylene. Write the structure of the polymer of 2-methylpropene in two ways.
(a) One that shows four repeating units, one after the other
(b) The condensed structure

10.34 Polyvinyl acetate is a soft adhesive that is modified (by reactions that we have yet to study) into a material (Butvar) that bonds two glass sheets together in safety glass. Thus when safety glass breaks, the broken pieces cannot fly around. Using four vinyl acetate units, write part of the structure of a molecule of polyvinyl acetate. Also write its condensed structure. The structure of vinyl acetate is

$$
\begin{array}{c}
\text{O} \\
\parallel \\
\text{OCCH}_3 \\
| \\
\text{CH}_2\!=\!\text{CH} \quad \text{Vinyl acetate}
\end{array}
$$

Aromatic Properties

10.35 Dipentene has a very pleasant, lemon-like fragrance, but it is not classified as an aromatic compound. Why?

$$
\text{CH}_2\!=\!\underset{\underset{\text{CH}_3}{|}}{\text{C}}\!-\!\!\!\bigcirc\!\!\!-\text{CH}_3
$$

Dipentene

10.36 Sulfanilamide, one of the sulfa drugs, has no odor at all, but it is still classified as an aromatic compound. Explain.

$$
\text{NH}_2\!-\!\!\!\bigcirc\!\!\!-\underset{\underset{\text{O}}{\parallel}}{\overset{\overset{\text{O}}{\parallel}}{\text{S}}}\!\text{NH}_2
$$

Sulfanilamide

10.37 Write equations for the reactions, if any, of benzene with the following compounds.
(a) Chlorine (FeCl$_3$ present)
(b) Hot sodium hydroxide solution
(c) Hydrochloric acid

10.38 Explain why benzene strongly resists addition reactions and gives substitution reactions instead.

Predicting Reactions

10.39 Write the structures of the products to be expected in the following situations. If no reaction is to be expected, write "no reaction." To work this kind of exercise, you have to be able to do three things.

1. *Classify* a specific organic reactant into its proper family. Do this first.
2. *Recall* the short list of chemical facts about the family. (If there is no matchup between this list and the reactants and conditions specified by a given problem, assume that there is no reaction.)

3. *Apply* the recalled chemical fact, which might be some "map sign" associated with a functional group, to the specific situation.

Study the next two examples before continuing.

EXAMPLE 10.5   Predicting Reactions

What is the product, if any, of the following?

$$\text{CH}_3\text{CH}_2\text{CH}_2\text{CH}_3 + \text{H}_2\text{SO}_4 \rightarrow ?$$

ANALYSIS We note first that the organic reactant is an alkane, so we next turn to the list of chemical properties about all alkanes that we learned. With this family, of course, the list is very short. Except for combustion, we have learned no reactions for alkanes, and we therefore, assume, that there aren't any others, not even with sulfuric acid. Hence, the answer is "no reaction." (It should be said that there are a few reactions of alkanes that we have ignored.)

EXAMPLE 10.6   Predicting Reactions

What is the product, if any, in the following situation?

$$\text{CH}_3\text{CH}\!=\!\text{CHCH}_3 + \text{H}_2\text{O} \xrightarrow{\text{acid catalyst}} ?$$

ANALYSIS We first note that the organic reactant is an alkene, so we review our mental "file" of reactions of the carbon–carbon double bond.

1. Alkenes add hydrogen (in the presence of a metal catalyst) to form alkanes.
2. They add water in the presence of an acid catalyst to give alcohols.
3. They polymerize.

These are the chief chemical facts, the principal "map signs," about the carbon–carbon double bond that we have studied, and we see that the list includes a reaction with water in the presence of an acid catalyst. We remember that in all addition reactions the double bond changes to a single bond and the pieces of the adding molecule end up on the carbons at ends of the double bond. We also have to remember Markovnikov's rule to tell us which pieces of the water molecule go to each carbon. In this specific example, however, Markovnikov's rule does not apply because the alkene is symmetrical.

SOLUTION

$$CH_3CH_2CHCH_3$$
$$\mid$$
$$OH$$

Now work the following parts. (Remember that C_6H_6 stands for benzene and that C_6H_5 is the phenyl group.)

(a) $CH_3CH_2CH=CHCH_2CH_3 + H_2O \xrightarrow{\text{acid catalyst}}$

(b)
$$CH_3$$
$$\mid$$
$$CH_3CHCH=CH_2 + H_2 \xrightarrow{\text{Ni catalyst}}$$

(c) $C_6H_6 + Br_2 \xrightarrow{FeBr_3}$

(d) $CH_3CH_2CH_2CH_2CH_2CH_3 + H_2SO_4(\text{concd}) \rightarrow$

(e) $+ H_2O \xrightarrow{\text{acid catalyst}}$

(f) $CH_2=CHCH_2CH_2CH_3 + H_2 \xrightarrow{\text{Ni catalyst}}$

(g) $CH_2CH_3 + H_2SO_4(\text{concd}) \longrightarrow$

(h) $C_6H_5CH=CHC_6H_5 + H_2O \xrightarrow{\text{acid catalyst}}$

(i) $CH_3 + H_2O \xrightarrow{\text{Ni catalyst}}$

(j) $CH_3CH_2CH(CH_3)_2 + O_2 \xrightarrow[\substack{\text{(balance the} \\ \text{equation)}}]{\text{complete combustion}}$

10.40 Write the structures of the products in the following situations. If no reaction is to be expected, write "no reaction."

(a) $CH_2=CHCH_2CH_2CH=CH_2 + 2H_2 \xrightarrow{\text{Ni catalyst}}$

(b) $+ H_2O \xrightarrow{\text{acid catalyst}}$

(c) $+ H_2SO_4(\text{concd}) \longrightarrow$

(d) $\xrightarrow[\substack{\text{(balance the} \\ \text{equation)}}]{\text{complete combustion}}$

(e) $C_6H_6 + Cl_2 \xrightarrow{FeCl_3}$

(f) $C_6H_6 + NaOH(aq) \longrightarrow$

Fossil Fuels (Special Topic 10.1)

10.41 What are the three principal fossil fuels being used today?

10.42 What is the difference between petroleum, crude oil, and natural gas?

10.43 In general terms, describe what happened to change peat into lignite, then into soft coal, and then into hard coal.

10.44 What does *fraction* mean at an oil refinery?

10.45 What kinds of compounds predominate in the crude oil fractions that boil below 200 °C?

10.46 How do petroleum refineries increase the supply of gasoline?

Ozone in Smog (Special Topic 10.2)

10.47 What event in a vehicle engine launches the production of ozone in smog? (Write an equation.)

10.48 How is NO_2 formed in smog? (Write an equation.)

10.49 How is NO_2 involved in the production of ozone in smog? (Write equations.)

10.50 Why is ozone dangerous?

10.51 In areas with severe smog problems, air quality authorities seek to reduce emissions of hydrocarbons, even those arising from the use of power lawn mowers and similar machines. What is the connection between these uses of fuels and the ozone in smog?

Acid Catalyzed Addition of Water to a Double Bond (Special Topic 10.3)

10.52 What forms when a molecule of ethene accepts a proton from H_3O^+? (Write its name and structure.)

10.53 Write the structure of the carbocation that would form if 2-butene accepted a proton from H_3O^+.

10.54 What is the structure of the product that forms when a water molecule attaches itself to the ethyl carbocation? What kind of reaction must occur for this product to become a molecule of ethyl alcohol?

10.55 What is the structure of the product that forms when a water molecule attaches itself to the carbocation that forms in the reaction described in Review Exercise 10.53? When this product loses a proton from oxygen, what is the structure of the stable compound that forms?

The Ozone Shield (Special Topic 10.4)

10.56 What reaction makes possible the *startup* of an ozone cycle?

10.57 What two reactions constitute the ozone cycle?

10.58 What natural reaction can stop an ozone cycle?

10.59 How does CFC-11 interfere with an ozone cycle?

10.60 What is the net effect of the ozone cycle?

10.61 Why is the ozone shield important to us?

10.62 What specifically is the Antarctic "ozone hole" and what helps to confine it to the Antarctic?

10.63 What makes the development of the ozone hole a matter of concern to people living far from the Antarctic?

11

ALCOHOLS, ETHERS, THIOALCOHOLS, AND AMINES

Occurrence, Types, and Names of Alcohols

Physical Properties of Alcohols

Chemical Properties of Alcohols

Ethers

Thioalcohols and Disulfides

Occurrence, Names, and Physical Properties of Amines

Chemical Properties of Amines

Grape juice becomes wine when the natural sugar in grapes ferments and changes to ethyl alcohol, one of a family of thousands of alcohols introduced in this chapter. Carbohydrates, proteins, and nucleic acids all have the alcohol group.

11.1 OCCURRENCE, TYPES, AND NAMES OF ALCOHOLS

Molecules of the alcohol family have an OH group attached to a saturated carbon.

The OH group of alcohols is one of the most widely occurring in nature. The molecules in carbohydrates, proteins, and nucleic acids as well as in several vitamins and hormones carry this group. Rubbing alcohol, beverage alcohol, and antifreezes are also members of the alcohol family.

Many Families of Organic Compounds Have the OH Group. In the **alcohols,** the OH group is bound to a carbon atom from which only single C—C and C—H bonds extend. Only in this situation, illustrated by several simple alcohols in Table 11.1, can the OH group be called the **alcohol group.**

In the family of the *phenols* the OH group is attached to a benzene ring, and in the *carboxylic acids* it is held by a carbon that also has a double bond to oxygen.

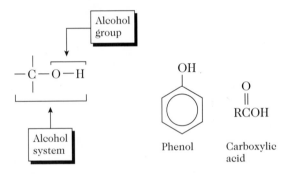

■ PRACTICE EXERCISE 1   Classify the following as alcohols, phenols, or carboxylic acids.

(a) CH_3—⬡—CH_2OH (b) CH_3—⬡—OH

(c) CH_3—⬡—$\overset{\overset{O}{\|}}{C}OH$ (d) $CH_2{=}CHCH_2OH$

(e) $CH_3CH_2CH_2CH_2OH$ (f) ⬡—OH

Some Chemical Properties Vary with the Subclasses of Alcohols. We'll see in Section 11.3 that the condition of the carbon holding the OH group, whether it has zero, one, or two H atoms, affects how alcohols are oxidized. **Primary alcohols** (1° alcohols) are those in which the OH group is held by a 1° carbon. In **secondary alcohols,** the OH group is held by a secondary (2°) carbon. When the OH group is on a 3° carbon, the alcohol is a **tertiary alcohol.**

■ Pronounce 1° as *primary,* 2° as *secondary,* and 3° as *tertiary.* These designations were defined on page 217.

TABLE 11.1   Common Alcohols

Name[a]	Structure	BP (C°)
Methyl alcohol (methanol)	CH_3OH	65
Ethyl alcohol (ethanol)	CH_3CH_2OH	78.5
Propyl alcohol (1-propanol)	$CH_3CH_2CH_2OH$	97

Isopropyl alcohol (2-propanol)	CH_3CHCH_3
	    |
	    OH
		82

Butyl alcohol (1-butanol)	$CH_3CH_2CH_2CH_2OH$	117

sec-Butyl alcohol (2-butanol)	$CH_3CH_2CHCH_3$
	       |
	       OH
		100

Isobutyl alcohol (2-methyl-1-propanol)	CH_3 | CH_3CHCH_2OH	108

t-Butyl alcohol (2-methyl-2-propanol)	CH_3
	   |
	CH_3COH
	   |
	   CH_3
		83

Ethylene glycol (1,2-ethanediol)	$CH_2 - CH_2$ | | OH OH	197

Propylene glycol (1,2-propanediol)	$CH_3 - CH - CH_2$
	       | |
	     OH OH
		189

Glycerol (1,2,3-propanetriol)	$CH_2 - CH - CH_2$ | | | OH OH OH	290

[a]The common names are given with the IUPAC names in parentheses.

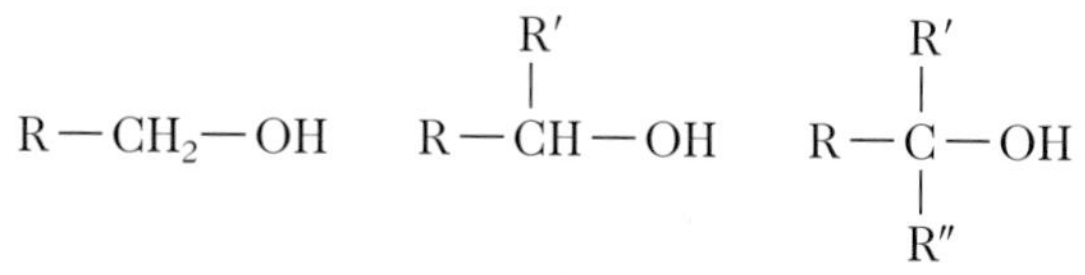

$$R-CH_2-OH \qquad \overset{\displaystyle R'}{\underset{}{R-CH-OH}} \qquad R-\overset{\displaystyle R'}{\underset{\displaystyle R''}{C}}-OH$$

Primary   Secondary   Tertiary
alcohol   alcohol   alcohol

■ The R groups in 2° and 3° alcohols don't have to be the same.

Monohydric alcohols, the "simple alcohols," have one OH per molecule. Some alcohols have more than one OH group. *Dihydric alcohols,* the **glycols,** have two. Ethylene glycol, an ingredient in some commercial antifreezes, is an example. **Trihydric alcohols,** like glycerol, have three alcohol groups per molecule. Many substances, particularly sugars, have several OH groups.

In di- and polyhydroxy alcohols, the OH groups are always on different carbon atoms. Almost no stable system is known in which one carbon holds two or three OH groups.

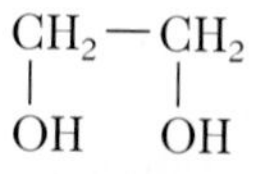

■ The following system is unstable: only rare examples are known.

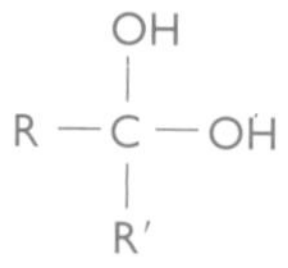

$CH_2 - CH_2$
 | |
OH OH
Ethylene glycol
(1,2-ethanediol,
a dihydric alcohol)

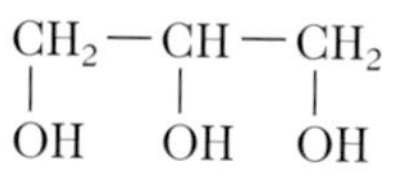

$CH_2 - CH - CH_2$
 | | |
OH OH OH
Glycerol
(1,2,3-propanetriol,
a trihydric alcohol)

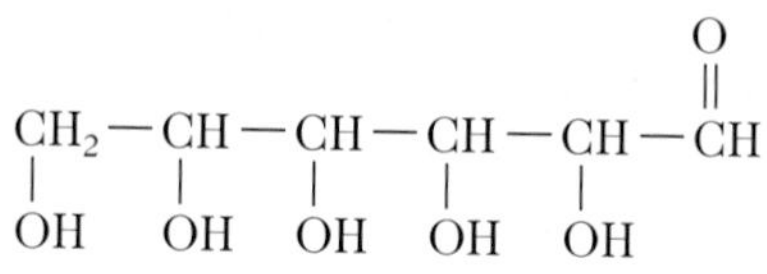

$$CH_2 - CH - CH - CH - CH - \overset{\displaystyle O}{\overset{\displaystyle \|}{CH}}$$
| | | | |
OH OH OH OH OH
Glucose, a sugar
(open form of molecule)

$$R - \overset{\displaystyle OH}{\underset{\displaystyle R'}{C}} - OH$$
1,1-Diols

■ **PRACTICE EXERCISE 2**   Classify each of the following as monohydric or dihydric. For each found to be monohydric, classify it further as 1°, 2°, or 3°. If the structure is too unstable to exist, state so.

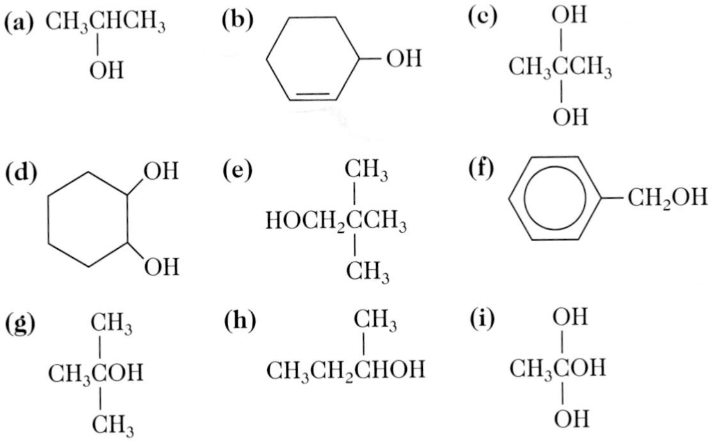

(a) CH_3CHCH_3
      |
      OH

(b) [cyclohexene ring]—OH

(c)   OH
        |
   CH_3CCH_3
        |
        OH

(d) [cyclohexane ring]—OH, OH

(e) CH_3
        |
  $HOCH_2CCH_3$
        |
      CH_3

(f) [benzene ring]—CH_2OH

(g) CH_3
      |
 CH_3COH
      |
   CH_3

(h) CH_3
        |
 CH_3CH_2CHOH

(i)   OH
      |
 CH_3COH
      |
   OH

The Simple Alcohols Are Usually Called by Their Common Names.   To write a common name, simply write the word *alcohol* after the name of the alkyl group present. For example:

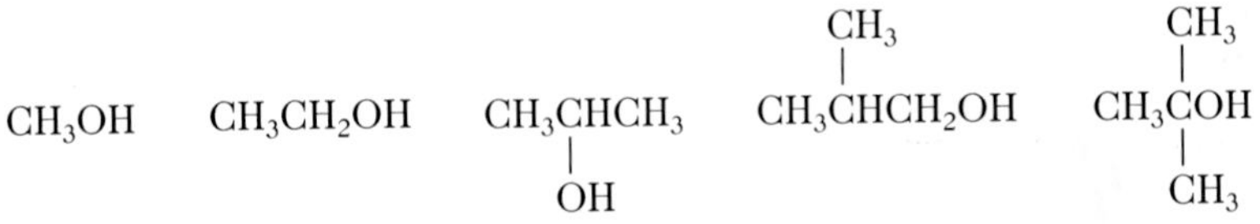

CH_3OH CH_3CH_2OH CH_3CHCH_3
                          |
                          OH

                                      CH_3 CH_3
                                        | |

CH_3CHCH_2OH CH_3COH
                           |
                          CH_3

Methyl alcohol   Ethyl alcohol   Isopropyl alcohol   Isobutyl alcohol   *t*-Butyl alcohol

■ These common names should be learned.

When the alkyl group has no common name, the IUPAC system is used (see Appendix III).

11.2   PHYSICAL PROPERTIES OF ALCOHOLS

Hydrogen bonding dominates the physical properties of alcohols.

■ The *donor* OH group has the H from which the hydrogen-bond (····) extends to the δ− site on the *acceptor*.

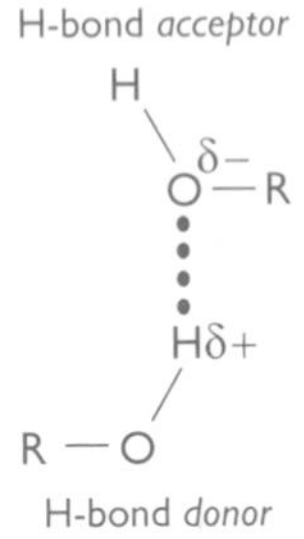

H-bond acceptor

H
 \
  δ−
  O—R
  ⋮
  ⋮
  Hδ+
  /
R—O

H-bond donor

Alcohol molecules are polar, and they can both donate and accept hydrogen bonds. These characteristics are behind an important molecular "map sign": *OH groups cause increases in boiling points and solubilities in water.* When we compare alkanes with alcohols of comparable formula mass (Table 11.2), you can see how greatly the OH group influences boiling point. Ethane and methyl alcohol, for example, boil 154 °C apart, which indicates how much greater is the attraction between the alcohol molecules compared with that between alkane molecules. The attraction stems from the occurrence of *two* hydrogen bonds between the alcohol molecules (Figure 11.1), whereas no hydrogen bonds exist between alkane molecules.

Water molecules experience three hydrogen bonds between neighbors (see Figure 11.1), so despite water's lower formula mass, it has a higher boiling point than methyl alcohol. Notice how the boiling point difference leaps when a dihydric alcohol, like ethylene glycol (Table 11.2), is compared with the alkane of comparable formula

TABLE 11.2   Influence of the Alcohol Group on Boiling Points

Name	Structure	Formula Mass	BP (°C)	Difference in BP
Ethane	CH_3CH_3	30	−89	
Methyl alcohol	CH_3OH	32	65	154
Propane	$CH_3CH_2CH_3$	44	−42	
Ethyl alcohol	CH_3CH_2OH	46	78	120
Butane	$CH_3CH_2CH_2CH_3$	58	0	
Ethylene glycol	$HOCH_2CH_2OH$	62	197	197

mass. *Four* hydrogen bonds can extend from a dihydric molecule, twice as many as occur between molecules of monohydric alcohols.

Methyl alcohol readily dissolves in water. Its molecules can donate and accept hydrogen bonds and so are able to slip into water's network of hydrogen bonds (Figure 11.2a). Methane, on the other hand, is insoluble in water because the tightly hydrogen-bonded water molecules cannot let the nonpolar methane molecules in (Figure 11.2b).

As the size of a monohydric alcohol molecule increases, it becomes more and more alkane-like. In decyl alcohol, for example, the small, water-like OH group is overwhelmed by the long hydrocarbon chain.

$$CH_3CH_2CH_2CH_2CH_2CH_2CH_2CH_2CH_2CH_2OH$$
Decyl alcohol

The flexings and twistings of long chains interfere too much with water's hydrogen bonding networks, and so water cannot let decyl alcohol molecules into solution. This alcohol and most with five or more carbons are insoluble in water. They do dissolve, however, in such nonpolar solvents as diethyl ether, benzene, and gasoline.

The more OH groups there are per molecule, the more soluble the compound is in water and the less soluble it is in nonpolar solvents. Glycerol, sugars, and even the dihydric alcohols do not dissolve in nonpolar solvents but dissolve in water.

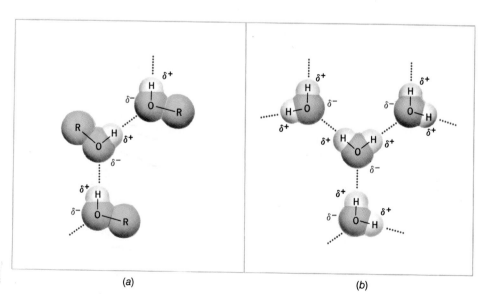

(a)   (b)

FIGURE 11.1
Hydrogen bonding (a) in alcohols and in water (b).

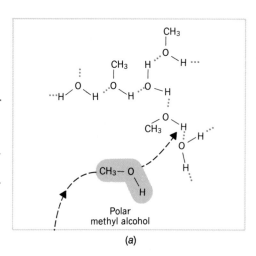 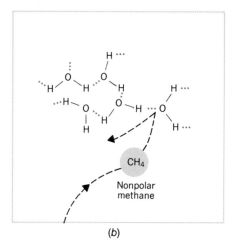

(a)   (b)

FIGURE 11.2
How a short-chain alcohol dissolves in water. (a) The alcohol molecule can take the place of a water molecule in the hydrogen bonding network of water. (b) An alkane molecule cannot break into the hydrogen bonding network in water, so the alkane cannot dissolve.

11.3 CHEMICAL PROPERTIES OF ALCOHOLS

The loss of water (dehydration) and the loss of hydrogen (oxidation) are two important reactions of alcohols.

Alcohols react with both inorganic and organic compounds, but we study only the inorganic reactants in this chapter. We first mention, however, a reaction that alcohols do not have.

Alcohols Do Not Ionize in Water.   The alcohols, like water, are exceptionally weak as either donors or acceptors of H^+ or OH^- ions. Alcohols are thus neither acids nor bases in the conventional sense, but are neutral compounds.

■ When an alcohol dissolves in water, it doesn't raise or lower the pH.

■ The opposite of *in vivo* is *in vitro,* meaning "in glass," that is, in lab vessels.

Alcohols Can Be Dehydrated to Alkenes.   Heat and an acid catalyst make an alcohol molecule lose H_2O leaving behind a carbon–carbon double bond. Called the *dehydration* of an alcohol, the reaction gives an alkene. The pieces of the water molecule, one H and one OH, come from *adjacent* carbons. Specific enzymes catalyze this reaction *in vivo*, meaning in living systems, where strongly acidic catalysts, of course, do not occur. In general:

$$-\underset{\underset{H}{|}}{C}-\underset{\underset{OH}{|}}{C}- \xrightarrow[\text{heat}]{H^+\text{catalyst}} \underset{}{C}=\underset{}{C} + H-OH$$

Alcohol   Alkene

Specific examples follow.

$$CH_3CH_2OH \xrightarrow[\text{heat}]{H^+\text{catalyst}} CH_2{=}CH_2 + H_2O$$
Ethanol   Ethene

$$\overset{\overset{\displaystyle OH}{|}}{CH_3CH_2CHCH_3} \xrightarrow[\text{heat}]{H^+\text{catalyst}} CH_3CH{=}CHCH_3 + CH_3CH_2CH{=}CH_2 + H_2O$$
2-Butanol   2-Butene   1-Butene
                          (chief product)   (minor product)

Sometimes, as in the second example, two or more alkenes seem to be possible products, because more than one adjacent carbon carries an H atom that might be removed to complete a molecule of H—OH. When, however, one of the possible alkenes has a double bond with more attached alkyl groups than any of the others, it is the chief product. Notice in the second example that the chief product, 2-butene, has two alkyl groups at the double bond (two CH_3 groups), but 1-butene has only one alkyl group (the CH_3CH_2 group). We aren't concerned about this for alcohol dehydrations *in vivo*, because enzymes are very specific in directing reactions.

$CH_3-CH=CH-CH_3$
2-Butene
(two alkyl groups)

$CH_2=CH-CH_2CH_3$
1-Butene
(one alkyl group)

EXAMPLE 11.1 Writing the Structure of the Alkene That Forms When an Alcohol Undergoes Dehydration

What is the product of the dehydration of isobutyl alcohol?

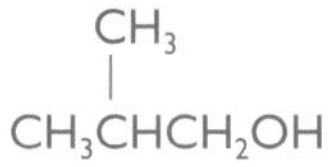

ANALYSIS Predicting the product of the dehydration of this alcohol involves rewriting the structure of the alcohol but leaving off *(a)* the OH group and *(b)* one H from a carbon adjacent to the carbon that holds the OH group. Then we write a double bond between these two carbon atoms.

SOLUTION

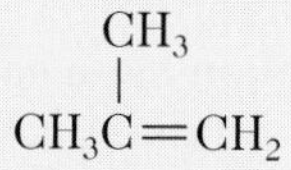

$$\begin{array}{c} CH_3 \\ | \\ CH_3C=CH_2 \end{array}$$
2-Methylpropene

CHECK Ask the following questions of the answer. Is it an *alkene?* (Alcohol dehydrations give alkenes.) Is its carbon skeleton the same as in the starting material? (Changes in the carbon skeleton do *not* occur in all of the examples that we will study.) Do four bonds extend from each carbon? (The rules of covalence must be obeyed.) By answering "yes" to each of these questions you have made a thorough check.

$$\begin{array}{c} CH_3 \\ | \\ CH_3CHCH_2OH \end{array}$$
Isobutyl alcohol

■ PRACTICE EXERCISE 3 Write the structures of the alkenes that can be made by the dehydration of the following alcohols.

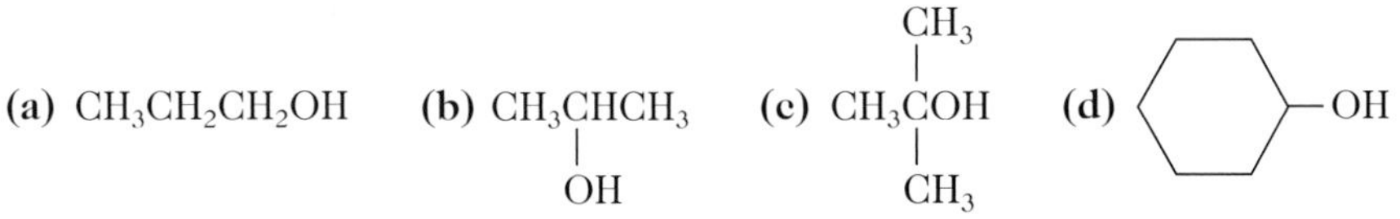

(a) $CH_3CH_2CH_2OH$ (b) CH_3CHCH_3 (c) $CH_3\overset{\displaystyle CH_3}{\underset{\displaystyle CH_3}{C}}OH$ (d) ⬡—OH
                                   |
                                  OH

Alcohols Can Be Oxidized. We learned in Section 3.2 that reactions in which electrons transfer are *redox reactions,* and these are common to many families of organic compounds. It is often tricky, however, to tell which atom in an organic molecule gains or loses electrons. Organic chemists therefore use alternative definitions of oxidation and reduction. An *oxidation* is the loss of H or the gain of O by a molecule. A *reduction* is the loss of O or the gain of H. The oxidations of alcohols are examples of losses of H atoms.

The goal of our study of alcohol oxidations is to be able to write the structures of the products. We will not learn how to balance equations for these reactions, so we will use unbalanced "reaction sequences," instead. We will simplify even further and

■ When $MnO_4^- = (O)$, its purple color gives way to a brown sludge of MnO_2 as the oxidation proceeds.

■ When $Cr_2O_7^- = (O)$, its orange color changes to the green of the Cr^{3+} ion as the oxidation proceeds.

use the symbol (O) for any oxidizing agent able to give the reaction. Typical oxidizing agents for *in vitro* oxidations, those done in the lab using glassware, are potassium permanganate, $KMnO_4$, and either sodium or potassium dichromate, $Na_2Cr_2O_7$ or $K_2Cr_2O_7$. The breath analyzer test for ethyl alcohol is an important application of dichromate oxidation (see Special Topic 11.1). *In vivo*, redox enzymes remove hydrogen from alcohols.

The hydrogen removed from an alcohol molecule by oxidation ends up in a water molecule. (The oxidizing agent furnishes the O atom for H_2O.) One H comes from the OH group of the alcohol, and the other H comes from the carbon that has been holding the OH group.

$$-\overset{\displaystyle |}{\underset{\displaystyle |}{\text{C}}}\overset{}{\underset{\displaystyle \text{H}}{}}\!\!-\!\text{O}\overset{}{\underset{\displaystyle \text{H}}{}} \;+\; (\text{O}) \longrightarrow \quad \overset{\diagdown}{\underset{\diagup}{\text{C}}}{=}\text{O} \;+\; \text{H}{-}\overset{\text{O}}{\underset{\diagdown\,\text{H}}{}}$$

$$[\text{H}{:}^- + \text{H}^+] \text{----} \quad (\text{O})$$

Only 1° and 2° alcohols can be oxidized this way, because 3° alcohols do not have an H atom on the carbon with the OH group.

The organic product of this oxidation has a carbon–oxygen double bond, but the actual family of this product depends on the subclass of the alcohol. The product can be an aldehyde, a carboxylic acid, or a ketone, all of which have carbon–oxygen double bonds.

$$\overset{\text{R}'}{\underset{\text{R}''}{\text{R}-\text{C}-\text{OH}}}$$

No H here

3° Alcohol system

$$\overset{\text{O}}{\underset{}{\overset{\|}{\text{R}-\text{C}-\text{H}}}}$$
Aldehyde

$$\overset{\text{O}}{\underset{}{\overset{\|}{\text{R}-\text{C}-\text{OH}}}}$$
Carboxylic acid

$$\overset{\text{O}}{\underset{}{\overset{\|}{\text{R}-\text{C}-\text{R}'}}}$$
Ketone

Primary Alcohols Are Oxidized to Aldehydes or Carboxylic Acids. When a conventional oxidizing agent is used, like potassium permanganate, 1° alcohols are oxidized first to aldehydes and then to carboxylic acids. It is difficult to stop at the aldehyde stage, because *aldehydes oxidize more easily than alcohols.* The oxidation of a 1° alcohol, therefore, is usually done with enough oxidizing agent to take the reaction all the way to the carboxylic acid. In body cells this is not a problem, because different enzymes are needed for each step. When the right enzyme catalyzes the reaction, an aldehyde forms.

$$\text{RCH}_2\text{OH} \xrightarrow[\text{(enzyme catalyzed)}]{(\text{O})} \overset{\text{O}}{\overset{\|}{\text{RCH}}} + \text{H}_2\text{O}$$
1° Alcohol   Aldehyde

A specific example is the oxidation of ethyl alcohol (in alcoholic beverages) to give acetaldehyde (which causes the hangover).

$$\text{CH}_3\text{CH}_2\text{OH} \xrightarrow[\text{(enzyme catalyzed)}]{(\text{O})} \overset{\text{O}}{\overset{\|}{\text{CH}_3\text{CH}}} + \text{H}_2\text{O}$$
Ethyl alcohol   Acetaldehyde

When potassium permanganate or sodium dichromate is used, then the intent normally is to carry the reaction through to the carboxylic acid, so enough oxidizing agent, as we said, is used at the start.

$$\text{RCH}_2\text{OH} \xrightarrow{(\text{O})} \left(\overset{\text{O}}{\overset{\|}{\text{RCH}}}\right) \xrightarrow{\text{more (O)}} \overset{\text{O}}{\overset{\|}{\text{RCOH}}}$$
1° Alcohol   Aldehyde   Carboxylic acid

A specific example is

$$CH_3CH_2CH_2OH \xrightarrow{(O)} CH_3CH_2\overset{\overset{\displaystyle O}{\|}}{C}OH + H_2O$$

Propyl alcohol   Propionic acid

EXAMPLE 11.2 Writing the Structure of the Product of the Oxidation of a Primary Alcohol

PROBLEM What aldehyde and what carboxylic acid could be made by the oxidation of butyl alcohol?

ANALYSIS I Butyl alcohol is a 1° alcohol, and every 1° alcohol has a CH_2OH group. The oxidation of a 1° alcohol strips the two H atoms from CH_2OH. One H comes from O and the second H is taken from C. This loss of two H atoms creates a double bond between C and O, and so changes CH_2OH to $CH=O$, an aldehyde group. (The two H atoms emerge in a molecule of water.) *The fundamental skeleton of all of the heavy atoms in the 1° alcohol, like C and O, remains intact.*

SOLUTION I The aldehyde that forms is

$$CH_3CH_2CH_2CH=O \quad \text{or} \quad CH_3CH_2CH_2\overset{\overset{\displaystyle O}{\|}}{C}H \quad \text{(butyraldehyde)}$$

ANALYSIS 2 Oxidation of an aldehyde group brings about the insertion of O between C and H.

$$CH_3CH_2CH_2\overset{\overset{\displaystyle O}{\|}\curvearrowright O}{C}H \xrightarrow{(O)} CH_3CH_2CH_2\overset{\overset{\displaystyle O}{\|}}{C}OH \quad \text{(butyric acid)}$$

SOLUTION 2 The product, as you see, is a carboxylic acid.

CHECKS Be sure that the aldehyde group is correct; a common error is to leave a second H atom on its C atom. Examine the functional group of the acid; its C atom must hold both OH and $=O$ besides *one* other group.

■ **PRACTICE EXERCISE 4** Write the structures of the aldehydes and carboxylic acids that can be made by oxidation of the following alcohols.

(a) $CH_3\overset{\overset{\displaystyle CH_3}{|}}{C}HCH_2OH$ (b) 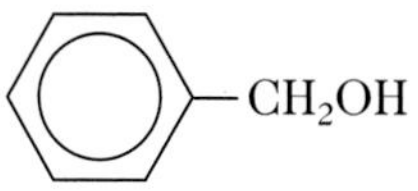 ⬡—CH_2OH

Secondary Alcohols Are Oxidized to Ketones. Ketones strongly resist further oxidation, so they are easily made by the oxidation of 2° alcohols using strong oxidizing agents like $KMnO_4$ and $Na_2Cr_2O_7$. *In vivo,* enzymes accomplish the identical overall reaction, the dehydrogenation of a 2° alcohol group to a keto group. In general, for 2° alcohols,

$$R\overset{\overset{\displaystyle OH}{|}}{C}HR' + (O) \longrightarrow R\overset{\overset{\displaystyle O}{\|}}{C}R' + H_2O$$

2° Alcohol   Ketone

Throughout all of human history people have drunk alcoholic beverages both for their pleasing tastes and their ability to alter moods. Because of the consequences of alcohol abuse, opposition to the use of such beverages is also widespread. One of the world's major religions, Islam, teaches total abstinence, but a very large percentage of Christendom uses wine in some rites. A fraction of Christendom, however, also practices total abstinence partly to demonstrate the possibility of having fun without alcohol.

With respect to alcohol abuse as an issue of public policy in the United States, there are roughly two camps. One camp sees alcohol itself as the culprit and so encourages abstinence. The other camp sees alcohol abusers at the heart of the problem and so espouses treatment centers and drug abuse education for them and temperance for the rest. We can do no more in this Special Topic than concentrate on facts to which both camps would agree.

Alcohol Beverages Pure ethyl alcohol is a colorless liquid with a taste so burning that no one drinks it pure. Users always drink dilute forms. Beer and ale are 3 to 6% alcohol; wine is 12 to 14%. (When the concentration reaches this point, the enzymes that catalyze the fermentation of fruit sugar, mostly glucose, are deactivated.) Fortified wines result from adding alcohol to wine to make the concentration about 20%. Stronger liquors are made by distilling wines or beers—brandy from wine and whiskey from a beer-like mixture. Such distilled beverages or "spirits" are 40 to 55% ethyl alcohol.

The "proof" of a beverage is always twice the percentage of alcohol. Pure alcohol is thus 200 proof. Historically, early distillers or their customers poured a small amount of the product onto a mound of gunpowder. If the powder would still burn, it was "proof" that the product had not been watered down. The lower limit on a positive test was called "100 proof" material, which we now know to be 50% alcohol.

Effects of Alcohol What is termed the *typical drink* contains 0.75 ounce of ethyl alcohol, which may be obtained by a pint of beer (16 oz of 4.5% alcohol), a 5-oz glass of wine (14% alcohol), a 3.5-oz glass of fortified wine, like sherry (20%), or a 1.5-oz "shot" of whiskey (50%). Observant friends of an adult, nonalcoholic, "standard," 150-lb male will notice changes in his behavior and mood at a blood alcohol level of 0.050% (39 mg/dL, 8.5 mmol/L)—the result of two typical drinks in succession. The mood changes of this state are largely what encourage moderate drinking.

The effects of drinking, however, depend on too many factors to warrant easy generalizations.

Body mass is only one factor in how quickly the blood alcohol level increases during drinking. The "standard" 120-lb adult, nonalcoholic female generally reaches a blood level of 0.050% with less drinking than the standard adult male. Food is another factor. The presence of food in the system slows down the absorption of alcohol while drinking. Ethyl alcohol is rather quickly metabolized; the rate of complete metabolism for the "standard" male is about one drink per hour.

As the blood alcohol level increases above 0.050%, the depressant effect of alcohol on the brain kicks in, and at a level of 0.10% ethyl alcohol (79 mg/dL, 17 mmol/L), voluntary motor actions, such as walking, hand and arm movements, and speech, become clumsy. Most states of the United States define "intoxication" in terms of a blood alcohol level of 0.10%. The driving laws of many states, however, specifically allow lower values—even as low as 0.05%—to be evidence for DWI, "driving while under the influence." At a level of 0.20% (157 mg/dL, 34 mmol/L) the typical nonalcoholic male's control over the entire motor area of the brain is impaired and the brain region controlling emotional behavior is affected. Staggering, boisterous talking, or anger (or weeping) appear; the person is a certified "drunk" to those who observe him. At a level of 0.30%, the user is very confused and may lapse into a stupor. With 0.40 to 0.50% blood alcohol, the individual will be in a coma and those parts of the brain that control the heart beat and breathing are so affected that death results.

Ethyl alcohol is absorbed directly into the bloodstream along any part of the intestinal tract. Enzymes in both the intestinal tract and the liver work to detoxify it, but the liver does most of such work. One enzyme takes ethyl alcohol to the aldehyde stage, producing acetaldehyde (CH_3CHO), which is chiefly responsible for cirrhosis. In time, the acetaldehyde is oxidized to the acetate ion, $CH_3CO_2^-$, which can be metabolized normally.

The Breathalyzer Test Law enforcement officers use a breath-analyzing device called a *breathalyzer* to determine the blood alcohol level of drivers suspected of being under the influence (Figure 1). It has been shown that the partial pressure of ethyl alcohol vapor in exhaled breath is proportional to the concentration of alcohol in the blood. The suspect, therefore, is asked to exhale through a tube that leads the air into aqueous sodium dichromate. The yellow dichromate ion changes to the green chromium(III) ion, Cr^{3+}, as it oxidizes ethyl alcohol to acetic acid.

$$3CH_3CH_2OH + 2Cr_2O_7^{2-} + 16H^+ \rightarrow$$
Ethyl alcohol   Dichromate
$$3CH_3CO_2H + 4Cr^{3+} + 11H_2O$$
Acetic acid

The instrument measures the amount of color change and translates the information directly into blood alcohol concentration.

Alcoholism   Ethyl alcohol is the chemical cause of *alcoholism*. Although "alcoholism" is not easily defined, most specialists agree that the definition must focus on the frequency of those symptoms that relate to drinking itself rather than something else. The essential features of alcoholism are a powerful craving for and a dependency on ethyl alcohol to the point that drinking or thinking about the next drink becomes the major preoccupation of life and withdrawal is very painful. Alcoholism is accompanied by disorders of the nervous system and muscles. In its early stages, it causes a fatty liver and in its later stages *cirrhosis* of the liver, an incurable and fatal disorganization of liver structure brought about by the development of nodules surrounded by fibrous tissue. One-third of heavy drinkers suffer heart damage and nearly half experience an increasing weakness of skeletal muscles. Alcoholics also have impaired immune systems and lowered resistance to pneumonia and other infectious diseases. You can see how all of these by-products of an alcoholic lifestyle would contribute immensely to a nation's medical costs.

Malnutrition, which is an insufficiency of proper nutrients (and not necessarily a lack of enough calories), complicates but does not cause the medical disorders of alcohol abuse. Two "fifths" of whiskey a day would supply enough calories for metabolism, but, without eating food, neither vitamins nor proteins are furnished. (A "fifth" equals one fifth of a gallon, or a bottle containing this amount. It's thus four-fifths of a quart or a little more than 750 mL.)

When taken in their totality over an entire population, individual human decisions concerning personal uses of alcoholic beverages are responsible for widespread misery and economic burdens on individuals, families, and society. In the late 1980s in urban areas, the third most frequent cause of death among those of 25 to 64 years of age was cirrhosis of the liver. The National Institute on Alcohol Abuse and Alcoholism estimates that the combined costs of health expenses and lost productivity are roughly \$120 billion annually in the United States.

FIGURE 1
The administration of the breathalyzer test along a highway.

Alcoholism Treatment   The conventional wisdom at alcoholism treatment centers is that alcoholics have a disease that arises from a strong genetic predisposition and is characterized by physical dependency. The conventional treatment is to get the alcoholic to stop drinking and henceforth to practice total abstinence. Less conventional views note that the long-term success of the orthodox treatment requires a step beyond abstinence. Success depends on changes in both outlook and values that confer a satisfying lifestyle not dependent on heavy drinking (or longing for it). Your genes may or may not dispose you to becoming addicted to ethyl alcohol. But you choose to take this or any other drug as an option in the quest for some desirable end, such as acceptance into a social group, release from stress, or a change in mood.

Alcohol and Pregnancy   *Fetal alcohol syndrome* appears to be caused directly by excessive ethyl alcohol itself, not by its oxidation products. The problem is that the fetus cannot detoxify ethyl alcohol, and the alcohol can stop cell division. The syndrome leads to babies with impaired mental abilities and many other disorders. The conventional wisdom today favors total abstinence for those who are pregnant, advice that can come too late for those unaware of their condition.

Some Additional Readings

1. S. Schenker and K. V. Speeg, "The Risk of Alcohol Intake in Men and Women," *The New England Journal of Medicine,* Jan. 11, 1990, page 127.
2. M. E. Charness, R. P. Simon, and D. A. Greenberg, "Ethanol and the Nervous System," *The New England Journal of Medicine,* Aug. 17 1989, page 442.
3. G. E. Vaillant, *The Natural History of Alcoholism,* Harvard University Press, 1983.
4. H. Fingarette, *Heavy Drinking: The Myth of Alcoholism as a Disease,* University of California Press, 1988.

A specific *in vitro* example of the oxidation of a 2° to a ketone is the following.

$$\underset{\text{\textit{sec}-Butyl alcohol}}{\overset{\overset{\text{OH}}{|}}{\text{CH}_3\text{CHCH}_2\text{CH}_3}} \xrightarrow{\text{(O)}} \underset{\text{Butanone}}{\overset{\overset{\text{O}}{||}}{\text{CH}_3\text{CCH}_2\text{CH}_3}} + \text{H}_2\text{O}$$

EXAMPLE 11.3   Writing the Structure of the Product of the Oxidation of a Secondary Alcohol

What ketone forms when isopropyl alcohol is oxidized?

ANALYSIS   Isopropyl alcohol is a 2° alcohol, and every 2° alcohol has a CHOH group. The oxidation of a 2° alcohol strips the two H atoms from CHOH, creates a double bond between C and O, and so changes CHOH to C=O, a keto group. (The two H atoms emerge in a molecule of water.) *The fundamental skeleton of all of the heavy atoms in the 2° alcohol, like C and O, remains intact.*

SOLUTION

$$\underset{\text{Isopropyl alcohol}}{\overset{\overset{\text{OH}}{|}}{\text{CH}_3\text{CHCH}_3}} \quad \text{is oxidized to} \quad \underset{\text{Acetone}}{\overset{\overset{\text{O}}{||}}{\text{CH}_3\text{CCH}_3}} \quad \text{plus H}_2\text{O}$$

CHECK   Is the product a *ketone?* (Secondary alcohols are oxidized to ketones.) Does the product have the same carbon skeleton as the starting material? Does each carbon have four bonds and each oxygen atom two?

■   PRACTICE EXERCISE 5   Write the structures of the ketones that can be made by oxidation of the following alcohols.

(a) $\overset{\overset{\text{OH}}{|}}{\text{CH}_3\text{CHCH}_2\text{CH}_3}$ **(b)** [benzene ring]—$\overset{\overset{\text{OH}}{|}}{\text{CHCH}_3}$ **(c)** [cyclopentane ring with OH]

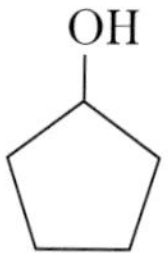

Phenols, Unlike Alcohols, Can Neutralize Hydroxide Ions.   In **phenols** the OH group is attached to a benzene ring. Many members of the phenol family are found both in nature and in commerce. The simplest member of the family is called phenol, and it is a raw material for making aspirin. Tyrosine, one of the amino acids that the body uses to make proteins, is also a phenol but it has other functional groups as well.

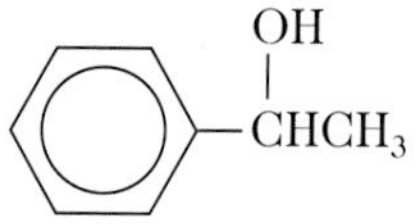

Unlike alcohols, phenols cannot be dehydrated. Although they are weak acids, phenols are strong enough to be able to neutralize sodium hydroxide. For example:

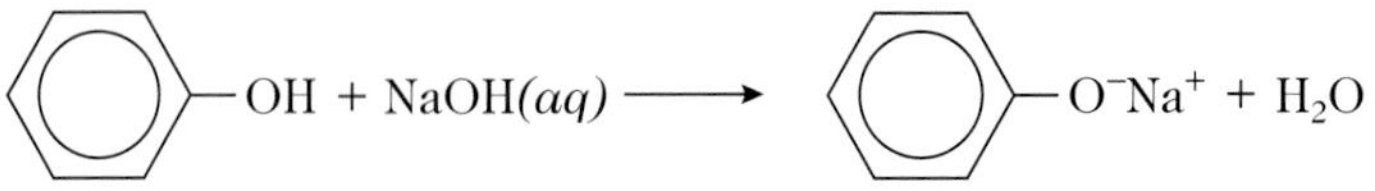

$$\underset{\text{Phenol}}{\text{[benzene ring]}-\text{OH}} + \text{NaOH}(aq) \longrightarrow \underset{\text{Sodium phenoxide}}{\text{[benzene ring]}-\text{O}^-\text{Na}^+} + \text{H}_2\text{O}$$

11.4 ETHERS

The ethers are almost as chemically unreactive as the alkanes.

Ethers are compounds in whose molecules two organic groups are joined to the same oxygen atom, and their family structure is R—O—R′ (see Table 11.3). The carbon joined to the bridging oxygen atom cannot be a carbonyl carbon, the one in CO. Thus the first three compounds given below are ethers, but methyl acetate is in the *ester* family, not the ether family, because the bridging oxygen is attached to a carbonyl carbon. (We study esters in a later chapter.)

CH₃CH₂—O—CH₂CH₃
Diethyl ether

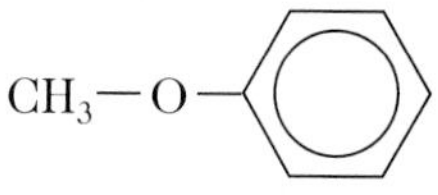

Methyl phenyl ether

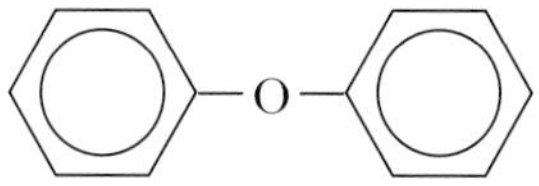

Diphenyl ether

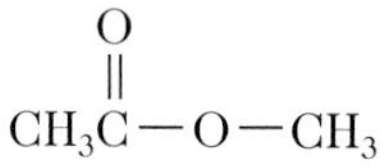

Methyl acetate
(an *ester*, not an ether)

The common names of ethers are made by naming the groups attached to the oxygen and adding the word *ether,* as illustrated in Table 11.3.

Ethers Cannot Donate Hydrogen Bonds. Because the ether group cannot donate hydrogen bonds, the boiling points of simple ethers are more like those of the alkanes of comparable formula masses than those of the alcohols. The oxygen of the ether group can accept hydrogen bonds, however, so ethers are more soluble in water than alkanes, as the data in the table in the margin show.

Ethers Have Few Chemical Reactions. At room temperature, ethers do not react with strong acids, bases, or strong oxidizing or reducing agents. Like all organic compounds, ethers burn. We will learn no other reactions of ethers, but we must be able to recognize the ether group and to remember that it is not very reactive toward anything.

Compound	BP (°C)	Solubility in Water (g/dL)
Pentane	36	0.036[a]
Diethyl ether	35	8.4[a]
1-Butanol	118	11[b]

[a]At 15 °C. [b]At 25 °C.

TABLE 11.3 Ethers

Common Name	Structure	BP (°C)
Dimethyl ether	CH₃OCH₃	−23
Methyl ethyl ether	CH₃OCH₂CH₃	11
Methyl *t*-butyl ether	CH₃OC(CH₃)₃	55.2
Diethyl ether	CH₃CH₂OCH₂CH₃	34.5
Dipropyl ether	CH₃CH₂CH₂OCH₂CH₂CH₃	91
Methyl phenyl ether	CH₃OC₆H₅	155
Diphenyl ether	C₆H₅OC₆H₅	259
Divinyl ether	CH₂=CHOCHCH₂	29

Ethers Can Be Prepared from Alcohols. We learned earlier in this chapter that alcohols can be dehydrated by the action of heat and an acid catalyst to give alkenes. The precise temperature that works best has to be discovered experimentally for each alcohol. If the right temperature is not used, a different path for dehydration occurs, one in which water splits out *between* two alcohol molecules, rather than from within one alcohol molecule, to give an ether. In general,

$$R—O—H + H—O—R \xrightarrow{\text{acid catalyst}} R—O—R + H_2O$$
Two alcohol molecules   Ether

A specific example is

■ Earlier we learned that concentrated H_2SO_4 acts on ethyl alcohol to give ethene when the temperature is 170 °C.

$$2CH_3CH_2OH \xrightarrow[\text{140 °C}]{H_2SO_4} CH_3CH_2—O—CH_2CH_3 + H_2O$$
Ethyl alcohol   Diethyl ether

The dehydration that produces an ether usually requires a temperature lower than that which gives an alkene, but we won't need the details. Our interest is simply in the *possibility* of making an ether from an alcohol as well as the structure of the ether that can be made.

EXAMPLE 11.4 Writing the Structure of an Ether That Can Form from an Alcohol

If the conditions are right, butyl alcohol can be converted to an ether. What is the structure of this ether?

ANALYSIS   The name of the starting material, *butyl* alcohol, gives us most of the answer. The ether that forms from butyl alcohol must receive its alkyl groups, so to write the ether's structure we write one O atom with two bonds from it and attach two butyl groups to it.

SOLUTION   The structure of the ether is

$$CH_3CH_2CH_2CH_2—O—CH_2CH_2CH_2CH_3$$

CHECK   Is the product truly an *ether?* (Is it of the form R—O—R?) Are the R groups identical to the R group of the parent alcohol? (In this example are they *butyl* groups?) Does each carbon atom have four bonds and each oxygen two?

■ PRACTICE EXERCISE 6   Write the structures of the ethers to which the following alcohols can be converted.

(a) CH_3OH **(b)** $CH_3CH_2CH_2OH$

(c) 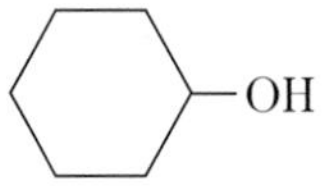 ⬡—OH

11.5 THIOALCOHOLS AND DISULFIDES

Both the SH group, an easily oxidized group, and the S—S system, an easily reduced group, are present in molecules of proteins.

Alcohols, ROH, can be viewed as alkyl derivatives of water, HOH. Similar derivatives of hydrogen sulfide, HSH, are also known and are in the family called the **thioalcohols,** commonly called the *mercaptans.*

$$R-S-H \qquad R-S-S-R'$$
Thioalcohols   Disulfides
(mercaptans)

The SH group itself is variously called the *thiol group,* the *mercaptan group,* or the *sulfhydryl group.*

The SH Group Is Easily Oxidized to the Disulfide Group We study the SH group to introduce one important property of proteins. Thioalcohols are *readily* oxidized to **disulfides,** compounds with two sulfur atoms joined by a covalent bond, R—S—S—R. (The R groups need not be identical.) In general,

$$R-S \quad H + H-S-R + (O) \rightarrow R-S-S-R + H_2O$$
Two molecules of   One molecule
a thioalcohol   of a disulfide

For example,

$$2CH_3SH + (O) \rightarrow CH_3-S-S-CH_3 + H_2O$$
Methanethiol   Dimethyl disulfide

■ *Mercaptan* is a contraction of *mercury capturer.* Compounds with SH groups form precipitates with mercury ions.

■ Lower-formula-mass thioalcohols are present in and are responsible for the respect usually accorded skunks.

EXAMPLE 11.5 Writing the Structure of the Product of the Oxidation of a Thioalcohol

What is the product of the oxidation of ethanethiol, CH_3CH_2SH?

ANALYSIS Thiols are oxidized to disulfides, so it's simply a matter of making a disulfide group, S—S, and attaching the hydrocarbon groups of the parent thioalcohol to the sulfur atoms.

SOLUTION The product is $CH_3CH_2S-SCH_2CH_3$ (diethyl disulfide).

Disulfides Are Easily Reduced to Thioalcohols The S—S bond in simple disulfides or in proteins is reduced by mild reducing agents to the thioalcohol parents of the disulfide. We will use the symbol (H) to represent any reducing agent that works, just as we used (O) for an oxidizing agent. In general,

$$RS-SR \quad + 2(H) \rightarrow RSH + HSR$$
One molecule of   Two molecules of
a disulfide   thioalcohol

For example,

$$CH_3CH_2S-SCH_2CH_2CH_3 + 2(H) \rightarrow CH_3CH_2SH + HSCH_2CH_2CH_3$$
Ethyl propyl disulfide   Ethanethiol   1-Propanethiol

Adrenergic Agents   Epinephrine and norepinephrine are two of the many hormones in our bodies. We will study the nature of hormones in a later chapter, but we can use a definition here. **Hormones** are compounds the body makes in special glands to serve as chemical messengers. In response to a stimulus somewhat unique for each hormone, such as fright, food odor, and sugar ingestion, the gland secretes its hormone into circulation in the bloodstream. The hormone then moves to some organ or tissue where it activates a particular metabolic series of reactions that constitute the biochemical response to the initial stimulus.

Maybe you have heard the expression, "I need to get my adrenaline flowing." Adrenaline—or epinephrine, its technical name—is made by the adrenal gland. If you ever experience a sudden fright, a trace amount of epinephrine immediately flows and the results include a strengthened heartbeat, an increase in blood pressure, and release of glucose into the circulation from storage—all of which get the body ready to respond to the threat.

Norepinephrine has similar effects, and because these two hormones are secreted by the adrenal gland, they are called *adrenergic agents*.

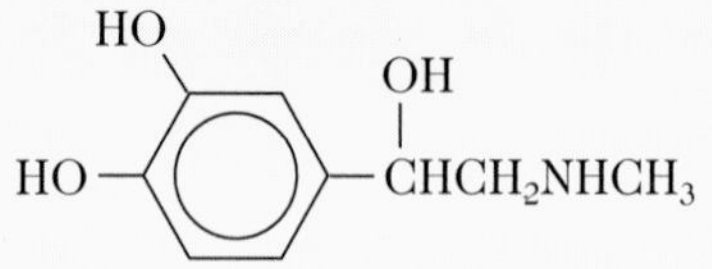

HO
OH
HO — CHCH$_2$NHCH$_3$

Epinephrine

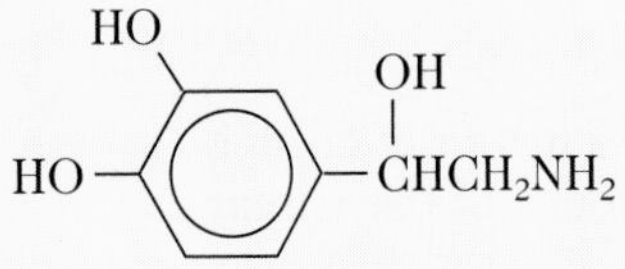

HO
OH
HO — CHCH$_2$NH$_2$

Norepinephrine

The β-phenylethylamines are another physiologically active family of amines. One example, dopamine, is the compound the body uses to make norepinephrine. (You can see how similar their structures are.) Synthetic dopamine is used to treat shock associated with severe congestive heart failure.

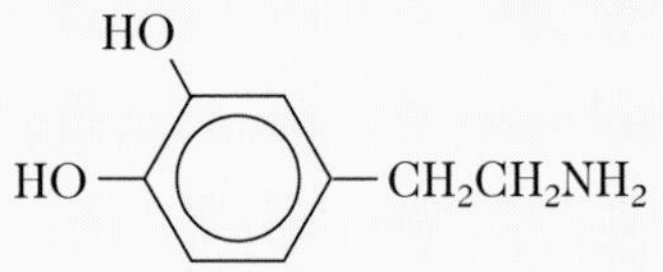

HO
HO — CH$_2$CH$_2$NH$_2$

Dopamine

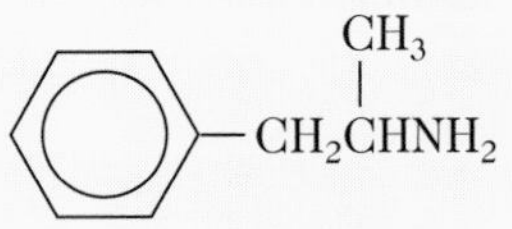

CH$_3$
CH$_2$CHNH$_2$

Dexedrine

Amphetamines   The *amphetamines* are a family of β-phenylethylamines that include Dexedrine ("speed"). Members of this family are legally prescribed as stimulants and antidepressants and sometimes for weight control programs. However, millions of amphetamine "pep pills" or "uppers" are sold illegally, and this use constitutes a serious drug abuse problem. The dangers of overuse include suicide, belligerence and hostility, paranoia, and hallucinations.

Cocaine ("Crack")   Cocaine is a psychoactive drug that comes in two forms, natural cocaine ("free base" cocaine, "crack" cocaine) and the salt of cocaine and hydrogen chloride. Free base cocaine has an amine system and so can form a salt with acids. The ways in which cocaine is used "on the street" reflect its properties. The free base has a relatively low melting point, 98 °C, and so is more volatile. The hydrogen chloride salt of cocaine is less volatile. It has a much higher melting point (197 °C) than the free base, but the salt is far more soluble in water. The salt is therefore used to prepare legal medications whenever this property is desirable. An aqueous solution of the salt is readily injected directly into the bloodstream.

■ PRACTICE EXERCISE 7   Complete the following equations by writing the structures of the products that form. If no reaction occurs, write "no reaction."

(a) CH$_3$SSCH$_3$ + (H) ⟶ ?   (b) CH$_3$CHCH$_3$ + (O) ⟶ ?
                                              |
                                              SH

(c) 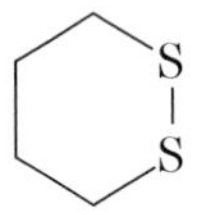 [ring with S and S] + (H) ⟶ ?   (d) 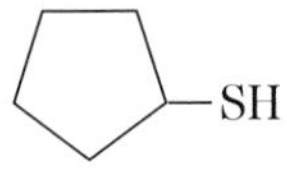 [ring]—SH + (O) ⟶ ?

Cocaine ("free base," "crack")

Cocaine hydrochloride

Cocaine in either form has two ester groups. We'll learn in Chapter 13 that esters react with water under acid or base catalysis or in the presence of the right enzyme. Because digestive reactions hydrolyze esters, cocaine is not taken orally (through the mouth). The free base ("crack"), having the much lower vapor pressure, can be inhaled. To make the vapor pressure higher and allow cocaine to be even more readily ab-

sorbed, it is heated first and then the vapors are inhaled or smoked. Cocaine migrates through the soft tissue of the nose directly into the bloodstream.

The value of "crack" over the salt to cocaine users is that "crack" can be introduced into circulation by inhalation, avoiding an injection, and the results desired by the user occur more rapidly and with greater intensity. The blood, however, normally carries an enzyme that catalyzes the hydrolysis of the ester groups of cocaine, so cocaine's effects by any entry route are quite shortlived. Individuals with atypically low levels of the enzyme experience effects of extreme clinical toxicity. Heavy users take cocaine by injection (it gets into circulation faster), and do so frequently. (Because the blood is slightly basic, the salt changes to the free base easily by deprotonation.)

Overdoses of cocaine cause tremors, convulsions, and delirium. Either respiratory failure or collapse of the cardiovascular system can cause death. Short of this, heavy users eventually experience hallucinations, paranoid delusions, and, frequently, dangerously violent behavior.

11.6 OCCURRENCE, NAMES, AND PHYSICAL PROPERTIES OF AMINES

The amino group, NH_2, has some of the properties of ammonia, including the ability to be involved in hydrogen bonding.

Both the *amino group* and its protonated form occur in living things in proteins, enzymes, and nucleic acids (the chemicals that carry our genes). When a carbonyl group is attached to nitrogen, the properties change sufficiently to warrant a different chemical family, the *amides*.

$$-NH_2 \qquad -NH_3^+ \qquad \overset{\displaystyle O}{\underset{\displaystyle |}{-\overset{||}{C}-N-}}$$

Amino   Protonated   Amide
group   amino group   system

Amines Are Ammonia-like Compounds. The **amines** are organic relatives of ammonia in which one, two, or all three of the H atoms on an ammonia molecule have been replaced by a hydrocarbon group (Table 11.4). Some examples follow.

CH_3NH_2 CH_3NHCH_3 CH_3NCH_3 $CH_3NHCH_2CH_3$ $-NH_2$

Methylamine   Dimethylamine   Trimethylamine   Methylethylamine   Aniline

■ These are the common, not the IUPAC names.

The hydrocarbon groups, as you can see, do not have to be alike, and they can be aromatic groups, as in aniline. The nitrogen atom can be a part of a ring, and such compounds, like pyridine, are classified as *heterocyclic amines*.

Pyridine

TABLE 11.4   Amines

Common Name	Structure	BP (°C)	Solubility in Water
Methylamine	CH_3NH_2	−8	Very soluble
Dimethylamine	$(CH_3)_2NH$	8	Very soluble
Trimethylamine	$(CH_3)_3N$	3	Very soluble
Ethylamine	$CH_3CH_2NH_2$	17	Very soluble
Diethylamine	$(CH_3CH_2)_2NH$	55	Very soluble
Triethylamine	$(CH_3CH_2)_3N$	89	14 g/dL
Propylamine	$CH_3CH_2CH_2NH_2$	49	Very soluble
Aniline	$C_6H_5NH_2$	184	4 g/dL

For a compound to be an amine, its molecules must have a nitrogen with three bonds, but none *directly* to a C=O group. If one bond is, the substance is an **amide**. Structure **1**, for example, is an amide, not an amine. Structure **2**, on the other hand, has two functional groups, a keto group and an amino group, because its N atom is not joined *directly* to C=O. One difference is that amines, like ammonia, are basic and amides are not.

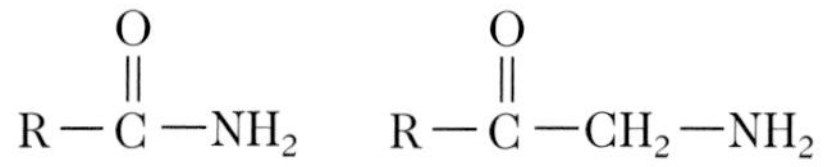

$$\underset{\mathbf{1}}{R-\overset{\overset{\displaystyle O}{\|}}{C}-NH_2} \qquad \underset{\mathbf{2}}{R-\overset{\overset{\displaystyle O}{\|}}{C}-CH_2-NH_2}$$

The common names of the simple, aliphatic amines are made by writing the names of the alkyl groups on nitrogen in front of the word *amine* (leaving no space). Table 11.4 gives examples. Many amines are physiologically active, including some hormones and drugs (see Special Topic 11.2).

Amines Form Weaker Hydrogen Bonds than Alcohols.   When the boiling points of compounds of about the same formula mass and chain branching are compared, amines of the type RNH_2 boil higher than alkanes but lower than alcohols. The size of $\delta-$ on N in RNH_2 is less than it is on O in ROH, because N is less electronegative than O. As a result, the size of $\delta+$ on H is less in the NH_2 group than in the OH group. The smaller sizes of $\delta+$ and $\delta-$ in amines still permit hydrogen bonding (Figure 11.3), but it is weaker than among the alcohols. Hydrogen bonding also helps amines to be much more soluble in water than alkanes, as Figure 11.3 also explains. Although hydrogen bonding in amines is weak, it is very important at the molecular level of life, because it stabilizes the special molecular shapes of proteins and nucleic acids.

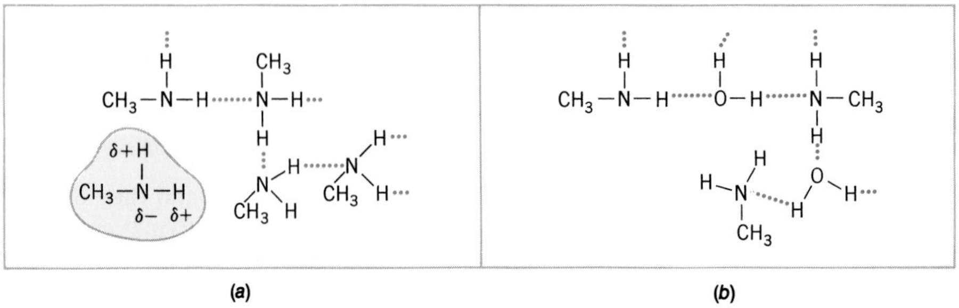

Compound	Formula Mass	BP (°C)
CH_3CH_3	30	−89
CH_3NH_2	31	−6
CH_3OH	32	65

FIGURE 11.3
Hydrogen bonds *(a)* in amines and *(b)* in aqueous solutions of amines.

(a)   (b)

11.7 CHEMICAL PROPERTIES OF AMINES

The amino group is a proton acceptor, and the protonated amino group is a proton donor.

We examine two chemical properties of amines that will be particularly important to our study of biochemicals: the basicity of amines (this section) and their conversion to amides (Chapter 13).

Amines Can Neutralize Strong Acids. Although amines, like ammonia, are *weak* bases, they are able to establish an equilibrium in water in which a small excess of $[OH^-]$ over $[H^+]$ exists. The solution is therefore basic.

$$R-NH_2 + H_2O \rightleftharpoons R-NH_3^+ + OH^-$$

$$\text{Amine} \qquad\qquad\qquad \substack{\text{Protonated} \\ \text{amine}}$$

Notice that when R = H, R—NH$_2$ is ammonia, NH$_3$, and R—NH$_3^+$ is the ammonium ion, NH$_4^+$. This equilibrium is exactly analogous to the behavior of ammonia in water. The equilibrium and the analogy to aqueous ammonia tell us that biochemicals with amino groups can increase the pH of a solution, a property of considerable importance at the molecular level of life, as we have often emphasized.

$$R-\overset{\displaystyle H}{\underset{\displaystyle H}{N:}} \;+\; H-\overset{\displaystyle H}{\underset{\displaystyle H}{\overset{+}{O}:}} \longrightarrow R-\overset{\displaystyle \overset{+}{H}}{\underset{\displaystyle H}{N}}-H \;+\; :\overset{\displaystyle H}{\underset{\displaystyle H}{O}:}$$

Amine (or   Hydronium   Protonated amine
ammonia,   ion   (or the ammonium
when R = H)   ion when R = H)

For example, hydrochloric acid is neutralized by methylamine as follows.

$$CH_3NH_2 \;+\; H_3O^+ + Cl^- \longrightarrow CH_2NH_3^+Cl^- \;+\; H_2O$$

$$\substack{\text{Methylamine}} \quad \substack{\text{Hydrochloric} \\ \text{acid}} \qquad \substack{\text{Methylammonium} \\ \text{chloride}}$$

It doesn't matter if the nitrogen atom in an amine bears one, two, or three hydrocarbon groups. The amine can still neutralize strong acids, because the reaction involves only the unshared pair of electrons on N, not any of the bonds to the other groups.

> The previously unshared electron pair on N now holds the H atom to N

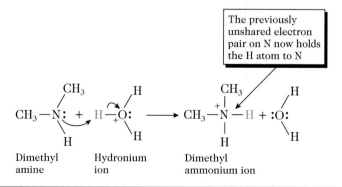

Dimethyl   Hydronium   Dimethyl
amine   ion   ammonium ion

■ PRACTICE EXERCISE 8   What are the structures of the cations that form when the following amines react completely with hydrochloric acid?
(a) aniline   (b) trimethylamine   (c) NH$_2$CH$_2$CH$_2$NH$_2$

Protonated Amines Can Neutralize Strong Bases.   A combination of a protonated amine and an anion make up an organic salt called an **amine salt.** Like all salts, amine salts are crystalline solids at room temperature. In addition, like the salts of the ammonium ion, nearly all amine salts of strong acids are soluble in water, *even when the parent amine is not.* Amine salts are much more soluble in water than amines because the *full* charges carried by the ions of an amine salt can be much better hydrated by water molecules than the amine itself, where only the small, partial charges of polar bonds occur.

■ Remember that *cations* are positively charged ions. *Anions* are negatively charged ions.

Protonated amine cations also neutralize the hydroxide ion and revert to amines in the following manner (where we show only skeletal structures).

$$-\overset{+}{\underset{|}{\text{N}}}-\text{H} + :\overset{..}{\underset{..}{\text{O}}}-\text{H} \longrightarrow -\text{N}: + \quad \overset{\text{H}}{\underset{..}{\overset{/}{\underset{..}{\text{O}}}}}-\text{H}$$

Protonated   Hydroxide   Amine
amine   ion

■ Amino acids, the monomers for proteins, have the NH_3^+ group in acidic and neutral solutions.

For example

$$CH_3NH_3^+ + OH^- \longrightarrow CH_3NH_2 + H_2O$$

$$CH_3CH_2\overset{+}{N}H_2CH_3 + OH^- \longrightarrow CH_3CH_2NHCH_3 + H_2O$$

The Amino Group Is a Solubility Switch.   We have just learned that putting a proton on an amino group and taking it off are easily done at room temperature simply by adding an acid and then a base. We have also learned that the protonated amine is more soluble in water than the amine. This makes the amino group an excellent "solubility switch" when it's on a molecule.

The solubility of an amine can be switched on simply by adding enough strong acid to protonate the amine molecules. Triethylamine, for example, is insoluble in water, but we can switch on its solubility by adding a strong acid, like hydrochloric acid. The amine dissolves as its protonated form is produced.

■ The amino group is a solubility switch.

$$OH^- \quad \underset{\underset{\text{(more soluble)}}{RNH_3^+}}{\overset{\overset{\text{RNH}_2}{\text{(less soluble)}}}{\LARGE \circlearrowleft}} \quad H^+$$

$$(CH_3CH_2)_3N: + HCl(aq) \longrightarrow (CH_3CH_2)_3\overset{+}{N}HCl^-(aq)$$
Triethylamine   Hydrochloric   Triethylammonium
(water insoluble)   acid   chloride (water soluble)

Being an *ionic* compound, triethylammonium chloride is water soluble.

We can just as quickly and easily bring the amine back out of solution by adding a strong base, like the hydroxide ion. It takes the proton off the protonated amine and gives the less soluble form.

$$(CH_3CH_2)_3\overset{+}{N}HCl^-(aq) + OH^-(aq) \longrightarrow (CH_3CH_2)_3N: + H_2O$$
Triethylammonium   (supplied by   Triethylamine
chloride (water-soluble)   NaOH, for   (water-insoluble)
                    example)

The significance of this "switching" relationship is that the solubilities of complex compounds that have the amine function can be changed almost instantly simply by adjusting the pH of the medium. One application of this property involves medicinals.

Alkaloid Medicinals Are Complex Amines.   A number of amines obtained from the bark, roots, leaves, flowers, or fruit of various plants are useful drugs. These naturally occurring, acid-neutralizing, physiologically active amines are called **alkaloids,** and morphine and quinine are just two examples.

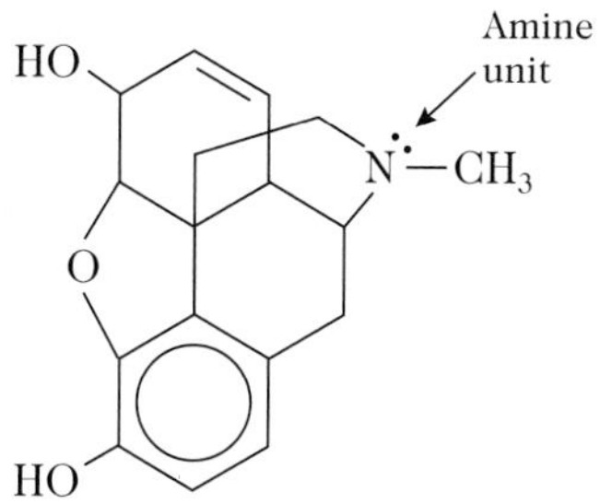

Morphine

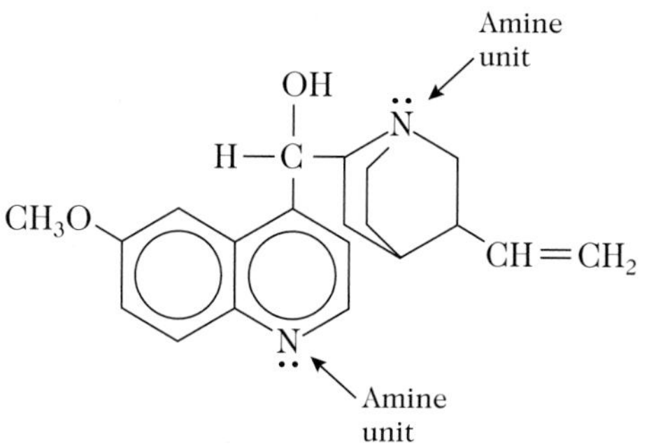

Quinine

To make it easier to administer alkaloidal drugs in the dissolved state, they are often prepared as their water-soluble amine salts. Morphine, for example, a potent sedative and painkiller, is given as morphine sulfate, the salt of morphine and sulfuric acid. Quinine, an antimalarial drug, is available as quinine sulfate. Special Topic 11.1 tells about a few other physiologically active amines, including cocaine, whose physical properties change as they are in the forms of salts.

EXAMPLE 11.6 Writing the Structure of the Product of the Deprotonation of the Cation of an Amine Salt

The protonated form of amphetamine is shown below. What is the structure of the product of its reaction with OH^-?

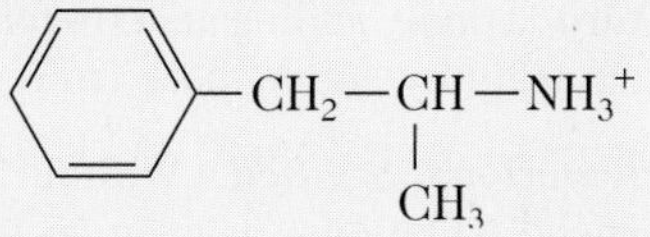

ANALYSIS Because OH^- removes just one H^+ from the nitrogen atom of a protonated amine's cation, all we have to do is reduce the number of H atoms on the nitrogen by one and cancel the positive charge.

SOLUTION Amphetamine is

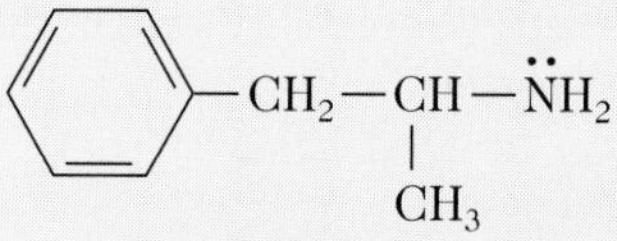

Amphetamine

■ PRACTICE EXERCISE 9   Write the structures of the products after the following protonated amines have reacted with OH^- in a 1:1 mole ratio.

(a)

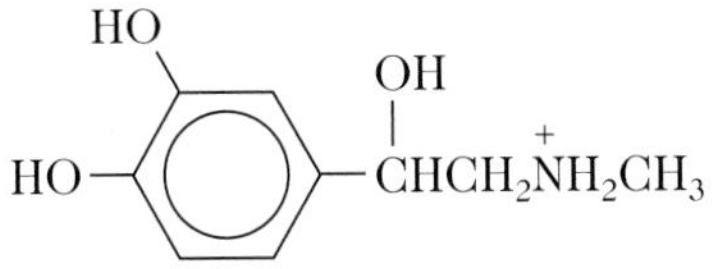

Epinephrine (adrenaline), a hormone is given here in its protonated form. Its chloride salt in a 0.1% solution, is injected in some cardiac failure emergencies. (See also Special Topic 11.2.)

(b)

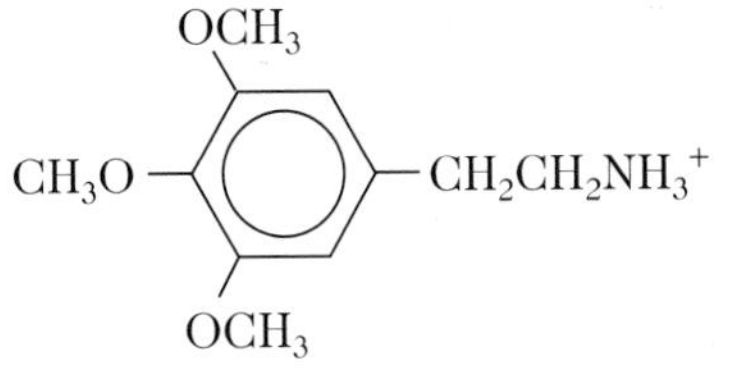

$$CH_3O \underset{OCH_3}{\overset{OCH_3}{\bigcirc}} CH_2CH_2NH_3^+$$

Mescaline, a mind-altering hallucinogen is shown here in its protonated form. It is isolated from the mescal button, a growth on top of the peyote cactus. Indians in the southwestern United States have used it in religious ceremonies.

SUMMARY

Alcohols   The alcohol system has an OH group attached to a saturated carbon. The common names have the word *alcohol* following the name of the alkyl group.

Alcohol molecules hydrogen bond to each other and to water molecules. By the action of heat and an acid catalyst they can be dehydrated internally to give carbon–carbon double bonds or externally to give ethers. Primary alcohols can be oxidized to aldehydes and to carboxylic acids. Secondary alcohols can be oxidized to ketones. Tertiary alcohols cannot be oxidized (without breaking up the carbon chain). The OH group in alcohols does not function as either an acid or a base in the ordinary sense.

Phenols   When the OH group is attached to a benzene ring, the system is the phenol system, and it is now acidic enough to neutralize strong bases.

Ethers   The ether system, R—O—R′, does not react at room temperature or body temperature with strong acids, bases, oxidizing agents, or reducing agents. It can accept hydrogen bonds but cannot donate them.

Thioalcohols   The thioalcohols, RSH, are easily oxidized to disulfides, RS—SR, and these are easily reduced back to the original thioalcohols.

Amines and Protonated Amines   When one, two, or three of the hydrogen atoms in ammonia are replaced by an organic group (other than a carbonyl group), the result is an amine. The nitrogen atom can be part of a ring, as in heterocyclic amines. Like ammonia, the amines are weak bases, but all can form salts with strong acids. The cations in these salts are protonated amines. Amine salts are far more soluble in water than their parent amines. Protonated amines are easily deprotonated by any strong base to give back the original and usually far less soluble amine. Thus, any compound with the amine function has a "solubility switch," because its solubility in an aqueous system can be turned on by adding acid (to form the amine salt) and turned off again by adding base (to recover the amine).

Reactions Studied   Without attempting to present balanced equations, or even all of the inorganic products, we can summarize the reactions studied in this chapter as follows. We have learned two "map signs" for alcohols, dehydration (either to alkenes or ethers) and oxidation. We have studied just one map sign apiece for thioalcohols, disulfides, and phenols. We also found that ethers are as unreactive as alkanes. Finally, we studied the chemistry of the amino group as a solubility switch. Here then is a summary of these properties.

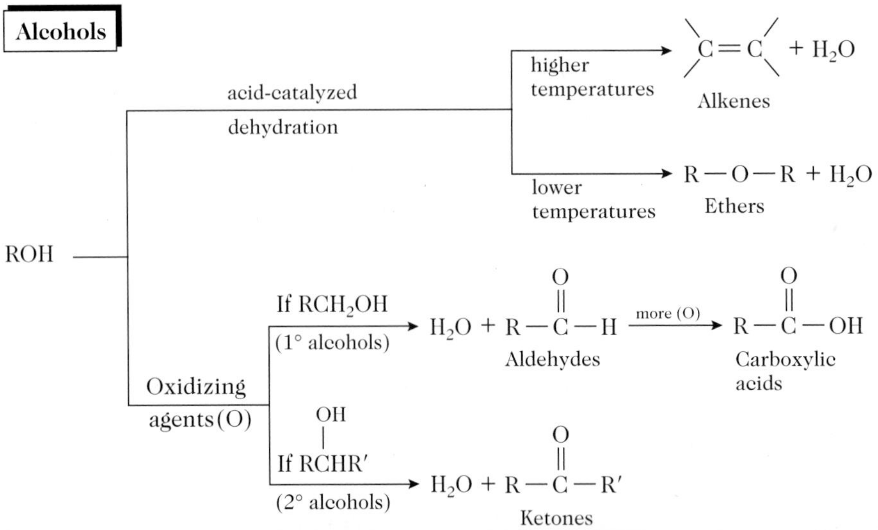

```
┌──────────┐
│ Alcohols │
└──────────┘
                              acid-catalyzed          higher          \      /
                              dehydration             temperatures     C = C    + H₂O
                                                                      /      \
                                                                        Alkenes

                                                      lower
                                                      temperatures    R—O—R + H₂O
                                                                        Ethers

  ROH
                                                          O                     O
                           If RCH₂OH                      ‖         more (O)     ‖
                           (1° alcohols)    H₂O + R—C—H  ─────────→  R—C—OH
                                                      Aldehydes               Carboxylic
                                                                             acids
         Oxidizing
         agents(O)         OH
                           │                              O
                           If RCHR′                       ‖
                           (2° alcohols)    H₂O + R—C—R′
                                                      Ketones
```

Phenols

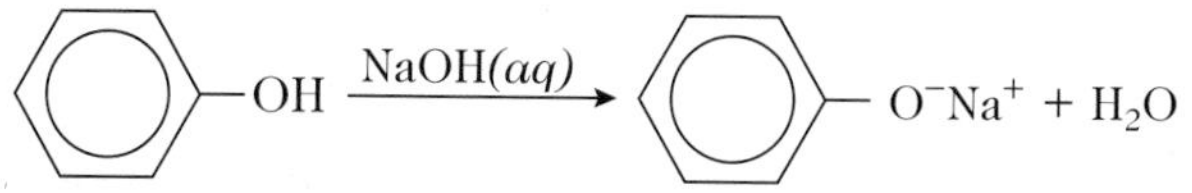

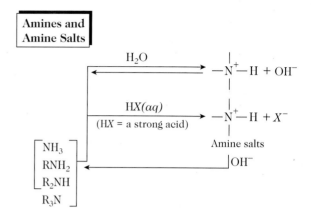

Thioalcohols and Disulfides

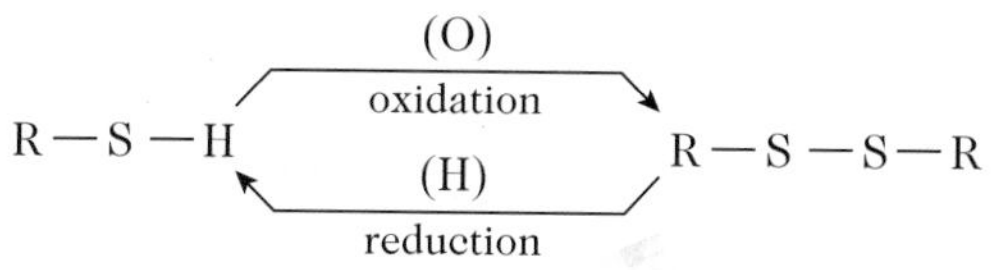

$$R-S-H \overset{(O)}{\underset{(H)}{\rightleftarrows}} R-S-S-R$$

oxidation / reduction

Amines and Amine Salts

REVIEW EXERCISES

The answers to Review Exercises whose numbers are in color are found in Appendix V. The answers to the other Review Exercises are found in the Study Guide that accompanies this text. The more challenging questions are marked with asterisks.

Functional Groups

11.1 Estriol is a hormone produced in the placenta during pregnancy. Identify by name the functional groups associated with the numbers given by its structure. Distinguish among 1°, 2°, and 3° alcohols and the phenol system.

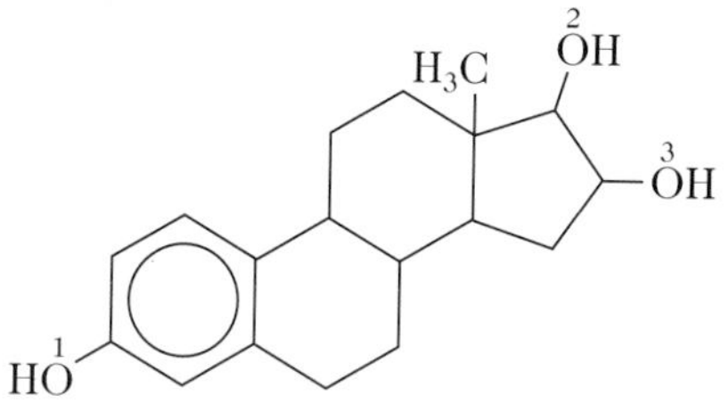

Estriol

11.2 What are the functional groups identified by the numbers in the structure of geranial, a constituent of oil of lemon grass?

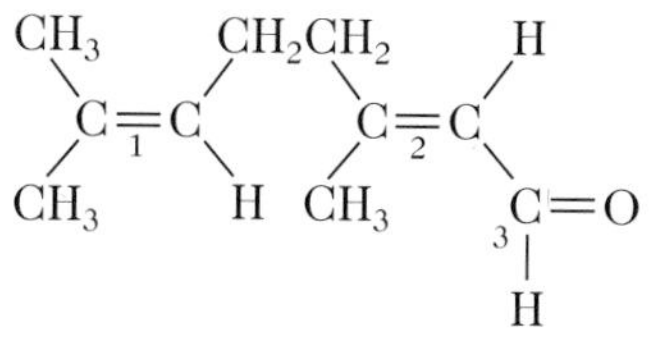

Geranial

11.3 Name the functional groups identified by the numbers in the structure of cortisone, a drug used in treating certain forms of arthritis. If a group is an alcohol, state if it is a 1°, 2°, or 3° alcohol.

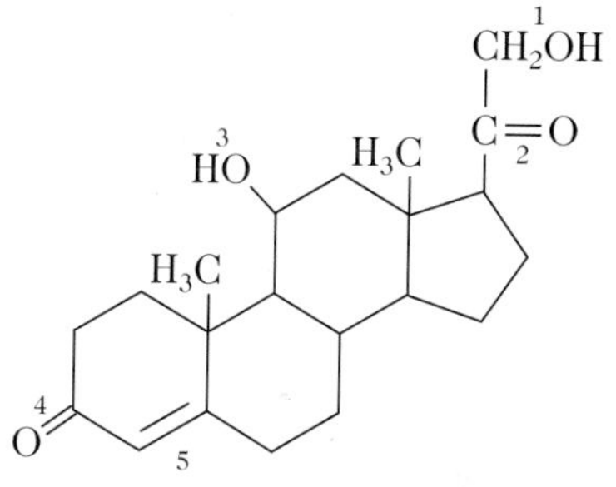

Cortisone

11.4 Give the names of the functional groups identified by the numbers in prostaglandin E_1, one of a family of compounds that are smooth muscle stimulants.

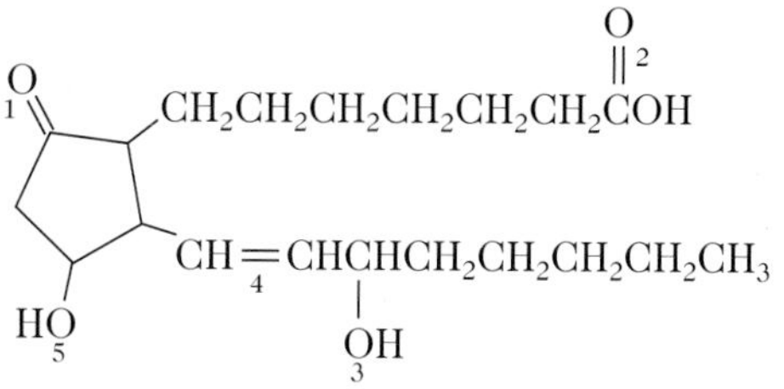

Prostaglandin E_1

*****11.5** The following compounds are all very active physiological agents. Name the numbered functional groups that are present in each.

(a)

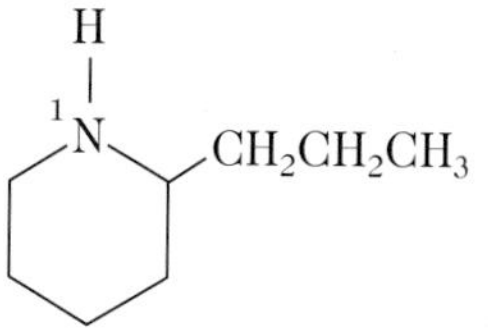

Coniine, the poison in the extract of hemlock that was used to execute the Greek philosopher Socrates

(b)

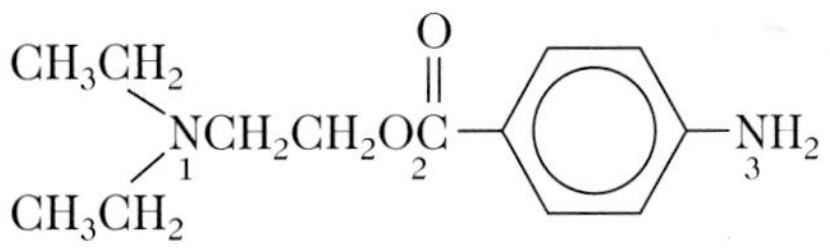

Novocaine, a local anesthetic

(c)

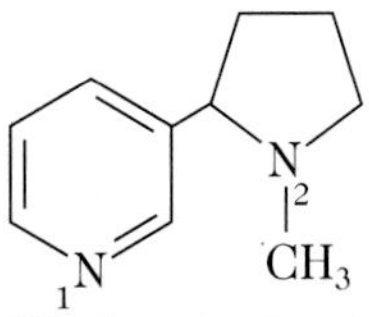

Nicotine, a poison in tobacco leaves

(d)

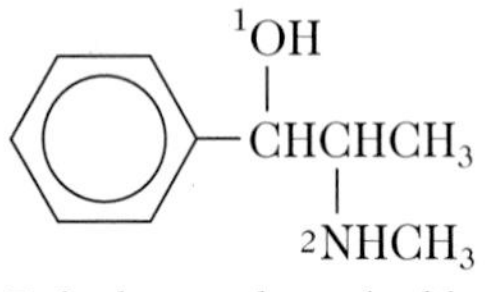

Ephedrine, a bronchodilator

*11.6 Some extremely potent, physiologically active compounds are in the following list. Name the functional groups that they have.

(a)

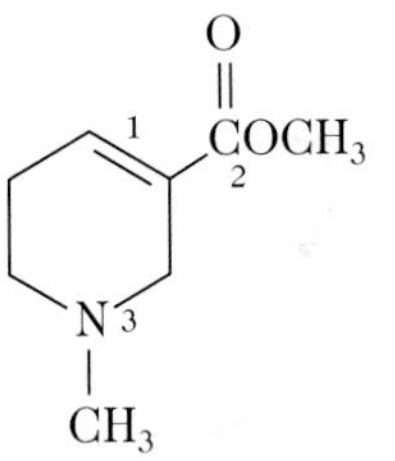

Arecoline, the most active component in the nut of the betel palm (This nut is chewed daily as a narcotic by millions of inhabitants of parts of Asia and the Pacific islands.)

(b)

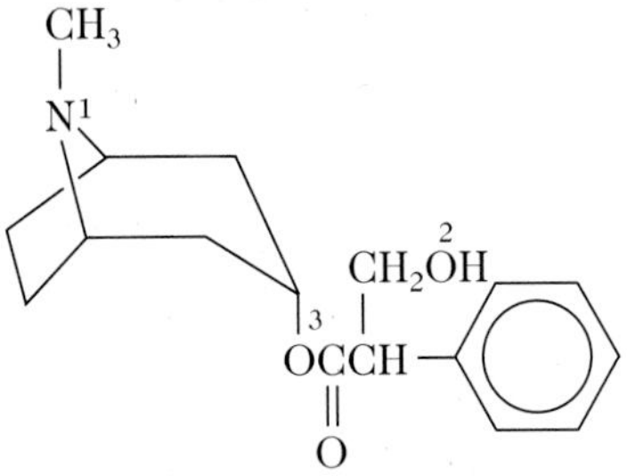

Hyoscyamine, a constituent of the seeds and leaves of henbane, and a smooth muscle relaxant (A similar form is called atropine, a drug used to counteract nerve poisons.)

(c)

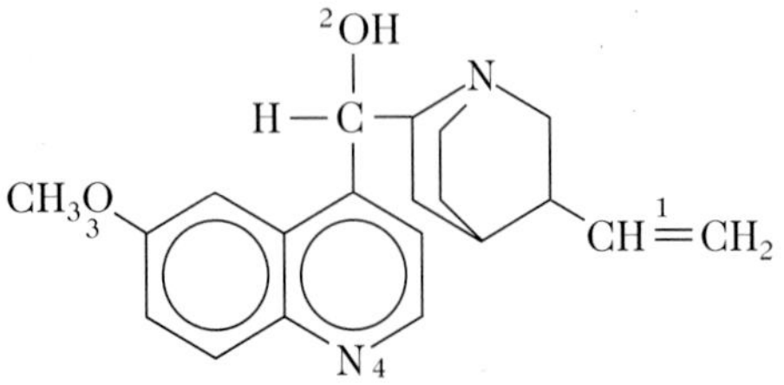

Quinine, a constituent of the bark of the chinchona tree in South America used to treat malaria

(d)

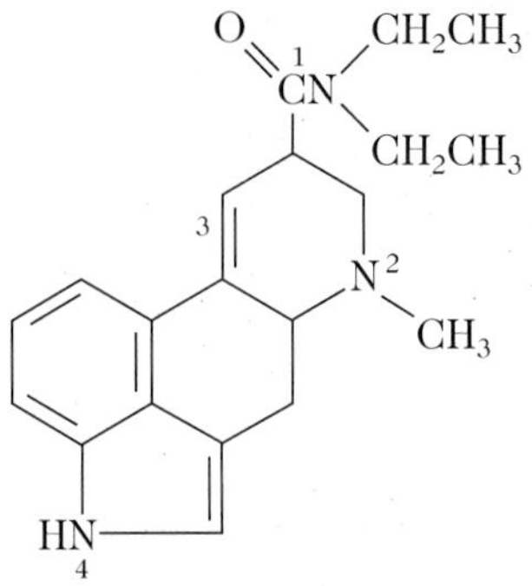

Lysergic acid diethylamide (LSD), a constituent of diseased rye grain and a notorious hallucinogen

Structures and Names of Alcohols

11.7 Write the structure of each compound.
(a) isobutyl alcohol   (b) isopropyl alcohol
(c) propyl alcohol   (d) glycerol

11.8 Write the structures of the following compounds.
(a) methyl alcohol   (b) *t*-butyl alcohol
(c) ethyl alcohol   (d) butyl alcohol

11.9 Give the common names of the following compounds.
(a) CH_3CH_2OH (b) $HOCH_2CH_2OH$

(c) $HOCH_2\overset{\underset{|}{OH}}{C}HCH_2OH$ (d) $CH_3\overset{\underset{|}{OH}}{C}HCH_3$

11.10 What are the common names of the following compounds?

(a) $CH_3CH_2CH_2OH$ (b) $HOCH_2CH_2CH_2CH_3$

(c) $HO\overset{\underset{|}{CH_3}}{\underset{\underset{CH_3}{|}}{C}}CH_3$ (d) $HOCH_2\overset{\underset{|}{CH_3}}{C}H\diagdown CH_3$

11.11 What is the structure and the common name of the simplest, *stable* dihydric alcohol?

11.12 Give the structure and the common name of the simplest, *stable* trihydric alcohol.

Physical Properties of Alcohols

11.13 Draw a figure that illustrates a hydrogen bond between two molecules of ethyl alcohol. Use a dotted line to represent this bond, and write in the $\delta+$ and the $\delta-$ symbols where they belong.

11.14 When ethyl alcohol dissolves in water, its molecules slip into the hydrogen bonding network in water. Draw a figure that illustrates this. Use dotted lines for hydrogen bonds, and place the $\delta+$ and $\delta-$ symbols where they belong.

*11.15 Arrange the following compounds in order of increasing boiling points. Place the letter symbol of the compound that has the lowest boiling point on the left end of the series.

CH$_3$CH$_2$CH$_2$OH   CH$_3$CH$_3$ HOCH$_2$CH$_2$OH   CH$_3$CH$_2$OCH$_3$
      A   B   C   D

         <   <   <

Lowest BP   Highest BP

*11.16 Arrange the following compounds in their order of increasing solubility in water. Place the letter symbol of the compound that has the least solubility on the left end of the series.

CH$_3$CH$_2$CH$_2$CHCH$_2$CH$_3$ HOCH$_2$CH$_2$CHCH$_2$OH   CH$_3$CH$_2$CH$_2$CH$_2$CH$_3$
           | |
          OH   OH

      A   **B**   **C**

         <   <

Lowest solubility   Highest solubility

Chemical Properties of Alcohols

11.17 Write the structures of the alkenes that form when the following alcohols undergo acid-catalyzed dehydration. Where more than one alkene is possible, identify which most likely forms in the greatest relative amount.

(a) CH$_3$CHCH$_2$OH   (b) CH$_3$CHCH$_2$CH$_3$
       | |
      CH$_3$ OH

(c) HO   CH$_3$ (d) [benzene ring]—CHCH(CH$_3$)$_2$
                                         |
    [cyclopentane]   OH

(e)   CH$_3$ (f)
              |
   CH$_3$CH$_2$CH$_2$CCH$_3$ HO—[cyclohexane]—CH$_3$
             |
            OH

11.18 Write the structures of the products of the oxidation of the alcohols given in Review Exercise 11.17. If the alcohol is a 1° alcohol, give the structures of both the aldehyde and the carboxylic acid that could be made by varying the quantities of the oxidizing agent.

*11.19 Write the structures of any alcohols that could be dehydrated to give each of the following alkenes. In some instances, more than one alcohol would work.

(a) CH$_2$=CHCH$_3$ (b) [cyclopentene]

           CH$_2$
           ||
(c) CH$_3$CCH$_3$ (d) [cyclohexene]—CH$_3$

*11.20 Write the structure of any alcohol that could be used to prepare each of the following compounds by an oxidation.

       O   O
       || ||
(a) HOCCH$_2$CH$_2$CH$_3$ (b) CH$_3$CH$_2$CH$_2$CCH(CH$_3$)$_2$

(c)   O   (d)
        || O
   H—C—[cyclopentane]   [benzene]—||
                                    COH

11.21 An unknown alcohol is either **A** or **B**.

   HO   CH$_3$ OH
   [cyclopentane]   [cyclopentane]

      A   **B**

When the unknown is shaken with a few drops of aqueous potassium permanganate, the purple color disappears and a brown precipitate forms. Which alcohol, **A** or **B**, is it? Explain how the given information led you to your conclusion.

11.22 An unknown liquid is either **A** or **B**.

     OH   CH$_3$
   [cyclopentane]   [cyclopentane]

     A   **B**

When it was shaken with aqueous Na$_2$Cr$_2$O$_7$ in acid, the orange color of the mixture did not change. What is the unknown? Explain. What would have been *seen* if the other compound had been the liquid?

Phenols

*11.23 What is one difference in *chemical* properties between **A** and **B**?

     OH   OH

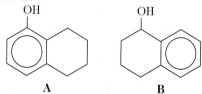

     A   **B**

11.24 A compound was either **A** or **B**.

      Br
   [benzene]—OH   [cyclohexane]—CH$_2$OH

     A   **B**

The compound dissolved in aqueous sodium hydroxide but not in water. Which compound was it? How can you tell? (Write an equation.)

Ethers

11.25 Which of the following compounds contain the ether function?

(a) $CH_3CH_2OCH_2CH_2CH_2OCH_3$

(b) $CH_3\overset{\displaystyle O}{\overset{\|}{C}}OCH_3$

(c) $CH_3O\overset{\displaystyle O}{\overset{\|}{C}}OCH_3$

(d)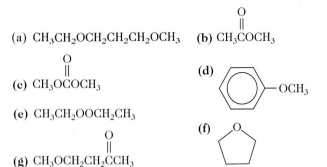

(e) $CH_3CH_2OOCH_2CH_3$

(f)

(g) $CH_3OCH_2CH_2\overset{\displaystyle O}{\overset{\|}{C}}CH_3$

11.26 Write the structures of the ethers that can be made from the following alcohols.

(a) $HOCH_2CH_3$

(b) CH_3CHCH_3
         $|$
         OH

(c) $CH_3OCH_2CH_2OH$

(d)

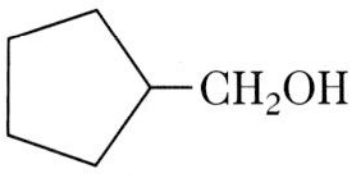

*11.27 Write the structures of the alcohols that could serve as the starting materials to prepare each of the following ethers.

(a) $CH_3CH_2CH_2OCH_2CH_2CH_3$

(b) $CH_3CHCH_2OCH_2CHCH_3$
         $|$ $|$
         CH_3 CH_3

(c)

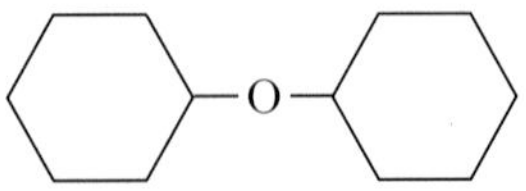

(d)

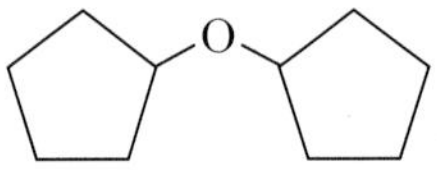

*11.28 Suppose that a mixture of 0.50 mol of ethanol, 0.50 mol of methanol, and a catalytic amount of sulfuric acid is heated under conditions that favor only ether formation. What organic products will be obtained? Write their structures.

11.29 What happens chemically when the following compound is heated with aqueous sodium hydroxide: $CH_3CH_2CH_2OCH_2CH_2CH_3$

Thioalcohols and Disulfides

11.30 We did not study rules for naming thioalcohols, but the patterns of the names used in Section 11.5 make these rules obvious. Write the structures of the following compounds.
(a) diethyl disulfide   (b) 1,2-propanedithiol
(c) isopropyl mercaptan   (d) 1-propanethiol

11.31 Complete the following reaction sequences by writing the structures of the organic products that form.
(a) $CH_3CH_2CH_2SH + (O) \rightarrow$
(b) $(CH_3)_2CHCH_2-S-S-CH_2CH(CH_3)_2 + (H) \rightarrow$

(c) 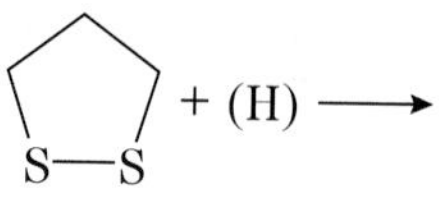 $+ (H) \longrightarrow$

(d) $HSCH(CH_3)_2 + (O) \rightarrow$

11.32 Ethanol, methanethiol, and propane have nearly the same formula masses, but ethanol boils at 78 °C, methanethiol at 6 °C, and propane at −42 °C. What do the boiling points suggest about the possibility of hydrogen bonding in the thioalcohol family? Does hydrogen bonding occur at all? Are the hydrogen bonds as strong as those in the alcohol family?

The Amine System

11.33 Classify the following as aliphatic, aromatic, or heterocyclic amines or amides, and name any other functional groups, too.

(a) $CH_3OCH_2\overset{\displaystyle O}{\overset{\|}{C}}NH_2$

(b) $CH_3O\overset{\displaystyle O}{\overset{\|}{C}}CH_2NH_2$

(c)

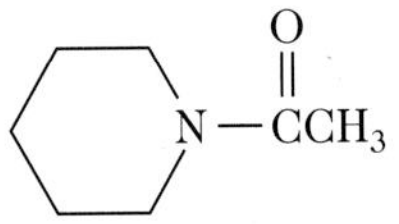

(d)

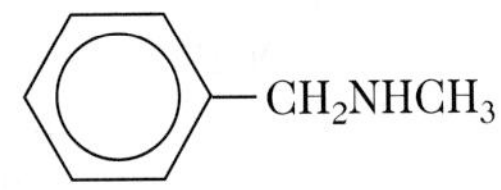

11.34 Classify each of the following as aliphatic, aromatic, or heterocyclic amines or amides. Name any other functional groups that are present.

(a)

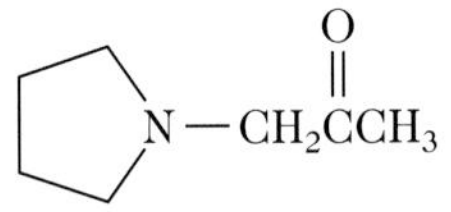

(b)

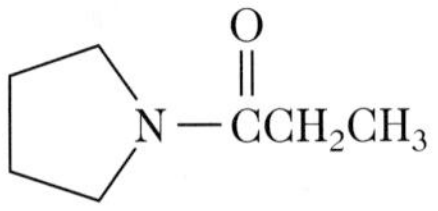

(c)

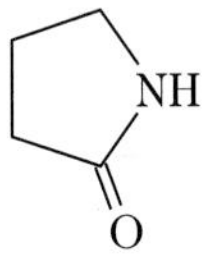

(d)

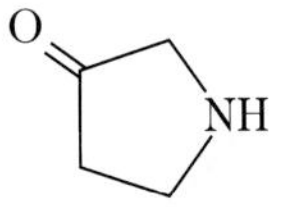

Names of Amines

11.35 Give the common names of the following compounds or ions.

(a) $CH_3CH_2CH_2NHCHCH_3$
                     $|$
                     CH_3

(b) $CH_3CH_2CH_2NCH_2CH_3$
                   $|$
                   CH_3

(c) $(CH_3)_3CNHCH_2CH(CH_3)_2$

(d) $CH_3CH_2CH_2NHCH_2CH_2$
                               $|$
                               CH_3

11.36 What are the common names of the following compounds?

(a) $CH_3CHCH_2CH_3$
         $|$
         NH_2

(b)

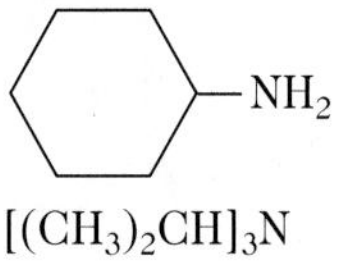

(c) $C_6H_5NH_2$

(d) $[(CH_3)_2CH]_3N$

Chemical Properties of Amines and Amine Salts

11.37 Complete the following reaction sequences by writing the structures of the organic products. If no reaction occurs, write "no reaction."
(a) $CH_3CH_2CH_2NH_2 + HCl(aq) \longrightarrow$
(b) $CH_3CH_2CH_2NH_3^+ \ Cl^- + NaOH(aq) \longrightarrow$
(c) $CH_3CH_2CH_2NH_3^+ \ Cl^- + HCl(aq) \longrightarrow$
(d) $CH_3CH_2CH_2NH_2 + NaOH(aq) \longrightarrow$

11.38 Write the structures of the organic products that form in each situation. Assume that all reactions occur at room temperature. (Some of the named compounds are described in Review Exercises 11.5 and 11.6.) If no reaction occurs, write "no reaction."

(a)

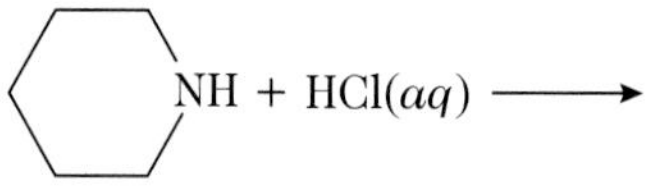

(b)

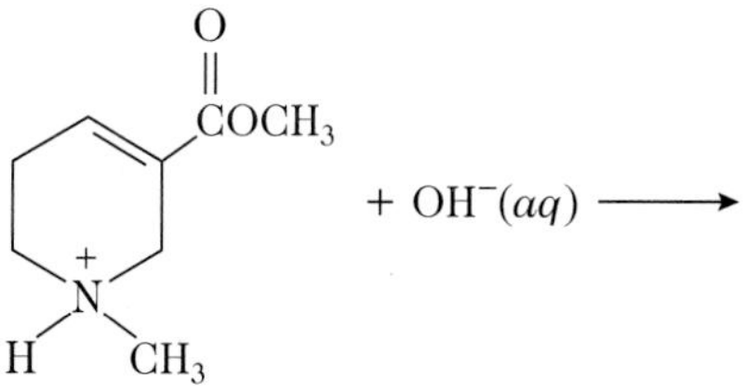

Protonated form
of arecoline

(c)

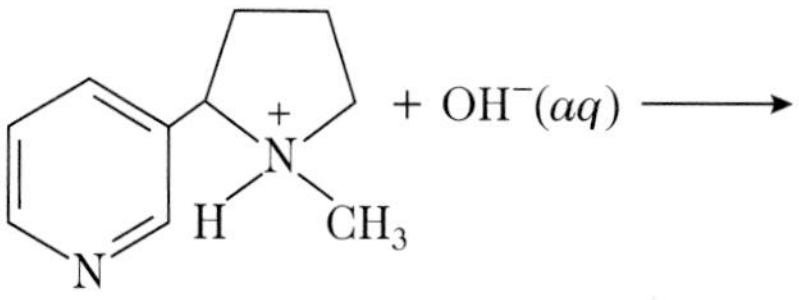

Protonated form
of nicotine

(d)

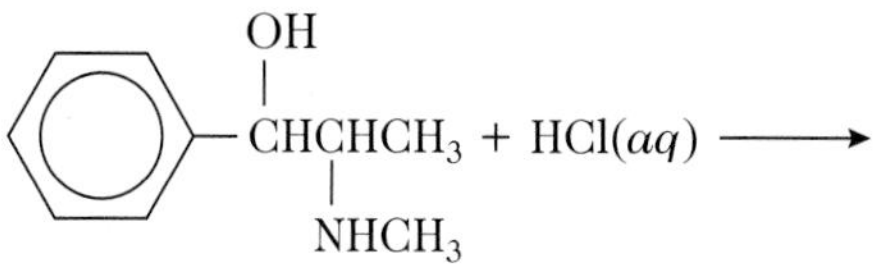

Ephedrine

(e)

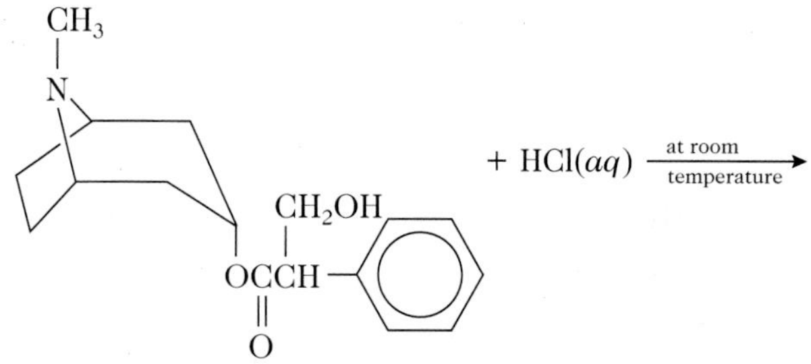

Hyoscyamine

11.39 Which is the stronger proton acceptor or base, **A** or **B**? Explain.

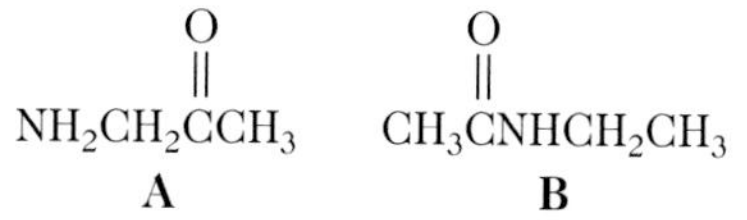

$$NH_2CH_2CCH_3 \qquad CH_3CNHCH_2CH_3$$
$$\text{A} \qquad\qquad\qquad \text{B}$$

11.40 Which is the stronger proton acceptor, **A** or **B**? Explain.

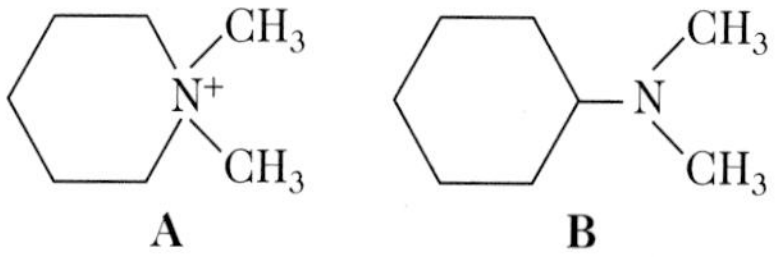

A   B

Organic Reactions

*11.41** Examine each of the following sets of reactants and conditions and decide if a reaction occurs. If one does, write the structures of the organic products. If no reaction occurs, write "no reaction." Some of the parts involve *alcohols* and their reactions. If the reaction is an oxidation of a 1° alcohol, write the structure of the *aldehyde* that can form, not the carboxylic acid.

When an alcohol is in the presence of an acid catalyst, we have learned that the alcohol might be dehydrated to an *alkene* or to an *ether*. To differentiate between these, use the following guide. When the alcohol structure has no coefficient, write the structure of the alkene that can form. When the alcohol structure has a coefficient of 2, give the structure of the ether that is possible. (This violates our rule that balancing an equation is the *last* step in writing an equation, but we need a signal here to tell what kind of reaction is intended.)

(a)

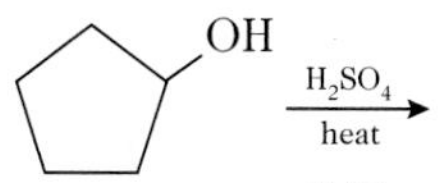

(b) $2HOCH_2CH_3 \xrightarrow[\text{heat}]{H_2SO_4}$

(c) $CH_3\overset{\displaystyle OH}{\underset{\displaystyle |}{C}}HCH_2CH_3 + (O) \longrightarrow$

(d)

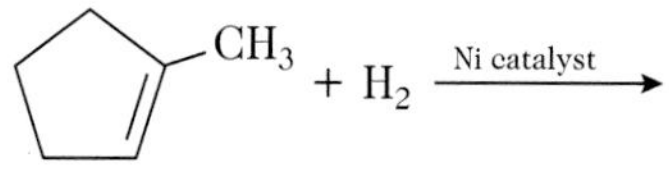

(e) $CH_3CH_2OH + NaOH(aq) \longrightarrow$

(f) $CH_3\overset{\displaystyle OH}{\underset{\displaystyle |}{C}}CH_2CH_2CH_3 + (O) \longrightarrow$
$\underset{\displaystyle CH_3}{}$

(g) $CH_3CH_2\overset{}{\underset{\displaystyle |}{C}}HCH_3 \xrightarrow[\text{heat}]{H_2SO_4}$
$\underset{\displaystyle CH_3}{}$

(h) $CH_3CH=CCH_3 + H_2 \xrightarrow{\text{Ni catalyst}}$
$\underset{\displaystyle CH_3}{}$

(i)

$-CH_3 + H_2O \xrightarrow{H^+}$

(j) HO

CH_3

$+ (O) \longrightarrow$

(k) $CH_3CH_2CH_2NH_2(aq) + HCl(aq) \longrightarrow$

(l) $C_6H_5NH_3^+(aq) + OH^-(aq) \longrightarrow$

(m) $(CH_3)_2CHSH + (O) \longrightarrow$

*11.42 Write the structure of the principal organic product that would be expected in the following situations. Follow the directions given for Review Exercise 11.41. If no reaction occurs, write "no reaction."

(a) CH$_3$CH$_2$CHCH$_3$ + NaOH(aq) $\longrightarrow$
    |
    CH$_2$CH$_3$

(b)

S—S

+ (H) $\longrightarrow$

(c)   CH$_3$
        |
    CH$_3$CHCHCH$_3$ + (O) $\longrightarrow$
            |
            OH

(d) 2CH$_3$CHOH $\xrightarrow[\text{heat}]{H_2SO_4}$
        |
        CH$_3$

(e) HOCH$_2$CH$_2$CH$_3$ + (O) $\longrightarrow$

(f) H$_3$C

CH$_3$

+ H$_2$O $\xrightarrow{H^+}$

(g)

+ NaOH(aq) $\longrightarrow$

(h)

CH$_3$—

—OH + (O) $\longrightarrow$

(i) CH$_3$CHCH$_3$ $\xrightarrow[\text{heat}]{H_2SO_4}$
        |
        OH

(j)   HO
        |
    CH$_3$CHCH$_2$OCH$_3$ + (O) $\longrightarrow$

*11.43 Isopropyl alcohol (C_3H_8O) can be oxidized to acetone (C_3H_6O) by potassium permanganate according to the following equation.

$$3C_3H_8O + 2KMnO_4 \rightarrow 3C_3H_6O + 2MnO_2 + 2KOH + 2H_2O$$

(a) How many moles of acetone can be prepared from 2.50 mol of isopropyl alcohol?

(b) How many moles of potassium permanganate are needed to oxidize 0.180 mol of isopropyl alcohol?

(c) A student began with 12.6 g of isopropyl alcohol. What is the minimum number of grams of potassium permanganate needed for this oxidation?

(d) Refering to part (c), what is the maximum number of grams of acetone that could be made? How many grams of MnO$_2$ are also produced?

Ethyl Alcohol and Alcoholism (Special Topic 11.1)

11.44 What ranges of percentages of ethyl alcohol occur in beer, wine, and "spirits?"

11.45 A vodka rated as 110 proof has what percentage of ethyl alcohol?

11.46 At what level of blood alcohol do the depressant effects become noticeable in most nonalcoholic drinkers?

11.47 Most states define intoxication as corresponding to what level of blood alcohol?

11.48 Describe in detail the connection between the concentration of ethyl alcohol in the blood and what happens (chemically and physically) in the breathalyzer test.

11.49 If the "standard," nonalcoholic adult male consumes two "typical drinks" over a 2-hour period, what will his blood alcohol level most likely be 3 hours from the start of the drinking?

11.50 Name some of the reasons why alcoholism contributes so much to a nation's health bill.

Physiologically Active Amines (Special Topic 11.2)

11.51 What are hormones and, in very broad terms, what is their function?

11.52 Hormones secreted by the adrenal gland are called what kinds of agents?

11.53 Name two hormones secreted by the adrenal gland.

11.54 When dopamine is called a "β-phenylethylamine," what do you suppose the *beta* (β) refers to? What is a use of dopamine in the body?

11.55 In what general family of the physiologically active amines are the amphetamines found?

11.56 What is the structural *difference* between "free base" or "crack" cocaine and the more water-soluble form of cocaine?

11.57 What *chemical* fact is behind the common methods of introducing cocaine into the blood, that is, by inhalation or injection, not by taking it orally?

11.58 Why is "free base" cocaine more volatile than the compound between cocaine and hydrogen chloride?

12

ALDEHYDES AND KETONES

Structural Features and Names of Aldehydes and Ketones

Oxidation of Aldehydes

Reduction of Aldehydes and Ketones

Reactions of Aldehydes and Ketones with Alcohols

Architects design reflecting surfaces to add drama and interest to buildings, as seen here. Depositing a thin film of silver on glass, one of the many ways to make such surfaces, uses a chemical property of aldehydes studied in this chapter. The aldehyde group occurs in carbohydrates.

12.1 STRUCTURAL FEATURES AND NAMES OF ALDEHYDES AND KETONES

Molecules of both aldehydes and ketones contain the carbonyl group.

We need to study aldehydes and ketones to understand carbohydrates. All simple sugars are either polyhydroxy aldehydes or polyhydroxy ketones, as illustrated by the structures of glucose and fructose.

■ We show here only one of the molecular forms of glucose and fructose molecules. Both are present in honey, and glucose is the chief carbohydrate in blood.

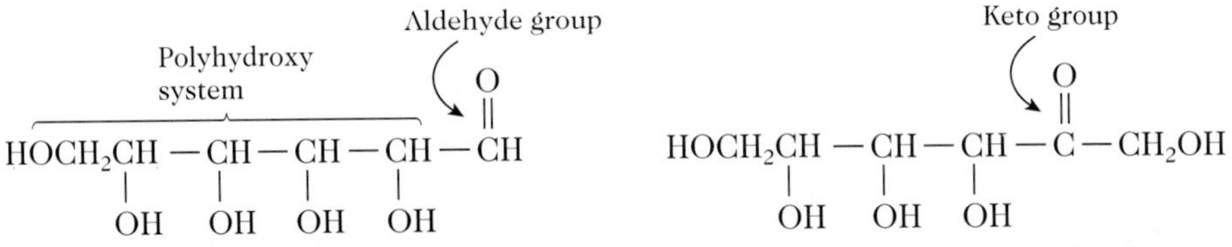

Aldehyde group

Keto group

Polyhydroxy system

$$HOCH_2CH-CH-CH-CH-CH$$
$$\quad\quad\;|\;\;\;\;|\;\;\;\;|\;\;\;\;|$$
$$\quad\quad OH\;\;OH\;\;OH\;\;OH$$

Glucose (open-chain form)

$$HOCH_2CH-CH-CH-C-CH_2OH$$
$$\quad\quad\;|\;\;\;\;|\;\;\;\;|$$
$$\quad\quad OH\;\;OH\;\;OH$$

Fructose (open-chain form)

Many intermediates in metabolism are also aldehydes or ketones.

Aldehydes and Ketones Have the *Carbonyl Group*. Both aldehydes and ketones contain the carbon–oxygen double bond, which is called the **carbonyl group** (pronounced car-bon-EEL group).

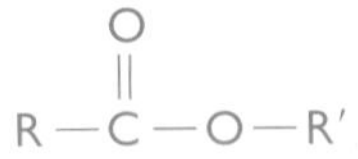

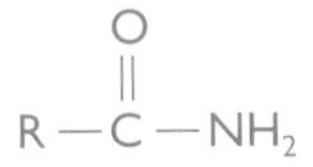

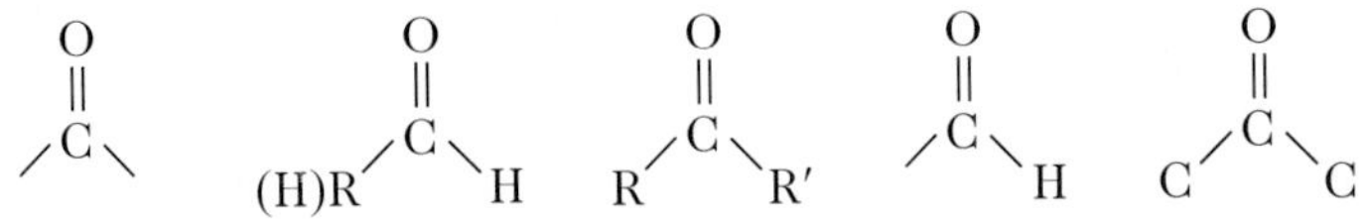

■ Compounds with carbonyl groups to be studied in the next chapter are

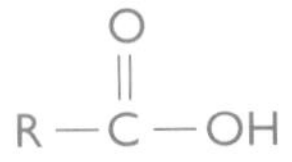

$$R-\overset{\displaystyle O}{\overset{\displaystyle \|}{C}}-OH$$
Carboxylic acids

$$R-\overset{\displaystyle O}{\overset{\displaystyle \|}{C}}-O-R'$$
Esters

$$R-\overset{\displaystyle O}{\overset{\displaystyle \|}{C}}-NH_2$$
Amides

Carbonyl group

Aldehydes

Ketones

Aldehyde group

Ketone system

Notice that the carbonyl group in aldehydes must hold at least one H atom (Table 12.1). In aldehydes this combination is called the **aldehyde group,** which is often condensed as CH=O or as CHO. To be an **aldehyde** there must be attached to the CH=O group either C or H, never N, O, or some other atom. Formaldehyde, the simplest aldehyde, H_2C=O, has a second H attached to CH=O.

In **ketones,** the carbonyl carbon must be joined on both sides by C atoms only (Table 12.2). When present in this combination, the carbonyl group is called the **keto group.**

The Carbonyl Group Makes a Molecule Moderately Polar. When compounds of nearly the same formula mass and chain branching are compared, aldehydes and ketones have boiling points between those of alkanes and alcohols (Table 12.3). We

TABLE 12.1 Aldehydes

Name	Structure	Formula Mass	BP (°C)	Solubility in Water
Formaldehyde	CH_2=O	30.0	−21	Very soluble
Acetaldehyde	CH_3CH=O	44.0	21	Very soluble
Propionaldehyde	CH_3CH_2CH=O	58.1	49	16 g/dL (25 °C)
Butyraldehyde	$CH_3CH_2CH_2CH$=O	72.1	76	4 g/dL
Benzaldehyde	C_6H_5CH=O	106.0	178	0.3 g/dL

TABLE 12.2 Ketones

Name	Structure	Formula Mass	BP (°C)	Solubility in Water
Acetone	O ‖ CH₃CCH₃	58.1	57	Very soluble
Ethyl methyl ketone	O ‖ CH₃CCH₂CH₃	72.1	80	33 g/dL (25 °C)
Methyl propyl ketone	O ‖ CH₃CCH₂CH₂CH₃	86.1	102	6 g/dL
Diethyl ketone	O ‖ CH₃CH₂CCH₂CH₃	86.1	102	5 g/dL
Cyclopentanone	(cyclopentane ring)=O	84.1	129	Slightly soluble
Cyclohexanone	(cyclohexane ring)=O	98.1	156	Slightly soluble

have learned that an alkane is almost nonpolar and that an alcohol is very polar. Therefore, we can justifiably infer that aldehydes and ketones are moderately polar.

Molecules of an alcohol, of course, are able both to donate and to accept hydrogen bonds. The carbonyl group cannot donate hydrogen bonds, but it can accept them, so both the aldehyde and keto groups help bring compounds into solution in water. As you can see in Tables 12.1 and 12.2, the aldehydes and ketones of low formula mass are relatively soluble in water.

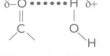

Hydrogen bond (••••••) between a water molecule and a carbonyl group

Aldehydes Have Common Names Based on Their Corresponding Acids. What is easy about the common names of aldehydes is that they all (well, nearly all) end in *-aldehyde*. The prefixes to this are the same as the prefixes of the carboxylic acids, discussed below, to which the aldehydes are easily oxidized. Even exposure to air causes the slow oxidation of an aldehyde, RCHO, to the corresponding carboxylic acid, RCO₂H. This is, therefore, a good place to study the common names of both carboxylic acids and aldehydes.

TABLE 12.3 Boiling Point versus Structure

Name	Structure	Formula Mass	BP (°C)
Butane	CH₃CH₂CH₂CH₃	58.2	0
Propionaldehyde	CH₃CH₂CH=O	58.1	49
Acetone	O ‖ CH₃CCH₃	58.1	57
Propyl alcohol	CH₃CH₂CH₂OH	60.1	98
Ethylene glycol	HOCH₂CH₂OH	62.1	198

The common names of the simple carboxylic acids are based on some natural sources of the acids. The one-carbon acid, for example, is called formic acid because it is present in the stinging fluid of ants, and the Latin root for ants is *formica*. The prefix in formic acid is *form-*, so the one-carbon aldehyde is called *formaldehyde*. Here are the four simplest carboxylic acids and their common names together with the structures and names, of their corresponding aldehydes.

■ Formic acid also appears to have an aldehyde group, but its second bond from C is to another O, not to H (or C), and it is classified as a carboxylic acid.

$$
\begin{array}{llll}
\overset{\displaystyle O}{\underset{\displaystyle \parallel}{}} & \overset{\displaystyle O}{\underset{\displaystyle \parallel}{}} & \overset{\displaystyle O}{\underset{\displaystyle \parallel}{}} & \overset{\displaystyle O}{\underset{\displaystyle \parallel}{}} \\
\mathrm{HCOH} & \mathrm{CH_3COH} & \mathrm{CH_3CH_2COH} & \mathrm{CH_3CH_2CH_2COH} \\
\text{Formic acid} & \text{Acetic acid} & \text{Propionic acid} & \text{Butyric acid}
\end{array}
$$

$$
\begin{array}{llll}
\overset{\displaystyle O}{\underset{\displaystyle \parallel}{}} & \overset{\displaystyle O}{\underset{\displaystyle \parallel}{}} & \overset{\displaystyle O}{\underset{\displaystyle \parallel}{}} & \overset{\displaystyle O}{\underset{\displaystyle \parallel}{}} \\
\mathrm{HCH} & \mathrm{CH_3CH} & \mathrm{CH_3CH_2CH} & \mathrm{CH_3CH_2CH_2CH} \\
\text{Formaldehyde} & \text{Acetaldehyde} & \text{Propionaldehyde} & \text{Butyraldehyde} \\
& \text{From the Latin} & \text{From the Greek} & \text{From the Latin } bu\text{-} \\
& acetum, \text{``vinegar.''} & proto, \text{``first,'' and} & tyrum, \text{``butter.''} \\
& & pion, \text{``fat.''}
\end{array}
$$

In the aromatic series we have the following examples.

$$
\begin{array}{ll}
\mathrm{C_6H_5CO_2H} & \mathrm{C_6H_5CHO} \\
\text{Benzoic acid} & \text{Benzaldehyde}
\end{array}
$$

Common Names of Simple Ketones Include the Names of Their Hydrocarbon Groups. Quite often the simpler ketones are given common names that are made by naming the two alkyl groups attached to the carbon atom of the carbonyl group and then following these names by the word *ketone*. For example,

■ Pronounce *-one* in ketone as *own.*

■ The name *acetone* stems from the fact that this ketone can be made by heating the calcium salt of *acetic* acid.

$$
\begin{array}{lll}
\overset{\displaystyle O}{\underset{\displaystyle \parallel}{}} & \overset{\displaystyle O}{\underset{\displaystyle \parallel}{}} & \overset{\displaystyle O}{\underset{\displaystyle \parallel}{}} \\
\mathrm{CH_3CH_2CCH_3} & \mathrm{CH_3CH_2CCH_2CH_3} & \mathrm{CH_3CCH_3} \\
\text{Methyl ethyl ketone} & \text{Diethyl ketone} & \text{(Dimethyl ketone)} \\
& & \text{Acetone}
\end{array}
$$

The name *acetone* is almost always used for dimethyl ketone.

■ **PRACTICE EXERCISE 1** Write the structures of the following ketones.

(a) ethyl isopropyl ketone   (b) methyl phenyl ketone

(c) dipropyl ketone   (d) di-*t*-butyl ketone

The IUPAC rules for naming aldehydes and ketones are similar to those for naming alkanes and alcohols (see Appendix III).

12.2 OXIDATION OF ALDEHYDES

The aldehyde group is easily oxidized to the carboxylic acid group, but the keto group is difficult to oxidize.

We learned in Chapter 11 that the oxidation of a 1° alcohol to an aldehyde requires special reagents, because aldehydes are themselves easily oxidized. We also learned that much less care is needed to oxidize a 2° alcohol to a ketone, because ketones resist further oxidation.

$$RCH_2OH + (O) \longrightarrow \overset{\overset{\displaystyle O}{\|}}{R}CH + H_2O$$

1° Alcohol   Oxidizing   Aldehyde
agent

$$\overset{\overset{\displaystyle OH}{|}}{R}CHR' + (O) \longrightarrow \overset{\overset{\displaystyle O}{\|}}{R}CR' + H_2O$$

2° Alcohol   Ketone

The ease with which the aldehyde group is oxidized by even mild reactants has led to some simple test tube tests for aldehydes.

Mild Oxidizing Agents in Benedict's and Tollens' Reagents React with Aldehydes. The aldehyde group is far more easily oxidized than 1° or 2° alcohols, so readily oxidized that aldehydes do not even store well when exposed to the oxygen of air. Certain heavy metal ions, like Ag^+ and Cu^{2+}, are also able to oxidize aldehydes if the solutions are slightly basic. The metal ions are themselves reduced in ways that adapt to test tube tests for the aldehyde group in general and carbohydrates in particular.

Heavy metal ions generally are insoluble in basic media (they form precipitates of hydroxides or oxides), but they can be made soluble by certain additional solutes. The citrate ion, for example, is a large, multicharged organic ion that enfolds the Cu^{2+} ion while not denying its availability to other reactants. An alkaline solution of copper(II) sulfate in sodium citrate, called *Benedict's reagent,* is commonly used to test a solution for the presence of glucose or other easily oxidized sugars, like fructose (see structures on page 272). **Benedict's test** consists of adding some of the solution to be tested to a few milliliters of Benedict's reagent and warming the mixture. Because of the Cu^{2+} ion, Benedict's reagent is bright blue, but the reaction changes the ion to Cu^+, which the citrate ion cannot hold in solution. Cu^+ reacts as it forms with the base present to form a brick-red precipitate of copper(I) oxide, Cu_2O. A positive Benedict's test for glucose, therefore, is the change from a bright blue solution to a brick-red precipitate. The extent of the color change depends on the concentration of the glucose. Very dilute solutions produce a greenish result.

Benedict's reagent has been a common test for glucose in the urine, a feature of untreated diabetes. Clinitest tablets, for example, contain all of the solid ingredients of Benedict's reagent in tablet form. When a tablet is added to a urine sample, the necessary heat is generated as the tablet dissolves. The glucose level of the urine is estimated by comparing the color change with a color code on the tablet bottle. The concentration of glucose in the urine of a diabetic person, however, is not the best indication of the glucose status of the blood. Diabetes specialists encourage the use of other methods to give blood glucose levels, but not all patients are willing to use them.

The silver ion is reduced by glucose or by other aldehydes to metallic silver. Ag^+, like Cu^{2+}, is insoluble in a basic solution, but ammonia molecules combine with it to give the base-soluble and colorless silver diammine ion, $[Ag(NH_3)_2]^+$. The ion is present in *Tollens' reagent,* a solution of silver nitrate in alkaline ammonia, and the corresponding test tube test is **Tollens' test.** When a small amount of an aldehyde is added to a few milliliters of Tollens' reagent and the mixture is warmed, the formation of a mirror of silver metal is obvious after a short while. It's beautiful to see and is basically the chemistry by which silver mirrors are made. If the glass is not grease free, a finely divided, gray precipitate of silver forms, but this is also a positive test.

Benedict's test. Benedict's reagent is bright blue. In a positive Benedict's test, a finely divided, brick-red precipitate of copper(I) oxide, Cu_2O slowly settles.

$$\begin{array}{c} CH_2CO_2^- \\ | \\ HOCCO_2^- \\ | \\ CH_2CO_2^- \end{array}$$

Citrate ion

■ All compounds with OH groups *alpha* (α) to aldehyde or keto groups give positive Benedict's tests.

$$\underset{\underset{\displaystyle OH}{|}}{-\,CHCR(H)}\ \overset{\overset{\displaystyle O}{\|}}{}$$

α position

■ Tollens' reagent deteriorates on standing, so it is freshly made before each use.

■ The test tube must be very clean with no soapy or greasy film on its inside surface.

12.3   REDUCTION OF ALDEHYDES AND KETONES

Aldehydes and ketones are reduced to alcohols when hydrogen adds to their carbonyl groups.

Reduction converts aldehydes to 1° alcohols and ketones to 2° alcohols. We study two methods of reduction: the direct addition of hydrogen and a reaction with hydride ion that we call *hydride ion transfer*. The overall results are the same.

Hydrogen Adds Catalytically to the Carbonyl Groups of Aldehydes and Ketones. Under heat and pressure and in the presence of a finely divided metal catalyst (like powdered nickel), aldehydes and ketones react with hydrogen as follows.

$$\underset{\text{Aldehyde}}{\overset{\overset{\text{O}}{\|}}{\text{RCH}}} \ + \ H_2 \xrightarrow[\text{heat, pressure}]{\text{Ni}} \underset{\text{1° Alcohol}}{\text{RCH}_2\text{OH}}$$

$$\underset{\text{Ketone}}{\overset{\overset{\text{O}}{\|}}{\text{RCR}'}} \ + \ H_2 \xrightarrow[\text{heat, pressure}]{\text{Ni}} \underset{\text{2° Alcohol}}{\overset{\overset{\text{OH}}{|}}{\text{RCHR}'}}$$

■ H_2 does not add as easily to the C=O double bond as it does to the C=C double bond.

The experimental *conditions* for these catalytic hydrogenations are impossible in living systems, of course, but they do show the *net effect* of the reductions of aldehyde and keto groups that is also accomplished in living cells. *The aldehyde group is reduced to a 1° alcohol and the keto group to a 2° alcohol.*

The Aldehyde or Keto Group Is Reduced by Acceptance of the Hydride Ion.   The hydride ion, $H:^-$, a powerful reducing agent, is supplied in the body by a small number of complex organic *hydride ion donors* whose structures are severely condensed to NADH, NADPH, $FMNH_2$, and $FADH_2$. (The structures of these, as we indicated, are complex, and we will not look at them until Chapter 17.)

■ The N in NADH and NADPH refers to nicotinamide, a B vitamin. The F in $FMNH_2$ or $FADH_2$ refers to riboflavin, another B vitamin.

The carbonyl groups of aldehydes and ketones can accept $H:^-$ as shown in the next equation, where we use NADH as a typical hydride ion donor *in vivo*. We represent NADH as NAD:H, however, to emphasize that an electron pair goes with H. As you can see, the overall result, the reduction of an aldehyde or a ketone to an alcohol by NADH, gives the same result as catalytic hydrogenation.

■ $H:^-$ is never free in water but always passes directly from donor to acceptor.

$$\underset{\substack{\text{Hydride} \\ \text{donor}}}{\text{NAD}:\text{H}} + \underset{\substack{\text{Aldehyde} \\ \text{or ketone}}}{\overset{}{\text{C}=\overset{..}{\underset{..}{\text{O}}}:}} \longrightarrow \text{NAD}^+ + \underset{\substack{\text{Anion of an} \\ \text{alcohol}}}{\text{H}-\overset{|}{\underset{|}{\text{C}}}-\overset{..}{\underset{..}{\text{O}}}:^-} \longrightarrow \underset{\text{Alcohol}}{\text{H}-\overset{|}{\underset{|}{\text{C}}}-\overset{..}{\underset{..}{\text{O}}}-\text{H}}$$

H$^+$ ← Supplied by the buffer

One of the many examples in cells of reduction by hydride ion donation is the reduction of the keto group in the pyruvate ion to the OH group of the lactate ion, a step in the metabolism of glucose.

$$\underset{\text{Pyruvate ion}}{\overset{\overset{:\text{O}:}{\|}}{\text{CH}_3\text{CCO}_2^-}} + \text{NAD}:\text{H}+\text{H}^+ \longrightarrow \underset{\text{Lactate ion}}{\overset{\overset{\text{OH}}{|}}{\text{CH}_3\text{CHCO}_2^-}} + \text{NAD}^+$$

With respect to the reductions of aldehydes and ketones, our needs center on what these reactions produce *in vivo* rather than *in vitro* . Therefore, we study only the *net effects* of such reactions and not the specific reagents and conditions commonly used *in vitro*. What we need to learn is how to write the products when we know the reactants.

EXAMPLE 12.1   Writing the Structure of the Product of the Reduction of an Aldehyde or Ketone

What is the product of the reduction of propionaldehyde?

ANALYSIS   All the action is at the carbonyl group; it changes to an alcohol group. Therefore all we have to do is copy over the structure of the given compound, change the double bond to a single bond, and supply the two hydrogen atoms—one to the oxygen atom of the original carbonyl group and one to the carbon atom.

SOLUTION   The product of the reduction of propionaldehyde is propyl alcohol.

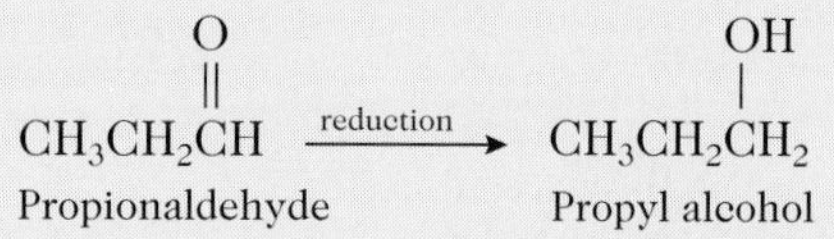

$$
\underset{\text{Propionaldehyde}}{\overset{\overset{\displaystyle O}{\overset{\|}{}}}{CH_3CH_2CH}} \quad \xrightarrow{\text{reduction}} \quad \underset{\text{Propyl alcohol}}{\overset{\overset{\displaystyle OH}{\overset{|}{}}}{CH_3CH_2CH_2}}
$$

CHECK   One common error is to change the structural skeleton, so be sure to check that the *sequence* of all of the atoms heavier than H, like C and O, *has not changed*. Another common error is to write a structure that includes a violation of the covalences of the heavy atoms—4 for C and 2 for O in neutral species. So go down the chain in the answer, atom by atom, to see that each carbon has four bonds and each oxygen has two. If you find an error, fix it.

■   PRACTICE EXERCISE 2   Write the structures of the products that form when the following aldehydes and ketones are reduced.

(a) 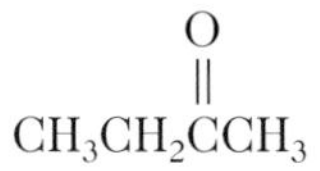
$$\overset{\overset{\displaystyle O}{\|}}{CH_3CH_2CCH_3}$$

(b) 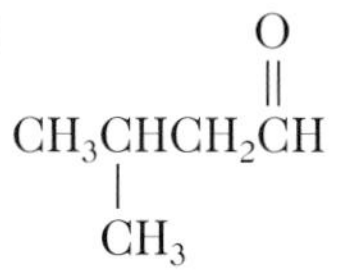
$$\underset{\overset{|}{CH_3}}{\overset{\overset{\displaystyle O}{\|}}{CH_3CHCH_2CH}}$$

(c)

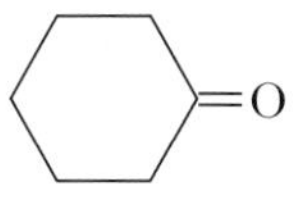

12.4   REACTIONS OF ALDEHYDES AND KETONES WITH ALCOHOLS

1,1-Diethers—acetals or ketals—form when aldehydes or ketones react with alcohols in the presence of an acid or enzyme catalyst.

This section is background for the study of carbohydrates whose molecules have the functional groups introduced here.

Alcohols Add to the Carbonyl Groups of Aldehydes and Ketones. When a solution of an aldehyde in an alcohol is prepared, molecules of the alcohol add to molecules of the aldehyde and the following equilibrium mixture forms.

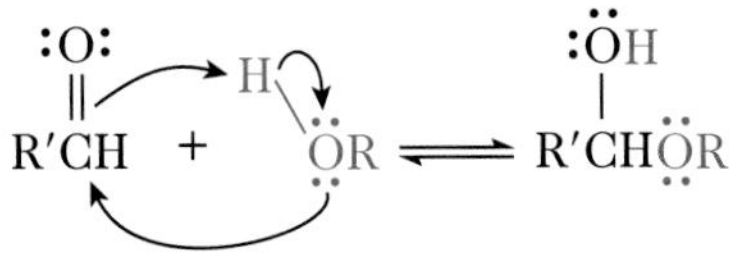

$$
\underset{\text{Aldehyde}}{\overset{\overset{\displaystyle :O:}{\parallel}}{R'CH}} \;+\; \underset{\text{Alcohol}}{\overset{\text{H}}{\ddot{O}R}} \;\rightleftharpoons\; \underset{\text{Hemiacetal}}{\overset{\overset{\displaystyle :\ddot{O}H}{|}}{R'CH\ddot{O}R}}
$$

■ The hemiacetal system:

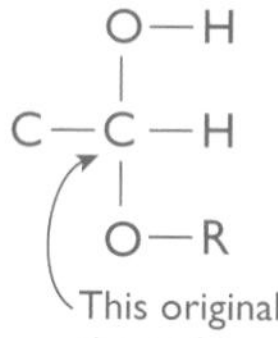

$$
\begin{array}{c}
\overset{\displaystyle O-H}{\underset{\displaystyle O-R}{C-\overset{|}{\underset{|}{C}}-H}}
\end{array}
$$

This originally was the carbon atom of an aldehyde group

■ The hemiketal system:

$$
\begin{array}{c}
\overset{\displaystyle O-H}{\underset{\displaystyle O-R}{C-\overset{|}{\underset{|}{C}}-C}}
\end{array}
$$

This originally was the carbon atom of a keto group

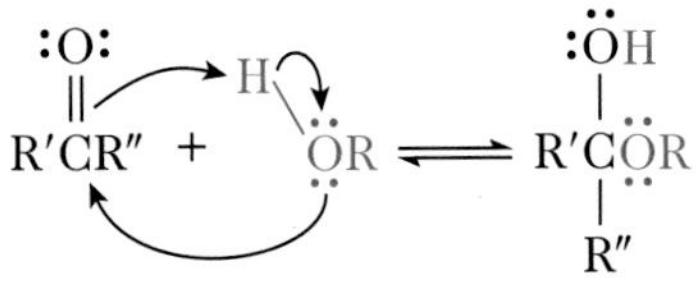

■ One form of the fructose molecule is a *cyclic* hemiketal (see Chapter 14).

The product, a **hemiacetal,** has molecules with a carbon atom holding both an OH group and an OR group. When both OH and OR are on the same carbon, they so modify each other's chemical properties that we have to create a separate family for the system. A hemiacetal is not an ordinary alcohol plus an ordinary ether in one molecule. For example, hemiacetals very readily break down to their parent aldehydes and alcohols. In contrast, ordinary ethers strongly resist reactions that break up their molecules. Only among the carbohydrates do we find relatively stable hemiacetal systems.

When a ketone is dissolved in an alcohol, a similar reaction occurs to give an equilibrium in which the product is called a **hemiketal** to signify its origin from a ketone.

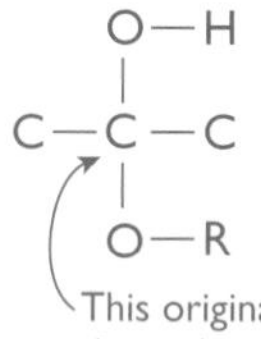

$$
\underset{\text{Ketone}}{\overset{\overset{\displaystyle :O:}{\parallel}}{R'CR''}} \;+\; \underset{\text{Alcohol}}{\overset{\text{H}}{\ddot{O}R}} \;\rightleftharpoons\; \underset{\text{Hemiketal}}{\overset{\overset{\displaystyle :\ddot{O}H}{|}}{\underset{\underset{\displaystyle R''}{|}}{R'C\ddot{O}R}}}
$$

Hemiketals are even less stable than hemiacetals. Yet, the hemiketal system does occur among carbohydrates. As we continue our study of these systems, however, we will deal almost entirely with hemiacetals because the extension of the principles to hemiketals is straightforward.

Except among carbohydrates, the hemiacetal system is almost always too unstable to exist in a pure compound. If we were to try to isolate and purify an ordinary hemiacetal, it would break up entirely and give only the original aldehyde and alcohol. Hemiacetals, in other words, generally exist only in the equilibrium that includes their parents. The relative ease of this breakup means that the hemiacetal system is a site of structural weakness, *even among molecules of carbohydrates.* For this reason, we must learn to recognize the system when it occurs in a structure.

EXAMPLE 12.2   Identifying the Hemiacetal System

Which of the following structures has the hemiacetal system? Place an asterisk by any carbon atoms that were initially those of aldehyde groups.

$$
\underset{\textbf{1}}{CH_3-O-CH_2-CH_2-OH} \qquad \underset{\textbf{2}}{CH_3-O-CH_2-OH} \qquad \underset{\textbf{3}}{\begin{array}{c} CH_2-O \\ \diagup \qquad\quad \diagdown \\ CH_2 \qquad\quad CH-OH \\ \diagdown \qquad\quad \diagup \\ CH_2-CH_2 \end{array}}
$$

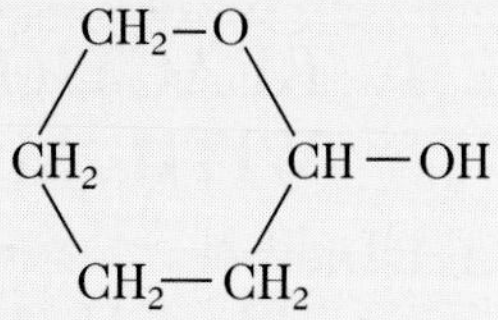

ANALYSIS   To have the hemiacetal system, the molecule must have a carbon to which are attached one OH group and one —O—C unit.

SOLUTION In structure **1** there is an OH group and an —O—C unit, but they are not joined to the *same* carbon. Therefore, **1** is not a hemiacetal. It has only an ordinary ether group plus an alcohol group.

In structure **2**, the OH and the —O—C are joined to the same carbon, so **2** is a hemiacetal. Similarly, in structure **3**, the carbon on the far right corner of the ring holds both OH and —O—C units, and **3** is also a hemiacetal, a cyclic hemiacetal. Structure **3** illustrates the way in which the hemiacetal system occurs in many carbohydrates, namely, as a cyclic hemiacetal. The asterisks in the following structures identify carbon atoms that were originally carbonyl carbons.

$$CH_3-O-\overset{*}{C}H_2-OH$$

$$\begin{array}{c} CH_2-O \\ \diagup \qquad \diagdown \\ CH_2 \qquad \overset{*}{C}H-OH \\ \diagdown \qquad \diagup \\ CH_2-CH_2 \end{array}$$

2.   3.

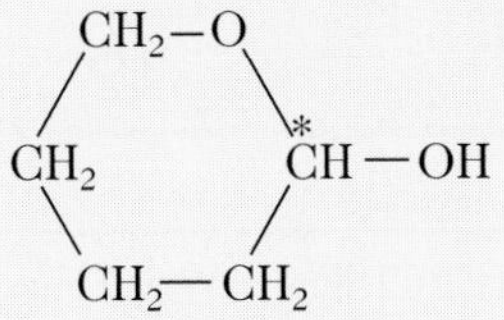

■ The ring system of **3** also occurs in glucose.

CHECK Make sure that any structure identified as a hemiacetal has at least one carbon attached to *two* O atoms by single bonds, that *one* of the O atoms is part of the OH group, and that the other is joined by its second single bond to C.

■ **PRACTICE EXERCISE 3** Identify the hemiacetals or hemiketals among the following structures, and place asterisks by the carbon atoms that initially were part of the carbonyl groups of parent aldehydes or ketones.

(a) 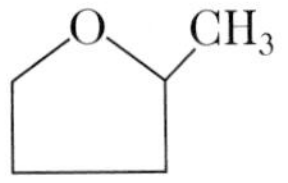   (b) 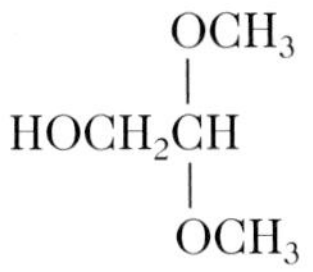 $\begin{array}{c} OCH_3 \\ | \\ HOCH_2CH \\ | \\ OCH_3 \end{array}$ (c) $HOCH_2OCH_2CH_3$ (d)

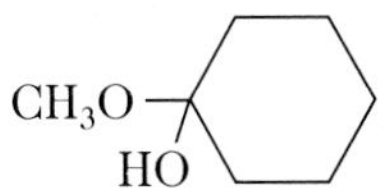

Another skill that will be useful in our study of carbohydrates is the ability to write the structure of a hemiacetal that could be made from a given aldehyde and alcohol.

EXAMPLE 12.3 Writing the Structure of a Hemiacetal Given Its Parent Aldehyde and Alcohol

Write the structure of the hemiacetal that is present at equilibrium in a solution of propionaldehyde in ethyl alcohol.

ANALYSIS A hemiacetal must have a carbon atom to which both an OH group and an OR group are attached. This carbon atom *is provided by the aldehyde.* The structure of the alcohol, CH_3CH_2OH, tells us that the R group in OR of the hemiacetal is CH_2CH_3.

SOLUTION The structure of the hemiacetal formed from propionaldehyde and ethyl alcohol is

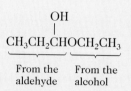

$$\underset{\text{From the}}{\underbrace{\overset{\overset{\displaystyle \text{OH}}{|}}{CH_3CH_2CH}}}\underset{\text{aldehyde}}{}O\underset{\substack{\text{From the}\\\text{alcohol}}}{\underbrace{CH_2CH_3}}$$

CHECK Find the carbon holding *two* O atoms and check the other two atoms or groups that it also holds *against the original aldehyde* (or ketone). These two atoms or groups—here, H and CH_3CH_2—must match those of the aldehyde (or ketone). Finally, check what else the two O atoms are holding; one must hold H (to make it an OH group) and the other must hold the alkyl group from the original alcohol.

■ **PRACTICE EXERCISE 4** Write the structures of the hemiacetals present in the equilibria that involve the following pairs of compounds.

(a) acetaldehyde and methyl alcohol   (b) butyraldehyde and ethyl alcohol

(c) benzaldehyde and propyl alcohol   (d) formaldehyde and methyl alcohol

---

Still another skill that will be useful in studying carbohydrates is the ability to write the structures of the aldehyde and alcohol that are liberated by the breakdown of a hemiacetal.

EXAMPLE 12.4 Writing the Breakdown Products of a Hemiacetal

What aldehyde and alcohol form when the following hemiacetal breaks down?

$$CH_3CH_2CH_2\overset{\overset{\displaystyle \text{OH}}{|}}{CH}-O-CH_2CH_3$$

ANALYSIS The key to solving this problem lies in analyzing the given structure. First, pick out the carbon atom of the original carbonyl group; it's the one holding *both* an OH group and an OR unit. Anything else this carbon holds—usually one H and one hydrocarbon group—completes what we need to write the structure of the aldehyde. The R group of OR is the hydrocarbon group of the original alcohol. Our analysis thus gives us:

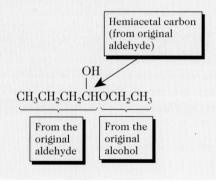

The original aldehyde is thus the unbranched four-carbon aldehyde, butyraldehyde, and the original alcohol is seen to be the two-carbon alcohol, ethyl alcohol.

SOLUTION The products of the breakdown of the given hemiacetal are

$$
\underset{\text{Butyraldehyde}}{CH_3CH_2CH_2\overset{\displaystyle O}{\overset{\displaystyle \|}{C}H}} \quad \text{and} \quad \underset{\text{Ethyl alcohol}}{HOCH_2CH_3}
$$

CHECK It is essential to notice that *only two bonds* in the hemiacetal are affected when we disassemble it, *the C—O bond of the hemiacetal carbon and the O—H bond*, not any other C—O bond, nor a C—C bond, and not a C—H bond. Failure to learn this fact is the most common error that students make in working problems like this. Examine your answer closely, therefore, to see that the original molecule is ruptured *only* at the C—O and O—H bonds of the hemiacetal carbon. *Break no other bonds*.

■ PRACTICE EXERCISE 5   Write the structures of the breakdown products of the following hemiacetals.

$$
\text{(a)} \; CH_3CH_2\overset{\displaystyle OH}{\overset{\displaystyle |}{C}}HOCH_3 \qquad \text{(b)} \; CH_3CH_2O\overset{\displaystyle OH}{\overset{\displaystyle |}{C}}HCH_2CH_3
$$

Acetals and Ketals Form When Alcohols React Further with Hemiacetals and Hemiketals. Hemiacetals and hemiketals are special kinds of alcohols, and they resemble alcohols in one important property. They can undergo a reaction that looks like the formation of an ether. A ordinary ether does not form, however, but a special kind, a 1,1-diether called an **acetal** or a **ketal**. The overall change that leads to an acetal is as follows.

$$
\underset{\text{Hemiacetal}}{R'C\overset{\displaystyle OH}{\overset{\displaystyle |}{H}}OR} + H-OR \xrightarrow{\text{acid catalyst}} \underset{\text{Acetal}}{R'C\overset{\displaystyle OR}{\overset{\displaystyle |}{H}}OR} + H_2O
$$

Hemiketals give the identical kind of reaction, but the products are called *ketals*. Unlike hemiacetals and hemiketals, both acetals and ketals are stable compounds that can be isolated and stored.

The difference between the formation of an acetal and the formation of an ordinary ether is that *acetals form and break more readily than ethers*. As a rule, when two functional groups are very close to each other in a molecule, each modifies the properties of the other in some way. Here, the OR group makes the OH group attached to the same carbon much more reactive toward the splitting out of water with an alcohol.

In a structural sense, an acetal is a 1,1-diether, but "1,1" does not refer to the numbering of the chain and the "ether" part of 1,1-diether does not connote "resistance to breaking up." The "1,1" means only that the two OR groups come to the *same* carbon. In this sense, ketals are also 1,1-diethers.

Ordinary ethers, R—O—R, do not break up in dilute acid or base, but *acetals and ketals are stable only if they are kept out of contact with aqueous acids*. Acids

■ The acetal system:

$$
C-\overset{\displaystyle O-R}{\underset{\displaystyle O-R}{\overset{\displaystyle |}{\underset{\displaystyle |}{C}}}}-H
$$

This originally was the carbon atom of a aldehyde group

■ The ketal system:

$$
C-\overset{\displaystyle O-R}{\underset{\displaystyle O-R}{\overset{\displaystyle |}{\underset{\displaystyle |}{C}}}}-C
$$

This originally was the carbon atom of a keto group

■ Sucrose (table sugar), lactose (milk sugar), and starch have the 1,1-diether system.

■ The hydrolysis of acetal and ketal systems in carbohydrates is the chemistry of the digestion of sugars and starch.

(or enzymes) catalyze the hydrolysis of acetals and ketals to their parent alcohols and aldehydes (or ketones). In aqueous *base,* however, the acetal (ketal) system is stable. Hydrolysis is the only chemical reaction of acetals and ketals that we need to study.

EXAMPLE 12.5   Writing the Structures of the Products of the Hydrolysis of Acetals or Ketals

What are the products of the following reaction?

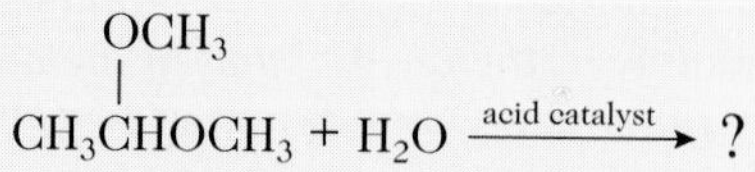

$$\begin{array}{c} OCH_3 \\ | \\ CH_3CHOCH_3 \end{array} + H_2O \xrightarrow{\text{acid catalyst}} \text{?}$$

ANALYSIS   The best way to proceed is to find the carbon atom in the structure that holds *two* oxygen atoms. This carbon is the carbonyl carbon atom of the parent aldehyde or ketone (see the asterisk in the structure below). Break both of its bonds to these oxygen atoms. *Do not break any other bonds.* Then at the carbon that once held two oxygens, make a carbonyl group. The other groups, those of the OR type (here, OCH_3), become alcohols.

SOLUTION   The final products of the hydrolysis of the given acetal are acetaldehyde and methyl alcohol.

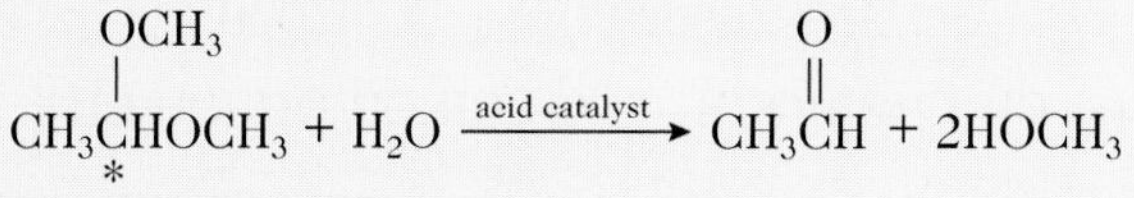

$$\begin{array}{c} OCH_3 \\ | \\ CH_3\underset{*}{C}HOCH_3 \end{array} + H_2O \xrightarrow{\text{acid catalyst}} \begin{array}{c} O \\ || \\ CH_3CH \end{array} + 2HOCH_3$$

■   PRACTICE EXERCISE 6   Write the structures of the aldehydes (or ketones) and the alcohols that are obtained by hydrolyzing the following compounds. If they do not hydrolyze like acetals or ketals, write "no reaction."

(a) $CH_3OCH_2OCH_3$ (b) $CH_3OCH_2CH_2OCH_2CH_3$ (c) $\begin{array}{c} H_3C\ \ OCH_3 \\ |\ \ \ | \\ CH_3CHCOCH_3 \\ | \\ CH_3 \end{array}$

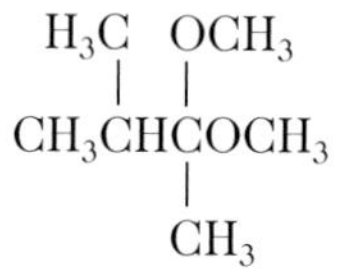

SUMMARY

Physical Properties of Aldehydes and Ketones   The carbonyl group confers moderate polarity, which gives aldehydes and ketones higher boiling points and solubilities in water than hydrocarbons but lower boiling points and solubilities in water than alcohols (that have comparable formula masses).

Chemical Properties of Aldehydes and Ketones   Aldehydes are easily oxidized to carboxylic acids, but ketones resist oxidation. Aldehydes give a positive Tollens' test and ketones do not. Glucose and similar easily oxidized carbohydrates give the Benedict's test.

When an aldehyde or a ketone is dissolved in an alcohol, some of the alcohol adds to the carbonyl group of the aldehyde or ketone. An equilibrium forms that includes molecules of a hemiacetal (or hemiketal). The chart at the end of this summary outlines the chemical properties of the aldehydes and ketones we have studied.

Hemiacetals and Hemiketals   Hemiacetals and hemiketals are usually unstable compounds that exist only in an equilibrium involving the parent carbonyl compound and the parent alcohol (which generally is the solvent). Hemiac-

etals and hemiketals readily break back down to their parent carbonyl compounds and alcohols. When an acid catalyst is added to the equilibrium, a hemiacetal or hemiketal reacts with more alcohol to form an acetal or ketal.

Acetals and Ketals   Acetals and ketals are 1,1-diethers that are stable in aqueous base or in water but not in aqueous acid. Acids catalyze the hydrolysis of acetals and ketals, and the final products are the parent aldehydes (or ketones) and alcohols.

Summary of Reactions

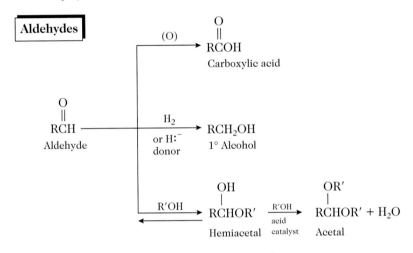

Acetals and
Ketals

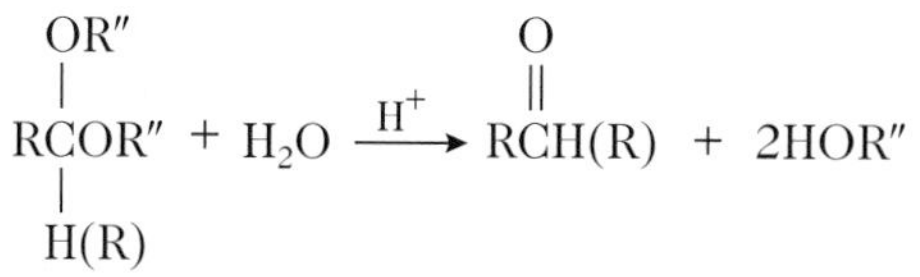

$$\underset{\substack{| \\ H(R)}}{\overset{\substack{OR'' \\ |}}{RCOR''}} + H_2O \xrightarrow{H^+} \underset{\text{Aldehyde or Ketone}}{\overset{\overset{O}{\parallel}}{RCH(R)}} + 2HOR''$$

Acetal or ketal   Aldehyde or   Alcohol
                         Ketone

REVIEW EXERCISES

The answers to Review Exercises whose numbers are in color are found in Appendix V. The answers to the other Review Exercises are found in the Study Guide that accompanies this text. The more challenging questions are marked with asterisks.

Names and Structures

12.1 Give the names of the functional groups present in the following structural formulas.

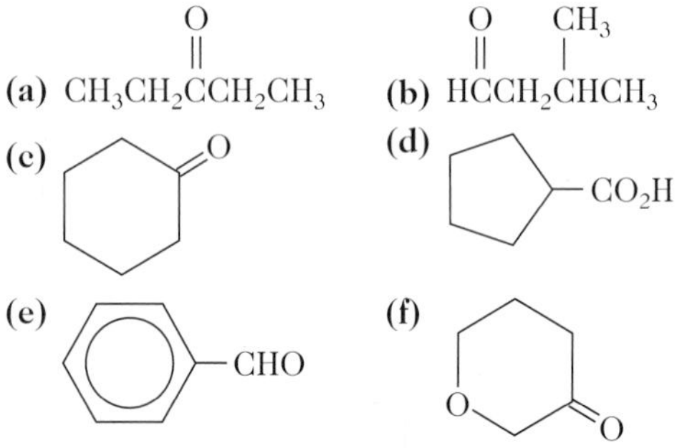

(a) $CH_3CH_2CCH_2CH_3$ (with O double bonded)

(b) $HCCH_2CHCH_3$ (with O double bonded and CH$_3$)

(c) cyclohexanone structure

(d) cyclopentane with CO_2H

(e) benzene with CHO

(f) tetrahydropyran ring with O

12.2 To display the *structural* differences among aldehydes, ketones, carboxylic acids, and esters, write the structure of one example of each using three carbons per molecule.

12.3 What are the structures of the following compounds?
(a) butyraldehyde   (b) acetaldehyde
(c) propionic acid   (d) formic acid

12.4 Write the structures of the following compounds.
(a) acetic acid   (b) propionaldehyde
(c) formaldehyde   (d) butyric acid

12.5 Give the common names of the following.
(a) C_6H_5CHO (b) CH_3CHO
(c) $CH_3CH_2CH_2CO_2H$ (e) CH_3CH_2CHO

12.6 Give the common names of the following.
(a) HCO_2H (b) $C_6H_5CO_2H$
(c) $CH_3CH_2CO_2H$ (d) $HCHO$

12.7 If the common name of $CH_3CH_2CH_2CH_2CO_2H$ is valeric acid, then what is the most likely common name of the following compound?

$$CH_3CH_2CH_2CH_2CHO$$

12.8 If the common name of **E** is glyceraldehyde, what is most the likely common name of **F**?

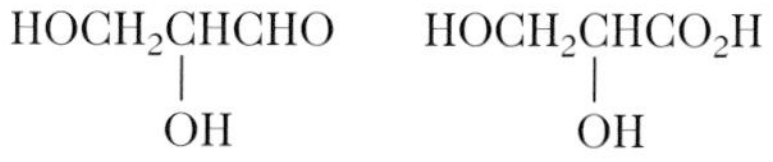

$$\underset{\substack{| \\ OH}}{HOCH_2CHCHO} \qquad \underset{\substack{| \\ OH}}{HOCH_2CHCO_2H}$$

$$\quad\;\; E \qquad\qquad\qquad F$$

Physical Properties of Aldehydes and Ketones

12.9 Arrange the following compounds in order of increasing boiling points. Do this by placing the letters that identify them in the correct order, starting with the lowest-boiling compound and moving in order to the highest-boiling compound. (They have about the same formula masses.)

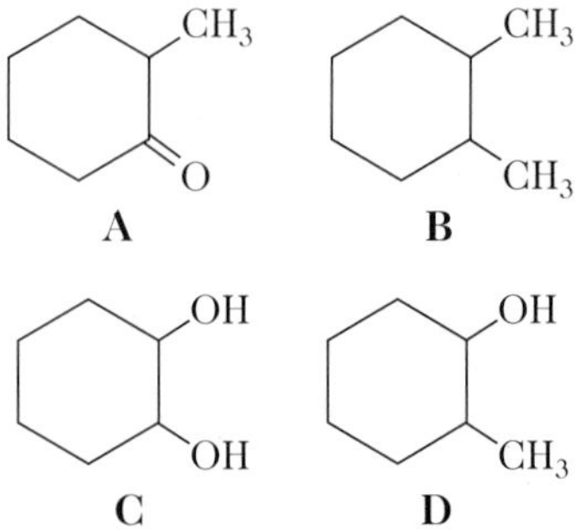

A: cyclohexane ring with CH_3 and O
B: cyclohexane ring with two CH_3 groups
C: cyclohexane ring with two OH groups
D: cyclohexane ring with OH and CH_3

12.10 Arrange the following compounds in order of increasing boiling points. Do this by placing the letters that identify them in the correct order, starting with the lowest-boiling compound and moving to the highest-boiling compound. (They all have about the same formula mass.)

$$
\begin{array}{cc}
\underset{\underset{\text{A}}{}}{\text{HOCH}_2\text{CH}_2\text{CH}_2\overset{\overset{\displaystyle\text{CH}_3}{|}}{\text{C}}\text{HOH}} & \underset{\underset{\text{B}}{}}{\text{CH}_3\text{CH}_2\text{CH}_2\text{CH}_2\overset{\overset{\displaystyle\text{O}}{\|}}{\text{C}}\text{CH}_3}
\end{array}
$$

$$
\begin{array}{cc}
\underset{\underset{\text{C}}{}}{\text{CH}_3\text{CH}_2\text{CH}_2\text{CH}_2\overset{\overset{\displaystyle\text{CH}_3}{|}}{\text{C}}\text{HCH}_3} & \underset{\underset{\text{D}}{}}{\text{HOCH}_2\text{CH}_2\text{CH}_2\overset{\overset{\displaystyle\text{CH}_3}{|}}{\text{C}}\text{HCH}_3}
\end{array}
$$

12.11 Reexamine the compounds in Review Exercise 12.9, and arrange them in order of increasing solubility in water.

12.12 Arrange the compounds in Review Exercise 12.10 in order of increasing solubility in water.

12.13 Draw the structure of a water molecule and a molecule of propionaldehyde and align them on the page to show how the propionaldehyde molecule can accept a hydrogen bond from the water molecule. Use a dotted line to represent this hydrogen bond and place $\delta+$ and $\delta-$ symbols where they are appropriate.

12.14 Draw the structures of molecules of methyl alcohol and acetone, and align them on the page to show how a hydrogen bond (which you are to indicate with a dotted line) can exist between the two. Place $\delta+$ and $\delta-$ symbols where they are appropriate.

Oxidation of Alcohols and Aldehydes

12.15 What are the structures of the aldehydes and ketones to which the following compounds can be oxidized?

(a) $\text{CH}_3\text{CH}_2\overset{\overset{\displaystyle\text{CH}_3}{|}}{\text{C}}\text{HOH}$

(b) $\text{CH}_3\overset{\overset{\displaystyle\text{CH}_3}{|}}{\text{C}}\text{HCH}_2\text{CH}_2\text{CH}_2\text{OH}$

(c) H_3C [cyclopentane ring] OH

(d) $\text{C}_6\text{H}_5\text{CH}_2\overset{\overset{\displaystyle\text{CH}_3}{|}}{\text{C}}\text{HCH}_2\text{OH}$

12.16 Examine each of the following compounds to see whether it can be oxidized to an aldehyde or to a ketone. If it can, write the structure of the aldehyde or ketone.

(a) $\text{HO}\overset{\overset{\displaystyle\text{CH}_3}{|}}{\text{C}}\text{HCH}_2\text{CH}_3$

(b) $\text{HOCH}_2\overset{\overset{\displaystyle\text{CH}_3}{|}}{\text{C}}\text{HCH}_3$

(c) [cyclohexane ring with CH_3 and OH]

(d) $\text{CH}_3\text{CH}_2\overset{\overset{\displaystyle\text{O}}{\|}}{\text{C}}\text{OH}$

(e) $\text{CH}_3\text{OCH}_2\text{CH}_2\overset{\overset{\displaystyle\text{CH}_3}{|}}{\text{C}}\text{HOH}$

(f) HOCH_2 [benzene ring] CO_2H

*12.17 An unknown compound, $\text{C}_3\text{H}_6\text{O}$, reacted with potassium permanganate to give $\text{C}_3\text{H}_6\text{O}_2$, and the same unknown also gave a positive Tollens' test. Write the structures of $\text{C}_3\text{H}_6\text{O}$ and $\text{C}_3\text{H}_6\text{O}_2$.

*12.18 An unknown compound, $\text{C}_3\text{H}_6\text{O}_2$, could be oxidized easily by potassium permanganate to $\text{C}_3\text{H}_4\text{O}_3$, and it gave a positive Benedict's test. Write structures for $\text{C}_3\text{H}_6\text{O}_2$ and $\text{C}_3\text{H}_4\text{O}_3$.

*12.19 Which of the following compounds can be expected to give a positive Tollens' test? All are intermediates in metabolism.

(a) $\text{HOCH}_2\overset{\overset{\displaystyle\text{O}}{\|}}{\text{C}}\text{CH}_2\text{OH}$

(b) $\text{HOCH}_2\overset{\overset{\displaystyle\text{OH}}{|}}{\text{C}}\text{HCHO}$

(c) $\text{CH}_3\overset{\overset{\displaystyle\text{O}}{\|}}{\text{C}}\text{CH}_2\text{CO}_2\text{H}$

(d) $\text{CH}_3\overset{\overset{\displaystyle\text{OH}}{|}}{\text{C}}\text{HCH}_2\text{CO}_2\text{H}$

*12.20 Which of the following compounds give a positive Benedict's test? (Most are intermediates in metabolism.)

(a) $\text{CH}_3\overset{\overset{\displaystyle\text{O}}{\|}}{\text{C}}\text{CO}_2\text{H}$

(b) $\text{CH}_3\overset{\overset{\displaystyle\text{O}}{\|}}{\text{C}}\text{CHO}$

(c) $\text{HOCH}_2\text{CH}_2\overset{\overset{\displaystyle\text{O}}{\|}}{\text{C}}\text{CH}_3$

(d) $\text{HOCH}_2\overset{\overset{\displaystyle\text{HO}}{|}}{\text{C}}\text{H}\overset{\overset{\displaystyle\text{OH}}{|}}{\text{C}}\text{HCHO}$

12.21 In what form does silver occur in Tollens' reagent?

12.22 What is the function of the citrate ion in Benedict's reagent?

12.23 What is the formula of the precipitate that forms in a positive Benedict's test?

12.24 Clinitest tablets are used for what?

12.25 What is one practical commercial application of Tollens' test?

12.26 One of the steps in the metabolism of fats and oils in the diet is the oxidation of the following compound:

$$\text{CH}_3\overset{\overset{\displaystyle\text{OH}}{|}}{\text{C}}\text{HCH}_2\text{CO}_2^-$$

Write the structure of the product of this oxidation.

*12.27 One of the important series of reactions in metabolism is called the *citric acid cycle*. Structures **A** and **B** are of compounds (actually, anions) that participate in this cycle. One of them, isocitric acid, is oxidized to a ketone. The other is not. Which one is isocitric acid, **A** or **B**? Write the structure of its corresponding ketone.

$$
\begin{array}{cc}
\underset{\underset{\text{A}}{}}{\text{HO}-\overset{\overset{\displaystyle\text{CH}_2\text{CO}_2^-}{|}}{\underset{\underset{\displaystyle\text{CH}_2\text{CO}_2^-}{|}}{\text{C}}}\text{CO}_2^-} & \underset{\underset{\text{B}}{}}{\overset{\overset{\displaystyle\text{HO}-\text{CHCO}_2^-}{|}}{\underset{\underset{\displaystyle\text{CH}_2\text{CO}_2^-}{|}}{\text{CHCO}_2^-}}}
\end{array}
$$

Reduction of Aldehydes and Ketones

12.28 Consider the reaction that occurs when a hydride ion transfers from its donor (which we can write as *Mtb*:H) to acetaldehyde.

(a) Write the structure of the organic anion that forms when the hydride ion is transferred to acetaldehyde.

(b) What is the net ionic equation of the reaction of the anion formed in (a) with water?

(c) What is the name of the organic product of the reaction of part (b)?

12.29 A donor of a hydride ion (*Mtb*:H) transfers H:$^-$ to a molecule of acetone.

(a) What is the structure of the organic ion that forms?

(b) What happens to this anion in the presence of water? (Write a net ionic equation.)

(c) What is the name of the organic product of this reaction with water?

*12.30 Metabolism of aspartic acid, an amino acid, occurs by a series of steps. A portion of this series is indicated below, where NAD:H is a reducing agent that becomes NAD$^+$ as it transfers hydride ion.

$$\underset{\text{Aspartate ion}}{\overset{+}{H_3}NCHCO_2^- \atop | \atop CH_2CO_2^-} \xrightarrow{\text{two steps}} \overset{+}{H_3}NCHCO_2^- \atop | \atop CH_2CHO} \xrightarrow{\text{NAD:H}}$$

$$NAD^+ + \overset{+}{H_3}NCHCO_2^- \atop | \atop \underset{A}{\overset{?}{\bigcirc}} \xrightarrow{\text{H}_2\text{O}} B + OH^-$$

Complete the structure of **A**, and write the structure of **B**.

*12.31 One of the steps the body uses to make long-chain carboxylic acids is a reaction similar to the following reaction.

$$\underset{}{\overset{O \quad\;\; O}{\overset{||\quad\;\; ||}{CH_3CCH_2CS}}}\!\!-\!\!\overset{}{\boxed{\text{enzyme}}} + \text{NAD:H} \longrightarrow$$

$$\overset{?}{\bigcirc}-CH_2CS \atop \overset{||}{O}-\boxed{\text{enzyme}} + NAD^+$$
$$\underset{A}{}$$
$$\xrightarrow{\text{H}_2\text{O}} B + OH^-$$

Complete the structure of **A** and write the structure of **B**.

12.32 Write the structures of the aldehydes or ketones that could be used to make the following compounds by reduction (hydrogenation).

(a) $\underset{\text{OH}}{\overset{|}{CH_3CHCH_2CH_3}}$

(b) $\underset{\text{CH}_3}{\overset{|}{HOCH_2CHCH_2CH_3}}$

(c) ⬡—OH

(d) CH_3—⬡—CH_2OH

Hemiacetals and Acetals, Hemiketals and Ketals

12.33 Examine each structure and decide whether it represents a hemiacetal, hemiketal, acetal, ketal, or something else.

(a) $\underset{}{\overset{CH_3}{\overset{|}{CH_3OCHOH}}}$

(b) $\underset{}{\overset{OCH_3}{\overset{|}{CH_3CHOCH_3}}}$

(c) $\underset{}{\overset{CH_3}{\overset{|}{CH_3OCHCH_2OCH_3}}}$

(d) $\underset{\overset{|}{CH_3}}{\overset{CH_3}{\overset{|}{CH_3OCOCH_3}}}$

12.34 Examine each structure and decide whether it represents a hemiacetal, hemiketal, acetal, ketal, or something else.

(a) $\underset{}{\overset{OCH_2CH_3}{\overset{|}{HOCH_2CH_2CHOCH_3}}}$

(b) ⬡ with OH and O in ring

(c) $CH_3CH_2OCH_2OH$

(d) $HOCH_2CH_2OCH_2CH_2CH_3$

12.35 Write the structures of the hemiacetals and the acetals that can form between propionaldehyde and the following two alcohols.

(a) methyl alcohol

(b) ethyl alcohol

12.36 What are the structures of the hemiketals and the ketals that can form between acetone and these two alcohols?

(a) methyl alcohol

(b) ethyl alcohol

*12.37 Write the structure of the hydroxyaldehyde (a compound having both the alcohol group and the aldehyde group in the same molecule) from which the following hemiacetal forms in a ring-closing reaction. (You may leave the chain of the open-chain compound somewhat coiled.)

$$\begin{array}{c} CH_3 \\ \; \\ CH\!-\!O \\ / \qquad\qquad \backslash \\ H_2C \qquad\quad CH\!-\!OH \\ \backslash \qquad\qquad / \\ CH_2\!-\!CH_2 \end{array}$$

*12.38 One form in which a glucose molecule exists is given by the following structure. (*Note:* The atoms and groups that are attached to the carbon atoms of the six-membered ring must be seen as projecting above or *below* the ring.)

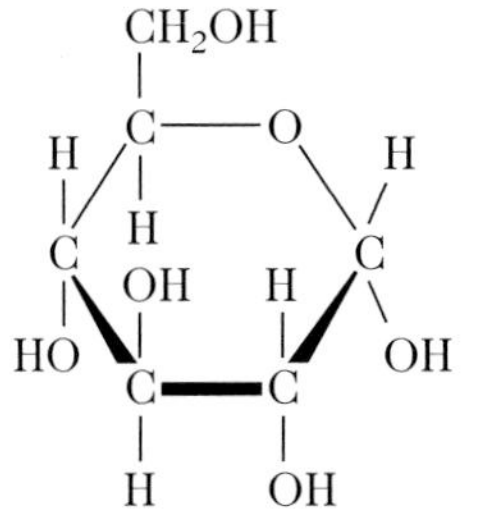

(a) Place an asterisk by the hemiacetal carbon.
(b) Write the structure of the open-chain form that has a free aldehyde group. (You may leave the chain coiled.)

*12.39 Write the structure of a hydroxyketone (a molecule that has both the OH group and the keto group) from which the following hemiketal forms in a ring-closing reaction. (You may leave the chain of the open-chain compound somewhat coiled.)

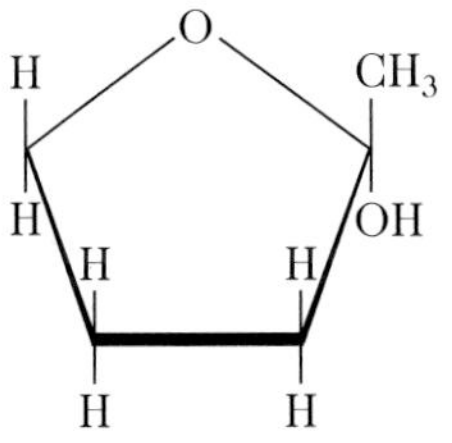

*12.40 Fructose occurs together with glucose in honey, and it is sweeter to the taste than table sugar. One form in which a fructose molecule can exist is given by the following structure.

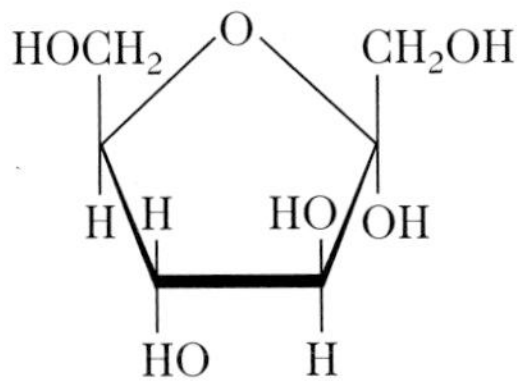

(a) Place an asterisk by the carbon of the hemiketal system that came initially from the carbon atom of a keto group.
(b) In water, fructose exists in equilibrium with an open-chain form of the given structure. This form has a keto group in the same molecule as five OH groups. Draw the structure of this open-chain form (leaving the chain coiled somewhat as it was in the structure that was given).

12.41 The digestion of some carbohydrates is simply their hydrolysis catalyzed by enzymes. Acids catalyze the same kind of hydrolysis of acetals and ketals. Write the structures of the products, if any, that form by the action of water and an acid catalyst on the following compounds.

(a) 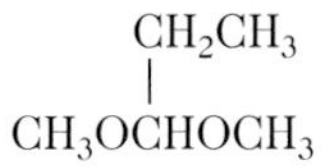
CH_2CH_3
|
$CH_3OCHOCH_3$

(b) 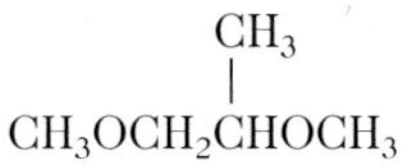
CH_3
|
$CH_3OCH_2CHOCH_3$

(c) 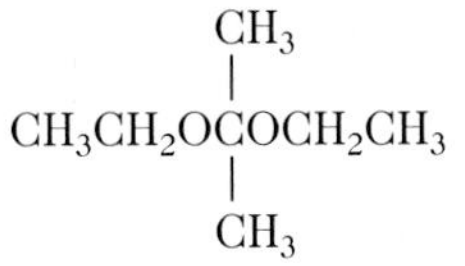
CH_3
|
$CH_3CH_2OCOCH_2CH_3$
|
CH_3

(d)

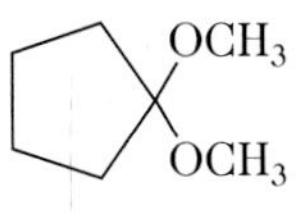

12.42 What are the structures of the products, if any, that form by the acid-catalyzed reaction of water with the following compounds?

(a) 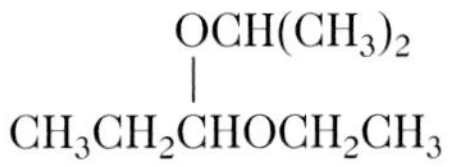
$OCH(CH_3)_2$
|
$CH_3CH_2CHOCH_2CH_3$

(b) 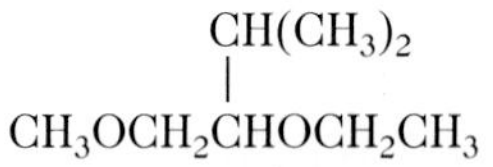
$CH(CH_3)_2$
|
$CH_3OCH_2CHOCH_2CH_3$

(c)

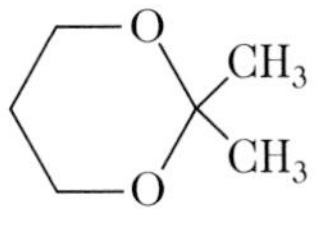

(d) 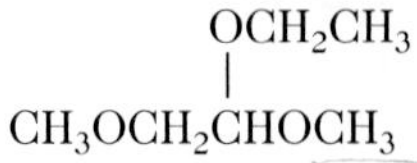
OCH_2CH_3
|
$CH_3OCH_2CHOCH_3$

Additional Exercises

*12.43 Complete the following reaction sequences by writing the structures of the organic products that form. If no reaction occurs, write "no reaction." Reviewed here too are some reactions of earlier chapters. Remember that (O) stands for an oxidizing agent like $Cr_2O_7^{2-}$ or MnO_4^-.

(a) 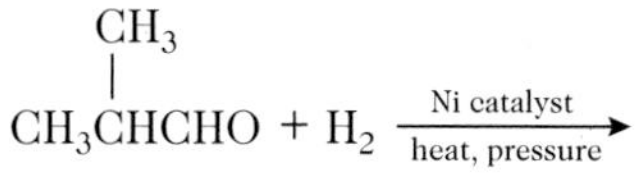
CH_3
|
$CH_3CHCHO + H_2 \xrightarrow[\text{heat, pressure}]{\text{Ni catalyst}}$

(b) 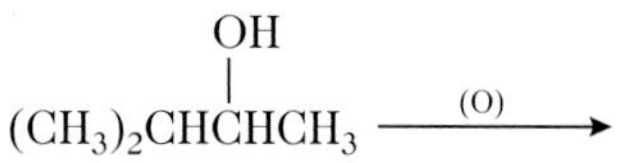
OH
|
$(CH_3)_2CHCHCH_3 \xrightarrow{(O)}$

(c)

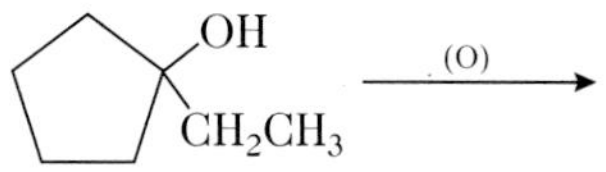

(d) $CH_3CH = CHCH_2CH_3 + H_2 \xrightarrow{\text{Ni catalyst}}$

(e) $CH_3OH + CH_3CH_2CHO \rightleftharpoons$

(f) $CH_3CHO + Mtb:H \xrightarrow[\text{by H}^+]{\text{(followed}}$
(where $Mtb:H$ is a metabolite able to donate hydride ion)

(g) $CH_3CHO + 2CH_3CH_2OH \xrightarrow[\text{catalyst}]{\text{acid}}$

(h) 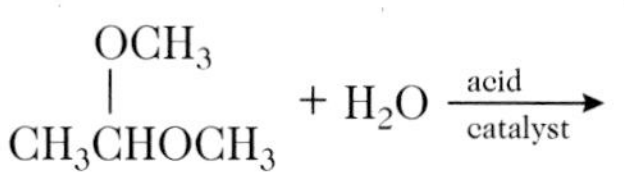
OCH_3
|
CH_3CHOCH_3 $+ H_2O \xrightarrow[\text{catalyst}]{\text{acid}}$

(i)

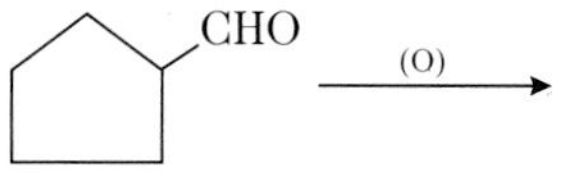

(j)

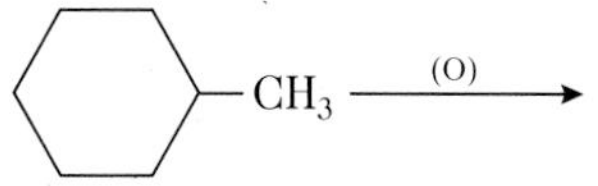

(k) $C_6H_5CH_2NH_2 + HCl(aq) \longrightarrow$

*12.44   Write the structures of the organic products that form in each of the following situations. If no reaction occurs, write "no reaction." (Some of the situations constitute a review of reactions in earlier chapters.)

(a)

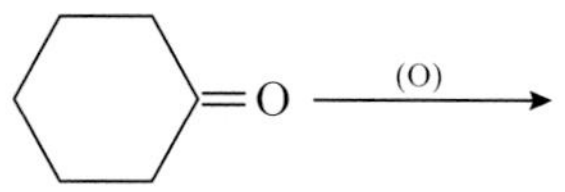

$=O \xrightarrow{\text{(O)}}$

(b)

$$CH_3CH_2\underset{\underset{CH_3CH_2}{|}}{\overset{\overset{CH_3}{|}}{C}}OH \xrightarrow{\text{(O)}}$$

(c) $CH_3CH_2CH_2OH + CH_3CHO \rightleftharpoons$

(d)

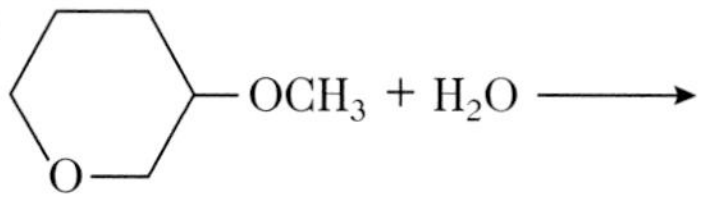

$-OCH_3 + H_2O \longrightarrow$

(e)   OH

$$CH_3\underset{\underset{CH_3}{|}}{\overset{\overset{OH}{|}}{C}}OCH_2CH_3 \rightleftharpoons$$

(f) $CH_3OCH_2CH_2CHO + Mtb\!:\!H \xrightarrow[\text{by } H^+]{\text{(followed}}$
(where $Mtb\!:\!H$ is a metabolite able to donate hydride ion)

(g) OCH_2CH_3

$$CH_3CH_2\underset{\underset{OCH_2CH_3}{|}}{\overset{\overset{OCH_2CH_3}{|}}{C}}CH_3 + H_2O \xrightarrow[\text{catalyst}]{\text{acid}}$$

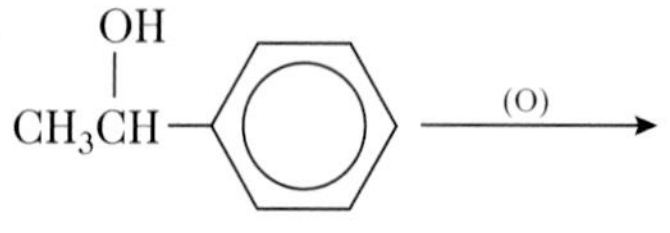

(h)   OH

$CH_3CH- \xrightarrow{\text{(O)}}$

(i)

$OCH_3 + H_2 \xrightarrow{\text{Ni catalyst}}$

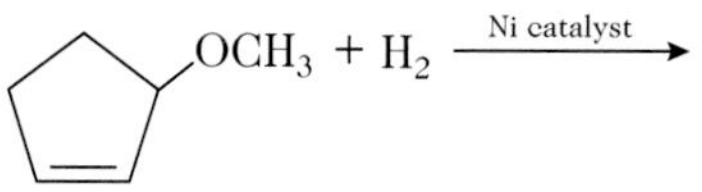

(j)

$-CHO + 2CH_3OH \xrightarrow[\text{catalyst}]{\text{acid}}$

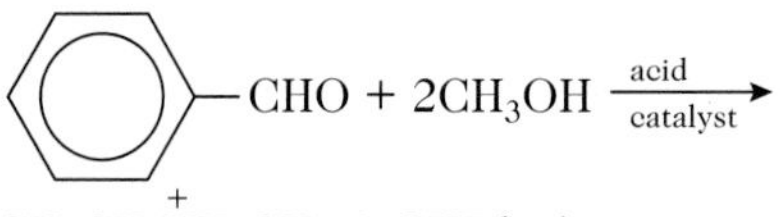

(k) $CH_3CH_2\overset{+}{N}H_2CH_3 + OH^-(aq) \longrightarrow$

*12.45   Catalytic hydrogenation of compound **A** (C_3H_6O) gave **B** (C_3H_8O). When **B** was heated strongly in the presence of sulfuric acid, it changed to compound **C** (C_3H_6). The acid-catalyzed addition of water to **C** gave compound **D** (C_3H_8O); and when **D** was oxidized, it changed to **E** (C_3H_6O). Compounds **A** and **E** are isomers, and compounds **B** and **D** are isomers. Write the structures of compounds **A** through **E**.

*12.46   When compound **F** ($C_4H_{10}O$) was gently oxidized, it changed to compound **G** (C_4H_8O), but vigorous oxidation changed **F** (or **G**) to compound **H** ($C_4H_8O_2$). Action of hot sulfuric acid on **F** changed it to compound **I** (C_4H_8). The addition of water to **I** (in the presence of an acid catalyst) gave compound **J** ($C_4H_{10}O$), a compound that could not be oxidized. Compounds **F** and **J** are isomers. Write the structures of compounds **F** through **J**.

*12.47   A student was assigned the preparation of the dimethyl acetal of butyraldehyde for which the equation is

$$CH_3CH_2CH_2CH{=}O + 2CH_3OH \xrightarrow[\text{catalyst}]{\text{acid}}$$

$$CH_3CH_2CH_2CH(OCH_3)_2 + H_2O$$

A solution of 12.5 g of butyraldehyde in 50.0 mL of methyl alcohol was used for this reaction. The density of methyl alcohol is 0.787 g/mL.

(a) How many moles of butyraldehyde were taken?

(b) How many moles of methyl alcohol were used?

(c) Was sufficient methanol taken? (Calculate the minimum number of grams of methyl alcohol that would be required.)

(d) How many grams of water would be obtained?

(e) Offer a reason for using an excess quantity of methyl alcohol.

13

CARBOXYLIC ACIDS AND THEIR DERIVATIVES

Occurrence and Structural Features

Chemical Properties of Carboxylic Acids and Their Salts

Esters of Carboxylic Acids

Organophosphate Esters and Anhydrides

Amides of Carboxylic Acids

The great strengths of fibers of Dacron and nylon, respectively a polyester and a polyamide, offer some security to those who must hurl themselves from cliffs. Esters and amides, groups studied in this chapter, are parts of the molecules of fats, oils, cell membranes, and proteins.

13.1   OCCURRENCE AND STRUCTURAL FEATURES

The functional groups studied in this chapter occur throughout all metabolism.

■ *Carboxyl comes from carbonyl + hydroxyl.*

Anywhere we care to look in the living world we find compounds with the functional groups studied in this chapter. The *carboxyl group* characterizes the **carboxylic acids.** This group or its negatively charged ion, the *carboxylate ion,* is in all of the amino acids and proteins, all *fatty acids* of fats and oils, and many of the metabolic intermediates of carbohydrates. The *ester group* is the chief functional group in all fats and oils. The *amide group* contributes every third bond to the "backbones" of all proteins.

■ Common abbreviations of the carboxyl group are CO_2H and $COOH$. Abbreviations of the carboxylate ion group include CO_2^- and COO^-.

$$\underset{\text{Carboxyl group}}{-\overset{\displaystyle O}{\overset{\|}{C}}-O-H} \qquad \underset{\text{Carboxylate ion group}}{-\overset{\displaystyle O}{\overset{\|}{C}}-O^-} \qquad \underset{\text{Ester group}}{\overset{\boxed{\text{Ester bond}}}{-\overset{\displaystyle O}{\overset{\|}{C}}-O-\overset{\displaystyle |}{\underset{|}{C}}-}} \qquad \underset{\text{Amide group}}{\overset{\boxed{\text{Amide bond}}}{-\overset{\displaystyle O}{\overset{\|}{C}}-\overset{\displaystyle |}{N}-}}$$

The Names of All Acid Derivatives Are Derived from the Names of the Acids.   Salts of carboxylic acids, esters, and amides are called *derivatives of carboxylic acids* because they can be made from the acids and changed back to them by relatively simple reactions. The names of the derivatives are related, and all are based on characteristic prefixes and suffixes that should now be learned (Table 13.1).

■ All of the acids with 3 to 10 carbon atoms have vile odors, the odors of long unwashed athletic socks or stale locker rooms.

The common names of the carboxylic acids originated in their early sources, as we noted in Chapter 12. Thus, valeric acid gets its name from the Latin *valerum,* meaning "to be strong," and what is strong about valeric acid is its odor. Among the common carboxylic acids (Table 13.2) are several obtained by the hydrolysis (digestion) of fats and oils, so the simple acids with 4 to 20 or more carbon atoms are often called the **fatty acids.**

Molecules of Carboxylic Acids Form Hydrogen Bonds between Them.   The *effective* formula mass of a carboxylic acid is higher than its calculated formula mass because its molecules form strongly hydrogen-bonded pairs.

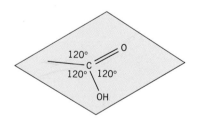

The carboxyl group has a planar geometry at the carbonyl group.

Hydrogen bonds (●●●) hold two molecules of a carboxylic acid together.

Carboxylic acids therefore have higher boiling points than even alcohols of comparable formula masses. Formic acid (BP 101 °C), for example, has about the same formula mass as ethyl alcohol (BP 78.5 °C), but its boiling point is 21.5 °C higher.

The carboxylic acids of lower formula mass (the C_1–C_4 acids) are soluble in water largely because the carboxyl group has *two* oxygen atoms that can accept hydrogen bonds from water molecules. In addition, the carboxyl group has the OH group that can donate hydrogen bonds. When their chains are longer than five carbon atoms, carboxylic acids are insoluble in water.

TABLE 13.1   Writing Common Names for Carboxylic Acids and Their Derivatives

Class	Name Ending of Class	Characteristic Prefix in the Acid Name			
		C_1 Form-	C_2 Acet-	C_3 Propion-	C_4 Butyr-
Acids	-ic acid	Formic acid	Acetic acid	Propionic acid	Butyric acid
Negative ions	-ate (ion)	Formate ion	Acetate ion	Propionate ion	Butyrate ion
Esters	-ate	(Alkyl) formate[a]	(Alkyl) acetate	(Alkyl) propionate	(Alkyl) butyrate
Amides	-amide	Formamide	Acetamide	Propionamide	Butyramide

[a] The name of the specific alkyl group attached to oxygen would be used in the name of a specific ester.

TABLE 13.2   Carboxylic Acids

n	Structure	Name[a]	Origin of Name	MP (°C)	BP (°C)	Solubility in Water[b]
1	HCO_2H	Formic acid (methanoic acid)	L. *formica*, "ant"	8	101	∞
2	CH_3CO_2H	Acetic acid (ethanoic acid)	L. *acetum*, "vinegar"	17	118	∞
3	$CH_3CH_2CO_2H$	Propionic acid (propanoic acid)	L. *proto, pion*, "first," "fat"	−21	141	∞
4	$CH_3(CH_2)_2CO_2H$	Butyric acid (butanoic acid)	L. *butyrum*, "butter"	−6	164	∞
5	$CH_3(CH_2)_3CO_2H$	Valeric acid (pentanoic acid)	L. *valere*, valerian root	−35	186	4.97
6	$CH_3(CH_2)_4CO_2H$	Caproic acid (hexanoic acid)	L. *caper*, "goat"	−3	205	1.08
7	$CH_3(CH_2)_5CO_2H$	Enanthic acid (heptanoic acid)	Gr. *oenanthe*, "vine blossom"	−9	223	0.26
8	$CH_3(CH_2)_6CO_2H$	Caprylic acid (octanoic acid)	L. *caper*, "goat"	16	238	0.07
9	$CH_3(CH_2)_7CO_2H$	Pelargonic acid (nonanoic acid)	*Pelargonium*, geranium	15	254	0.03
10	$CH_3(CH_2)_8CO_2H$	Capric acid (decanoic acid)	L. *caper*, "goat"	32	270	0.015
12	$CH_3(CH_2)_{10}CO_2H$	Lauric acid (dodecanoic acid)	Laurel	44	—	0.006
14	$CH_3(CH_2)_{12}CO_2H$	Myristic acid (tetradecanoic acid)	*Myristica*, nutmeg	54	—	0.002
16	$CH_3(CH_2)_{14}CO_2H$	Palmitic acid (hexadecanoic acid)	Palm oil	63	—	0.0007
18	$CH_3(CH_2)_{16}CO_2H$	Stearic acid (octadecanoic acid)	Gr. *stear*, "solid"	70	—	0.0003

Miscellaneous Carboxylic Acids

	$C_6H_5CO_2H$	Benzoic acid	Gum benzoin	122	249	0.34 (25 °C)
	$C_6H_5CH\!=\!CHCO_2H$	Cinnamic acid (trans isomer)	Cinnamon	132	—	0.04
	$CH_2\!=\!CHCO_2H$	Acrylic acid	L. *acer*, "sharp"	13	141	Soluble

[a] In parentheses below each common name is the IUPAC name.
[b] In grams of acid per 100 g water at 20 °C (except where noted).

13.2 CHEMICAL PROPERTIES OF CARBOXYLIC ACIDS AND THEIR SALTS

Although carboxylic acids are weak acids, they can neutralize bases and they react with alcohols to form esters.

■ Common abbreviations of carboxylate ions are RCO_2^- and $RCOO^-$.

The properties of carboxylic acids and their negatively charged ions most needed as background for biochemistry are their acid–base relationships and the formation of esters and amides. Both the CO_2H and CO_2^- systems, strongly resist both oxidation and reduction.

Carboxylic Acids Are Weak Acids.   Toward a water molecule, a very weak proton acceptor, a carboxylic acid is a weak proton donor or acid. In 1 M acetic acid, for example, only about 0.5% of the solute molecules are ionized according to the following equilibrium.

$$CH_3CO_2H + H_2O \rightleftharpoons CH_3CO_2^- + H_3O^+$$

■ A serious decrease in the blood's pH occurs in many medical emergencies, but a similar outcome from strenuous sports activities is prevented simply by the body's saying, "Take a break."

The hydroxide ion, the carbonate ion, and the bicarbonate ion, on the other hand, are strong enough proton acceptors to neutralize carboxylic acids entirely. This reaction is important at the molecular level of life, because the carboxylic acids that we normally produce by metabolism must be neutralized. Otherwise, the pH of body fluids, like the blood, would fall too low to sustain life. The acids of metabolism must also be made more soluble in water, which they are in the form of carboxylate ions, RCO_2^-.

With hydroxide ion, the equation for the neutralization reaction is as follows.

$$RCO_2H + OH^- \longrightarrow RCO_2^- + H—OH$$

With bicarbonate ion, the base of the blood's carbonate buffer, the reaction occurs by the following equation.

$$RCO_2H + HCO_3^- \longrightarrow RCO_2^- + H_2O + CO_2$$

Some specific examples are as follows.

$$CH_3CO_2H + OH^- \longrightarrow CH_3CO_2^- + H_2O$$
Acetic acid   Acetate ion

$$CH_3(CH_2)_{16}CO_2H + OH^- \longrightarrow CH_3(CH_2)_{16}CO_2^- + H_2O$$
Stearic acid   Stearate ion
(insoluble in water)   (soluble in water)

■ The stearate ion is one of several organic ions in soap.

■ PRACTICE EXERCISE 1   Write the structures of the carboxylate ions that form when the following carboxylic acids are neutralized.

(a) $CH_3CH_2CO_2H$ (b) 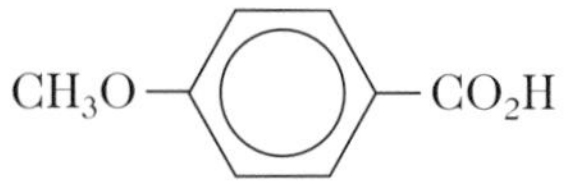 CH_3O—⬡—CO_2H (c) $CH_3CH=CHCO_2H$

The Sodium and Potassium Salts of Carboxylic Acids Are Soluble in Water.   The salts of carboxylic acids are true salts, assemblies of oppositely charged ions. All, therefore, are solids at room temperature. They follow the "like dissolves like rule" by being insoluble in nonpolar solvents, like ether and hydrocarbons. However, the *salts* of carboxylic acids with Group IA metals (but not with Group IIA metals) are soluble in water. Several salts are used as decay inhibitors in foods and bottled beverages.

The names of the ions of carboxylic acids are made by dropping *-ic acid* from the acid's name and substituting *-ate ion*. Thus, the negatively charged ion of ace*tic acid* is the ace*tate ion*.

Carboxylate Ions Can Neutralize Hydronium Ions. Because carboxylate ions are the ions of *weak* acids, they themselves must be relatively good proton acceptors or bases, especially toward the hydronium ion, a strong proton donor. At room temperature, the following neutralization of a strong acid by a carboxylate ion occurs virtually instantaneously. It is the most important reaction of the carboxylate ion that we will study, because it makes the carboxylate ion group a neutralizer of excess acid at the molecular level of life.

$$R-\overset{\overset{\displaystyle O}{\|}}{C}-\overset{..}{\underset{..}{O}}:^- \;+\; \overset{H\;\;\;\;H}{\underset{\underset{H}{|}_+}{\overset{..}{O}}} \longrightarrow R-\overset{\overset{\displaystyle O}{\|}}{C}-\overset{..}{\underset{..}{O}}\underset{H}{} \;+\; \overset{H\;\;\;\;H}{\underset{..}{O}}$$

Carboxylate ion   Hydronium ion   Carboxylic acid   Water

■ These rapid proton transfers don't require any heating either.

For example (and assuming that the carboxylate ions are salts involving Na^+ or K^+),

$$C_6H_5CO_2^- + H_3O^+ \longrightarrow C_6H_5CO_2H + H_2O$$

Benzoate ion   Benzoic acid
(soluble in water)   (insoluble in water)

$$CH_3(CH_2)_{16}CO_2^- + H_3O^+ \longrightarrow CH_3(CH_2)_{16}CO_2H + H_2O$$

Stearate ion   Stearic acid
(soluble in water)   (insoluble in water)

■ These carboxylate ions are soluble in water only when the associated positive ions are from group IA, like Na^+ and K^+ (or NH_4^+), not from other groups.

The Carboxylic Acid Group Is a Solubility "Switch." In the reactions just studied, we noted the solubilities in water of several species to draw attention to a very important property given to a molecule by the carboxyl group. The CO_2H group is an "on–off switch" for the solubility in water of any substance that contains it. When we try to increase the pH of a solution—by adding a strong base (like NaOH or KOH)—a water-insoluble carboxylic acid almost instantly dissolves, because it changes to its carboxylate ion. Similarly, when we try to decrease the pH of a solution—by adding a strong acid—a water-soluble carboxylate ion instantly changes to its much less soluble, free carboxylic acid form. In other words, by suitably adjusting the pH of an aqueous solution, we can make a substance with a carboxyl group more soluble or less soluble in water.

■ The carboxylic acid solubility switch is

$$H^+ \;\;\underset{RCO_2^-\ (\text{more soluble})}{\overset{RCO_2H\ (\text{less soluble})}{\rightleftharpoons}}\;\; OH^-$$

■ The other solubility "on–off" switch that we have studied involves amines (page 262).

───────────────────────────────

■ **PRACTICE EXERCISE 2** Write the structures of the organic products of the reactions of the following compounds with dilute hydrochloric acid at room temperature.

(a) $CH_3O-\langle\!\bigcirc\!\rangle-CO_2^-K^+$ **(b)** $CH_3CH_2CO_2^-Li^+$ **(c)** $(CH_3CH{=}CHCO_2^-)_2Ca^{2+}$

───────────────────────────────

Carboxylic Acids React with Alcohols to Form Esters. When a solution of a carboxylic acid in an alcohol is heated in the presence of a strong acid catalyst, the following species become involved in an equilibrium.

■ The ester *group* is

$$-\overset{\overset{\displaystyle O}{\|}}{C}-O-\overset{|}{C}-$$

$$\underset{\text{Carboxylic acid}}{R-\overset{\overset{\displaystyle O}{\|}}{C}-O-H} + \underset{\text{Alcohol}}{H-O-R'} \xrightarrow{H^+} \underset{\text{Ester}}{R-\overset{\overset{\displaystyle O}{\|}}{C}-O-R'} + H-O-H$$

When the alcohol is in *excess*, the equilibrium shifts so much to the right (in accordance with Le Châtelier's principle) that the reaction is a good method for making an ester in the lab. This synthesis of an ester is called **esterification.** Some specific examples are as follows.

$$\underset{\text{Acetic acid}}{CH_3-\overset{\overset{\displaystyle O}{\|}}{C}-O-H} + \underset{\substack{\text{Ethyl alcohol}\\(\text{large excess})}}{H-O-CH_2CH_3} \xrightarrow{H^+} \underset{\text{Ethyl acetate}}{CH_3-\overset{\overset{\displaystyle O}{\|}}{C}-O-CH_2CH_3} + H_2O$$

■ These equations are actually forward reactions of equilibria, and the excesses of alcohols force them to the right.

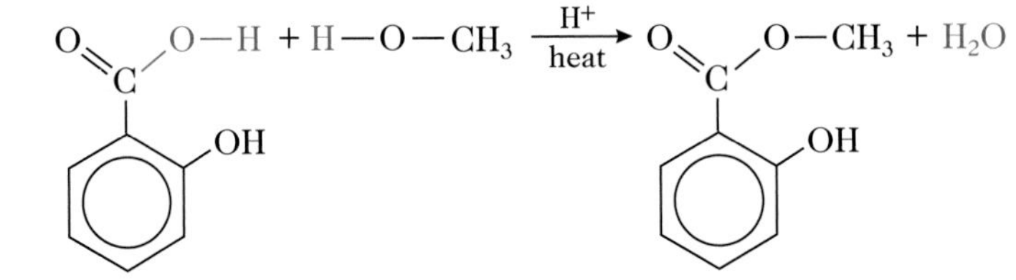

Salicylic acid   Methyl alcohol (large excess)   →   Methyl salicylate (oil of wintergreen)

EXAMPLE 13.1   Writing the Structure of a Product of Esterification

What is the structure of the ester that can be made from benzoic acid and methyl alcohol?

ANALYSIS   The ester structure *must* include the ester group,

$$-\overset{\overset{\displaystyle O}{\|}}{C}-O-C$$

The *names* of the starting materials usually give away the rest of the ester structure (provided we know the structures corresponding to the names). The carboxylic acid *always* furnishes the group joined to the carbon of the CO group, the C_6H_5 or phenyl group in our example. The parent alcohol *always* supplies the hydrocarbon group attached to O in the ester, the CH_3 group in our example. All that remains is to join C_6H_5 and CH_3 to the correct sites of the ester group's skeleton.

■ The structure of an ester can be condensed to RCO_2R'.

SOLUTION: The ester that forms is methyl benzoate, which we may write as $C_6H_5CO_2CH_3$, or as

$$C_6H_5-\overset{\overset{\displaystyle O}{\|}}{C}-O-CH_3$$

■   PRACTICE EXERCISE 3   Write the structures of the esters that form by the esterification of acetic acid by the following alcohols.

(a) methyl alcohol   (b) propyl alcohol   (c) isopropyl alcohol

■ PRACTICE EXERCISE 4   Write the structures of the esters that can be made by the esterification of ethyl alcohol by the following acids.
(a) formic acid   (b) propionic acid   (c) benzoic acid

13.3 ESTERS OF CARBOXYLIC ACIDS

The ester group is split apart by water in the presence of either acid or base.

Esters are compounds with the carbonyl–oxygen–carbon network (Table 13.3). Interestingly, the esters of vile-smelling acids generally have pleasant fragrances. Special Topic 13.1 describes some important esters in more detail.

The ester group cannot donate hydrogen bonds, because it has no HO (or HN) group, but it can *accept* them. This allows the esters of lower formula mass to be relatively soluble in water.

Esters React with Water to Give the Parent Carboxylic Acid and Alcohol.   The hydrolysis of an ester, its reaction with water, is very slow unless an acid catalyst is used. (In the body, we hydrolyze ester groups when we digest fats and oils, only here an enzyme acts as the catalyst.) In general,

■ $HCO_2CH_2CH(CH_3)_2$
Isobutyl formate (raspberries)
$CH_3CO_2(CH_2)_4CH_3$
Pentyl acetate (bananas)
$CH_3CH_2CH_2CO_2CH_2CH_3$
Ethyl butyrate (pineapples)

$$R-\overset{\overset{\displaystyle O}{\|}}{C}-O-R' + H-O-H \xrightarrow[\text{heat}]{H^+} R-\overset{\overset{\displaystyle O}{\|}}{C}-O-H + H-O-R'$$

Ester   Carboxylic acid   Alcohol

TABLE 13.3   Esters of Carboxylic Acids

Name[a]	Structure	MP (°C)	BP (°C)	Solubility in Water[b]
Ethyl Esters of Straight-Chain Carboxylic Acids, $RCO_2C_2H_5$				
Ethyl formate (ethyl methanoate)	$HCO_2C_2H_5$	−79	54	Soluble
Ethyl acetate (ethyl ethanoate)	$CH_3CO_2C_2H_5$	−82	77	7.35 (25 °C)
Ethyl propionate (ethyl propanoate)	$CH_3CH_2CO_2C_2H_5$	−73	99	1.75
Ethyl butyrate (ethyl butanoate)	$CH_3(CH_2)_2CO_2C_2H_5$	−93	120	0.51 (22 °C)
Miscellaneous Esters				
Methyl acrylate	$CH_2{=}CHCO_2CH_3$		80	5.2
Methyl benzoate	$C_6H_5CO_2CH_3$	−12	199	Insoluble
Natural waxes	$CH_3(CH_2)_nCO_2(CH_2)_nCH_3$			
	$n = 23–33$, carnauba wax			
	$= 25–27$, beeswax			
	$= 14–15$, spermaceti			

[a]Common names; IUPAC names are in parentheses.
[b]In grams of ester per 100 g water at 20 °C (unless otherwise specified).

ESTERS IN FOOD ADDITIVES, MEDICINALS, AND SYNTHETIC FILMS AND FIBERS

Esters of *p*-Hydroxybenzoic Acid—The Parabens
Several alkyl esters of *p*-hydroxybenzoic acid—referred to as *parabens* on ingredient labels—are used to inhibit molds and yeasts in cosmetics, pharmaceuticals, and food.

Salicylates Certain esters and salts of salicylic acid are analgesics (pain suppressants) and antipyretics (fever reducers). The parent acid, salicylic acid, is itself too irritating to the stomach for these uses, but sodium salicylate and acetylsalicylic acid (aspirin) are commonly used. Methyl salicylate, a pleasant-smelling oil, is used in liniments, for it readily migrates through the skin.

FIGURE I

The knitted tubing for this aortic heart valve is made of Dacron fibers.

Sodium salicylate   Acetylsalicylic acid (aspirin)   Methyl salicylate (oil of wintergreen)

Dacron Dacron, a polyester of exceptional strength, is widely used to make fabrics and film backing for recording tapes. (Actually, the name *Dacron* applies just to the fiber form of this polyester. When it is cast as a thin film, its name is *Mylar.*) Dacron fabrics have been used to repair or replace segments of blood vessels (Figure 1).

The formation of Dacron and many other polyfunctional polymers starts with two difunctional monomers, *aAa* and *bBb*. Their functional groups are able to react with each other to split out a small molecule, *ab*. The monomer fragments, *A* and *B*, join end-

to-end to make a very long polymer molecule. In principle, the polymerization can be represented as follows:

$$aAa + bBb + aAa + bBb + aAa + bBb + \text{etc.} \rightarrow$$
$$-A-B-A-B-A-B-\text{ etc. } + n(ab)$$
A copolymer

Because two monomers are used, the reaction is called *copolymerization.*

One monomer used to make Dacron is ethylene glycol, which has two alcohol OH groups. The other monomer is dimethyl terephthalate, which has two methyl ester groups. The copolymerization of these two monomers depends on a reaction of esters with alcohols that we will not study; the *ab* molecule that splits out is methyl alcohol. The copolymerization proceeds as follows.

Ethylene glycol

Dimethyl terephthalate

(repeating unit)
Dacron/Mylar

Specific examples are

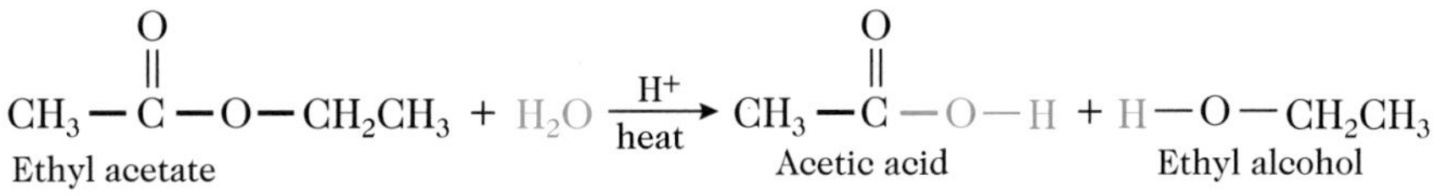

$$
\underset{\text{Ethyl acetate}}{CH_3 - \overset{\overset{\displaystyle O}{\|}}{C} - O - CH_2CH_3} + H_2O \xrightarrow[\text{heat}]{H^+} \underset{\text{Acetic acid}}{CH_3 - \overset{\overset{\displaystyle O}{\|}}{C} - O - H} + \underset{\text{Ethyl alcohol}}{H - O - CH_2CH_3}
$$

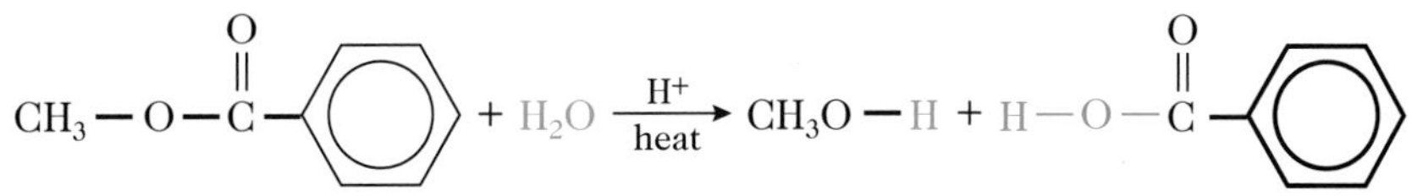

$$
\underset{\text{Methyl benzoate}}{CH_3 - O - \overset{\overset{\displaystyle O}{\|}}{C} - \bigcirc} + H_2O \xrightarrow[\text{heat}]{H^+} \underset{\text{Methyl alcohol}}{CH_3O - H} + \underset{\text{Benzoic acid}}{H - O - \overset{\overset{\displaystyle O}{\|}}{C} - \bigcirc}
$$

To avoid a mistake that students often make, notice that the *only* bond that breaks in ester hydrolysis is the bond joining the carbonyl group to the oxygen atom, the "ester bond." Notice also that the products are always the "parents" of the ester, so the *names* of the parents are available from the name of the ester itself. Thus, methyl benzoate hydrolyzes to *methyl* alcohol and *benzoic* acid. Let's work an example that involves only the ester structure, not its name.

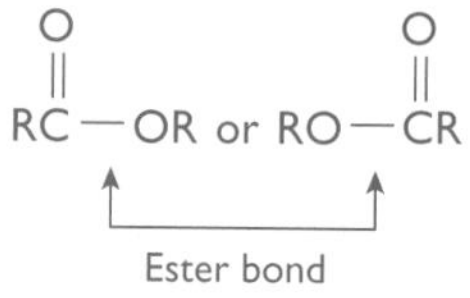

$$
\underset{\text{Ester bond}}{\overset{\overset{\displaystyle O}{\|}}{RC} - OR \ \text{or} \ RO - \overset{\overset{\displaystyle O}{\|}}{CR}}
$$

EXAMPLE 13.2 Predicting the Products of an Ester Hydrolysis

What are the products of the hydrolysis of the following ester?

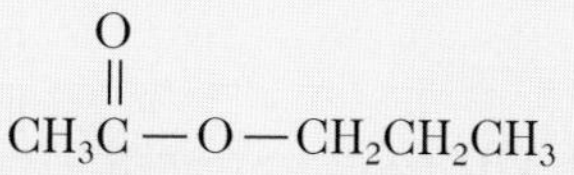

$$
\overset{\overset{\displaystyle O}{\|}}{CH_3C} - O - CH_2CH_2CH_3
$$

ANALYSIS Finding the ester bond, the carbonyl–to–oxygen bond, is the crucial step because this is the bond that is broken when an ester hydrolyzes. It doesn't matter in which direction this bond happens to point on the page; it is the *only* bond that breaks.

$$
\overset{\overset{\displaystyle O}{\|}}{CH_3C} - O - CH_2CH_2CH_3 \quad \text{or} \quad CH_3CH_2CH_2 - O - \overset{\overset{\displaystyle O}{\|}}{CCH_3}
$$

These are identical compounds.

Carbonyl-to-oxygen bond, the ester bond

Break the carbonyl–oxygen bond. Erase it and separate the fragments. (We'll continue with the structure on the left, above.)

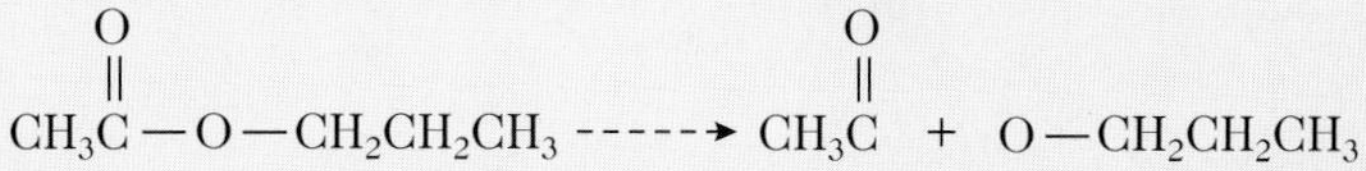

$$
\overset{\overset{\displaystyle O}{\|}}{CH_3C} - O - CH_2CH_2CH_3 \ \text{-----} \blacktriangleright \ \overset{\overset{\displaystyle O}{\|}}{CH_3C} \ + \ O - CH_2CH_2CH_3
$$

Next, attach the pieces of the water molecule to make the "parents" of the ester. Attach OH to the carbonyl carbon and put H on the oxygen atom of the other fragment.

SOLUTION The products are therefore acetic acid and propyl alcohol.

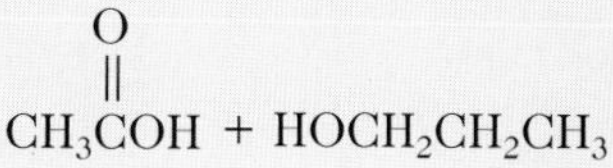

$$
\overset{\overset{\displaystyle O}{\|}}{CH_3COH} + HOCH_2CH_2CH_3
$$

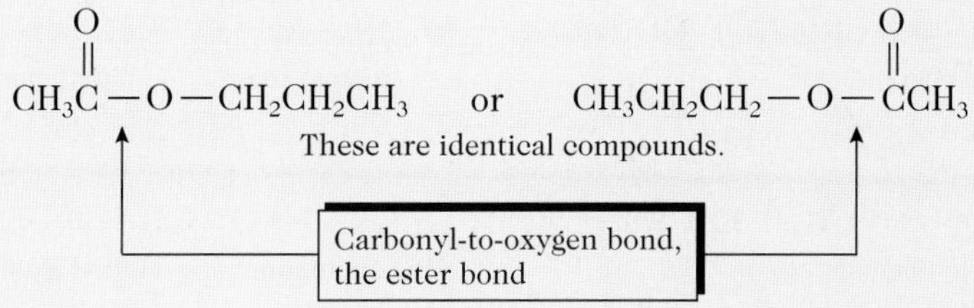

■ The most common mistakes made by students are breaking the wrong bond and breaking too many bonds.

CHECK   Reexamine the structure of the original ester. Its alcohol portion, the R group on O, has three carbons in a straight chain, so the alcohol must be like this. The acid portion of the ester has two carbons, so the acid must be a two-carbon acid (Double-check each C and O for the correct number of bonds.)

■ **PRACTICE EXERCISE 5**   Write the structures of the products of the hydrolysis of the following esters.

$$
\begin{array}{ccc}
\underset{\text{(a) CH}_3\text{OCCH}_3}{\overset{\displaystyle O}{\overset{\|}{}}} &
\underset{\text{(b) CH}_3\text{CH}_2\text{C}-\text{O}-\text{CHCH}_3}{\overset{\displaystyle O \qquad\quad CH_3}{\overset{\|\qquad\quad |}{}}} &
\underset{\text{(c) CH}_3\text{CH}-\text{C}-\text{OCH}_2\text{CH}_2\text{CH}_3}{\overset{\displaystyle CH_3\; O}{\overset{|\quad\;\|}{}}}
\end{array}
$$

■ From the Latin *sapo,* "soap," and *onis,* "to make." Ordinary soap is made by saponification of the ester groups in fats and oils.

Saponification of an Ester Gives the Salt of the Parent Acid and the Parent Alcohol. Bases, like OH^- and $CO_3{}^{2-}$, also promote the breakup of an ester at its ester bond. One product is the parent alcohol, but the other is the *negatively charged ion* of the parent carboxylic acid. The reaction of an ester with aqueous base is called **saponification,** and it requires a full mole (not just a catalytic trace) of base for each mole of ester bonds. Esters can be saponified using group IA carbonates, like Na_2CO_3, or hydroxides, like NaOH and KOH. We'll use OH^- as the active saponifying agent in our illustrations; in general,

$$
\underset{\text{Ester}}{\overset{\displaystyle O}{\overset{\|}{\text{R}-\text{C}-\text{O}-\text{R}'}}} + OH^-(aq) \xrightarrow{\text{heat}} \underset{\substack{\text{Carboxylate}\\ \text{anion}}}{\overset{\displaystyle O}{\overset{\|}{\text{R}-\text{C}-\text{O}^-}}} + \underset{\text{Alcohol}}{\text{H}-\text{O}-\text{R}'}
$$

Specific examples are as follows.

$$
\underset{\text{Ethyl acetate}}{\overset{\displaystyle O}{\overset{\|}{\text{CH}_3-\text{C}-\text{O}-\text{CH}_2\text{CH}_3}}} + OH^-(aq) \xrightarrow{\text{heat}} \underset{\text{Acetate ion}}{\overset{\displaystyle O}{\overset{\|}{\text{CH}_3-\text{C}-\text{O}^-}}} + \underset{\text{Ethyl alcohol}}{\text{H}-\text{O}-\text{CH}_2\text{CH}_3}
$$

$$
\underset{\text{Methyl benzoate}}{\overset{\displaystyle O}{\overset{\|}{\text{C}_6\text{H}_5-\text{C}-\text{O}-\text{CH}_3}}} + OH^-(aq) \xrightarrow{\text{heat}} \underset{\text{Benzoate ion}}{\overset{\displaystyle O}{\overset{\|}{\text{C}_6\text{H}_5-\text{C}-\text{O}^-}}} + \underset{\text{Methyl alcohol}}{\text{H}-\text{O}-\text{CH}_3}
$$

Notice that the *names* of the esters are strong clues to the names *and structures* of the products of saponification. In the next example, however, we'll start with the ester *structure,* not its name.

EXAMPLE 13.3   Writing the Structures of the Products of Saponification

What are the products of saponification of the following ester?

$$
\overset{\displaystyle O}{\overset{\|}{\text{CH}_3\text{CH}_2\text{C}-\text{O}-\text{CH}_3}}
$$

ANALYSIS Saponification is very similar to ester hydrolysis. The ester bond is broken. *Break only this bond.* Separate the fragments:

$$\underset{\text{CH}_3\text{CH}_2\overset{\displaystyle O}{\overset{\displaystyle \|}{\text{C}}}-\text{O}-\text{CH}_3} \ \ \text{------->} \ \ \text{CH}_3\text{CH}_2\overset{\displaystyle O}{\overset{\displaystyle \|}{\text{C}}} \ + \ -\text{O}-\text{CH}_3$$

To change the fragment with the carbonyl group into the *negatively charged ion* of a carboxylic acid, attach O^- to the carbonyl carbon atom. Then attach H to the oxygen atom of the other fragment to make the alcohol molecule.

SOLUTION The products are the propionate ion and methyl alcohol.

$$\text{CH}_3\text{CH}_2\overset{\displaystyle O}{\overset{\displaystyle \|}{\text{C}}}-\text{O}^- \ + \ \text{H}-\text{O}-\text{CH}_3$$

■ **PRACTICE EXERCISE 6** Write the structures of the products of saponification of the following esters.

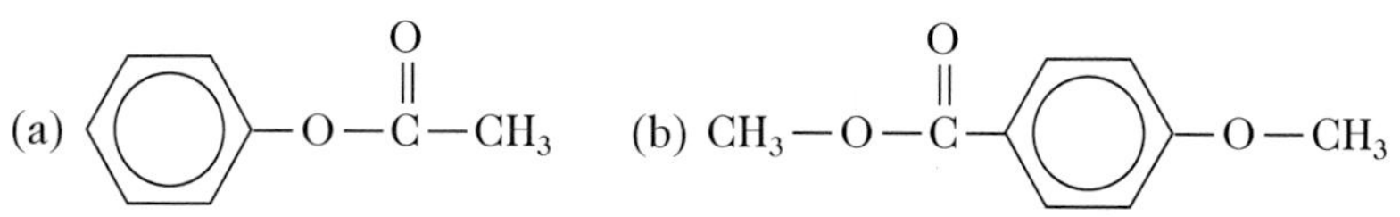

(a) C₆H₅—O—C(=O)—CH₃   (b) CH₃—O—C(=O)—C₆H₄—O—CH₃

13.4   ORGANOPHOSPHATE ESTERS AND ANHYDRIDES

Some of the most widely distributed esters in living organisms are those of phosphoric acid, diphosphoric acid, and triphosphoric acid.

Phosphoric acid appears in several forms and ions in the body, but the three fundamental parents are *phosphoric acid, diphosphoric acid,* and *triphosphoric acid.* We study the last two and their esters particularly because the P—O—P network in their structures is a major storehouse of chemical energy in cells.

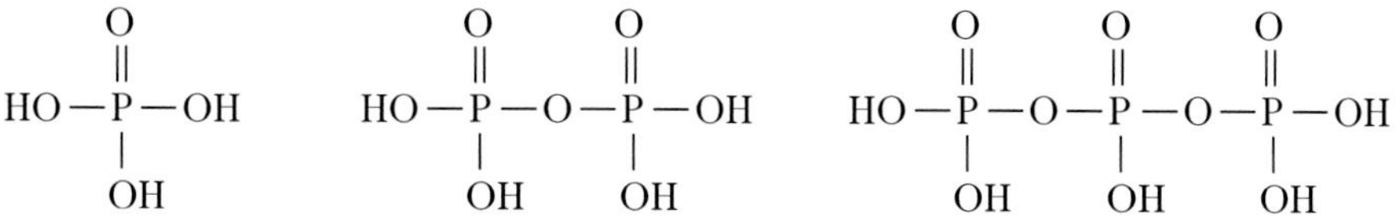

$$\underset{\text{Phosphoric acid}}{\text{HO}-\overset{\displaystyle O}{\overset{\displaystyle \|}{\underset{\displaystyle |}{\underset{\displaystyle \text{OH}}{\text{P}}}}}-\text{OH}} \qquad \underset{\text{Diphosphoric acid}}{\text{HO}-\overset{\displaystyle O}{\overset{\displaystyle \|}{\underset{\displaystyle |}{\underset{\displaystyle \text{OH}}{\text{P}}}}}-\text{O}-\overset{\displaystyle O}{\overset{\displaystyle \|}{\underset{\displaystyle |}{\underset{\displaystyle \text{OH}}{\text{P}}}}}-\text{OH}} \qquad \underset{\text{Triphosphoric acid}}{\text{HO}-\overset{\displaystyle O}{\overset{\displaystyle \|}{\underset{\displaystyle |}{\underset{\displaystyle \text{OH}}{\text{P}}}}}-\text{O}-\overset{\displaystyle O}{\overset{\displaystyle \|}{\underset{\displaystyle |}{\underset{\displaystyle \text{OH}}{\text{P}}}}}-\text{O}-\overset{\displaystyle O}{\overset{\displaystyle \|}{\underset{\displaystyle |}{\underset{\displaystyle \text{OH}}{\text{P}}}}}-\text{OH}}$$

These three acids are all *polyprotic,* but in the slightly alkaline media of body fluids, they exist not as free acids but as a mixture of negatively charged ions instead. The ions' net charges and their relative concentrations depend on pH.

Esters of Alcohols and Phosphoric Acid Are Monophosphate Esters. If you look closely at the structure of phosphoric acid, you can see that part of it resembles a carboxyl group.

$$HO-\overset{\displaystyle O}{\overset{\|}{P}}- \qquad HO-\overset{\displaystyle O}{\overset{\|}{C}}-$$

Network in a molecule   Network in a molecule
of phosphoric acid   of a carboxylic acid

■ Each of the two OH groups in a phosphate ester can be converted into an ester. Nucleic acid molecules, for example, have the phosphate diester system.

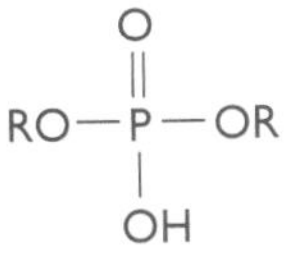

$$RO-\overset{\displaystyle O}{\overset{\|}{\underset{\underset{\displaystyle OH}{|}}{P}}}-OR$$

A phosphate diester

Therefore, it isn't surprising that *esters* of phosphoric acid exist and that they are structurally similar to esters of carboxylic acids.

$$R'O-\overset{\displaystyle O}{\overset{\|}{P}}- \qquad R'O-\overset{\displaystyle O}{\overset{\|}{C}}-$$

Network in a molecule   Network in a molecule
of phosphate ester   of a carboxylic acid ester

One large difference between a phosphate ester and a carboxylate ester is that *a phosphate ester is still a diprotic acid*. Its molecules carry two proton-donating OH groups. Depending on the pH of the medium, therefore, a phosphate ester can exist in any one of three forms, and usually there is an equilibrium mixture of all three.

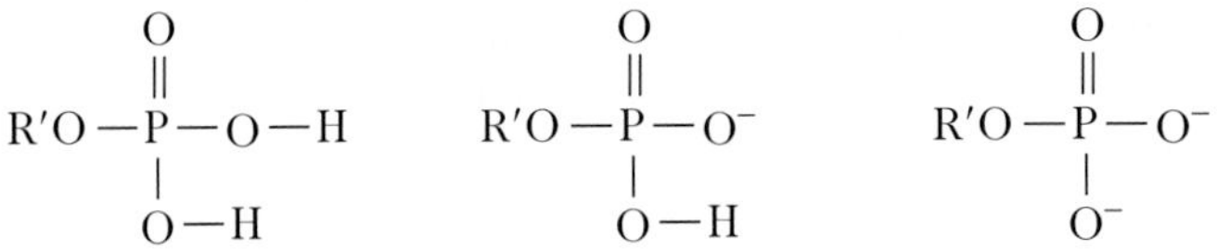

$$R'O-\overset{\displaystyle O}{\overset{\|}{\underset{\underset{\displaystyle O-H}{|}}{P}}}-O-H \qquad R'O-\overset{\displaystyle O}{\overset{\|}{\underset{\underset{\displaystyle O-H}{|}}{P}}}-O^- \qquad R'O-\overset{\displaystyle O}{\overset{\|}{\underset{\underset{\displaystyle O^-}{|}}{P}}}-O^-$$

Phosphate ester   Phosphate ester   Phosphate ester
(as a diprotic acid)   (as a singly ionized   (as a doubly ionized
                         species)   species)

Favored at low pH   Favored at a pH   Favored at pH
                         just below 7   values above 7

■ The pH of most body fluids is slightly greater (more basic) than pH 7.0.

At the pH of most body fluids, phosphate esters exist mostly as the doubly ionized species—as the di-negative ion. All forms, however, are generally soluble in water, and one reason that the body converts so many substances into their phosphate esters may be to improve their solubilities in water. One of the important *monophosphates* in the body is adenosine monophosphate, or AMP, shown here in the fully ionized form in which it mostly exists in cells.

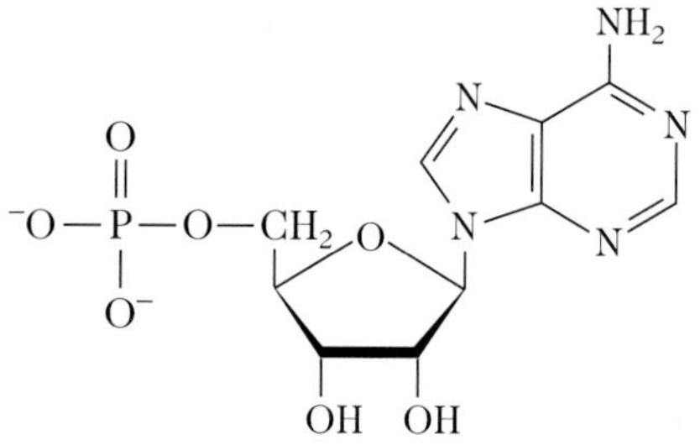

Adenosine monophosphate, AMP
(fully ionized form)

Alcohols and Diphosphoric Acid Form Diphosphate Esters. A diphosphate ester actually has three functional groups: a phosphate ester group, proton-donating OH groups, and a system new to our study, the *phosphoric anhydride* system.

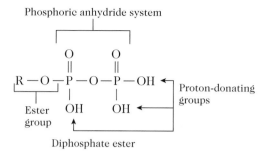

Phosphoric anhydride system

Ester group

Proton-donating groups

Diphosphate ester

A portion of this structure is called the phosphoric *anhydride* system because it can be viewed as having formed by a dehydration—the removal of H_2O—between two molecules of phosphoric acid.

$$
\begin{array}{c}
O \\ \| \\ -P-O-H \\ | \end{array}
+
\begin{array}{c}
O \\ \| \\ H-O-P- \\ | \end{array}
\longrightarrow
\begin{array}{c}
O \quad\quad O \\ \| \quad\quad \| \\ -P-O-P- \\ | \quad\quad | \end{array}
+ H_2O
$$

Portions of two molecules of phosphoric acid

Part of the diphosphate system—the phosphoric anhydride system

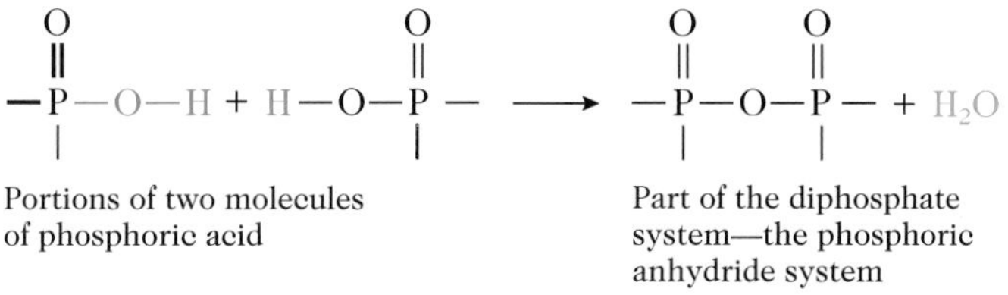

(Do not regard this equation as the way by which cells make the phosphoric anhydride system. View the equation as simply a device for seeing structural relationships and as one part of the explanation for the word *anhydride*.)

Adenosine diphosphate, or ADP, is one of many *diphosphate esters* in the body, where it exists largely as its triply charged ion.

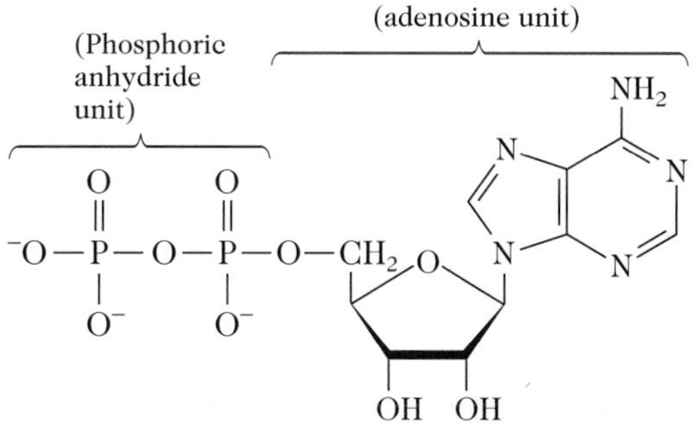

(adenosine unit)

(Phosphoric anhydride unit)

NH_2

Adenosine diphosphate, ADP (fully ionized form)

The Phosphoric Anhydride Group Is a Storehouse of Chemical Energy in Living Systems. The phosphoric anhydride group can be broken up by water or alcohols. Alcohols, for example, react as follows, where the curving arrows indicate where the attack occurs and how bonds break and form.

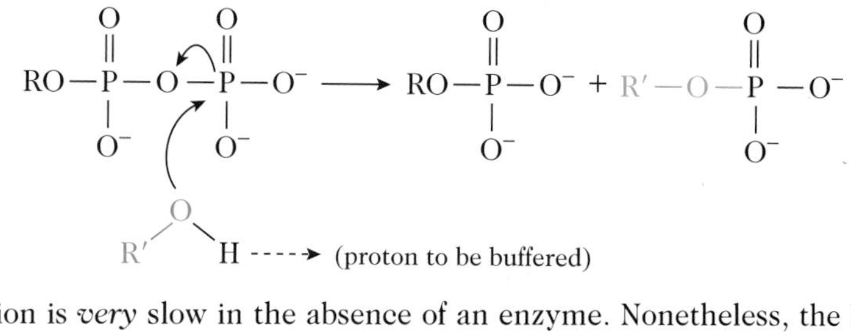

$$
RO-P-O-P-O^- \longrightarrow RO-P-O^- + R'-O-P-O^-
$$

(proton to be buffered)

The reaction is *very* slow in the absence of an enzyme. Nonetheless, the breakup of the phosphoric anhydride system generates considerable energy per mole, which tells

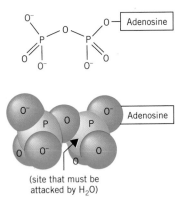

FIGURE 13.1

The oxygen atoms in the phosphoric anhydride system of ADP screen the phosphorus atoms. The negative charges on these oxygen atoms deflect incoming, electron-rich particles such as molecules of an alcohol or of water. The phosphoric anhydride system therefore reacts very slowly in cells with alcohols or water, unless a special enzyme is present.

us that this system *contains* much chemical energy. For this reason, compounds with phosphoric anhydride systems are designated *energy-rich* compounds. The phosphoric anhydride group in ADP and similar compounds (like ATP, below) turns out to be the chief means for storing chemical energy in cells. Let's see why.

The source of the internal energy in the triply charged ion of ADP is the tension up and down the anhydride chain. The central chain bears oxygen atoms with full negative charges, and these charges repel each other. *This internal repulsion primes the phosphoric anhydride system for breaking apart exothermically when it is attacked by a suitable reactant.* The system is like a spring under compression and awaiting release. Yet an enzyme catalyst for the release is essential because the phosphorus atoms in the chain are buried within a clutch of negatively charged oxygen atoms. These would actually *repel* alcohol molecules (Figure 13.1). Yet, if an alcohol molecule is to attack, break up the phosphoric anhydride system, and release energy, the electron-rich O atom of ROH must reach a phosphorus atom, as we visualized above. So the internal tension in the diphosphate cannot be relieved unless an enzyme for the reaction is present. *The body exerts control over the energy-releasing reactions of diphosphates by its control of the enzymes for these reactions.*

Water could make the same kind of exothermic attack on a diphosphate ester as an alcohol, but *the body has no enzymes inside cells that catalyze this reaction.* Hence, energy-rich diphosphates are able to exist in cells despite the abundance of a potential reactant, water.

Alcohols and Triphosphoric Acid Form Triphosphate Esters. Adenosine triphosphate, or ATP, is the most common and widely occurring member of a small family of energy-rich *triphosphate esters*. Because the triphosphates have *two* (overlapping) phosphoric anhydride systems in each molecule, on a mole-for-mole basis the triphosphates are among the most energy-rich substances in the body.

Adenosine triphosphate, ATP
(fully ionized form)

Triphosphates are much more widely used in cells as sources of energy than are diphosphates. The overall reaction for the contraction of a muscle, for example, can be written as follows. We now introduce the symbol P_i to stand for the set of inorganic phosphate ions, mostly $H_2PO_4^-$ and HPO_4^{2-}, produced in the breakup of ATP and present at equilibrium at body pH.

$$\text{Relaxed muscle} + \text{ATP} \xrightarrow{\text{enzyme}} \text{contracted muscle} + \text{ADP} + P_i$$

Although it is extremely simplified, the "equation" shows that muscular work requires ATP. If the body's supply of ATP were consumed *with no way to remake it,* we'd soon lose all capacity for such work. The resynthesis of ATP from ADP and inorganic phosphate ion is one of the major uses of the chemical energy in the food we eat.

ATP is resynthesized only as needed. As soon as ADP and phosphate ion appear following the use of ATP, they trigger the resynthesis of ATP. There is thus a rapid turnover of ATP in the body. The chemical energy in the food we eat drives the synthesis of roughly 40 kg (88 lb) of ATP every 24 hours! We'll study how the body does this remarkable task in Chapter 19.

13.5 AMIDES OF CARBOXYLIC ACIDS

Amides are neutral nitrogen compounds that can be hydrolyzed to carboxylic acids and ammonia (or amines).

We study the *amide* system because all proteins are essentially polyamides, polymers whose molecules have regularly spaced *amide bonds*. Nylon is a synthetic polymer with repeating amide groups (see Special Topic 13.2).

The carbonyl–nitrogen bond is called the **amide bond** because it is the bond that forms when amides are made, and it is the bond that breaks when amides are hydrolyzed. As the following general structures show, an amide can be derived either from ammonia or from amines. Those derived from ammonia itself are referred to as *simple* amides (Table 13.4).

■ The amide bond is called the *peptide bond* in protein chemistry.

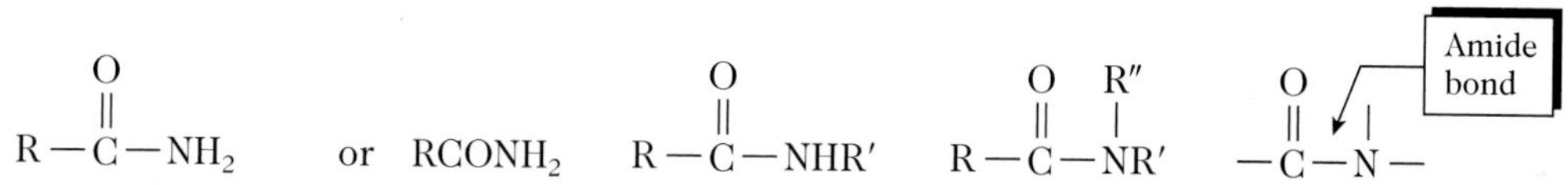

$$
\begin{array}{cc}
\underset{\substack{\| \\ R-C-NH_2}}{O} & \text{or} \quad RCONH_2 \qquad \underset{\substack{\| \\ R-C-NHR'}}{O} \qquad \underset{\substack{\| \ | \\ R-C-NR'}}{O \ R''} \qquad \underset{\substack{\| \ | \\ -C-N-}}{O}
\end{array}
$$

Amides of ammonia (simple amides)   Amides of amines   Amide group   Amide bond

Hydrogen Bonding Is Strong in Amides. Amide molecules are quite polar, and when they have at least one H atom attached to N, they can both donate and accept hydrogen bonds. These forces of attraction add up so much in simple amides that all except formamide are solids at room temperature. Simple amides have considerably higher boiling points than alkanes, alcohols, or even carboxylic acids of comparable formula mass, as the data in the margin show. When we study proteins, we'll see how hydrogen bonding is involved in stabilizing the shapes of protein molecules, shapes that are as important to the functions of proteins as anything else about their structures.

Compound	Formula Mass	BP (°C)
$CH_3CH_2CH_2CH_3$	58	−42
$CH_3CH_2CH_2OH$	60	97
CH_3CO_2H	60	118
CH_3CONH_2	59	222

TABLE 13.4 Amides of Carboxylic Acids

IUPAC Name	Structure	MP (°C)
Formamide	$HCONH_2$	3
N-Methylformamide	$HCONHCH_3$	−5
N,N-Dimethylformamide	$HCON(CH_3)_2$	−61
Acetamide	CH_3CONH_2	82
N-Methylacetamide	$CH_3CONHCH_3$	28
N,N-Dimethylacetamide	$CH_3CON(CH_3)_2$	−20
Propionamide	$CH_3CH_2CONH_2$	79
Butyramide	$CH_3CH_2CH_2CONH_2$	115
Benzamide	$C_6H_5CONH_2$	133

SPECIAL TOPIC 13.2
NYLON, A POLYAMIDE

The term *nylon* is a coined name that applies to any synthetic, long-chain, fiber-forming polymer with repeating amide linkages. One of the most common members of the nylon family, nylon-66, is made from 1,6-hexanediamine and hexanedioic acid.

$$NH_2(CH_2)_6NH_2 \qquad HOC(CH_2)_4COH$$

1,6-Hexanediamine   Hexanedioic acid

$$\text{etc.} \left[C(CH_2)_4CNH(CH_2)_6NH \right]_n \text{etc.}$$

Repeating unit in nylon-66

(The "66" means that each monomer has six carbon atoms.) To be useful as a fiber-forming polymer, each nylon-66 molecule should contain from 50 to 90 of each of the monomer units. Shorter molecules form weak or brittle fibers.

When molten nylon resin is being drawn into fibers, newly emerging strands are caught up on drums and stretched as they cool. Under this tension, the long polymer molecules within the fiber line up side by side, overlapping each other, to give a finished fiber of unusual strength and beauty (Figure 1). Part of nylon's strength comes from the innumerable hydrogen bonds that extend between the polymer molecules and that involve their many regularly spaced amide groups.

Nylon is more resistant to combustion than wool, rayon, cotton, or silk, and it is as immune to insect attack as fiberglass. Molds and fungi do not attack ny-

FIGURE I
This woman's life depends of the strength of nylon when she is parasailing.

lon molecules either. In medicine, nylon is used in specialized tubing, and as velour for blood contact surfaces. Nylon sutures were the first synthetic sutures and are still commonly used.

Amides Are Made from Amines by Acyl Group Transfer Reactions. In living systems, amides are never made directly from carboxylic acids and amines (or ammonia). Instead, special molecules serve as carriers for **acyl groups,** which are units present in carboxylic acids and their derivatives but without independent existence.

■ Acyl groups are carbonyl counterparts of alkyl groups.

■ To name an acyl group, change *-ic acid* to *-yl* in the name of the parent acid.

$$\underset{\text{Acyl group}}{R-C-} \qquad \text{For example:} \quad \underset{\text{Acetyl group}}{CH_3C-} \qquad \underset{\text{Propionyl group}}{CH_3CH_2C-} \qquad \underset{\substack{\text{Amino acetyl group} \\ \text{(an aminoacyl group)}}}{NH_2CH_2C-}$$

In the presence of the correct enzyme, an acyl group transfers from its carrier molecule to the amino group of another molecule to make an amide bond.

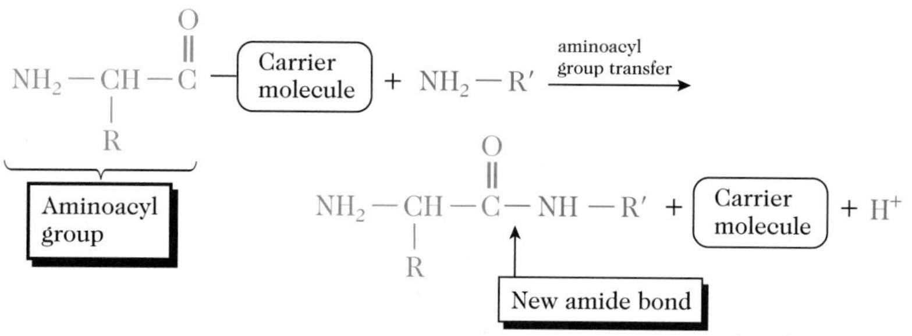

$$NH_2-CH-\overset{\overset{\displaystyle O}{\|}}{C}-\boxed{\begin{array}{c}\text{Carrier}\\\text{molecule}\end{array}}\ +\ NH_2-R'\ \xrightarrow{\ \overset{\text{aminoacyl}}{\text{group transfer}}\ }$$

$$\underbrace{}_{\boxed{\begin{array}{c}\textbf{Aminoacyl}\\\textbf{group}\end{array}}}$$

with R below CH.

$$NH_2-CH-\overset{\overset{\displaystyle O}{\|}}{C}-NH-R'\ +\ \boxed{\begin{array}{c}\text{Carrier}\\\text{molecule}\end{array}}\ +\ H^+$$

R below CH; New amide bond points to C—NH.

■ R and R′ are organic groups but not necessarily alkyl groups.

What we need to learn now is how to figure out the structure of an amide, given a parent carboxylic acid and either ammonia or an amine.

EXAMPLE 13.4 Writing the Structure of an Amide That Can Be Made from the Acyl Group of an Acid and an Amine

---

What amide can be made from the following two substances, assuming that a suitable acyl group transfer process is available?

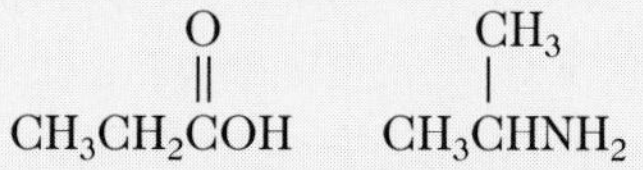

$$\overset{\overset{\displaystyle O}{\|}}{CH_3CH_2COH}\qquad \overset{\overset{\displaystyle CH_3}{|}}{CH_3CHNH_2}$$

ANALYSIS The amide system must be part of the structure we seek, so the best way to proceed is to write the skeleton of the amide system and then build on it. It doesn't matter how we orient this skeleton, left-to-right or right-to-left, as we'll demonstrate by showing both approaches.

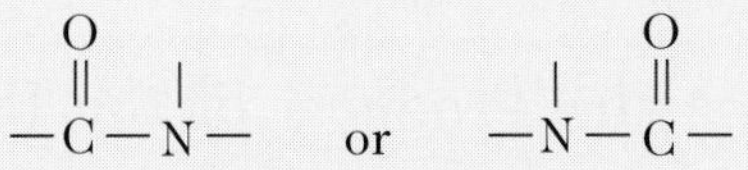

$$-\overset{\overset{\displaystyle O}{\|}}{C}-\overset{|}{N}-\qquad \text{or}\qquad -\overset{|}{N}-\overset{\overset{\displaystyle O}{\|}}{C}-\qquad \text{(incomplete)}$$

Then we look at the structure of the *acid* to see what group must be on the carbon atom of this skeleton, and we see that it's an *ethyl* group. We look at the amine to see what group(s) it carries, and it's an *isopropyl* group. So we attach the ethyl group to the carbonyl carbon atom and the isopropyl group to N. (If there had been *two* organic groups on N in the amine, we would attach both, of course.)

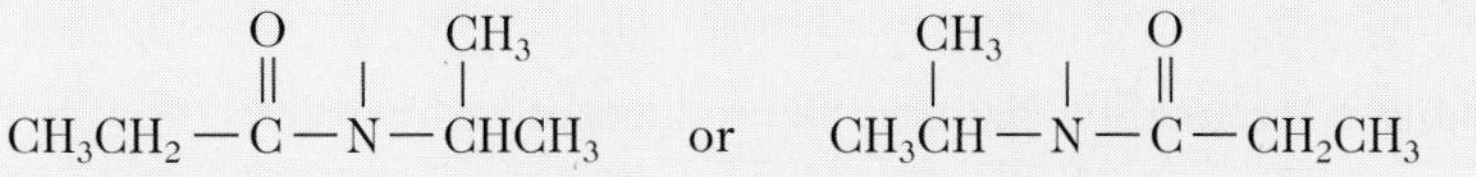

$$CH_3CH_2-\overset{\overset{\displaystyle O}{\|}}{C}-\overset{|}{N}-\overset{\overset{\displaystyle CH_3}{|}}{CHCH_3}\qquad \text{or}\qquad \overset{\overset{\displaystyle CH_3}{|}}{CH_3CH}-\overset{|}{N}-\overset{\overset{\displaystyle O}{\|}}{C}-CH_2CH_3\qquad \text{(incomplete)}$$

Finally, of the two H atoms on N in the parent amine, one H survives, so our last step is to write it in. (Recall that N needs three bonds in a neutral species.)

SOLUTION The amide is

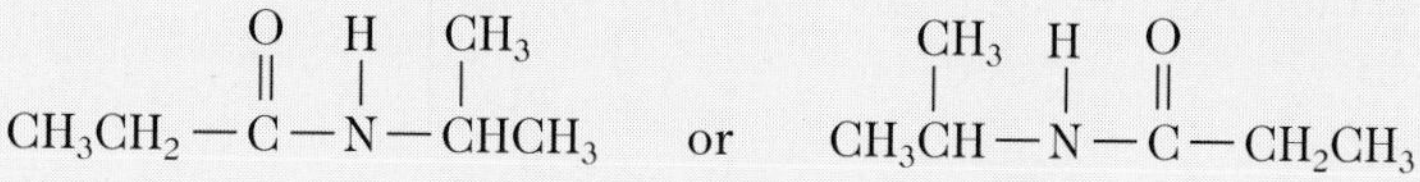

$$CH_3CH_2-\overset{\overset{\displaystyle O}{\|}}{C}-\overset{\overset{\displaystyle H}{|}}{N}-\overset{\overset{\displaystyle CH_3}{|}}{CHCH_3}\qquad \text{or}\qquad \overset{\overset{\displaystyle CH_3}{|}}{CH_3CH}-\overset{\overset{\displaystyle H}{|}}{N}-\overset{\overset{\displaystyle O}{\|}}{C}-CH_2CH_3$$

The two structures, of course, are identical.

■   PRACTICE EXERCISE 7   What amides, if any, could be made by suitable acyl group transfer reactions from the following pairs of compounds?

(a) CH_3NH_2 and CH_3CHCO_2H (b) $NH_2C_6H_5$ and CH_3CO_2H
$\qquad\qquad\qquad\qquad |$
$\qquad\qquad\qquad\quad CH_3$

$\qquad\qquad O$
$\qquad\qquad ||$
(c) $CH_3CCH_2NH_2$ and CH_3NH_2 (d) CH_3CO_2H and CH_3NCH_3
$\qquad\qquad\qquad\qquad\qquad\qquad\qquad\qquad\qquad |$
$\qquad\qquad\qquad\qquad\qquad\qquad\qquad\qquad CH_3$

Amides Can Be Hydrolyzed to Carboxylic Acids and Amines (or Ammonia).   Acids, bases, and enzymes all promote the hydrolysis of amides, the reaction by which proteins are digested. The actual products formed when an amide is hydrolyzed depend on the pH of the medium. If the medium is alkaline, the carboxylate ion and the amine (or ammonia) form. If the medium is acidic, the carboxylic acid and the protonated amine (or NH_4^+) are the products. Of course, the hydrolysis of an amide can occur without any catalyst or promoter, but it's much slower.

■ What specific products form in amide hydrolysis depend on the pH much in the way that the products of ester hydrolysis do.

To simplify, we'll write amide hydrolysis as a simple reaction with water to give the free carboxylic acid and the free amine. Here are some examples.

$$R-\overset{\overset{\displaystyle O}{||}}{C}-NH_2 \;+\; H_2O \xrightarrow{\text{enzyme}} R-\overset{\overset{\displaystyle O}{||}}{C}-OH \;+\; NH_3$$

$$R-\overset{\overset{\displaystyle O}{||}}{C}-NHR' \;+\; H_2O \xrightarrow{\text{enzyme}} R-\overset{\overset{\displaystyle O}{||}}{C}-OH \;+\; NH_2-R'$$

■ The digestive tract provides several protein-digesting enzymes called *proteases*.

$$R-\overset{\overset{\displaystyle O}{||}}{C}-\overset{\overset{\displaystyle R''}{|}}{N}R' \;+\; H_2O \xrightarrow{\text{enzyme}} R-\overset{\overset{\displaystyle O}{||}}{C}-OH \;+\; H-\overset{\overset{\displaystyle R''}{|}}{N}-R'$$

EXAMPLE 13.5   Writing the Products of the Hydrolysis of an Amide

Acetophenetidin (phenacetin) has been used in some headache remedies.

$$CH_3CH_2-O-\!\!\!\left\langle\!\!\bigcirc\!\!\right\rangle\!\!-NH-\overset{\overset{\displaystyle O}{||}}{C}-CH_3$$

Acetophenetidin

■ An old medicine, APC, for example, consisted of aspirin, phenacetin, and caffeine.

If this compound is an amide, what are the products of its hydrolysis?

ANALYSIS   Acetophenetidin does have the amide bond, NH to carbonyl, so it can be hydrolyzed. (The functional group on the left side of this structure is an *ether*, and *ethers do not react with water.*) Because the amide bond breaks when an amide is hydrolyzed, simply erase this bond from the structure and separate the parts. *Do not break any other bond.*

$$CH_3CH_2-O-\!\!\!\left\langle\!\!\bigcirc\!\!\right\rangle\!\!-NH\!\!\!\lesssim\!\!\overset{\overset{\displaystyle O}{||}}{C}-CH_3 \;\text{------->}$$

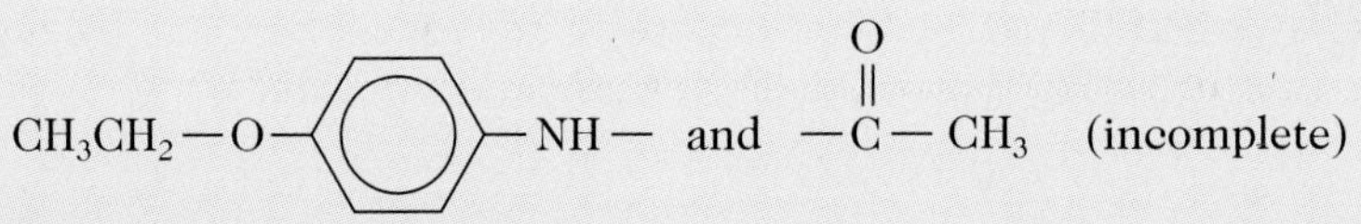

$$CH_3CH_2-O-\text{⟨benzene⟩}-NH- \quad \text{and} \quad -\overset{\overset{\displaystyle O}{\|}}{C}-CH_3 \quad \text{(incomplete)}$$

We know that the hydrolysis uses HO—H to give a *carboxylic acid* and an *amine*, so we put a HO group on the carbonyl group and we put H on the nitrogen of the other fragment.

SOLUTION The products of the hydrolysis of acetophenetidin are

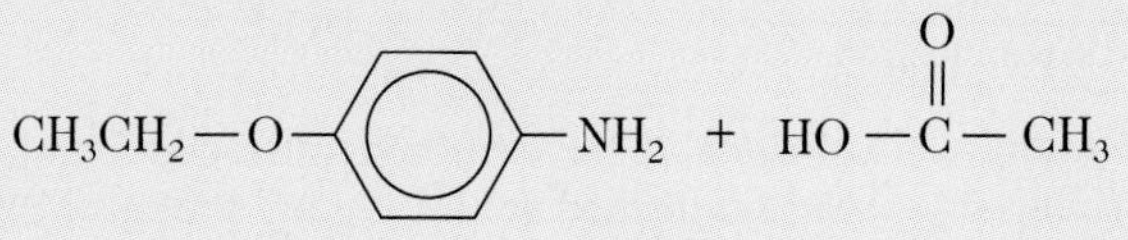

$$CH_3CH_2-O-\text{⟨benzene⟩}-NH_2 \;+\; HO-\overset{\overset{\displaystyle O}{\|}}{C}-CH_3$$

■ PRACTICE EXERCISE 8   For all compounds in the following list that are amides, write the products of their hydrolysis.

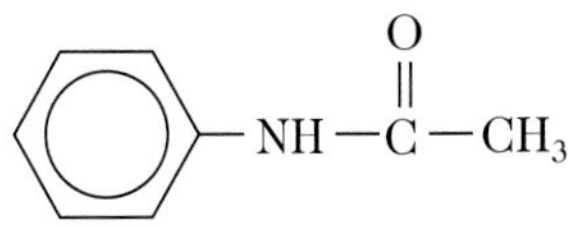

(a) 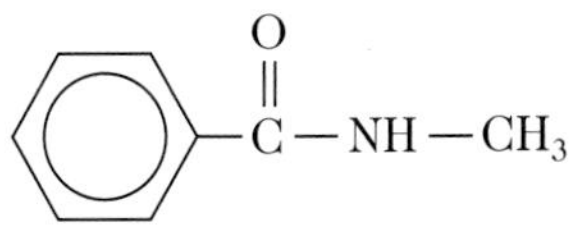 $\text{⟨benzene⟩}-\overset{\overset{\displaystyle O}{\|}}{C}-NH-CH_3$ (b) $\text{⟨benzene⟩}-\overset{\overset{\displaystyle O}{\|}}{C}-CH_2-NH_2$

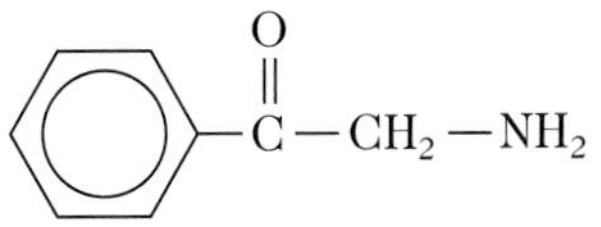

(c) $\text{⟨benzene⟩}-NH-\overset{\overset{\displaystyle O}{\|}}{C}-CH_3$ (d) 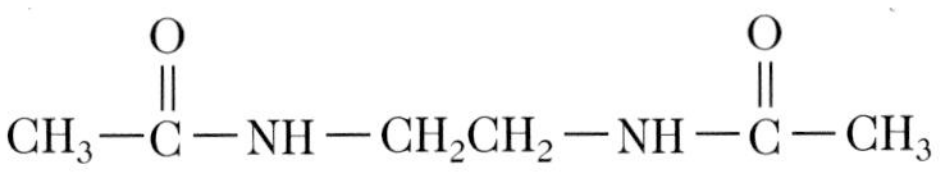 $CH_3-\overset{\overset{\displaystyle O}{\|}}{C}-NH-CH_2CH_2-NH-\overset{\overset{\displaystyle O}{\|}}{C}-CH_3$

SUMMARY

Acids and Their Salts The carboxyl group, CO_2H, is a polar group that confers moderate water solubility to a molecule without preventing its solubility in nonpolar solvents. This group is resistant to oxidation and reduction. Carboxylic acids are strong proton donors toward hydroxide ions. Toward water, carboxylic acids are weak acids. Carboxylate ions, therefore, are good proton acceptors.

Salts of carboxylic acids are ionic compounds, and the potassium and sodium salts are soluble in water. Hence, the carboxyl group is one of nature's important "solubility switches." An insoluble acid becomes soluble in base, but it is thrown out of solution again by the addition of acid.

The derivatives of acids—salts, esters, and amides—can be made from the acids and are converted back to the acids. We can organize the reactions we have studied for the carboxylic acids and esters as follows.

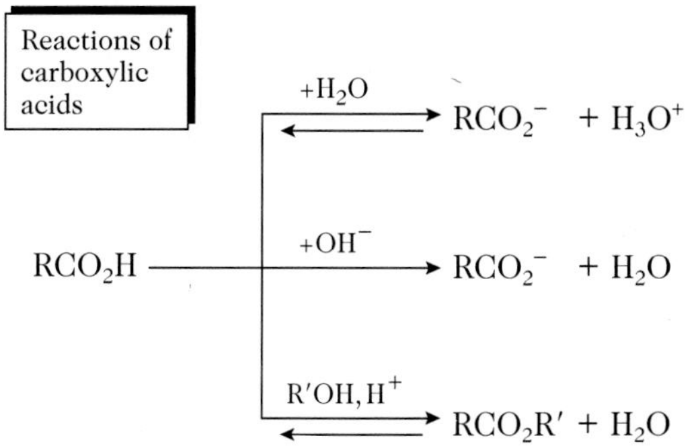

Reactions of carboxylic acids

$$RCO_2H \begin{cases} \xrightarrow{+H_2O} RCO_2^- + H_3O^+ \\ \xrightarrow{+OH^-} RCO_2^- + H_2O \\ \xrightarrow{R'OH, H^+} RCO_2R' + H_2O \end{cases}$$

Esters   Esters can be hydrolyzed and saponified.

Reactions of esters

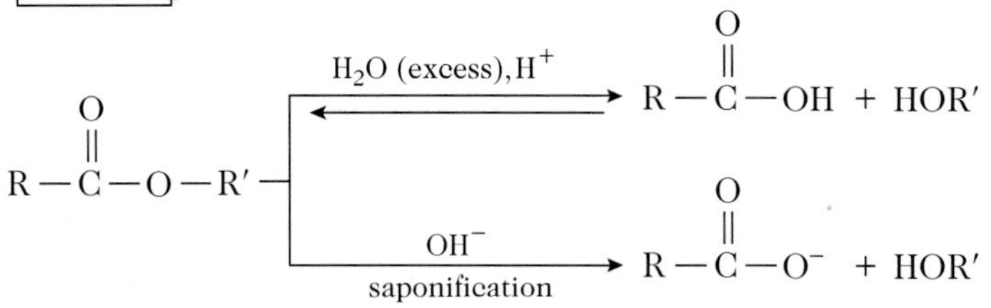

$$R-\overset{\overset{\displaystyle O}{\|}}{C}-O-R' \quad \xrightarrow[\text{saponification}]{\substack{H_2O\ (\text{excess}),\,H^+ \\[6pt] OH^-}} \quad \begin{array}{l} R-\overset{\overset{\displaystyle O}{\|}}{C}-OH \;+\; HOR' \\[14pt] R-\overset{\overset{\displaystyle O}{\|}}{C}-O^- \;+\; HOR' \end{array}$$

Esters and Anhydrides of the Phosphoric Acid System
Esters of phosphoric acid, diphosphoric acid, and triphosphoric acid occur in living systems largely as negatively charged ions, because these esters are also polyprotic acids. In addition, esters of di- and triphosphoric acid are phosphoric anhydrides, which are energy-rich compounds.

Amides   The carbonyl–nitrogen bond, the amide bond, can be formed by letting an amine or ammonia react with anything that can transfer an acyl group. Amides are neither basic nor acidic, but are neutral. They can be hydrolyzed to their parent acids and amines. We can summarize the reactions involving amides as follows:

Amide formation and hydrolysis

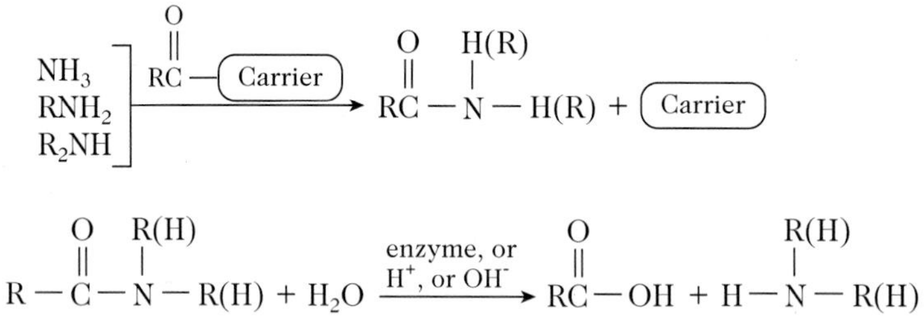

$$\left.\begin{array}{c} NH_3 \\ RNH_2 \\ R_2NH \end{array}\right] \quad RC-\boxed{\text{Carrier}} \quad \longrightarrow \quad RC-N-H(R) \;+\; \boxed{\text{Carrier}}$$

$$R-\overset{\overset{\displaystyle O}{\|}}{C}-\overset{\overset{\displaystyle R(H)}{|}}{N}-R(H) \;+\; H_2O \quad \xrightarrow[H^+,\ or\ OH^-]{\text{enzyme, or}} \quad RC-OH \;+\; H-\overset{\overset{\displaystyle R(H)}{|}}{N}-R(H)$$

REVIEW EXERCISES

The answers to Review Exercises whose numbers are in color are found in Appendix V. The answers to the other Review Exercises are found in the Study Guide that accompanies this text. The more challenging questions are marked with asterisks.

Structures and Names of Carboxylic Acids and Their Salts

13.1 What is the structure of the carboxyl group, and in what way does it differ from the functional group in an alcohol? In a ketone?

13.2 Fatty acids are carboxylic acids obtained from what substances?

13.3 Write the structures of the following substances.
(a) propionic acid   (b) benzoic acid
(c) acetic acid   (d) formic acid

13.4 Write the structures of the following.
(a) sodium acetate   (b) butyric acid
(c) sodium benzoate   (d) potassium butyrate

Physical Properties of Carboxylic Acids

13.5 Draw a figure that shows how two acetic acid molecules can pair in a hydrogen-bonded form.

***13.6** The hydrogen bond system in formic acid includes an array of molecules, one after the other, each carbonyl oxygen of one molecule attracted to the HO group of the next molecule in line. Represent this linear array of hydrogen-bonded molecules of formic acid by a drawing.

13.7 Give the following compounds in their order of increasing solubility in water. Do this by arranging their identifying letters in a row in the correct order, placing the letter of the least soluble on the left.

CH₃CH₂CH₂CH₂CO₂H   HCO₂H
A   **B**

CH₃CH₂CH=CHCH₂CH₃
C

*13.8 Give the following compounds in their order of increasing boiling points by arranging their identifying letters in a row in the correct order. Place the letter of the lowest-boiling compound on the left.

HO₂CCH₂CH₂CO₂H   CH₃CH₂CH₂CH₃
A   **B**

CH₃CH₂OH   CH₃CO₂H
C   **D**

13.9 The acid in sour milk is lactic acid. What is the name of its anion?

Carboxylic Acids as Weak Acids

13.10 Arrange the following compounds in order of increasing acidity by placing their identifying letters in a row in the correct sequence. (Place the letter of the least acidic compound on the left.)

CH₃CH₂OH   H₂SO₄   CH₃—〇—OH   CH₃CO₂H

A   **B**   **C**   **D**

*13.11 Give the order of increasing acidity of the following compounds. Arrange their identifying letters in the order that corresponds to their acidity, with the letter of the least acidic compound on the left.

CH₃—〇—OH   CH₃—〇—CH₂OH

A   **B**

CH₃—〇—CO₂H   HNO₃

C   **D**

13.12 Write the net ionic equation for the complete reaction, *if any,* of aqueous sodium hydroxide with each of the compounds in Review Exercise 13.11 at room temperature.

13.13 What are the net ionic equations for the reactions of the following compounds with aqueous potassium hydroxide at room temperature?

(a) HO₂CCH₂CH₂CO₂H   **(b)** HOCH₂CH₂CH₂CO₂H

(c)
    O
    ‖
HCCH₂CH₂CH₂CO₂H

(d) O=〇—CO₂H

Salts of Carboxylic Acids

13.14 Which compound, **A** or **B**, is more soluble in ether? Explain.

CH₃(CH₂)₆CO₂Na   CH₃(CH₂)₆CO₂H
A   **B**

13.15 Which compound, **A** or **B**, is more soluble in water? Explain.

CH₃CH₂—〇—ONa   CH₃CH₂—〇—OH

A   **B**

*13.16 Suppose that you add 0.1 mol of hydrochloric acid to an aqueous solution that contains 0.1 mol of the compound given in each of the following parts. If any reaction occurs rapidly at room temperature, write its net ionic equation.
(a) CH₃CH₂CO₂⁻
(b) ⁻O₂CCH₂CH₂CH₂CO₂⁻
(c) NH₃

*13.17 Suppose that you have each of the following compounds in a solution in water. What reaction, if any, will occur rapidly at room temperature if an equimolar quantity of hydrochloric acid is added? Write net ionic equations.
(a) HOCH₂CH₂CO₂⁻   **(b)** HOCH₂CH₂CO₂H

(c)
〇—O⁻

13.18 What are the structures of the products of the esterification by ethyl alcohol of each compound? In part (d), show the esterification of both carboxyl groups.

(a) CH₃CH₂CO₂H   CH₃
                                            |
(b) CH₃CHCO₂H

(c) NO₂—〇—CO₂H   **(d)** HO₂C—〇—CO₂H

13.19 When acetic acid is esterified by each of the following compounds, what are the structures of the esters that form?
(a) methyl alcohol
(b) isopropyl alcohol
(c) phenol
(d) HOCH₂CH₂OH (ethylene glycol) (Show the esterification of both alcohol groups.)

13.20 Write the structures of the following compounds.

(a) isopropyl formate   **(b)** propyl benzoate

13.21 What are the structures of the following compounds?

(a) ethyl propionate   (b) isobutyl acetate

*13.22 Arrange the following compounds in order of increasing boiling points. Do this by placing their identifying letters in a row, starting with the lowest-boiling compound on the left.

$$CH_3CO_2CH_2CH_2CH_3 \qquad CH_3CH_2CH_2CH_2CO_2H$$
$$\textbf{A} \qquad\qquad\qquad\qquad \textbf{B}$$
$$CH_3CH_2CH_2CH_2CH_3 \qquad HO_2CCH_2CH_2CH_2CH_2CH_3$$
$$\textbf{C} \qquad\qquad\qquad\qquad \textbf{D}$$

*13.23 Arrange the following compounds in order of increasing solubility in water by placing their identifying letters in the correct sequence, beginning with the least soluble on the left.

$$CH_3CH_2CH_2CH_2CO_2Na \qquad CH_3CH_2CH_2CH_2CO_2CH_2CH_3$$
$$\textbf{A} \qquad\qquad\qquad\qquad \textbf{B}$$
$$CH_3CH_2CH_2CH_2CO_2H \qquad CH_3CH_2CH_2CH_2CH_3$$
$$\textbf{C} \qquad\qquad\qquad\qquad \textbf{D}$$

Reactions of Esters

13.24 Write the equation for the acid-catalyzed hydrolysis of each compound. If no reaction occurs, write "no reaction."

(a)
$$\underset{CH_3COCH_2CHCH_3}{\overset{\overset{\displaystyle O}{\|}\qquad\overset{\displaystyle CH_3}{|}}{}}$$

(b)
$$CH_3CH_2O\overset{\overset{\displaystyle O}{\|}}{C}-\bigcirc$$

(c)
$$\underset{CH_3CCH_2OCH_3}{\overset{\overset{\displaystyle O}{\|}}{}}$$

(d) $CH_3CH_2OCH_2CH_2CH_3$

13.25 What are the equations for the acid-catalyzed hydrolyses of the following compounds? If no reaction occurs, write "no reaction."

(a)
$$\bigcirc-\underset{\overset{|}{CH_3}}{O\overset{\overset{\displaystyle O}{\|}}{C}CHCH_3}$$

(b)
$$\bigcirc-\underset{\overset{|}{CH_3}}{\overset{\overset{\displaystyle O}{\|}}{C}OCHCH_3}$$

(c)
$$\bigcirc-\overset{\overset{\displaystyle O}{\|}}{C}CH_2OCH_3$$

(d)
$$CH_3CH_2O\overset{\overset{\displaystyle O}{\|}}{C}CH_2CH_2\overset{\overset{\displaystyle O}{\|}}{C}OCH_2CH_3$$

13.26 The digestion of fats and oils involves the complete hydrolysis of molecules such as the following. What are the structures of its hydrolysis products?

$$CH_3(CH_2)_{12}\overset{\overset{\displaystyle O}{\|}}{C}OCH_2CHCH_2O\overset{\overset{\displaystyle O}{\|}}{C}(CH_2)_{14}CH_3$$
$$\underset{\overset{\displaystyle \|}{\underset{\displaystyle O}{}}}{\underset{OC(CH_2)_{10}CH_3}{|}}$$

*13.27 Cyclic esters are known compounds. What is the structure of the product when the following compound is hydrolyzed?

$$\overset{\displaystyle O}{\underset{\displaystyle}{\|}}$$

13.28 What are the structures of the products of saponification (by NaOH) of the compounds in Review Exercise 13.24?

13.29 What forms, if anything, when the compounds of Review Exercise 13.25 are subjected to saponification by aqueous KOH? Write their structures.

13.30 What are the products of saponification of the compound given in Review Exercise 13.26? (Assume that aqueous NaOH is used.)

13.31 Write the structure of the organic ion that forms when the compound in Review Exercise 13.27 is saponified.

Phosphate Esters and Anhydrides

13.32 Write the structures of the following compounds.
(a) monomethyl phosphate
(b) monoethyl diphosphate
(c) monopropyl triphosphate

13.33 State one apparent advantage to the body of its converting many compounds into phosphate esters.

13.34 What part of the structure of ATP is particularly responsible for its being described as an *energy-rich* compound? Explain.

Synthesis of Amides

*13.35 Examine the following acyl group transfer reaction.

$$NH_2CH_2\overset{\overset{\displaystyle O}{\|}}{C}NH\underset{\overset{|}{CH_3}}{CH}\overset{\overset{\displaystyle O}{\|}}{C}\boxed{\begin{array}{c}\text{Carrier}\\\text{molecule}\end{array}}_1 + NH_2\underset{\overset{|}{CH_2C_6H_5}}{CH}\overset{\overset{\displaystyle O}{\|}}{C}\boxed{\begin{array}{c}\text{Carrier}\\\text{molecule}\end{array}}_2 \longrightarrow$$

$$NH_2CH_2\overset{\overset{\displaystyle O}{\|}}{C}NH\underset{\overset{|}{CH_3}}{CH}\overset{\overset{\displaystyle O}{\|}}{C}NH\underset{\overset{|}{CH_2C_6H_5}}{CH}\overset{\overset{\displaystyle O}{\|}}{C}\boxed{\begin{array}{c}\text{Carrier}\\\text{molecule}\end{array}}_2 + \boxed{\begin{array}{c}\text{Carrier}\\\text{molecule}\end{array}}_1$$

(a) Which specific acyl group transferred? (Write its structure.)
(b) How many amide bonds are showing (or implied) in the product?

13.36 Write the structures of the amides that can be made from the following pairs of compounds.
(a) propionic acid and ammonia
(b) acetic acid and ethyl amine
(c) benzoic acid and dimethylamine
(d) butyric acid and ammonia

13.37 What are the structures of the amides that can be made from the following pairs of compounds?
(a) acetic acid and aniline
(b) propionic acid and methylamine
(c) butyric acid and diethylamine
(d) formic acid and ammonia

Reactions of Amides

13.38 What are the products of the hydrolysis of the following compounds? (If no hydrolysis occurs, state so.)

(a) $CH_3CH_2NHCCH_3$ (with O double bond)

(b) $CH_3CHNHCCH_2CH_3$ (with CH_3 and O)

(c) $CH_3NHCCHCH_3$ (with OCH_3 and O)

(d) $CH_3CCH_2NHCH_3$ (with O)

13.39 Write the structures of the products of the hydrolysis of the following compounds. If no reaction occurs, state so. If more than one bond is subject to hydrolysis, be sure to hydrolyze all of them.

(a) $NH_2CHCH_2CNHCH_2COH$ (with CH_3 and two O)

(b) $NH_2CCH_2CHCNH_2$ (with H_3C and two O)

(c) [ring structure with NH and O and CH_3]

(d) NH_2CNH_2 (with O)

Additional Exercises

*13.40** Complete the following reaction sequences by writing the structures of the organic products. If no reaction occurs, state so. (These constitute a review of this and earlier chapters on organic chemistry.)

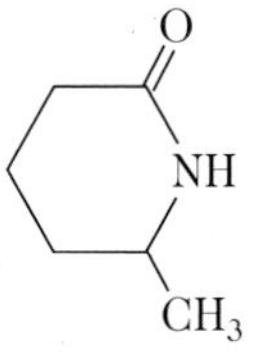

(a) $CH_3CHCO_2H + NaOH(aq) \longrightarrow$
   with CH_3O

(b) $CH_3CH_2COCH_3 + H_2O \underset{catalyst}{\overset{acid}{\rightleftharpoons}}$ (with O)

(c) $CH_3CHO \xrightarrow{K_2Cr_2O_7(aq)}$

(d) 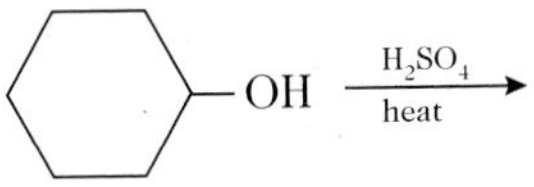 [cyclohexane]—OH $\xrightarrow[heat]{H_2SO_4}$

(e) $CH_3CHCH_2COCH_3 + NaOH(aq) \longrightarrow$
   with CH_3 and O

(f) $CH_3CHCHCH_3 \xrightarrow{KMnO_4(aq)}$
   with H_3C and OH

(g) $CH_3CH_2COH + CH_3CH_2OH \underset{heat}{\overset{H^+}{\rightleftharpoons}}$ (with O)

(h) $CH_3CH_2CHOCH_3 + H_2O \xrightarrow[catalyst]{acid}$
   with OCH_3

(i) 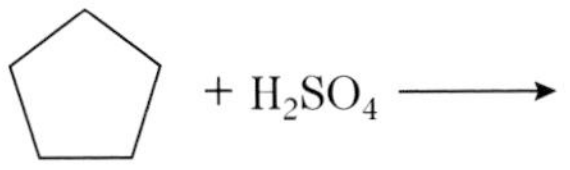 [cyclopentane] $+ H_2SO_4 \longrightarrow$

(j) $CH_3CH_2OH + $ [benzene ring]$-CO_2H \underset{catalyst}{\overset{acid}{\rightleftharpoons}}$

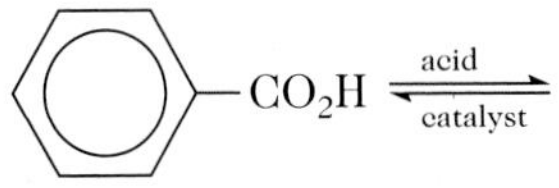

(k) $H_2C{=}CHCH_2CH_3 + HCl(g) \longrightarrow$

(l) $CH_3CH_2OCH_2CH_2CCH_3 + H_2O \longrightarrow$ (with O)

*13.41** Write the structures of the organic products, if any, that form in the following situations. If no reaction occurs, state so. (Some of these constitute a review of the reactions of earlier chapters.)

(a) 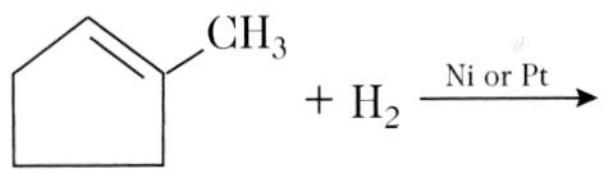 [cyclopentene with CH_3] $+ H_2 \xrightarrow{Ni\ or\ Pt}$

(b) $CH_3COH + HOCH_2CH_2CH_3 \underset{heat}{\overset{H^+}{\rightleftharpoons}}$ (with O)

(c) $CH_3CHCH_2CO_2^- + HCl(aq) \longrightarrow$
   with CH_3

(d) 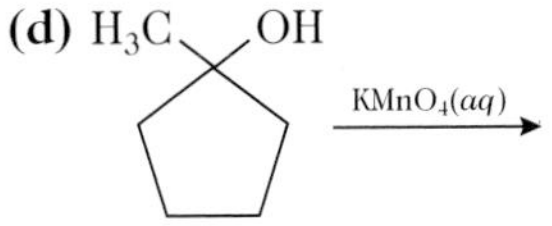 [cyclopentane with H_3C and OH] $\xrightarrow{KMnO_4(aq)}$

(e) $CH_3CH_2OCCH_2CH_2COCH_3 + NaOH(aq) \longrightarrow$
   (with two O) (excess)

(f) 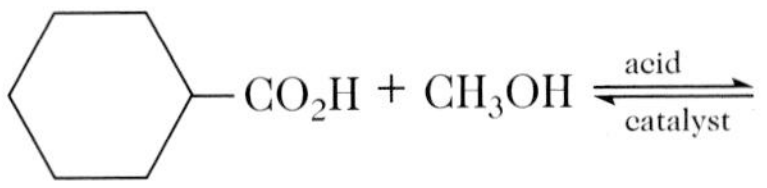 [cyclohexane ring]$-CO_2H + CH_3OH \underset{catalyst}{\overset{acid}{\rightleftharpoons}}$

(g) $CH_3CH_2OCCH_2CH_2OCCH_2CH_3 + NaOH(aq) \longrightarrow$
   (with two O) (excess)

(h) 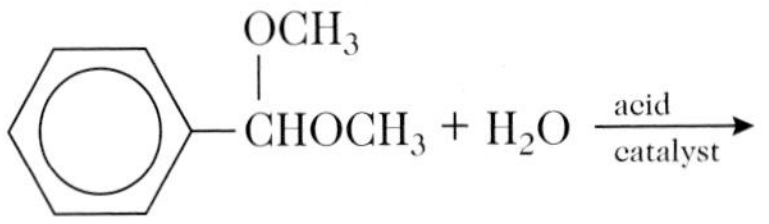 [benzene ring]$-CHOCH_3 + H_2O \xrightarrow[catalyst]{acid}$
   with OCH_3

(i) $HO_2CCH_2CH_2CH_2CH_3 + NaOH(aq) \longrightarrow$

(j) $CH_3CH_2CH_2OCCH_2CH_2OCCH_3 + H_2O \xrightarrow[\text{catalyst}]{\text{acid}}$
(with two C=O groups shown above)
(excess)

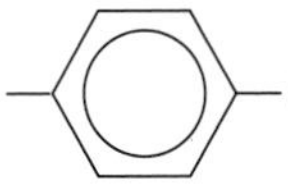

(k) $HO_2C-\langle\bigcirc\rangle-CO_2H + CH_3OH \underset{\text{catalyst}}{\overset{\text{acid}}{\rightleftharpoons}}$
(excess)

(l) $CH_3OCH_2CH_2CO_2^- + HCl(aq) \longrightarrow$

*13.42 What are all the functional groups we have studied that can be changed by each of the following reactants? Write the equations for the reactions, using general symbols such as ROH, RCO_2H, and so forth to illustrate these reactions, and name the organic families to which the reactants and products belong.
(a) water, either with an acid or an enzyme catalyst
(b) hydrogen (or a hydride ion donor) and any needed catalysts and special conditions
(c) an oxidizing agent represented by (O), such as $Cr_2O_7^{2-}$ or MnO_4^-, but not oxygen as used in combustion

13.43 We have described three functional groups that typify those involved in the chemistry of the digestion of carbohydrates, fats and oils, and proteins. What are the names of these groups and to which type of food does each belong?

*13.44 A student performed an experiment that hydrolyzed 1.65 g of benzamide.
(a) What is the maximum number of grams of benzoic acid that could be obtained?
(b) How many milliliters of 0.482 M HCl would be needed to convert into ammonium chloride all of the ammonia that can form from the hydrolysis of the sample of benzamide?

*13.45 Write the structures of the organic products that would form in the following situations. If no reaction occurs, state so. These constitute a review of nearly all the organic reactions we have studied, beginning with Chapter 9.

(a) $CH_3COCH_3 + H_2O \xrightarrow[\text{catalyst}]{\text{acid}}$ (with C=O)

(b) $CH_3CCH_2CH_3 + H_2 \xrightarrow[\text{pressure}]{\text{Ni or Pt}}$ (with C=O)

(c) $CH_3CH_2CH_2CH_2CH_3 + MnO_4^-(aq) \longrightarrow$

(d) $C_6H_5CO_2H + Na_2CO_3(aq) \longrightarrow$

(e) 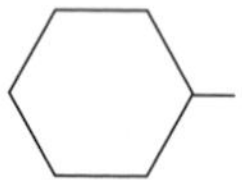 $\langle\bigcirc\rangle-CH_2CH_3 + Cr_2O_7^{2-}(aq) \longrightarrow$

(f) $CH_3CHCH_2CH_3 + NaOH(aq) \longrightarrow$ (with CH_3 substituent)

(g) $CH_3CHOCH_2CH_3 + H_2O \underset{\text{catalyst}}{\overset{\text{acid}}{\rightleftharpoons}}$ (with OCH_2CH_3 substituent)

(h) $CH_3CH_2CH + Cr_2O_7^{2-}(aq) \longrightarrow$ (with C=O)

(i) $CH_3CH_2CCH_3 + Cr_2O_7^{2-}(aq) \longrightarrow$ (with C=O)

(j) $CH_3OH + CH_3CH_2COH \overset{H^+}{\rightleftharpoons}$ (with C=O)

(k) $CH_3CH_2CNH_2 + NaOH(aq) \xrightarrow{\text{heat}}$ (with C=O)

(l) $CH_3CH_2CH + 2CH_3OH \xrightarrow[\text{catalyst}]{\text{acid}}$ (with C=O)

(m) $CH_3CH_2SH + (O) \longrightarrow$

(n) $C_6H_5COCH_2CHCH_3 + NaOH(aq) \longrightarrow$ (with C=O and CH_3 substituent)

(o) $NH_2CH_2CH_2CHCH_3 + HCl(aq) \longrightarrow$ (with CH_3 substituent)

(p) $CH_3CH=CH_2CH_2OCH_3 + H_2 \xrightarrow{\text{Ni or Pt}}$

*13.46 What are the structures of the organic products that form in the following situations? (If there is no reaction, state so.) These reactions review most of the chemical properties of functional groups we have studied, beginning with Chapter 9.

(a) $HOCCH_2CH_2CH_3 + NaOH(aq) \longrightarrow$ (with C=O)

(b) 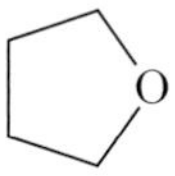 (cyclopentane ring with) $O + NaOH(aq) \longrightarrow$

(c) $CH_3CH_2OCH_2CH_2CH + MnO_4^-(aq) \longrightarrow$ (with C=O)

(d) $NH_2CH_2CH_2NH_2 + HCl(aq) \xrightarrow{} \text{excess}$

(e) 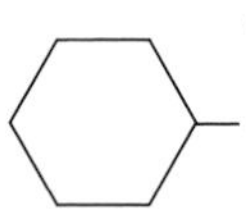 (cyclohexane ring)$-CCH_2CH_3 + H_2 \xrightarrow[\text{pressure}]{\text{Ni or Pt}}$ (with C=O)

(f) $CH_3(CH_2)_5CH_3 + Cr_2O_7^{2-}(aq) \longrightarrow$

(g) $CH_3-\langle\bigcirc\rangle-CH + 2CH_3OH \xrightarrow[\text{catalyst}]{\text{acid}}$ (with C=O)

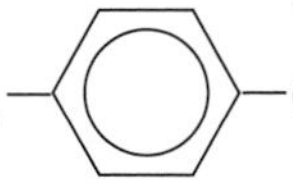

(h) $CH_3-\langle\bigcirc\rangle-COCH_2CHCH_3 + NaOH(aq) \longrightarrow$ (with C=O and CH_3 substituent)

(i)

$$
\underset{\substack{|\\\text{OH}}}{CH_3CHCH_2CH_3} + MnO_4^-\,(aq) \longrightarrow
$$

(j)

$$
\underset{\substack{|\\CH_3}}{\overset{\substack{OCH_2CH_3\\|}}{CH_3COCH_2CH_3}} + H_2O \xrightarrow[\text{catalyst}]{\text{acid}}
$$

(k)

$$
\underset{}{CH_3CH_2O\overset{\substack{O\\||}}{C}(CH_2)_3\overset{\substack{O\\||}}{C}OCH_3CH_3} + H_2O \xrightarrow[\text{catalyst}]{\text{acid}}
$$

(l)

$$
\underset{\substack{|\\ }}{CH_3O\overset{\substack{CH_2CH_3\\|}}{CH}OCH_3} + H_2O \xrightarrow[\text{catalyst}]{\text{acid}}
$$

(m) $CH_3CH_2CO_2H + HCO_3^- \longrightarrow$

(n) $CH_3SSCH_3 \xrightarrow{\text{reduction (2H)}}$

(o) $CH_3(CH_2)_5CO_2H + CH_3OH \underset{}{\overset{H^+}{\rightleftharpoons}}$

(p)

$$
CH_3NH\overset{\substack{O\\||}}{C}CH_2CH_2NH\overset{\substack{O\\||}}{C}CH_2CH_3 + \underset{\text{excess}}{H_2O} \xrightarrow{\text{enzyme}}
$$

13.47 Identify by letter which of the following compounds would be more soluble in water at pH 12 than at pH 7. Explain.

$$
\underset{\textbf{A}}{CH_3(CH_2)_6CO_2CH_3} \qquad \underset{\textbf{B}}{CH_3(CH_2)_6CO_2H}
$$

$$
\underset{\textbf{C}}{CH_3(CH_2)_6CH_2NH_2}
$$

13.48 Identify by letter which of the following compounds would be more soluble in water at pH 2 than at pH 7. Explain.

$$
\underset{\textbf{A}}{C_6H_5CH_2CH_2CONH_2} \qquad \underset{\textbf{B}}{C_6H_5CH_2COCH_2NH_2}
$$

$$
\underset{\textbf{C}}{C_6H_5CH_2CH_2CO_2H}
$$

Common Esters (Special Topic 13.1)

13.49 Esters of *p*-hydroxybenzoic acid are referred to by what common name? How are these esters used in commerce?

13.50 What is meant by a *copolymer*?

13.51 What polymer has been used in surgical grafts?

13.52 Salicylates are described as *analgesics* and *antipyretics*. What do these terms mean?

13.53 Why is salicylic acid, the parent of the salicylates, structurally modified for medicinal uses?

13.54 Concerning salicylic acid,
(a) What two functional groups does it have?
(b) Which functional group is esterified in acetylsalicylic acid?
(c) Which group is esterified in methyl salicylate?

Nylon (Special Topic 13.2)

13.55 What functional group is present in nylon-66?

13.56 The strength of a nylon fiber is attributed in part to what relatively weak bond?

14

CARBOHYDRATES

The flow of solar energy from sun to grass to cows to milk to people (in the farmhouse) is all here in this photo. The primary product of photosynthesis is glucose, an important carbohydrate at the molecular level of all living things.

14.1 BIOCHEMISTRY—AN OVERVIEW

Building materials, information, and energy are basic essentials for life.

Biochemistry is the systematic study of the chemicals of living systems, their organization, and the principles of their participation in the processes of life.

The Cell Is the Smallest Unit That Lives.   The molecules of living systems are lifeless, yet life has a molecular basis. Whether studied in cells or when isolated from them, the chemicals at the foundation of life obey all of the known laws of chemistry and physics. Yet, in isolation, not one compound of a cell has life. The cell's intricate *organization* of compounds in a cell is as important to life as the chemicals themselves. Thus, the cell is the smallest unit of matter that lives and that, in the proper environment, can make a new cell.

The Life of a Cell Requires Materials, Information, and Energy.   Our purpose in the remainder of this book is to study the molecular basis of meeting the three basic needs of a living system, its needs for materials, information, and energy. Without the daily satisfaction of these, life at any of the many loftier levels, like creativity, relationships, and love, would be severely constrained. Our focus will be on the molecular basis of life in the human body.

In this chapter and the next two, we will study the materials of life, starting with the three main classes of foodstuffs: carbohydrates, lipids, and proteins. We use their molecules to build and run our bodies and to try to stay in some state of repair. Plants rely on carbohydrates for cell walls, and animals obtain energy from carbohydrates by eating plants. Fats and oils (Chapter 15) serve many purposes but mainly as materials for animal cell membranes and as sources of chemical energy. Proteins (Chapter 16) are particularly important to the structures and functions of both plant and animal cells. Enzymes (Chapter 17), a particular family of proteins, have been mentioned often as the catalysts in living systems.

■ Cornstarch, potato starch, table sugar, and cotton are all carbohydrates.

■ Butter, lard, margarine, and corn oil are all lipids.

■ Meat is rich in protein.

The Circulatory System Delivers Needed Compounds and Carries Wastes Away. Carbohydrates, lipids, and proteins must be broken down (hydrolyzed) to much smaller molecules before our cells can use them. The enzymes of our digestive juices and the chemical reactions of digestion (Chapter 18) handle this work. The bloodstream delivers the small molecular products of digestion together with oxygen from the lungs to all parts of the body. One of the special emphases of the entire book is the molecular basis for using oxygen and releasing carbon dioxide during metabolism. The blood also carries hormones from endocrine glands and the many components of the immune system while managing to deliver wastes to the kidneys and the lungs.

Some Materials Are Used Mainly for Their Chemical Energy.   The molecular basis of energy for life (Chapter 19) is another broad topic of our study. One of the questions that we will address is: How can the body extract from a sandwich the energy for running, skipping, and laughing? As we study biochemical energetics, we'll also have opportunities to study the molecular basis of some disorders and diseases. The molecules in the diet are broken down, of course, not only to be used for energy but also to be built back up into the larger molecules needed by cells (Chapter 20).

Every Cell Has an Information System.   Enzymes and hormones are components of the intricate information system in our bodies. Materials and energy without infor-

Calvin and Hobbes

by Bill Watterson

FIGURE 14.1

Calvin has not quite discovered the molecular basis of life, but he's onto something—more likely the molecular basis of self-destruction (and Hobbes says "no thanks, not with this driver"). Used by permission of Bill Watterson and Universal Press Syndicate. All rights reserved.

mation—plans or blueprints—would produce only rubble and rubbish. Monkeys swinging hammers would reduce a stack of lumber only to splinters. Carpenters, using the same materials and expending no more raw energy, can build a building, because they possess information both as plans and as remembered experience.

Although enzymes are elements in the cell's information system, enzymes do not *originate* the blueprints. They only help to carry them out. The blueprint for any one member of a species is encoded in the molecular structures of its nucleic acids (Chapter 21), which direct the syntheses of enzymes. Hormones and neurotransmitters, other elements of cellular information, depend on enzymes for their own existence. Hundreds of unwanted conditions are caused by deficiencies of enzymes and neurotransmitters, problems traced to faults in nucleic acids. Thus, a study of the enzyme-makers is included in our study of the molecular basis of life. Our look at nucleic acids will help us see how *different* species can use raw materials from essentially the same pool of amino acids, fatty acids, and carbohydrates and yet synthesize species-unique enzymes. Camels, after all, have camel babies, not petunias or butterflies.

Our nucleic acids, while wondrous molecules, also render us vulnerable to atomic radiation and to chemicals that mimic their effects (Chapter 22). At the same time, atomic radiation can be used for medical good. We will close our study of the molecular basis of life and disease with the molecular effects of radiation on cellular materials.

■ Individual genes are sections of molecules of a polymer called DNA, which is one kind of nucleic acid.

We Launch a New Beginning with the Study of Carbohydrates. As Calvin indicates in a remark to Hobbes (Figure 14.1), we've an exciting trip ahead as we push on to an understanding of the "functioning of this complex, fragile and miraculous chunk of meat that is (our) body." In the preceding chapters we have slowly and carefully built a solid foundation of chemical principles. It's been like a mountain climbing trip where the route for a large part of the trek is through country with few grand vistas and yet with a beauty of its own. Now we're moving to elevations where the vistas begin to open. It'll not be a "WAHOOOOoooo" trip into the abyss with eyes shut, but more like a grand excursion. We're now at a new beginning, and we start it with a study of the first of the three chief classes of food materials, the carbohydrates.

14.2 MONOSACCHARIDES

The structure of glucose is the key to the structures of most carbohydrates.

Carbohydrates are polyhydroxy aldehydes, polyhydroxy ketones, or substances that yield these by simple hydrolysis (by reaction with water).

The Monosaccharides Are the Monomers for the Hydrolyzable Carbohydrates. Carbohydrates that do not react with water are called **monosaccharides** or sometimes **simple sugars.** Virtually all have names ending in *-ose*, and they have the general formula, $(CH_2O)_n$. Subclasses based on total carbon content and functional group are devised as follows.

■ The oxidized and reduced forms of polyhydroxy aldehydes and ketones as well as certain amino derivatives are also in the family of carbohydrates.

1. The number of carbons in one molecule, or the value of n in $(CH_2O)_n$.

If the number of carbons is	The monosaccharide is a
3	triose
4	tetrose
5	pentose
6	hexose, etc.

2. The nature of the carbonyl group. If an aldehyde group is present, the monosaccharide is an **aldose.** If a keto group is present, the monosaccharide is a **ketose.**

Combinations of these terms are often used. *Glucose,* for example, is a hexose with an aldehyde group, so it's an **aldohexose.** *Fructose,* a hexose with a keto group, is a **ketohexose.**

Disaccharides and Polysaccharides Are Hydrolyzable Carbohydrates. The molecules of **disaccharides,** such as the nutritionally important *sucrose, maltose,* and *lactose,* can be hydrolyzed to two monosaccharide molecules. *Starch* and *cellulose* are called **polysaccharides,** because each of their molecules can be hydrolyzed to hundreds of monosaccharide molecules, those of glucose.

■ *Oligosaccharide* molecules yield three to a few dozen monosaccharide molecules when they are hydrolyzed.

Most Mono- and Disaccharides Are Reducing Sugars. Carbohydrates that give positive Tollens' or Benedict's tests (page 275) are called **reducing sugars** because something in the reagent, Ag^+ or Cu^{2+}, is reduced in a positive test (to Ag or Cu_2O). All monosaccharides and nearly all disaccharides—sucrose is an exception—are reducing sugars, but polysaccharides are not.

Glucose Is the Centerpiece of Carbohydrate Chemistry. The widespread distribution of cellulose and starch in the enormous numbers and kinds of plants very likely make glucose the most abundant organic species on earth. Cellulose, a polymer of glucose, is present in the walls of virtually all plant cells, and starch, another glucose polymer, is the way plants store chemical energy. Special Topic 14.1 describes in general terms how glucose units are made in plants by photosynthesis.

Glucose is also by far the most common carbohydrate in blood and is often called **blood sugar.** The structure and properties of glucose, as you can see, are thus central to carbohydrate chemistry.

Glucose is $C_6H_{12}O_6$ and many of its properties can be understood in terms of one of its forms, an open-chain pentahydroxy aldehyde.

■ Glucose is also called corn sugar, because it can be made by the hydrolysis of cornstarch.

The energy released when a piece of wood burns came originally from the sun. The wood, of course, isn't just bottled sunlight. It's a complex, highly organized mixture of compounds, mostly organic. The solar energy needed to make these compounds is temporarily stored in wood in the form of distinctive arrangements of electrons and nuclei that characterize energy-rich molecules.

They are made from very simple, energy-poor substances such as carbon dioxide, water, and soil minerals. In the living world only plants have the ability to use solar energy to convert energy-poor substances into complex, energy-rich, organic compounds. The overall process by which plants do this is called **photosynthesis.**

The simplest statement of photosynthesis in equation form is

$$n\text{CO}_2 + n\text{H}_2\text{O} + \underset{\text{energy}}{\text{solar}} \xrightarrow[\text{plant enzymes}]{\text{chlorophyll}} (\text{CH}_2\text{O})n + n\text{O}_2$$

To make glucose, a hexose, n must equal 6. The symbol (CH_2O) stands for a molecular unit in carbohydrates, but plants can use the energy of carbohydrates (which came from the sun) to make other substances as well—proteins, lipids, and many others. In the final analysis, the synthesis of all the materials in our bodies consumes solar energy, and all our activities that use energy ultimately depend on a steady flow of solar energy through plants to the plant materials we eat. The meat and dairy products in our diets also depend on the consumption of plants by animals.

Chlorophyll is the green pigment in the systems of plants, usually their leaves, that can absorb and use solar energy. Chlorophyll molecules absorb solar en-

ergy and, in their energized states, trigger the subsequent reactions leading to carbohydrates. A large number of steps and several enzymes are involved. The rate of photosynthesis increases as the air temperature increases and as the concentration of CO_2 in air increases.

Notice that another product of photosynthesis is oxygen, and this process continuously regenerates the world's oxygen supply. Roughly 400 billion tons of oxygen is set free by photosynthesis each year, and about 200 billion tons of carbon (as CO_2) is converted into compounds in plants. Of all of this activity, only about 10 to 20% occurs in land plants. The rest is done by tiny phytoplankton and algae in the earth's oceans. In principle, it would be possible to dump so much poison into the oceans that the cycle of photosynthesis would be gravely affected. It is quite clear that the nations of the world must see that this does not happen.

When plants die and decay, their carbon atoms end up eventually in carbon dioxide again, and the reactions of decay consume oxygen. The combustion of fuels such as petroleum, coal, and wood also uses oxygen and animals consume oxygen during respiration. Thus there exists a grand cycle in nature in which atoms of carbon, hydrogen, and oxygen move from CO_2 and H_2O into complex forms plus molecular O_2. The latter then interact in various ways to regenerate CO_2 and H_2O.

Someone has estimated that all the oxygen in the earth's atmosphere is renewed by this cycle once in about 20 centuries and that all the CO_2 in the atmosphere and the earth's waters goes through this cycle every three centuries.

$$\text{HOCH}_2\text{CH}-\overset{\displaystyle |}{\underset{\displaystyle \text{OH}}{\text{CH}}}-\overset{\displaystyle |}{\underset{\displaystyle \text{OH}}{\text{CH}}}-\overset{\displaystyle |}{\underset{\displaystyle \text{OH}}{\text{CH}}}-\overset{\displaystyle \overset{\text{O}}{\|}}{\underset{\displaystyle \text{OH}}{\text{C}}}-\text{H}$$

Basic structure of all aldohexoses, including glucose

This structure, however, turns out to be just one of three forms in which glucose (or other aldohexoses) exists in solution. The other two are ring structures, as we'll study next.

Glucose Exists Almost Entirely in Cyclic Forms with Easily Opened Rings. Figure 14.2 gives the structures of the three forms of glucose, including the open form we just looked at. (It's shown in Figure 14.2 in a coiled configuration to make it easier to see its relationship to the cyclic forms.) In an aqueous solution, all three forms are in a dynamic equilibrium.

$$\alpha\text{-Glucose} \rightleftharpoons \text{open form of glucose} \rightleftharpoons \beta\text{-glucose}$$

The key to this equilibrium is the instability of the hemiacetal system.

The ordinary form of crystalline glucose is the cyclic α-form. Its hemiacetal system is at ring carbon 1, where there is both an ether linkage and an OH group. Notice for future reference that the OH group at ring carbon 1 (C-1) in α-glucose points to the *opposite side* of the ring from the CH₂OH group. Because the hemiacetal system is an integral part of the ring of α-glucose, the ring is not as stable as an ordinary six-membered ring. The spontaneous breaking up and reforming of the hemiacetal group cause the ring to open and close. The breakup opens the ring and the hemiacetal's parent aldehyde group emerges. Its parent alcohol group also appears, but *it is on the same molecule,* only down the chain. What is now important to the existence of three forms of glucose is that *groups are free to rotate about single bonds in open-chain systems.*

When a rotation of one-half turn takes place at the bond from C-1 to C-2 before the ring recloses, the OH group at C-1 of the reformed ring points on the *same* side of the ring as the C-6 CH₂OH group. This is a different orientation from that of α-glucose. Thus we have another form, β-glucose, also a cyclic hemiacetal and therefore also somewhat unstable. A β-glucose molecule can open up again, experience chain rotations, and then reclose to give either the same β form or, after a rotation,

■ We studied hemiacetals in Section 12.4 where we learned that their breakdown gives a parent aldehyde and alcohol group

■ The H's at C-2 and C-5 and the OH at C-3 do not stick *inside* the ring. They stick *above or below the plane* of the ring.

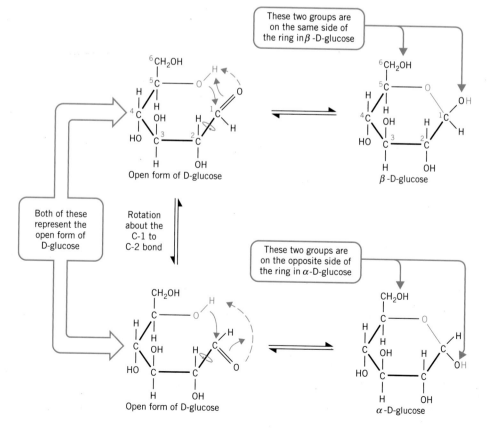

FIGURE 14.2
The α and β forms of D-glucose arise from the same intermediate, the open-chain form. Depending on how the aldehyde group, CHO, is pointing when the ring closes, one form of the ring or the other results.

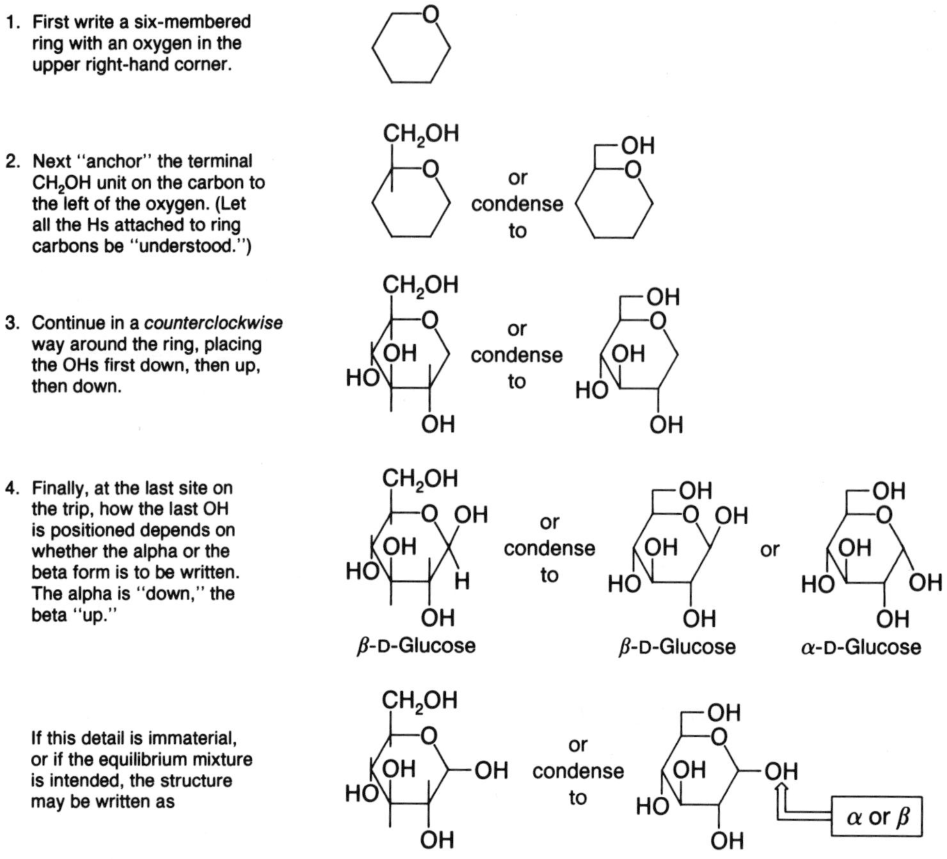

1. First write a six-membered ring with an oxygen in the upper right-hand corner.
O
2. Next "anchor" the terminal CH_2OH unit on the carbon to the left of the oxygen. (Let all the Hs attached to ring carbons be "understood.")
CH_2OH
O
or condense to
OH
O
3. Continue in a *counterclockwise* way around the ring, placing the OHs first down, then up, then down.
CH_2OH
O
OH
HO
OH
or condense to
OH
O
OH
HO
OH
4. Finally, at the last site on the trip, how the last OH is positioned depends on whether the alpha or the beta form is to be written. The alpha is "down," the beta "up."
CH_2OH
O
OH
OH
HO
H
OH
β-D-Glucose
or condense to
OH
O
OH
OH
HO
OH
β-D-Glucose
or
OH
O
OH
HO
OH
OH
α-D-Glucose
If this detail is immaterial, or if the equilibrium mixture is intended, the structure may be written as
CH_2OH
O
OH
OH
HO
OH
or condense to
OH
O
OH
OH
HO
OH
α or β

FIGURE 14.3
How to draw the cyclic forms of D-glucose in a highly condensed way.

the α form. These ring openings and closings are what take place in the dynamic equilibrium of aqueous glucose.

■ The two cyclic forms of glucose are isomers with the special name of *anomers.*

The two cyclic forms of glucose differ only in the orientation of the OH group at C-1, upward in β-glucose and downward in α-glucose. *The projections of all of the other OH groups do not change.* At equilibrium the proportions are 36% α-glucose, 0.02% open-chain glucose, and 64% β-glucose. Figure 14.3 shows how to write these structures easily and quickly.

■ This is another example of Le Châtelier's principle in action.

The open-chain form of glucose occurs only in solution, but it is the form attacked by Tollens' or Benedict's reagents. As the open-form molecules react and are thus removed from the equilibrium, they are replaced by a steady shifting of the equilibrium from closed forms to the open form. This is why glucose, despite the fact that it exists almost entirely in one cyclic form or the other, still gives the chemical properties of a pentahydroxy aldehyde (and why it's still all right to define a monosaccharide as a polyhydroxy aldehyde—or ketone—rather than as a cyclic hemiacetal).

■ Lactose is milk sugar.

Galactose, an Isomer of Glucose, Also Exists in Cyclic Hemiacetal Forms. Galactose, an aldohexose and an isomer of glucose, occurs in nature mostly as structural units in larger molecules such as the disaccharide lactose. Galactose differs from glucose only in the orientation of the C-4 OH group. Like glucose, it is a reducing sugar, and it exists in solution in three forms: α, β, and open. The arrows show how bonds and one H atom become relocated when ring closure occurs. Notice how similar these equilibria are to those of glucose (see Figure 14.2).

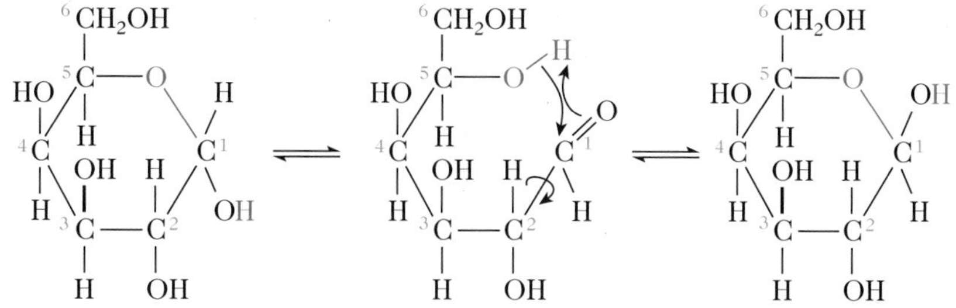

1
α-Galactose

2
Open form of galactose

3
β-Galactose

Fructose Is an Important Ketohexose. Fructose, a reducing sugar, is found together with glucose and sucrose in honey and fruit juices. It, too, can exist in more than one form, including cyclic hemiketals. We show only one cyclic form and leave the other to a Review Exercise.

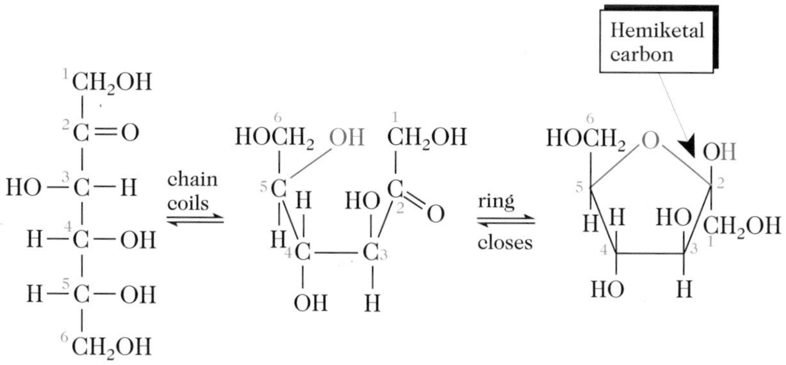

Fructose, open forms

β-Fructose

We obtain fructose when we digest table sugar, and various phosphate esters of fructose are important compounds in metabolism.

Deoxycarbohydrates Have Fewer OH Groups. A deoxycarbohydrate is one lacking an OH group where normally this group is expected. 2-Deoxyribose, for example, is the same as ribose except that there is no OH group at C-2, just two H's instead. Ribose and 2-deoxyribose are building blocks of nucleic acids, ribose for the ribonucleic acids or RNAs and deoxyribose for deoxyribonucleic acid or DNA. Each of these aldopentoses can exist in three forms, two cyclic hemiacetals and an open form. We show only one of the closed forms of each.

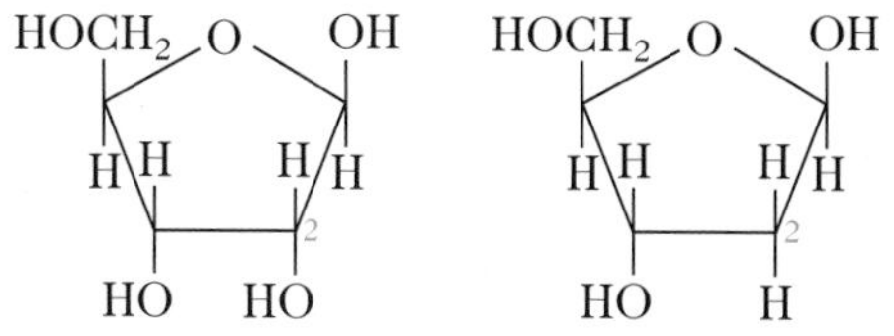

β-Ribose

β-2-Deoxyribose

14.3   OPTICAL ISOMERISM AMONG THE CARBOHYDRATES[1]

Carbohydrate molecules possess a unique handedness that adapts them to the enzymes that catalyze their reactions.

■ Galactose is one of these isomers of glucose.

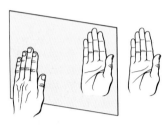

The image of the left hand is just
like the right hand in the relative
orientations of fingers, thumbs, and
palms.

Glucose is one of 16 isomers, not counting those that differ only in the easily convertible α- and β-forms. (Each of the 16 isomers exists in these cyclic forms.) We refer here to a different kind of isomerism, one new to our study, one in which molecules are as closely related as a left hand is to the right hand.

Molecules Can Be as Nearly Alike as an Object and Its Mirror Image and Still Be Different.   The molecules of one of the 16 isomers of glucose is an exact mirror image to those of naturally occurring glucose, and yet this isomer is useless to the life of a cell. Figure 14.4 shows two structures with a mirror between them. To the left of the mirror is naturally occurring α-glucose. If you put this molecular model in front of a real mirror, and then carefully made a model of what you saw in the mirror as the image, you would have the structure shown on the right. It's as if you put your left hand in front of a mirror and made a model of what you see as an image; it would be of your *right* hand.

The structures in Figure 14.4 are so closely alike that we want to name them both α-glucose, so to designate which is which, we call the structure of natural glucose α-D-glucose and that of its mirror image α-L-glucose. Thus we say that one glucose is in the D family of carbohydrates, and the other is in the L family.

The designations D and L actually are the "names" of two entire families of carbohydrates. A molecule in the D family always has a mirror-image counterpart in the L family, although very few members of the L family occur in nature. The essential structural feature of **D family** carbohydrates is that the CH_2OH group is *above* the plane of the ring when the ring's O atom is to its *right* and in the *upper right-hand corner* (see Figure 14.4). Members of the **L family** also have the CH_2OH group topside, but over on the *right* side of the ring with the ring's oxygen atom to its *left*. Thus, if you look back, the cyclic forms of galactose are also in the D family.

[1]Where time is short, this section can be omitted without causing problems in later chapters.

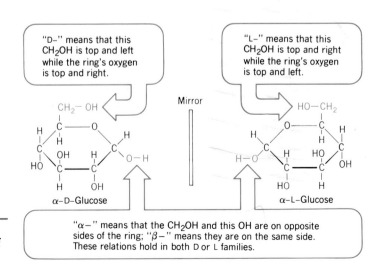

FIGURE 14.4
The D and L forms of
α-glucose.

"D–" means that this
CH_2OH is top and left
while the ring's oxygen
is top and right.

"L–" means that this
CH_2OH is top and right
while the ring's oxygen
is top and left.

Mirror

α-D-Glucose

α-L-Glucose

"α–" means that the CH_2OH and this OH are on opposite
sides of the ring; "β–" means they are on the same side.
These relations hold in both D or L families.

The aldohexoses differ in the ways in which their OH groups project, up or down, around the ring. As we said, the aldohexoses exist as 16 isomers, 8 in the D family and 8 in the L family. So the 16 occur as 8 pairs, each *pair* consisting of molecules related as object to mirror image. D-glucose and D-galactose are but 2 of the 16. Like glucose and galactose, each of the others exists in solution in equilibrium with three forms—α, β, and open.

■ Almost without exception, the naturally occurring carbohydrates are in the D family.

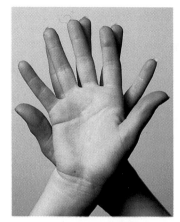

The two hands are related as an object to its mirror image, but they cannot be superimposed.

Two Molecules Are Truly Identical Only If They Superimpose. If, as we have said, α-D- and α-L-glucose are *isomers,* then they are truly *different* substances. One striking difference, as we noted, is the uselessness of α-L-glucose as a nutrient to living things. These two isomers are no more identical than a left-hand and a right-hand glove. The right-hand glove is relatively useless to the left hand. It doesn't fit properly. (This is an advance hint that the ability of an enzyme to fit to a particular molecule is important.)

If the two gloves were identical, we should be able to *superimpose* them. We should be able to perform an imaginary blending and merging of one with its mirror image, and do this so successfully that they would coincide identically and simultaneously in every part.

Try this with our left and right hands. They're related as an object to its own mirror image (ignoring wrinkles, fingerprints, scars, and jewelry). Line up your hands palms facing and imagine one blending into and through the other. It may seem that everything is merging identically—thumb to thumb, little finger to little finger, and so forth—but the palms are coming out on opposite sides. If you turn one hand over to get the palms to come out right, then the fingers won't superimpose. Your left and right hands, although as closely related as an object to its mirror image, do not superimpose. This is how structurally close D and L isomers are.

■ The test of superimposability is the single most definitive test for judging if two molecular models are of the same compound.

Molecules That Do Not Superimpose with Their Mirror Images Are *Chiral*. A large number of substances have molecules that are different in the way that two gloves differ. These molecules, like gloves, also have *handedness.* The technical term is **chirality.** A molecule with chirality does not superimpose with its mirror image and is called a **chiral molecule.** Any molecule, like one of H_2, H_2O, or CH_4, that does superimpose with its mirror image lacks chirality or handedness and is called an *achiral molecule.* Achiral molecules are always symmetrical in some way.

■ *Chiral* is from the Greek *cheir,* meaning "hand."

Isomers whose molecules are related as object to mirror image that cannot be superimposed are called **enantiomers.** α-D-Glucose and α-L-glucose are thus enantiomers of each other.

Enantiomers Have Optical Activity. Using your hands as a pair of enantiomers, think for a moment about how similar they are. They have identical "formula masses" and "molecular formulas." All "bond angles" are identical, like the angle between the thumb and the little finger. All "bond distances" are also identical, like the distance between the tips of the thumb and little finger in each hand. If your fingertips were electronegative atoms joined to a heavy central atom (the palm area), the "molecules" of your two hands would *have* to have identical polarities.

■ Notice that α-D-glucose and β-D-glucose are *not* enantiomers because their molecules are not related as object to mirror image. The *enantiomer* of α-D-glucose is α-L-glucose.

The result is that we would expect enantiomers to have identical physical properties, for example, boiling points, melting points, solubilities, and densities. Such is exactly observed among pairs of enantiomers. There is only one physical difference between two enantiomers. (Luckily; otherwise, how would we tell them apart?)

The physical difference concerns the behavior of two enantiomers to plane-polarized light. When plane-polarized light is sent through a solution of α-D-glucose, the solute molecules force the plane of the plane-polarized light to rotate to the right a certain number of degrees. You would have to twist your head the same way if,

■ Light reflected from a highway, a snow field, or a lake is rich in polarized components, and these are filtered out by Polaroid sunglasses.

■ All of the 16 isomers of the aldohexoses are optically active to different extents.

■ Nicknames for glucose and fructose, *dextrose* and *levulose*, respectively, reflect their optical rotatory powers.

■ [Ag(NH₃)₂]⁺ is the key component of Tollens' reagent (page 275).

wearing Polaroid sunglasses, you wanted to filter out the polarized light *emerging from the solution*. (In the lab, a special instrument called a *polarimeter* is used to measure the number of degrees of twist.) Any compound that is able to rotate the plane of plane polarized light is said to have **optical activity.** All enantiomers are optically active.

The number of degrees of rotation depends on concentration and distance traveled by the light through the solution, *but the direction, right or left, is a unique property of each enantiomer*. α-D-Glucose twists the plane of plane-polarized light to the right. In the same container at the identical concentration, α-L-glucose twists the plane the identical number of degrees in the opposite direction, to the left. This one property, the *direction*, is the physical difference between enantiomers. We say that α-D-glucose is *dextrorotatory* after a Latin root, *dexter*, meaning "to the right." α-L-Glucose is *levorotatory*, where *levo* signifies "to the left." Because of their effects on polarized light, isomers related as enantiomers have long been called **optical isomers,** and this kind of isomerism is called **optical isomerism.**

Remember that the designations D and L are no more than the names of two families. No automatic association exists between the names D and L and the properties of being dextro- and levorotatory. Thus, naturally occurring fructose is in the D family but is strongly levorotatory. Naturally occurring glucose, also in the D family, is strongly dextrorotatory. Signs of rotation are sometimes placed within the name of an enantimer. You may see a bottle with (+)-glucose listed as an ingredient. The (+) means dextrorotatory, just as the (−), meaning levorotatory, might appear on a label of fructose.

Toward *Achiral* Substances, Enantiomers Have Identical Chemical Properties. When either D- or L-glucose is mixed with a reactant whose molecules or ions are achiral, like those of H₂ or [Ag(NH₃)₂]⁺, the enantiomers have identical reactions. This shouldn't be too surprising; by analogy, either of your chiral hands "reacts"

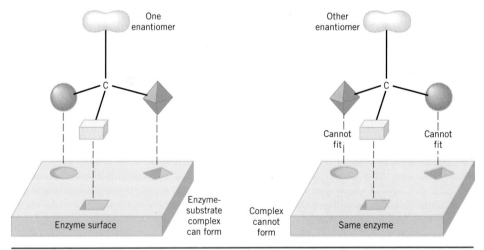

FIGURE 14.5

Because an enzyme is chiral, it can accept the substrate molecules of only one member of a pair of enantiomers. Simple geometric forms are used here to explain why. The enzyme is chiral, but only one form is depicted. Notice that the objects labeled as enantiomers are in an object-to-mirror image relationship, but they cannot be superimposed. On the left, the enzyme can accept the molecule of one enantiomer. On the right, the same enzyme cannot accept a molecule of the other enantiomer, because the shapes don't match.

identically toward achiral objects, like broom handles and water glasses. The tremendous "trifle" in nature occurs when the reactants are themselves chiral. Then enantiomers display dramatic differences in reactivity.

Molecules of All Enzymes Are Chiral. Virtually all reactions of living systems require enzyme catalysts, and *all of nature's enzymes are chiral*. We use the term **substrate** for any compound undergoing an enzyme-catalyzed reaction. The "moment of truth" occurs when a substrate molecule approaches the surface of an enzyme molecule. Unless the former can physically fit to the surface of the latter, no catalysis occurs. Figure 14.5 uses simple models of enantiomers to show how one but not the other can fit to a chiral surface. This is why L-glucose is useless in cells. Its molecules are like left hands rummaging in a box of right-hand gloves. Nothing fits. Thus, subtle differences in molecular geometries are as much matters of life and death as gross structural features.

■ Proteins are also in optical families, and not only are all enzymes chiral, they are in the same optical family.

■ The fit of molecules of substrate and enzyme is as dependent on complementary shapes as the fit of a key to its lock.

14.4 DISACCHARIDES

The three nutritionally important disaccharides are maltose, lactose, and sucrose.

The relationships of the disaccharides that we will study to the monosaccharides just discussed are seen in the following "word equations."

$$\text{Maltose} + \text{H}_2\text{O} \xrightarrow[\text{enzyme (maltase)}]{\text{H}^+ \text{ or}} \text{glucose} + \text{glucose}$$

$$\text{Lactose} + \text{H}_2\text{O} \xrightarrow[\text{enzyme (lactase)}]{\text{H}^+ \text{ or}} \text{glucose} + \text{galactose}$$

$$\text{Sucrose} + \text{H}_2\text{O} \xrightarrow[\text{enzyme (sucrase)}]{\text{H}^+ \text{ or}} \text{glucose} + \text{fructose}$$

■ You can see why glucose is of such central interest to carbohydrate chemists.

Maltose Is Made from Two Glucose Units. Maltose or malt sugar is not found widely in nature, although it is present in germinating grain and in corn syrup, which is made from cornstarch by partial hydrolysis. The two glucose units of maltose are linked by an *acetal oxygen bridge* from C-1 of one unit to C-4 of the other.

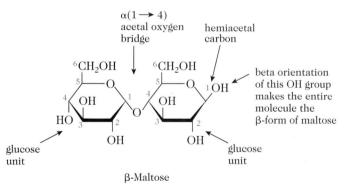

β-Maltose

■ If the OH group on the far right projected upward instead of downward, the structure would be that of α-D-maltose.

Because acetals can be hydrolyzed, maltose reacts with water. Either acids or the enzyme *maltase* catalyzes the reaction. The acetal oxygen bridge breaks and two molecules of glucose form. The *geometry* of the bridge bears on this reaction, as we'll next see, so we need a way to describe it.

■ We studied the hydrolysis of acetals on page 282.

The Acetal Oxygen Bridge in Maltose Is α(1→4).

Look at the glucose unit on the left in maltose, specifically at its O atom at C-1, the O atom furnished for the oxygen bridge. Notice that this O atom points *downward* and so has an *alpha* orientation. Therefore, the bridge to C-4 of the glucose unit on the right in maltose is called an *alpha* bridge and is given the symbol α(1→4). Had the O atom at C-1 (left side glucose unit) pointed upward in the *beta* direction, the bridge to C-4 would have been designated as β(1→4), *but the resulting molecule would not have been maltose.* Another disaccharide, *cellobiose,* has a β(1→4) bridge between two glucose units. This is not a trifle; we can digest maltose but not cellobiose. Humans have the enzyme maltase to catalyzes the digestion of maltose, but we have no enzyme to hydrolyze cellobiose.

■ Microorganisms living in the guts of ruminants, like cows, have an enzyme that catalyzes the digestion of β(1→4) oxygen bridges.

Maltose Also Has a Hemiacetal Group.

Notice that the glucose unit on the right side in maltose, the C-4-side of the α(1→4) bridge, has a hemiacetal group. This part of the maltose molecule, therefore, can open and close just like glucose. Thus maltose, like glucose, can exist in three forms, its own α, β, and open forms. In all three, the α(1→4) bridge holds. The ring opening and closing action occurs only at the hemiacetal section of the right-hand glucose unit. The availability of the open, aldehyde form makes maltose a reducing sugar. It gives positive Tollens' and Benedict's tests.

Lactose Is Made from a Galactose and a Glucose Unit.

Lactose or milk sugar occurs in the milk of mammals, 4 to 6% in cow's milk and 5 to 8% in human milk. It is obtained commercially as a by-product in the manufacture of cheese. Like maltose, lactose has an acetal oxygen bridge, only it's a *beta* bridge. From C-1 of its galactose unit there is a β(1→4) bridge to C-4 of a glucose unit. The glucose unit therefore still has a free hemiacetal system, so lactose is a reducing sugar, and it exists in its own two ring forms plus the aldehyde form in which one ring is open.

■ The OH group located the farthest right in the lactose structure is the hemiacetal OH group. Shown here is its *beta* orientation, making the given structure specifically β-lactose.

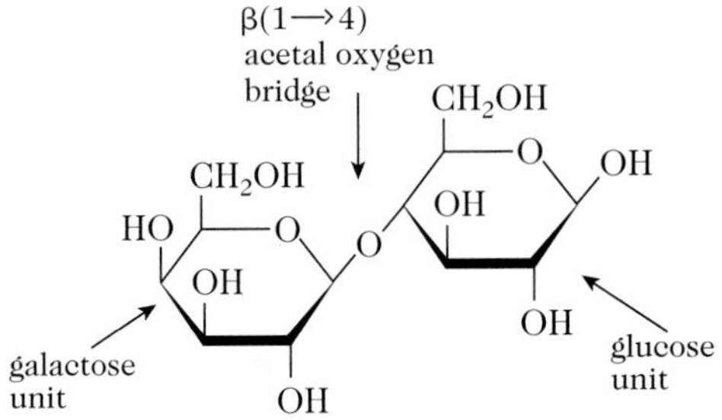

Lactose

■ Beet sugar and cane sugar are identical compounds, sucrose.

Sucrose Is Made from a Glucose and a Fructose Unit.

Sucrose, our familiar table sugar, is obtained from sugar cane or from sugar beets. The 50:50 mixture of glucose and fructose that forms when sucrose is hydrolyzed is called *invert sugar,* and it makes up the bulk of the carbohydrate in honey. Structurally, as you can see below, fructose has no hemiacetal or hemiketal group, so neither ring in sucrose can open and close spontaneously in water. Hence, sucrose cannot give positive tests with Tollens' or Benedict's reagents. It's our only common nonreducing sugar.

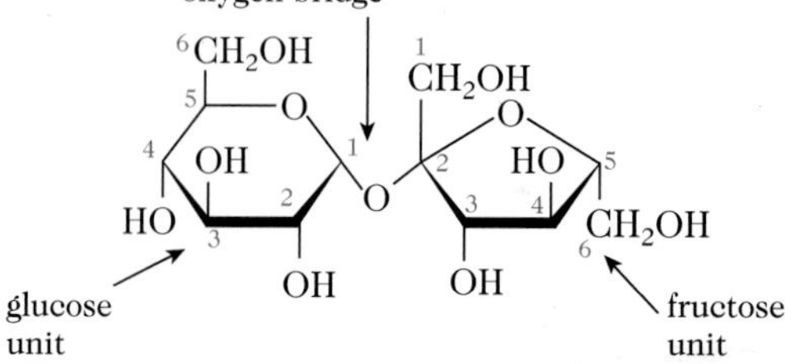

Sucrose

14.5 POLYSACCHARIDES

Polysaccharides are polymers of monosaccharides held together by oxygen bridges similar to those in disaccharides.

Much of the glucose a plant makes by photosynthesis (Special Topic 14.1) goes to make cellulose and other substances that it needs to build its cell walls and its rigid fibers. *Cotton,* for example, is about 98% cellulose. Some glucose is also stored for the food needs of the plant. Free glucose, however, is too soluble in water for storage, so most is converted to a much less soluble form, *starch,* a polymer of glucose particularly abundant in plant seeds. Plants use the glucose units in starch as nutrients much as we do when we harvest the seeds of plants for our own food.

Starch Is a Polymer of Glucose Units Linked by Acetal Oxygen Bridges. Starch is actually a mixture of two kinds of polymers of α-glucose (see Figure 14.6). One, *amylose,* is linear, and has α(1→4) oxygen bridges like those of maltose. Its long molecules are coiled, with hydrogen bonds between OH groups stabilizing each coil.

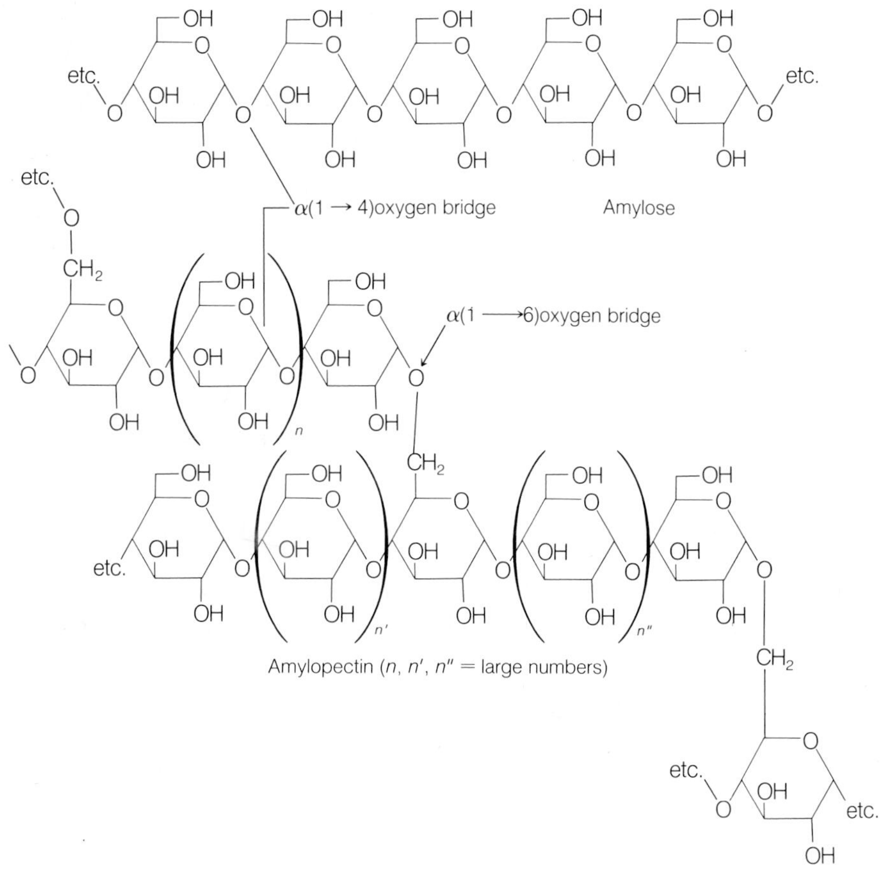

■ We could also view amylose as consisting of repeating *maltose* units linked by α(1→4) bridges. You can see why the *partial* hydrolysis of starch gives maltose.

FIGURE 14.6

Amylose and amylopectin, the glucose polymers in starch. Depending on the origin of the starch, the formula masses vary from 50,000 to several million. (A formula mass of 1 million corresponds to about 6000 glucose units per polymer molecule.)

FIGURE 14.7

Cellulose, a polymer of β-D-glucose. The value of *n* varies from 1000 to 13,000 (or 2000 to 26,000 glucose units) in different varieties of cotton. The strength of a cotton fiber comes in part from the thousands of hydrogen bonds that can exist between parallel and overlapping cellulose molecules.

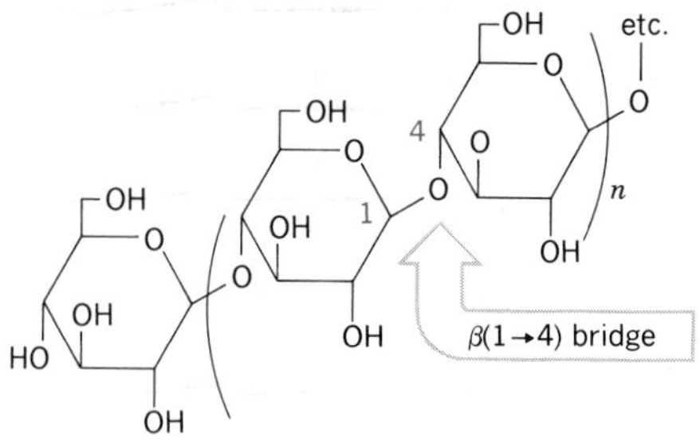

The other component of starch, *amylopectin,* is branched and has both α(1→4) and α(1→6) bridges. The latter link the C-1 ends of long amylose molecules to C-6 positions of other long amylose chains (see Figure 14.6). Hundreds of such α(1→6) links occur per amylopectin molecule, so amylopectin is heavily branched. The branching makes coiling impossible, so the OH groups of amylopectin are more exposed, making it more soluble in cold water than amylose. Neither dissolves well, however. The "solution" is actually a colloidal dispersion, because it gives the Tyndall effect.

Natural starches are about 10 to 20% amylose and 80 to 90% amylopectin. The acetal bridges in both are easily hydrolyzed in the presence of acids or the enzyme *amylase,* which we have. The complete digestion of starch gives glucose.

Neither amylose nor amylopectin is a reducing sugar and neither gives a positive Tollens' or Benedict's test. One unique test that starch does give is called the *iodine test* for starch, and it can detect extremely minute traces of starch[2] in water. When a drop of iodine reagent is added to starch, an intensely purple color develops as the iodine molecules become trapped within the vast network of starch molecules.

Glycogen Is the Animal Form of Amylopectin.   Molecules of *glycogen* consist of glucose units ranging in number from 1700 to 600,000 and are essentially like those of amylopectin, but even more branched.

Glycogen is the way that we store glucose, being found chiefly in the liver and in muscles. A network of enzymes handles its formation when glucose is in plentiful supply after a meal rich in carbohydrates. Another set of enzymes handles the debranching and hydrolysis of glycogen when glucose is needed between meals or in other periods of fasting.

Cellulose Is a Linear Polymer of Glucose Joined by β(1→4) Bridges.   Unlike starch or glycogen, cellulose is a polymer of the β form of glucose. Its molecules (Figure 14.7) are unbranched, so they bear a resemblance to amylose, but all of the bridges are β(1→4) instead of α(1→4). In cotton, the typical molecule has from 2000 to 9000 glucose units.

Humans have no enzyme that can catalyze the hydrolysis of the beta bridge in cellulose, so none of the huge supply of cellulose in the world's plants, or the cellobiose that could be made from it, is nutritionally useful to us. Many bacteria have this enzyme, however, and some strains dwell in the stomachs of cattle and other animals. Bacterial action converts cellulose in hay and other animal feed into other molecules that the larger animals then use.

■ So-called *soluble* starch is partially hydrolyzed starch, and its smaller molecules more easily dissolve in water.

■ Partial hydrolysis gives the smaller polymer molecules that make up *dextrin,* used to make mucilage and paste.

■ Although each starch molecule has at least one hemiacetal group (at a terminal glucose unit), the concentration of such groups is too low to give positive Tollens' or Benedict's tests.

■ Glycogen is sometimes called animal starch.

■ Just two of the units in cellulose constitute cellobiose.

[2]The starch–iodine reagent is made by dissolving iodine, I_2, in aqueous potassium iodide, KI. Iodine by itself is very insoluble in water, but iodine molecules combine with iodide ions to form the triiodide ion, I_3^-. Molecular iodine is readily available from this ion if some reactant is able to react with it.

SUMMARY

Carbohydrates Carbohydrates are polyhydroxy aldehydes, or ketones, or substances that can be hydrolyzed to these compounds. Those that can't be hydrolyzed are the monosaccharides, which in pure forms exist as cyclic hemiacetals or cyclic hemiketals and are reducing sugars.

Monosaccharides The three nutritionally important monosaccharides are glucose, galactose, and fructose. Glucose is the chief carbohydrate in blood. In water, three forms are present. Two are cyclic hemiacetals designated as the α and β forms, and the third is an open-chain polyhydroxy aldehyde. Galactose, which differs from glucose only in the orientation of the OH at C-4, is obtained (together with glucose) from the hydrolysis of lactose. Fructose is a reducing ketohexose. Deoxysugars such as 2-deoxyribose lack an OH group at one available position.

Disaccharides The disaccharides break up into two monosaccharides when they react with water. Maltose is made of two glucose units joined by an α(1→4) acetal bridge. In a molecule of lactose (milk sugar), a galactose unit joins a glucose unit by a β(1→4) oxygen bridge. In sucrose (cane or beet sugar), there is an oxygen bridge from C-1 of a glucose unit to C-2 of a fructose unit. Both maltose and lactose retain hemiacetal systems, so both are reducing sugars and exist in solution in three interconvertible forms, like the monosaccharides. They also exist in α and β forms. Sucrose is a nonreducing disaccharide. The digestion of maltose gives only glucose; that of lactose gives glucose plus galactose; and that of sucrose gives glucose plus fructose.

Polysaccharides Three important polysaccharides of glucose are starch (a plant product), glycogen (an animal product), and cellulose (a plant fiber). In molecules of each, (1→4) acetal bridges occur. They're alpha bridges in starch and glycogen and beta bridges in cellulose. In the amylopectin portion of starch as well as in glycogen, numerous α(1→6) bridges also occur. No polysaccharide gives a positive test with Tollens' or Benedict's reagents. Starch gives a positive iodine test. As starch is hydrolyzed its molecules successively break down to smaller pieces and, finally, glucose. Humans have enzymes that catalyze the hydrolysis of α(1→4) bridges but not polysaccharide β(1→4) bridges.

Optical Activity Molecules of carbohydrates possess chirality (handedness), and each carbohydrate is optically active. Molecules of carbohydrates exist as mirror-image pairs called enantiomers, and enantiomer molecules do not superimpose. Enantiomers have identical physical properties except in the direction they rotate the plane of plane-polarized light. They also have identical chemical properties toward all achiral reactants. Toward chiral substances, including enzymes, enantiomers react differently, and the reactions of carbohydrates in living cells are catalyzed only by enzymes to which their molecules can physically fit.

REVIEW EXERCISES

The answers to Review Exercises whose numbers are in color are found in Appendix V. The answers to the other Review Exercises are found in the Study Guide that accompanies this text. The more challenging questions are marked with asterisks.

Biochemistry

14.1 Substances in the diet must provide raw materials for what three essentials for life?

14.2 What are the three broad classes of foods?

14.3 What kind of compound carries the genetic "blueprints" of a cell?

14.4 What is as important to the life of a cell as the chemicals that make it up or that it receives?

Carbohydrate Terminology

14.5 Examine the following structures and identify by letters which structures fit each of the labels. If a particular label is not illustrated by any structure, state so.

$$
\begin{array}{cccc}
\text{CH}{=}\text{O} & & \text{CH}_2\text{OH} & \\
| & & | & \\
\text{CHOH} & \text{CH}{=}\text{O} & \text{C}{=}\text{O} & \text{CH}{=}\text{O} \\
| & | & | & | \\
\text{CHOH} & \text{CHOH} & \text{CHOH} & \text{CHOH} \\
| & | & | & | \\
\text{CHOH} & \text{CHOH} & \text{CHOH} & \text{CHOH} \\
| & | & | & | \\
\text{CHOH} & \text{CHOH} & \text{CHOH} & \text{CH}_2 \\
| & | & | & | \\
\text{CH}_2\text{OH} & \text{CH}_2\text{OH} & \text{CH}_2\text{OH} & \text{CH}_2\text{OH} \\
\mathbf{A} & \mathbf{B} & \mathbf{C} & \mathbf{D}
\end{array}
$$

(a) ketoses (b) deoxy sugars
(c) aldohexoses (d) aldopentoses

14.6 Write the structure (open chain form) that illustrates (a) any ketopentose, (b) any aldotetrose

14.7 What is the structure of the simplest aldose?

14.8 What is the structure of the simplest ketose?

14.9 A sample of 0.0001 mol of a carbohydrate reacted with water in the presence of a catalyst and 1 mol of glucose was produced. Classify this carbohydrate as a mono-, di-, or polysaccharide.

14.10 An unknown carbohydrate gives a positive Benedict's test. Classify it as a reducing or a nonreducing carbohydrate.

Monosaccharides

14.11 What is the name of the most abundant aldohexose in nature?

14.12 What is the name of the most abundant carbohydrate in (a) blood and in (b) corn syrup?

14.13 What is the name of the ketose present in honey?

Cyclic Forms of Carbohydrates

*14.14 Consider the following cyclic hemiacetal.

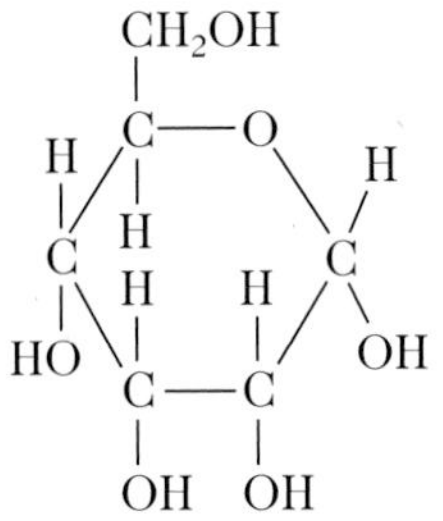

What is the structure of its open form? (Write the open form with its chain coiled in the same way it is coiled in the closed form, above.)

14.15 Examine the following structure. If you judge that it is either a cyclic hemiketal or a cyclic hemiacetal, write the structure of the open chain form (coiled in like manner as the chain of the ring).

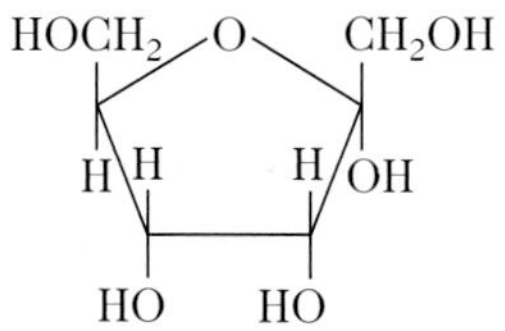

14.16 Mannose, an isomer of glucose, is identical to glucose except that in the cyclic structures the OH at C-2 in mannose projects on the same side of the ring as the CH$_2$OH group. Write the structures of the three forms of mannose that are in equilibrium in a solution in water. Identify which corresponds to α-mannose and which to β-mannose.

14.17 Allose is identical to glucose except that in its cyclic form the OH group at C-3 projects on the opposite side of the ring from the CH$_2$OH group. Write the structures of the three forms of allose that are in equilibrium in an aqueous solution. Which structures are α- and β-allose?

14.18 If less than 0.05% of all galactose molecules are in their open chain form at equilibrium in water, how can galactose give a strong, positive Tollens' test, a test good for the aldehyde group?

14.19 At equilibrium in an aqueous solution, there are three forms of glucose, with one being just a trace of the open form. Suppose that in some enzyme-catalyzed process the β-form is removed from this equilibrium. What becomes of the other forms?

14.20 Study the cyclic form of fructose on page 321 again. If its designation as β-fructose signifies a particular relationship between the CH$_2$OH group at C-5 and the OH group at C-2, what must be the cyclic formula of α-fructose?

14.21 Is the following structure that of α-fructose, β-fructose, or something else? Explain.

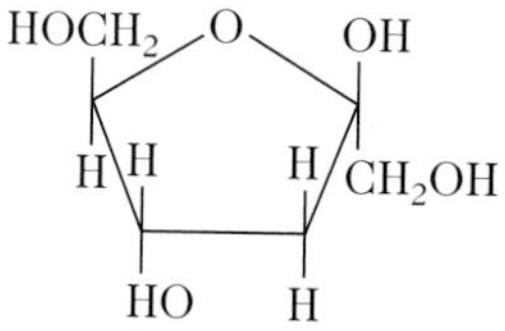

14.22 Using the structure of β-2-deoxyribose given on page 321, write the cyclic structure of α-3-deoxyribose and mark its hemiacetal carbon with an asterisk.

14.23 Could 4-deoxyribose exist as a cyclic hemiacetal with a five-membered ring (one of whose atoms is O)? Explain.

Optical Activity

14.24 When we say that α-D-glucose has *chirality,* what does this mean?

14.25 Would β-L-glucose give a positive Tollens' test? Explain.

14.26 Among the cyclic forms of the aldohexoses, what specifically is true about those in the D family? In the L family?

14.27 Suppose a 1 *M* solution of an aldohexose in a given container causes the plane of plane-polarized light to rotate 26° to the right.
(a) Is this compound dextrorotatory or levorotatory?
(b) What will a 1 *M* solution of the enantiomer of this compound do to plane polarized light in the identical container?

14.28 In what *structural* way are α-D-fructose and α-L-fructose related? (You need not write actual structures to answer this.)

14.29 Why do D-glucose and L-glucose respond so differently to enzymes?

Disaccharides

14.30 What are the names of the three nutritionally important disaccharides?

14.31 What is invert sugar?

14.32 Why isn't sucrose a reducing sugar?

*14.33 Examine the following structure and answer the questions about it.

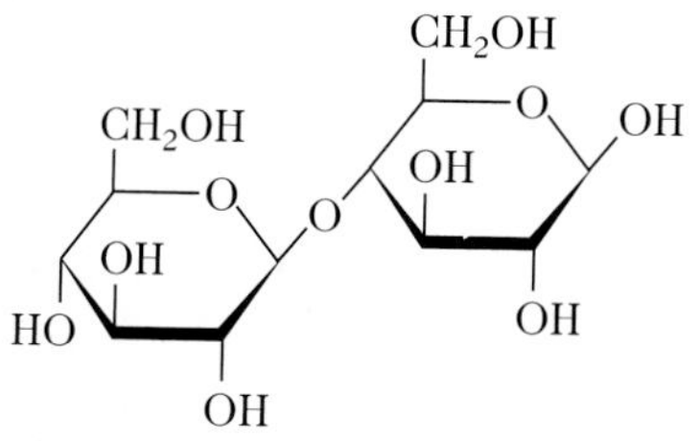

(a) Does it have a hemiacetal system? Where? (Draw an arrow to it or circle it.)

(b) Does it have an acetal system? Where? (Circle it.)

(c) Does this substance give a positive Benedict's test? Explain.

(d) In what specific structural way does it differ from maltose?

(e) What are the names of the products of the acid-catalyzed hydrolysis of this compound?

14.34 Trehalose is a disaccharide found in young mushrooms and yeast, and it is the chief carbohydrate in the hemolymph of certain insects. On the basis of its structural features, answer the following questions.

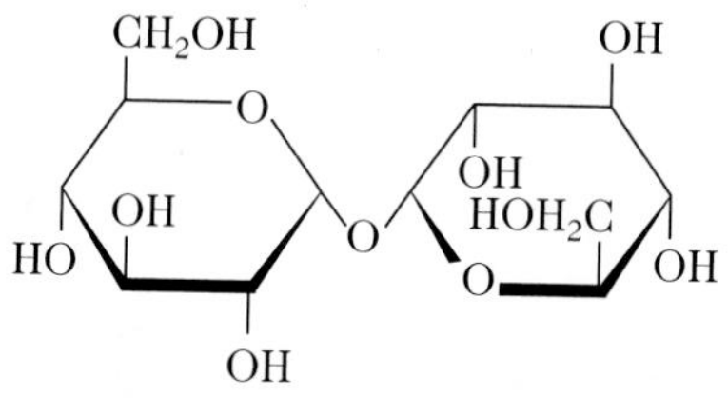

(a) Is trehalose a reducing sugar? Explain.

(b) Identify, by name only, the products of the hydrolysis of trehalose.

*14.35 Maltose has a hemiacetal system. Write the structure of maltose in which this group has changed to the open form.

14.36 Write the open form of lactose.

Polysaccharides

14.37 Name the polysaccharides that give only glucose when they are completely hydrolyzed.

14.38 What is the main structural difference between amylose and cellulose?

14.39 How are amylose and amylopectin alike structurally?

14.40 How are amylose and amylopectin different structurally?

14.41 Why can't humans digest cellulose?

14.42 What is the iodine test? Describe the reagent and state what it is used to test for and what is seen in a positive test.

14.43 How do amylopectin and glycogen compare structurally?

14.44 How does the body use glycogen?

Additional Exercises

*14.45 A freshly prepared aqueous solution of sucrose gives a negative Tollens' test (as expected), but when the solution has stood at room temperature for about a week it gives this test. Explain.

*14.46 A freshly prepared solution (actually a dispersion) of starch in water gives a positive iodine test. If this solution is warmed with a trace of human saliva, however, the ability of the solution to give this test gradually disappears. How might this observation be explained?

Photosynthesis (Special Topic 14.1)

14.47 The energy available in glucose originated in the sun. Explain in general terms how this happened.

14.48 Write the simple, overall equation for photosynthesis.

14.49 What is the name and color of the energy-absorbing pigment in plants?

14.50 In what general region of the planet earth is most of the photosynthesis carried out? By what organisms?

14.51 Describe in general terms the oxygen cycle of our planet including the function of photosynthesis in it.

15

LIPIDS

What Lipids Are

Chemical Properties of
Triacylglycerols

Phospholipids

Steroids

Cell Membranes—Their
Lipid Components

The oils that are pressed from peanuts, olives, and corn are the polyunsaturated oils recommended over saturated fats for heart-friendly diets. Fats and oils are but one group in the family of lipids, studied in this chapter.

15.1 WHAT LIPIDS ARE

The lipids include the edible fats and oils whose molecules consist of esters of long chain fatty acids and glycerol.

When undecomposed plant or animal material is crushed and ground with a nonpolar solvent, like ether, whatever dissolves is classified as a **lipid.** The operation catches a large variety of relatively nonpolar substances, and all are lipids. Thus this broad family is defined not by one common structure but by the technique used to isolate its members, *solvent extraction.* Among the many substances that won't dissolve in nonpolar solvents are carbohydrates, proteins, other very polar organic substances, inorganic salts, and water. The chart in Figure 15.1 outlines the many kinds of lipids.

Lipids Are Broadly Subdivided According to the Presence of Hydrolyzable Groups. One of the major classes of lipids, the **hydrolyzable lipids,** consists of compounds with one or more groups that can be hydrolyzed. In nearly all examples, these are *ester groups.* A number of families hydrolyzable lipids including neutral fats, waxes, phospholipids, and glycolipids. The *neutral fats* include such familiar food products as butterfat, lard (pork fat), tallow (beef fat), olive oil, corn oil, and peanut oil. Thus, some neutral fats are solids and others are liquids at room temperature. The solid neutral fats are generally from animals and so are called the *animal fats.* The neutral fats from plants are liquids and are called the *vegetable oils.* Fats and oils are the high-calorie components of the diet. The conversion factor used by the National Academy of Sciences is 9.0 kcal/g for food fat, which can be compared with 4.0 kcal/g for proteins or carbohydrates.

The **nonhydrolyzable lipids** lack groups that can be hydrolyzed and include the steroids, the family to which cholesterol and many sex hormones belong (see Section 15.4). Many plants produce another group of nonhydrolyzable lipids called the *terpenes,* which are responsible for the pleasant odors of many plant oils. Oil of rose, for example, is 40 to 60% geraniol, a terpene alcohol with a sizable hydrocarbon unit.

The *Fatty Acids* Are Mostly Long-Chain, Unbranched Monocarboxylic Acids. When hydrolyzable lipids react with water, carboxylic acids—the **fatty acids**—or their anions are among the products. Fatty acids from most plants and animals share the following features.

■ *Lipid* is from the Greek *lipos,* "fat."

■ Extraction means to shake or stir a mixture with a solvent that dissolves just part of the mixture.

■ Hydrolysis and saponification of esters were discussed in Section 13.3.

■ What is *neutral* about the neutral fats is the absence of electrical charges on their molecules.

■ Cholesterol and several sex hormones are *steroids.*

Geraniol, a terpene and a component of rose oil

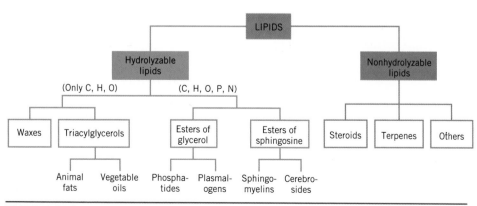

FIGURE 15.1
Lipid families.

STRUCTURAL FEATURES OF THE COMMON FATTY ACIDS

1. They are usually *mono*carboxylic acids, RCO_2H.
2. The R group is usually a long *unbranched* chain.
3. The number of carbon atoms is almost always *even*.
4. The R group can be saturated, or it can have one or more double bonds, which are *cis*.

Model of a typical saturated fatty acid

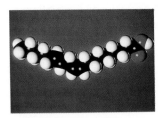

Model of linoleic acid, a fatty acid with one *cis* alkene group

Thus, just two functional groups are present in the fatty acids, often both in the same molecule, namely, the alkene double bond and the carboxyl group.

The most abundant *saturated* fatty acids are palmitic acid, $CH_3(CH_2)_{14}CO_2H$, and stearic acid, $CH_3(CH_2)_{16}CO_2H$, which have 16 and 18 carbons, respectively. Refer back to Table 13.1 for the other saturated fatty acids obtainable from lipids—the acids with more carbon atoms than acetic acid but with *even* numbers of carbons, like butyric (C_4), caproic (C_6), caprylic (C_8) and capric (C_{10}) acids. Fatty acids with fewer than 16 carbons, however, are relatively rare in nature.

The *unsaturated* fatty acids most commonly obtained from lipids are listed in Table 15.1 and include palmitoleic acid (C_{16}) and the C_{18} acids—oleic, linoleic, and linolenic acids. The double bonds in the unsaturated fatty acids of Table 15.1 are *cis*. Oleic acid is the most abundant and most widely distributed fatty acid in nature. Linoleic acid is rated as an *essential fatty acid,* meaning that it is almost like a vitamin, being particularly needed for health and growth in infants but poorly supplied unless the diet includes vegetable oils. In adults, dietary linoleic acid acts to lower the serum level of "bad" cholesterol, that which is associated with a lipid transport complex called the *low-density lipoprotein complex* or LDL (see Special Topic 20.3). The body also uses linoleic acid to synthesize arachidonic acid (Table 15.1), which is then used to make members of the family of *prostaglandins.* These make up an unusual family of fatty acids with 20 carbons, five-membered rings, and a wide variety of effects in the body (see Special Topic 15.1).

Linolenic acid is an example of an ω-3 fatty acid, those with alkene double bonds that spring from the third carbon in from the omega (ω) end of the chain, opposite the carboxyl end (the α end). Interest in two other ω-3 fatty acids (below) was aroused in the late 1980s when it was noticed that Eskimos, despite high cholesterol levels, have a low incidence of heart disease and receive considerable ω-3 fatty acids from the fish oils of their diets.

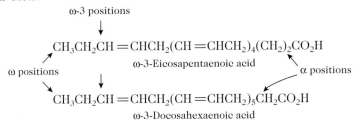

ω-3 positions

$$CH_3CH_2CH=CHCH_2(CH=CHCH_2)_4(CH_2)_2CO_2H$$
ω-3-Eicosapentaenoic acid

ω positions   α positions

$$CH_3CH_2CH=CHCH_2(CH=CHCH_2)_5CH_2CO_2H$$
ω-3-Docosahexaenoic acid

Human milk, but not infant formulas, contains ω-3-docosahexaenoic acid (DHA), which the infant's retina and brain uses. Studies using monkeys have found that a deficiency of DHA in infancy can cause functional changes that might be permanent.

The *cis*-alkene groups of the unsaturated fatty acids are kinks in their long and otherwise flexible hydrocarbon groups. These kinks inhibit the molecules from packing closely together in the solid state. Without such closeness, there can only be relatively weak forces of attraction between molecules. Therefore, as the number of alkene groups per molecule increases, the melting points of the fatty acids decrease (see Table 15.1).

The prostaglandins were discovered in the mid-1930s by a Swedish scientist, Ulf von Euler (Nobel prize, 1970), but they didn't arouse much interest in medical circles until the late 1960s, largely through the work of Sune Bergstrom. It became apparent that these compounds, which occur widely in the body, affect a large number of processes. Their general name comes from an organ, the prostate gland, from which they were first obtained. About 20 are known, and all have five-membered rings, carboxyl groups, and alkene double bonds. In some there are keto groups. Alcohol groups are also prevalent, as in the one example given below, prostaglandin PGF$_{2\alpha}$.

Prostaglandins are made in the body from 20-carbon fatty acids, like arachidonic acid. By coiling a molecule of this acid, as shown, you can see how its structure needs only a ring closure (suggested by the dashed arrow) and three more oxygen atoms to become PGF$_{2\alpha}$. The oxygen atoms are all provided by molecular oxygen itself.

Prostaglandins as Chemical Messengers The prostaglandins are like hormones in many ways, except that they do not act globally, that is, over the entire body. They do their work within the cells where they are made or in nearby cells, so they are sometimes called *local hormones*. This is perhaps why the prostaglandins have such varied functions; they occur and express their roles in such varied tissues. They work together with hormones to modify the chemical messages that hormones bring to cells. In some cells, the prostaglandins inhibit enzymes and in others they activate them. In some organs, the prostaglandins help to regulate the flow of blood within them. In others, they affect the transmission of nerve impulses.

Some prostaglandins enhance inflammation in a tissue, and it is interesting that aspirin, an inflammation reducer, does exactly the opposite. This effect is caused by aspirin's ability to inhibit the work of an enzyme needed in the synthesis of prostaglandins.

Prostaglandins as Pharmaceuticals In experiments that use prostaglandins as pharmaceuticals, they have been found to have an astonishing variety of effects. One prostaglandin induces labor at the end of a pregnancy. Another stops the flow of gastric juice while the body heals an ulcer. Other possible uses are in the treatment of high blood pressure, rheumatoid arthritis, asthma, nasal congestion, and certain viral diseases.

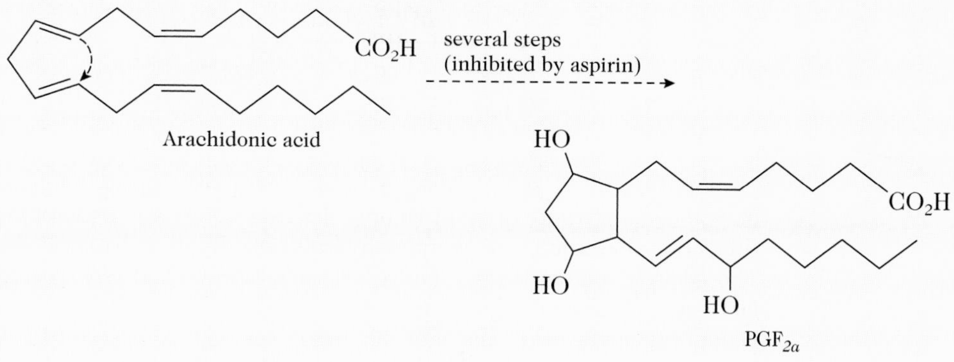

Arachidonic acid

several steps
(inhibited by aspirin)

PGF$_{2\alpha}$

TABLE 15.1 Common Unsaturated Fatty Acids

Name	Double Bonds	Total Carbons	Structure	MP (°C)
Palmitoleic acid	1	16	$CH_3(CH_2)_5CH{=}CH(CH_2)_7CO_2H$	32
Oleic acid	1	18	$CH_3(CH_2)_7CH{=}CH(CH_2)_7CO_2H$	4
Linoleic acid	2	18	$CH_3(CH_2)_4CH{=}CHCH_2CH{=}CH(CH_2)_7CO_2H$	−5
Linolenic acid	3	18	$CH_3CH_2CH{=}CHCH_2CH{=}CHCH_2CH{=}CH(CH_2)_7CO_2H$	−11
Arachidonic acid	4	20	$CH_3(CH_2)_4(CH{=}CHCH_2)_4(CH_2)_2CO_2H$	−50

The relationship between double bonds per molecule and melting point carries over to the neutral fats. Animal fats have fewer alkene groups per molecule than vegetable oils and so are likely to be solids at room temperature, while the vegetable oils tend to be liquids.

■ PRACTICE EXERCISE 1   To visualize how a *cis* double bond introduces a kink into a molecule, write the structure of oleic acid in a way that correctly shows the *cis* geometry of the alkene group. (Without the double bond, the entire side chain can stretch out into a perfect zigzag conformation, as in stearic acid. This makes it easy for two side chains to nestle very close to each other.)

The properties of the fatty acids are those to be expected of compounds with carboxyl groups, double bonds (where present), and long hydrocarbon chains. Thus, they are insoluble in water and soluble in nonpolar solvents. The fatty acids are neutralized by bases to form salts, and they can be esterified by reacting with alcohols (see Section 13.3). Fatty acids with alkene groups also react with hydrogen in the presence of a catalyst.

■ The name *triglycerides* is common in the older scientific literature on triacylglycerols.

The Triacylglycerols Are Triesters of Glycerol and Fatty Acids.   The molecules of the most abundant lipids are the **triacylglycerols**, often called the **triglycerides.** They are triesters between glycerol and three fatty acids.

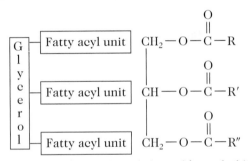

Components of triacylglycerols (neutral fats and oils)

As you can see, there are no (+) or (−) charges on triacylglycerol molecules, and so they are unlike the more complex hydrolyzable lipids to be studied in Section 15.3. The triacylglycerols include lard (pork fat), tallow (beef fat), butterfat—all animal fats—as well as such plant oils (vegetable oils) as olive oil, cottonseed oil, corn oil, peanut oil, soybean oil, coconut oil, and linseed oil.

In a particular fat or oil, certain fatty acids predominate, others either are absent or are present in trace amounts, and virtually all of the molecules are triacylglycerols. Data on the fatty acid compositions of several fats and oils are listed in Table 15.2. Oleic acid (C_{18}, one alkene group) is very common among both the fats and oils. Notice particularly, however, that the vegetable oils tend to incorporate more of the acyl groups of the unsaturated fatty acids, like those of oleic and linoleic acid, than do the animal fats. Thus, vegetable oils have more double bonds per molecule and so are often described as *polyunsaturated.* The saturated fatty acyl units of palmitic and stearic acids are far more common in animal fats, which are thus sometimes called the *saturated fats.*

■ An acyl group has the general structure:

$$RC- \quad \overset{\displaystyle O}{\overset{\displaystyle \|}{}}$$

The three acyl units in any given molecule found among the more common triacylglycerols are contributed by two or three *different* fatty acids, and not all molecules are identical. Fats and oils are thus mixtures of different molecules that share common structural features. Although we cannot give, for example, *one* struc-

TABLE 15.2   Fatty Acids Obtained from Neutral Fats and Oils

Type of Lipid	Fat or Oil	Average Composition of Fatty Acids (%)					
		Myristic acid	Palmitic acid	Stearic acid	Oleic acid	Linoleic acid	Others
Animal fats	Butter	8–15	25–29	9–12	8–33	2–4	a
	Lard	1–2	25–30	12–18	48–60	6–12	b
	Beef tallow	2–5	24–34	15–30	35–45	1–3	b
Vegetable oils	Olive	0–1	5–15	1–4	67–84	8–12	
	Peanut	—	7–12	2–6	30–60	20–38	
	Corn	1–2	7–11	3–4	25–35	50–60	
	Cottonseed	1–2	6–10	2–4	20–30	50–58	
	Soybean	1–2	6–10	2–4	20–30	50–58	c
	Linseed	—	4–7	2–4	14–30	14–25	d
Marine oils	Whale	5–10	10–20	2–5	33–40	—	e
	Fish	6–8	10–25	1–3	—	—	e

[a]Also, 3–4% butyric acid, 1–2% caprylic acid, 2–3% capric acid, 2–5% lauric acid.

[b]Also, 1% linolenic acid.

[c]Also, 5–10% linolenic acid.

[d]Also, 45–60% linolenic acid.

[e]Large percentages of other highly unsaturated fatty acids.

ture for cottonseed oil, we can describe what is probably a fairly typical molecule, like that of structure 1.

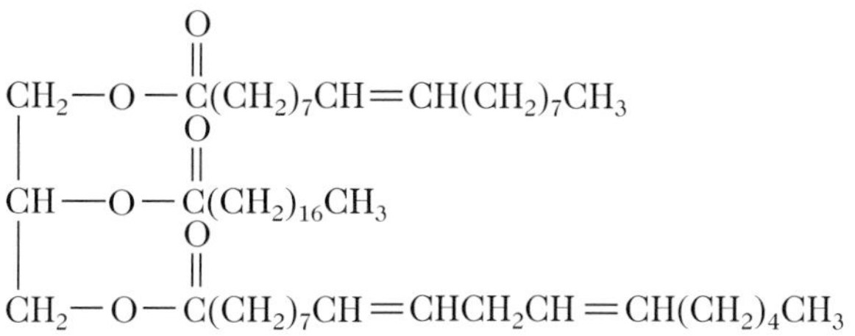

$$CH_2-O-\overset{\overset{\displaystyle O}{\|}}{C}(CH_2)_7CH=CH(CH_2)_7CH_3$$
$$CH-O-\overset{\overset{\displaystyle O}{\|}}{C}(CH_2)_{16}CH_3$$
$$CH_2-O-\overset{\overset{\displaystyle O}{\|}}{C}(CH_2)_7CH=CHCH_2CH=CH(CH_2)_4CH_3$$

1

Model of structure **1**

Plant Waxes Are Simple Esters with Long Hydrocarbon Chains.   The waxes occur as protective coatings on fruit and leaves as well as on fur, feathers, and skin. Nearly all **waxes** are esters of long-chain monohydric alcohols and long-chain monocarboxylic acids in both of which there are an *even* number of carbons. As many as 26 to 34 carbon atoms can be incorporated into *each* of the alcohol and acid units, which makes the waxes almost totally hydrocarbon-like.

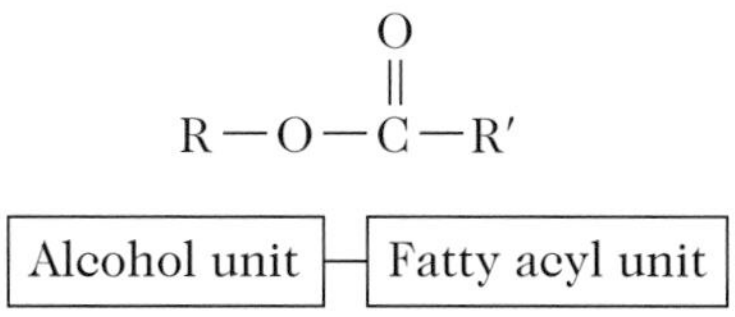

$$R-O-\overset{\overset{\displaystyle O}{\|}}{C}-R'$$

| Alcohol unit | Fatty acyl unit |

Components of waxes

Any particular wax, like beeswax, consists of a mixture of similar compounds that share the kind of structure shown above. In molecules of lanolin (wool fat), however, the alcohol portion is contributed by steroid alcohols, which have large ring systems that we'll study in Section 15.4. Waxes exist in sebum, a secretion of human skin that helps to keep the skin supple.

■ Lanolin is used to make cosmetic skin lotions.

■ **PRACTICE EXERCISE 2** One particular ester in beeswax can be hydrolyzed to give a straight-chain primary alcohol with 26 carbons and a straight-chain carboxylic acid with 28 carbons. Write the structure of this ester.

15.2 CHEMICAL PROPERTIES OF TRIACYLGLYCEROLS

Triacylglycerols can be hydrolyzed, saponified, and hydrogenated.

Triacylglycerols Can Be Hydrolyzed. When we need the chemical energy of the triacylglycerols stored in our fat tissue, a special enzyme (a lipase) catalyzes their complete hydrolysis. The fatty acid molecules are then sent to the liver. In general,

■ The solubility of free fatty acids in water is extremely low, only about 10^{-6} mol/L, so to be transported in blood, fatty acids must be carried by protein molecules in lipoprotein complexes (see Section 20.4).

$$
\begin{array}{c}
\text{CH}_2\text{—O—C(=O)—R} \\
| \\
\text{CH—O—C(=O)—R}' \\
| \\
\text{CH}_2\text{—O—C(=O)—R}'' \\
\text{Triacylglycerol}
\end{array}
+ 3\text{H}_2\text{O}
\xrightarrow[\text{lipase)}]{\text{enzyme (triacyl-glycerol}}
\begin{array}{c}
\text{CH}_2\text{OH} + \text{HO—C(=O)—R} \\
| \\
\text{CHOH} + \text{HO—C(=O)—R}' \\
| \\
\text{CH}_2\text{OH} + \text{HO—C(=O)—R}'' \\
\text{Glycerol} \quad \text{Fatty acids}
\end{array}
$$

$$
\begin{array}{c}
\text{CH}_2\text{O—C(=O)(CH}_2)_7\text{CH}=\text{CH(CH}_2)_7\text{CH}_3 \\
| \\
\text{CHO—C(=O)(CH}_2)_{16}\text{CH}_3 \\
| \\
\text{CH}_2\text{O—C(=O)(CH}_2)_7\text{CH}=\text{CHCH}_2\text{CH}=\text{CH(CH}_2)_4\text{CH}_3
\end{array}
+ 3\text{H}_2\text{O}
\xrightarrow[\text{lipase)}]{\text{enzyme (triacyl-glycerol}}
\begin{array}{c}
\text{CH}_2\text{OH} \\
| \\
\text{CHOH} + \\
| \\
\text{CH}_2\text{OH} \\
\text{Glycerol}
\end{array}
$$

$$
\text{HOC(=O)(CH}_2)_7\text{CH}=\text{CH(CH}_2)_7\text{CH}_3 + \text{HOC(=O)(CH}_2)_{16}\text{CH}_3
$$
Oleic acid   Stearic acid

$$
+\text{HOC(=O)(CH}_2)_7\text{CH}=\text{CHCH}_2\text{CH}=\text{CH(CH}_2)_4\text{CH}_3
$$
Linoleic acid

When we *digest* triacylglycerols, hydrolysis is not complete. The digestive enzyme (pancreatic lipase) takes the hydrolysis only to monoacylglycerols, fatty acids, and some diacylglycerols.

Soaps Are Made by the Saponification of Triacylglycerols. The saponification of the ester links in triacylglycerols by the action of a base (e.g., NaOH, KOH, or Na_2CO_3) gives glycerol and a mixture of the salts of fatty acids. These salts are soaps, and how they exert their detergent action is described in Special Topic 15.2. In general,

SPECIAL TOPIC 15.2
HOW DETERGENTS WORK

Soap Water is a very poor cleansing agent because it can't penetrate greasy substances, the "glues" that bind soil to skin and fabrics. When just a little soap is present, however, water, especially warm water, cleans very well. Soap is a simple chemical, a mixture of the sodium or potassium salts of the long-chain fatty acids obtained by saponification of fats or oils.

Detergents Soap is just one kind of detergent. All detergents are surface-active agents that lower the surface tension of water. All consist of ions or molecules that have long hydrocarbon portions plus ionic or very polar sections at one end. The accompanying structures illustrate these features and show the varieties of detergents that are available.

Although soap is manufactured, it is not called a synthetic detergent. This term is limited to detergents that are not soap, that is, not the salts of naturally occurring fatty acids obtained by saponification of lipids. Most synthetic detergents are salts of sulfonic acids, but others have different kinds of ionic or polar sites. The great advantage of synthetic detergents is that they work in hard water and are not precipitated by the hardness ions—Mg^{2+}, Ca^{2+}, and the two ions of iron. These ions form messy precipitates ("bathtub ring") with the anions of the fatty acids present in soap. The anions of synthetic detergents do not form such precipitates.

Figure 1 shows how detergents work. In Figure 1a we see the hydrocarbon tails of the detergent work their way into the hydrocarbon environment of the grease layer. ("Like dissolves like" is the principle at work here.) The ionic heads stay in the water phase, and the grease layer becomes pincushioned with electrically charged sites. In Figure 1b we see the grease layer breaking up, aided with some agitation or scrubbing. Figure 1c is a magnified view of grease globules

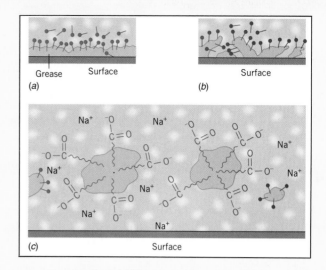

Grease   Surface
(a)

Surface
(b)

(c)   Surface

studded with ionic groups, and, being like charged, these globules repel each other. They also tend to dissolve in water, so they are ready to be washed down the drain.

$$CH_3(CH_2)_{14}CO_2^-Na^+$$
Soap—an anionic deteragent

$$CH_3(CH_2)_{13}OSO_3^-Na^+$$
Sodium alkyl sulfate—an anionic detergent

$$CH_3(CH_2)_8-\bigcirc-SO_3^-Na^+$$

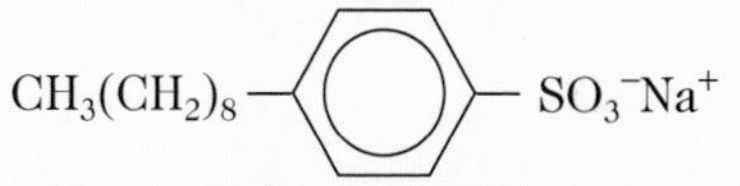

A sodium alkylbenzenesulfonate—
an anionic detergent

$$CH_3(CH_2)_{11}\overset{+}{N}(CH_3)_3Cl^-$$
A triethylalkylammonium ion—
a cationic detergent

$$CH_3(CH_2)_8O(CH_2CH_2O)_nH$$
A nonionic detergent

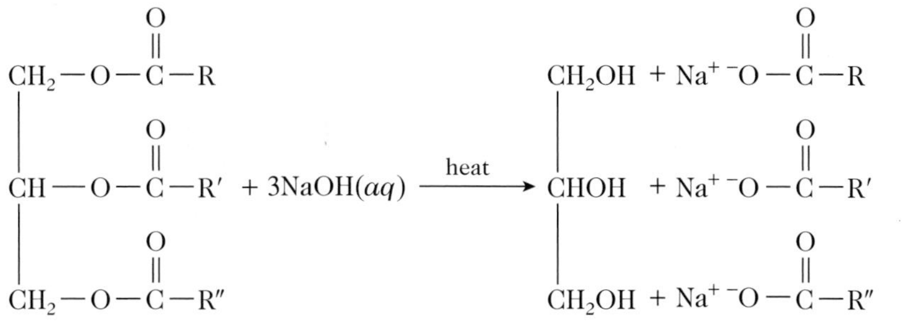

Triacylglycerol $+ 3NaOH(aq) \xrightarrow{heat}$ Glycerol   Mixture of sodium salts of fatty acids (soap)

■ PRACTICE EXERCISE 3 Write a balanced equation for the saponification of **1** with sodium hydroxide.

The Hydrogenation of Vegetable Oils Gives Solid Shortenings and Margarine. When hydrogen is made to add to some of the double bonds in vegetable oils, the oils become like animal fats, both physically and structurally. One very practical consequence of such partial hydrogenation is that the oils change from being liquids to solids at room temperature. Many people prefer solid, lardlike shortening for cooking, instead of a liquid oil. Therefore, the manufacturers of such "hydrogenated vegetable oils" as Crisco and Spry take inexpensive, readily available vegetable oils, like corn oil and cottonseed oil, and catalytically add hydrogen to some (not all) of the alkene groups in their molecules. We say that the double bonds become *saturated*. Unlike natural lard, the vegetable shortenings have no cholesterol.

■ Except for the absence of cholesterol, hydrogenated vegetable oils are chemically and nutritionally identical to animal fats.

■ PRACTICE EXERCISE 4 Write the balanced equation for the complete hydrogenation of the alkene links in structure **1**.

■ If *all* the alkene groups in a vegetable oil were hydrogenated, the product would be just like beef or mutton fat and would not melt on the tongue.

The chief lipid material in margarine is produced from vegetable oils in the same way. The hydrogenation is done with special care so that the final product can melt on the tongue, one property that makes butterfat so pleasant.

The popular brands of peanut butter, those with peanut oils that do not separate, are made by the partial hydrogenation of the oil in real peanut butter. The peanut oil changes to a solid, hydrogenated form at room temperature and therefore it cannot separate.

15.3 PHOSPHOLIPIDS

Phospholipid molecules have very polar or ionic sites in addition to long hydrocarbon chains.

Phospholipids are esters either of glycerol or of sphingosine, which is a long-chain, dihydric amino alcohol with one double bond. Note its long hydrocarbon chain.

■ *Phosphoglycerides* is an older name for the glycerophospholipids.

The Glycerophospholipids Have Phosphate Units plus Two Acyl Units. The **glycerophospholipids** occur in two broad types, the *phosphatides* and the *plasmalogens*. Both are esters of glycerol. Molecules of the **phosphatides** have two ester bonds from glycerol to fatty acids plus one ester bond to phosphoric acid. The phosphoric acid unit, in turn, is joined by a phosphate ester link to a small alcohol molecule. Without this link, the compound is called *phosphatidic acid*.

$$CH_3(CH_2)_{12}CH = CHCHCHCH_2OH$$
$$\underset{\displaystyle HO \quad NH_2}{\phantom{CH_3(CH_2)_{12}CH = CHCH}||}$$

Sphingosine

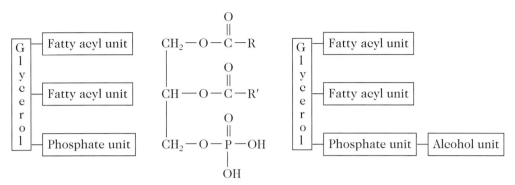

Phosphatidic acid components　　Phosphatidic acid　　　Phosphatide components

The phospholipids all have small, very polar molecular parts that are important to the formation of cell membranes.

Three particularly important phosphatides have phosphate ester links either to choline, to ethanolamine, or to serine, forming, respectively, *phosphatidylcholine* (lecithin), **2**, *phosphatidylethanolamine* (cephalin), **3**, and *phosphatidylserine,* **4**.

■ *Cephalin* is from the Greek *kephale,* "head." Cephalin is found in brain tissue.

$$\overset{+}{\text{HOCH}_2\text{CH}_2\text{N}(\text{CH}_3)_3} \qquad \text{HOCH}_2\text{CH}_2\text{NH}_2 \qquad \text{HOCH}_2\overset{|}{\underset{\underset{\overset{+}{\text{NH}_3}}{|}}{\text{CH}}}\text{CO}_2^-$$

Choline　　　　　　　　　Ethanolamine　　　　　　　Serine

As the structures of **2, 3,** and **4** given below show, one part of each phosphatide molecule is very polar because it carries full electrical charges. These charges are partly responsible for the phosphatides being somewhat more soluble in water than triacylglycerols. The remainder of a phosphatide molecule is nonpolar and hydrocarbon-like, so phosphatides can be extracted from animal matter by nonpolar solvents.

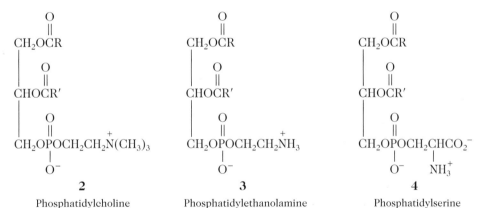

2
Phosphatidylcholine (lecithin)

3
Phosphatidylethanolamine (cephalin)

4
Phosphatidylserine

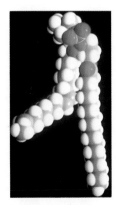

A typical phosphatide

■ *Lecithin* is from the Greek *lekitos,* "egg yolk," a rich source of this phospholipid.

These three are the most common lipids used to make animal cell membranes.

When pure, lecithin is a clear, very hygroscopic, waxy solid. In air, it is quickly attacked by oxygen, which makes it turn brown in a few minutes. Lecithin is a powerful emulsifying agent for triacylglycerols; this is why egg yolks, which contain lecithin, are used to make the emulsions found in mayonnaise, ice cream, custards, candies, and cake dough.

The Plasmalogens Have Both Ether and Ester Groups. The **plasmalogens,** as we said, make up another family of glycerophospholipids, and they occur widely in the membranes of both nerve and muscle cells. They differ from the phosphatides by the presence of an unsaturated *ether* group instead of an acyl group at one end of the glycerol unit.

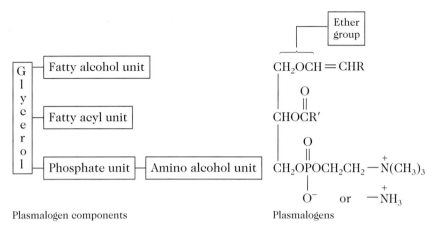

Plasmalogen components   Plasmalogens

Plasmalogen molecules, like phosphatides, also carry electrically charged positions as well as long hydrocarbon chains.

The Sphingolipids Are Based on Sphingosine, Not Glycerol. The two types of sphingosine-based lipids or **sphingolipids** are the *sphingomyelins* and the *cerebrosides*. They are also important constituents of cell membranes, particularly those of nerve cells. The sphingomyelins are phosphate diesters of sphingosine. Their acyl units occur as acylamido parts, and they come from unusual fatty acids that are not found in neutral fats. Like the molecules of the phosphatides, those of the sphingolipids have two nonpolar tails.

The cerebrosides are not phospholipids but are **glycolipids** instead, lipids with a carbohydrate unit and not a phosphate ester system. The sugar segment, usually a galactose or a glucose unit, provides many OH groups and so furnishes a strongly polar region to the molecule.

■ *Glyco-* is from *glycose*, a generic name for any monosaccharide.

A sphingomyelin

Sphingolipid components   Sphingomyelins   Cerebrosides

The cerebrosides are particularly prevalent in the membranes of brain cells.

15.4 STEROIDS

Cholesterol and other steroids are nonhydrolyzable lipids.

Steroids are high-formula-mass aliphatic compounds whose molecules include a characteristic four-ring feature called the *steroid nucleus.* It consists of three six-membered rings and one five-membered ring, as seen in structure **5.** Several steroids are physiologically active (Table 15.3).

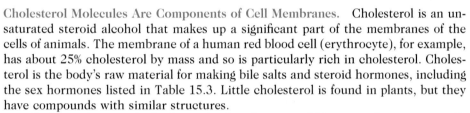

5
Steroid nucleus

Cholesterol

Cholesterol

Some steroids or steroidlike synthetics, the *anabolic steroids,* have been illicitly used by athletes striving for better performances in competition (see Special Topic 15.3).

Cholesterol Molecules Are Components of Cell Membranes. Cholesterol is an unsaturated steroid alcohol that makes up a significant part of the membranes of the cells of animals. The membrane of a human red blood cell (erythrocyte), for example, has about 25% cholesterol by mass and so is particularly rich in cholesterol. Cholesterol is the body's raw material for making bile salts and steroid hormones, including the sex hormones listed in Table 15.3. Little cholesterol is found in plants, but they have compounds with similar structures.

■ Steroid alcohols are called *sterols.*

Cholesterol enters the body via the diet, but up to 800 mg per day can normally be synthesized in the liver from two-carbon acetate units. Cholesterol made in the liver is used to make bile salts like sodium cholate or is converted to esters of cholesterol.

■ Cholesterol is the chief constituent in gallstones.

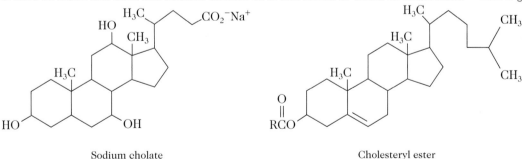

Sodium cholate
(a bile salt)

Cholesteryl ester
(R is long-chain)

Bile salts are secreted into the intestinal tract where they function as powerful surface-active agents (detergents). They aid both in the digestion of dietary lipids and in the absorption of fat-soluble vitamins and fatty acids from the digestive tract (eventually) into circulation. Cholesterol itself is put into circulation in the bloodstream as a component of *very low density lipoproteins* (VLDLs). Cholesterol esters, on the other hand, occur in higher-density complexes. The relationships among cholesterol, cholesteryl esters, the various kinds of lipoproteins, and the risk of heart disease will be discussed when we study the metabolism of lipids. We need to know more about proteins, genes, and enzymes first.

■ *Lipoprotein* molecules are combinations of lipids and proteins.

TABLE 15.3 Important Steroids

Vitamin D₃ Precursor

Irradation of this derivative of cholesterol by ultraviolet light opens one of the rings to produce vitamin D₃. Meat products are sources of this compound.

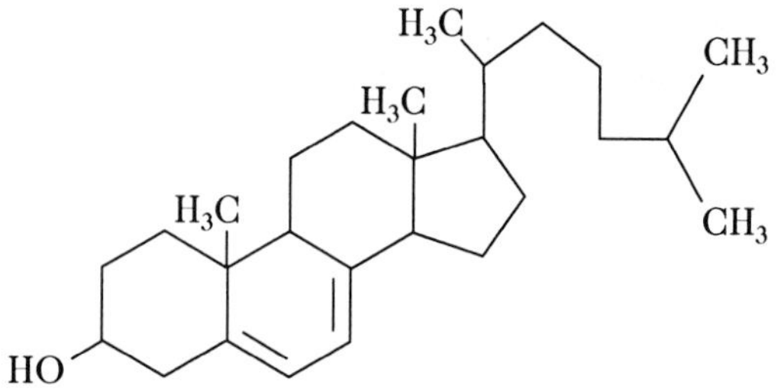

7-Dehydrocholesterol

ultraviolet radiation

Vitamin D₃ is an antirachitic factor. Its absence leads to rickets, an infant and childhood disease characterized by faulty deposition of calcium phosphate and poor bone growth.

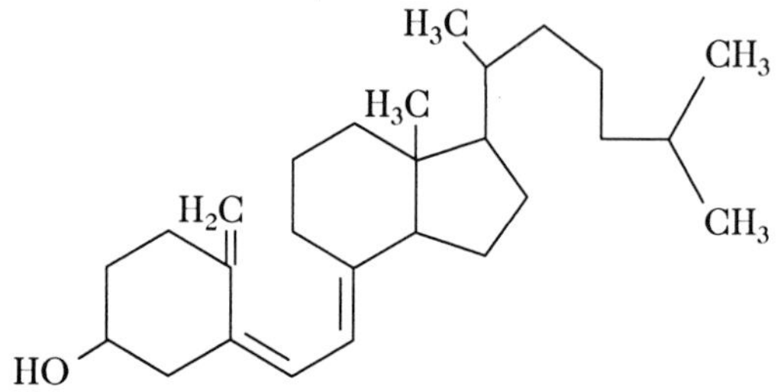

Vitamin D₃

Adrenocortical Hormone

Cortisol is one of the 28 hormones secreted by the cortex of the adrenal gland. Cortisone, very similar to cortisol, is another such hormone. When cortisone is used to treat arthritis, the body changes much of it to cortisol by reducing a keto group to the 2° alcohol group that you see in the structure of cortisol

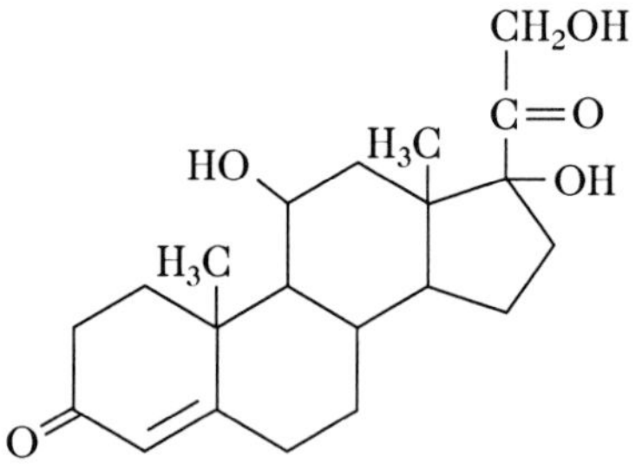

Cortisol

Sex Hormones

Estradiol is a human estrogenic hormone.

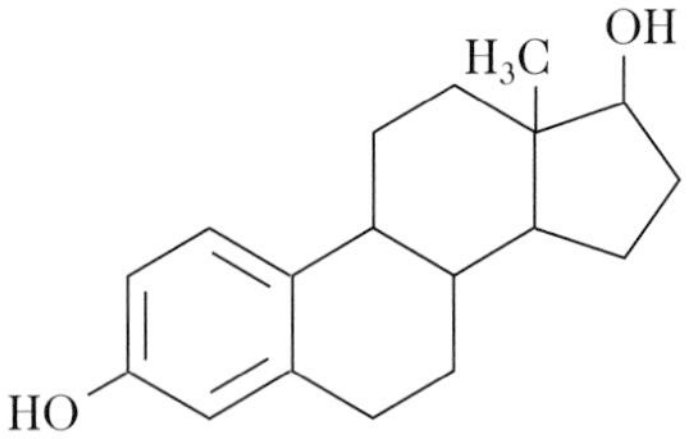

Estradiol

Progesterone, a human pregnacy hormone, is secreted by the corpus luteum.

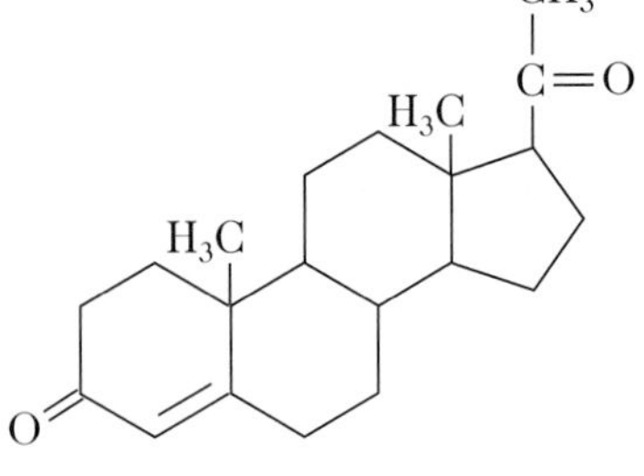

Progesterone

TABLE 15.3 (continued)

Testosterone, a male sex hormone, regulates the development of reproductive organs and secondary sex characteristics.

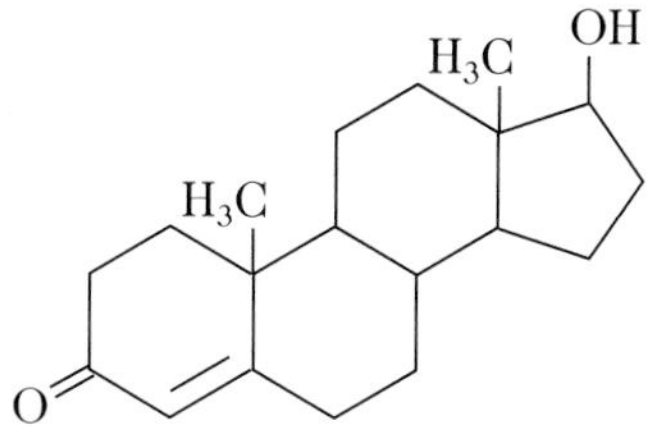

H₃C OH

H₃C

O

Testosterone

Androsterone is another male sex hormone.

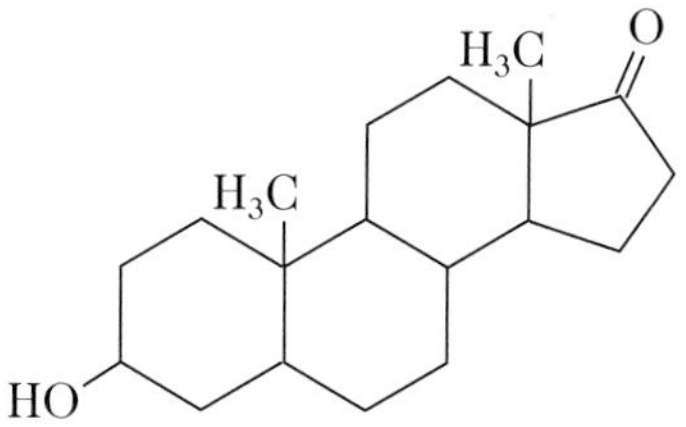

H₃C O

H₃C

HO

Androsterone

Synthetic Hormones in Fertility Control

Most oral contraceptive pills contain one or two synthetic, hormone-like compounds. (Synthetics must be used because the real hormones are broken down in the body.)

Synthetic Estrogens

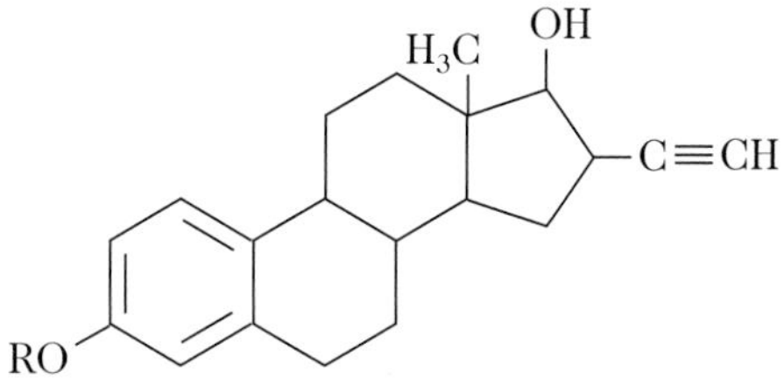

H₃C OH

$C \equiv CH$

RO

R = H, ethinylestradiol
R = CH₃, mestranol

Synthetic Progestin

The most widely used pills have a combination of an estrogen (20 to 100 μg if mestranol and 20 to 50 μg if ethinylestradiol) plus a progestin (0.35 to 2.5 mg depending on the compound).

A relatively new birth control technology, one that prevents implantation of a fertilized ovum in the uterus, has been developed. The compound is an antiprogesterone called mifepristone, or RU 486, and how it works is described in Special Topic 16.2.

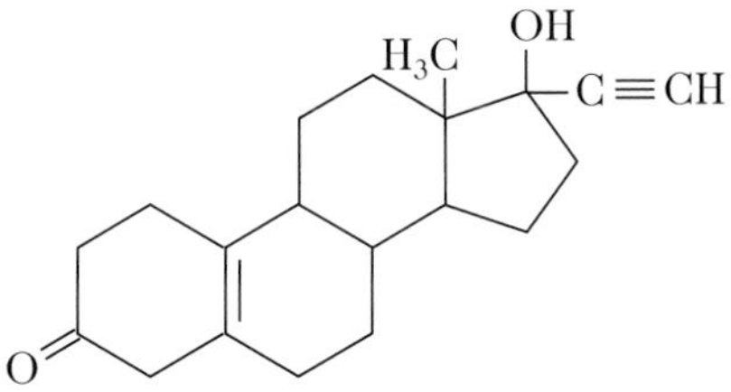

H₃C OH

$C \equiv CH$

O

Norethynodrel

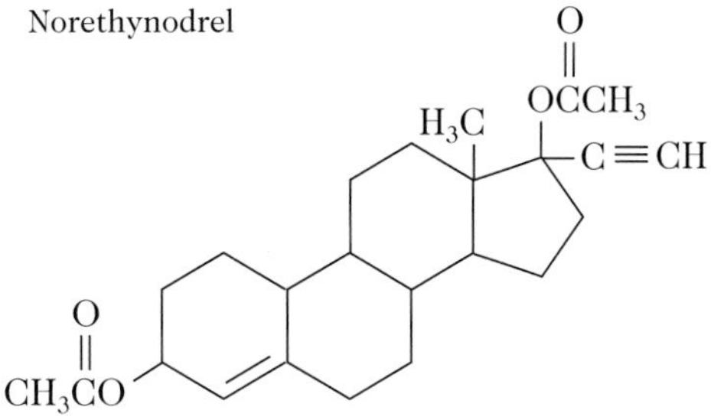

O
‖
OCCH₃

H₃C $C \equiv CH$

O
‖
CH₃CO

Ethynodiol diacetate

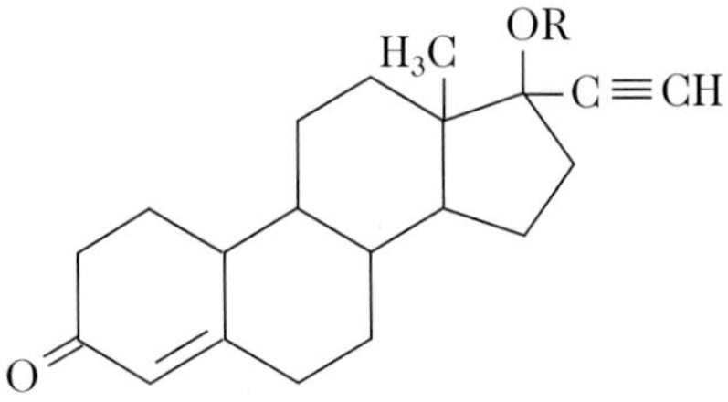

H₃C OR

$C \equiv CH$

O

R = H, norethindrone
R = COCH₃, norethindrone acetate

345

Athletes seek the winning edge largely through intensive training. The sexes compete separately, however, because male athletes generally have an edge in brute strength given them simply by the steroid hormone testosterone. Some athletes have actually taken testosterone to improve their prospects of winning. However, the side effects of excessive testosterone, including secondary sexual characteristics, are sometimes serious, and the hormone is not well absorbed from the intestinal tract. It has to be taken by injection, not orally.

Testosterone is called an *anabolic* steroid because it has an anabolic effect, meaning a building-up effect in metabolism. It promotes the formation of larger muscles, for example, from their building blocks, the amino acids. To counter the side effects of testosterone, synthetic analogs have been made, some of which can be taken orally. A few are given next.

Methenolone and methandrostenolone can be taken orally. Nandrolone decanoate must be taken by injection. (Why?) The side effects seriously outweigh any advantages of the anabolic steroids. They include damage to the kidneys and the liver, impotence and infertility, and aggressive behavior in the extreme. When taken by women athletes, the steroids can cause menstrual problems and both facial hair and baldness.

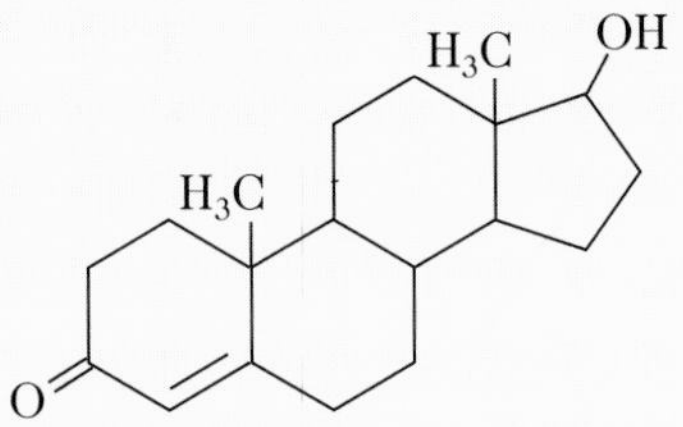

Testosterone

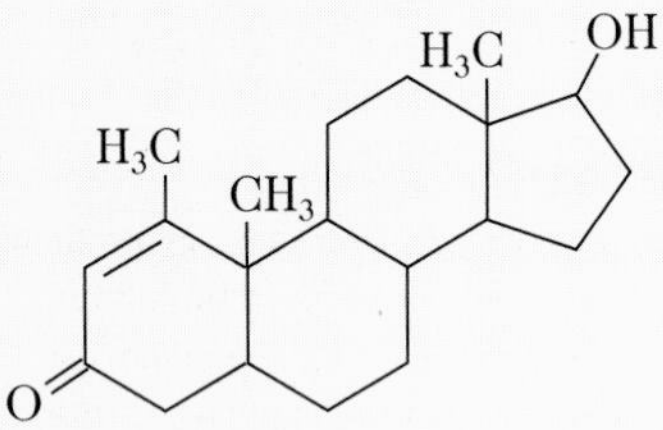

Methenolone
(The acetate ester is
Primonabol.)

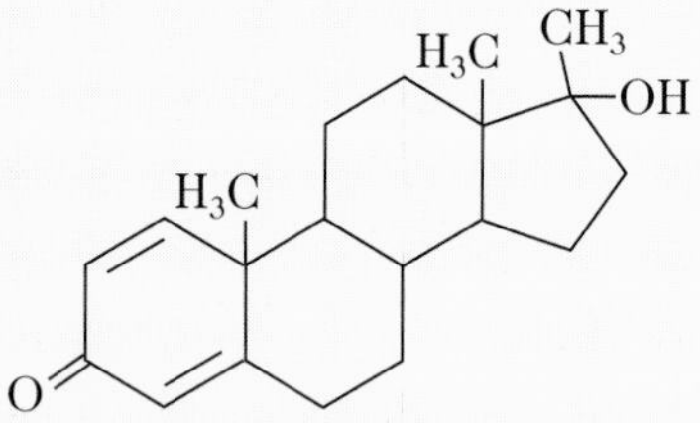

Methandrostenolone
(Danabol, Nerobol, Nabolin,
Stenolon, Dianabol)

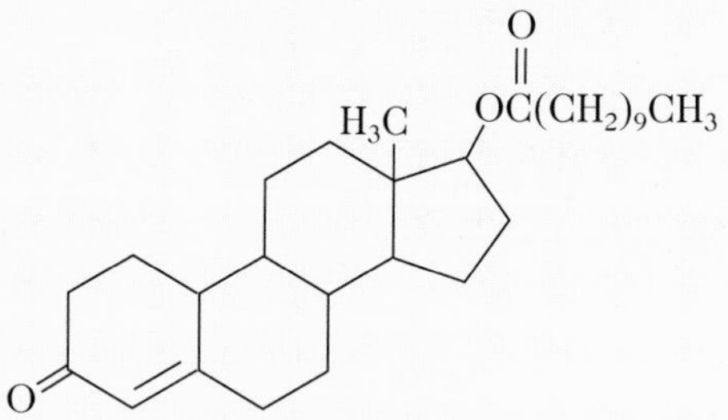

Nandrolone decanoate (Deca-
Durabolin, Deca-Durabol, Deca-
Hybolin, Retabolil)

15.5 CELL MEMBRANES— THEIR LIPID COMPONENTS

Cell membranes consist mainly of a *lipid bilayer* plus molecules of proteins and cholesterol.

Animal cell membranes are made of both lipids and proteins, but the lipid molecules actually hold the cell together. The protein components have a variety of other functions. Some are *receptors* that bind hormone molecules and relay them or their "messages" inside the cell. Other proteins provide passages—closable molecular channels—for small ions and molecules. Some membrane protein assemblies act as "pumps" to move solutes across the membrane. We focus here on the lipids of the membrane and leave the proteins to Chapter 16.

Both Hydrophilic and Hydrophobic Groups Are Necessary for Cell Membranes. The principal lipids of animal cell membranes are not triacylglycerols but more complex lipids, like phospholipids and glycolipids, as well as cholesterol. Molecules of phospho- or glycolipids have two nonpolar "tails" plus sections that are either very polar or fully ionic. Such sections are called **hydrophilic groups,** because they attract water molecules. In phospholipids, for example, the hydrophilic groups are the phosphate diester units, which have ionic sites. In a glycolipid, the sugar unit with its many OH groups is the hydrophilic group. Hydrophilic groups force molecules of membrane lipids to take up positions that expose such groups to maximum contact with water molecules both inside and outside the cell.

The nonpolar, hydrocarbon sections of membrane lipids are called **hydrophobic groups** because they are water avoiding. Hydrophobic groups tend to force membrane lipid molecules to become positioned *within* the membrane so that these groups remain out of contact with water as much as possible.

Substances with both hydrophilic and hydrophobic groups, like phospholipids and glycolipids, are called **amphipathic compounds.** When mixed in the right proportion with water, their molecules spontaneously group together into *micelles.* A **micelle** is a globular aggregation in which hydrophobic contacts are minimized and hydrophilic interactions are maximized. Figure 15.2 shows how a micelle forms when the amphipathic molecules have a *single* hydrocarbon "tail," like a detergent or soap molecule.

In the Lipid Bilayer of Cell Membranes, Hydrophobic Groups Intermingle between the Membrane Surfaces. As we said, molecules of phospholipids and glycolipids have two hydrocarbon "tails." These force a micelle made of such lipids to take up an extended disklike shape (Figure 15.3). Further extension of the shape shown in Figure 15.3 would produce two rows of molecules or a **lipid bilayer** arrangement, a sheetlike array that consists of two layers of lipid molecules aligned side by side. This is the basic architecture of an animal cell membrane (Figure 15.4). The hydrophobic "tails" of the lipid molecules intermingle in the center of the bilayer away from water molecules. In a sense, these "tails" dissolve in each other, following the "like dissolves like" rule. The hydrophilic "heads" stick out into the aqueous phase and so are in contact with water. Water-avoiding and water-attracting properties, not covalent bonds, are thus the major "forces" that stabilize an animal cell membrane.

Cholesterol Molecules Also Help to Stabilize Membranes. The ring system of cholesterol makes its molecules somewhat long, flat, and rigid. In the lipid bilayer, they occur with their long axes lined up side by side with the hydrocarbon chains of the other lipids. The cholesterol OH groups are hydrogen-bonded to O atoms of ester groups in the membrane lipid molecules. Because the cholesterol units are relatively rigid, much more so than the fatty acid chains, cholesterol molecules help to keep a membrane from being too fluidlike.

■ *Hydrophilic* from the Greek *hydor,* "water," and *philos,* "loving." *Hydrophobic* from the Greek *hydor,* "water," and *phobikos,* "fearing."

■ Soaps and detergents are also examples of amphipathic compounds.

(a)   (b)

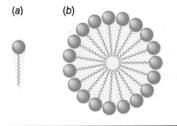

FIGURE 15.2
Detergent micelle. *(a)* Space-filling requirements of an amphipathic detergent or soap molecule with one hydrophobic tail (wavy line) and a hydrophilic head (blue sphere). *(b)* Micelle in water. The hydrophobic tails gather together as the hydrophilic heads have maximum exposure to the aqueous medium.

(a)

(b)

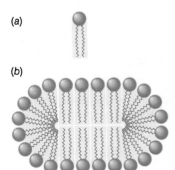

FIGURE 15.3
Glycerophospholipid micelle in water. *(a)* Space-filling requirements for an amphipathic glycerophospholipid, which has two hydrophobic tails (wavy lines) and a hydrophilic head (blue sphere). *(b)* A disklike micelle whose "wall" for the most part (top and bottom segments) is a lipid bilayer.

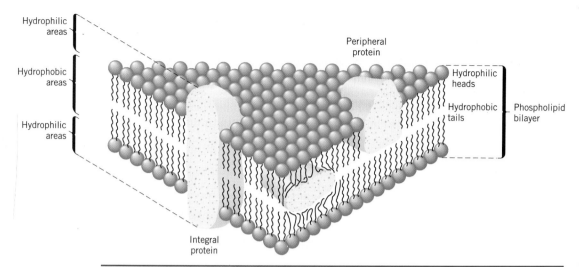

FIGURE 15.4
Cell membrane.

The Lipid Bilayer Is Self-Sealing. If a pin were stuck through a cell membrane and then pulled out, the lipid bilayer would close back spontaneously. Such flexibility is allowed because, as we said, no covalent bonds hold neighboring lipid molecules to each other. Only the net forces of attraction implied by the terms *hydrophobic* and *hydrophilic* are at work. Yet the bilayer is strong enough to hold a cell together, and it is flexible enough to let things in and out. Water molecules move back and forth easily, but other molecules and ions are much less free to move. Their migrations depend on the protein components of the membrane, also indicated in Figure 15.4. We must therefore postpone further discussion of membranes until we have learned more about proteins, particularly such families of proteins as enzymes and receptors.

SUMMARY

Lipids Lipids are ether-extractable substances in animals and plants, and they include hydrolyzable esters and non-hydrolyzable compounds. The esters are generally of glycerol or sphingosine, with their acyl portions contributed by long-chain carboxylic acids called fatty acids. The fatty acids obtained from lipids by hydrolysis generally have long chains of even numbers of carbons, seldom are branched, and often have one or more alkene groups. The alkene groups are *cis*. Because molecules of all lipids are mostly hydrocarbon-like, lipids are soluble in nonpolar solvents but not in water.

Triacylglycerols Molecules of neutral fats, those without electrically charged sites or sites that are similarly polar, are esters of glycerol and a variety of fatty acids, both saturated and unsaturated. Vegetable oils have more double bonds per molecule than animal fats. The triacylglycerols can be hydrogenated, hydrolyzed, and saponified.

Waxes Molecules of the waxy coatings on leaves and fruit, or in beeswax or sebum, are simple esters between long-chain alcohols and fatty acids.

Glycerophospholipids Molecules of the glycerophospholipids are esters both of glycerol and of phosphoric acid. A second ester bond from the phosphate unit goes to a small alcohol molecule that can also have a positively charged group. Thus, this part of a glycerophospholipid is strongly hydrophilic. The two types of glycerophospholipids are the phosphatides and the plasmalogens. Both are vital to animal cell membranes. In phosphatide molecules there are two fatty acyl ester units besides the phosphate system. In plasmalogens, there is one fatty acyl unit and a long-chain, unsaturated ether unit in addition to the phosphate system.

Sphingomyelins Sphingomyelins are esters of sphingosine, a dihydric amino alcohol. They also have a strongly hydrophilic phosphate system.

Glycolipids   Also sphingosine based, the glycolipids use a monosaccharide instead of the phosphate-to-small alcohol unit to provide the hydrophilic section. Otherwise, they resemble the sphingomyelins.

Steroids   Steroids are nonhydrolyzable lipids with the steroid nucleus of four fused rings (three being C-6 rings and one a C-5 ring). Several steroids are sex hormones, and oral fertility control drugs mimic their structure and functions. Cholesterol, the raw material used by the body to make bile salts and other steroids, is manufactured in the liver. Cho-

lesterol is carried in circulation in lipoprotein complexes. Cholesterol molecules are essential components of animal cell membranes.

Animal Cell Membranes   A double layer of phospholipids or glycolipids plus cholesterol and assemblies of protein molecules make up the lipid bilayer part of a cell membrane. The hydrophobic tails of the amphipathic lipids intermingle within the bilayer, away from the aqueous phase. The hydrophilic heads are in contact with the aqueous medium. Cholesterol molecules help to stiffen the membrane.

REVIEW PROBLEMS

The answers to Review Exercises whose numbers are in color are found in Appendix V. The answers to the other Review Exercises are found in the Study Guide that accompanies this text. The more challenging questions are marked with asterisks.

Lipids in General

15.1   Crude oil is soluble in ether, yet it isn't classified as a lipid. Explain.

15.2   Cholesterol has no ester group, yet we classify it as a lipid. Why?

15.3   Ethyl acetate has an ester group, but it isn't classified as a lipid. Explain.

15.4   What are the criteria for deciding if a substance is a lipid?

Fatty Acids

15.5   What are the structures and the names of the two most abundant *saturated* fatty acids?

15.6   Write the structures and names of the *unsaturated* fatty acids that have 18 carbons each and that have no more than three double bonds. Show the correct geometry at each double bond.

15.7   Write the equations for the reactions of palmitic acid with (a) NaOH(*aq*) (b) CH$_3$OH (when heated in the presence of an acid catalyst).

*15.8   What are the equations for the reactions of oleic acid with each substance?
(a) KOH(*aq*)
(b) H$_2$ (in the presence of a catalyst)
(c) CH$_3$CH$_2$OH (heated with an acid catalyst)

15.9   Which of the following acids, **A** or **B**, is more likely to be obtained by the hydrolysis of a lipid? Explain.

$$CH_3(CH_2)_{12}CO_2H \qquad CH_3\overset{\overset{\displaystyle CH_3}{|}}{C}H(CH_2)_{11}CO_2H$$
$$\textbf{A} \qquad\qquad\qquad \textbf{B}$$

15.10   Without writing structures, state what kinds of chemicals the prostaglandins are.

Triacylglycerols

*15.11   Write the structure of a triacylglycerol that involves linolenic acid, linoleic acid, and palmitic acid, besides glycerol.

*15.12   What is the structure of a triacylglycerol made from glycerol, stearic acid, oleic acid, and palmitic acid?

*15.13   Write the structures of all the products that would form from the complete hydrolysis of the following lipid. (Show the free carboxylic acids, not their anions.)

$$
\begin{array}{l}
\quad\quad\quad\quad\quad \overset{\displaystyle O}{\overset{\displaystyle \|}{}} \\
CH_2OC(CH_2)_7CH{=}CHCH_2CH{=}CH(CH_2)_4CH_3 \\
\quad\quad\quad \overset{\displaystyle O}{\overset{\displaystyle \|}{}} \\
CHOC(CH_2)_{12}CH_3 \\
\quad\quad\quad \overset{\displaystyle O}{\overset{\displaystyle \|}{}} \\
CH_2OC(CH_2)_7CH{=}CH(CH_2)_7CH_3
\end{array}
$$

*15.14   Write the structures of the products that are produced by the saponification (by NaOH) of the triacylglycerol whose structure was given in Review Exercise 15.13.

*15.15   Hydrolysis of a lipid produced glycerol, lauric acid (Table 13.1), linoleic acid, and oleic acid in equimolar amounts. Write a structure that is consistent with these results. Is there more than one structure that can be written? Explain.

*15.16   Hydrolysis of 1 mol of a lipid gives 1 mol each of glycerol and oleic acid and 2 mol of lauric acid (Table 13.1). Write a structure consistent with these facts. Is more than one constitution (structure) possible?.

15.17   What is the structural difference between the triacylglycerols of the animal fats and the vegetable oils?

15.18   Products such as corn oil are advertised as being "polyunsaturated." What does this mean in terms of the structures of the molecules that are present? Corn oil is "more polyunsaturated" than what?

15.19 What chemical reaction is used in the manufacture of margarine?

15.20 Lard and butter are chemically almost the same substances, so what is it about butter that makes it so much more desirable a spread for bread than, say, lard or tallow?

Waxes

15.21 One component of beeswax has the formula $C_{36}H_{72}O_2$. When it is hydrolyzed, it gives $C_{18}H_{36}O_2$ and $C_{18}H_{38}O$. Write the most likely structure of this compound.

*15.22 When all the waxes from the leaves of a certain shrub are separated, one has the formula $C_{60}H_{120}O_2$. Its structure is **A**, **B**, or **C**. Which is it most likely to be? Explain why the others can be ruled out.

$$CH_3(CH_2)_{56}CO_2CH_2CH_3 \qquad CH_3(CH_2)_{29}CO_2(CH_2)_{28}CH_3$$
$$\textbf{A} \qquad\qquad\qquad \textbf{B}$$
$$CH_3(CH_2)_{28}CO_2(CH_2)_{29}CH_3$$
$$\textbf{C}$$

Phospholipids

15.23 Why are the phosphatides and plasmalogens both called glycerophospholipids?

15.24 What site in a glycerophospholipid carries a negative charge? What atom carries a positive charge?

15.25 In general terms, how are the sphingomyelins and cerebrosides structurally alike? How are they structurally different?

15.26 What structural unit provides the most polar groups in a molecule of a glycolipid? (Name it.)

15.27 Phospholipids are not classified as neutral fats. Explain.

15.28 Phospholipids are particularly common in what part of a cell?

15.29 What are the names of the two types of sphingosine-based lipids?

15.30 Are the sugar units that are incorporated into the cerebrosides bound by acetal-like links or by ordinary ether links? How can one tell? Which kind of link is more easily hydrolyzed (assuming an acid catalyst)?

*15.31 Complete hydrolysis of 1 mol of a phospholipid gave 1 mol each of the following compounds: glycerol, linolenic acid, oleic acid, phosphoric acid, and the cation, $HOCH_2CH_2N(CH_3)_3{}^+$.
(a) Write a structure of this phospholipid that is consistent with the information given.
(b) Is the substance a glycerophospholipid or a sphingolipid? Explain.
(c) Is it an example of a lecithin or a cephalin? How can you tell?

*15.32 When 1 mol of a certain phospholipid was hydrolyzed, there was obtained 1 mol each of lauric acid, oleic acid, phosphoric acid, glycerol, and $HOCH_2CH_2NH_3{}^+$.

(a) What is a possible structure for this phospholipid?
(b) Is it a sphingolipid or a glycerophospholipid? Explain.
(c) Is it a cephalin or a lecithin? Explain.

Steroids

15.33 What is the name of the steroid that occurs as a detergent in our bodies?

15.34 What is the name of a vitamin that is made in our bodies from a dietary steroid by the action of sunlight on the skin?

15.35 Give the names of three steroid sex hormones.

15.36 What is the name of a steroid that is part of the cell membranes in animal tissues?

15.37 What is the raw material used by the body to make bile salts? How does the body use the bile salts?

15.38 How does the body carry cholesterol in circulation in the bloodstream?

Cell Membranes

15.39 Describe in your own words what is meant by the *lipid bilayer* structure of cell membranes.

15.40 How do the hydrophobic parts of phospholipid molecules avoid water in a lipid bilayer?

15.41 Besides lipids, what kinds of substances are present in a cell membrane?

15.42 What kinds of forces are at work to hold a cell membrane together?

15.43 Name the functions that the proteins of a cell membrane can serve.

The Prostaglandins (Special Topic 15.1)

15.44 Name the fatty acid that is used to make the prostaglandins.

15.45 What effect is aspirin believed to have on prostaglandins, and how is this related to aspirin's medicinal value?

Detergent Action (Special Topic 15.2)

15.46 Which is the more general term, soap or detergent? Explain.

15.47 What kind of chemical is soap?

15.48 For household laundry work, which product is generally preferred, a synthetic detergent or soap? Why?

15.49 Why are soap and sodium alkyl sulfates called *anionic* detergents?

15.50 Explain in your own words how a detergent can loosen oils and greases from fabrics.

Illicit Steroids (Special Topic 15.3)

15.51 What do the anabolic steroids supposedly do that is sought by their users?

15.52 Suggest a reason why nandrolone decanoate cannot be taken orally.

15.53 Why is the use of the anabolic steroids by athletes banned?

Additional Exercises

15.54 Examine the following structure and answer the questions about it.

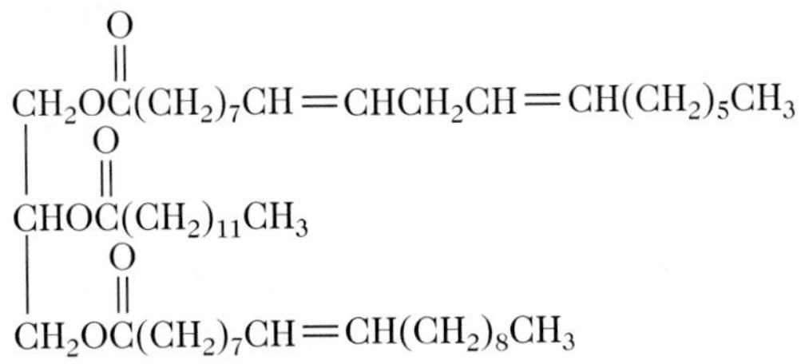

$$
\begin{array}{l}
\quad\quad\ \ \overset{\displaystyle O}{\overset{\displaystyle \|}{}} \\
CH_2OC(CH_2)_7CH\!=\!CHCH_2CH\!=\!CH(CH_2)_5CH_3 \\
|\quad\ \ \overset{\displaystyle O}{\overset{\displaystyle \|}{}} \\
CHOC(CH_2)_{11}CH_3 \\
|\quad\ \ \overset{\displaystyle O}{\overset{\displaystyle \|}{}} \\
CH_2OC(CH_2)_7CH\!=\!CH(CH_2)_8CH_3
\end{array}
$$

(a) Is it a triester of glycerol?
(b) Does it have hydrophobic groups?

(c) What are its hydrophilic functional groups?
(d) What would form if all of its alkene groups were hydrogenated?
(e) Is this molecule likely to be found among naturally occurring triacylglycerols? Explain.

15.55 Examine the following structure and answer the questions that follow.

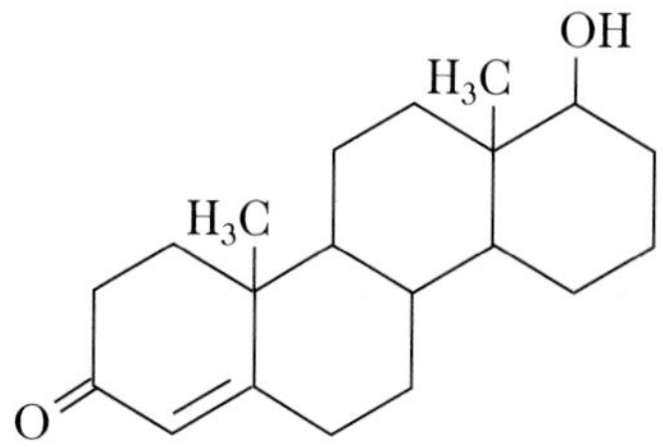

(a) Can this compound be described as amphipathic? Explain?
(b) Is it a member of the steroid family? How can one tell?

16

PROTEINS

Cables of proteins enable the giraffe to support its long yet flexible neck. How the natural polymers of proteins are structured is explained in this chapter.

16.1 AMINO ACIDS—THE BUILDING BLOCKS OF PROTEINS

Living things select from among the molecules of about 20 α-amino acids to make the polypeptides in proteins.

Proteins, found in all cells and in virtually all parts of cells, constitute about half of the body's dry mass. They give strength and elasticity to skin and blood vessels. In muscles and tendons, they work as the cables that move our arms and legs. Proteins reinforce teeth and bones much as steel rods strengthen concrete. The proteins in blood—antibodies, hemoglobin, and other forms—serve as protectors and as the long-distance haulers of oxygen and lipids. Other proteins form parts of the communications network of our nervous system. Nearly all enzymes, neurotransmitters, and membrane receptors, as well as some hormones, are proteins that direct the work of repair, construction, communication, and energy conversion in the body. No other class of compounds is involved in such a variety of functions, all essential to life. They deserve the name *protein,* taken from the Greek *proteios,* "of the first rank."

■ The chief elements in proteins are C, H, O, N, and S.

Polypeptides Are Made from α-Amino Acids.   The dominant structural units of **proteins** are polymers called **polypeptides.** A metal ion or a small organic molecule or ion is also often bound together with a polypeptide to make up a complete protein molecule (Figure 16.1). Many proteins, however, are made entirely of polypeptides.

The monomer units of polypeptides are **α-amino acids,** which have the general structure given by **1**. These provide the **amino acid residues** or **peptide** units for polypeptide molecules.

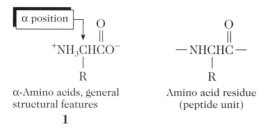

α-Amino acids, general structural features   Amino acid residue (peptide unit)

1

■ With few exceptions, the amino acids not in the "standard" set consist of modifications of "standard" amino acid molecules made *after* a polypeptide has been put together.

Twenty α-amino acids make up the "standard set" of polypeptide monomers (Table 16.1), the same set being used by all species of plants and animals. In any given polypeptide some monomers are used many times as hundreds of amino acid residues are joined together to make a single molecule. However, before we can study how polypeptides are put together, we must learn more about their α-amino acid building blocks.

■ A few nonstandard amino acids are used in certain tissues as well as in some strains of bacteria.

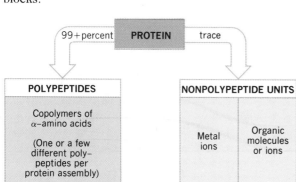

FIGURE 16.1
Components of proteins. Some proteins consist exclusively of polypeptide molecules, but most also have nonpolypeptide units such as small organic molecules, metal ions, or both.

TABLE 16.1   Amino Acids: $^+NH_3CHCO_2^-$
$\qquad\qquad\qquad\qquad\qquad\qquad |$
$\qquad\qquad\qquad\qquad\qquad\quad\ R$

Type	Side Chain, R	Name	Three-Letter Symbol	pI
Side chain is nonpolar	—H	Glycine	Gly	5.97
	—CH$_3$	Alanine	Ala	6.00
	—CH(CH$_3$)$_2$	Valine	Val	5.96
	—CH$_2$CH(CH$_3$)$_2$	Leucine	Leu	5.98
	—CHCH$_2$CH$_3$ 　　| 　CH$_3$	Isoleucine	Ile	6.02
	—CH$_2$C$_6$H$_5$	Phenylalanine	Phe	5.48
	—CH$_2$ 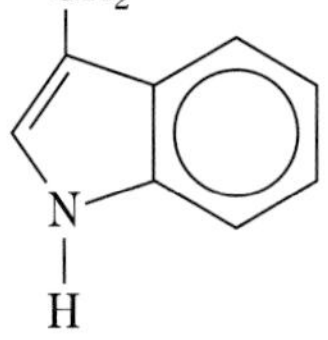	Tryptophan	Trp	5.89
	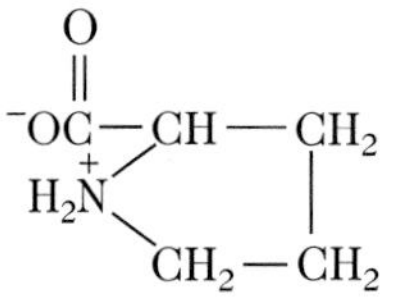 (complete structure, proline)	Proline	Pro	6.30
Side chain has a hydroxyl group	—CH$_2$OH	Serine	Ser	5.68
	—CHOH 　| 　CH$_3$	Threonine	Thr	5.64
	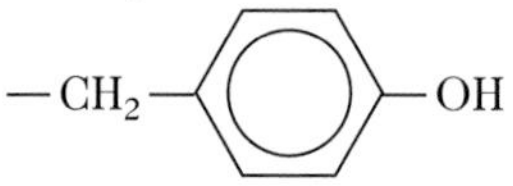 —CH$_2$—⟨ ⟩—OH	Tyrosine	Tyr	5.66
Side chain has a carboxyl group (or amide group)	—CH$_2$CO$_2$H	Aspartic acid	Asp	2.77
	—CH$_2$CH$_2$CO$_2$H	Glutamic acid	Glu	3.22
	—CH$_2$CONH$_2$	Asparagine	Asn	5.41
	—CH$_2$CH$_2$CONH$_2$	Glutamine	Gln	5.65
Side chain has a basic amino group	—CH$_2$CH$_2$CH$_2$CH$_2$NH$_2$	Lysine	Lys	9.74
	—CH$_2$CH$_2$CH$_2$NHCNH$_2$ 　　　　　　　　|| 　　　　　　　　NH	Arginine	Arg	10.76
	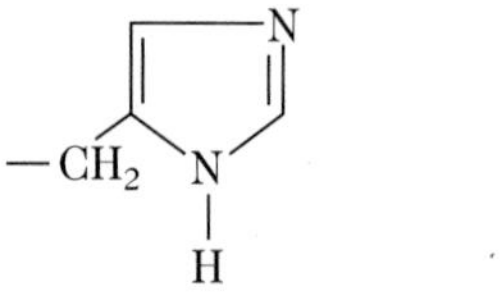 —CH$_2$ (imidazole)	Histidine	His	7.59
Side chain contains sulfur	—CH$_2$SH	Cysteine	Cys	5.07
	—CH$_2$CH$_2$SCH$_3$	Methionine	Met	5.74

α-Amino Acids Have the Same *Main Chain* as α-Amino Acetic Acid but Have Different *Side Chains*.   As you can see in Table 16.1, the amino acids differ in the **R** group, called a **side chain,** located at the α-position in **1.** All amino acids (except proline) have the same *main chain:* $^+NH_3CHCO_2^-$. *Be sure to notice this.* The only differences are in the side chains attached to the α position (the CH group) of the main chain. Also be careful to notice that the second column of Table 16.1 gives only the structures of the R groups, not the complete amino acid structure (except for proline, in which the R group bends back and is joined to the N atom to form a ring).

In the solid state, amino acids exist entirely in the form shown by **1,** which is called a **dipolar** ion or a **zwitterion.** It is an electrically neutral particle, but it has a positive and negative charge and so is exceedingly polar. Amino acids, like salts, therefore have melting points that are considerably higher than those of most molecular compounds. Amino acids also tend to be much more soluble in water than in nonpolar solvents.

Structure **1** is actually an *internally neutralized molecule.* We can imagine that **1** begins its life with a regular amino group, NH_2, and an ordinary carboxyl group, CO_2H. If it were glycine (R = H), for example, the structure could be written as $NH_2CH_2CO_2H$. But the amino group, a proton acceptor, would then "do its thing." It would take a proton from the carboxyl group, a proton donor *close by in the same molecule,* and change the molecule to the dipolar ion, **1.** Of course, a dipolar ion has its own (weaker) proton-donating group, NH_3^+, and its own (also weaker) proton-accepting group, CO_2^-, so *dipolar ions can neutralize acids and bases of sufficient strength,* like H_3O^+ and OH^-. Each amino acid, in other words, is a buffer system all by itself. A *pair* of separate components, one a base and the other an acid, is not required. One dipolar ion serves the two functions of a buffer; its NH_3^+ is a base neutralizer and its CO_2^- is an acid neutralizer. We will soon see that all protein molecules likewise have at least one NH_3^+ group and one CO_2^- group, so *proteins are also able to serve as buffers.*

Structure **1** is an **isoelectric molecule** because it has an equal number of positive and negative charges. Because isoelectric molecules are electrically neutral, overall, they cannot migrate to either electrode in an electrolysis experiment (page 149). However, in an amino acid solution, the following equilibrium mixture always exists. It includes a protonated form of **1,** namely **2,** and a deprotonated species, **3.** The proportions of **1, 2,** and **3** depend altogether on the pH of the medium.

$$^+NH_3CHCO_2^-$$
$$|$$
$$R \quad \mathbf{1}$$

$$OH^- \quad\quad\quad\quad OH^-$$
$$H^+ \quad\quad\quad\quad H^+$$

$$^+NH_3CHCO_2H \quad\quad\quad NH_2CHCO_2^-$$
$$| \quad\quad\quad\quad\quad\quad |$$
$$R \quad\quad\quad\quad\quad\quad R$$
$$\mathbf{2} \quad\quad\quad\quad\quad\quad \mathbf{3}$$

Provided an amino acid has neither a carboxyl group nor an amino group on its side chain (like glutamic acid or lysine), it can exist in water as an isoelectric molecule, **1,** only if the pH is between 6 and 7. The exact value depends on the specific amino acid. As the pH is made less than 6 (more acidic), the CO_2^- groups of isoelectric molecules, **1,** accept protons in increasing numbers and change to CO_2H groups. The resulting species—its structure is given by **2**—is positively charged and can now migrate toward the negative electrode in electrolysis. On the other hand, if the pH is made more than 7 (more basic), the NH_3^+ groups of isoelectric molecules, **1,** give up

■ Hereafter, when we say "amino acid," we'll mean α-amino acid.

■ α-Amino acids melt around 300 °C but their simple esters, which cannot be dipolar ions, generally melt around 100 °C.

$$O$$
$$\|$$
$$NH_2CHCOCH_3$$
$$|$$
$$R$$

Methyl ester of an α-amino acid

■ We can still call dipolar ions *molecules* because, overall, they are electrically neutral particles.

■ These shifts of H^+ ions illustrate Le Châtelier's principle at work.

protons to the base and become NH_2 groups. The new species—its structure is shown by **3**—is negatively charged and so can migrate to the positive electrode in electrolysis.

Each amino acid has a particular value of pH, its **isoelectric point** or **p*I***, at which its molecules all behave as if they are isoelectric and give no net movements in an electrolysis experiment. The isoelectric points of the standard amino acids are given in the last column of Table 16.1. Now let's see what all this has to do with proteins.

Proteins, like Amino Acids, Have Isoelectric Points.   As we have mentioned, all proteins have NH_3^+ and CO_2^- groups or can acquire them by a change in the pH of the surrounding medium. Whole protein molecules, therefore, can be made isoelectric with the right adjustment of the pH. *Each protein has its own isoelectric point.*

Now think of what happens when the pH of the protein's aqueous medium is changed to some value *other than* the protein's p*I* value. Initially *isoelectric* protein molecules will engage in proton transfers and pick up net charges, exactly like amino acids in the same circumstances. By adding a strong acid or a strong base—by changing the pH of the protein's medium—*the entire electrical condition of a huge protein molecule can be made either positively or negatively charged almost instantly.* ("Instantly" suggests "possible medical emergency," of course.)

Changes in the net electrical charges of protein molecules have serious consequences at the molecular level of life. We cannot overemphasize the fact that an organism must control the pH values of its fluids. Nearly all medical emergencies and all strenuous sports activities, when carried to extremes, threaten the pH of the blood and so the forms in which the blood carries its proteins. The electrical charge condition of a particular protein can dramatically affect its chemical reactions or greatly alter its solubility. Enzyme molecules, for example, work best as catalysts only over narrow ranges of pH.

To facilitate the further study of proteins, you should now memorize the structures of a minimum of five amino acids: glycine, alanine, cysteine, lysine, and glutamic acid. These are suggested because they illustrate important features. How to use Table 16.1 to write their structures is described in the following example. Be sure to study this example, because the most common difficulty students have with Table 16.1 is converting its information into the structure of a specific amino acid.

■ Casein, the protein in milk, precipitates when milk turns sour because a change in pH causes the casein molecules to become isoelectric.

■ The amino acid proline is the only one of the standard 20 in which the α-amino group is itself joined to one end of the side chain.

Proline

EXAMPLE 16.1   Writing the Structure of an Amino Acid

What is the structure of cysteine?

ANALYSIS   Regard any α-amino acid as having two structural features, a main chain common to *all* α-amino acids plus a unique side chain at the α position. So write the common unit first and then add the side chain.

SOLUTION   The common unit, the main chain, is

$$^+NH_3CHCO_2^-$$
$$|$$

The correct side chain, either from memory or by looking it up, is CH_2SH, so simply attach this group to the α carbon. Cysteine is

$$^+NH_3CHCO_2^-$$
$$|$$
$$CH_2SH$$

That's all there is to it; so be sure to learn the main chain and then the side chains of the five amino acids.

■ PRACTICE EXERCISE 1 Write the structures of the dipolar ionic forms of glycine, alanine, lysine, and glutamic acid.

Several Amino Acids Have Hydrophobic Side Chains. The first amino acids in Table 16.1, including alanine, have essentially nonpolar, hydrophobic side chains. When a long polypeptide molecule folds into its distinctive shape, these hydrophobic groups tend to be folded next to each other as much as possible rather than next to highly polar groups or to water molecules in the solution. This water avoidance by a nonpolar side chain is called the **hydrophobic interaction** of the side chain. Water molecules have a strong tendency to form hydrogen bonds between each other and thus "reject" the presence of molecules or groups that cannot themselves "offer" hydrogen bonds to them. Water molecules, in a sense, "club together," forcing nonpolar groups to stay away and be by themselves. We are beginning to see how the nature of the side chains is a major factor in the final shape of a protein.

Some Amino Acids Have Hydrophilic OH Groups on Their Side Chains. The second set of amino acids in Table 16.1 consists of those whose side chains carry alcohol or phenol OH groups, which are polar and hydrophilic and can form hydrogen bonds. The way in which a long polypeptide chain folds into its final most stable shape will thus be influenced by the water-attracting effects of side chains with OH groups. As much as possible, these groups will stick out into the surrounding aqueous phase to which they are attracted by hydrogen bonds.

Two Amino Acids Have Carboxyl Groups on Their Side Chains. The side chains of aspartic acid and glutamic acid carry proton-donating CO_2H groups. Because body fluids are generally slightly basic, the protons available from side-chain CO_2H groups have been neutralized, so these groups actually occur mostly as CO_2^- groups on protein molecules in body fluids.

Aspartic acid and glutamic acid often occur as asparagine and glutamine in which their side chain CO_2H groups have become amide groups, $CONH_2$, instead. These are also polar, hydrophilic groups, *but they are not electrically charged.* They are neither acidic nor basic groups. They can form hydrogen bonds, however, and both the $CONH_2$ and CO_2^- groups influence the final overall shape taken by a polypeptide.

■ PRACTICE EXERCISE 2 Write the structure of aspartic acid (in the manner of **1**) with the side chain carboxyl (a) in its carboxylate form and (b) in its amide form.

Lysine, Arginine, and Histidine Have Basic Groups on Their Side Chains. The extra NH_2 group on lysine makes its side chain hydrophilic, basic, and proton-accepting. Arginine and histidine have similarly basic side chains. Proteins whose molecules have such side chains tend to become positively charged in acid because protons go onto NH_2 groups to change them to NH_3^+ groups. Both NH_2 and NH_3^+ groups can form hydrogen bonds and so influence polypeptide shapes.

■ Aspartame, a popular artificial sweetener, has an aspartic acid residue.

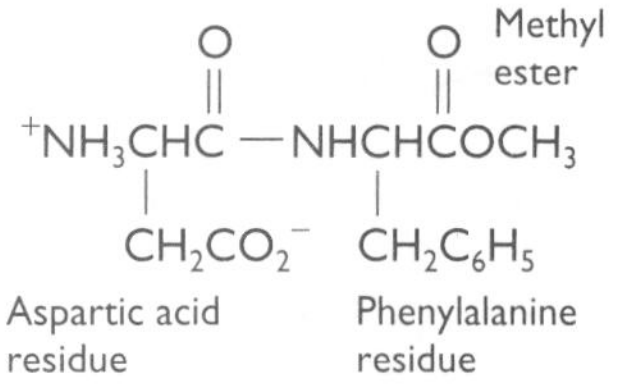

$$^+NH_3CHC-NHCHCOCH_3$$

Aspartic acid residue   Phenylalanine residue

Aspartame (Nutrasweet)

■ PRACTICE EXERCISE 3   Write the structure of arginine in the manner of **1**, but with its side-chain amino group in its protonated form. (Put the extra proton on the $=NH$ unit, not the NH_2 unit of the side chain.)

■ PRACTICE EXERCISE 4   Is the side chain of the following amino acid hydrophilic or hydrophobic? Does the side chain have an acidic, basic, or neutral group?

$$^+NH_3CHCO_2^-$$
$$|$$
$$CH_2CH_2CONH_2$$

Cysteine and Methionine Have Sulfur-Containing Side Chains.   The side chain in cysteine has an SH group. As we studied in Section 11.5, molecules with this group are easily oxidized to disulfide systems, and disulfides are easily reduced to SH groups:

$$2RSH + (O) \xrightarrow{\text{oxidation}} RSSR + H_2O$$

$$RSSR + 2(H) \xrightarrow{\text{reduction}} 2RSH$$

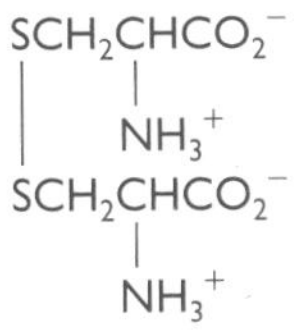

$$SCH_2CHCO_2^-$$
$$|$$
$$NH_3^+$$
$$SCH_2CHCO_2^-$$
$$|$$
$$NH_3^+$$
Cystine

Cysteine and its oxidized form, cystine, are therefore interconvertible by oxidation and reduction, a property of far-reaching importance in some proteins. Two polypeptides, for example, can be joined by *disulfide links,* such as seen in cystine, to make a much larger molecule. The disulfide link is especially prevalent in the proteins that have a protective function, such as those in hair, fingernails, and the shells of certain crustaceans, but many simpler proteins, like insulin, have them as well.

16.2   OVERVIEW OF PROTEIN STRUCTURE

Protein molecules can have four levels of complexity.

Protein structures are more complicated by far than those of carbohydrates or lipids, and every aspect of their structures is vital at the molecular level of life. We'll begin, therefore, with a broad overview of protein architecture.

■ We'll study denaturing agents in Section 16.7.

Protein Structure Involves Four Features.   A protein with the structure and overall shape that it normally has and that permit it to function biologically in a living system is called a **native protein**. The same protein after losing its molecular shape, even while retaining its original molecular constitution (covalent structure), is a **denatured protein.** There are four possible levels of complexity to the native structures of proteins. Disarray at any level almost always denatures proteins and renders them biologically useless.

The first and most fundamental level of protein structure, the **primary structure,** concerns only the *sequence of amino acid residues* in the polypeptide(s) of the protein. However, it is this sequence that ultimately determines the three-dimensional structure of a protein and, therefore how a protein can function.

The next level of protein complexity, the **secondary structure,** also concerns only individual polypeptides. It consists of the particular way in which a long polypeptide strand has coiled or in which strands have intertwined or lined up side to side to give helices or sheets of molecules. Secondary structures are stabilized largely by dipole–dipole attractions, particularly the hydrogen bond.

The **tertiary structure** of a polypeptide concerns the further coiling, bending, kinking, or twisting of secondary structures. If you've ever played with a coiled door spring or a "slinky," you know that the coil (secondary structure) can be bent and twisted (tertiary structure). By and large, noncovalent forces such as hydrogen bonds and hydrophobic interactions stabilize these shapes. When polypeptides have disulfide bonds, they form from SH groups *after* the polypeptide has been synthesized in the cell, so the S—S covalent bond is classified as a feature of tertiary, not primary structure.

Finally, we deal with those proteins having still another complexity, **quaternary structure**. A protein's quaternary structure forms when two or more polypeptides, often with additional small molecules or ions, aggregate in a precise manner and form one grand whole. Each polypeptide unit—now called a *subunit* of the protein—has all previous levels of structure.

■ In many polypeptides, attractions and repulsions between electrically charged sites on side chains are also involved in tertiary structure.

16.3 PRIMARY STRUCTURES OF PROTEINS

The backbones of all polypeptides of all plants and animals have a repeating series of N—C—C(=O) units.

The Peptide Bond Joins Amino Acid Residues Together in a Polypeptide. The **peptide bond** is the covalent bond that forms when amino acids are put together in a cell to make a polypeptide. It's nothing more than an amide system, carbonyl–nitrogen. To illustrate it and to show how polypeptides acquire their primary structure, we will begin by simply putting together only two amino acids to make a *dipeptide.*

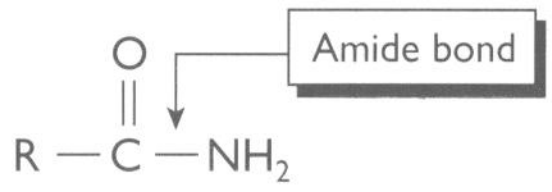

Suppose that glycine acts at its carboxyl end and alanine acts at its amino end such that, by a series of steps (not given in detail), a molecule of water splits out and a carbonyl-to-nitrogen bond, a peptide bond, is created.

■ *Simple* amides have the following structure (see also Section 17.3):

$$
\underset{R}{R} - \overset{\overset{\displaystyle O}{\|}}{C} - NH_2 \quad \boxed{\text{Amide bond}}
$$

■ How a cell causes a peptide bond to form involves a cell's nucleic acids and is described in Chapter 21.

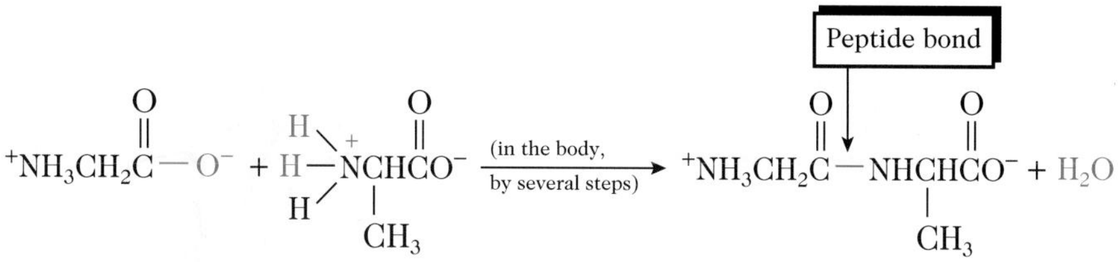

$$
\overset{O}{\underset{\|}{^+NH_3CH_2C}} - O^- + \underset{\underset{H}{\diagup}}{\overset{H}{\diagdown}}\overset{+}{N}\overset{O}{\underset{\underset{CH_3}{|}}{\overset{\|}{CHCO^-}}} \xrightarrow[\text{by several steps}]{\text{(in the body,}} \boxed{\text{Peptide bond}} \; {^+NH_3CH_2}\overset{O}{\underset{\|}{C}} - NH\overset{O}{\underset{\underset{CH_3}{|}}{\overset{\|}{CHCO^-}}} + H_2O
$$

Glycine (Gly)　　　　　Alanine (Ala)　　　　　　　　Glycylalanine (Gly-Ala)

4

Of course, there is no reason why we could not picture the roles reversed so that alanine acts at its carboxyl group and glycine at its amino group. The result is a different dipeptide (but an isomer of the first).

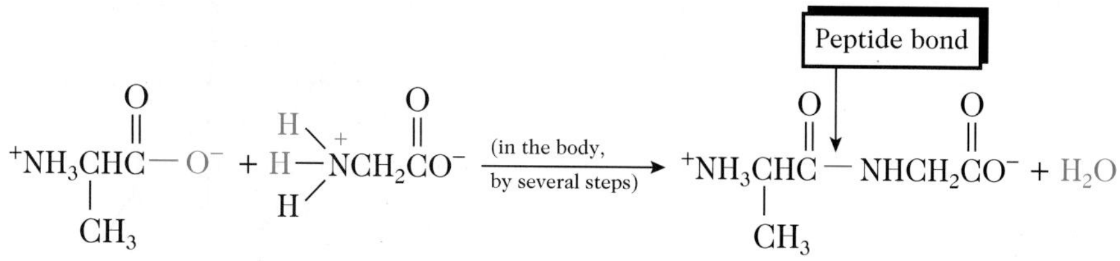

$$
\overset{O}{\underset{\underset{CH_3}{|}}{\underset{\|}{^+NH_3CHC}}} - O^- + \underset{\underset{H}{\diagup}}{\overset{H}{\diagdown}}\overset{+}{N}\overset{O}{\underset{\|}{CH_2CO^-}} \xrightarrow[\text{by several steps}]{\text{(in the body,}} \boxed{\text{Peptide bond}} \; {^+NH_3}\overset{O}{\underset{\underset{CH_3}{|}}{\overset{\|}{CHC}}} - NH\overset{O}{\underset{\|}{CH_2CO^-}} + H_2O
$$

Alanine (Ala)　　　　　Glycine (Gly)　　　　　　　　Alanylglycine (Ala-Gly)

5

■ The *di* in *dipeptide* signifies that *two* amino acid residues are present, not the number of peptide bonds.

The product of the union of any two amino acid residues by a peptide bond is called a **dipeptide**, and all dipeptides have the following features:

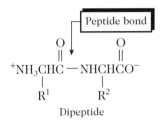

Dipeptide

Structures **4** and **5** differ only in the *sequence* in which the side chains, H and CH_3, occur on α-carbons. This is fundamentally how polypeptides also differ—in the sequences of their side chains. The amino acid residue with the free NH_3^+ group is called the *N-terminal* residue, which is Gly in **4** and Ala in **5**. The amino acid residue with the free CO_2^- group at the other end of the main chain is called the *C-terminal* residue; it's Ala in **4** and Gly in **5**. By convention, the N-terminal residue is always written at the left end of the main chain.

EXAMPLE 16.2 Writing the Structure of a Dipeptide

What are the two possible dipeptides that can be put together from alanine and cysteine?

ANALYSIS   Both dipeptides must have the same backbone, so we write two of these first. We follow the convention that such backbones are always written in the N to C—left to right—direction.

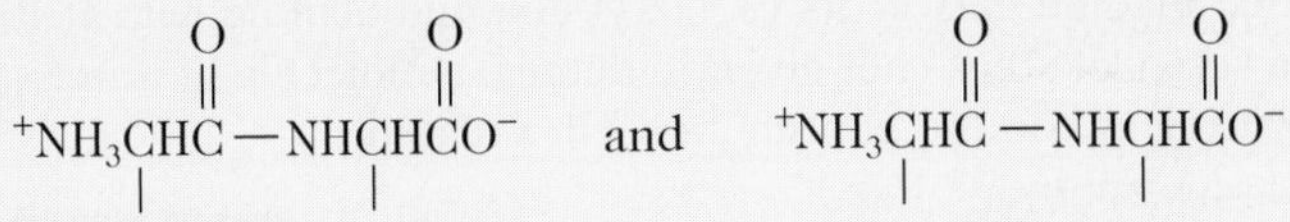

$$\overset{O}{\overset{\|}{{}^+NH_3CHC}}-\overset{O}{\overset{\|}{NHCHCO^-}} \quad and \quad \overset{O}{\overset{\|}{{}^+NH_3CHC}}-\overset{O}{\overset{\|}{NHCHCO^-}}$$

Then, either from memory or by using a table, we know that the two side chains are CH_3 for alanine and CH_2SH for cysteine. We simply attach these in their two possible orders to make the finished structures.

SOLUTION

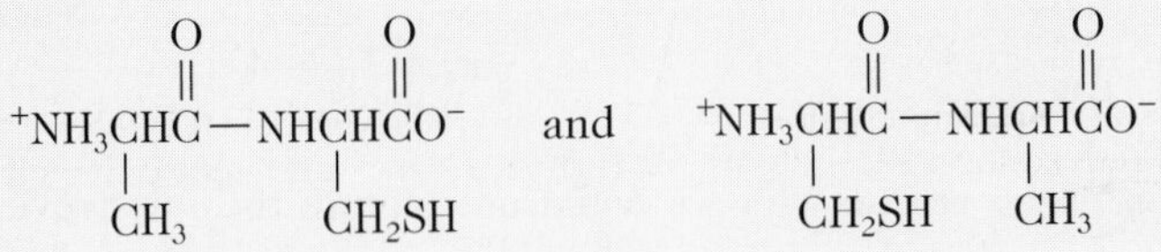

$$\underset{\underset{CH_3}{|}}{\overset{O}{\overset{\|}{{}^+NH_3CHC}}}-\underset{\underset{CH_2SH}{|}}{\overset{O}{\overset{\|}{NHCHCO^-}}} \quad and \quad \underset{\underset{CH_2SH}{|}}{\overset{O}{\overset{\|}{{}^+NH_3CHC}}}-\underset{\underset{CH_3}{|}}{\overset{O}{\overset{\|}{NHCHCO^-}}}$$

Ala-Cys   Cys-Ala

It would be worthwhile at this time simply to memorize the easy repeating sequence in a dipeptide, because it carries forward to higher peptides.

nitrogen–carbon–carbonyl–nitrogen–carbon–carbonyl-etc.

Sequence common to all polypeptides

(Remember that the "carbon" in "nitrogen–carbon–carbonyl" is the *alpha* carbon. Remember also that the direction—left to right means nitrogen–carbon–carbonyl—is a convention, not a law of nature.)

■ PRACTICE EXERCISE 5   Write the structures of the two dipeptides that can be made from alanine and glutamic acid.

Three-Letter Symbols for Amino Acid Residues Simplify the Writing of Polypeptide Structures.   Each amino acid has a three-letter symbol, given in the second from the last column in Table 16.1. To use them to write a polypeptide structure, separate each symbol by hyphens and remember the convention that the N-terminal reside is on the left. Above, we wrote the structure of the dipeptide **4**, for example, as Gly-Ala, and that of **5** as Ala-Gly.

■ The backbones or main chains of **4** and **5** are identical.

Dipeptides still have NH_3^+ and CO_2^- groups, so a third amino acid can react at either end. In general,

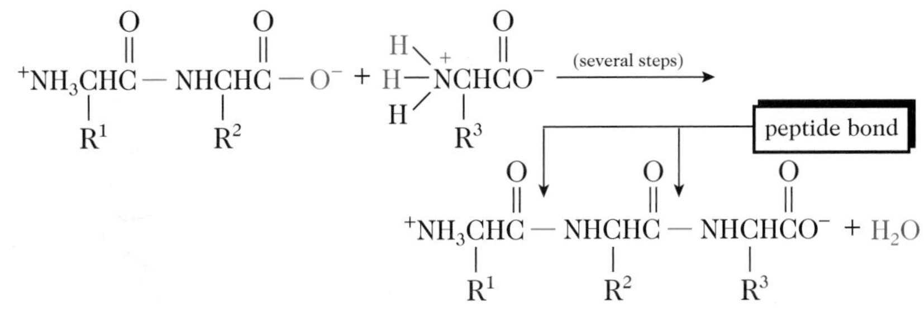

$$^+NH_3CHC-NHCHC-O^- + H-\overset{H}{\underset{H}{N}}{}^+CHCO^- \xrightarrow{\text{(several steps)}}$$

(with R^1, R^2 below first two carbons, R^3 below third, and "peptide bond" boxed)

$$^+NH_3CHC-NHCHC-NHCHCO^- + H_2O$$

(with R^1, R^2, R^3 below the carbons)

A tripeptide

When R^1 is H, R^2 is CH_3, and R^3 is $CH_2CH_2CO_2H$, the tripeptide is Gly-Ala-Glu, but it is only one of six possible tripeptides that involve these three different amino acids. The set of all possible sequences for a tripeptide made from glycine, alanine, and glutamic acid is as follows:

Gly-Ala-Glu   Ala-Gly-Glu   Glu-Gly-Ala
Gly-Glu-Ala   Ala-Glu-Gly   Glu-Ala-Gly

Each tripeptide has main chain groups, NH_3^+ and CO_2^-, that can interact at either end with a fourth amino acid to make a tetrapeptide. The product, of course, would itself carry the end groups from which the chain could be further extended. You can see how, by chain extensions repeated many hundreds of times, a long polypeptide can form. You can also see how, with even a few different amino acids, a large number of isomeric polypeptides are possible.

■ From four *different* amino acids, 24 tetrapeptides are possible ($1 \times 2 \times 3 \times 4$ or 4 factorial, 4!, for statistics' buffs).

The Sequence of Side Chains on the Repeating N—C—C(=O) Backbone Is the *Primary* **Structure of All Polypeptides.**   All polypeptides have the following skeleton in common. They differ in length *(n)* and in the kinds and sequences of side chains.

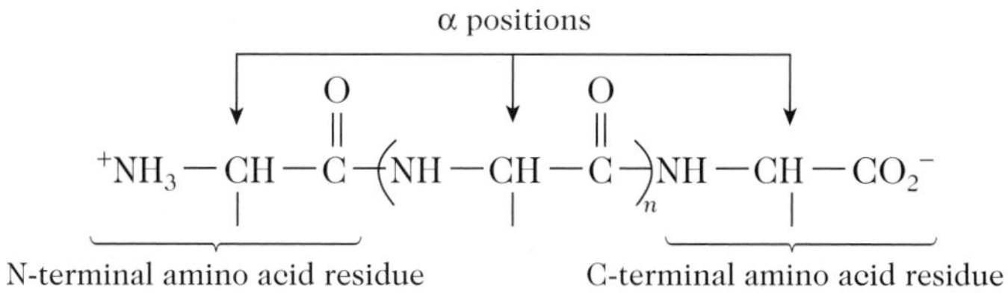

α positions

$$^+NH_3-CH-\overset{O}{\overset{\|}{C}}\left(NH-CH-\overset{O}{\overset{\|}{C}}\right)_n NH-CH-CO_2^-$$

N-terminal amino acid residue   C-terminal amino acid residue
Polypeptide "backbone" (*n* can equal several thousand)

As the number of amino acid residues increases, some used several times, the number of possible polypeptides increases rapidly. For example, if 20 different amino acids are incorporated, each used only once, there are 2.4×10^{18} possible isomeric polypeptides!

■ A polypeptide with only 20 amino acid residues in its molecule is a *very* small polypeptide.

16.4   SECONDARY STRUCTURES OF PROTEINS

The α-helix and the β-pleated sheet are two important kinds of secondary protein structures.

Once a cell puts together a polypeptide, largely noncovalent forces, like the hydrogen bond and hydrophobic interactions, help to determine how a polypeptide twists into a particular native shape. The hydrophobic interactions are largely what "drive" the formation of an overall shape, and hydrogen bonds help to stabilize it.

The α-Helix Is a Major Secondary Structure of Polypeptides.   One of the most common configurations is the *α-helix*, a coiled configuration of a polypeptide strand (Figure 16.2). In the **α-helix,** the polypeptide backbone coils as a right-handed screw, which permits all of its side chains to stick to the outside of the coil.

Hydrogen Bonds Stabilize α-Helices   In α-helices, hydrogen bonds extend from the H atoms of NH units in the main chain to oxygen atoms of carbonyl units farther along the chain. Individually, hydrogen bonds are weak forces of attraction, but they add up much like the individual forces in a zipper. Generally, only *segments* of long polypeptide molecules, not entire lengths, are in α-helix configuration. Coils about 11 residues long are common, but as many as 53 make up the a-helix section of one polypep-

■ The turns of a right-handed helix or screw are in the same direction taken as the fingers of your right hand curl when your thumb points along the axis of the helix in the direction with which the helix advances. Wood screws have right-handed helices.

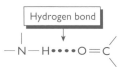

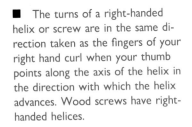

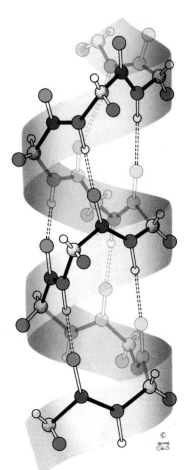

FIGURE 16.2

α-Helix. The polypeptide backbone follows the spiraling ribbon as a right-handed helix. Oxygen atoms of carbonyl groups are in red; N atoms are in dark blue; H atoms of NH units are in white. Side-chain groups, R, are represented here only by simple spheres (purple). Note how they project to the *outside* of the ribbon. Dashed lines show how hydrogen bonds extend from each NH group's H atom to a carbonyl group's O atom four residues along the backbone.

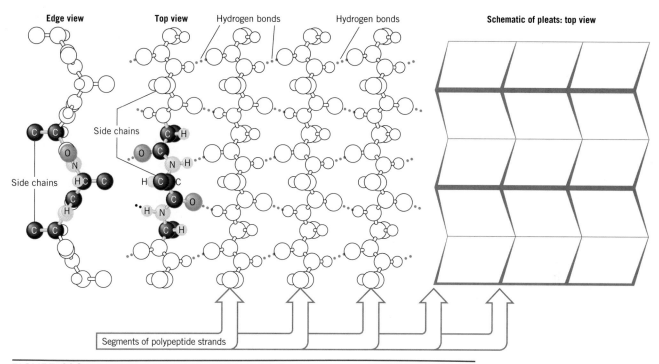

FIGURE 16.3
β-Pleated sheet.

tide. Much of the remaining uncoiled portions often occur *between* coiled segments, but sometimes they are themselves involved in another secondary structure, a *pleated sheet* array (to be studied soon.)

A Left-Handed Helix Characterizes the Individual Polypeptide Strands in Collagen. The *collagens,* the most abundant proteins in vertebrates, are a family of extracellular proteins that give strength to bone, teeth, cartilage, tendons, skin, blood vessels, and certain ligaments. Glycine contributes a third of the amino acid residues in collagen's subunits. Another 15 to 30% are furnished by proline and its hydroxy derivatives, all of which, because of their rings, limit the flexibility of a collagen molecule and force it to coil as a *left-handed* helix.

Vitamin C Is Essential to the Synthesis of Collagen. The hydroxylated derivatives of proline residues in collagen are made *after* the initial polypeptide is formed. One needed enzyme requires ascorbic acid (vitamin C) units in its structure, which is why vitamin C is essential to growing children. They need it to make the collagen required for strong bones and teeth as well as all other tissues that depend on collagen. When an adult's diet is deficient in ascorbic acid, wounds do not heal well, blood vessels become fragile, and an overall vitamin deficiency condition called *scurvy* can result.

The β-Pleated Sheet Is a Side-by-Side Array of Polypeptide Units. Adjacent segments in some polypeptides line up side by side to form a somewhat pleated sheetlike array called the β-pleated sheet (Figure 16.3). Stabilized by hydrogen bonds, the β-pleated sheet is a feature of *portions* of many polypeptides. The side chains project above and below the surface of the sheet. As few as 2 segments of a polypeptide strand and as many as 15 have been found aligned side by side in the same pleated sheet,

■ The α-helix was discovered by Linus Pauling and R. B. Corey using X-ray technology.

■ The protein in the cornea of the eye is also a member of the collagen family.

■ Linus Pauling won the 1954 Nobel prize in chemistry for discovering the α-helix and β-sheet configurations.

FIGURE 16.4

Segments of polypeptide strands can exist as β-sheets (adjacent to and parallel with each other) in more than one way. *(a)* A hairpin loop brings the next segment into an antiparallel arrangement. *(b)* A back-and-over loop allows the next segment to have a parallel alignment. *(c)* A left-handed crossover loop also permits a parallel alignment. The loops are made of segments of the polypeptide chain.

with each strand ranging from 6 to 15 residues long. When a hairpin turn carries the polypeptide chain from one segment of a pleated sheet to the next, the strands run in opposite direction. Other kinds of turns are possible, however, but are less common (Figure 16.4).

16.5   TERTIARY STRUCTURES OF PROTEINS

Tertiary structures are the results of folding, bending, and twisting of secondary structures.

■ Special proteins called *chaperonines* (after "chaperone") guide the formation of the final shapes of newly formed polypeptides.

Once primary and secondary structures are in place, the final shaping of a protein occurs. These changes occur spontaneously in cells, sometimes in a matter of seconds after the polypeptide molecule has been made and sometimes only after several minutes.

Tertiary Protein Structure Involves the Folding and Bending of Secondary Structure. As we have said, when α-helices take shape, their side chains tend to project outward, where they can be in contact with water. Even in water-soluble proteins, however, as many as 40% of the side chains are hydrophobic and cannot break up the hydrogen bonding networks among water molecules. An entire α-helix or β-sheet, therefore, undergoes further twisting and folding until, as much as possible, hydrophobic groups are tucked to the inside, away from water, and hydrophilic groups stay exposed to water. The final shape of a polypeptide, its **tertiary structure,** thus emerges in response to simple molecular forces set up by the water-avoiding and the water attracting properties of side chains.

■ The three-letter symbol for cystine is

Cys
|
Cys

Disulfide Bonds Give Loops in Polypeptides or Join Two Strands Together. Polypeptides that are to have disulfide bonds receive them by the oxidation of SH groups during the development of tertiary structure. If the SH group on the side chain of cysteine appears on two neighboring polypeptide molecules, then mild oxidation is all it takes to link the two molecules by a disulfide bond. Such cross-linking can also occur between parts of the same polypeptide molecule, in which case a closed loop results.

Ionic Bonds Also Stabilize Tertiary Structures. Another force that can stabilize a tertiary structure is the attraction between a full positive and a full negative charge, each occurring on a particular side chain. At the pH of body fluids, the side chains of

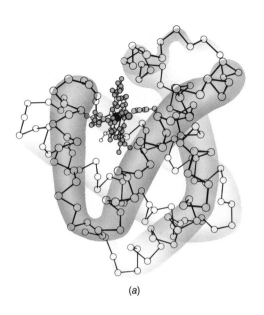

(a) (b)

FIGURE 16.5

(a) Myoglobin (sperm whale), a polypeptide with 153 amino acid residues. The tubelike forms outline the eight segments that are in an α-helix. The flat, purple structure is the heme unit, and the red circle is an oxygen molecule. Only the atoms that make up the backbone of the chain are indicated (by circles). The side chains have been omitted. *(b)* Heme molecule with its Fe^{2+} ion.

both aspartic acid and glutamic acid carry CO_2^- groups. The side chains of lysine and arginine carry NH_3^+ groups. The oppositely charged groups attract each other like the attraction of oppositely charged ions in an ionic crystal, so this feature of polypeptide shape is sometimes called a **salt bridge.**

Many Polypeptides Carry *Prosthetic Groups.* In myoglobin, the oxygen-holding protein of muscle tissue, about 75% of the polypeptide consists of α-helix segments (Figure 16.5*a*). This allows virtually all of the hydrophobic groups to be folded inside where they avoid water, while the hydrophilic groups project to the outside. A non-protein molecule, heme (Figure 16.5*b*), completes the native structure of myoglobin. The heme unit is held in the folded globin molecule by electrical attractions between two electrically charged side chains and the Fe^{2+} ion in heme.

A nonprotein, organic compound that associates with a polypeptide, like heme in myoglobin, is called a **prosthetic group** and is often the focus of the protein's biological purpose. Heme, for example, is the actual oxygen holder in myoglobin, and it serves the same function in **hemoglobin,** the oxygen carrier in blood.

■ *Prosthetic* is from the Greek *prosthesis,* "an addition."

16.6 QUATERNARY STRUCTURES OF PROTEINS

For many proteins, the native form emerges only as two or more polypeptides assemble into a quaternary structure.

Proteins, like myoglobin, have finished shapes at the tertiary level. They are made up of single polypeptide molecules, sometimes with prosthetic groups. Many proteins, however, are aggregations of two or more polypeptides, and these aggregations constitute **quaternary structures.** Individual molecules of polypeptides that make up an intact protein molecule are called the protein's *subunits.* If the subunits of a native protein become separated, even if nothing happens to their primary, secondary, or tertiary structural features, the protein is rendered biologically useless.

Hemoglobin has four subunits, two of one kind (designated α subunits) and two of another (the β subunits), each subunit supporting a heme molecule (Figure 16.6). Each subunit has specific primary, secondary, and tertiary structural features, and a

SICKLE-CELL ANEMIA AND ALTERED HEMOGLOBIN

The decisive importance of the primary structure to all other structural features of a polypeptide or its associated protein is illustrated by the grim story of sickle-cell anemia. This inherited disease is widespread among those whose roots are in equatorial regions of central and western Africa.

In its mild form, where only one parent carries the genetic trait, the symptoms of sickle-cell anemia are seldom noticed except when the environment has a low partial pressure of oxygen, as at high altitudes. In the severe form, when both parents carry the trait, the infant usually dies by the age of 2 unless treatment is begun early. The problem is *an impairment in blood circulation traceable to the altered shape of hemoglobin in sickle-cell anemia.* The altered shape is particularly a problem after the hemoglobin has delivered oxygen and is on its way back to the heart and lungs for more.

The fault at the molecular level lies in a β subunit of hemoglobin. One of the amino acid residues should be glutamic acid, but is valine instead. Thus, instead of a side-chain CO_2^- group, which is electrically charged and hydrophilic, there is an isopropyl side chain, which is neutral and hydrophobic. Normal hemoglobin, symbolized as HbA, and sickle-cell hemoglobin, HbS, therefore have different patterns of electrical charges. Both have about the same solubility in well-oxygenated blood, but oxygen-free molecules of HbS clump together inside red cells and precipitate. This deforms the cells into a telltale sickle shape (Figure 1). The distorted cells are harder to pump through capillaries, where the cells often create plugs. Sometimes the red cells split open. Any of these events places a strain on the heart. The error in one side chain seems insignificant, but it is far from small in human terms.

The sickle cell trait offers some resistance to malaria, which almost certainly explains why the trait survives largely where this tropical disease is most common. Normally, the mosquito-borne parasite that causes malaria resides within a red blood cell. However, the parasite cannot survive very long inside a sickled cell. The parasite has a high need for potassium ion, but the membrane of a sickled cell allows too much potassium ion to get through and escape. Thus, people with the sickle cell trait are statistically more likely than individuals unprotected from malaria to live long enough to bear children and so pass the trait to their offspring.

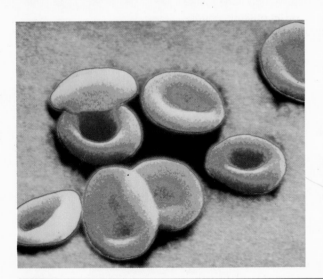

FIGURE 1
Electron micrographs of a normal red blood cell, left, and a sickle cell, right.

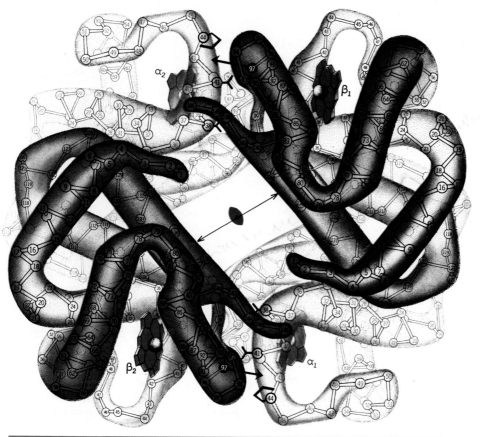

FIGURE 16.6

Hemoglobin. Four polypeptide chains, each with one heme molecule represented here by the flat colored plates that contain spheres (Fe^{2+} ions), are nestled together.

combination of hydrophobic and dipole–dipole attractions, like hydrogen bonds, hold the subunits together. If even one amino acid residue is wrong in one subunit, the results can be very serious, as in the example of sickle-cell anemia (see Special Topic 16.1).

Covalent Cross Links Occur in Collagen. The polypeptide units in collagen, each with about 1000 amino acid residues and each in a left-handed helix, assemble in units of three molecules each. The three left-handed helices wrap around each other in a relatively open *right*-handed helix of helices to form the **triple helix**, cablelike system called *tropocollagen* (Figure 16.7). Between the polypeptide strands of the triple helix, *covalently bonded* molecular bridges are erected by a series of reactions that cause lysine side chains to link together. *Covalent* crosslinks are better able than hydrogen bonds or hydrophobic interactions to resist forces that could work to undo the tertiary structure of collagen and so weaken bones and ligaments. With aging, however, additional covalent crosslinks develop between collagen strands, and this leads to less flexibility and agility.

A microfiber or *fibril* of collagen of considerable strength forms when individual tropocollagen cables overlap lengthwise. The mineral deposits in bones and teeth become tied into the protein at the gaps between the heads and tails of tropocollagen molecules.

■ Meat from old animals is tougher because of their more highly cross-linked collagen.

■ A collagen fibril only 1 mm in diameter can hold a suspended mass as large as 10 kg (22 lb).

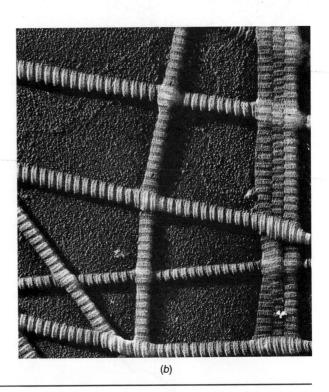

(a)   (b)

FIGURE 16.7
Triple helix of collagen. *(a)* Schematic drawing. *(b)* Electron micrograph of collagen fibrils from the skin. A fibril is an orderly aggregation of collagen molecules aligned side by side but overlapping each other in a regularly repeating manner that produces the banded appearance. (Micrograph courtesy of Jerome Gross, Massachusetts General Hospital.)

16.7   COMMON PROPERTIES OF PROTEINS

Even small changes in the pH of a solution can affect a protein's solubility and its physiological properties.

Although proteins come in many diverse biological types, they generally have similar chemical properties toward ordinary substances because they have similar functional groups.

Protein Digestion Is Hydrolysis.   The digestion of a protein is nothing more than the hydrolysis of its peptide bonds to give a mixture of amino acids. Different digestive enzymes handle the cleavage of peptide bonds according to the nature of the side chains nearby. To illustrate digestion, we may use the hydrolysis of a tripeptide.

■ Enzymes that catalyze the hydrolysis of proteins are called *proteases*.

$$^{+}NH_3CH_2\overset{\overset{\displaystyle O}{\|}}{C} - NH\underset{\underset{\displaystyle CH_3}{|}}{C}HC\overset{\overset{\displaystyle O}{\|}}{} - NH\underset{\underset{\displaystyle CH_2C_6H_5}{|}}{C}HC\overset{\overset{\displaystyle O}{\|}}{}O^{-} + 2H_2O \xrightarrow[\text{(enzyme catalyzed)}]{\text{digestion}}$$

Hydrolyzable peptide bonds

Glycylalanylphenylalanine (Gly-Ala-Phe)

TABLE 16.2   Denaturing Agents for Proteins

Denaturing Agent	How the Agent May Operate
Heat	Disrupts hydrophobic interactions and hydrogen bonds by making molecules vibrate too violently. Produces coagulation, as in the frying of an egg.
Microwave radiation	Causes violent vibrations of molecules that disrupt hydrogen bonds and hydrophobic interactions.
Ultraviolet radiation	Probably operates much like the action of heat (e.g., sunburning).
Violent whipping or shaking	Causes molecules in globular shapes to extend to longer lengths and then entangle (e.g., beating egg white into meringue).
Soaps	Probably affect hydrogen bonds and salt bridges.
Organic solvents (e.g., ethyl alcohol and isopropyl alcohol)	May interfere with hydrogen bonds because these solvents can also form hydrogen bonds or can disrupt hydrophobic interactions. Quickly denature proteins in bacteria, killing them (e.g., disinfectant action of 70% ethyl alcohol).
Strong acids and bases	Disrupt hydrogen bonds and salt bridges. Prolonged action leads to actual hydrolysis of peptide bonds.
Salts of heavy metals (e.g., salts of Hg^{2+}, Ag^+, Pb^{2+})	Cations combine with SH groups and form precipitates. (These salts are all poisons.)

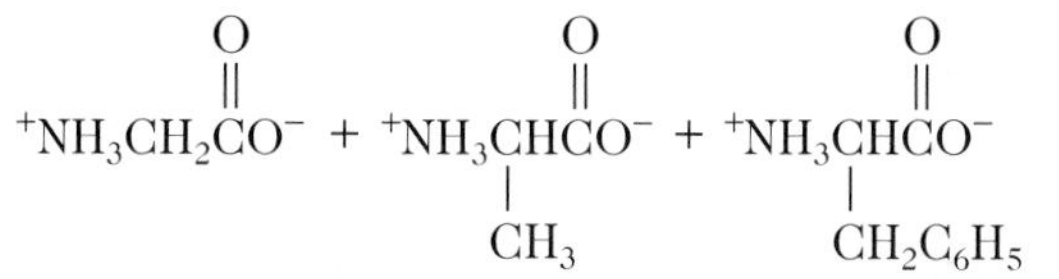

$$^+NH_3CH_2CO^- \quad + \quad ^+NH_3CHCO^- \quad + \quad ^+NH_3CHCO^-$$

Glycine (Gly) Alanine (Ala) Phenylalanine (Phe)

(with O double-bonds above carbonyls; CH_3 on Ala, $CH_2C_6H_5$ on Phe)

Protein *Denaturation* Is the Loss of Protein Shape.   Peptide bonds are not necessarily hydrolyzed when a protein is denatured. All that needs to happen is some disruption of secondary or higher structural features. The disorganization of the overall molecular shape of a native protein is called **denaturation.** It can occur as an unfolding or uncoiling of helices or as the separation of subunits. Because native proteins have their overall shapes both within and in response to an aqueous environment, even the removal of water can cause the denaturation of many proteins.

Usually, denaturation is accompanied by a major loss of solubility in water. The nearly transparent protein in raw egg white, for example, is a somewhat soluble albumin whose molecules are tightly folded. When egg white is whipped or heated, however, the protein denatures as its molecules unfold and become entangled among themselves. Denatured egg albumin does not blend with water and is not transparent.

Table 16.2 lists several reagents or physical forces that cause denaturation, with brief explanations of how they work. The effectiveness of a given denaturing agent depends on the protein. Those of hair and skin or of fur and feathers quite strongly resist denaturation because they are rich in disulfide crosslinks.

Protein Solubility Depends Greatly on pH.   As we learned on page 356, an entire polypeptide molecule can bear a net charge that is easily altered by changing the pH. For example, CO_2^- groups become electrically neutral CO_2H groups as they pick up

■ There are a few proteins that can be *renatured* after they've been denatured.

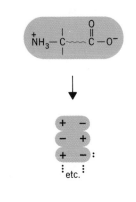

FIGURE 16.8
Several isoelectric protein molecules (*top*) can aggregate into very large clusters that no longer dissolve in water.

■ When buffers break down, the body declares a medical emergency.

■ Oligosaccharide units of the glycolipids of membranes are involved in the actions of toxins, viruses, and bacteria.

protons when a strong acid is added. Suppose that the net charge on a polypeptide is $1-$ and that one extra CO_2^- is responsible for it. When acid is added, we might imagine the following change, where the elongated shape is the polypeptide system.

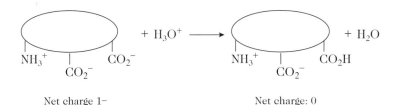

Net charge $1-$ Net charge: 0

The simple transfer of one proton makes the polypeptide isoelectric and neutral. On the other hand, a polypeptide might have a net charge of $1+$, caused by an excess of one NH_3^+ group. Now the addition of base, e.g., OH^-, can make the polypeptide isoelectric and neutral by removing H^+ from the NH_3^+ group and changing it into an amino group, NH_2.

At a characteristic pH, its isoelectric point, each protein has a net charge of zero. At any different pH, most if not all of the protein's molecules carry *identical* charges, either all positive or all negative (depending on the pH). *Identically charged protein molecules repel each other and so cannot clump together.* Proteins, therefore, have their *greatest* solubilities in water when the pH is *not* the isoelectric pH.

Electrically *neutral* polypeptide molecules, on the other hand, can aggregate. They can clump together provided they have the freedom to become aligned so that oppositely charged sites on neighboring molecules are able to get together (Figure 16.8). By aggregating, protein molecules make particles of such great size that they simply drop out of solution. *A protein is least soluble in water when the pH equals the protein's isoelectric point.* You can see that whenever a protein must be *in solution* to work, as is true for many enzymes, the pH of the medium must *not* be that of the protein's isoelectric point. Buffers in body fluids ensure the necessary control of pH.

An example of the effect of pH on solubility is illustrated by casein, the protein in milk whose p*I* value is 4.7. As milk turns sour, the pH drops from its normal value of 6.3–6.6 to 4.7. As a result, more and more casein molecules become isoelectric, clump together, and separate as curds. As long as the pH of milk is something *other* than the p*I* for casein, the protein remains colloidally dispersed.

16.8   CELL MEMBRANES REVISITED— GLYCOPROTEIN COMPONENTS

Glycoproteins provide "recognition sites" on the surfaces of cell membranes.

In Section 15.5 we introduced the general features of cell membranes, giving particular attention to their lipid components. With the knowledge about protein structure gained in this chapter and about carbohydrates learned in Chapter 14, we can take a second, deeper look at cell membranes.

Membrane Proteins Are Glycoproteins.   Most cells are essentially "sugar coated." The coatings are not the ordinary sugars of nutrition but *oligosaccharides* derived from them, carbohydrates that can be hydrolyzed to give three to a few dozen

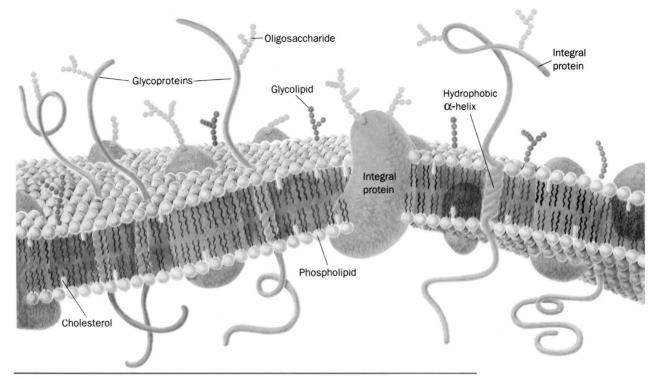

FIGURE 16.9

Glycoproteins as structural units in a cell membrane. The blue spheres, each with two wavy tails, are phospholipid molecules. Shown in yellow are cholesterol molecules. Chains of green beads represent glycolipids. Chains of yellow beads attached to polypeptides (in brown) are oligosaccharide units.

monosaccharide molecules. Some oligosaccharides are bound to proteins to form the **glycoproteins** of membranes (Figure 16.9). Most proteins, in fact, are glycoproteins; several thousand have been identified and the list is growing rapidly. Recall that some carbohydrates are covalently joined to lipids to form the *glycolipids* of the membrane.

■ We learned in Section 15.3 that cerebrosides are glycolipids.

Membrane Proteins Help to Maintain Concentration Gradients. If the cell membrane were an ordinary dialyzing membrane, any kind of small molecule or ion could move freely back and forth between a cell and its surrounding fluid. The workings of cells, however, require that only *some* things be let in and that others be let out. Concentration *gradients* develop, therefore, between a cell's interior and whatever fluid is outside.

■ A gradient in sugar concentration exists in unstirred coffee just after you add sugar.

A **gradient** is an unevenness in the value of some physical property throughout a system. A *concentration gradient* exists in a solution, for example, when one region of the solution has a higher concentration of solute than another. Gradients are generally unstable systems compared to those with the same components but more thoroughly mixed up. In liquids and gases, the natural fate of a gradient is eventually to disappear because of the random motions of ions and molecules. Thus, solute concentrations eventually become uniform. Despite this tendency, living cells are able to maintain a number of gradients through the use of energy-consuming events. For example, as the data in the margin show, both sodium ions and potassium ions are in sharp concentration gradients between the fluids inside a cell and the fluids outside. *These gradients must be maintained against nature's spontaneous tendency to remove concentration gradients.* Assemblies of protein molecules in cell membranes carry out this vital function.

Ion	Concentration (mmol/L)	
	Plasma	Cells
Na^+	135–145	10
K^+	3.5–5.0	125

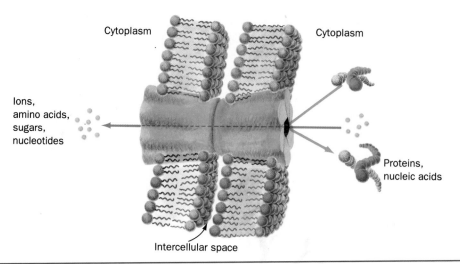

Cytoplasm   Cytoplasm

Ions,
amino acids,
sugars,
nucleotides

Proteins,
nucleic acids

Intercellular space

FIGURE 16.10
Gap junctions. A channel made of protein molecules between two cells enables some small particles to pass directly from one cell to another. The blue spheres, each with two wavy tails, are
phospholipid molecules of the cell membranes.

One kind of assembly, the *sodium–potassium pump,* can move sodium ions
against their gradient from the less concentrated to the more concentrated regions.
When too many sodium ions manage to get inside a cell, they are "pumped" back into
the external fluid by this assembly. The same pump can move potassium ions back
inside a cell. The movement of a solute through a membrane against a concentration
gradient requires chemical energy and is an example of the **active transport** of the
species through the membrane.

■ Other reactions in cells supply
the chemical energy that lets the
pump work.

Gap Junctions Enable Substances to Move Directly from One Cell to Another.   In
the cells of most tissues of multicelled organisms, membrane proteins provide a *direct*
route for the movements of ions and molecules from one cell to another. Such routes
are through **gap junctions,** which are tubules fashioned from proteins that "rivet" cells
together (Figure 16.10).

So many gap junctions occur in some tissues that the entire tissue is intercon-
nected from within. Gap junctions in bone tissue, for example, enable bone cells at
some distance from capillaries to receive nourishment and to remove wastes. Heart
muscle is able to contract *synchronously* because gap junctions allow ions to move
easily between cells. The gaps are large enough to allow certain ions, like Ca^{2+}, and
certain relatively small molecules to pass, but are not large enough for macromolecules like proteins and nucleic acids to get through.

■ Molecules with formula masses
up to about 1200 can negotiate
some gap junctions.

Some Proteins in Cell Membranes Are Receptors for Hormones.   A **receptor molecule** is one whose unique shape enables it to fit only to the molecule of a compound
that it is supposed to receive, its *substrate.* Thus, a receptor is able to "recognize" the
molecules of just one compound from among the hundreds whose molecules bump
against it. This is roughly how specific hormones are able to find and stop only at the
cells where they are meant to stop, by being able to fit to unique receptors. Some
medications work the same way. For example, an antiprogesterone birth control
agent, RU 486, acting as a hormone mimic, binds to a receptor protein (see Special
Topic 16.2).

SPECIAL TOPIC 16.2
MIFEPRISTONE (RU 486)—RECEPTOR BINDING OF A SYNTHETIC ANTIPREGNANCY COMPOUND

When released from an ovarian follicle, the natural female hormone progesterone acts to prepare the system for pregnancy both by inhibiting further production of ova (egg cells) and by preparing the uterus for the implantation of the fertilized ovum. The action of progesterone involves the binding of its molecules to protein receptors within cells of the lining of the uterus (the endometrium).

Antiprogestins (see Table 15.3) mimic the work of progesterone in that they cause a pseudopregnant state ("false pregnancy state") and so suppress the production of ova. Fertilization cannot occur without an ovum, of course, and so the synthetic progestin-containing medications are birth control pills.

RU 486 (RU for Roussel-Uclaf, a French pharmaceutical company) was first prepared in 1980. It was soon discovered to be a strong binder to the progesterone binding sites on the receptor protein for progesterone. This action blocks the normal action of progesterone, and RU 486 is thus an *antiprogesterone*. If RU 486 is taken during the 5- to 6-day postcoital period (the period immediately following intercourse), its blocking action prevents pregnancy by suppressing the implantation of a fertilized ovum in the uterus. If used within 72 hours of unprotected intercourse, its failure rate is very low. It can thus be used as a "morning after" pill.

RU 486, followed by the use of two prostaglandin-like compounds, also induces abortion. Thus, RU 486 is also an abortifacient (an abortion-inducing agent). Under medical supervision, the use of RU 486 for this purpose has a 96% success record. The failures include continued pregnancy, only partial expulsion of the fetus, and the need for procedures to stem uterine bleeding.

That RU 486 "prevents pregnancy" is a controversial statement, because some view the onset of pregnancy as occurring at the moment of fertilization. Others regard pregnancy as not starting until the fertilized ovum has become implanted in the uterus. The controversy thus involves the questions, When does *pregnancy* begin? and the not identical question, When does *human life* begin? Around these questions have surged some of the stormiest waters of the pro-life–prochoice controversy.

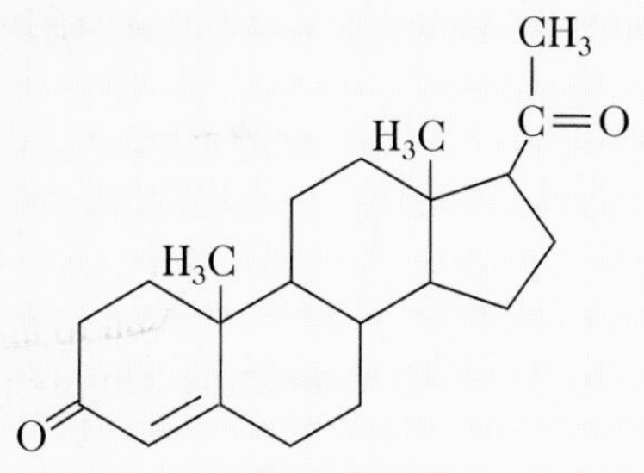

Progesterone

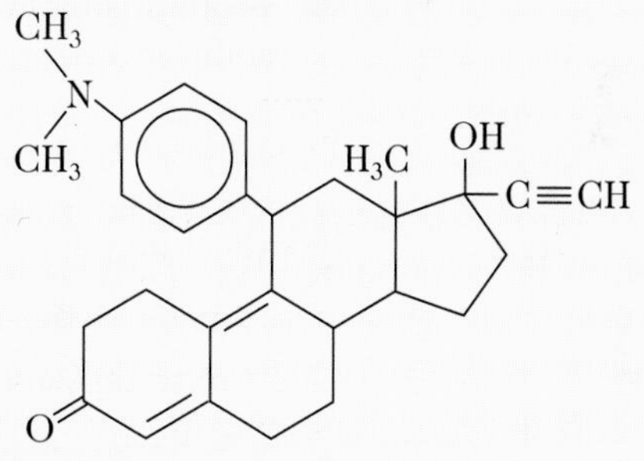

Mifepristone (RU 486)

16.9 CLASSES OF PROTEINS

Two criteria for classifying proteins are their solubility in aqueous systems and their biological function.

We began this chapter with hints about the wide diversities of the kinds and uses of proteins. Now that we know about their structures, we can better understand how so many classes of proteins with so many functions are possible.

Proteins Can Be Classified according to Solubility.   Two families, *fibrous proteins* and *globular proteins,* are based on solubility in water.

Fibrous proteins are insoluble in water and include the following types.

■ When meat is cooked, some of its collagen changes to gelatin, which makes the meat easier to digest.

1. **Collagens** occur in bone, teeth, tendons, skin, blood capillaries, cartilage, and some ligaments. When such tissue is boiled with water, the portion of its collagen that dissolves is called *gelatin.*

2. **Elastins,** which have elastic, rubberlike qualities, are also in cartilage and are found in stretchable ligaments, such as the walls of large blood vessels like the aorta, as well as in the lungs and in the necks of grazing animals. Elastin, like collagen, is rich in glycine and proline residues. Elastin chains are crosslinked by covalently bonded units that are largely responsible for elastin's elasticity.

■ Elastin is not changed to gelatin by hot water.

3. **Keratins** occur in hair, wool, animal hooves, horns, nails, porcupine quills, and feathers. The keratins are rich in disulfide links, which contribute to the unusual stabilities of these proteins to environmental stresses.

4. **Myosins** are the proteins in contractile muscle.

5. **Fibrin** is the protein of a blood clot. During clotting, fibrin forms from its precursor, fibrinogen, by an exceedingly complex series of reactions.

Globular proteins are soluble in water or in water that contains certain salts. There are two broad types.

1. **Albumins** are present in egg white and in blood. In the blood, the albumins are buffers, transporters of water-insoluble molecules of lipids or fatty acids, and carriers of metal ions, like Cu^{2+} ions, that are insoluble in aqueous media at pH values higher than 7.

2. **Globulins** include antibodies, factors of the body's defenses against diseases. In addition, enzymes, many transport proteins, and receptor proteins are globulins.

Proteins Can Be Classified according to Biological Function.   Perhaps no other system more clearly dramatizes the importance of proteins than classification by biological function.

1. Enzymes, the biological catalysts.

2. Contractile muscle, with stationary filaments, myosin, and moving filaments, actin.

3. Hormones, such as growth hormone, insulin, and others.

4. Neurotransmitters, such as the enkephalins and endorphins.

5. Storage proteins—proteins that store nutrients that the organism will need, such as seed proteins in grains, casein in milk, ovalbumin in egg white, and ferritin, the iron-storing protein in human spleen.

6. Transport proteins—proteins that carry things from one place to another, such as hemoglobin, the serum albumins, and ceruloplasmin, a copper-carrying protein.

7. Structural proteins—proteins that hold a body structure together, such as collagen, elastin, keratin, and glycoproteins in cell membranes.

8. Protective proteins—proteins that help the body to defend itself, such as antibodies and fibrinogen.

9. Toxins—poisonous proteins, such as snake venom, diphtheria toxin, and *Clostridium botulinum* toxin (a toxic substance that causes some types of food poisoning).

SUMMARY

Amino Acids About 20 α-amino acids supply the amino acid residues that make up a polypeptide. In the solid state or in water at a pH of roughly 6 to 7, amino acids exist as dipolar ions or zwitterions. Isoelectric points or p*I* values are the pH values of solutions in which amino acid (or protein) molecules are electrically neutral (isoelectric). Several amino acids have hydrophobic side chains, but the side chains in others are strongly hydrophilic. The SH group of cysteine opens the possibility of disulfide crosslinks between or within polypeptide units.

Polypeptides Amino acid residues are held together by peptide (amide) bonds, so the repeating unit in polypeptides is —NH—CH—CO—. Each amino acid residue has its own side chain. This repeating system with a unique sequence of side chains constitutes the primary structure of a polypeptide.

Once the primary structure is fashioned, the polypeptide coils and folds into higher features—secondary and tertiary—that are stabilized largely by hydrophobic interactions and hydrogen bonds. The most prominent secondary structures are the α-helix—a right-handed helix—and the β-pleated sheet. Individual polypeptides in collagen, which has an abundance of glycine, proline, and hydroxylated proline residues, are in a left-handed helix. Disulfide bonds form from SH groups on cysteine residues as many proteins assume their tertiary structure.

Proteins Many proteins consist just of one kind of polypeptide. Many others have nonprotein, organic groups—prosthetic groups—or metal ions. And still other proteins—those with quaternary structure—involve two or more polypeptides whose molecules aggregate in definite ways, stabilized by hydrophobic interactions, hydrogen bonds, and salt bridges. Thus the terms *protein* and *polypeptide* are not synonyms, although for some specific proteins they turn out to be.

Because of their higher levels of structure, proteins can be denatured by agents that do nothing to peptide bonds. A few denatured proteins can be renatured, but this is uncommon. The acidic and basic side chains of polypeptides affect protein solubility, and when a protein is in a medium whose pH equals the protein's isoelectric point, the substance is least soluble. The amide bonds (peptide bonds) of proteins are hydrolyzed during digestion.

Membrane Proteins—Glycoproteins Incorporated into the lipid bilayer membranes of cells are proteins (and lipids) with attached oligosaccharide units. Some of the proteins of a cell membrane provide conduits by which active transport processes can maintain concentration gradients. Other proteins provide gap junctions for direct movements, cell to cell, for certain dissolved species. The oligosaccharides of the membrane proteins serve as cell-recognition features for molecules, like hormones, moving near the cell.

REVIEW EXERCISES

The answers to Review Exercises whose numbers are in color are found in Appendix V. The answers to the other Review Exercises are found in the Study Guide that accompanies this text. The more challenging questions are marked with asterisks.

Amino Acids

16.1 One of the following structures is not of an amino acid on the list of the standard 20. Which one is not on the list? How can you tell without looking at Table 16.1?

$$NH_2CH_2CH_2CH_2CH_2CHCO_2^-$$
$$|$$
$$NH_3^+$$
A

$$^+NH_3CH_2CHCO_2^-$$
$$|$$
$$CH_3$$
B

$$^+NH_3CHCO_2^-$$
$$|$$
$$CH_2CO_2H$$
C

16.2 The following amino acid is one of the standard list of 20.

$$^+NH_3CHCO_2^-$$
$$|$$
$$CH_2CH(CH_3)_2$$

(a) What part of its structure would be its amino acid *residue* in the structure of a polypeptide? (Write the structure of this residue.)
(b) With the aid of Table 16.1, write the name and the three-letter symbol of this amino acid.
(c) Is its side chain hydrophobic or hydrophilic?

16.3 What structure will nearly all the molecules of glycine have at a pH of about 1?

16.4 What structure will most of the molecules of alanine have at a pH of about 12?

16.5 Pure alanine does not melt, but at 290 °C it begins to char and decompose. However, the ethyl ester of alanine, which has a free NH_2 group, has a low melting point, 87 °C.

Write the structure of this ethyl ester, and explain this large difference in melting point.

16.6   Which of the following amino acids has the more hydrophilic side chain? Explain.

$$^+NH_3CHCO_2^-$$
$$|$$
$$CH_2CH_2CH_2NHCNH_2$$
A

$$NH$$
$$||$$

$$^+NH_3CHCO_2^-$$
$$|$$
$$CH_3CHCH_2CH_3$$
B

16.7   Which of the following amino acids has the more hydrophobic side chain? Explain.

$$^+NH_3CHCO_2^-$$
$$|$$
$$CH_2OH$$
A

$$^+NH_3CHCO_2^-$$
$$|$$
$$CH_2C_6H_5$$
B

16.8   One of the possible forms for histidine, one of the standard 20 amino acids, is

$$^+NH_3CHCO_2^-$$
$$|$$
$$CH_2$$

(a) Is histidine most likely to be in this form at pH 1 or at pH 11? Explain.
(b) Would histidine in this form migrate to the positive electrode or to the negative electrode, or would it not migrate at all in an electrolysis apparatus?
(c) Is the side chain in histidine hydrophobic or hydrophilic?

16.9   Glutamic acid can exist in the following form.

$$NH_2CHCO_2^-$$
$$|$$
$$CH_2CH_2CO_2^-$$

(a) Would this form predominate at a pH of 2 or a pH of 10? Explain.
(b) To which electrode—negative, positive, or neither— would aspartic acid in this form migrate in an electrolysis apparatus?

16.10   When it is said that a substance is poorly soluble in water because of a *hydrophobic interaction*, what does "hydrophobic interaction" mean?

16.11   What kind of a reactant is required to convert cysteine into cystine, an acid, base, oxidizing agent, or reducing agent?

*16.12   Write two equilibrium equations that show how glycine, in its isoelectric form, can serve as a buffer.

Polypeptides

*16.13   Each of the following structures has an amide linkage. Each can be hydrolyzed to glycine and lysine. The amide linkage in one of the two structures, however, cannot properly be called a *peptide bond*. This is true of which structure? Why?

$$^+NH_3CH(CH_2)_4NHCCH_2$$
$$| \qquad\qquad\quad |$$
$$CO_2^- \qquad\quad NH_3^+$$

$$O$$
$$||$$

A

$$O$$
$$||$$
$$^+NH_3CH_2CNHCHCO_2^-$$
$$|$$
$$(CH_2)_4NH_2$$
B

16.14   Write both the conventional and the condensed structure (three-letter symbols) of the dipeptides that can be made from lysine and cysteine.

16.15   What are the condensed structures of the dipeptides that can be made from glycine and glutamic acid? (Do not use the three-letter symbols.)

16.16   Using three-letter symbols, write the structures of all of the tripeptides that can be made from lysine, glutamic acid, and cysteine.

16.17   Write the structures in three-letter symbols of all of the tripeptides that can be made from glycine, cysteine, and alanine.

*16.18   What is the conventional structure of Val-Ile-Phe?

*16.19   Write the conventional structure for Val-Phe-Ala-Gly-Leu.

*16.20   Write the conventional structure for Asp-Lys-Glu-Thr-Tyr.

*16.21   Compare the side chains in the pentapeptide of Review Exercise 16.19 (call it **A**) with those in the following, which we can call **B**.

Lys-Glu-Asp-Thr-Ser

(a) Which of the two, **A** or **B**, is the more hydrocarbon-like?
(b) Which is probably more soluble in water? Explain.

*16.22   Compare the side chains in the pentapeptide of Review Exercise 16.20, which we'll label **C**, with those in Phe-Leu-Gly-Ala-Val, which we can label **D**. Which of the two would tend to be less soluble in water? Explain.

Higher Levels of Protein Structure

16.23   Which *level* of polypeptide complexity concerns the molecular "backbone" and the sequence of side chains?

16.24   What is meant by *native* protein?

16.25   To what level of protein complexity is the disulfide bond normally assigned?

16.26   An enzyme consists of two polypeptide chains associated together in a unique manner. To what level of protein structure is this detail assigned?

16.27 Describe the specific geometrical features of an α-helix structure. What force of attraction stabilizes it? Between what two kinds of sites in the α-helix does this force operate? How do the side chains become positioned in the α-helix?

16.28 Give a brief description of the secondary structure of an individual polypeptide strand in collagen.

16.29 What function does ascorbic acid (vitamin C) perform in the formation of strong bones?

16.30 What specific force of attraction stabilizes a β-pleated sheet? Where do the side chains take up positions?

16.31 Does an α-helix or a β-sheet describe the *entire* secondary structure of a polypeptide? If not, how do these features occur?

16.32 What factors affect the bending and folding of α-helices in the presence of an aqueous medium?

16.33 What is meant by a salt bridge?

16.34 When is the disulfide bond normally put into place during the formation of a protein?

16.35 In what way does hemoglobin represent a protein with quaternary structure (in general terms only)?

16.36 How do myoglobin and hemoglobin compare (in general terms only)?
(a) Structurally, at the quaternary level?
(b) Where they are found in the body?
(c) In terms of their prosthetic group(s)?
(d) In terms of their functions in the body?

Properties of Proteins

*16.37 What products form when the following polypeptide is completely digested?

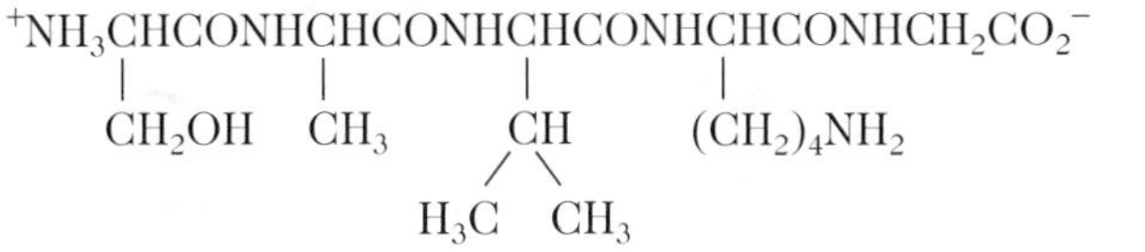

16.38 Explain why a protein is least soluble in an aqueous medium that has a pH equal to the protein's p*I* value.

16.39 What is the difference between the *digestion* and the *denaturation* of a protein?

Cell Membranes

16.40 What is the most common kind of protein in cell membranes?

16.41 What is meant by "gradient" in the term *concentration gradient?*

16.42 Which has the higher level of Na^+, plasma or cell fluid?

16.43 Does cell fluid or plasma have the higher level of K^+?

16.44 In which fluid, plasma or cell fluid, would the level of sodium ion increase if the sodium ion gradient could not be maintained?

16.45 What does the sodium–potassium pump do?

16.46 What does "active" refer to in the term *active transport?*

16.47 What is a *gap junction* and what services does it perform?

Types of Proteins

16.48 What experimental criterion distinguishes between fibrous and globular proteins?

16.49 What is the relationship between collagen and gelatin?

16.50 How are collagen and elastin alike? How are they different?

16.51 What experimental criterion distinguishes between the albumins and the globulins?

16.52 What is fibrin and how is it related to fibrinogen?

Sickle-Cell Anemia (Special Topic 16.1)

16.53 What is the primary *structural* fault in the hemoglobin of sickle-cell anemia?

16.54 What happens in blood cells in sickle-cell anemia that causes their shapes to become distorted?

16.55 What problems are caused by the distorted shapes of the red cells?

Mifepristone (RU 486) (Special Topic 16.2)

16.56 What is meant by a "receptor protein?"

16.57 In general terms only, how does RU 486 work in the early postcoital period?

Additional Exercises

16.58 Write the structure of a pentapeptide that would hydrolyze to give only alanine.

*16.59 Consider the following structure.

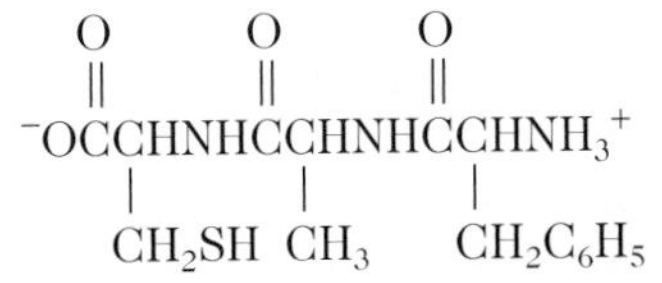

(a) If a polypeptide were *partially* hydrolyzed, could a molecule of this structure possibly form in theory? Explain.
(b) What is the three-letter symbol of the N-terminal residue?
(c) How would the structure of this compound be represented using the three-letter symbols and following the rules for writing such a structure?
(d) Would a mild reducing agent have any effect on this compound? If so, write the structure of the product.

17

ENZYMES, HORMONES, AND NEUROTRANSMITTERS

The enzymes and neurotransmitters that try to keep this woman from being killed must function with blitzing speed. Your system has this speed, too, although your brain probably says, "Don't ever do this."

17.1 ENZYMES

The catalytic abilities of enzymes often depend on cofactors that are made from B vitamins.

Nearly all of the body's thousands of catalysts, its *enzymes,* are proteins. A few, like ribonuclease P, have enzymatic activity caused solely by a nucleic acid, but we will not study them. We begin our study with some general properties of enzymes.

Enzymes Are Unusually Specific for Substrates.   Enzymatic activity begins when molecules of enzyme and its reactant, a **substrate**, physically fit together. Because each substrate has molecules with shapes unique to it, *each reaction needs its own specific enzyme.*

   Few enzymes are so specific that they catalyze only one particular reaction of one compound. Most enzymes possess *relative specificity.* Enzymes that catalyze the hydrolysis of esters, for example, usually handle a variety of esters, not just one, but they usually do not catalyze the hydrolysis of other functional groups.

   ■ All enzyme molecules and most substrate molecules are chiral; they have handedness.

Enzymes Display Remarkable Rate Enhancements.   Enzymes, like all catalysts, affect the *rates* of reactions by providing a reaction pathway with a lower energy of activation than the uncatalyzed reaction can take (see page 117). Even small reductions of energy barriers can cause spectacular increases in rates as illustrated by *carbonic anhydrase.* This enzyme catalyzes the equilibration of bicarbonate ion and protons, on the one hand, with carbon dioxide and water, on the other:

   ■ Many enzymes have been isolated for use in causing chemical reactions *in vitro.*

$$CO_2 + H_2O \xrightleftharpoons[\text{carbonic anhydrase}]{} HCO_3^- + H^+$$

In actively metabolizing cells, where the supply of CO_2 is relatively high, the equilibrium shifts to the right. In blood circulating through the lungs, where exhaling keeps the supply of CO_2 low, the equilibrium must shift to the left, *and the same enzyme participates in this change.* Each molecule of carbonic anhydrase aids in the conversion of 600,000 molecules of CO_2 *each second!* This is ten million times faster than the uncatalyzed reaction, which makes the speed of action of carbonic anhydrase among the highest of all known enzymes.

   ■ The equilibrium *must* shift to the right to make HCO_3^- and H^+ when the supply of CO_2 is high—a consequence of Le Châtelier's principle.

Enzymes Establish Equilibria Rapidly.   It is important to remember that a catalyst speeds up *equilibration,* meaning the reestablishment of an equilibrium that has somehow been upset. A catalyst accelerates *both* the forward and the reverse reactions. Whether the equilibrium shifts to the right or to the left doesn't depend on the catalyst at all. It depends strictly on other factors, namely, the temperature, the nature of the equilibrium, the relative concentrations of reactants and products, and how other reactions might feed substances into the equilibrium or continuously remove them. All the catalyst does is make whatever *shift* in equilibrium is mandated by changes in these conditions to occur very rapidly.

Most Enzymes Consist of Polypeptides plus Cofactors.   The molecules of most enzymes include a nonpolypeptide component called a **cofactor**. The polypeptide portion itself is called the **apoenzyme,** but without the cofactor there is no enzymatic activity.

   ■ Cofactors are *prosthetic groups* (page 365).

   The cofactors of many enzymes are simply the trace metal ions of nutrition. Zn^{2+} is the metal ion in carbonic anhydrase, for example. Fe^{2+} occurs in the cytochromes, a family of enzymes involved in biological oxidations. Ions of copper, cobalt, manganese, and chromium are also cofactors. In other enzymes the cofactor is an organic molecule or ion called a **coenzyme.** Some enzymes have both a metal ion cofactor and a coenzyme. Coenzymes are usually made out of vitamin molecules. **Vitamins** are organic compounds that our bodies must have but cannot make. They *must* be in the diet because their absence causes *vitamin deficiency diseases,* like scurvy and beriberi. Vitamins occur only in trace concentrations in foods and are not classified as carbohydrates, lipids, or proteins.

B Vitamins Are Used to Make Coenzymes.   Thiamine diphosphate, a coenzyme with structure **1**, is a diphosphate ester of thiamine, a B vitamin, shown here in its fully ionized form.

■ When the diet is deficient in thiamine, often called vitamin B$_1$, a disease called *beriberi* results.

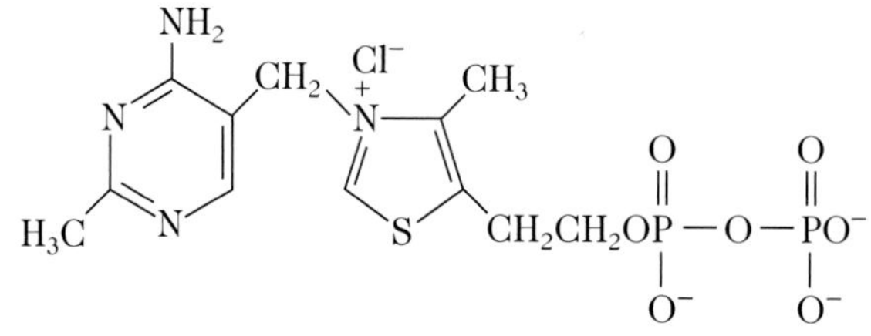

1
Thiamine diphosphate

■ Nicotinamide's other name is *niacin.* A deficiency of this vitamin leads to *pellegra.*

When pure, this coenzyme is a triprotic acid, but at the pH of body fluids it is ionized approximately as shown.

Nicotinamide, another B vitamin, is part of the structure of nicotinamide adenine dinucleotide, **2a,** another important coenzyme. Mercifully, its long name is usually shortened to NAD$^+$ (or, sometimes, just NAD). It is the nicotinamide unit that provides the positive charge in NAD$^+$.

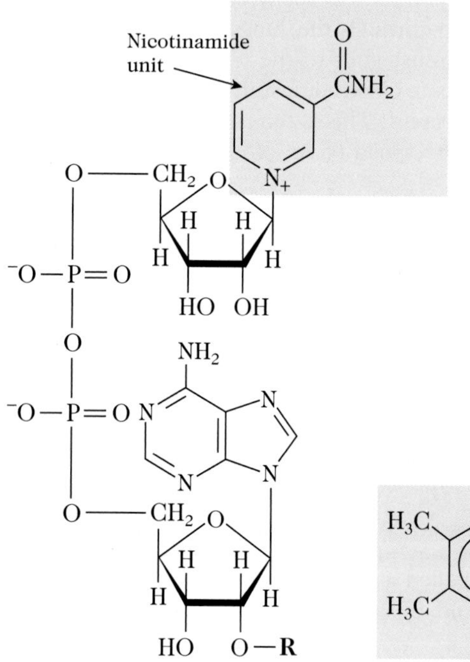

2

2a   NAD$^+$ **R** = H
2b   NADP$^+$ **R** = PO$_3{}^{2-}$

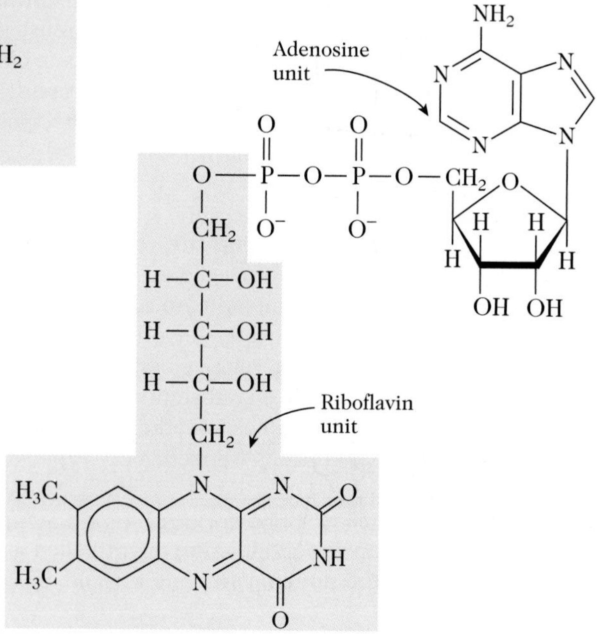

3

FAD

Quite often equations that involve enzymes with recognized coenzymes are written with the symbol of the coenzyme as a reactant or product. NAD^+, for example, is the cofactor for the enzyme that catalyzes the body's oxidation of ethyl alcohol to acetaldehyde. It serves, in fact, as the actual acceptor of the hydride ion, $H:^-$, given up by ethyl alcohol. The overall equation is written simply as follows.

$$CH_3CH_2OH + NAD^+ \rightarrow CH_3CH{=}O + NAD:H + H^+$$

Ethyl alcohol   Acetaldehyde   Reduced form of NAD^+ Hydrogen ion (buffered)

When the symbol of a coenzyme is used in an equation, remember that it stands for the entire enzyme that bears the coenzyme. In this reaction, the NAD^+ unit in the enzyme accepts $H:^-$ from the alcohol, and we can write this part of the reaction by the following equation.

$$NAD^+ + H:^- \rightarrow NAD:H$$

By accepting the *pair of electrons* in $H:^-$, NAD^+ is reduced to $NAD:H$ (usually written as NADH), called the *reduced form* of NAD^+. $NADP^+$ can also accept hydride ion, and its reduced form is written as NADPH.

■ The *P* in $NADP^+$ refers to the extra phosphate ester unit.

Notice that the oxidation of ethyl alcohol by NAD^+ also gives H^+, which is neutralized by the buffers in the surrounding fluids. When NADH is a *reactant*, the proton is needed again and is readily taken back from the buffer system. NADH and H^+, in other words, make a pair either as products or as reactants, so to simplify matters, the symbol $NADH/H^+$ is sometimes used in the place of NADH + H^+.

We learned earlier that we may not call something a catalyst unless it undergoes no *permanent* change, which the foregoing examples seem to contradict. In the body, however, a reaction that alters an enzyme is followed at once by one that regenerates it. Thus, the $NADH/H^+$ that forms in the oxidation of ethyl alcohol by NAD^+ is changed back to NAD^+ in the next step, a step whose enzyme has the cofactor FAD (for flavin adenine dinucleotide), **3**. FAD incorporates still another B vitamin, *riboflavin*. FAD, for example, can accept hydrogen from $NADH/H^+$, change to $FADH_2$, and so regenerate NAD^+. The overall reaction is

■ Riboflavin is vitamin B_2.

$$NADH/H^+ + FAD \rightarrow NAD^+ + FADH_2$$

The FAD-containing enzyme is, of course, now in its reduced form, $FADH_2$. $FADH_2$ hands on its load of hydrogen and electrons in yet another step and so is reoxidized and restored to FAD. The steps continue, but we'll stop here. The main points are that B vitamins are key parts of coenzymes and that the *catalytic sites* of the associated enzymes are contributed by these vitamins.

Flavin mononucleotide or FMN, a near relative of FAD, also contains riboflavin. (FMN is FAD minus the adenosine unit.) The reduced form of FMN is $FMNH_2$, and FMN is also involved in biological oxidations.

Enzymes Are Named after Their Substrates or Reaction Types. Nearly all enzymes have names that end in *-ase*. The prefix is either from the name of the substrate or from the kind of reaction. For example a **hydrolase** catalyzes hydrolysis reactions. An *esterase* is a hydrolase that aids the hydrolysis of esters. A *lipase* works on the hydrolysis of lipids. A *peptidase* or a *protease* catalyzes the hydrolysis of peptide bonds.

■ Whenever we see *-ase* as a suffix in the name of any substance or type of reaction, the word is the name of an enzyme.

Enzymes with NAD^+, $NADP^+$, FAD, or FMN are examples of **oxidoreductases** because they catalyze redox equilibria. Sometimes an oxidoreductase is called an *oxidase* when the favored reaction is an oxidation and a *reductase* when the reaction is a reduction. A **transferase** catalyzes the transfer of a group from one molecule to another, and a *kinase* is a special transferase that handles phosphate groups. Other

■ The digestive enzymes *trypsin*, *chymotrypsin*, and *pepsin*, all peptidases, have old (nonsystematic) names that do not end in *-ase*.

■ ATP is adenosine triphosphate, an important, high-energy triphosphate ester (see page 302).

broad categories of enzymes are the **lyases,** which catalyze elimination reactions that form double bonds; **isomerases,** which cause the conversion of a compound into an isomer; and **ligases,** which cause the formation of bonds at the expense of chemical energy in triphosphates, like ATP.

Enzymes Often Occur as a Family of Similar Compounds Called *Isoenzymes* **with Identical Functions.**   Identical reactions are often catalyzed by enzymes with identical cofactors but slightly different apoenzymes. Biological catalysts that vary slightly in structure but have the same function are called **isoenzymes.**

■ Here, *iso-* signifies the same catalytic function, not identical formulas. Some references refer to isoenzymes as *isozymes.*

Creatine kinase, abbreviated CK, for example, consists of two polypeptide chains labeled *M* (for skeletal muscle) and *B* (for brain). CK occurs as three isoenzymes. All catalyze the transfer of a phosphate group (to a compound called creatine). One CK isoenzyme, called CK(MM), has two *M* units and occurs in skeletal muscle. Another, CK(BB), has two *B* units and occurs in brain tissue. The third, CK(MB), has one *M* and one *B* polypeptide, and it is present almost exclusively in heart muscle, where it accounts for 15 to 20% of the total CK activity. The rest is contributed by CK(MM).

We have given this much detail about creatine kinase because it and similar sets of isoenzymes have an important function in clinical analysis and medical diagnosis, as we'll see later in the chapter.

17.2   ENZYME–SUBSTRATE COMPLEX

The chirality and flexibility of an enzyme and the side chains of its amino acid residues permit only the enzyme's substrate to fit.

When an enzyme catalyzes a reaction of its substrate, molecules of each must momentarily fit to each other and form a temporary combination called an **enzyme–substrate complex.** The complex is one component of a series of chemical equilibria that carry the substrate through a number of changes until the final products form.

$$E + S \rightleftharpoons E\text{–}S \rightleftharpoons E\text{–}S^* \rightleftharpoons E\text{–}P \rightleftharpoons E + P$$

Enzyme Substrate   Enzyme–substrate complex, *ES*   Activated *ES* complex   Enzyme–product complex   Enzyme (recovered) Product

■ A hormone uses a flexible lock-and-key kind of recognition to ensure that it is taken up only by its intended target cells.

According to current theory, the first step is the binding of the enzyme to the substrate. Because this step resembles the fitting of a key (substrate) to a tumbler lock (enzyme), the theory is often called the **lock-and-key theory** of enzyme action. Shaped pieces that fit together are said to have *complementary shapes,* or we say that there is *complementarity* between the two shapes (Figure 17.1). Special Topic 17.1 discusses the importance of molecular complementarity to such diverse functions as the immune system and the ABO blood groups.

Two kinds of complementarity are required for an enzyme–substrate complex to form. The first is *geometrical complementarity:* a square peg fits a square hole better than a round hole, for example. The other is *physical complementarity,* which concerns factors other than shape—hydrophobic interactions, hydrogen bonds, and electrical charges of *opposite* nature nestling *nearest* each other as the complex forms.

To get the substrate to fit to the enzyme depends on some flexibility in the enzyme molecule, much as the tumblers inside a lock adjust when a key is inserted. When the substrate molecule nestles onto the enzyme, the molecular groups of the substrate induce the enzyme molecule to adjust its shape to achieve the best fit (Figure 17.2). The initial contact with substrate and enzyme may cause changes in

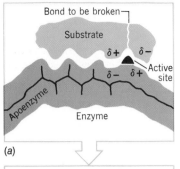

(a)

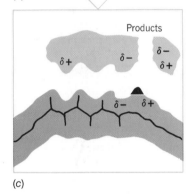

(b)

(c)

FIGURE 17.1
Lock-and-key model for enzyme action. *(a)* The enzyme and its substrate fit together to form an enzyme–substrate complex. *(b)* A reaction, such as the breaking of a chemical bond, occurs. *(c)* The product molecules separate from the enzyme.

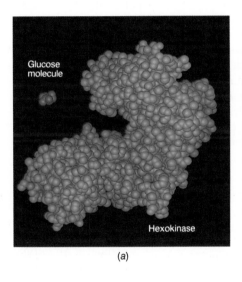

(a)

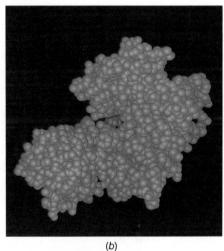

(b)

FIGURE 17.2
Induced-fit theory. *(a)* A molecule of an enzyme, hexokinase, has a gap into which a molecule of its substrate, glucose, can fit. *(b)* The entry of the glucose molecule induces a change in the shape of the enzyme molecule, which now surrounds the substrate entirely. (Courtesy of T.A. Steitz, Department of Molecular Biophysics and Biochemistry, Yale University, New Haven, CT.)

tertiary structure in the polypeptide of the protein. Such changes, which induce stress in the polypeptide, force the enzyme to modify its shape further. This **induced-fit** mechanism of enzyme action is now viewed as a modification of the lock-and-key theory, because induced fitting appears to be a general characteristic of how enzymes work. Once a proper fit is achieved, a reactive portion of the substrate molecule is poised over the enzyme's *catalytic site,* that part of the enzyme that is directly involved in catalysis.

17.3 REGULATION OF ENZYMES

Enzymes are switched on and off by inhibitors, antimetabolites, genes, poisons, hormones, and neurotransmitters.

A cell cannot be doing everything at once. Some of its possible reactions have to be shut down while others occur. One way to keep a reaction switched off is to prevent its enzyme from forming, as we will learn when we study how genes can be controlled (see Chapter 21). Hormones and neurotransmitters (see Section 17.5) also regulate enzymes. Still other regulatory mechanisms operate, which we'll study here.

■ The prevention of enzyme synthesis often involves the regulation of genes and their work of directing the synthesis of polypeptides.

The *immune system,* as large and complex as the nervous system, is the body's array of defenses against *pathogens,* disease-causing microorganisms and viruses. We cannot in a brief Special Topic do justice to the immune system, of course, but we can take note of some of the ways in which it shares basic operating principles with the enzyme–substrate reaction. The concept of the fitting of a substrate to an enzyme by means of geometrical and physical complementarity is at the molecular base of the body's immune system, as well as the existence of blood type groups.

When a pathogen has penetrated the first line of defense, the physical barriers of skin and mucous membranes, white blood cells known as *lymphocytes* go into action. They all begin life in bone marrow, but not all mature there. Two kinds of immunity involving two kinds of lymphocytes are recognized. One is *cellular immunity,* and it is handled by *T lymphocytes* or *T cells* (after *t*hymus tissue, where T cells mature). Cellular immunity handles viruses that have gotten inside cells, as well as parasites, fungi, and foreign tissue.

AIDS   The human immunodeficiency virus (HIV) is able to destroy certain kinds of T cells, the *helper T cells.* This renders the immune system deficient in its ability to handle infections and results in AIDS, acquired immune deficiency syndrome. Relatively nonlethal diseases normally handled routinely by the body thus become lethal in persons with AIDS.

Antigen–Antibody Reaction   The second kind of immunity is *humoral immunity* (after an old word for fluid, *humor*). Humoral immunity is the responsibility of the *B lymphocytes* or *B cells* (because they mature in *b*one marrow). B cells act mostly against bacterial infections but also against those workings of viral infections that occur outside of cells. We'll limit the continuing discussion to the work of B cells.

B cells carry and manufacture *antibodies,* glyco-

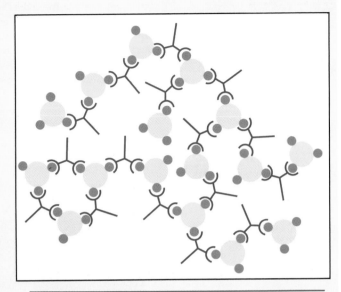

FIGURE 1
Molecules of the antibody crosslink with antigen particles to form a mass resembling a huge copolymer.

proteins that are able, by an interaction like that between substrate and enzyme, to attract and take antigens out of circulation and defeat the spread of the pathogen. An *antigen* is any molecular species or any pathogen that induces the immune system to make antibodies as well as gives the immune system a molecular-cellular memory for the antigen. Thus, at a later invasion of the same antigen, the immune system is poised for a far more rapid defensive response than it initially had. A *vaccine* is able to start the initial defensive response leading to the molecular memory for the antigen without causing the disease itself.

Figure 1 represents an antibody that is *dipolar;* it has two crosslinking molecular groups. The antigen in Figure 1 is represented by a unit that can become bound to at least three antibody binding sites. The

Enzymes Can Be Activated by the Chemical Removal of an Inhibitor Molecule. Some enzymes are kept in an inactive form by the work of an **inhibitor,** a substance that combines with the enzyme to cover the catalytic site or to change the enzyme's shape. Inhibitors whose molecules compete with substrate molecules for fitting to the catalytic site are called *competitive inhibitors. Noncompetitive inhibitors* bind to the enzyme somewhere other than its catalytic site, causing a distortion of the enzyme's overall shape and so making the fit of substrate to enzyme impossible. Enzyme inhi-

TABLE 1 Blood Type Acceptor–Donor Options

If Your Blood Type Is	You Can Accept Blood from One of This Type	You Can Donate Blood to One of This Type
O	O	O, A, B, AB
A	A or O	A or AB
B	B or O	B or AB
AB	AB, A, B, O	AB

antibody protein is specific for just one antigen. As you can see, the interaction of antibody with its antigen essentially "polymerizes" the entire system into one vast "copolymer." The product is now in a far less soluble form and it bears molecular markings that are recognized by other white cells (phagocytes) that engulf the "polymer" and destroy it. There are some antigen–antibody complexes that are destroyed by a series of interacting proteins called the *complement system*. By tying up the antigens, the antibodies prevent the spread of the infection and thus allow the system the time needed to destroy the antigens.

The ABO Blood Groups The surfaces of red blood cells carry projecting carbohydrate units of glycolipids. There are differences, however, among individuals in the structures of these sugar residues. One of the consequences of these differences is the existence of blood group systems, one being the *ABO system*. You might have type A, type B, or type O blood. Some have a combination type, AB blood.

If you are of type A and by a transfusion are given blood from a type B person, the red cells from the type B blood will clump together (agglutinate), likely causing a blockage of blood capillaries that could be fatal. Thus your type A blood is able to "see" something in type B blood as a foreign material. Type A blood contains in the serum portion (the liquid minus the cells) an antibody against type B blood. What specifically is the antigen in the type B blood is the molecular unit at the tip of a multiring carbohydrate unit joined to a glycolipid of the (type B) red cell. This is why each kind of red cell is described as carrying an *antigen*, one of three types, A, B, and H.

If you are of type A, your serum includes anti-B antibodies. People with type B blood have anti-A antibodies. AB-type people with AB blood have neither anti-A nor anti-B antibodies and so are able to accept transfusions from people of any blood type. (However, in all but emergency situations, transfusions are normally done using type AB blood.) Type O people carry the H "antigen" on their red cells and their blood has *both* anti-A and anti-B antibodies. Type O people, therefore, can *receive* transfusions only from individuals with type O blood. At the same time, type O people can *give* transfusions to all types—they are universal donors—because the antigen on the red cells in type O people actually has no "enemies" in other types of blood, no antibodies that can attack and agglutinate type O red cells when they are transfused into people of other blood types.

The H antigen in type O blood is given the name antigen because it is the precursor to the A and B antigens of other blood types. Type A individuals make type A antigen by adding a modified glucose residue to the tip of a glycolipid on the red cell. Type B people make type B antigen by adding a galactose residue to the same glycolipid. Type O individuals simply lack the enzymes needed for these transformations. The differences among these enzymes are thought to involve single amino acid residue substitutions in the enzymes' polypeptides. Table 1 summarizes donor–acceptor relationships for the blood types.

bition is normally used only for temporary control. In such *reversible inhibition,* inhibitors are removed when no longer needed. Some poisons, however, cause *irreversible inhibition.*

An inhibitor can be a *product* of the very reaction that the enzyme catalyzes or a product of a reaction farther down a successive series of reactions. The action is called **feedback inhibition** because, as the inhibitor's concentration increases, its molecules more and more react with ("feed back" to) and inhibit the molecules of ar enzyme needed to make them. The amino acid isoleucine, for example, is made from another amino acid, threonine, by a series of steps, each with its own enzyme. Molecules of isoleucine are able to act as reversible inhibitors of the enzyme for the first step, E_1.

$$^+NH_3CHCO_2^- \xrightarrow{E_1} \xrightarrow{E_2} \xrightarrow{E_3} \xrightarrow{E_4} \xrightarrow{E_5} {}^+NH_3CHCO_2^-$$

$$\underset{\text{Threonine}}{\overset{|}{\underset{CH_3}{\overset{|}{CHOH}}}} \qquad \begin{array}{c} \textit{Inhibition of } E_1 \\ \textit{by molecules} \\ \textit{of isoleucine} \end{array} \qquad \underset{\text{Isoleucine}}{\overset{|}{\underset{CH_2CH_3}{\overset{|}{CHCH_3}}}}$$

When sufficient quantities of isoleucine have been made, their further synthesis stops, because all of E_1 is inactivated.

The beautiful feature of feedback inhibition is that the system making a product shuts down automatically when enough is made. Then, as the cell consumes this product, even those of its molecules serving as inhibitors are consumed. Now the enzyme is released to help make more product The phenomenon, very common in nature, is sometimes called **homeostasis,** the response of an organism to a stimulus that starts a series of events that restore the system to the original state.

An inhibitor doesn't have to be a product of the enzyme's own work. It can be something else the cell makes or it could be a medication. What its molecules must do is bind, without reacting, somewhere on the enzyme, and they must be removable as needed.

Substrate Molecules Can Activate Enzymes.   Some enzymes have more than one active site. When no substrate is around, the enzyme adopts a configuration in which the shapes of *both* active sites are poorly matched to substrate molecules. When substrate concentration builds up, however, their molecules, beating against the enzyme, manage to force entry onto one active site. As this successfully happens, *the whole enzyme alters in shape, and the other active site becomes activated.* Now *both* active sites are well matched to the substrate, and they continue to work as long as substrate is present. Here, then, is another example of how the activity of an enzyme is related to the concentration of a substrate.

Nerve Signals Can Cause Nonsubstrate Molecules to Activate Enzymes.   Nonsubstrate molecules that activate enzymes are called **effectors,** and *calmodulin* and *troponin* are two important examples. Both are proteins, with calmodulin present in most cells and troponin largely in muscle cells, like heart muscle. They are able to activate enzymes needed for chemical work initiated by nerve signals, like muscle contraction.

For calmodulin or troponin to be effectors, they must first combine with calcium ion. The problem is that this ion is normally almost entirely absent from the fluid of the cytosol. Calcium ion cannot occur in this fluid at a concentration higher than about $10^{-7} M$, because the fluid has a level of phosphate ion of about $10^{-3} M$. This is sufficiently high to make all Ca^{2+} ion insoluble as $Ca_3(PO_4)_2$, except for that in a concentration of $10^{-7} M$. So calcium ions have to be sequestered within storage compartments of the cytoplasm until the instant they are needed. Calcium ions in the bloodstream constitute another supply that cells draw on as needed.

In electrically excited cells (e.g., nerve and muscle cells), a nerve signal establishes the need for Ca^{2+} ion in the cytosol. The signal opens a *calcium channel* through a membrane, releasing Ca^{2+} ions for travel, and they bind to calmodulin or troponin. The binding of Ca^{2+} activates calmodulin or troponin, which become *effectors* and able to activate other enzymes. Once the other enzymes do their work,

■ Many enzymes consist of two or more polypeptide subunits *each of which has binding and catalytic sites and all of which become involved in the overall reaction.*

■ The *cytosol* is the *solution* in the cytoplasm and does not include the organelles in the cytoplasm. The *cytoplasm* consists of the entire contents of the cell outside the cell nucleus.

■ The normal range of concentration of Ca^{2+} ion in blood is 1.14 to 1.30 × 10^{-3} mol/L.

and functionally this is completed when the nerve signal is over, the calcium channels close. Then an active transport mechanism pipes calcium ions out of the cytosol, and the effector is inactivated.

To summarize, a nerve signal opens calcium channels and releases calcium ions inside a cell where they bind to and activate calmodulin or troponin. Now effectors, they activate other enzymes. When the nerve signal stops, calcium ions are removed to storage or pumped out of the cell, deactivating the effectors and their associated enzymes.

Some Enzymes Are Initially in Inactive Forms Called Zymogens. Several enzymes are first made as **zymogens,** polypeptides with molecular portions folded over active sites. When the active enzyme is needed, an activator (another enzyme) clips off these portions to expose the active sites. Circulating in the blood, for example, is a zymogen called *plasminogen.* In the next section we'll see how its conversion to the enzyme *plasmin* is important following the formation of a blood clot. Several digestive enzymes are also made as zymogens.

■ Zymogens are sometimes called *proenzymes* and other kinds of "proproteins" are known, like *proinsulin,* the precursor to the hormone *insulin,* a blood sugar regulator. The conversion of proinsulin to insulin entails the removal of a 33-residue polypeptide unit.

Antimetabolites Inhibit Bacterial Enzymes. *Antibiotics* are members of a broad family of compounds called **antimetabolites,** substances that inhibit or prevent the normal metabolism of a disease-causing bacterial system. Some antimetabolites work by inhibiting an enzyme that the bacterium needs for its own growth. Both the sulfa drugs and penicillin work this way.

■ An antimetabolite is called an *antibiotic* when it is the product of the growth of a fungus or a natural strain of bacteria.

Poisons Can Permanently Inhibit Key Enzymes. The most dangerous **poisons** are effective even at very low concentrations because they are powerful, irreversible inhibitors of enzymes. The cyanide ion is a dangerous poison, for example, that forms a strong complex with one of the metal ion cofactors in an enzyme needed for our use of oxygen. Cyanide ion is thus able to shut down cellular respiration, and death follows promptly.

Enzymes that have SH groups are denatured and deactivated by such heavy metal ions as Hg^{2+}, Pb^{2+}, Cu^{2+}, and Ag^+, all poisons. Nerve gas poisons and their weaker cousins, the organophosphate insecticides, inactivate enzymes of the nervous system.

17.4 ENZYMES IN MEDICINE

The specificity of an enzyme for its substrate provides several methods of clinical analysis and medical diagnosis.

Enzymes that normally work only inside cells are not found in the blood except at extremely low concentrations. When cells are diseased or injured, however, their enzymes spill into the bloodstream. Much can be learned about the disease or injury by detecting such enzymes and measuring their levels. In this section, we learn some examples of such medical technology.

Enzyme Assays of Blood Use Substrates as Chemical "Tweezers." Despite the enormous complexity of blood and the very low levels of enzymes in it, enzyme assays are relatively easy (at least in principle). The substrate for the enzyme is used to find its own enzyme; the specificity of the enzyme–substrate reaction ensures that the substrate finds nothing else. If no enzyme is present to match the substrate, nothing happens. Otherwise, the extent of the substrate's reaction, catalyzed by the enzyme, measures the concentration of the enzyme.

388

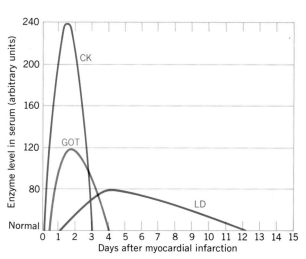

FIGURE 17.3

The concentrations of three enzymes in blood serum increase after a myocardial infarction. CK, creatine kinase; GOT, glutamate:oxaloacetate aminotransferase; LD, lactate dehydrogenase.

Viral Hepatitis Is Detected by the Appearance of GPT and GOT in Blood. Heart, muscle, kidney, and liver tissue all contain the enzyme *glutamate:pyruvate aminotransferase,* mercifully abbreviated GPT. The liver, however, has about three times as much GPT as any other tissue, so the appearance of GPT in the blood generally indicates liver damage or a virus infection of the liver, like viral hepatitis.

The level of another enzyme, *glutamate:oxaloacetate aminotransferase* or GOT, also increases in viral hepatitis, but the GPT level goes much higher than the GOT level. The ratio of GPT to GOT in the serum of someone with viral hepatitis is typically 1.6, compared with a level of 0.7 to 0.8 in healthy individuals.

■ Notice that we speak here of *ratios* of GPT to GOT, not absolute amounts, which are normally very low.

Heart Attacks Cause Increased Levels of Three Enzymes in Blood Serum. A *myocardial infarction* is the blockage by deposits, by hardening, or by a clot of one or more blood capillaries that supply heart muscle tissue with oxygen and nutrients. Withering of a portion of the heart muscle occurs, and if the patient survives, the withered muscle becomes scar tissue. If treatment is started very promptly, the outlook for survival is good. An exceptionally reliable diagnosis of an infarction can be made by the analysis of the serum for several enzymes and isoenzymes.

■ The popular term for this set of events is *heart attack*.

When a myocardial infarction occurs, the serum levels of three enzymes normally confined inside heart muscle cells—CK (page 382), GOT (just described), and LD (lactate dehydrogenase)—begin to increase (Figure 17.3). As we learned on page 382, the CK enzyme occurs as three isoenzymes, CK*(MM)*, CK*(BB)*, and CK*(MB)*, so a special technique is used to determine specifically the concentration of the heart tissue isoenzyme, CK*(MB)*. The same technique is used to separate the five isoenzymes of LD, because the relative concentrations of two of them, LD_1 and LD_2, change when an infarction occurs (Figure 17.4). Normally the LD_1 level is *less* than the LD_2 level, but following a myocardial infarction, an "LD_1–LD_2 flip" occurs, and their concentrations reverse. When both the CK*(MB)* band and the LD_1–LD_2 flip are found, the diagnosis of a myocardial infarction is essentially 100% certain.

In the CHEMSTRIP Matchmaker device, the blood glucose level is measured by the intensity of a dye produced enzymatically and converted into milligrams of glucose per deciliter (100 mL) of blood.

The Blood Glucose Level Can Be Determined Enzymatically. The regular determination of the status of glucose in blood is important to people with diabetes, because when the blood glucose level is poorly managed, several complications can occur. One commercially available test uses a combination of chemicals, including enzymes, that react with blood glucose to generate a dye on a test surface. The intensity of the resulting dye is proportional to the blood glucose level.

Similar enzyme-based tests are available to measure the serum levels of urea, triacylglycerols, bilirubin (a breakdown product of hemoglobin), and other compounds.

■ The technology for measuring the glucose status of the blood has been changing rapidly.

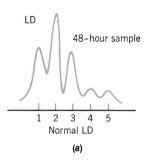

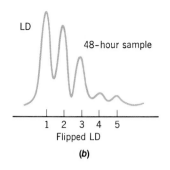

FIGURE 17.4
Lactate dehydrogenase isoenzymes. *(a)* The normal pattern of the relative concentrations of the five isoenzymes. *(b)* The pattern after a myocardial infarction. Notice the LD_1–LD_2 flip, the reversal in relative concentration between LD_1 and LD_2.

A Natural Blood Clot-Dissolving Enzyme Can Be Activated by Other Enzymes. If you cut yourself, there is set in motion a huge cascade of enzyme-catalyzed reactions that bring about the formation of *fibrin* from a circulating polypeptide, *fibrinogen*. The long, stringy fibrin molecules form a brush-heap mat that entraps water and puts a seal, a blood clot, on the cut. After the wound heals, the clot must be dissolved. No part of a blood clot must break loose and circulate to the heart, because it will be stopped by tiny capillaries in heart muscle tissue. Clots in the lungs are also very serious.

To dissolve the fibrin of a clot, the body uses a special enzyme, *plasmin*, which normally circulates in an inactive form called *plasminogen*, a zymogen (see page 387). Even as a clot forms, circulating plasminogen starts being absorbed by it. Thus, the conversion of plasminogen to plasmin will occur exactly where plasmin is needed to catalyze the hydrolysis of fibrin and so make the clot "dissolve." The activation of plasminogen to plasmin involves two factors, the binding to fibrin itself plus the work of a circulating material called *tissue plasminogen activator* (TPA).

As we mentioned, the formation of a clot in a heart muscle capillary is one cause of a myocardial infarction. The most widely used treatment is to speed up the activation of plasminogen to plasmin by injecting something that will more quickly launch the work of TPA. Such *plasminogen activation therapy* is intended to open blocked capillaries as rapidly as possible before the oxygen starvation of surrounding heart muscle tissue spreads the damage. The sooner this therapy is applied following an infarction, the better the chances of survival and of only slight damage to the heart muscle.

Two enzymes are most frequently used for plasminogen activation therapy, streptokinase and synthetic TPA. (Aspirin and heparin are sometimes given concurrently.) TPA acts more rapidly than streptokinase, so given the urgency of reopening blocked heart vessels, TPA has a higher percentage of infarct survivors than streptokinase. There is a huge difference in cost, however; in the early 1990s, TPA cost about $2200 per dose, but the cost of streptokinase was just under $200 per dose.

■ About 150,000 people a year die from clots that start in their lungs.

■ Tissue plasminogen activator made by genetic engineering (see Section 21.5) is referred to as *recombinant* tissue plasminogen activator.

■ This therapy is called *thrombolytic therapy* because it *lyses* (breaks down) *thrombi* (blood clots).

■ Heparin is an *anticoagulant,* a substance that interferes with the formation of a blood clot.

■ Aspirin increases the effectiveness of plasminogen activation.

17.5 CHEMICAL COMMUNICATION— AN INTRODUCTION

Hormones and neurotransmitters are the chief methods by which cells communicate with one another.

Like any complex organism, the body is made up of highly specialized parts. Information, therefore, must flow among the parts to maintain a well-coordinated system. This flow is handled by chemical messengers, *hormones* or *neurotransmitters,* sent in response to a variety of signals.

■ From the Greek *hormon,* "arousing."

Hormones are compounds made in specialized organs, the endocrine glands, and secreted into the bloodstream and usually sent some distance away where they launch responses in their particular **target tissue** or **target cells.** The distance might be as close as another neighboring cell or as far away as 15 to 20 cm. The signal for releasing a hormone might be something conveyed by one of our senses, like light or an odor, or it might be a stress or a variation in the level of a particular substance in the blood or in another fluid. Insulin, for example, is released when the level of glucose in blood increases.

Neurotransmitters are chemicals made in nerve cells, called *neurons,* and sent to the next nerve cells. Thus, the distinction between hormones and neurotransmitters largely concerns a difference in how far they go to exert their action. How these two kinds of messengers cause what they do when they arrive at target cells, however, bears many similarities.

■ We can imagine the unique long-chain carbohydrate units of the membrane's glycoproteins as fishing lines dangling unique lures at their tips that are attractive only to one kind of hormone or neurotransmitter molecule (refer back to Figure 16.9, page 371).

Target Cell Receptors Identify Chemical Messengers.   At a target cell, a hormone or neurotransmitter delivers its messages by first binding to a cell **receptor.** Each receptor, a unique glycoprotein, has molecules so structured that they normally accept only molecules of the messenger intended for them. A lock-and-key mechanism is at work. Sometimes, however, molecules of toxic substances, or virus particles, or even dangerous bacteria "recognize" a glycoprotein on the cells of one particular tissue and cause wholly unwanted changes.

Receptors sometimes serve other purposes. Receptor-like proteins, for example, recognize complementary molecular units on neighboring cells, lock to them, and so cause cells to bind together. Cell binding helps to control cell division. In cancerous tissue, cell-to-cell binding is weakened, and cancer cells more easily proliferate and even break off. They then enter circulation in blood or lymph, and so spread the cancer.

Receptor-like glycoproteins on sperm cells are able to lock to molecular units on only the ova of the same species, thanks to a lock-and-key mechanism. Antibodies "recognize" antigens in the body's immune response by a lock-and-key mechanism as we described in Special Topic 17.1. As evidence that such responses involve cellular *surfaces,* sometimes even killed bacteria or deactivated virus particles cause the body's immune system to develop antibodies. Immunity, once initiated by any mechanism, is usually retained by the body for a long time.

■ When immunity is not life-lasting, as with the influenza virus, it is because the disease-causing material undergoes mutations.

Chemical Messengers Enter Cells by Four Major Mechanisms.   The formation of a receptor—messenger complex changes the receptor structure, so that now it is activated to do something. It might be to activate a gene or an enzyme, or to alter the

FIGURE 17.5
Ways by which hormones get chemical messages into cells.

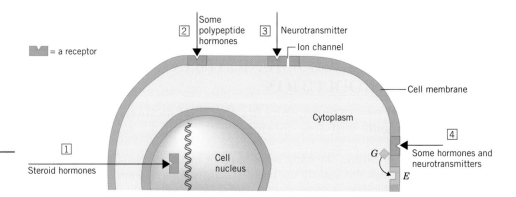

permeability of a cell membrane so that certain ions or small molecules can move across it. In nerve cells, the activation of a receptor sends the nerve signal on. We'll consider specific examples later. We offer only a broad overview here.

Figure 17.5 outlines the principal ways by which signals enter cells. Some hormones, once "recognized" by the target cell, move directly through the cell membrane, enter the cytosol, and find a receptor inside the cell. Steroid hormones work in this way. See ☐1☐ in Figure 17.5. They bind to receptors close to or inside the cell nucleus, where they induce changes in the way the cell uses DNA.

Polypeptide hormones cannot migrate directly through cell membranes, so they bind to receptors that are integral parts of the membrane (see ☐2☐ in Figure 17.5). Insulin and growth hormone work in this way.

Neurotransmitters also bind to membrane-bound receptors, and this opens channels through the membrane for metal ions, ☐3☐. We have already seen how movements of the calcium ion can affect calmodulin or troponin and so activate a series of enzyme-catalyzed reactions.

Some receptors, after they accept their neurotransmitters or hormones, are able to activate a small polypeptide called the *G-protein* (see ☐4☐ in Figure 17.5). The G-protein then activates an enzyme, *E* (see Figure 17.5). Remarkably, a large variety of cells share just a few mechanisms for taking advantage of the action of the G-protein. We'll study only one, the cyclic AMP system, to illustrate general principles.

■ The *G* in *G-protein* stands for *guanyl nucleotide-binding protein.*

The Formation and Hydrolysis of Cyclic AMP Constitute a Major Mechanism by Which Many Cells Pass On Messages. Cyclic nucleotides, particularly 3′,5′-cyclic AMP, are important *secondary chemical messengers.* How cyclic AMP works is sketched in Figure 17.6.

At the top of the figure we see a hormone—it could just as well be a neurotransmitter—that combines with an outward-facing receptor at the surface of a cell. The complex that forms activates the G-protein, which, in turn, activates the enzyme

■ *Cyclic* refers to the *extra* ring of the phosphate diester system.

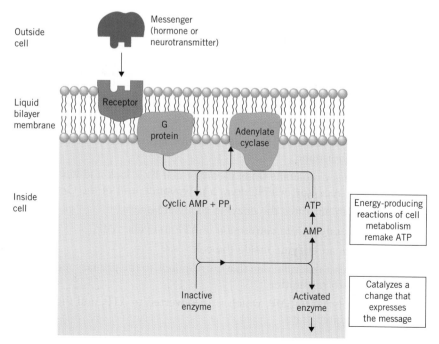

FIGURE 17.6

Activation of the enzyme adenylate cyclase by a hormone (or a neurotransmitter). The hormone–receptor complex alters a unit of the G-protein, which activates this enzyme. It then catalyzes the formation of cyclic AMP, which, in turn, activates an enzyme inside the cell.

adenylate cyclase, an inward facing protein bound on the cytosol side of the lipid bilayer. The receptor and adenylate cyclase do not touch; the G-protein mediates the hormone's signal from one to the other.

Once adenylate cyclase is activated, the "message" is on the inside of the cell membrane. Now the enzyme promptly catalyzes the conversion of ATP into cyclic AMP and diphosphate ion, PP_i.

■ The complete structure of ATP is shown on page 302.

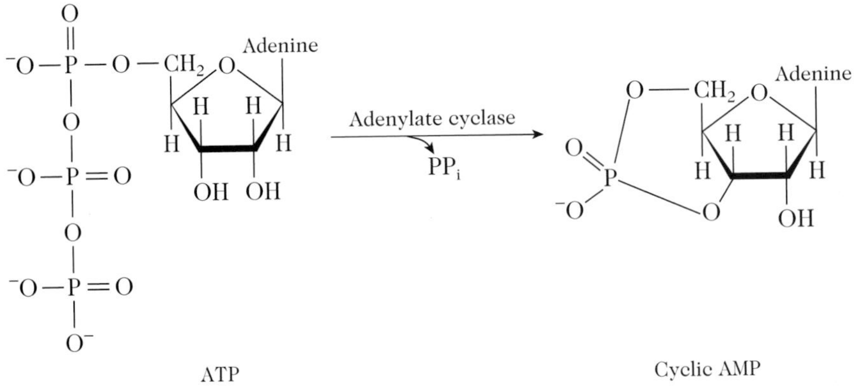

ATP   Cyclic AMP

The newly formed cyclic AMP now activates a particular enzyme that, in turn, catalyzes a specific reaction. This last event is what the original message was all about. Finally, the cycle of events is switched off by the work of *phosphodiesterase,* an enzyme that catalyzes the hydrolysis of cyclic AMP to AMP.

■ E. W. Sutherland, Jr., an American scientist, won the 1971 Nobel prize in physiology and medicine for his work on cyclic AMP.

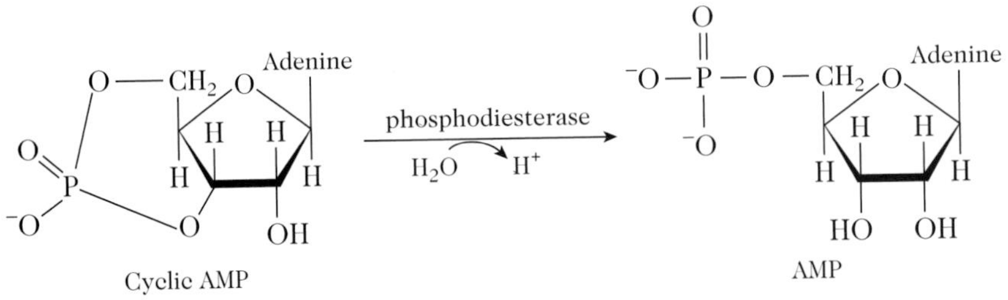

Cyclic AMP   AMP

■ Hormones that work through the cyclic AMP cascade that we will encounter later include epinephrine, glucagon, norepinephrine, and vasopressin.

Other reactions in the cell will now convert AMP to ATP and so put the cell in readiness for another outside signal. Let's summarize the steps in this remarkable chemical "cascade."

1. A signal releases the hormone or neurotransmitter.
2. It travels to its target cell, next door for a neurotransmitter but some farther distance away for a hormone.
3. The primary messenger molecule finds its target cell by a lock-and-key mechanism and binds to a receptor, which alters a polypeptide unit of the G-protein.

■ The cholera toxin is able to lock the G-protein into its active form, so adenylate cyclase cannot be shut off. This stimulates so much active transport of Na^+ ion into the gut, along with water, that massive diarrhea kills the victim.

4. The altered G-protein activates the enzyme adenylate cyclase
5. Adenylate cyclase catalyzes the conversion of ATP to cyclic AMP.
6. Cyclic AMP activates an enzyme inside the cell.
7. The enzyme catalyzes a reaction, one that corresponds to the primary message of the hormone or neurotransmitter.
8. Cyclic AMP is hydrolyzed to AMP, which is reconverted to ATP, and the system returns to the preexcited state.

17.6 HORMONES AND NEUROTRANSMITTERS

Interventions in the work of hormones and neurotransmitters are the bases of the action of a number of drugs, both licit and illicit.

Structurally, Hormones Come in Four Broad Types. It is impossible to do justice to a subject as vast as hormones in one section of one chapter, so what follows is a very broad sketch of a few chemical aspects of hormone action. We will mention specific hormones in later chapters where they are relevant to some particular metabolic activity. Let's next consider some features of hormone molecules, which come in four general types.

The steroid hormones are made from cholesterol and so have largely hydrocarbon-like molecules. This feature enables steroid molecules to slip easily through the lipid bilayers of their target cells. Inside they find their final receptors, and the hormone–receptor complexes move to DNA molecules, where they bind and affect the transcriptions of genetic messages. The sex hormones like estradiol, progesterone, and testosterone work in this way.

Many *growth factors* as well as insulin, oxytocin, and thyroid-stimulating hormone consist of polypeptides or proteins. These are able to alter the permeabilities of their target cells to the migrations of small molecules. Growth factors, for example, aid in getting amino acids inside cells where they are needed for growth. Insulin helps to get glucose inside its target cells. Either the absence of insulin or the absence (or inactivity) of insulin receptors results in the disease diabetes mellitus ("diabetes"). Several neurotransmitters are also polypeptides, and they alter the permeability of a neuron membrane to Ca^{2+} and Na^+. The cross-membrane movements of these ions are involved in the electrical signal that flows down a neuron.

The prostaglandins (see Special Topic 15.1) are classified as hormones, but as *local hormones* because they work where they are made.

Finally, a number of hormones are relatively simple amino compounds made from amino acids. These include epinephrine (page 258) and thyroxin. Some of the amines are also neurotransmitters, as we'll see next.

Neurotransmitters Move across the Narrow Synaptic Gap from One Neuron to the Next. Each nerve cell has a fiberlike part called an *axon* that reaches to the face of the next neuron or to one of its filament-like extensions called *dendrites*. A nerve impulse consists of a traveling wave of electrical charge that sweeps down the axon as small ions migrate at different rates between the inside and the outside of the neuron. The problem is how to get this impulse launched into the next neuron so that it can continue along the length of the nerve fiber. This is solved by *chemical* communication from one neuron to the next, made possible by neurotransmitters. They are made from amino acids within the neuron and stored in sacs, called *vesicles*, located near the ends of the axons (Figure 17.7). Some neurotransmitters are nothing more than simple amino acids, for example, the α-amino acids glycine and glutamic acid and γ-aminobutyric acid. Other neurotransmitters are small polypeptides and some are even monoamines, like those mentioned in Special Topic 11.4. Some substances, for example, norepinephrine, are able to function as either a neurotransmitter or a hormone, depending on the tissue.

Between the terminal of an axon and the end of the next neuron, there is a very narrow, fluid-filled gap called the *synapse*. Neurotransmitters move across the synapse when the electrical wave causes them to be released from their vesicles. When neurotransmitter molecules lock to their receptors on the other side of the synapse,

■ Growth factors stimulate cell division, so amino acids are needed for building the proteins of new cells.

■ The enkephalins (page 396) are pentapeptide neurotransmitters.

■ The traveling wave of electrical charge moves rapidly, but still not as rapidly as electricity moves in electrical wires.

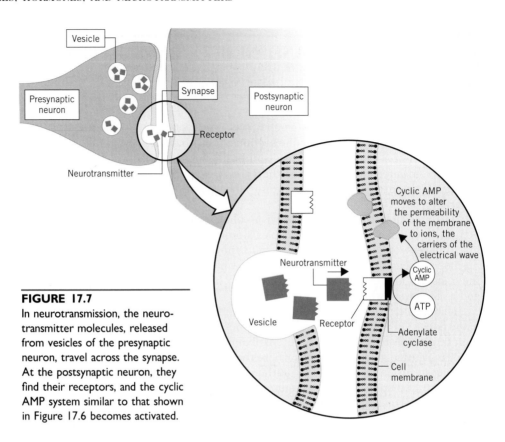

FIGURE 17.7
In neurotransmission, the neuro-transmitter molecules, released from vesicles of the presynaptic neuron, travel across the synapse. At the postsynaptic neuron, they find their receptors, and the cyclic AMP system similar to that shown in Figure 17.6 becomes activated.

adenylate cyclase (or something like it) is activated (see Figure 17.7). The formation of cyclic AMP is catalyzed, which then initiates whatever change is programmed by the chemicals in the target neuron. Unless the neurotransmitter is to continue to act without interruption, it must be removed or deactivated. How this is accomplished depends on the cell and the neurotransmitter.

■ The ANS nerves handle the signals that run the organs that have to work autonomously (without conscious effort), such as the heart and the lungs.

The Neurotransmitter Acetylcholine Is Swiftly Hydrolyzed.   One method used to detach a neurotransmitter is to break it up by a chemical reaction. Acetylcholine, for example, a neurotransmitter in the cholinergic nerves of the autonomic nervous system or ANS, is catalytically hydrolyzed to choline and acetic acid. The enzyme is choline acetyltransferase.

$$(CH_3)_3\overset{+}{N}CH_2CH_2O\overset{\overset{\displaystyle O}{\displaystyle \|}}{C}CH_3 + H_2O \underset{\text{acetyltransferase}}{\overset{\text{Choline}}{\rightleftharpoons}} (CH_3)_3\overset{+}{N}CH_2CH_2OH + HO\overset{\overset{\displaystyle O}{\displaystyle \|}}{C}CH_3$$

Acetylcholine   Choline   Acetic acid

■ A medication used to treat Alzheimer's disease, Cognex (Warner-Lambert), inhibits choline acetyltransferase where, in the Alzheimer victim's brain, acetylcholine is itself in short supply.

Within 2 milliseconds (2×10^{-3} s) of the release of acetylcholine in the synapse, all its molecules are broken down. The synapse is now cleared for a fresh release of acetylcholine from the presynaptic neuron if the signal for its release continues. If the signal does not come, then the action is shut down.

The Botulinum Toxin Prevents a Neuron from Making Acetylcholine.   The botulinum toxin is an extremely powerful toxic agent made by the food-poisoning botulinum bacterium and works by preventing the *synthesis* of acetylcholine. Without this neurotransmitter, the cholinergic nerves of the ANS can't work.

Local Anesthetics Block Receptors for Acetylcholine. Nupercaine, procaine, and tetracaine, common local anesthetics, work competitively against acetylcholine by blocking the receptor protein. Compounds that prevent the action of a neurotransmitter are called **antagonists** to the neurotransmitter. The neurotransmitter itself is sometimes referred to as an **agonist.**

Some Neurotransmitters Are Reabsorbed by the Presynaptic Neuron. Norepinephrine, another neurotransmitter, is deactivated by being reabsorbed by the neuron that released it, where it is then degraded. (Some is also deactivated right within the synapse.) The degradation of norepinephrine and similarly reabsorbed neurotransmitters is catalyzed by enzymes called the **monoamine oxidases** or MAOs.

Drugs That Inactivate the Monoamine Oxidases Are Used to Treat Depression. One place where norepinephrine works is in the brain stem, where mood regulation is centered. If for any reason the monoamine oxidases are inactivated, then an excess of norepinephrine builds up in brain stem cells, and some spills back into the synapse and sends signals on. In some mental states, like depression, an abnormally *low* level of norepinephrine develops, so now one would want to *inactivate* the monoamine oxidases. This would leave what norepinephrine there is to carry on its work. Thus, some of the antidepressant drugs, like iproniazid, work by inhibiting monoamine oxidases.

Other antidepressants, like amitriptyline (Elavil) and imipramine (Tofranil), inhibit the reabsorption of norepinephrine by the presynaptic neuron. Without reabsorption into the degradative hands of the monoamine oxidases, the level of norepinephrine and its signal-sending work stays high. Perhaps the most widely used antidepressant, fluoxetine hydrochloride (Prozac), inhibits the reabsorption of serotonin, another neurotransmitter.

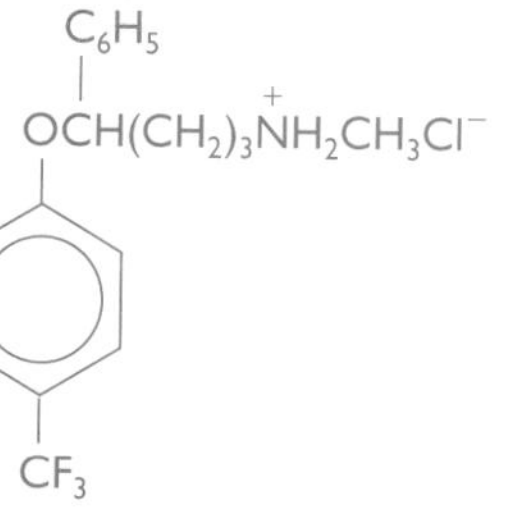

■ The nerves that use norepinephrine are called the adrenergic nerves (after an earlier name for norepinephrine, noradrenaline).

$$OCH(CH_2)_3 \overset{+}{N}H_2CH_3Cl^-$$

Fluoxetine hydrochloride (Prozac)

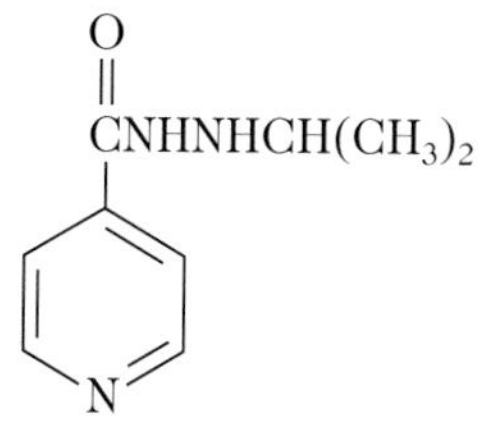

Iproniazid

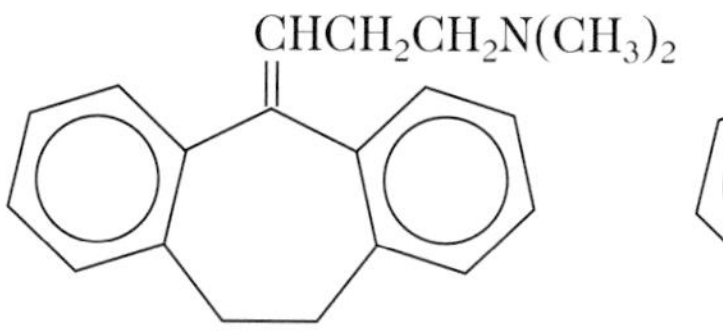

Amitriptyline (Elavil)

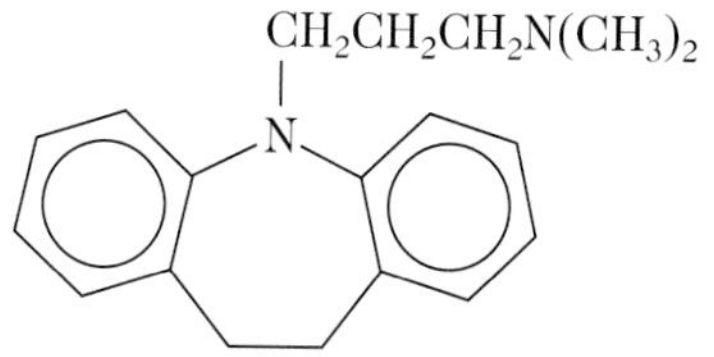

Imipramine (Trofranil)

Dopamine Excesses Occur in Schizophrenia. Dopamine, like norepinephrine, is also a monoamine neurotransmitter. It occurs in the midbrain in neurons involved with feelings of pleasure and arousal as well as with the control of certain movements. (The structure of dopamine is in Special Topic 11.2, page 258.)

In schizophrenia, neurons using dopamine are *overstimulated,* because either the releasing mechanism or the receptor mechanism is overactive. Drugs commonly used to treat schizophrenia, like chlorpromazine (Thorazine) and haloperidol (Haldol), competitively bind to dopamine receptors and thus inhibit the signal-sending work of dopamine.

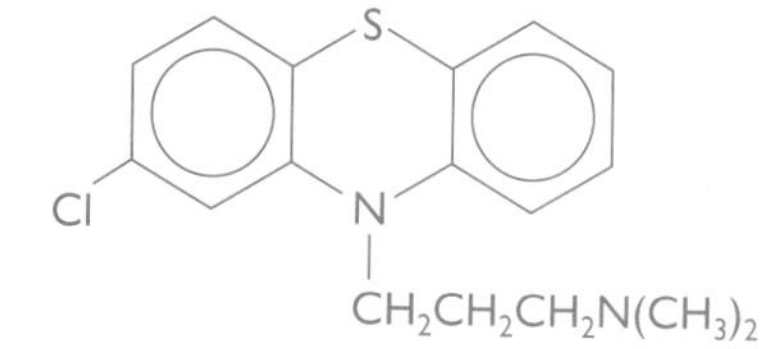

Chlorpromazine (Thorazine)

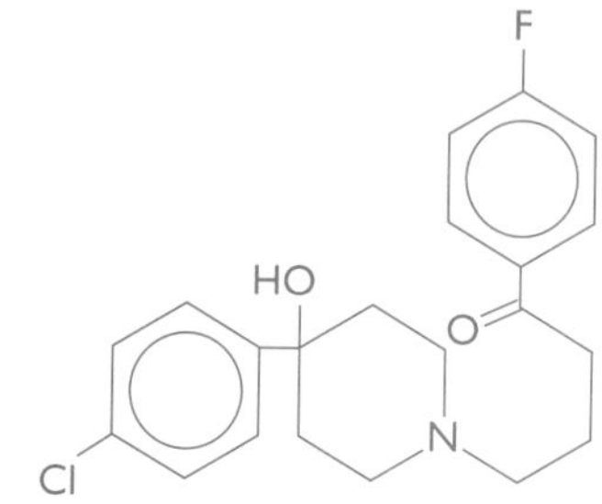

Amphetamine Abuse Causes Schizophrenia-like Symptoms. Stimulants like the amphetamines (cf. Special Topic 11.2) work by triggering the release of dopamine into the arousal and pleasure centers of the brain. The effect is a "high," but it's easy to abuse amphetamines and so cause the same kind of overstimulation associated with schizophrenia and the same results—delusions of persecution, hallucinations, and other disturbances of the thought processes.

Haloperidol (Haldol)

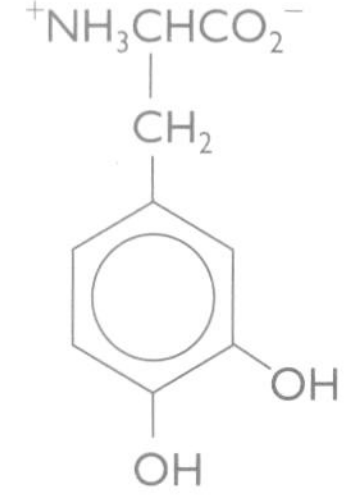

$^+NH_3CHCO_2^-$

CH$_2$

OH

OH

L-DOPA

MPTP   MPP$^+$

■ $^+NH_3CH_2CH_2CH_2CO_2^-$
GABA

Dopamine-Releasing Neurons Have Degenerated in Parkinson's Disease. When the dopamine-using neurons in the brain degenerate, as in Parkinson's disease, an extra supply of dopamine itself is then needed to compensate. A compound called L-DOPA (*levorotatory dihydroxyphenylalanine*) is commonly administered because affected neurons that still work can use it to make extra dopamine.

In the mid-1980s, by an accidental discovery, it was found that a contaminant in street heroin, called MPTP for short, rapidly destroys the same cells that degenerate in Parkinson's disease. The active agent is actually a metabolic breakdown product called MPP$^+$. The cells affected by Parkinson's disease apparently do not die at once but first go into a dormant state. The rejuvenation of such cells has long been sought. Scientists, working in the early 1990s with experimental rats, found that BDNF (for brain-derived neurotrophic factor) protects the dormant cells, stimulates them into recovery, and even protects susceptible cells. In 1993, another neurotrophic factor, GDNF (for glial cell line-derived neurotrophic factor), was also found to promote the survival of dopamine-releasing neurons.

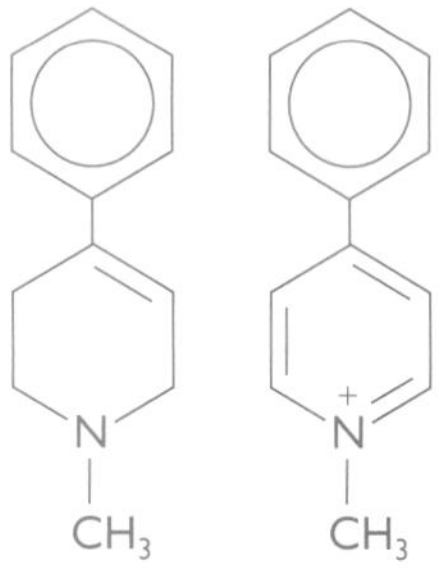

Another line of research into treating Parkinson's disease involves transplants of healthy fetal brain tissue. Preliminary results of the use of this technique on human patients, made public in 1992, were promising, but much research remains. The use of fetal tissue for medical research has been a source of great controversy.

GABA Inhibits Nerve Signals. The normal function of some neurotransmitters is to *inhibit* signals instead of to initiate them. Gamma-aminobutyric acid (GABA) is an example, and as many as a third of the synapses in the brain are supplied with GABA.

The inhibiting work of GABA can be made even greater by mild tricyclic tranquilizers such as diazepam (Valium) and chlordiazepoxide hydrochloride (Librium), as well as by ethyl alcohol. The augmented inhibition of signals reduces anxiety, affects judgment, and induces sleep.

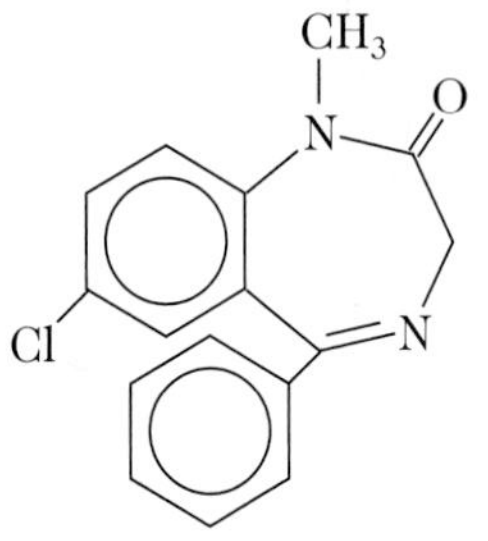

Diazepam (Valium)

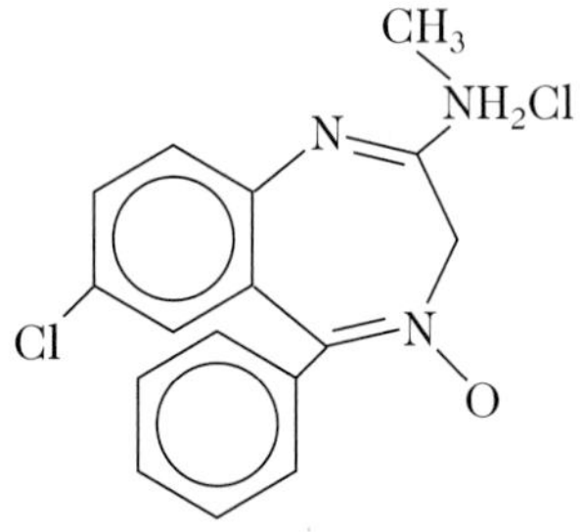

Chlordiazepoxide
hydrochloride (Librium)

■ From the Greek *chorea*, "dance."

GABA Is Deficient in Huntington's Chorea. The victims of Huntington's chorea, a hereditary neurological disorder, suffer from speech disturbances, irregular movements, and a steady mental deterioration, all related to a deficiency in GABA. Unhappily, GABA can't be administered in this disease, because it can't move out of circulation and into the regions of the brain where it works.

■ *En-* or *end-*, within; *kephale*, brain; *-orph-*, from morphine.

Several Polypeptides Act as Painkilling Neurotransmitters. As we said, some neurotransmitters are relatively small polypeptides. One type includes the *enkephalins*.

Tyr-Gly-Gly-Phe-Met   Tyr-Gly-Gly-Phe-Leu
Met-enkephalin   Leu-enkephalin

The *endorphins* are larger polypeptides, and both enkephalins and endorphins are powerful pain inhibitors. One compound, *dynorphin,* is the most potent painkiller yet discovered, being 200 times stronger than morphine, an opium alkaloid that is widely used to relieve severe pain. Sites in the brain that strongly bind molecules of morphine also bind those of the enkephalins, so these natural painkillers are now often referred to as the body's natural opiates.

Enkephalins Inhibit the Release of Substance P.   Substance P is a pain-signaling, polypeptide neurotransmitter. According to one theory, when a pain-transmitting neuron is activated to send a pain signal, the transmission is accomplished by the release of Substance P into the synapse. However, butting against the pain signal-sending neuron are other neurons that can release enkephalins. When released, they *inhibit* the work of Substance P and so tone down the intensity of the pain signal. The action of the enkephalin might explain the delay of pain that sometimes occurs during an emergency, when the brain and the body must continue to function to escape the emergency.

Many Neurotransmitters Exert More than One Effect.   Several neurotransmitters can be received by more than one kind of receptor. For example, at least three types of receptors for the opiates have been identified thus far. Receptor multiplicity may explain how the same neurotransmitters can have multiple effects. Thus not only do opiates reduce pain, but they affect emotions, induce sleep, and disturb the appetite, with each of the opiate receptors handling a different one of these functions.

Calcium Channel Blockers **Are Drugs That Reduce the Vigor of Heart Muscle Contractions.**   As we learned earlier, calcium ions are secondary chemical messengers, and neurotransmitters are able to open channels for calcium ions through cell membranes. Heart muscle tissue receives such signals at a rate that paces the heart as its muscles contract and relax during the heartbeat. Calcium ions are what finally deliver the message to contract. Then the cell pumps them back out and the muscle relaxes until another cycle starts.

Drugs like nifedipine (Procardia, Adalat), diltiazem (Cardizem), and verapamil (Isoptin) find calcium ion channels in heart muscle and block them. Not all are blocked, of course, so the effect is to reduce the migrations of Ca^{2+} through cell membranes. These *calcium channel blockers* thus make each heart muscle contraction less vigorous. This reduces the risk of heart attacks in people known to be at risk, such as those who experience angina pectoris and cardiac arrhythmias (heartbeat irregularities).

■ Calcium channel blockers are also called *calcium antagonists* or *slow channel blocking agents*.

Nitric Oxide Is a Retrograde Messenger That Might Be Involved in the Storage of Memory.   In the early 1990s, scientists were surprised to discover that two very simple compounds, nitrogen monoxide (NO) and carbon monoxide (CO), are messenger molecules involved at the molecular level of learning and memory. Learning involves putting experiences, information, and data into memory in such a way that they can be retrieved and applied in new situations. According to one broad model of how some types of learning take place, a nerve cell that receives a signal associated with something to be remembered manufactures a *retrograde messenger*. This moves back to the signal-sending cell and strengthens the connection between the two cells. Each additional time that the same signal is sent, the signal is further strengthened and memory improved.

■ *Retrograde* means moving or directed backward.

One of the retrograde messengers is nitric oxide, NO. Its tiny molecules are able to slip easily through cell membranes, which is an essential property for any retrograde messenger. It is known that cultured brain cells can make nitric oxide when certain receptors are stimulated, but when a binder of NO is present, experimental rats are unable to learn certain spatial tasks.

■ Abnormal NO systems have been found in some people with hypertension and in others with high blood pressure.

The discovery of a neurotransmitter role for an otherwise noxious gas, NO, means that not all chemical messenger work is done by a common mechanism. NO is a gas and can diffuse, as we said, from cell to cell relatively easily in several directions. The

■ One enzyme that NO is known to stimulate directly is guanylate cyclase, an enzyme similar to adenylate cyclase.

"classical" neurotransmitters must be stored and then released on signal. Next, they must find a receptor in a membrane to influence events within the cell. NO is able to slip directly through a membrane and so needs no membrane-bound receptor.

Carbon Monoxide Is Probably a Retrograde Messenger. In the early 1990s, evidence was found that carbon monoxide, another noxious gas, might also be a brain messenger and possibly involved in a long-term learning-linked process. (One is certainly tempted to say, "Of all things!") In relatively large concentrations, when carried into the body in inhaled air, CO is a poison. It binds so strongly to heme units in hemoglobin that the latter cannot carry oxygen, causing death. Yet in small concentrations manufactured within the brain, CO activates an enzyme involved in learning. This is an area of rapidly moving research driven in part by intense curiosity about *how* we remember.

What we have done in this section is look at some *molecular* connections between conditions of the nervous system and particular chemical substances. This whole field is one of the most rapidly moving areas of scientific investigation today, and during the next several years we may expect to see a number of dramatic advances both in our understanding of what is happening and in the strategies of treating mental and heart diseases.

SUMMARY

Enzymes Enzymes are the catalysts in cells. Some consist wholly of one or more polypeptides, and other enzymes include a cofactor besides the polypeptides. The cofactor can be an organic coenzyme, a metal ion, or both. Some coenzymes are phosphate esters of B vitamins, and in these examples, the vitamin unit usually furnishes the enzyme's active site. Because they are mostly polypeptide in nature, enzymes are vulnerable to all of the conditions that denature proteins. The name of an enzyme, which almost always ends in -*ase,* usually discloses either the identity of its substrate or the kind of reaction it catalyzes.

Some enzymes occur as small families called isoenzymes in which the polypeptide components vary slightly from tissue to tissue in the body. An enzyme is very specific both in the kind of reaction it catalyzes and in its substrate. Enzymes make possible reaction rates that are substantially higher than the rates of uncatalyzed reactions.

Induced Fitting When an enzyme–substrate complex forms, the active site is brought together with the part of the substrate that is to react. Binding sites on the enzyme guide the substrate molecule in and induce a change in the conformation of the enzyme molecule to produce the best fit of the substrate. The recognition of the enzyme by the substrate occurs as a flexible, lock-and-key model that involves complementary shapes and electrical charges.

Regulation of Enzymes Enzymes can be activated either by their own substrates or by effectors. Some enzymes are

activated by genes. Other enzymes, parts of cell membranes (or of small bodies within cells), are activated by the interaction between a hormone or a neurotransmitter and its receptor protein. The work of many of these is to cause changes in the calcium ion level in a cell.

Other enzymes, such as certain digestive enzymes, exist as zymogens and are activated when some agent removes a small part that blocks the active site.

Enzymes can be reversibly inhibited by a nonproduct inhibitor or by competitive feedback that involves a product of the enzyme's action. Competitive inhibition and noncompetitive inhibition are two mechanisms. Some of the most dangerous poisons bind to active sites and irreversibly inhibit the work of an enzyme, or they carry enzymes out of solution by a denaturant action. Many antibiotics and other antimetabolites work by inhibiting enzymes in pathogenic bacteria.

Medical Uses of Enzymes The serum levels of many enzymes increase when the tissues or organs that hold these enzymes are injured or diseased. By monitoring these serum levels and by looking for certain isoenzymes, we can diagnose diseases, for example, viral hepatitis and myocardial infarctions. Enzymes are also used in analytical systems that measure concentrations of substrates, such as in tests for glucose.

When a blood clot threatens or causes a heart attack, streptokinase or tissue plasminogen activator (TPA) can be used to initiate the hydrolysis of the fibrin of the clot.

***Chemical Communication with Hormones and Neuro-
transmitters*** Glycoprotein molecules that are parts of
membrane surfaces provide recognition sites for chemical
messengers. The messenger molecule and the receptor protein form a complex. One common consequence is the alteration of the G-protein that, in the adenylate cyclase cascade, activates adenylate cyclase, which then triggers the
formation of cyclic AMP. In turn, cyclic AMP sets off other
events, such as the activation of an enzyme that catalyzes a
reaction, one that is ultimately what the "signal" of the neurotransmitter was all about.

Hormones Endocrine glands secrete hormones, and these
primary chemical messengers travel to their target cells in
the blood, where they activate a gene, or an enzyme, or
affect the permeability of a cell membrane. They recognize
their own target cells by binding to specific receptor proteins.

The steroid hormones can move into a cell to its nucleus and there find a receptor. Polypeptide hormones bind
to membrane-bound receptors to initiate their action.

Neurotransmitters In response to an electrical signal, vesicles in an axon release a neurotransmitter that moves
across the synapse. Its molecules bind to a receptor protein
on the next neuron, and then the pattern is much like that
of hormones. The result, however, is to open channels
through the cell membrane for the migration of ions.

Neurotransmitters include amino acids, monoamines,
and polypeptides. Some neurotransmitters *activate* a response in the next neuron, whereas others *deactivate* an
activity. A number of medications work by interfering with
neurotransmitters or with the opening of calcium ion channels. Retrograde chemical messengers, like NO and CO, are
involved in learning and memory.

REVIEW EXERCISES

*The answers to Review Exercises whose numbers are in
color are found in Appendix V. The answers to the other
Review Exercises are found in the Study Guide that accompanies this text. The more challenging questions are
marked with asterisks.*

Nature of Enzymes

17.1 What are (a) the function and (b) the composition, in
general terms only, of an enzyme?

17.2 To what does *specificity* refer in enzyme chemistry?

17.3 Define and distinguish among the following terms.

(a) apoenzyme   (b) cofactor   (c) coenzyme

17.4 Write the equation for the equilibrium catalyzed by
carbonic anhydrase. What is particularly remarkable about
the enzyme?

17.5 What in general does an enzyme do to an equilibrium?

Coenzymes

17.6 What B vitamin is involved in the NAD^+/NADH system?

17.7 The active part of either FAD or FMN is furnished by
which vitamin?

17.8 Complete and balance the following equation.

$$\underset{\substack{| \\ \text{CH}_3\text{CHCH}_3}}{\text{OH}} + \text{NAD}^+ \longrightarrow \underset{\substack{\| \\ \text{CH}_3\text{CCH}_3}}{\text{O}} + \underline{\quad} + \underline{\quad}$$

17.9 Complete and balance the following equation.

$$\underline{\quad} + \text{NADH} + \text{FAD} \longrightarrow \text{NAD}^+ + \underline{\quad}$$

17.10 In what structural way do NAD^+ and $NADP^+$ differ?
What formula can be used for the reduced form of $NADP^+$?

Kinds of Enzymes

17.11 What *kind* of reaction does each of the following
enzymes catalyze?
(a) an oxidase   (b) transmethylase
(c) hydrolase   (d) oxidoreductase

17.12 What is the difference between lactose and lactase?

17.13 What is the difference between a hydrolase and hydrolysis?

17.14 What are isoenzymes (in general terms)?

17.15 What are the three isoenzymes of creatine kinase?
Give their symbols and state where they are principally
found.

Theory of How Enzymes Work

17.16 What name is given to the part of an enzyme where
the catalytic work is carried out?

17.17 How is enzyme specificity explained?

17.18 What is the induced-fit theory?

Enzyme Activation and Inhibition

17.19 How does competitive inhibition of an enzyme
work?

17.20 Feedback inhibition of an enzyme works in what way?

17.21 Why is feedback inhibition an example of a homeostatic mechanism?

17.22 How does a rising concentration of a substrate sometimes work to activate the enzyme for the substrate when the enzyme has two active sites?

17.23 What, in general terms, is an *effector*?

17.24 What are the names of two important effectors? Which one is used in muscle cells?

17.25 What are the approximate concentrations of calcium ion in the cytosol and the fluid just outside a cell? Why doesn't simple diffusion wipe out this concentration gradient?

17.26 Why must the concentration of Ca^{2+} be so low in the cytosol?

17.27 Does the concentration of Ca^{2+} in the cytosol measure the amount of Ca^{2+} in the whole cytoplasm? Explain.

17.28 What does Ca^{2+} do to calmodulin or troponin?

17.29 When Ca^{2+} combines with troponin, what happens with respect to other proteins in the cell? What then happens to Ca^{2+}?

17.30 What is the relationship of a zymogen to its corresponding enzyme? Give an example of an enzyme that has a zymogen.

17.31 What is plasmin and in what form does it normally circulate in the blood?

17.32 How do the following poisons work?
(a) CN^- (b) Hg^{2+}
(c) nerve gases or organophosphate insecticides

17.33 What are antimetabolites, and how are they related to antibiotics?

17.34 In broad terms, how does penicillin work?

Enzymes in Medicine

17.35 What viral infection might be indicated by an increase in the serum levels of the enzymes GPT and GOT?

17.36 What chemical fact about an enzyme is at the heart of the ability of a clinical chemist, medical technologist, or medical technician to analyze for the enzyme in the presence of so many other solutes in blood, including many other enzymes?

17.37 The serum levels of what three enzyme systems increase in a myocardial infarction?

17.38 What is the significance of CK(*MB*) in serum in trying to find out whether a person has had a heart attack and not just some painful injury in the chest region?

17.39 What CK isoenzyme would increase if the injury in Review Exercise 17.38 were to skeletal muscle?

17.40 What is the LD_1–LD_2 flip, and how is it used in diagnosis?

17.41 What are two enzymes that are available to help dissolve a blood clot? Which one occurs in human blood and how is it obtained for therapeutic uses?

17.42 What substance makes up most of a blood clot, and what happens to it when TPA works?

Chemical Communication

17.43 What are the names of the sites of the synthesis of (a) hormones and (b) neurotransmitters?

17.44 What do the lock-and-key and induced fit concepts have to do with the work of hormones and neurotransmitters?

17.45 In what general ways do hormones and neurotransmitters resemble each other?

17.46 What general name is given to the substance on a target cell that recognizes a hormone or neurotransmitter?

17.47 What function does adenylate cyclase have in the work of at least some hormones?

17.48 How is cyclic AMP involved in the work of some hormones and neurotransmitters?

17.49 After cyclic AMP has caused the activation of an enzyme inside a cell, what happens to the cyclic AMP that stops its action until more is made?

Hormones

17.50 What are the four broad types of hormones?

17.51 What structural fact about the steroid hormones makes it easy for them to get through a cell membrane?

17.52 In each case, what substance (or kind of substance) can enter a target cell more readily following the action of the hormone?
(a) insulin   (b) growth hormone   (c) a neurotransmitter

Neurotransmitters

17.53 What happens to acetylcholine after it has worked as a neurotransmitter? What is the name of the enzyme that catalyzes this change?

17.54 A competitive inhibitor of a neurotransmitter receptor can be called an *antagonist*. What does this mean?

17.55 How does a local anesthetic such as procaine affect the functioning of acetylcholine as a neurotransmitter?

17.56 How does the botulinum toxin work?

17.57 What, in general terms, are the monoamine oxidases, and in what way are they important?

17.58 What does iproniazid do chemically in the neuron signaling that is carried out by norepinephrine?

17.59 In general terms, how do antidepressants such as amitriptyline and imipramine work?

17.60 Which neurotransmitter is also a hormone, and what is the significance of this dual character to the body?

17.61 The overactivity of which neurotransmitter is thought to be one biochemical problem in schizophrenia?

17.62 How do the schizophrenia-control drugs chlorpromazine and haloperidol work?

17.63 How can the amphetamines, when abused, produce schizophrenia-like symptoms?

17.64 How does L-DOPA work in treating Parkinson's disease?

17.65 How does a neurotrophic factor like BDNF work in treating Parkinson's disease?

17.66 Which common neurotransmitter in the brain is a signal inhibitor? How do such tranquilizers as Valium and Librium affect it?

17.67 Why is enkephalin called one of the body's own opiates? How does it appear to work?

17.68 What does substance P do?

17.69 What is meant by the term *retrograde messenger?*

17.70 What physical property of nitric oxide or carbon monoxide enables them to act in a retrograde manner?

Complementarity and Immune Responses (Special Topic 17.1)

17.71 What is a pathogen? (in general terms)

17.72 What two kinds of immunity are recognized and how do they differ?

17.73 The HIV particle attacks what kind of immunity and in what way?

17.74 What is an antibody? An antigen?

17.75 At the molecular level, what aspects of molecular structure explain the high specificity of the immune response?

17.76 In what kind of immunity are the B cells operative, and what kind of substance do B cells eventually make to counter an alien material?

17.77 Why is a glycolipid on a red blood cell of an A-type individual referred to as an antigen and not an antibody?

17.78 At the molecular level involving their red blood cells, in what specific ways do the A, B, and O types differ?

17.79 What is present in the blood of an A-type person that makes receiving blood from a B-type person dangerous?

17.80 Why is it that O-types can donate blood to people of any type but can receive blood only from other O-types?

Additional Exercises

*****17.81** If you drink enough methyl alcohol, you will become blind or die. One strategy to counteract methyl alcohol poisoning is to give the victim a nearly intoxicating drink of dilute ethyl alcohol. As the ethyl alcohol floods the same enzyme that attacks the methyl alcohol, the methyl alcohol gets a lessened opportunity to react and it is slowly and relatively harmlessly excreted. Otherwise, it is oxidized to formaldehyde, the actual poison from an overdose of methyl alcohol:

$$CH_3OH \xrightarrow{\text{dehydrogenase}} CH_2O$$

Methyl alcohol   Formaldehyde

What kind of enzyme inhibition might ethyl alcohol be achieving here? (Name it.)

*****17.82** Truffles are edible, potato-shaped fungi that grow underground in certain parts of France, and are highly prized by gourmet cooks and gourmands. Pigs are used to locate truffles buried as much as one meter below the surface because truffles carry traces of a steroid, androsten-16-en-3-ol, which is a powerful sex attractant for pigs. A sex attractant is a species-specific chemical compound made and released in trace amounts by a female member of a species toward which a male member experiences a powerful sexual response. A male pig, of course, does not initially know that it is a truffle emitting the attractant. (It appears that all sex attractants for humans are *non*chemical, being better understood, perhaps, as public relations activities.)

Androsten-16-en-3-ol

Androsterone

Notice the structural similarity between androsten-16-en-3-ol and a human sex hormone, androsterone.

(a) In terms of what general theory would we explain how androsten-16-en-3-ol has a particular specificity in pigs and not in humans but androsterone has a specificity in humans and not in pigs?

(b) To convert androsten-16-en-3-ol into androsterone *in vitro,* what specific *series* of changes in functional groups would have to be carried out? Your answer would begin with something like "First, we have to change such and such a group into. . . . Then this new group would have to be changed. . . ." (All needed changes involve one-step reactions we have studied.)

*****17.83** Referring to the equation given on page 386 for the conversion of threonine to isoleucine, what is the maximum number of milligrams of isoleucine that could be made from 150.0 mg of threonine?

18

EXTRACELLULAR FLUIDS OF THE BODY

Digestive Juices

Blood and the
Absorption of
Nutrients by Cells

Chemistry of the
Exchange of
Respiratory Gases

Acid–Base Balance of
the Blood

Acid–Base Balance and
Some Chemistry of
Kidney Functions

High-altitude work, like climbing along the southeast buttress of Fitzroy in Patagonia (Argentina), would cause the unacclimated to overbreathe and so remove carbon dioxide too rapidly from the body. This loss of a base-neutralizer leads to alkalosis and possibly fatal altitude sickness. One of the major topics of the entire book, the acid–base status of the blood, is studied in this chapter.

18.1 DIGESTIVE JUICES

The end products of the complete digestion of the nutritionally important carbohydrates, lipids, and proteins are monosaccharides, fatty acids, monoacylglycerols, and amino acids.

Life engages two environments, the outside environment, of which we are so aware, and the *internal environment,* which we usually take for granted. When healthy, our bodies have nearly perfect control over the internal environment. We are able to handle large external changes fairly well, for example, large temperature fluctuations, chilling winds, stifling humidity, and a fluctuating tide of atmospheric dust and pollutants.

Cells Exist in Contact with Interstitial Fluids and Blood. The **internal environment** consists of all the **extracellular fluids,** meaning those not actually inside cells. About three-quarters of the extracellular fluids consists of the **interstitial fluid,** which fills the spaces or interstices *between* cells. The blood makes up nearly all the rest. The lymph, cerebrospinal fluid, digestive juices, and synovial fluids are also extracellular fluids.

The chemistry occurring inside cells, in the *intracellular fluid,* has been and will continue to be a major topic of our study. Here we focus on two of the extracellular fluids, digestive juices and blood.

The Digestive Tract Is a Convoluted Tube Running through the Body with Access to Several Solutions of Hydrolytic Enzymes. The **digestive juices** are dilute solutions of electrolytes and hydrolytic enzymes (or their zymogens) either in the cells lining the intestinal tract or in solutions entering the tract from organs (Figure 18.1).

■ The fluids of the internal environment make up about 20% of the mass of the body.

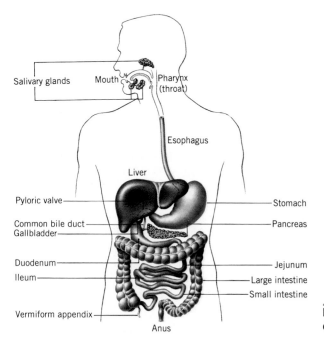

FIGURE 18.1
Organs of the digestive tract.

Food Stimulates the Release of Gastric Juice. When food begins its trip into the stomach, gastric cells in the gastric lining are stimulated to produce the neurotransmitter *acetylcholine.* Also launched is the synthesis of *histamine,* being made from the α-amino acid histidine by the loss of a molecule of carbon dioxide (by decarboxylation).

$$^{+}NH_3CHCO_2^{-} \xrightarrow[\text{decarboxylation}]{CO_2} NH_2CH_2$$

Histidine   Histamine

When food actually arrives in the stomach, the stomach becomes distended (enlarged dowward), and this stimulates the gastric cells to release the hormone *gastrin.* These three compounds, gastrin, acetylcholine, and histamine, now stimulate the release of the fluids that combine to give **gastric juice.** One kind of

gastric cell secretes mucin, which coats the stomach to protect it against its own digestive enzymes and its acid. Mucin is continuously produced and only slowly digested. If for any reason its protection of the stomach is hindered, part of the stomach itself could be digested, which would lead to a gastric ulcer (one found in the stomach).

Histamine Stimulates the K⁺–H⁺ Ion Pump. Another kind of gastric cell (parietal cells) secretes hydrogen ion at a concentration of roughly 0.15 mol/L (pH 0.8 or more than a million times more acidic than blood). This is accomplished by the operation of a K^+–H^+ ion "pump," an assemblage of protein molecules that exists in the membranes of the parietal cells. The K^+–H^+ pump, which uses the chemical energy of ATP to operate, takes K^+ ion out of stomach fluids and puts H^+ in. K^+ ions thus are moved one way (out of the stomach) through the pump as H^+ ions move the other way (into the stomach). The exchange of positive ions ensures overall electrical neutrality. The K^+ ions are then returned to the stomach fluids, but to ensure electrical neutrality, a negatively

■ Dextrins are partial breakdown products of amylopectin, a component of starch.

Saliva Provides α-Amylase, a Starch-Splitting Enzyme. The flow of **saliva** is stimulated by the sight, smell, taste, and even thought of food. Besides water (99.5%), saliva includes a food lubricant called *mucin* (a glycoprotein) and an enzyme, α-*amylase.* This enzyme catalyzes the partial hydrolysis of starch to dextrins and maltose, and it works best at the pH of saliva, 5.8 to 7.1. Proteins and lipids pass through the mouth essentially unchanged.

Gastric Juice Starts the Digestion of Proteins with Pepsin. When food arrives in the stomach, the cells of the gastric glands are stimulated by hormones to release the fluids that together make up **gastric juice.** One kind of gastric gland secretes mucin, which coats the stomach to protect it against its own digestive enzymes and its acid. Mucin is continuously produced and only slowly digested. If for any reason mucin's protection of the stomach is hindered, part of the stomach itself could be digested, and a gastric ulcer would form (see Special Topic 18.1).

■ A *protease* is an enzyme that catalyzes the digestion of proteins.

Another gastric gland secretes hydrochloric acid at a concentration of roughly 0.1 mol/L, about a million times more acidic than blood. The acid of the gastric juice coagulates proteins and activates a protease. Protein coagulation retains protein longer in the stomach for exposure to the protease. The gastric juice of an infant is less acidic than that of the adult. To compensate for the protein-coagulating work normally done in the adult by acid, infant gastric juice contains *rennin,* a powerful protein coagulator.

charged chloride ion goes along. Because the Cl^- ion is the chief anion in gastric juice, *the overall effect of this coming and going is the secretion of hydrochloric acid into the stomach.* Histamine is the specific stimulator of the K^+-H^+ ion pump.

Histidine Mimics Inhibit the Pump. Two of the most common and successful medications for the treatment of stomach ulcers are cimetidine (Tagamet) and ranitidine (Zantac). They are classified as antisecretory agents because they inhibit the secretion of gastric juice. Their molecules are able to mimic histamine molecules and bind to the receptor protein of the K^+-H^+ pump without stimulating the pump to work.

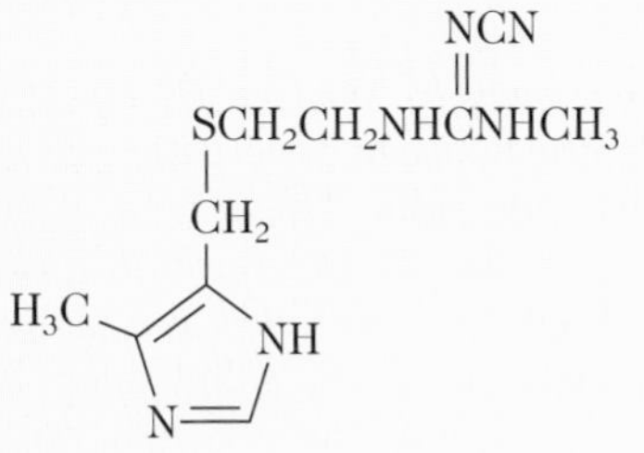

Cimetidine (Tagamet)

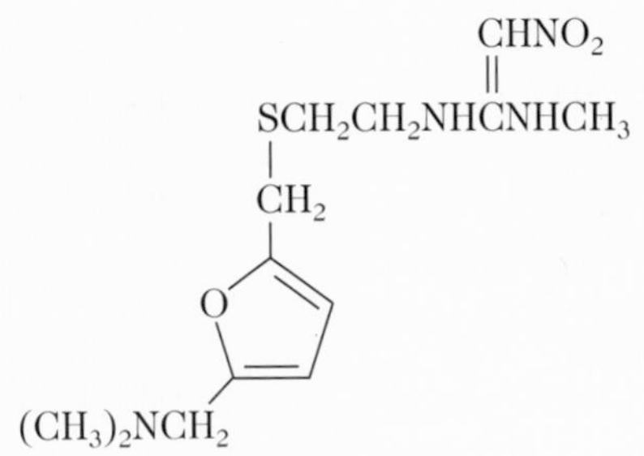

Ranitidine (Zantac)

When molecules of these medications are attached to the receptor sites of the K^+-H^+ pump, histamine molecules have nowhere to bind. By preventing histamine binding, the K^+-H^+ pump shuts down. Now the secretion of acid into the stomach is severely inhibited, giving time for the stomach to heal the ulcer.

The Ultimate Cause of Ulcers Is a Bacterium. The conventional wisdom is that high stress and poor diets cause ulcers. Throughout the 1980s, however, evidence accumulated that ulcers are caused by the bacterium *Helicobacter pylori.* The proof became sufficiently convincing to the National Institutes of Health that in early 1994 it recommended that antimicrobial agents, including antibiotics, be made a part of ulcer treatment along with antisecretory agents.

Another gastric gland secretes the zymogen *pepsinogen.* Pepsinogen is changed to *pepsin,* a protease, by the action of hydrochloric acid and traces of pepsin. The optimum pH of pepsin is in the range 1 to 1.5, which is found in the stomach fluid. Pepsin catalyzes the only important digestive work in the stomach, the hydrolysis of some of the peptide bonds of proteins to make shorter polypeptides. Adult gastric juice also has a lipase, but it does not start its work until it passes with the stomach contents into the higher-pH medium of the upper intestinal tract. Because the pH of an infant's gastric juice is higher than that of an adult, its lipase gets an early start on lipid digestion in the stomach itself.

The churning and digesting activities in the stomach produce a liquid mixture called *chyme.* This is released in portions through the pyloric valve into the duodenum, the first 12 inches of the upper intestinal tract.

Pancreatic Juice Furnishes Several Zymogens and Enzymes. As soon as chyme appears in the duodenum, hormones are released that circulate to the pancreas and induce this organ to release two juices. One is almost entirely dilute sodium bicarbonate, which neutralizes the acid in chyme. The other is the one usually called **pancreatic juice.** It carries enzymes or zymogens that become involved in the digestion of practically everything in food. It contributes an *α-amylase* similar to that present in saliva, a *lipase, nucleases,* and zymogens for protein-digesting enzymes.

The conversion of the proteolytic zymogens to active enzymes begins with a

■ The nucleases include ribonuclease (RNase) and deoxyribonuclease (DNase).

"master switch" enzyme called *enteropeptidase*. It is released from cells that line the duodenum when chyme arrives, and it then catalyzes the formation of trypsin from its zymogen, trypsinogen.

$$\text{Trypsinogen} \xrightarrow{\text{enteropeptidase}} \text{trypsin}$$

Trypsin then catalyzes the change of the other zymogens into their active enzymes.

$$\text{Procarboxypeptidase} \xrightarrow{\text{trypsin}} \text{carboxypeptidase}$$

$$\text{Chymotrypsinogen} \xrightarrow{\text{trypsin}} \text{chymotrypsin}$$

$$\text{Proelastase} \xrightarrow{\text{trypsin}} \text{elastase}$$

Trypsin, chymotrypsin, and *elastase* catalyze the hydrolysis of large polypeptides to smaller ones. *Carboxypeptidase,* working in from C-terminal ends of small polypeptides, carries the action further to amino acids and di- or tripeptides.

■ These proteases must exist as zymogens first or they will catalyze the self-digestion of the pancreas, which does happen in acute pancreatitis.

Bile Salts Are Powerful Surfactants Necessary to Manage Dietary Lipids and Fat-Soluble Vitamins.   To digest most lipids, the lipase in pancreatic juice needs the help of the powerful detergents in bile, called **bile salts.** These help to emulsify water-insoluble fatty materials and so greatly increase the exposure of lipids to water and lipase. Triacylglycerols are hydrolyzed chiefly to fatty acids and monoacylglycerols.

■ The structure of a typical bile salt was given on page 343.

   Bile is a juice that enters the duodenum from the gallbladder. Its secretion is stimulated by a hormone released when chyme contains fatty material. Bile is also an avenue of excretion because it carries cholesterol and breakdown products of hemoglobin. These and further breakdown products constitute the bile pigments, which give color to feces.

   The bile salts also assist in the absorption of the fat-soluble vitamins (A, D, E, and K) from the digestive tract into the blood. This work reabsorbs some bile pigments, some of which eventually leave the body via the urine. Thus the bile pigments are responsible for the color of both feces and urine.

Cells of the Intestines Carry Several Digestive Enzymes.   The term **intestinal juice** embraces not only a secretion but also the enzyme-rich fluids found inside certain kinds of cells that line the duodenum and jejunum. The secretion of some of these cells delivers an amylase and enteropeptidase, which we just described.

■ These intestinal cells last only about 2 days before they self-digest. They are constantly being replaced.

   The other enzymes supplied by this region of the digestive tract work within their own cells on digestible compounds already being absorbed. An *aminopeptidase,* working inward from the N-terminal ends of small polypeptides, digests them to amino acids. The enzymes *sucrase, lactase,* and *maltase* handle the digestion of disaccharides: sucrose to fructose and glucose; lactose to galactose and glucose; and maltose to glucose. An *intestinal lipase* and enzymes for the hydrolysis of nucleic acids are also present.

   As fatty acids and monoacylglycerols migrate through the cells of the duodenal lining, much is reconstituted into triacylglycerols, which are taken up by the lymph system rather than the blood.

Some Vitamins and Essential Amino Acids Are Made in the Large Intestine.   No digestive functions are performed in the large intestine. Microorganisms in residence there, however, make vitamins K and B, plus some essential amino acids. These are absorbed by the body, but their contribution to overall nutrition in humans is not large. Water and sodium chloride are reabsorbed from the ileum, and undigested matter (including fiber) and some water make up the feces.

18.2 BLOOD AND THE ABSORPTION OF NUTRIENTS BY CELLS

The balance between the blood's pumping pressure and its colloidal osmotic pressure tips at capillary loops.

The circulatory system is one of our two main lines of chemical communication between the external and internal environments (Figure 18.2). All of the veins and arteries together are called the *vascular compartment*. The *cardiovascular compartment* includes this plus the heart.

The Blood Moves Nutrients, Oxygen, Messengers, Wastes, and Disease Fighters throughout the Body. Blood in the pulmonary branches moves through the lungs where waste carbon dioxide in the blood is exchanged for oxygen from freshly inhaled air. The oxygenated blood then moves to the rest of the system via the systemic branches. At endocrine glands, the blood picks up and circulates hormones whose secretions are often in response to something present in the blood.

■ The nervous system with its neurotransmitters is the other line of communication.

■ About 8% of the body's mass is blood. In the adult, the blood volume is 5 to 6 L.

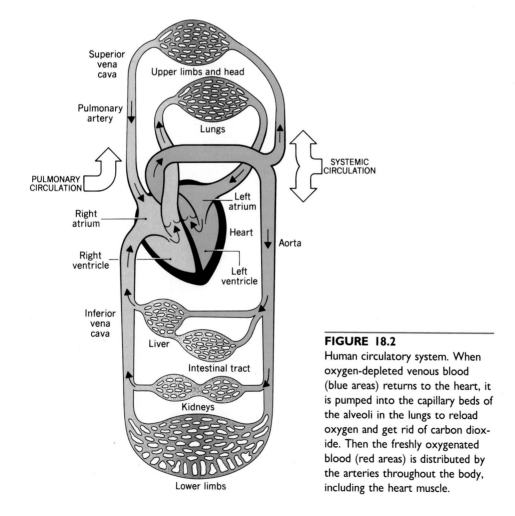

FIGURE 18.2
Human circulatory system. When oxygen-depleted venous blood (blue areas) returns to the heart, it is pumped into the capillary beds of the alveoli in the lungs to reload oxygen and get rid of carbon dioxide. Then the freshly oxygenated blood (red areas) is distributed by the arteries throughout the body, including the heart muscle.

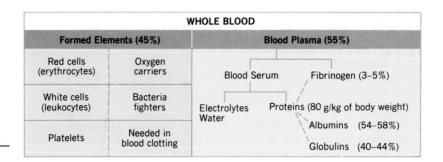

FIGURE 18.3
Major components of blood.

At the intestinal tract, the blood picks up the products of digestion. Most are immediately monitored at the liver, and many alien chemicals are modified so they can be eliminated. In the kidneys, the blood is purified of nitrogen wastes, particularly urea, and kidney cells replenish the supply of bicarbonate ion for the carbonate buffer of the blood. The pH of blood and its electrolyte balance thus depend hugely on the chemical work of the kidneys.

White cells in blood give protection against bacteria; red cells or *erythrocytes* carry oxygen and a waste product, bicarbonate ion; and the platelets in blood are needed for blood clotting and other purposes. The blood also carries several zymogens needed for the blood clotting mechanism.

The Proteins in Blood Are Vital to Its Colloidal Osmotic Pressure.   The principal types of substances in whole blood are summarized in Figure 18.3. One kilogram of blood plasma contains 80 g of proteins: albumins (54–58%), globulins (40–44%), and fibrinogen (3–5%). *Albumins* help carry hydrophobic molecules such as fatty acids, other lipids, and steroid hormones, and they contribute 75–80% of the osmotic effect of the blood. Some globulins carry ions (e.g., Fe^{2+} and Cu^{2+}) that otherwise could not be soluble in a fluid with a pH slightly greater than 7. Other globulins help to protect the body against infectious disease. Circulating *fibrinogen* is converted to an insoluble form, *fibrin,* when a blood clot forms.

Figure 18.4 shows the quantities of various components of the major body fluids. The **electrolytes** of blood, the inorganic ions, dominate. The sodium ion is the chief cation in both blood and interstitial fluid, and the potassium ion is the major cation inside cells. A sodium–potassium pump, a special protein complex that uses energy, maintains these gradients (see page 372). Both Na^+ and K^+ ions are needed to maintain osmotic pressure relationships, and both are a part of the regulatory system for acid-base balance.

The protein content of blood plasma is a major difference between plasma and interstitial fluid. Plasma proteins are the principal reason why blood has a higher osmotic pressure than interstitial fluid.[1] The *total* osmotic pressure of blood is caused by all of the dissolved and colloidally dispersed solutes: electrolytes, organic compounds and ions, and proteins. However, the small ions and molecules can dialyze back and forth between the blood and the interstitial compartment. The large protein molecules can't do this, so it is their presence that gives to blood the higher effective osmotic pressure. This contribution to the blood's osmotic pressure made by colloidally dispersed substances is called the **colloidal osmotic pressure** of blood.

■ About a quarter of the plasma proteins are replaced each day.

■ The osmolarity of plasma is about 290 mOsm/L.

[1]As a reminder and a useful memory aid, high solute concentration means high osmotic pressure; and in osmosis or dialysis, solvent flows from a region where the solute is dilute to a region where it is concentrated. The "goal" of this flow is to even out the concentrations everywhere.

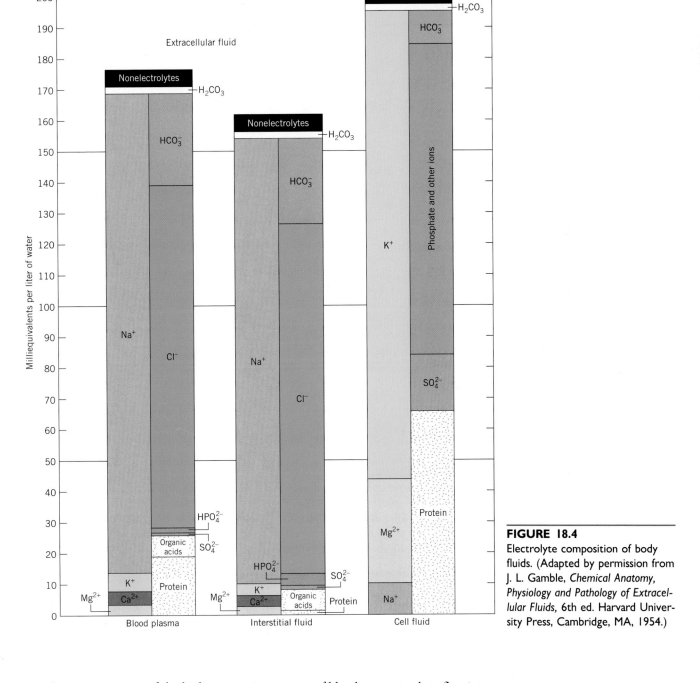

FIGURE 18.4

Electrolyte composition of body fluids. (Adapted by permission from J. L. Gamble, *Chemical Anatomy, Physiology and Pathology of Extracellular Fluids*, 6th ed. Harvard University Press, Cambridge, MA, 1954.)

As a consequence of the higher osmotic pressure of blood, water tends to flow into the blood from the interstitial compartment. Of course, this can't be allowed to happen everywhere and continually, or the interstitial spaces and then the cells would eventually become too dehydrated to maintain life.

Fluids That Leave the Blood Must Return in Equal Volume. The blood vessels undergo extensive branching until the narrowest tubes called the *capillaries* are reached. Blood enters a capillary loop as arterial blood, but it leaves on the other side of the loop as venous blood (Figure 18.5). During the switch, fluids and nutrients leave the blood and move into the interstitial fluids and then into the tissue cells them-

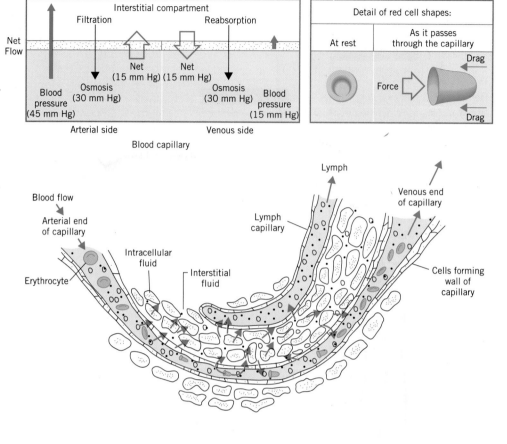

FIGURE 18.5

Exchange of nutrients and wastes at capillaries. As indicated at the top, on the arterial side of a capillary loop the blood pressure counteracts the pressure from dialysis and osmosis, and fluids are forced to leave the bloodstream. On the venous side of the loop, the blood pressure has decreased below that of dialysis and osmosis, so fluids flow back into the bloodstream. On the top right and the bottom is shown how a normal red cell distorts as it squeezes through a capillary loop. Red cells in sickle-cell anemia do not pass through as smoothly. The bottom drawing also shows how some fluids enter the lymph system.

■ The lymph system makes antibodies, and it has white cells that help defend the body against infectious diseases.

selves. *In the same volume* the fluids must return to the blood, but now they must carry the wastes of metabolism.

The rate of this diffusion of fluids throughout the body is sizable, about 25 to 30 L per second! Some fluids return to circulation by way of the lymph ducts, which are thin-walled, closed-end capillaries that bed in soft tissue.

Blood Pressure Overcomes Osmotic Pressure on the Arterial Side of a Capillary Loop. On the arterial side of a capillary loop, the blood pressure is high enough to overcome the natural tendency of fluids to move *into* the blood. Water and dissolved solutes are instead forced out of the blood and into the surrounding tissue, where exchanges of chemicals occur.

Osmotic Pressure Overcomes Blood Pressure on the Venous Side of a Capillary Loop. As blood emerges from the thin constriction of a capillary into the venous side, its pressure drops. It is now too low to prevent the natural diffusion of fluids back into the bloodstream. By this time, of course, the fluids are carrying waste products. The colloidal osmotic pressure contributed by the macromolecules in blood, particularly the albumins, make the difference in determining the direction of diffusion.

Blood Loses Albumins in the Shock Syndrome. When the capillaries become more permeable to blood proteins, as they do in such trauma as sudden severe injuries, major surgery, and extensive burns, the proteins migrate out of the blood. Unfortunately, this protein loss also means the loss of the colloidal osmotic pressure that helps fluids to return from the tissue areas to the bloodstream. As a result, the total volume of circulating blood drops quickly, which drastically reduces the blood's ability to carry oxygen and to remove carbon dioxide. The drop in blood volume and the resulting loss of oxygen supply to the brain send the victim into *traumatic shock.*

Blood Also Loses Proteins in Kidney Disease and Starvation.   Sometimes the proteins in the blood are lost at malfunctioning kidneys. The effect, although gradual, is a slow but unremitting decrease in the blood's colloidal osmotic pressure. Fluids begin to accumulate in interstitial regions. Because this takes place relatively slowly and because water continues to be ingested, no sudden drop in blood volume occurs, as in shock. The victim appears puffy and waterlogged, however, a condition called *edema.* Edema can also appear at one stage of starvation, when the body has metabolized its circulating proteins to make up for the absence of dietary proteins.

Any obstruction in the veins can also cause edema, as in varicose veins and certain forms of cancer. Now it is the venous blood pressure that increases, creating a back pressure that reduces the rate at which fluids can return to circulation from the tissue areas. The localized swelling that results from a blow is a temporary form of edema caused by injuries to the capillaries.

■ The prompt restoration of blood volume is mandatory in the treatment of shock.

■ From the Greek *oidema,* "swelling."

18.3   THE CHEMISTRY OF THE EXCHANGE OF RESPIRATORY GASES

The binding of oxygen to hemoglobin is cooperative, and it is affected by the pH, *PCO_2*, and *PO_2* of the blood.

The carrier of oxygen in blood is *hemoglobin,* a complex protein found inside red blood cells (erythrocytes). When hemoglobin is oxygen free, it is sometimes called *deoxyhemoglobin,* and when it carries oxygen it is *oxyhemoglobin.* The discussion that follows frequently invokes Le Châtelier's principle, so it would be well to recall it: an equilibrium, when upset by some stress, will shift in whatever direction most directly absorbs the stress. Changes in concentrations and gas pressures are the major stresses that we'll encounter.

The First Oxygen Molecule to Bind to Hemoglobin Activates the Binding of Three More Oxygen Molecules.   Hemoglobin, recall, consists of four polypeptide subunits, each with an oxygen-binding heme unit. Thus, a fully oxygenated hemoglobin molecule carries *four* oxygen molecules. For maximum efficiency in moving oxygen out of the lungs, all hemoglobin molecules should leave the lungs fully loaded. None should leave only partially oxygenated.

What ensures full oxygenation are changes in the shapes of the hemoglobin subunits caused by the binding of the first oxygen molecule. This is very similar to the way a substrate can activate an enzyme with more than one catalytic site (see Section 17.3) The first molecule to force its way into one subunit induces configurational changes in the other three to make all other oxygen-binding sites far more receptive than initially. Such changes to hemoglobin's remaining three subunits, caused by the binding of oxygen to the first, make it easier for the next three oxygen molecules to bind. They flood into the partially oxygenated hemoglobin molecule more easily than into an empty molecule. Thus the oxygenation is *cooperative,* meaning that changes in hemoglobin subunit shapes cooperate with oxygenation.

■ Each red cell carries about 2.8 × 10^8 molecules of hemoglobin.

■ The structure in Figure 16.6, page 367, is actually deoxyhemoglobin.

■ Carbon monoxide binds 150 to 200 times more strongly to hemoglobin than does oxygen and thus prevents the oxygenation of hemoglobin and causes internal suffocation.

■ About 20% of a smoker's hemoglobin is more or less permanently tied up by carbon monoxide.

The Relatively High PO_2 in the Lungs Aids the Oxygenation of Hemoglobin. The partial pressure of oxygen (PO_2) is higher in the lungs than anywhere else in the body, being 100 mm Hg in freshly inhaled air in alveoli and only about 40 mm Hg in oxygen-depleted tissues. Oxygen, therefore, naturally migrates from the lungs into the bloodstream. It's as if the higher partial pressure *pushes* oxygen into the blood.

The Formation of Deoxyhemoglobin Pulls Oxygen into the Lungs. For the sake of the remaining discussion, we'll simplify what we have just described by letting the symbol H*Hb* represent an entire hemoglobin molecule. The first H in H*Hb* stands for a potential hydrogen ion, and we overlook the fact that more than one is actually present in hemoglobin. (We're now also overlooking the fact that *four* molecules of O_2 bind to one of hemoglobin.) With this in mind, we can represent the oxygenation of hemoglobin as the *forward* reaction in the following equilibrium where oxyhemoglobin is represented as the anion HbO_2^-:

$$HHB(aq) + O_2(g) \rightleftharpoons HbO_2^-(aq) + H^+(aq) \qquad (18.1)$$
$$\underset{\text{Hemoglobin}}{} \qquad\qquad \underset{\text{Oxyhemoglobin}}{}$$

Two facts indicated by Equilibrium 18.1 are that H*Hb* is a weak acid and that it becomes a stronger acid as it becomes oxygenated. Being stronger, it is produced in its ionized state as HbO_2^- and H^+. *The presence of H^+ in Equilibrium 18.1 means that the equilibrium can be shifted one way or another simply by changing the pH,* a fact of enormous importance at the molecular level of life, as we'll see.

To understand the oxygenation of hemoglobin, we have to see how various stresses shift Equilibrium 18.1 to the *right* in the lungs. One stress, as we've already noted, is the relatively high value of PO_2 (100 mm Hg) in the alveoli. This stress acts on the left side of Equilibrium 18.1, so it helps to shift the equilibrium to the right.

- The stimulation of H*Hb* to bind O_2 caused by the removal of H^+ is called the **Bohr effect** after Christian Bohr, a Danish scientist (and the father of nuclear physicist Niels Bohr).

$$HHb(aq) + O_2(g) \rightarrow HbO_2^-(aq) + H^+(aq) \qquad (18.2)$$
$$\underset{\substack{\text{In the}\\\text{red cell}}}{} \quad \underset{\substack{\text{From}\\\text{air}}}{} \quad \underset{\substack{\text{In the}\\\text{red cell}}}{} \quad \underset{\substack{\text{In the}\\\text{red cell}}}{}$$

Another stress that isn't evident from Equilibrium 18.1 is the removal of H^+ *as it forms* by a second reaction, that of H^+ with HCO_3^-. A red cell entering the lungs carries waste HCO_3^-, so the newly formed H^+ ions promptly equilibrate with HCO_3^-, CO_2, and H_2O with the help of carbonic anhydrase.

- The equilibrium managed by carbonic anhydrase was discussed on page 379.

$$H^+(aq) + HCO_3^-(aq) \xrightleftharpoons{\text{carbonic anhydrase}} CO_2(aq) + H_2O \qquad (18.3)$$

The *forward* reaction of Equilibrium 18.3, in other words, occurs preferentially because this uses up H^+ (made by Equation 18.2) and because of the ample supply of HCO_3^-. Both stresses cause Equilibrium 18.3 to shift to the right, a shift made very rapid by carbonic anhydrase. We'll rewrite this forward shift as Equation 18.4.

- Carbonic anhydrase, one of the body's fastest-working enzymes, has to work rapidly because the red cells are always on the move.

$$H^+(aq) + HCO_3^-(aq) \xrightarrow{\text{carbonic anhydrase}} CO_2(aq) + H_2O \qquad (18.4)$$
$$\underset{\substack{\text{In the}\\\text{red cell}}}{} \quad \underset{\substack{\text{In the}\\\text{red cell}}}{} \quad\qquad \underset{\substack{\text{In the}\\\text{red cell}}}{}$$

The switch from the appearance of H^+ as a product (of Equation 18.2) to its disappearance as a *reactant* (in Equation 18.4) is called the **isohydric shift.** We see here one of the beautiful examples of coordinated activity in the body, the coupling of the uptake of oxygen to the release of the carbon dioxide to be exhaled.

The isohydric shift produces dissolved carbon dioxide, $CO_2(aq)$. But $CO_2(aq)$ changes to gaseous carbon dioxide, $CO_2(g)$, in the lungs and *the loss of CO_2 from the red cell by exhaling forces both Equations 18.4 and 18.2 to occur in the lungs.* The lower partial pressure of carbon dioxide in the lungs helps to draw $CO_2(g)$ from $CO_2(aq)$.

$$CO_2(aq) \xrightarrow{\text{exhaling}} CO_2(g) \qquad (18.5)$$

In the red cell   In exhaled air

Thus, the uptake of O_2 as hemoglobin oxygenates, which *pushes* Equilibrium 18.1 to the right, simultaneously produces a chemical (CO_2) whose loss pulls the same equilibrium to the right.

Let's now see how waste CO_2 is picked up at working cells and is carried to the lungs, and let's also see how the pickup is done cooperatively with the *release* of oxygen at the same cells now needing it.

Circulating Oxyhemoglobin Gives Up Oxygen Only Where Cells Have Made Waste Carbon Dioxide. Consider a tissue that has done chemical work, used up oxygen, and made waste carbon dioxide. When a fully oxygenated red cell arrives in such tissue, *the events that we just described reverse themselves.*

We can think of the reversal as beginning with the diffusion of waste CO_2 from the working tissue into the blood. An impetus for such diffusion is the higher PCO_2 (50 mm Hg) in active tissue versus its value in blood (40 mm Hg). Once the CO_2 arrives in the blood, it moves inside a red cell where it encounters carbonic anhydrase. Equilibrium 18.3 must now run in reverse so that the arriving CO_2 can be consumed and changed to $HCO_3{}^-(aq)$.

■ CO_2 molecules diffuse in body fluids 30 times more easily than O_2 molecules, so the partial pressure gradient for CO_2 need not be as steep as that for O_2.

$$H_2O + CO_2(aq) \xrightarrow{\text{carbonic anhydrase}} HCO_3{}^-(aq) + H^+(aq) \quad (18.3\text{—reverse})$$

From the working tissue   In the red cell

Notice that hydrogen ions are also produced. If you'll look back to Equilibrium 18.1, you will see that an increase in the level of H^+ (caused by the influx of waste CO_2) can only make Equation 18.1 run in reverse. Moreover, oxygen is drawn into the tissue needing it by the partial pressure gradient; PO_2 is lower in such tissue than in blood.

$$H^+(aq) + HbO_2{}^-(aq) \rightarrow HHb(aq) + O_2(g) \qquad (18.1\text{—reverse})$$

Just made in the red cells at working tissue   In the red cell (just arrived at tissue)   In the red cell   Will diffuse into the tissue needing oxygen

This reaction, *another isohydric shift,* not only neutralizes the acid generated by the arrival of waste CO_2, but also helps to force oxyhemoglobin to give up its oxygen. Notice the cooperation. The tissue that has used up oxygen has made CO_2 and so needs O_2 again. So the tissue has indirectly made the H^+ needed to release O_2 from newly arrived $HbO_2{}^-$.

The changes in shapes of the hemoglobin subunits now operate in reverse, and all oxygen molecules smoothly leave. It's all or nothing again, and the efficiency of the unloading of oxygen is so high that if one O_2 molecule leaves, the other three follow essentially at once. Partially deoxygenated hemoglobin units do not slip through and return to the lungs.

To summarize the reactions just studied, we can write the following equations. The cancel lines show how we arrive at the overall net results. (We'll omit the designations of physical state, like *aq* and *g.*)

■ The cancel lines help pinpoint isohydric shifts. CA is carbonic anhydrase.

Oxygenation:
(These reactions occur in the lungs.)

$$H\!Hb + O_2 \longrightarrow HbO_2{}^- + \cancel{H}^+$$

| In the red cell | From air | Will go in red cell to tissue | In the red cell |

$$\cancel{H}^+ + HCO_3{}^- \xrightarrow{\text{CA}} CO_2 + H_2O$$

| Just made | In red cell (but from tissues) | In red cell (but will be exhaled) |

■ **Net effect of oxygenating hemoglobin:**

$$H\!Hb + O_2 + HCO_3{}^- \longrightarrow HbO_2{}^- + CO_2 + H_2O \qquad (18.6)$$

| In the red cell | From air | In red cell (but from tissues) | Will go in red cell to tissue | Will leave the lungs in exhaled air |

Deoxygenation:
(These reactions occur wherever tissues are low in oxygen.)

$$CO_2 + H_2O \xrightarrow{\text{CA}} HCO_3{}^- + \cancel{H}^+$$

| Waste from tissues | | In red cell (in blood still within tissue but will go to the lungs) | In the red cell |

$$\cancel{H}^+ + HbO_2{}^- \longrightarrow H\!Hb + O_2$$

| Just made | In red cell (in blood within tissues) | In red cell (will return to the lungs) | Goes into tissue needing it |

■ **Net effect of deoxygenating oxyhemoglobin:**

$$CO_2 + H_2O + HbO_2{}^- \longrightarrow H\!Hb + HCO_3{}^- + O_2 \qquad (18.7)$$

| Waste from tissues | | In red cell (in blood with tissues) | In red cell (will go to the lungs) | Goes in blood to the lungs | Goes into tissue |

■ For those entering careers in nursing and respiratory therapy and sports medicine, there is no other single topic of such career-lasting importance than the chemistry of respiration and its associated electrolyte balance.

Some Waste CO_2 Is Carried to the Lungs on Hemoglobin.   Not all of the waste CO_2 made by working cells winds up as $HCO_3{}^-$. Some reacts with the hemoglobin just freed by deoxygenation and causes the following equilibrium to shift to the right.

$$CO_2(aq) + H\!Hb(aq) \rightleftharpoons Hb\!-\!CO_2{}^-(aq) + H^+(aq) \qquad (18.8)$$

| Waste from tissue | Just released by deoxygenation of $HbO_2{}^-$ | Carbamino-hemoglobin (in red cells) |

In working tissue where CO_2 is made, the forward reaction of Equilibrium 18.8 consumes some of the freshly made CO_2. The product, $Hb\!-\!CO_2{}^-$, is called **carbamino-hemoglobin,** and it is one form in which some waste CO_2 travels in the blood back to the lungs.

Notice in Equilibrium 18.8 that the forward reaction also produces H^+, just as did the reaction of water and waste CO_2 (Equation 18.3—reverse). Thus, whether waste CO_2 is changed to $HCO_3{}^-$ and H^+ or to $Hb\!-\!CO_2{}^-$ and H^+, hydrogen ions are made that are needed to react with $HbO_2{}^-$ and make it unload O_2 where O_2 is needed.

When the red cell reaches the lungs, $Hb\!-\!CO_2{}^-$ must release CO_2 (to be exhaled). In other words, Equilibrium 18.8 must shift to the left. The shift occurs in the lungs, not elsewhere, because the H^+ needed for the shift will be *generated* by oxygen uptake.

$$HHb(aq) + O_2(g) \longrightarrow HbO_2^-(aq) + H^+(aq) \qquad \text{(18.2, again)}$$

Bicarbonate Ion Travels Largely Outside the Red Cell. As we said, most of the waste CO_2 is carried as HCO_3^-. As it forms inside red cells, HCO_3^- migrates into the serum. The cause of its migration is the movement of chloride ion into the red cell. The Cl^- ion forms a weak bond to hemoglobin, which helps to attract Cl^- into the red cell. To keep everything electrically neutral, for every chloride ion that moves into the red cell, a like-charged ion must leave, and HCO_3^- is the ion that does this. The switch—Cl^- *into* the red cell and HCO_3^- *out* of it—is called the **chloride shift.**

At the lungs, the chloride shift reverses because HCO_3^- ions cannot help but be drawn back into the red cells. The reason is the HCO_3^- must be changed swiftly to CO_2, and the enabling enzyme (carbonic anhydrase) is only *inside* the red cell, not in the serum. The newly formed CO_2 leaves the lungs. In accordance with Le Châtelier's principle, the equilibrium shifts as much as possible to replace it. Because this shift is the forward reaction of Equilibrium 18.3, the red cell needs HCO_3^- ion.

$$HCO_3^- + H^+ \xrightarrow{\text{carbonic anhydrase}} CO_2 + H_2O \qquad \text{(18.2—reverse)}$$

Moving into red cell (with Cl^- moving out)

Leaves the lungs (drawing HCO_3^- in)

Before continuing, it would be a good idea to study this much of Section 18.3 again. It's not easy going, but the reward is understanding something vital at the molecular level of life. Think back and reflect on how the chemical events of respiration so crucial to our lives occur *automatically,* partly because equilibria shift quickly in accordance with Le Châtelier's principle at precisely those locations in the body where the shifts *must* occur. Reflect also on the role of the hydrogen ion in the equilibria. We are now in a better position to understand why the body must maintain a proper acid–base balance of the blood. Our very ability to breathe is at stake, whether in medical emergencies or in strenuous sports activities.

Myoglobin Binds Oxygen More Strongly than Hemoglobin. Myoglobin (H*Mb*) is a heme-containing protein in red muscle tissue, like heart muscle. Myoglobin's function is to bind and store oxygen for the needs of such tissue. Unlike hemoglobin, the myoglobin molecule has only one polypeptide unit and only one heme unit. Myoglobin, however, binds oxygen more strongly than hemoglobin, so *myoglobin is able to take oxygen from oxyhemoglobin.*

$$HbO_2^- + HMb \longrightarrow HHb + MbO_2^-$$

This ability is vital to heart muscle which, as much as the brain, must have an assuredly continuous supply of oxygen. When oxymyoglobin, MbO_2^-, gives up its oxygen for the cell's needs, it can at once get a fresh supply from the circulating blood. Not only does a working muscle cell now have CO_2 and H^+ available to deoxygenate HbO_2^-, it also has the superior oxygen affinity of its own myoglobin to draw more O_2 into the cell.

Fetal Hemoglobin Binds Oxygen Better than Adult Hemoglobin. The hemoglobin in a fetus is slightly different from that in an adult, and it also binds oxygen more strongly than does adult hemoglobin. This ensures that the fetus can successfully pull oxygen from the mother's oxyhemoglobin and satisfy its own needs. Soon after birth, fetal hemoglobin changes to adult hemoglobin.

■ For those who simply want to understand the respiratory demands (and limitations) of active sports, including skiing and trekking at altitude, the chemistry of respiration is the key.

■ H*Mb* = myoglobin.

18.4   ACID–BASE BALANCE OF THE BLOOD

The proper treatment of acidosis or of alkalosis depends on knowing if the underlying cause is a metabolic or a respiratory disorder.

■ Acidosis is sometimes called *acidemia,* and alkalosis is called *alkalemia.*

Acid–base balance in the blood exists when the pH of blood is in the range 7.35 to 7.45. A decrease in pH, *acidosis,* or an increase, *alkalosis,* is serious and requires prompt attention, because all the equilibria that involve H^+ in the oxygenation or the deoxygenation of blood are sensitive to pH. If the pH falls below 6.8 or rises above 7.8, life is not possible.

Disturbances in Either Metabolism or Respiration Upset the Blood's Acid–Base Balance.   In general, acidosis results from either the retention of acid or the loss of base by the body, and these can be induced by disturbances in either metabolism or respiration. Similarly, alkalosis results either from the loss of acid or the retention of base, and some disorder in either metabolism or respiration can be the underlying cause.

A malfunction in respiration can be caused by an injury or disease of the lungs or by any injury to the *respiratory centers.* These are units in the brain that sense changes in the pH and PCO_2 of the blood and instruct the lungs to breathe either more rapidly or more slowly.

■ Normal values (arterial blood):
pH = 7.35–7.45
PCO_2 = 35–45 mm Hg
$[HCO_3^-]$ = 19–24 mmol/L
(1.0 mmol HCO_3^- =
                       61 mg HCO_3^-)
1 mmol = 10^{-3} mol)

What we'll do in this section is study four situations, metabolic and respiratory acidosis and metabolic and respiratory alkalosis. We will learn how the values of pH, PCO_2, and serum $[HCO_3^-]$ change in each situation.

Metabolic Acidosis Receives a Respiratory Compensation, Hyperventilation.   In **metabolic acidosis,** the lungs and the respiratory centers are working, and the problem is metabolic. Acids are being produced faster than they are neutralized, or they are being exported too slowly.

Excessive loss of base, such as from severe diarrhea, can also result in metabolic acidosis. In diarrhea, the alkaline fluids of the duodenum leave the body, and as base migrates to replace them, there will be a depletion of base somewhere else, such as in the blood, at least for a period of time.

As the pH of the blood decreases and the molar concentration of H^+ increases, there are parallel *but momentary* increases in the values of PCO_2 and $[HCO_3^-]$. The value of PCO_2 starts to increase because the carbonate buffer, working hard to neutralize the extra H^+, manufactures CO_2.

$$H^+(aq) + HCO_3^-(aq) \rightarrow CO_2(aq) + H_2O \qquad (18.9)$$

Produced
by acidosis

The kidneys also work harder to try to replenish the supply of HCO_3^-.

The chief compensation for metabolic acidosis, however, involves the respiratory system. The respiratory centers, which are sensitive to changes in the value of PCO_2, instruct the lungs to blow CO_2 out of the body. The lungs, in other words, hyperventilate. As Equation 18.9 indicates, the loss of each molecule of CO_2 means a net neutralization of one H^+ ion.

Hyperventilation, however, is overdone. So much CO_2 is blown out that the value of PCO_2 actually decreases. A low arterial PCO_2 is called **hypocapnia.** Thus, as the blood pH decreases so, too, do the values of PCO_2 (from hyperventilation) and $[HCO_3^-]$ (from the reaction with H^+). We can summarize a number of situations that involve metabolic acidosis as follows.

Situations of Metabolic Acidosis

Lab results	pH ↓ (7.20); PCO_2 ↓ (30 mm Hg); $[HCO_3^-]$ ↓ (12 mmol/L).
Typical patient	An adult male comes to the clinic with a severe infection. He does not know that he has diabetes.
Range of causes	Diabetes mellitus; severe diarrhea (with loss of HCO_3^-); kidney failure (to export H^+ or to make HCO_3^-); prolonged starvation; severe infection; aspirin overdose; alcohol poisoning.
Symptoms	Hyperventilation (because the respiratory centers have told the lungs to remove excess CO_2 from the blood); increased urine output (to remove H^+ from the blood); thirst (to replace water lost as urine); drowsiness; headache; restlessness; disorientation.
Treatment	If the kidneys function, use isotonic HCO_3^- intravenously to restore HCO_3^- level, thereby neutralizing H^+ and raising PCO_2. In addition, restore water. In diabetes, use insulin therapy. If the kidneys do not function, hemodialysis must be tried.

■ (↓) means a decrease from a normal and (↑) means an increase. Some typical values are in parentheses. Note that some changes do not necessarily bring values outside the normal ranges.

Respiratory Acidosis Is Compensated by a Metabolic Response. In **respiratory acidosis**, either the respiratory centers or the lungs have failed, and the lungs are hypoventilating *because they cannot help it*. The blood now cannot help but retain CO_2. An increase in arterial PCO_2 is called **hypercapnia**. The retention of CO_2 functionally means the retention of acid, because CO_2 can neutralize base by the following equation.

$$OH^-(aq) + CO_2(aq) \longrightarrow HCO_3^-(aq) \qquad (18.10)$$

The *decrease* in the level of base *lowers* the pH of the blood and gives rise to respiratory acidosis.

Situations of Respiratory Acidosis

Lab results	pH ↓ (7.21); PCO_2 ↑ (70 mm Hg); $[HCO_3^-]$ ↑ (27 mmol/L).
Typical patient	Chain smoker with emphysema or has chronic obstructive pulmonary disease.
Range of causes	Emphysema; severe pneumonia; asthma; anterior poliomyelitis; or any cause of shallow breathing such as an overdose of narcotics, barbiturates, or general anesthesia; severe head injury.
Symptoms	Shallow breathing (which is involuntary).
Treatment	Underlying problem must be treated; possibly intravenous sodium bicarbonate; possibly hemodialysis.

The body responds metabolically as best it can to respiratory acidosis by using HCO_3^- to neutralize the acid, by making more HCO_3^- in the kidneys, and by exporting H^+ via the urine.

Metabolic Alkalosis Also Receives a Respiratory Compensation, Hypoventilation.
In **metabolic alkalosis**, the system has lost acid, or it has retained base (HCO_3^-), or it has been given an overdose of base (e.g., antacids). Metabolic alkalosis can also be caused by a kidney-associated decrease in the serum levels of K^+ or Cl^-. The loss of these ions means the retention of Na^+ and HCO_3^- ions, because this pair of ions works in tandem and oppositely to the K^+ and Cl^- pair. The loss of acid could be from prolonged vomiting, which removes gastric acid. This is followed by an effort to borrow serum H^+ to replace it, and the pH of the blood increases.

■ Improperly operated nasogastric suction can also remove too much gastric acid.

Whatever the cause, the respiratory centers sense an increase in the level of base in the blood (as the level of acid drops), and they instruct the lungs to retain the most readily available neutralizer of base it has, namely CO_2. (Recall that CO_2 removes OH^- according to Equation 18.10.) To help retain "acid" (CO_2) to neutralize base, the lungs hypoventilate and so remove CO_2 at a slower rate. Thus, hypoventilation compensates for metabolic alkalosis.

Notice carefully that hypoventilation alone cannot be used to diagnose if a patient has metabolic *alkalosis* or respiratory *acidosis*. Either condition means hypoventilation. One condition, respiratory acidosis, could be treated by intravenous sodium bicarbonate, a base, but this treatment would aggravate metabolic alkalosis. You can see that lab data on pH, PCO_2, and $[HCO_3^-]$ must be obtained to determine which condition is actually present. Otherwise, the treatment could be just the opposite of what should be done.

■ Compensation by hypoventilation is obviously limited by the fundamental need of the body for some oxygen.

Situations of Metabolic Alkalosis

Lab results	pH ↑ (7.53); PCO_2 ↑ (56 mm Hg); $[HCO_3^-]$ ↑ (45 mmol/L)
Typical patient	Postsurgery patient with persistent vomiting.
Range of causes	Prolonged loss of stomach contents (vomiting or nasogastric suction); overdose of bicarbonate or ulcer medications; severe exercise, or stress, or kidney disease (with loss of K^+ and Cl^-); overuse of a diuretic.
Symptoms	Hypoventilation (to retain CO_2); numbness, headache, tingling; possibly convulsions.
Treatment	Isotonic ammonium chloride (a mild acid), intravenously with great care; replace K^+ loss.

■ An overdose of "bicarb" ($NaHCO_3$), from a too aggressive use of this home remedy for "heartburn," can cause metabolic alkalosis.

■ Ammonium ion acts as a neutralizer as follows:

$$NH_4^+ + OH^- \rightarrow NH_3 + H_2O$$

Respiratory Alkalosis Is Compensated Metabolically by a Reduced Bicarbonate Level. In **respiratory alkalosis,** the body has lost acid usually by some involuntary hyperventilation such as hysterics, prolonged crying, overbreathing at high altitudes, or the mismanagement of a respirator. The respiratory centers have lost control, and the body expels CO_2 too rapidly. The loss of CO_2 means the loss of a base neutralizer from the blood. Hence, the level of base increases; the blood pH therefore increases. The body's compensation is metabolic; the kidneys excrete base, HCO_3^-, so the serum level of HCO_3^- decreases and the blood pH decreases.

Extreme respiratory alkalosis can occur in high mountain climbers, like climbers of Mount Everest (8848 m, 29,030 ft). At its summit, the barometric pressure is 253 mm Hg and the PO_2 of the air is only 43 mm Hg (as compared with 149 mm Hg at sea level). Hyperventilation brings a climber's arterial PCO_2 down to only 7.5 mm Hg (compared with a normal of 40 mm Hg) and the blood pH to above 7.7! Given a few days time, the body tries to adjust to high altitude by making more hemoglobin and more red blood cells.

■ No conditioning at a low altitude can get the cardiovascular system ready for a low pO_2 at a high altitude.

■ Tissue that gets too little O_2 is in a state of *hypoxia*. If it gets none at all, it is in a state of *anoxia*.

Situations of Respiratory Alkalosis

Lab results	pH ↑ (7.56); PCO_2 ↓ (23 mm Hg); $[HCO_3^-]$ ↓ (20 mmol/L).
Typical patient	Someone nearing surgery and experiencing anxiety.
Range of causes:	Prolonged crying; rapid breathing at high altitudes; hysterics; fever; disease of the central nervous system; improper management of a respirator.
Symptoms	Hyperventilation (that can't be helped); numbness, headache, tingling; possibly convulsions.
Treatment	Rebreathe one's own exhaled air (by breathing into a sack); administer carbon dioxide. Treat underlying causes.

Take careful notice that hyperventilation alone cannot be used to tell what the condition is. Either metabolic *acidosis* or respiratory *alkalosis* is accompanied by hyperventilation, but the treatments are opposite in nature.

Combinations of Primary Acid—Base Disorders Are Possible. We have just surveyed the *four primary acid—base disorders.* Combinations of these are often seen, and health care professionals have to be alert to the ways in which the lab data vary in such combinations. Someone with diabetes, for example, might also suffer from an obstructive pulmonary disease. Diabetes causes metabolic acidosis and a *decrease* in $[HCO_3^-]$. The pulmonary disease causes respiratory acidosis with an *increase* in $[HCO_3^-]$. In combination, then, the lab data on bicarbonate level will not be in the expected pattern for either. We will not carry the study of such complications further. We mention them only to let you know they exist. There are standard ways to recognize them.[2]

■ People working in emergency care situations obtain the requisite lab data rapidly, and they must be able to interpret the data on the spot.

18.5 ACID—BASE BALANCE AND SOME CHEMISTRY OF KIDNEY FUNCTIONS

Both filtration and chemical reactions in the kidneys help to regulate the electrolyte balance of the blood.

Diuresis is the formation of urine in the kidneys, and it is an integral part of the body's control of the electrolyte and buffer levels in blood.

Urea Is the Chief Nitrogen Waste Exported in the Urine. Huge quantities of fluids leave the blood by diffusion each day at the hundreds of thousands of filtering units (called *glomeruli*) in each kidney. Substances in solution but not those in colloidal dispersions (e.g., proteins) leave in these fluids. Then active transport processes in kidney cells pull all of any escaped glucose, any amino acids, and most of the fluids and electrolytes back into the blood. Most of the wastes are left in the urine being made.

■ The net urine production is 0.6 to 2.5 L/day.

Urea is the chief nitrogen waste (30 g/day), but creatinine (1 to 2 g/day), uric acid (0.7 g/day), and ammonia (0.5 g/day) are also excreted with the urine. If the kidneys are injured or diseased and cannot function, wastes build up in the blood, which leads to a condition known as *uremic poisoning.* When the kidneys fail, either hemodialysis must be tried (Special Topic 6.3) or a kidney transplant is necessary.

■ *Ur-*, "of the urine"; *-emia,* "of the blood." *Uremia* means substances of the urine present in the blood.

The Hormone Vasopressin Helps Control Water Loss. A nonapeptide hormone, *vasopressin,* instructs the kidneys to retain or excrete water and thus helps to regulate the overall levels of solutes in blood. The hypophysis, where vasopressin is made, releases it when the osmotic pressure of blood increases by as little as 2%. At the kidneys, vasopressin promotes the reabsorption of water, and therefore it is often called the *antidiuretic hormone* or ADH.

When the blood's osmotic pressure is higher than normal (hypertonicity), it means that there is a higher concentration of solutes and colloids in blood. The released vasopressin therefore helps the blood to retain water and thus keeps the solute levels from going still higher. In the meantime, the thirst mechanism is stimulated to bring in water to dilute the blood.

■ The monitors in the hypophysis of the blood's osmotic pressure are called *osmoreceptors.*

[2]See, for example, H. Valtin and F. J. Gennari, *Acid—Base Disorders, Basic Concepts and Clinical Management,* 1987. Little, Brown and Company, Boston.

Conversely, if the osmotic pressure of blood decreases (becomes hypotonic) by as little as 2%, the hypophysis retains vasopressin. None reaches the kidneys, so the water that has left the bloodstream at filtering units does not return as much. Remember that a low osmotic pressure means a low concentration of solutes, so the absence of vasopressin at the kidneys when the blood is hypotonic lets urine form. This reduces the amount of water in the blood and thereby raises the concentrations of its dissolved matter. You can see that with the help of vasopressin a normal individual can vary the intake of water widely and yet preserve a stable, overall concentration of substances in blood.

■ In *diabetes insipidus,* vasopressin secretion is blocked, and unchecked diuresis can make from 5 to 12 L of urine a day.

The Hormone Aldosterone Helps the Blood Retain Sodium Ion.   The adrenal cortex makes *aldosterone,* a hormone that works to stabilize the sodium ion level of the blood. A steroid hormone, aldosterone, is secreted if the blood's sodium ion level drops. When aldosterone arrives at the kidneys, it initiates reactions that make sodium ions that have left the blood return. Of course, to keep things isotonic in the blood, the return of sodium ions also requires the return of water.

Conversely, if the sodium ion level of the blood increases, then aldosterone is not secreted, and sodium ions that have migrated out of the blood at a filtering unit are permitted to stay out. They remain in the urine being made, together with some extra water.

The Kidneys Make HCO₃⁻ for the Blood's Buffer System.   We have seen that breathing is the body's most direct means of controlling *acid* as breathing removes or retains CO_2. The kidneys are the body's means of controlling *base,* as they make or remove HCO_3^-.

■ Urine taken after several hours of fasting normally has a pH of 5.5 to 6.5.

The kidneys also adjust the blood's levels of HPO_4^{2-} and $H_2PO_4^-$, the anions of the phosphate buffer. Moreover, when acidosis develops, the kidneys can put H^+ ions into the urine. Some neutralization of these ions by HPO_4^{2-} and by NH_3 takes place, but the urine becomes definitely more acidic as acidosis continues.

■ In severe acidosis, the pH of urine can go as low as 4.

Figure 18.6 shows the various reactions that take place in the kidneys, particularly during acidosis. (The numbers in the following boxes refer to this figure.) The breakdown of metabolites, [1], makes carbon dioxide, which enters the equilibrium whose formation is catalyzed by carbonic anhydrase, [2], to give both bicarbonate ion and hydrogen ion. The bicarbonate ion goes into the bloodstream, [3], but the hydrogen ion is put into the tubule, [4], where urine is accumulating. The urine already contains sodium ions and monohydrogen phosphate ions, but to make step [4] possible, *some* positive ion has to go with the HCO_3^-. Otherwise, there would be no net electrical balance. The kidneys have the ability to select Na^+ to go with HCO_3^- at [3]. The kidneys can make Na^+ travel one way and H^+ the other. Newly arrived H^+ can be buffered by HPO_4^{2-} in the developing urine, [5]. Moreover, the kidneys have an ability not generally found in other tissues to synthesize ammonia and use it to neutralize H^+, [6]. Thus the ammonium ion also appears in the urine.

The Kidneys Excrete Organic Anions.   When acidosis has a metabolic origin, the serum levels of the anions of organic acids increase. Organic acids are made at accelerated rates in metabolic disorders, like diabetes and starvation, and are the chief cause of the pH change in metabolic acidosis. The base in the blood buffer has to neutralize them.

The kidneys let organic anions stay in the urine, but only by letting increasing quantities of water stay, too. There is a limit to how concentrated the urine can become, so as solutes stay in the urine, water must also stay. Someone with metabolic acidosis, therefore, can experience a general dehydration as the system borrows water from other fluids to make urine. The thirst mechanism normally brings in replacement water, so the individual drinks copious amounts of fluids.

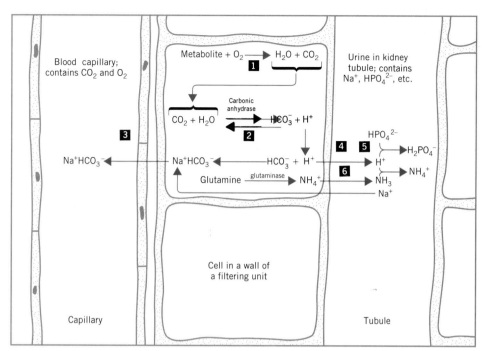

FIGURE 18.6
Acidification of the urine. The numbers refer to the text discussion.

The Kidneys Can Export HCO$_3^-$. In alkalosis, the kidneys can put bicarbonate ion into the urine and so use HCO$_3^-$, not HPO$_4^{2-}$ to neutralize H$^+$. Both actions raise the pH of the urine, and in severe alkalosis it can go over 8.

The Kidneys Also Help to Regulate Blood Pressure. If the blood pressure drops, as in hemorrhaging, the kidneys secrete a trace of *renin* into the blood. Renin is an enzyme that acts on one of the zymogens in blood, angiotensinogen, to convert it to the enzyme angiotensin I. This, in turn, helps to convert still another protein in blood to angiotensin II, a neurotransmitter.

Angiotensin II is the most potent vasoconstrictor known. When it makes blood capillaries constrict, the heart has to work harder, and this makes the blood pressure increase. This helps to ensure that some semblance of proper filtration continues at the kidneys.

Angiotensin II also triggers the release of aldosterone, which we've already learned helps the blood to retain water. This is important because the maintenance of the overall blood volume is needed to sustain a proper blood pressure.

SUMMARY

Digestion α-Amylase in saliva begins the digestion of starch. Pepsin in gastric juice starts the digestion of proteins. In the duodenum, trypsinogen (from the pancreas) is activated by enteropeptidase (from the intestinal juice) and becomes trypsin, which helps to digest proteins. Enteropeptidase also activates chymotrypsin (from chymotrypsinogen), carboxypeptidase (from procarboxypeptidase), and elastin (from proelastin). These also help to digest proteins.

The pancreas supplies an important lipase, which, with the help of the bile salts, catalyzes the digestion of hydrolyzable lipids. The bile salts also aid in the absorption of the fat-soluble vitamins, A, D, E, and K.

Intestinal juice supplies enzymes for the digestion of disaccharides, nucleic acids, small polypeptides, and lipids.

The end products of the digestion of proteins are amino acids; those of carbohydrates are glucose, fructose, and galactose; and those of the triacylglycerols are fatty acids and monoacylglycerols. Complex lipids are also hydrolyzed, and nucleic acids yield phosphate, pentoses, and heterocyclic amines.

Blood   Proteins in blood give it a colloidal osmotic pressure that assists in the exchange of nutrients at capillary loops. Albumins are carriers for hydrophobic molecules and serum-soluble metallic ions. Fibrinogen is the precursor of fibrin, the protein of a blood clot.

Among the electrolytes, anions of carbonic and phosphoric acid are involved in buffers, and all ions are involved in regulating the osmotic pressure of the blood. The chief cation in blood is Na^+, and the chief cation in cells is K^+.

The blood transports oxygen and products of digestion to all tissues. It carries nitrogen wastes to the kidneys. It unloads cholesterol and heme breakdown products at the gallbladder. And it transports hormones to their target cells. Lymph, another fluid, helps to return some substances to the blood from tissues.

Sudden failure to retain the protein in blood leads to an equally sudden loss in blood volume and a condition of shock. Slower losses of protein, as in kidney disease or starvation, lead to edema.

Respiration   The relatively high pressure in the lungs helps to force O_2 into HHb. This creates HbO_2^- and H^+. In an isohydric shift, the H^+ is neutralized by HCO_3^-, which is returning from working tissues that make CO_2, which leaves during exhaling. Some of the H^+ also converts $HbCO_2^-$ to HHb and CO_2.

In deoxygenating HbO_2^- at cells that need oxygen, the influx of CO_2 makes HCO_3^- and H^+. The H^+ then moves (isohydric shift) to HbO_2^- and breaks it down to HHb and O_2. Both oxygenation and deoxygenation of blood are done with the cooperative flexibility of the shapes of the subunits of hemoglobin.

In red muscle tissue, myoglobin's superior ability to bind oxygen ensures that such tissue obtains oxygen from the deoxygenation of oxyhemoglobin. Fetal hemoglobin also has a superior oxygen binding ability.

Acid–Base Balance   The body uses the bicarbonate ion of the carbonate buffer to inhibit acidosis by irreversibly removing H^+ when the lungs release CO_2. The HCO_3^- ion is replaced by the kidneys, which can also put excess H^+ into the urine. Dissolved CO_2 in the blood's carbonate buffer works to control alkalosis by neutralizing OH^-. Metabolic acidosis, with hyperventilation, and metabolic alkalosis, with hypoventilation, arise from dysfunctions in metabolism. Respiratory acidosis, with hypoventilation, and metabolic alkalosis, with hyperventilation, occur when the respiratory centers or the lungs are not working.

Diuresis   The kidneys, with the help of hormones and changes in blood pressure, blood osmotic pressure, and concentrations of ions, monitor and control the concentrations of solutes in blood. Vasopressin tells the kidneys to keep water in the bloodstream. Aldosterone tells the kidneys to keep sodium ion (and therefore water also) in the bloodstream. In acidosis, the kidneys transfer H^+ to the urine and replace some of the HCO_3^- lost from the blood. In alkalosis the kidneys put some HCO_3^- into urine.

REVIEW EXERCISES

The answers to Review Exercises whose numbers are in color are found in Appendix V. The answers to the other Review Exercises are found in the Study Guide that accompanies this text. The more challenging questions are marked with asterisks.

Digestion

18.1   What are the names of the two chief extracellular fluids?

18.2   Name the fluids that have digestive enzymes or digestive zymogens.

18.3   What enzymes or zymogens are there, if any, in each of the following?
(a) saliva   (b) gastric juice
(c) pancreatic juice (d) bile
(e) intestinal juice

18.4   Name the enzymes and the digestive juices that supply them (or their zymogens) that catalyze the digestion of each of the following.
(a) large polypeptides   (b) triacylglycerols
(c) amylose   (d) sucrose
(e) di- and tripeptides   (f) nucleic acids

18.5   What are the end products of the digestion of each of the following?
(a) proteins   (b) carbohydrates
(c) triacylglycerols

*18.6   What functional groups are hydrolyzed when each of the substances in Review Exercise 18.5 is digested? (Refer back to earlier chapters if necessary.)

18.7   In what way does enteropeptidase function as a "master switch" in digestion?

18.8   What would happen if the pancreatic zymogens were activated within the pancreas?

18.9   What services do the bile salts render in digestion?

18.10   What does mucin do (a) for food in the mouth and (b) for the stomach?

18.11   What is the catalyst for each of the following reactions?
(a) Pepsinogen $\rightarrow$ pepsin
(b) Trypsinogen $\rightarrow$ trypsin
(c) Chymotrypsinogen $\rightarrow$ chymotrypsin
(d) Procarboxypeptidase $\rightarrow$ carboxypeptidase
(e) Proelastase $\rightarrow$ elastase

18.12   Rennin does what for an infant?

18.13   Why is gastric lipase unimportant to digestive processes in the adult stomach but useful in the infant stomach?

18.14   In terms of where they work, what is different about intestinal juice and pancreatic juice?

18.15   What secretion neutralizes chyme, and why is this work important?

18.16   What happens to the molecules of the monoacylglycerols and fatty acids that form from digestion?

18.17   In a patient with a severe obstruction of the bile duct the feces appear clay-colored. Explain why the color is light.

Substances in Blood

18.18   In terms of their general composition, what is the greatest difference between blood plasma and interstitial fluid?

18.19   What is the largest contributor to the net osmotic pressure of the blood as compared with the interstitial fluid?

18.20   What is fibrinogen? Fibrin?

18.21   What services are performed by albumins in blood?

18.22   In what two different regions are Na^+ and K^+ ions mostly found? What are the chief functions of these ions?

Exchange of Nutrients at Capillary Loops

18.23   What two opposing forces are at work on the arterial side of a capillary loop? What is the net result of these forces, and what does the net force do?

18.24   On the venous side of a capillary loop there are two opposing forces. What are they, what is the net result, and what does this cause?

18.25   Explain how a sudden change in the permeability of the capillaries can lead to shock.

18.26   Explain how each of the following conditions leads to edema.
(a) kidney disease   (b) starvation   (c) a mechanical blow

Exchange of Respiratory Gases

18.27   What are the respiratory gases?

18.28   What compound is the chief carrier of oxygen to actively metabolizing tissues?

18.29   The binding of oxygen to hemoglobin is said to be *cooperative*. What does this mean, and why is it important?

*__18.30__   Write the equilibrium expression for the oxygenation of hemoglobin. In what direction does this equilibrium shift when
(a) The pH decreases?
(b) The PO_2 decreases?
(c) The red cell is in the lungs?
(d) The red cell is in a capillary loop of an actively metabolizing tissue?
(e) CO_2 comes into the red cell?
(f) HCO_3^- ions flood into the red cell?

*__18.31__   Using chemical equations, describe the isohydric shift when a red cell is (a) in actively metabolizing tissues and (b) in the lungs.

*__18.32__   In what two ways does the oxygenation of hemoglobin in red cells in alveoli help to release CO_2?

*__18.33__   In what way does waste CO_2 at active tissues help to release oxygen from the red cell?

*__18.34__   In what way does extra H^+ at active tissue help release oxygen from the red cell?

18.35   Where is carbonic anhydrase found in the blood, and what function does it have in the management of the respiratory gases in (a) an alveolus and (b) actively metabolizing tissues?

18.36   What are the two main forms in which waste CO_2 moves to the lungs?

18.37   How is the ability of oxygen to bind to hemoglobin affected by PCO_2, and how is this beneficial?

18.38   What is the chloride shift and how does it aid in the exchange of respiratory gases?

18.39   In what way is the superior ability of myoglobin over hemoglobin to bind oxygen important?

18.40   Fetal hemoglobin has a higher oxygen affinity than adult hemoglobin. Why is this important to the fetus?

Acid–Base Balance of the Blood

18.41   Construct a table using arrows ($\uparrow$ or $\downarrow$) and typical lab data that summarize the changes observed in respiratory and metabolic acidosis and alkalosis. The column headings should be as follows.

Condition	pH	PCO_2	$[HCO_3^-]$

*__18.42__   With respect to the *directions* of the changes in the values of pH, PCO_2, and $[HCO_3^-]$ in both respiratory acidosis and metabolic acidosis, in what way are the two types of acidosis the same? In what way are they different?

18.43 Hyperventilation is observed in what two conditions that relate to the acid–base balance of the blood? In one, administration of carbon dioxide is sometimes used, and in the other, administration of isotonic HCO_3^- can be a form of treatment. Which treatment goes with which condition and why?

18.44 In what two conditions that relate to the acid–base balance of the blood is hypoventilation observed? Isotonic ammonium chloride and isotonic sodium bicarbonate are possible treatments. Which treatment goes with which condition, and how do they work?

18.45 In which condition relating to acid–base balance does hyperventilation have a beneficial effect? Explain.

*18.46 Hyperventilation is part of the *cause* of the problem in which condition relating to the acid–base balance of the blood?

*18.47 Hypoventilation is the body's way of helping itself in which condition that relates to the acid–base balance of the blood?

*18.48 In which condition that concerns the acid–base balance of the blood is hypoventilation part of the *problem* rather than the cure?

18.49 How can a general dehydration develop in metabolic acidosis?

*18.50 Which condition, metabolic or respiratory acidosis or alkalosis, results from each of the following situations?

(a) hysterics (b) overdose of bicarbonate
(c) emphysema (d) narcotic overdose
(e) diabetes (f) overbreathing at a high altitude
(g) severe diarrhea (h) prolonged vomiting
(i) cardiopulmonary disease (j) barbiturate overdose

*18.51 Referring to Review Exercise 18.50, which is happening in each situation, hyperventilation or hypoventilation?

18.52 Why does hyperventilation in hysterics cause alkalosis?

18.53 Explain how emphysema leads to acidosis.

18.54 Prolonged vomiting leads to alkalosis. Explain.

18.55 Uncontrolled diarrhea can cause acidosis. Explain.

18.56 Respiratory alkalosis causes hypocapnia or hypercapnia?

18.57 For each 1 °C above normal of human body temperature, the rate of CO_2 production increases by 13 percent. If the rate of breathing does not increase, what results—hypocapnia or hypercapnia?

Blood Chemistry and the Kidneys

18.58 If the osmotic pressure of the blood has increased, what, in general terms, has changed to cause this?

18.59 How does the body respond to an increase in the osmotic pressure of the blood?

18.60 If the sodium ion level of the blood falls, how does the body respond?

18.61 What is the response of the kidneys to a decrease in blood pressure?

18.62 Alcohol in the blood suppresses the secretion of vasopressin. How does this affect diuresis?

18.63 In what ways do the kidneys help to reduce acidosis?

Gastric Flow and Ulcer Treatment (Special Topic 18.1)

18.64 Describe how the flow of the acid component of gastric juice is controlled.

18.65 Cimetidine is described as a *competitive enzyme inhibitor*. What do you suppose this means? How does cimetidine work to aid in the healing of an ulcer?

Additional Exercises

18.66 Monoacylglycerols are able to migrate through membranes of the cells of the intestinal tract that absorb them. Glycerol, however, is unable to accomplish this movement. How might we explain these relative abilities to migrate through a cell membrane?

*18.67 When the gallbladder is surgically removed, lipids of low formula mass are the only kinds that can be easily digested. Explain.

*18.68 It has been reported that some long-distance Olympic runners have trained at high altitudes and then had some of their blood withdrawn and frozen. Days or weeks after returning to lower altitudes and just prior to a long race, they have used some of this blood to replace an equal volume of what they are carrying. This is supposed to help them in the race. How would it work?

*18.69 Aquatic diving animals are known to have much larger concentrations of myoglobin in their red muscle tissue than humans. How is this important to their lives?

18.70 If the H^+ that forms when O_2 combines with hemoglobin is not neutralized, how is oxygen uptake by the red cell in the lungs affected? Explain.

18.71 If the bicarbonate ion carried by the blood is not neutralized as the blood moves through the lungs, what happens to oxygen uptake by the blood in the lungs?

*18.72 What happens to oxygen uptake in the lungs if you breathe air which has been enriched in carbon dioxide, *without affecting the partial pressure of its oxygen?* Explain. (Assume that the enrichment in CO_2 has been done at the expense of the nitrogen in air, not the oxygen.)

19

MOLECULAR BASIS OF ENERGY FOR LIVING

Overview of
  Biochemical
  Energetics

Energy from
  Carbohydrates

Citric Acid Cycle

Respiratory Chain

Energy from Fatty
  Acids

Energy from Amino
  Acids

Acidosis and Energy
  Problems

All of the body's energy delivering systems are surging in Norwegian speedskater Edel Hoiseth as she rounds a turn during the 1994 Olympics. The molecular basis of energy for living is the subject of this chapter.

19.1   OVERVIEW OF BIOCHEMICAL ENERGETICS

High-energy phosphates, like ATP, are the body's means of trapping chemical energy from the oxidation of the products of digestion.

■ *Catabolism* is from the Greek *cata-,* "down;" and *ballein,* "to throw or cast." Its opposite is *anabolism,* reactions that make larger molecules from smaller ones.

We cannot use solar energy, like plants, nor can we use steam energy, like locomotives. We need *chemical* energy for living, and we obtain it from food, which we use to make our own high-energy molecules. Their energy then drives the chemical processes behind the work of muscles, nerves, and the synthesis of compounds in various tissues. We tap into the chemical energy of foods largely by the **catabolism** (i.e., break down) of carbohydrates and fatty acids, although we can also use proteins for energy.

The Resynthesis of ATP from ADP and P_i *Is the Centerpiece of Biochemical Energetics.*   From the lowest to the highest forms of life, the principal carriers of chemical energy are high-energy phosphates, like *adenosine triphosphate* (ATP). ATP's chemical energy, stored in its phosphoric anhydride system (page 300), is used to drive the chemical reactions required to operate muscles and nerves and to make other compounds. As ATP delivers its energy, its molecules break down to adenosine diphosphate (ADP) and inorganic phosphate ion (P_i).

■ P_i is mostly $H_2PO_4^-$ and HPO_4^{2-}.

Once ATP has been consumed, more must quickly be made or no further work is possible. One of the major purposes of catabolism is to transfer chemical energy from molecules of carbohydrates and lipids in such a way that ATP is resynthesized from ADP and P_i. The resynthesis of ATP is the chief means by which living systems trap the energy available by the oxidation of metabolites. We'll now do a broad survey of all of the pathways for ATP synthesis in humans (Figure 19.1).

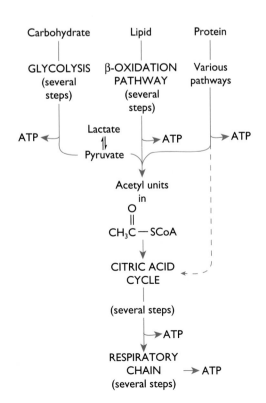

FIGURE 19.1
Major pathways for making ATP.

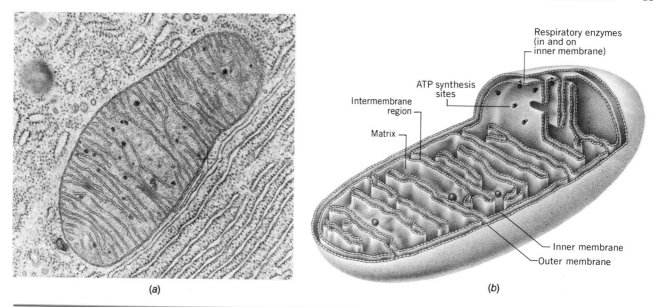

(a)   *(b)*

FIGURE 19.2

A mitochondrion. *(a)* Electron micrograph ($\times 53,000$) of a mitochondrion in a pancreas cell of a bat. *(b)* Perspective showing the interior. The respiratory enzymes are incorporated into the inner membrane. On the inside of this membrane are enzymes that catalyze the synthesis of ATP. (Micrograph courtesy of Dr. Keith R. Porter.)

The Appearance of ADP and P$_i$ Initiate the Respiratory Chain. The resynthesis of ATP from ADP and P$_i$ is under feedback control, meaning that when the supply of ATP is high, ATP resynthesis is shut down. A low supply of ATP means that its breakdown products, ADP and P$_i$, are necessarily present, and they are the species that launch the operation of the **respiratory chain**, the body's major supplier of ATP. So when ATP has to be made, the respiratory chain goes into action. Its operation requires both oxygen and organic suppliers of hydride ion, H:$^-$.

■ When the level of ATP drops and the levels of ADP and P$_i$ increase, the rate of breathing is also accelerated.

The Mitochondrion Is the Chief Site of ATP Synthesis. The cell's principal site of ATP synthesis, a *mitochondrion*, is often dubbed the powerhouse of the cell (Figure 19.2). Some tissues have thousands of them in each cell. Each mitochondrion has two important membranes, one outer and one inner. The inner membrane is very convoluted and has a surface area many times that of the outer. The region between the two membranes is called the *intermembrane space*. The space deep inside the inner membrane is filled with a gel-like material called the mitochondrial *matrix*. It's less than 50% water and rich in soluble enzymes, cofactors, inorganic ions, and substrates.

■ A cell in the flight muscle of a wasp has about a million mitochondria.

The Citric Acid Cycle Is a Major Supplier of Hydride Ion to the Respiratory Chain. When the respiratory chain starts, other cycles of metabolism stir to supply compounds required to run the chain. The purpose of the **citric acid cycle**, for example (next up from the bottom in Figure 19.1), is to supply the respiratory chain with donors of hydride ion, H:$^-$. Several intermediates of the citric acid cycle serve this purpose. The citric acid cycle itself, however, cannot operate without its own fuel, which is *acetyl coenzyme A*, often written as acetyl CoA.

Acetyl coenzyme A
(acetyl CoA)

Acetyl Coenzyme A Can Be Made by the Catabolism of Carbohydrates. All three food groups can supply chemical energy, as we said, but carbohydrates and fatty acids are most often used in a well-nourished person. Carbohydrates supply acetyl CoA

■ From the Greek *glykos,* "sugar or sweet;" and *-lysis,* dissolution.

either from starch or from glucose by a pathway called **glycolysis** (see also Figure 19.1). Its reactions break glucose to the pyruvate ion, and then a short pathway converts pyruvate into acetyl CoA.

Glycolysis also produces some ATP independently of the respiratory chain. A secondary source of ATP is vital when a cell is temporarily low on oxygen and the chain cannot operate rapidly enough. Glycolysis is able to make some ATP until the oxygen supply is reestablished.

The full sequence of oxygen-consuming reactions from glucose through glycolysis, acetyl CoA, and the respiratory chain is called the **aerobic sequence** of glucose catabolism. Glycolysis alone, when run without oxygen, is called the **anaerobic sequence** of glucose catabolism. Athletes "go anaerobic" during particularly strenuous efforts, like a 100-m dash. Such a pace cannot be sustained for a long-distance run, however, because there are limits to the abilities of tissues to tolerate both a low supply of oxygen and the wastes from anaerobic glycolysis. Rapid breathing and less strenuous efforts (or rest) sooner or later get the system back on the aerobic track.

■ *Aerobic* signifies the use of air. *Anaerobic,* stemming from "not air," means in the absence of the use of oxygen.

Acetyl CoA Is Also Made by the Catabolism of Fatty Acids. A great deal of ATP production can come from the catabolism of fatty acids, and sustained athletic efforts draw on them. A series of reactions called **β-oxidation** breaks fatty acids two carbons at a time into acetyl CoA, which is fed into the respiratory chain. Some of the intermediates in β-oxidation can also feed hydride ion directly to the chain.

Acetyl CoA Can Also Be Made by the Breakdown of Amino Acids. The bodies of undernourished or starving people, who obtain insufficient dietary carbohydrate and fat, draw on the proteins of their own tissues to provide ATP. Most amino acids can be catabolized to acetyl units or to intermediates of the citric acid cycle itself. In doing this, people not only waste away, they also upset osmotic pressure relationships internally. The osmolarity of the blood decreases as its albumins are slowly used up, which leads to water retention in the interstitial and intracellular compartments and general edema.

19.2 ENERGY FROM CARBOHYDRATES

The end product of glycolysis is pyruvate ion, when oxygen is available, and lactate ion under anaerobic conditions.

Glycolysis Begins with Either Glycogen or Glucose. As we noted earlier, *glycolysis* is a series of reactions that catabolize glucose while a small but important amount of ATP is made (Figure 19.3). The starting point can be either glucose or a glucose unit in glycogen. Glycolysis also accommodates galactose and fructose. Near the top of Figure 19.3, we see fructose entering the scheme, and galactose is changed (by reactions not shown) to an early intermediate in glycolysis. Thus, *glycolysis is central to the catabolism of all dietary monosaccharides.*

■ *Bis* in "bisphospho-" means that there are two *separate* phosphate ester groups, not one diphosphate unit with a phosphoric anhydride system (see page 300).

The initial steps of glycolysis that lead to fructose-1,6-bisphosphate are actually up an energy hill, because they consume ATP. But this is like pushing a sled or bike up the short backside of a long hill. It's an investment in energy more than repaid by the long, downhill slide that follows.

The connection between aerobic glycolysis and the citric acid cycle is a short series of reactions that convert pyruvate into acetyl CoA. The overall change is

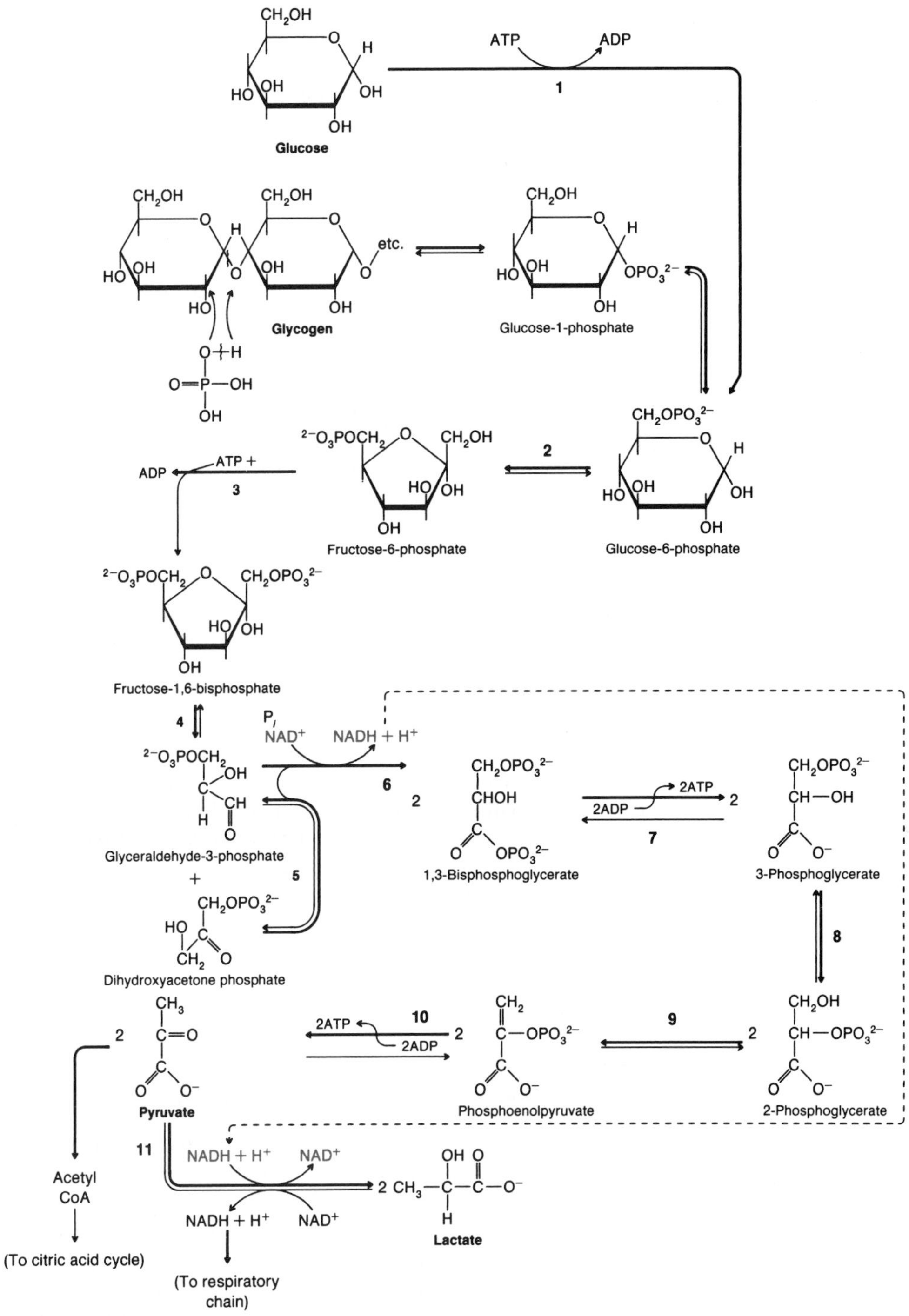

FIGURE 19.3

Glycolysis.

$$\overset{O}{\underset{\parallel}{CH_3CCO_2^-}} + CoASH + NAD^+ \xrightarrow{\quad \text{five steps} \quad} \overset{O}{\underset{\parallel}{CH_3CSCoA}} + CO_2 + NADH$$

Pyruvate   Coenzyme A   Acetyl coenzyme A

It is acetyl CoA that directly enters the citric acid cycle.

Anaerobic Glycolysis Ends in Lactate. Except during extensive exercise, glycolysis is operated with sufficient oxygen and so is *aerobic*. Sometimes a cell receives oxygen at a slower rate than necessary to sustain the work of making ATP by the respiratory chain. The cell or its tissue is then said to be running an **oxygen debt,** and glycolysis now ends with the lactate ion. The overall equation for *anaerobic glycolysis* is

$$\underset{\text{Glucose}}{C_6H_{12}O_6} + 2ADP + 2P_i \xrightarrow[\text{glycolysis}]{\text{anaerobic}} \underset{\text{Lactate}}{2CH_3\overset{\overset{\displaystyle OH}{|}}{C}HCO_2^-} + 2H^+ + 2ATP$$

Notice that acid (H^+) is one product. As lactate is produced, so is acid, and the body must neutralize the acid by its buffers. If anaerobic glycolysis taxes the buffers too much, a form of metabolic acidosis called *lactic acid acidosis* results. *The accumulation of lactate and acid during strenuous physical activities challenges the buffers severely and eventually limits athletic accomplishment.* Hyperventilation is initiated to blow out carbon dioxide and thus remove acid, but the athlete must eventually slow down, breathe deeply, and so allow the body to remove the lactate and to operate aerobically again. The conditioning that athletes endure is partly meant to make the heart muscle very strong and to strengthen the lungs and all aspects of respiration. Then the intake of oxygen, its delivery, and the removal of waste carbon dioxide operate at maximum efficiency. The removal of lactate, whose presence is commonly associated with sore muscles, requires the oxygen debt in affected tissues be repaid by rest and recuperation. Then lactate can be oxidized to pyruvate, which breaks up into acetyl CoA and so fuels the respiratory chain as we just studied.

Lactate Forms in Anaerobic Glycolysis in order to Regenerate an Enzyme. We will not go through glycolysis step by step, but one step is important to an understanding of the difference between the aerobic and anaerobic alternatives. This is the oxidation of glyceraldehyde-3-phosphate to 1,3-bisphosphoglycerate. It's roughly in the middle of Figure 19.3.

■ On page 381 we introduced the idea of using $NADH/H^+$ to represent $NADH + H^+$.

$$\underset{\substack{\text{Glyceraldehyde-}\\\text{3-phosphate}}}{^{-2}O_3POCH_2\overset{\overset{\displaystyle HO}{|}}{C}H\overset{\overset{\displaystyle O}{\|}}{C}H} + P_i + NAD^+ \longrightarrow \underset{\text{1,3-Bisphosphoglycerate}}{^{-2}O_3POCH_2\overset{\overset{\displaystyle HO}{|}}{C}H\overset{\overset{\displaystyle O}{\|}}{C}OPO_3^{2-}} + NADH/H^+$$

This step requires NAD^+ as an oxidizing agent, so its reduced form, $NADH/H^+$, is in the mixture of products. When oxygen is available, $NADH/H^+$ is reoxidized to NAD^+ by the respiratory chain, and the associated enzyme is restored to its initial form. In other words, aerobic conditions let glycolysis continue over and over because all of its enzymes, with the help of the chain, are routinely regenerated. When the cell lacks sufficient oxygen, however, another way to reoxidize $NADH/H^+$ to NAD^+ must be provided and so restore the associated enzyme. Otherwise, with a "plugged" enzyme, glycolysis would shut down, leaving the cell without a means to continue the synthesis of ATP.

■ No reaction in the body is truly complete until its enzyme is fully restored.

The reoxidation of $NADH/H^+$ in the absence of oxygen and without the help of the chain is done by using pyruvate as the agent for the oxidation of $NADH/H^+$. Normally, pyruvate is the end product of aerobic glycolysis, but its keto group, $C{=}O$, is able to accept $H{:}^-$ from $NADH/H^+$, change to the $CH{-}OH$ group of lactate, and thus restore the NAD^+ enzyme needed if anaerobic glycolysis is to continue. (See the long dashed line in Figure 19.3.) *Lactate serves to store H:$^-$ until the cell once again becomes aerobic,* and $H{:}^-$ can be fed into the respiratory chain. This is why lactate, not pyruvate, is the end product of anaerobic glycolysis. Of course, there are limits. The longer the cell operates anaerobically, the more lactate accumulates and the more lactic acid acidosis develops, as noted earlier.

The Pentose Phosphate Pathway of Glucose Catabolism Makes an Essential Reducing Agent, NADPH. Glycolysis for the purpose of making ATP is not the only use of glucose by the body. The **pentose phosphate pathway** of glucose catabolism is another use, and its purpose is to make NADPH, the reduced form of $NADP^+$. NADPH (or rather the NADPH/H^+ system) is a reducing agent, and the biosyntheses of substances like the fatty acids require such. Fatty acids are almost entirely alkane-like and, like alkanes, are the most reduced types of organic compounds.

The complicated series of reactions of the pentose phosphate pathway (which we'll not study in detail) is very active in adipose tissue, where fatty acid synthesis occurs. The balanced equation for the complete oxidation of one glucose molecule via the pentose phosphate pathway is as follows.

■ $NADP^+$ is a phosphate derivative of NAD^+.

$$6 \text{ Glucose-6-phosphate} + 12NADP^+ + 6H_2O \xrightarrow{\text{pentose phosphate pathway}}$$
$$5 \text{ glucose-6-phosphate} + 6CO_2 + 12NADPH + 12H^+ + P_i$$

Notice that the net effect of this overall reaction is the oxidation of *one* glucose unit.

19.3 CITRIC ACID CYCLE

Acetyl CoA is used to make the citrate ion that is then broken down bit by bit to CO_2 and units of $(H{:}^- + H^+)$, which funnel into the respiratory chain.

Figure 19.4 gives the steps of the *citric acid cycle,* a series of reactions for breaking down the acetyl groups obtained from acetyl CoA by earlier processes.[1] The two carbon atoms of the acetyl group end up in molecules of CO_2, and the hydrogen atoms are fed into the respiratory chain. The hydrogens eventually reduce oxygen and so become parts of water molecules. The reactions of the citric acid cycle occur in the innermost compartment of a mitochondrion, the matrix.

■ Hans Krebs won a share of the 1953 Nobel prize in medicine and physiology for his work on the citric acid cycle.

To launch the citric acid cycle, an acetyl group transfers from acetyl CoA to the oxaloacetate ion, giving the citrate ion. (The addition of one organic species to another is a type of reaction that we did not study in earlier chapters.) Now begins a series of reactions by which the citrate ion is degraded until another oxaloacetate ion is recovered. The numbers of the following steps match those in Figure 19.4. Notice particularly where the figure shows $(H{:}^- + H^+)$ units being passed to the respiratory chain. These are the pieces of the hydrogen molecule extracted from an intermediate of the citric acid cycle and taken up by an enzyme of the chain, either as NADH/H^+ or as $FADH_2$. In this way, the citric acid cycle, as we said, "fuels the respiratory chain."

■ At physiological pH, the acids in the cycle exist largely as their anions.

1. Citrate is dehydrated to give the double bond of *cis*-aconitate. (The dehydration of an alcohol was studied on page 248.)

2. Water adds to the double bond of *cis*-aconitate to give an isomer of citrate called isocitrate. (The addition of water to a double bond was studied on page 227.)

[1]In various references, the citric acid cycle also goes by two other names: the *tricarboxylic acid cycle* (or *TCA cycle*) and the *Krebs cycle*.

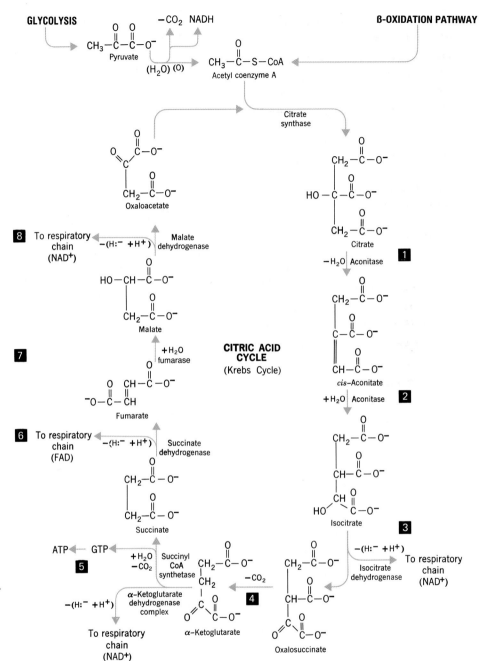

FIGURE 19.4
Citric acid cycle. The boxed numbers refer to the text discussion. The names of the enzymes for each step are given by the arrows.

■   The gain of electrons or of an electron carrier such as H:⁻ is *reduction*; the loss of e⁻ or of H:⁻ is *oxidation*.

The net effect of steps 1 and 2 is to switch the alcohol group in citrate to a different carbon atom, but this changes the alcohol from tertiary to secondary, from an alcohol that cannot be oxidized to one that can.

3. The secondary alcohol group in isocitrate is dehydrogenated (oxidized) to give oxalosuccinate. (We studied the oxidation of a 2° alcohol on page 251.) The oxidation is done by NAD^+, which is reduced to $NADH/H^+$. The latter passes an electron pair down the respiratory chain, so more ATP is made.

4. Oxalosuccinate loses its middle carboxyl group—*it decarboxylates*—to give α-ketoglutarate.

5. α-Ketoglutarate now changes to succinate. This is a very complicated series of reactions, all catalyzed by one team of enzymes that includes coenzyme A. At one stage, guanosine triphosphate or GTP, another high-energy triphosphate similar to ATP, is generated. GTP, however, is able to transfer one phosphate unit directly to ADP to make ATP, so making GTP functionally means that, besides succinate, ATP is the final outcome of step 5.

6. Succinate donates $H:^-$ to FAD (not to NAD^+), and both $FADH_2$ and the fumarate ion form. $FADH_2$ is also part of the respiratory chain, so it enables the synthesis of still more ATP.

7. Fumarate adds water to its double bond, and malate forms.

8. The secondary alcohol in malate is oxidized (dehydrogenated) by NAD^+ to give oxaloacetate; $NADH/H^+$ also forms and still more ATP can be made by the chain.

The citric acid cycle has now completed one turn, and one molecule of the initial carrier, oxaloacetate, has been regenerated. It can accept another acetyl group from acetyl coenzyme A to launch another turn if the demand for ATP is still high enough (meaning that considerable ADP and P_i remain).

A maximum of 12 molecules of ATP are generated from the degradation of each acetyl group by the citric acid cycle. Three are the result of step 3, four at step 5, two at step 6, and three at step 8. We next learn how the respiratory chain finally makes all of this ATP from ADP and P_i.

19.4 RESPIRATORY CHAIN

The flow of electrons to oxygen in the respiratory chain creates a proton gradient in mitochondria that drives the synthesis of ATP.

The term *respiration* refers to more than just breathing. It includes the chemical reactions that use oxygen in cells. Oxygen is reduced to water, and the following equation is the most basic statement we can write for what happens.

■ Remember in the following discussion that when any molecule loses electrons it is oxidized and the acceptor is reduced.

$$(:) + 2H^+ + \cdot \ddot{O} \cdot \longrightarrow H\!-\!\ddot{O}\!-\!H + energy$$

| Pair of electrons | Pair of protons | Atom of oxygen | Molecule of water |

The electrons and protons come mostly from intermediates in the citric acid cycle.

The Respiratory Enzymes Are Agents of Electron Transfers. The long series of oxidation–reduction reactions that make up the respiratory chain are catalyzed by the **respiratory enzymes**. The flow of electrons from the initial donor is irreversibly down an energy hill all the way to oxygen.

The cell's principal site of ATP synthesis, as we said, is the *inner* membrane of a mitochondrion (see Figure 19.2). One very important property of this membrane is that it is not permeable to protons, H^+, except at tiny channels that lead to an enzyme for the synthesis of ATP.

The respiratory chain involves four multienzyme complexes (Figure 19.5). NADH, which is part of complex I, is generated by the transfer of $H:^-$ from metabolites, like certain intermediates of the citric acid cycle. We'll use the general symbol

■ Think of MH_2 as $M\!:\!H$ which becomes M when $H:^-$ and H^+ leave it.

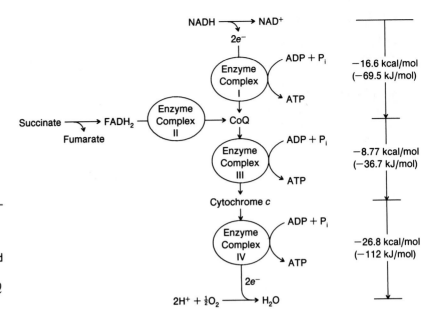

FIGURE 19.5

Enzyme complexes of the respiratory chain and the energy drop from NAD^+ to reduced oxygen (water). NADH is the enzyme with reduced nicotinamide as a cofactor. $FADH_2$ is the enzyme with reduced riboflavin as a cofactor. CoQ is another enzyme.

MH_2 for a metabolite. MH_2 is used to make NADH (or $NADH/H^+$) by the following equation.

$$MH_2 + NAD^+ \longrightarrow M + NADH/H^+$$

- Think of NADH as NAD:H, in which the dots represent the electron pair that moves along the respiratory chain.

(The proton produced by this reaction and symbolized as part of $NADH/H^+$ is buffered, as we discussed on page 381.) An electron pair has now moved from MH_2 to $NADH/H^+$, and the flow of electrons has started toward oxygen.

The next enzyme in the chain, FMN, is also a unit of enzyme complex I in Figure 19.5. (The coenzyme of FMN is riboflavin; see page 381.) The hydride unit in $NADH/H^+$ now transfers as $H:^-$ to FMN, so we can write the following equation.

- Think of $FMNH_2$ as FMN:H, with H above, in which the electron pair dots are those that move along the respiratory chain.

$$NADH/H^+ + FMN \longrightarrow FMNH_2 + NAD^+$$

This restores the NAD^+ enzyme, and moves the electron pair one more step down the respiratory chain.

As Electrons Flow in the Respiratory Chain, Protons Move across the Inner Mitochondrial Membrane. What happens next is the transfer of just the pair of electrons, not whole units of $H:^-$. The hydrogen *nucleus* of $H:^-$ leaves the carrier as acid, H^+, *but acid forces a change in the configuration of a membrane-bound protein.* One result of the configurational change is the outward ejection of protons through the inner membrane into the fluid of the intermembrane space. The pH of this fluid becomes 1.4 units less than (and so is more acidic than) the pH in the matrix (interior) side of the inner membrane. *The flow of electrons in the respiratory chain, in other words, is setting up a proton gradient across the inner membrane.*

- Visualize these enzymes and their reactions as occurring in the structure of the inner membrane itself.

Proton Channels of the Inner Mitochondrial Membrane Are Parts of the ATP-Making Enzyme. Nature abhors gradients; random motions always make things more mixed up and randomly distributed. Protons, however, cannot diffuse back just anywhere along the inner membrane and so destroy the proton gradient. Instead, embedded in the inner membrane are tubelike complexes of protein molecules that are components of a complex enzyme called **proton-pumping ATPase.** Each tube is a channel through the inner membrane into a cluster of polypeptide molecules that project like

(a)

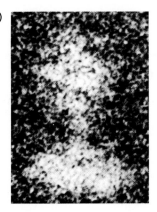

(b)

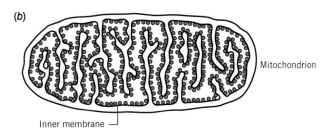

Mitochondrion

Inner membrane

(c)

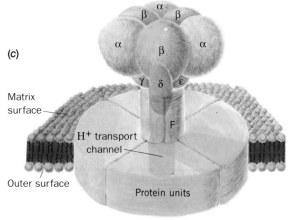

Matrix
surface

H^+ transport
channel

Outer surface

Protein units

FIGURE 19.6
Proton-pumping ATPase. *(a)* Electron micrograph of an intact inner mitochondrial membrane showing the "lollipop" projections of the enzyme into the matrix side. [From D. F., Parson, *Science,* **140,** 985 (1963). Copyright © 1963 American Association for the Advancement of Science. Used by permission.] *(b)* Interpretive drawing of the mitochondrion with the same projections shown in part *a. (c)* Interpretive drawing of one proton-pumping ATPase complex. The unit labeled δ comprises the proton channel beneath it (F_1) and the other subunits of the enzyme complex above it. The three subunits labled β are sites where ATP is made. The proton flow would be from bottom to top through the indicated channel and into the polypeptides labeled α and β. (Adapted with permission from D. Voet and J. G. Voet, *Biochemistry,* John Wiley & Sons, Inc., 1990. Used with permission.)

lollipops into the matrix (Figure 19.6). The flow of H^+ ions through a tube thus brings H^+ ions directly into the polypeptides of the lollipop region. As we have learned, a change in the acidity of a region holding a polypeptide can force the polypeptide to change its shape. Just such a change of shape now occurs to some of the lollipop polypeptides, *forcing the release of already-made ATP.* The flow of hydrogen ions is not what catalyzes the *formation* of ATP from ADP and P_i. Instead, the proton flow catalyzes the *release* of ATP *already formed* but held tightly to the enzyme. Some of the lollipop polypeptides are the ATP-forming enzymes, but the ATP so made cannot "let go." The arrival of H^+ ions is required to force the release of ATP.

Because a series of *oxidations* create the proton gradient, the overall synthesis of ATP by the chain is called **oxidative phosphorylation.** Each unit of NADH/H^+ that launches the chain at enzyme complex I can lead to a maximum of three ATPs. Each $FADH_2$ entering at complex II can cause two ATP molecules to form. The main outline of oxidative phosphorylation was first proposed in 1961 by Peter Mitchell of England, who called it the *chemiosmotic theory.* Let's now go back over its major steps.

■ Peter Mitchell won the 1978 Nobel prize in chemistry for his work.

■ Because chemical reactions create the gradients, we have the *chemi-* part of the term *chemiosmotic.*

■ This migration through a semipermeable membrane explains the *-osmotic* part of the term *chemiosmotic.*

OXIDATIVE PHOSPHORYLATION (CHEMIOSMOTIC THEORY)

1. The synthesis of ATP occurs at an enzyme located on the matrix side of the inner mitochondrial membrane.
2. ATP synthesis is driven by a flow of protons that occurs from the intermembrane side to the matrix side of the inner membrane.
3. The flow of protons is through special membrane channels and down a concentration gradient of protons created across the inner membrane.
4. The energy to create the proton gradient is provided by a flow of electrons through the respiratory chain enzymes packaged into four complexes located in the inner membrane.
5. The inner membrane is a closed envelope except for special channels some of which handle the flow of protons and others that let solutes move into or out of the matrix.

■ Rotenone is a naturally occurring insecticide.

Some Antibiotics and Poisons Inhibit Oxidative Phosphorylation. One of the barbiturates, amytal sodium, and the powerful insecticide, rotenone, block the respiratory chain by interfering at enzyme complex I. The antibiotic antimycin A stops the chain at complex III. The cyanide ion blocks the chain at its very end in complex IV. These agents, by inhibiting the chain, thus work to inhibit the creation of the proton gradient and in this manner inhibit ATP synthesis.

19.5 ENERGY FROM FATTY ACIDS

Acetyl groups are produced by the β-oxidation of fatty acids and are fed into the citric acid cycle and respiratory chain.

The degradation of fatty acids takes place by a repeating series of steps known as the **β-oxidation pathway** (Figure 19.7). The pathway has this name because the overall result of all but the last step is the oxidation of the CH_2 group beta to a keto group. (Long ago, chain positions in carbonyl compounds were lettered outward from the carbonyl group, using the Greek alphabet, the α position being adjacent to the carbonyl group.)

$$\underset{\text{Fatty acid}}{\overset{\beta \quad \alpha}{RCH_2CH_2CO_2H}} \xrightarrow[\text{three steps}]{\beta\text{-oxidation}} \underset{\text{A }\beta\text{-keto fatty acid}}{\overset{O}{\overset{\|}{RCCH_2CO_2H}}}$$

■ AMP is adenosine monophosphate. PP$_i$ is the diphosphate ion.

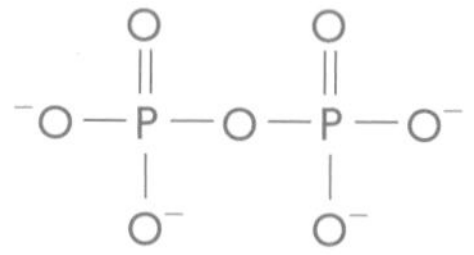

A Fatty Acid Is First Joined to Coenzyme A. The β-oxidation pathway occurs in the mitochondrial matrix, but before a fatty acid molecule can enter β-oxidation, it has to be joined to coenzyme A. The cost in chemical energy is paid by the breaking of *two* P—O bonds of the phosphoric anhydride system in ATP, but now the fatty acyl unit is activated for oxidation. The ATP itself breaks down to AMP and PP$_i$, the latter being further broken to two P$_i$ units.

$$\overset{O}{\overset{\|}{^-O-P}}-O-\overset{O}{\overset{\|}{P}}-O^-$$

Diphosphate ion, PP$_i$

Fatty Acyl CoA Is Catabolized Two Carbons at a Time. The repeating sequence of β-oxidation consists of four steps. The overall result of one series is one molecule of

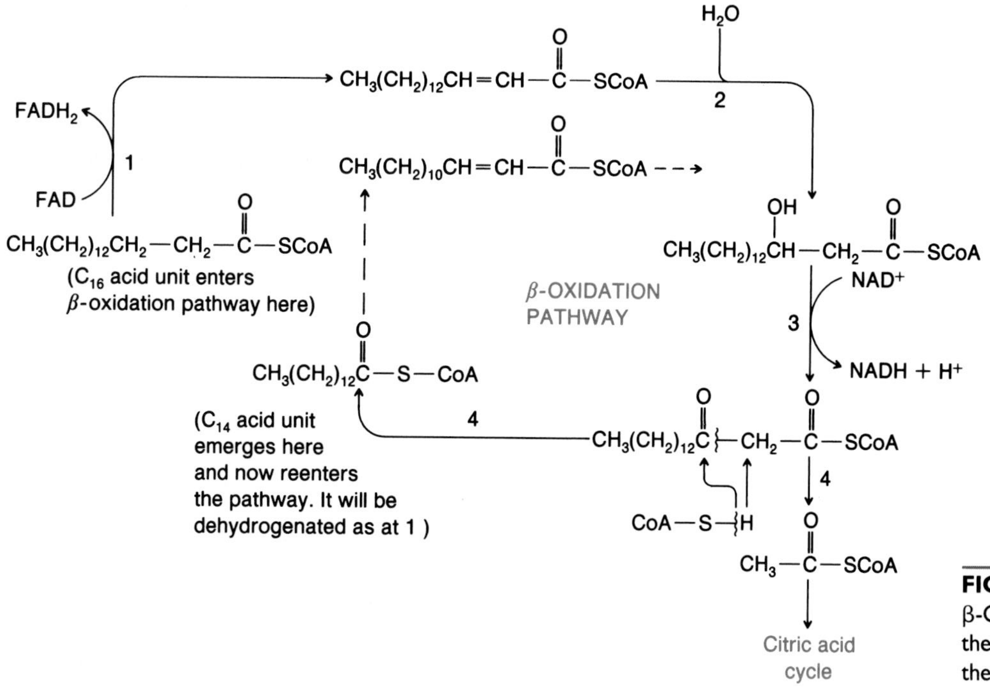

FIGURE 19.7
β-Oxidation. The numbers refer to the numbered steps discussed in the text.

FADH$_2$, one of NADH/H$^+$, one of acetyl coenzyme A, and a fatty acyl unit with two fewer carbons.

The now shortened fatty acyl unit is carried again through the four steps, and the process is repeated until no more two-carbon acetyl units can be made. The FADH$_2$ and the NADH/H$^+$ are parts of enzymes of the respiratory chain and so fuel the operation of the chain. The acetyl groups pass into the citric acid cycle, or they enter the general pool of acetyl coenzyme A on which the body draws to make other substances (e.g., cholesterol). Let's now look at the four steps in greater detail. The numbers that follow refer to Figure 19.7.

■ Franz Knoop directed much of the research on β-oxidation, so this pathway is sometimes called *Knoop oxidation.*

1. Dehydrogenation (oxidation) occurs. FAD accepts (H:$^-$ + H$^+$) from the α- and β-carbons of the fatty acyl unit of palmityl coenzyme A.

$$\underset{\text{Palmityl coenzyme A}}{CH_3(CH_2)_{12}CH_2 - CH_2 \overset{O}{\overset{\|}{C}}SCoA} + FAD \xrightarrow{\boxed{1}}$$

$$\underset{\substack{\text{An }\alpha,\beta\text{-unsaturated acyl derivative} \\ \text{of coenzyme A}}}{CH_3(CH_2)_{12}CH = CH\overset{O}{\overset{\|}{C}}SCoA} + FADH_2 \xrightarrow[\text{FAD}]{} \underset{\substack{\text{Transfers to} \\ \text{respiratory chain}}}{[H:^- + H^+]}$$

FADH$_2$ interacts at enzyme complex II of the respiratory chain (page 434).

2. The hydration of a double bond takes place. Water adds to the alkene group formed by step 1, and a 2° alcohol group forms.

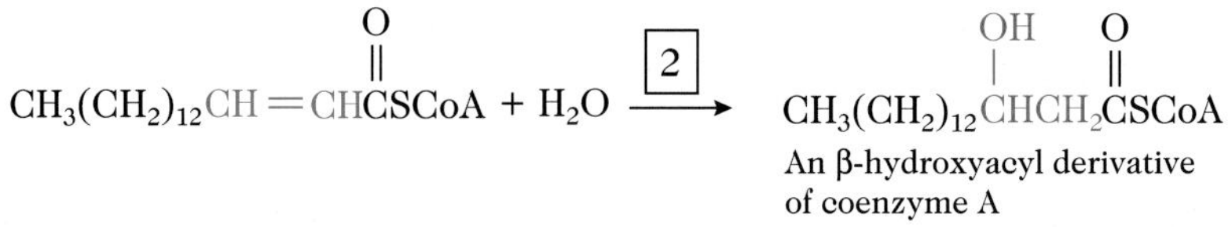

$$CH_3(CH_2)_{12}CH = CHCSCoA + H_2O \xrightarrow{\boxed{2}} CH_3(CH_2)_{12}CHCH_2CSCoA$$

An β-hydroxyacyl derivative
of coenzyme A

3. Dehydrogenation occurs once again. The loss of $[H:^- + H^+]$ oxidizes the 2° alcohol to a keto group. Notice that the overall result of steps 1 to 3 is to oxidize the CH_2 group at the β position of the original fatty acyl group to a keto group, which is why the pathway is called *beta* oxidation.

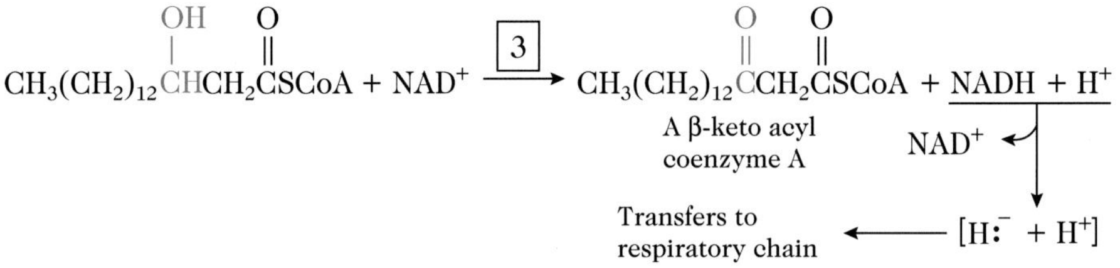

$$CH_3(CH_2)_{12}CHCH_2CSCoA + NAD^+ \xrightarrow{\boxed{3}} CH_3(CH_2)_{12}CCH_2CSCoA + \underline{NADH + H^+}$$

A β-keto acyl
coenzyme A

NAD$^+$

Transfers to
respiratory chain $\longleftarrow$ $[H:^- + H^+]$

4. The bond between the α-carbon and the β-carbon breaks as one unit of acetyl coenzyme A forms. This bond was weakened by the stepwise oxidation of the β-carbon, and now coenzyme A (CoASH) splits it.

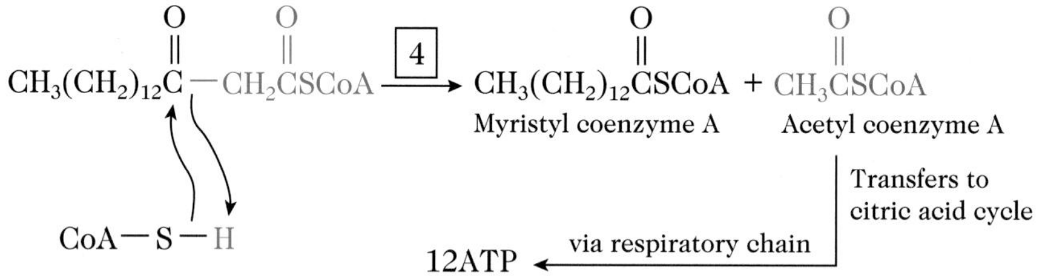

$$CH_3(CH_2)_{12}C - CH_2CSCoA \xrightarrow{\boxed{4}} CH_3(CH_2)_{12}CSCoA + CH_3CSCoA$$

Myristyl coenzyme A   Acetyl coenzyme A

CoA—S—H

Transfers to
citric acid cycle

12ATP $\longleftarrow$ via respiratory chain

The remaining acyl unit, the original shortened by two carbons, now goes through the cycle of steps again: dehydrogenation, hydration, dehydrogenation, and cleavage. After seven such cycles, one molecule of palmityl coenzyme A is broken into eight molecules of acetyl coenzyme A.

One Palmityl Unit Yields 129 ATP Molecules.   Table 19.1 shows how the maximum yield of ATP from the oxidation of one unit of palmityl coenzyme A adds up to 131 ATPs. The net from palmitic acid is 2 ATP fewer, or 129 ATP, because the activation of the palmityl unit—joining it to CoA—requires this initial investment, as we mentioned earlier.

TABLE 19.1   Maximum Yield of ATP from Palmityl CoA

Intermediates Produced by Seven Turns of the Fatty Acid Cycle	ATP Yield per Intermediate	Total ATP Yield
7 FADH$_2$	2	14
7 NADH	3	21
8 Acetyl coenzyme A	12	96
		131 ATP
Deduct two high-energy phosphate bonds for activating the acyl unit		−2
Net ATP yield per palmityl unit		129 ATP

19.6 ENERGY FROM AMINO ACIDS

When not needed for making proteins, amino acids are catabolized for energy or are converted into glucose or fatty acids.

Amino Acids Enter the Body's Nitrogen Pool. Amino acids, the end products of protein digestion, are rapidly taken up by circulation and enter the general **nitrogen pool,** the name given to the whole collection of nitrogen compounds wherever they are found (Figure 19.8).

Some amino acids are used to make replacements for worn out enzymes, hormones, and tissue proteins. Others are used to make nonprotein nitrogen compounds such as heme, certain neurotransmitters, and nucleic acids. Any amino acids left over are catabolized, and their chemical energy is used to make high-energy phosphates. The products of the complete catabolism of amino acids are carbon dioxide, water, and urea.

Amino Acids Are Sometimes Used to Make Glucose. Liver cells can use intermediates in the catabolism of many amino acids to make glucose. We'll say more about it in Chapter 20, but this synthesis of glucose from smaller molecules is called **gluconeogenesis** (*gluco-,* glucose; *neo-,* new; *-genesis,* creation), and it is indispensable to the brain. The brain's normal source of energy is the aerobic sequence starting with glucose taken from circulation. As the supply of glucose in the blood decreases between meals, during fasting, or in starvation, the liver makes glucose from molecular bits and pieces of amino acids and fatty acids.

When new glucose isn't needed and the supply of amino acids is still higher than needed to make proteins, then the molecular rummage from the partial catabolism of amino acids goes into making fatty acids. We'll learn how fatty acid synthesis occurs in Chapter 20.

The major point we make here is that the chemical energy in amino acids can be used directly to make high-energy phosphates or it can be put into the chemical energy of new glucose or fatty acids, which the body later and elsewhere taps. Excess amino acids are not excreted. You can gain weight on an all-protein diet, which places strains on the liver and the kidneys as they work to convert amino groups to urea and then export this nitrogen waste.

We won't study in detail how each amino acid is catabolized, because each requires its own particular scheme, usually quite complicated. Three general types of reactions occur, however, which we describe here: *oxidative deamination, transamination,* and *direct deamination.*

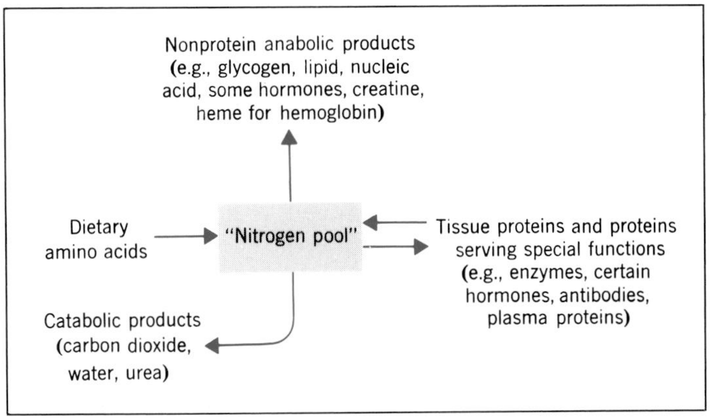

■ The polypeptides in enzymes have a particularly rapid turnover.

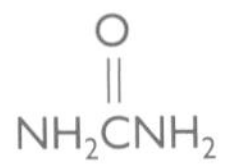

$$\overset{\text{O}}{\underset{\text{Urea}}{\underset{\|}{NH_2CNH_2}}}$$

FIGURE 19.8
Nitrogen pool.

Oxidative Deamination Removes an Amino Group.   The replacement of the amino group of an amino acid by a keto group is called **oxidative deamination**. The amino group leaves as an ammonium ion in a reaction that is largely used as a step in one of the body's mechanisms for moving NH_2 groups to urea. Two *coupled* steps are involved. The display that follows is an "equation" new to our study, but a kind often used when coupled reactions are described.

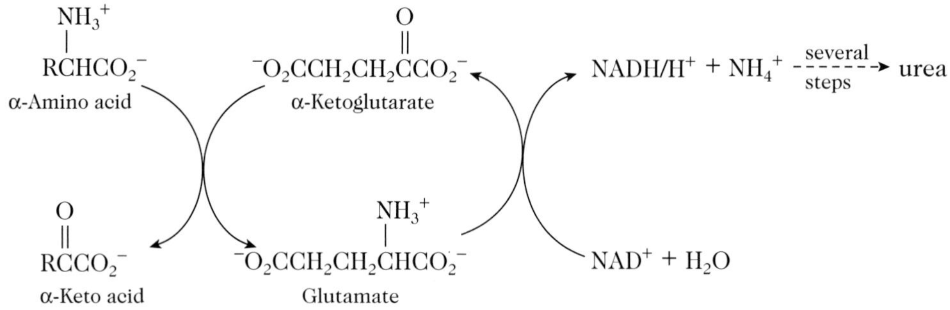

The step on the left, signified by the two downward pointing arrows is called **transamination**, because an amino group is transferred from one reactant, the α-amino acid, to the other, α-ketoglutarate. In the next step, indicated by the two upward pointing arrows, the nitrogen of the amino group (in glutamate) is detached and changed to the ammonium ion as α-ketoglutarate is regenerated for another cycle. The ammonium ion next enters a complex series of reactions called the *urea cycle* (which we will not study in detail) and is changed to urea. The α-ketoglutarate–glutamate pair thus provides a switching mechanism in catabolism. The pair switches nitrogen from amino acids to intermediates that can launch the insertion of nitrogen into waste urea.

Direct Deamination Occurs without an Oxidation.   Two amino acids, including serine, experience the simultaneous loss of both water and ammonia by **direct deamination**, a reaction that removes an amino group without consuming an oxidizing agent (like NAD^+). The reaction can occur to serine because it has an OH group on the carbon adjacent to the amino group's carbon atom. The alcohol group is dehydrated in the first step.

■ The imine group, C═NH, hydrolyzes easily because it can add water to give HO—C—NH_2, which then splits out NH_3 to leave O═C.

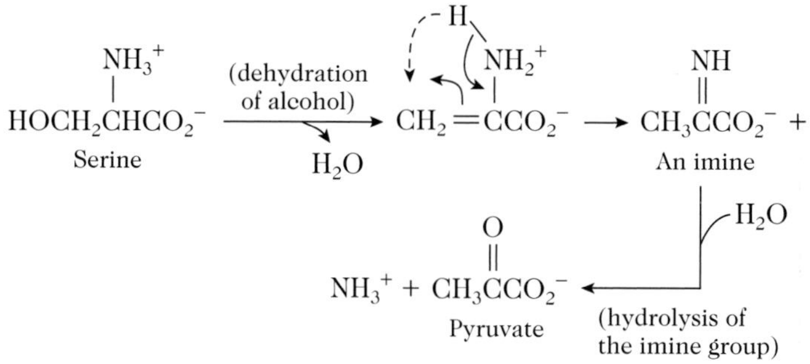

The first product, an unsaturated amine, spontaneously rearranges into an *imine*, a compound with a carbon–nitrogen double bond. Water adds to this double bond, and the product spontaneously breaks up, so the net effect is the hydrolysis of the imine group to a keto group and ammonia. Thus, serine breaks down to pyruvate, which, as we learned in connection with aerobic glycolysis, can send an acetyl group into the citric acid cycle. It can also be used to make new glucose by gluconeogenesis or new fatty acids.

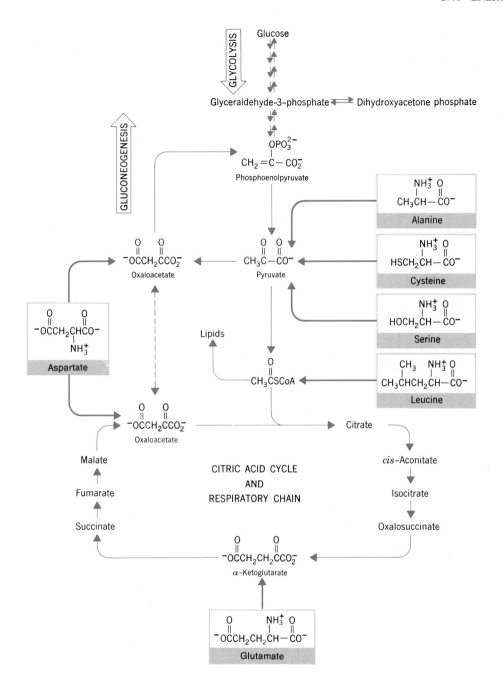

FIGURE 19.9
Catabolism of some amino acids.

Figure 19.9 gives an overview of the catabolism of several individual amino acids, placing them into the context of major metabolic pathways. We will study the catabolism of two amino acids in more detail.

Alanine Is a Source of Pyruvate.   The transamination of alanine gives pyruvate, which can enter the citric acid cycle, can go into gluconeogenesis, or can be used for the biosynthesis of fatty acids.

$$\underset{\text{Alanine}}{CH_3\overset{\overset{\displaystyle NH_3^+}{|}}{C}HCO_2^-} \xrightarrow{\text{transamination}} \underset{\text{Pyruvate}}{CH_3\overset{\overset{\displaystyle O}{\|}}{C}CO_2^-} \longrightarrow \text{Acetyl CoA} \longrightarrow \begin{array}{l}\text{citric}\\\text{acid}\\\text{cycle}\end{array}$$

$$\downarrow$$
gluconeogenesis $\longrightarrow$ fatty acid synthesis

Aspartic Acid Is a Source of Oxaloacetic Acid.   The transamination of aspartic acid gives oxaloacetate, an intermediate in both gluconeogenesis and the citric acid cycle.

$$\underset{\text{Aspartic acid}}{^-O_2CH_2\overset{\overset{\displaystyle NH_3^+}{|}}{C}HCO_2^-} \xrightarrow{\text{transamination}} \underset{\text{Oxaloacetate}}{^-O_2CCH_2\overset{\overset{\displaystyle O}{\|}}{C}CO_2^-} \longrightarrow \text{citric acid cycle}$$

$$\downarrow$$
gluconeogenesis

Notice in Figure 19.9 that oxaloacetate occurs in three places, namely, as an intermediate in the citric acid cycle, as the product of the oxidative deamination of aspartate, and as a raw material for making new glucose. For a *net gain* of glucose molecules via gluconeogenesis, the oxaloacetate in the citric acid cycle cannot be counted as available to make new glucose. Only oxaloacetate made from amino acids can give a net gain of glucose this way. This is why gluconeogenesis under conditions of starvation necessarily breaks down body proteins. It needs some of their amino acids to make new glucose via oxaloacetate.

19.7   ACIDOSIS AND ENERGY PROBLEMS

An acceleration of β-oxidation tips some equilibria in a direction that leads to ketoacidosis.

Cells of certain tissues must engage in gluconeogenesis in two serious conditions, starvation and uncontrolled diabetes mellitus. In starvation, the blood sugar level drops because of nutritional deficiencies, so the body (principally the liver) tries to compensate by making glucose. The long-term consequences are fatal unless the underlying causes are treated.

The Level of Acetyl CoA Increases When Gluconeogenesis Is Accelerated.   We have just learned that gluconeogenesis consumes oxaloacetate. When oxaloacetate supplies decline, so does the ability of acetyl coenzyme A to insert its acetyl group into the citric acid cycle. (Oxaloacetate is the acetyl group acceptor of this cycle.) *Yet acetyl coenzyme A continues to be made by β-oxidation.* The inevitable result is that *acetyl CoA levels increase.*

As the supply of acetyl CoA increases in the liver, the following equilibrium shifts to the right to make acetoacetyl CoA.

■ The stress on this equilibrium is an increase in the level of acetyl CoA, so by Le Châtelier's principle, the equilibrium shifts to the right to absorb this stress.

$$\underset{\text{Acetyl CoA}}{2CH_3\overset{\overset{\displaystyle O}{\|}}{C}SCoA} \rightleftharpoons \underset{\text{Acetoacetyl CoA}}{CH_3\overset{\overset{\displaystyle O}{\|}}{C}CH_2\overset{\overset{\displaystyle O}{\|}}{C}SCoA} + CoASH$$

As the level of acetoacetyl CoA increases, the following equilibrium shifts to the right. (The forward reaction is the same reaction that occurs early in the body's synthesis of cholesterol, for which (HMG)—CoA synthase is the catalyst.)

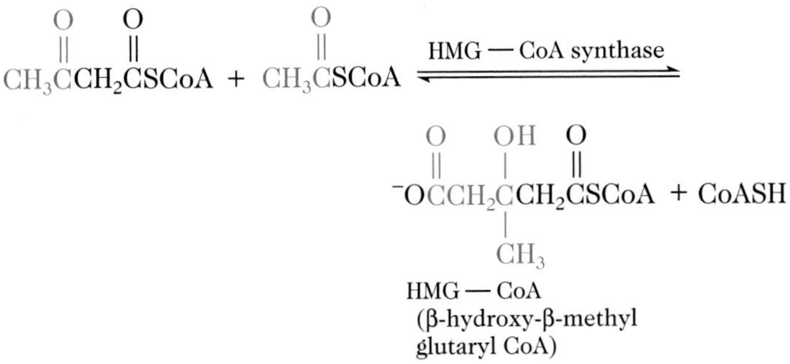

$$\underset{\text{O \quad O \qquad\qquad O}}{CH_3CCH_2CSCoA \ + \ CH_3CSCoA} \ \underset{\text{HMG — CoA synthase}}{\rightleftharpoons}$$

$$\underset{\substack{CH_3 \\ \text{HMG — CoA} \\ (\beta\text{-hydroxy-}\beta\text{-methyl} \\ \text{glutaryl CoA})}}{{}^-OCCH_2CCH_2CSCoA \ + \ CoASH}$$

As the level of HMG—SCoA increases, a liver enzyme splits it to acetoacetate ion and acetyl coenzyme A.

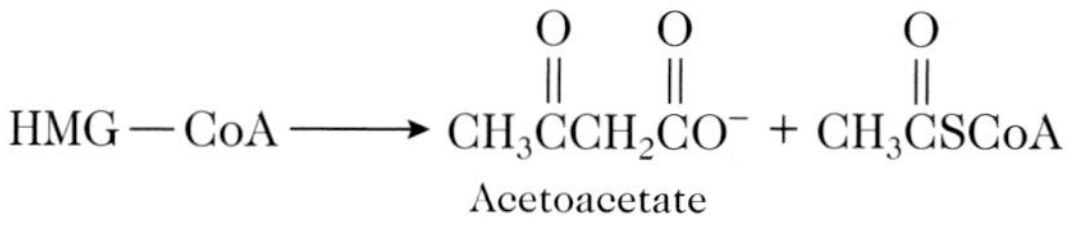

$$HMG — CoA \longrightarrow \underset{\text{Acetoacetate}}{CH_3CCH_2CO^-} + CH_3CSCoA$$

The net effect of these steps, starting from acetyl CoA, is the following:

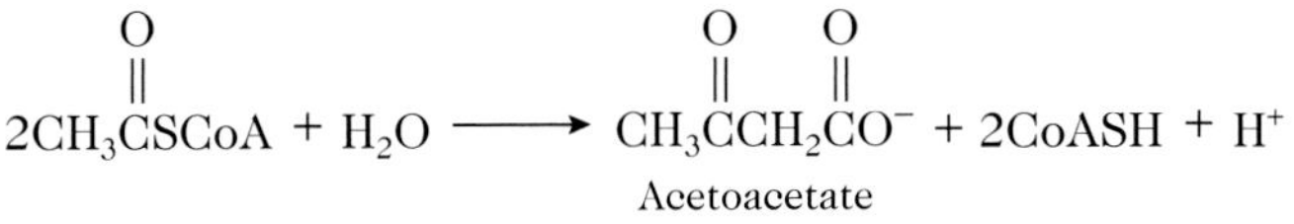

$$2CH_3CSCoA + H_2O \longrightarrow \underset{\text{Acetoacetate}}{CH_3CCH_2CO^-} + 2CoASH + H^+$$

Notice the hydrogen ion. It makes the situation dangerous, and an increased rate of synthesis of "new" glucose was the cause.

Accelerated Acetoacetate Production Leads to Acidosis. The acid produced by the formation of acetoacetate must be neutralized by the buffer. Under an increasingly rapid production of acetoacetate and hydrogen ion, the blood buffer slowly loses ground. A condition of *acidosis* sets in. It is *metabolic* acidosis, because the cause lies in a disorder of metabolism. Because the chief species responsible for this acidosis has a keto group, the condition is often called **ketoacidosis.**

Blood Levels of the Ketone Bodies Increase in Starvation and Diabetes. The *acetoacetate ion* is called one of the **ketone bodies.** The two others are *acetone* and the *β-hydroxybutyrate ion*. Both are produced from the acetoacetate ion. Acetone arises from acetoacetate by the loss of the carboxyl group:

■ β-Hydroxybutyrate is called a ketone body despite its not having a keto group.

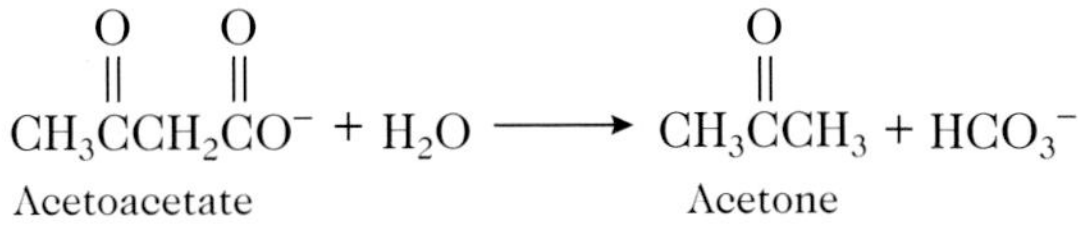

$$\underset{\text{Acetoacetate}}{CH_3CCH_2CO^-} + H_2O \longrightarrow \underset{\text{Acetone}}{CH_3CCH_3} + HCO_3^-$$

β-Hydroxybutyrate is produced when the keto group of acetoacetate is reduced by NADH:

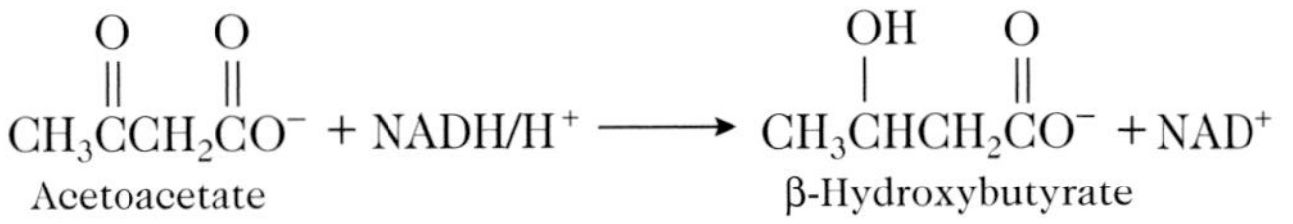

$$\underset{\text{Acetoacetate}}{CH_3CCH_2CO^-} + NADH/H^+ \longrightarrow \underset{\text{β-Hydroxybutyrate}}{CH_3CHCH_2CO^-} + NAD^+$$

■ The vapor pressure of acetone at body temperature is about 400 mm Hg, making its loss by evaporation from the blood in the lungs easy.

■ 1 μmol = 1 micromole
          = 10^{-6} mol

Condition	$[HCO_3^-]_{blood}$ (mmol/L)
Normal	22–30
Mild acidosis	16–20
Moderate acidosis	10–16
Severe acidosis	< 10

The ketone bodies enter general circulation. Because acetone is volatile, most of it leaves the body via the lungs, and individuals with severe ketoacidosis have "acetone breath," the noticeable odor of acetone on the breath.

Acetoacetate and β-hydroxybutyrate can be used in skeletal muscles to make ATP. Heart muscle actually uses these two for energy in preference to glucose. Even the brain, given time, can adapt to using acetoacetate and β-hydroxybutyrate for energy when the blood sugar level drops in starvation or prolonged fasting. Thus the ketone bodies are not in themselves abnormal constituents of blood. Only when they are produced at a rate faster than the blood buffer can handle the acid produced with them is there a problem.

The Conditions of Ketonemia, Ketonuria, and "Acetone Breath" Collectively Constitute *Ketosis*. Normally, the levels of acetoacetate and β-hydroxybutyrate in the blood are, respectively, 2 μmol/dL and 4 μmol/dL. In prolonged, undetected, and untreated diabetes, these values can increase as much as 200-fold. The condition of excessive levels of ketone bodies in the blood is called **ketonemia**.

As ketonemia becomes more and more advanced, the ketone bodies begin to appear in the urine, a condition called **ketonuria**. When there is a combination of ketonemia, ketonuria, and acetone breath, the overall state is called **ketosis**. The individual is described as *ketotic*. As unchecked ketosis becomes more severe, the associated ketoacidosis worsens and the pH of the blood continues its fatal descent.

The Urinary Removal of Organic Anions Means the Loss of Base from the Blood. To leave the negatively charged ketone bodies in the urine, the kidneys have to leave positive ions with them to keep everything electrically neutral. Na^+ ions, the most abundant cations in blood, are used. One Na^+ ion has to leave the blood and enter the urine with each acetoacetate ion, for example. This loss of Na^+ is often referred to as the "loss of base" from the blood, although Na^+ is not a base. However, the loss of one Na^+ stems from the appearance of one acetoacetate ion *plus one H^+ ion* that the blood had to neutralize. Thus, each Na^+ that leaves the body corresponds to the loss of one HCO_3^- ion, the true base, consumed in neutralizing one H^+. This explains why the loss of Na^+ is taken as an indicator of the loss of the true base.

The kidneys normally manufacture HCO_3^- ion, the true base, in order to replenish the blood buffer system. Another way, therefore, to understand the loss of Na^+ to the urine as the loss of *base* from the blood is that an Na^+ ion is needed to accompany a newly made bicarbonate ion when it goes from the kidneys into the blood. Therefore, the greater the number of Na^+ ions that must be left in the *urine* to clear ketone bodies from the blood, the less is the amount of true base, HCO_3^-, that can be put into the *blood*.

Diuresis Must Accelerate to Handle Ketosis. The solutes that are leaving the body in the urine cannot, of course, be allowed to make the urine too concentrated. Otherwise, osmotic pressure balances are upset. Therefore, increasing quantities of water must be excreted. To satisfy this need, the individual has a powerful thirst. Other wastes, such as urea, are also being produced at higher than normal rates, because amino acids are being sacrificed in gluconeogenesis. These wastes add to the demand for water to make urine.

■ *Polyuria* is the technical name for the overproduction of urine.

Internal Water Shortages in Ketosis Spell Dehydration of Critical Tissues. If, during a state of ketosis, insufficient water is drunk, then water is simply taken from extracellular fluids. The blood volume therefore tends to drop, and the blood becomes more concentrated. It also thickens and becomes more viscous, which makes the delivery of blood more difficult.

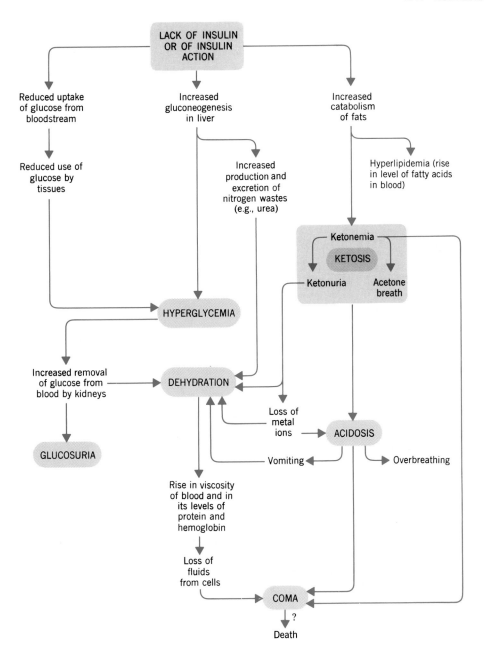

FIGURE 19.10
Principal sequence of events in untreated diabetes.

Because the brain has the highest priority for blood flow, some of this flow is diverted from the kidneys to try to ensure that the brain gets what it needs. This only worsens the situation in the kidneys, and they have an increasingly difficult time clearing wastes. As the water shortage worsens, some water is borrowed from the intracellular supply. This, in addition to a combination of other developments, leads to coma and eventually death.

Figure 19.10 outlines the succession of events in untreated type I diabetes. It is nothing short of remarkable how the absence of one chemical, insulin, can release such a vast train of biochemical events. At the molecular level of life, however, this kind of story occurs very often.

SUMMARY

High-Energy Compounds   Triphosphate esters such as adenosine triphosphate, ATP, are the chief storehouse of chemical energy in living systems. When they react and make some of their energy available to cellular processes, they break down to organic diphosphate (e.g., ADP) and inorganic phosphate ion. (Sometimes they break down to monophosphates, like AMP.) The chief task of biochemical energetics is to remake high-energy compounds using the chemical energy in foods.

Glycolysis   Glycolysis uses glucose or glycogen to make pyruvate, which can be converted to acetyl CoA, the fuel of the citric acid cycle. Glycolysis also generates some ATP. (Galactose and fructose enter glycolysis, too.) The glycolysis pathway has one enzyme that under aerobic conditions sends hydride directly to the respiratory chain.

Under anaerobic conditions, glycolysis ends at the lactate ion, not the pyruvate ion. Under a low oxygen supply, the keto group in pyruvate is reduced to lactate so as to store $H:^-$ and thereby regenerate one of the enzymes of glycolysis. This enables glycolysis to be a source of ATP when insufficient oxygen is available and the tissue is running an oxygen debt.

The sequence from glucose through acetyl CoA, the citric acid cycle, and the respiratory chain is called the aerobic sequence of glucose catabolism.

Pentose Phosphate Pathway   The body's need for NADPH to make fatty acids is met by catabolizing glucose through the pentose phosphate pathway.

Citric Acid Cycle   Hydride ion is provided to the respiratory chain by the citric acid cycle. Acetyl groups from acetyl coenzyme A are joined to a four-carbon carrier, oxaloacetate, to make citrate. This six-carbon salt of a tricarboxylic acid then is degraded bit by bit as $[H:^- + H^+]$ is fed to the chain. Each acetyl unit leads to the synthesis of a maximum of 12 ATP molecules.

Respiratory Chain   A series of electron-transfer enzymes called the respiratory enzymes occur together as groups called enzyme complexes in the inner membranes of mitochondria. These complexes process metabolites (MH_2), which often are obtained by the operation of the citric acid cycle. Either NAD^+ or FAD can accept $(H:^- + H^+)$ from a metabolite. Their reduced forms, NADH or $FADH_2$, then pass on the electrons into the enzyme complexes of the respiratory chain until finally the electrons are used (together with H^+ from the buffer) to reduce oxygen to water.

Oxidative Phosphorylation   According to the chemiosmotic theory, the flow of electrons in the respiratory chain forces protons across the inner membrane. This creates a proton gradient between the intermembrane region (higher H^+ level) and the mitochondrial matrix (lower H^+ level). The H^+ ions flow back through channels of the inner membrane and into the polypeptides that make up the proton-pumping ATPase. Bound to a portion of this enzyme is a molecule of ATP already made from ADP and P_i. With the arrival of the protons, a configurational change occurs in the enzyme that expels the ATP and opens the enzyme to receive more ADP and P_i and make more ATP. The new ATP's release awaits the arrival of more H^+ from the proton gradient. Various drugs and antibiotics can block the respiratory chain.

Catabolism of Fatty Acids   Fatty acyl groups, after being pinned to coenzyme A, are catabolized by a series of reactions called β-oxidation. By a succession of four steps—dehydrogenation, hydration of a double bond, oxidation of the resulting alcohol, and cleavage of the bond from the α to the β-carbon—one two-carbon acetyl group is clipped from a fatty acyl group. The series then repeats as the shortened fatty acyl group continues to be degraded. Each turn of the β-oxidation cycle produces one $FADH_2$ and one NADH, and these pass $H:^-$ to the respiratory chain for the synthesis of ATP. Each turn also sends one acetyl group into the citric acid cycle which, via the chain, leads to several more ATPs. The net ATP production is 129 ATPs per palmityl residue.

Amino Acid Distribution   The nitrogen pool receives amino acids from the diet, from the breakdown of proteins in body fluids or tissues, and from any synthesis of nonessential amino acids that occurs. Amino acids are used to build and repair tissue, replace proteins of body fluids, make nonprotein nitrogen compounds, provide chemical energy if needed, and supply molecular parts for gluconeogenesis or fatty acid synthesis.

Amino Acid Metabolism   By reactions of transamination and oxidative deamination, α-amino acids shuffle amino groups between themselves and intermediates of the citric acid cycle. Deaminated amino acids eventually become acetyl coenzyme A, acetoacetyl CoA, pyruvate, or an intermediate in the citric acid cycle. The skeletons of most amino acids can be used to make glucose, fatty acids, or the ketone bodies. Their nitrogen atoms become part of urea.

Ketoacidosis   Acetoacetate, β-hydroxybutyrate, and acetone build up in the blood—ketonemia—in starvation or in diabetes. The first two are normal sources of energy in some tissues. When made faster than metabolized, however, they cause a loss of bicarbonate ion from the blood buffer, which leads to ketoacidosis.

The kidneys try to leave the negatively charged ketone bodies in the urine, but this requires Na^+ (for electrical neutrality) and water (for osmotic pressure balances). The loss of Na^+ from the blood is called a loss of "base," because its loss means less Na^+ is available to accompany replacement HCO_3^- ions needed to replenish the blood buffer system.

Under developing ketoacidosis, the kidneys have extra nitrogen wastes and, in diabetes, extra glucose to be exported in the urine. For these, more water is needed, and unless it is brought in by the thirst mechanism, it has to be sought from within. But the brain has first call on blood flow, so the kidneys suffer more. Eventually, if these events continue unchecked, the victim goes into a coma and dies.

REVIEW EXERCISES

The answers to Review Exercises whose numbers are in color are found in Appendix V. The answers to the other Review Exercises are found in the Study Guide that accompanies this text. The more challenging questions are marked with asterisks.

Energy Sources

19.1 What products of the digestion of carbohydrates, triacylglycerols, and the polypeptides can be used as sources of biochemical energy to make ATP?

19.2 Complete the following structure of ATP.

$$\text{Adenosine} -\text{O}-\overset{\overset{\text{O}}{\|}}{\underset{\underset{\text{O}^-}{|}}{\text{P}}}-$$

19.3 Write the structures of ADP and of AMP in the manner started by Review Exercise 19.2.

19.4 At physiological pH, what does the term *inorganic phosphate* stand for? (Give formulas and names.)

19.5 What is the general overall purpose of biochemical energetics?

Overview of Metabolic Pathways

19.6 In the general area of biochemical energetics, what is the purpose of each of the following pathways?
(a) respiratory chain   (b) anaerobic glycolysis
(c) citric acid cycle   (d) fatty acid cycle

19.7 What prompts the respiratory chain to go into operation?

19.8 Which tends to increase the rate of breathing, an increase in the body's supply of ATP or an increase in its supply of ADP? Explain.

19.9 In general terms, the intermediates that send electrons down the respiratory chain come from what metabolic pathway that consumes acetyl groups?

19.10 Arrange the following sets of terms in sequence in the order in which they occur or take place. Place the identifying letter of the first sequence of a set to occur on the left of the row of letters.

(a) Citric acid cycle   pyruvate   acetyl CoA
   A   B   C
 respiratory chain   glycolysis
    D   E

(b) Citric acid cycle   β-oxidation acetyl CoA
   A   B   C
 respiratory chain
    D

19.11 The *aerobic sequence* begins with what metabolic pathway and ends with which pathway?

19.12 The β-oxidation pathway occurs to what kind of compound?

19.13 The anaerobic sequence begins and ends with what compounds?

19.14 What does it mean for the cell that anaerobic glycolysis is a "backup?"

Catabolism of Glucose

19.15 Fill in the missing substances and balance the following incomplete equation:

$$C_6H_{12}O_6 + 2ADP + \underline{\quad} \longrightarrow 2C_3H_5O_3^- + 2H^+ + \underline{\quad}$$
Glucose   Lactate

19.16 What particular significance does glycolysis have when a tissue is running an oxygen debt?

19.17 What happens to pyruvate (a) under aerobic conditions and (b) under anaerobic conditions?

19.18 What happens to lactate when an oxygen debt is repaid?

19.19 The pentose phosphate pathway uses $NADP^+$, not NAD^+. What forms *from* $NADP^+$, and how does the body use what forms (in general terms)?

Citric Acid Cycle

19.20 What makes the citric acid cycle start up?

19.21 What chemical unit is degraded by the citric acid cycle? Give its name and structure.

19.22 How many times is a secondary alcohol group oxidized in the citric acid cycle?

19.23 Water adds to a carbon–carbon double bond how many times in one turn of the citric acid cycle?

*19.24   The enzyme for the conversion of isocitrate to oxalosuccinate (see Figure 19.4) is stimulated by one of two substances, ATP or ADP. Which one is the more likely activator? Explain.

Respiratory Chain

19.25   Is the following (unbalanced) change a reduction or an oxidation? How can you tell?

$$\underset{\displaystyle \text{CH}_3\text{CHCH}_2\text{CO}_2^-}{\overset{\displaystyle \text{OH} \atop |}{}} \longrightarrow \underset{\displaystyle \text{CH}_3\text{CCH}_2\text{CO}_2^-}{\overset{\displaystyle \text{O} \atop ||}{}}$$

19.26   Which kind of enzyme would be more likely to cause the change given in Review Exercise 19.25, an enzyme like aconitase or an enzyme like isocitrate dehydrogenase? Explain.

19.27   What is missing in the following basic expression for what must happen in the respiratory chain?

$$\tfrac{1}{2}\text{O}_2 + 2\text{H}^+ \rightarrow \text{H}_2\text{O}$$

19.28   What general name is given to the set of enzymes involved in electron transport?

19.29   What does respiratory enzyme complex IV do?

19.30   What is FAD, and where is it involved in the respiratory chain?

19.31   Across which cellular membrane does the respiratory chain establish a gradient of H^+ ions? On which side of this membrane is the value of the pH lower?

19.32   According to the chemiosmotic theory, the flow of what particles most directly leads to the synthesis of ATP?

19.33   If the inner mitochondrial membrane is broken, the respiratory chain can still operate, but the phosphorylation of ADP that normally results stops. Explain this in general terms.

19.34   What does the flow of protons into the proton-pumping ATPase do, make ATP or cause the release of ATP?

19.35   Briefly describe the theory presented in this chapter that explains how a flow of protons across the inner mitochondrial membrane initiates the synthesis of ATP from ADP and P_i.

Catabolism of Fatty Acids

19.36   Arrange the following processes in the order in which they occur when the energy in storage in triacylglycerols is mobilized.

A   β-oxidation
B   oxidative phosphorylation
C   citric acid cycle
D   lipoprotein formation
E   lipolysis in adipose tissue

19.37   What specific function does β-oxidation have in obtaining energy from fatty acids?

19.38   What specific function does the citric acid cycle have in the use of fatty acids for energy?

19.39   How are long-chain fatty acids activated for β-oxidation?

*19.40   Write the equations for the four steps of β-oxidation as it operates on butyryl CoA. How many more cycles are possible after this one?

19.41   How is the FAD enzyme recovered from its reduced form, FADH_2, when β-oxidation operates?

19.42   How is the reduced form of the NAD^+-enzyme used in β-oxidation restored to its oxidized form?

19.43   Why is fatty acid catabolism called *beta* oxidation?

Nitrogen Pool

19.44   What is the nitrogen pool?

19.45   What are four ways in which amino acids are used in the body?

19.46   When the body retains more nitrogen than it excretes in all forms, the system is said to be on a *positive nitrogen balance*. Would this state characterize infancy or old age?

19.47   What happens to amino acids that are obtained in the diet but aren't needed to make any nitrogeneous compounds?

The Catabolism of Amino Acids[2]

*19.48   By means of two successive equations, one a transamination and the other an oxidative deamination, write the reactions that illustrate how the amino group of alanine can be removed as NH_4^+.

*19.49   When tyrosine undergoes decarboxylation, what forms? Write its structure.

*19.50   Write the structure of the product of the decarboxylation of tryptophan.

19.51   In the conditions of starvation and diabetes, what can the amino acids be used for?

*19.52   Write the structure of the keto acid that forms when phenylalanine undergoes transamination with α-ketoglutarate.

*19.53   When valine and α-ketoglutarate undergo transamination, what new keto acid forms? Write its structure.

19.54   Which nutrient becomes increasingly important as a source of energy (ATP) when the net effect either of starvation or of diabetes is the reduced availability of glucose as a source of ATP?

[2]Use Table 16.1, page 354, to determine the structures of the amino acids named in Review Exercises 19.48–19.53.

19.55 The catabolism of the nutrient of Review Exercise 19.54 produces what intermediate that is further catabolized by the citric acid cycle?

19.56 Two molecules of acetyl CoA can combine to give the coenzyme A derivative of what keto acid? Give its structure.

*19.57 Arrange the following compounds in the order in which they would be produced if the carbon skeleton of alanine were to appear in one of the ketone bodies.
1 Pyruvate
2 acetoacetyl CoA
3 acetoacetate
4 alanine
5 acetyl CoA

19.58 In two steps, the compound of Review Exercise 19.56 gives one unit of a ketone body and one other significant species (besides recovered CoA). What is it? Why is it a problem?

19.59 Give the names and structures of the ketone bodies.

19.60 What is ketonemia?

19.61 What is ketonuria?

19.62 What is meant by acetone breath?

19.63 Ketosis consists of what collection of conditions?

19.64 What is ketoacidosis? What form of acidosis is it, metabolic or respiratory?

19.65 The formation of which particular compound most lowers the supply of HCO_3^- in ketoacidosis?

19.66 What are the reasons for the increase in the volume of urine that is excreted in someone with untreated type I diabetes?

19.67 If the ketone bodies (other than acetone) can normally be used by heart and skeletal muscle, what makes them dangerous in starvation or in diabetes?

19.68 Why does the rate of urea production increase in untreated type I diabetes?

19.69 When a physician refers to the loss of Na^+ as the loss of *base*, what is actually meant?

*19.70 The conversion of pyruvate to an acetyl unit is both an oxidation and a decarboxylation.
(a) If *only* decarboxylation occurred, what would form from pyruvate? Write the structure of the other product in

$$CH_3\overset{O}{\overset{\|}{C}}CO_2^- + H^+ \longrightarrow \underline{\quad} + O{=}C{=}O$$

(b) If this product is oxidized, what is the name and the structure of the product of such oxidation?

(c) Referring to Figure 19.4, which specific compound undergoes an oxidative decarboxylation similar to that of pyruvate? (Give its name.)

*19.71 One cofactor in the enzyme assembly that catalyzes the oxidative decarboxylation of pyruvate requires thiamine, one of the B vitamins. Therefore in beriberi, the deficiency disease for this vitamin, the level of what substance can be expected to rise in blood serum (and for which an analysis can be made as part of the diagnosis of beriberi)?

*19.72 Complete the following equations for one cycle of β-oxidation by which a six-carbon fatty acyl group is catabolized.

$$\text{(a)}\ CH_3CH_2CH_2CH_2\overset{O}{\overset{\|}{C}}SCoA + FAD \longrightarrow \underline{\quad} + \underline{\quad}$$

(b) $\underline{\quad} + H_2O \longrightarrow \underline{\quad}$

(c) $\underline{\quad} + NAD^+ \longrightarrow \underline{\quad} + NADH/H^+$

(d) $\underline{\quad} + CoASH \longrightarrow \underline{\quad} + \underline{\quad}$

*19.73 Myristic acid, $CH_3(CH_2)_{12}CO_2H$, can be catabolized by β-oxidation just like palmitic acid.
(a) How many units of acetyl CoA can be made from it?
(b) In producing this much acetyl CoA, how many times does $FADH_2$ form and then deliver its hydrogen to the respiratory chain?
(c) Referring again to part (a), how many times does NADH form as acetyl CoA is produced and then deliver its hydrogen to the respiratory chain?
(d) Complete the following table by supplying the missing numbers of molecules that are involved in the catabolism of myristic acid to acetyl CoA.

Intermediate	Maximum Number of ATP from each	Total Number of ATP Possible from Each as Acetyl CoA Forms
— FADH$_2$	—	—
— NADH	—	—
— CH$_3$CO-SCoA	—	
		Sum = —
Deduct — high-energy phosphate bonds for activating the myristyl group		—
Net ATP produced for each myristyl group as it changes to acetyl CoA		—

20

METABOLISM AND MOLECULE BUILDING

Metabolic
  Interrelationships—
  An Overview

Glycogen Metabolism

Glucose Tolerance

Absorption,
  Distribution, and
  Synthesis of Lipids

Synthesis of Amino
  Acids

The bear does not know the luxury of obtaining salmon protein from a can, like pampered carnivores in people's homes. He's standing barefoot in frigid water at Brooks Falls, Alaska, and is no doubt as suprised as the salmon. We here survey some of the movements of nitrogen compounds at the molecular level of humans.

20.1 METABOLIC INTERRELATIONSHIPS—AN OVERVIEW

Acetyl CoA and pyruvate stand at major metabolic crossroads.

We will begin this chapter by using Figure 20.1 for a broad overview of metabolism. The figure places the pathways of *catabolism,* studied in the previous chapter, into the context of those of *anabolism,* or molecule building, studied in this chapter.

We start at the top of Figure 20.1 with products of digestion, namely, amino acids, monosaccharides, fatty acids, and monoacylglycerols. Amino acids enter the nitrogen pool, and they or their breakdown products become involved in nearly all of the other pathways shown in the figure.

The monosaccharides make up what is called **blood sugar,** but this is almost entirely glucose. Some glucose goes quickly to replenish glycogen reserves, chiefly in the liver and the muscles. Glycolysis makes some ATP and pyruvate, which itself leads to more ATP. Figure 20.1 seems to suggest that gluconeogenesis is simply the reverse of glycolysis, but it isn't quite that, as we will see.

Fatty acids might be used for energy or for building or repairing cell membranes, or they might be put into storage as fat in adipose tissue. Toward the right in Figure 20.1 we see β-oxidation feeding acetyl CoA toward the citric acid cycle and the respiratory chain. Acetyl CoA is also the raw material for making fatty acids, cholesterol, steroid hormones, bile salts, and the ketone bodies.

Figure 20.1 illustrates an old truism about nature that "everything is tied to everything else." Notice particularly how many pathways converge on either acetyl CoA or pyruvate. These two compounds are at the central metabolic crossroads of the body. No matter how complex are the compounds we take in, most of them can be broken down and their pieces used to make other compounds. Our study continues with the metabolism of glycogen, the chief storage form for glucose units.

■ Adipose tissue centers mostly around the middle of the body and cushions internal organs.

20.2 GLYCOGEN METABOLISM

Much of the body's control over the blood sugar level is handled by its regulation of the synthesis and breakdown of glycogen.

Glucose, delivered by general circulation, is the brain's favorite source of chemical energy, so the body must carefully manage the concentration of glucose in the blood. The blood glucose level is affected by several factors involved in its distribution, storage, and use.

A Special Vocabulary Exists to Describe Variations in the *Blood Sugar Level.* The concentration of monosaccharides in blood plasma has traditionally been called the **blood sugar level** but, more accurately, the **plasma sugar level.** It is very nearly the same as the glucose level, because glucose is overwhelmingly the major monosaccharide in plasma. When determined after several hours of fasting, the plasma sugar level, called the **normal fasting level,** is 70 to 110 mg/dL (3.9 to 6.1 mmol/L).[1] In a condition called **hypoglycemia,** the plasma sugar level is *below* normal, and in **hyperglycemia** it is *above* normal. The term **normoglycemia** is sometimes used for levels within the normal range.

■ mg/dL = milligrams per deciliter, where 1 dL = 100 mL.

---

[1]The normal reference laboratory values or "normals" used in this text are those published in *The New England Journal of Medicine,* **327,** page 718 (Sept. 3, 1992).

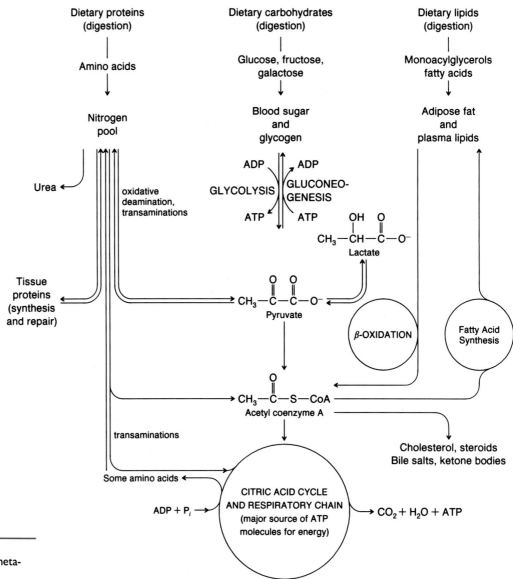

FIGURE 20.1

Interrelationships of major metabolic pathways.

■ -glyc-, "sugar"
-emia, "in blood"
hypo-, "under, below"
hyper-, "above, over"
renal, "of the kidneys"
-uria, "in urine"

Hypoglycemia Can Make You Faint. Your brain relies almost entirely on glucose for its chemical energy, consuming about 120 g of glucose per day. Brain cells do have the ability to switch over to other nutrients, but they cannot do this very rapidly. Thus, if hypoglycemia develops rapidly, you can become dizzy and may even faint.

Hyperglycemia Can Lead to Glucose in the Urine. When the blood sugar level becomes too hyperglycemic (becomes too high), the kidneys are unable to put back into the blood all of the glucose that temporarily leaves the blood as it circulates through the kidneys. Glucose then appears in the urine, a condition called **glucosuria.** The blood sugar level above which this happens is called the **renal threshold** for glucose, and it is in the range of roughly 160 to 180 mg/dL, and higher in some individuals.

Excess Blood Glucose Normally Is Withdrawn from Circulation. When more than enough glucose is in circulation to meet energy needs, the body does not simply eliminate the excess. Instead, the chemical energy of excess glucose is conserved either by making fatty acids (and then triacylglycerols in adipose tissue) or by making glycogen.

The synthesis of glycogen from glucose is called **glycogenesis** ("glycogen creation"), and most of this activity occurs in liver and muscle cells. The liver can hold 70 to 110 g of glycogen, and muscle cells, taken as a whole, contain from 170 to 250 g. When muscle cells need the chemical energy of glucose, they take it back out of glycogen by a series of reactions called **glycogenolysis** (lysis or hydrolysis of glycogen), a process controlled by several hormones. Liver glycogen serves largely to resupply the blood with glucose when the diet is not doing it. Thus, when the blood needs glucose because the blood sugar level has dropped too much, the liver hydrolyzes its glycogen reserves as needed and then puts the glucose into circulation.

■ *Lysis* means "breakup."

Epinephrine Stimulates Glycogenolysis. When muscular work is begun, the adrenal medulla secretes the hormone *epinephrine.* In muscle tissue and, to some extent, in the liver, epinephrine activates glycogenolysis by a long series of steps, all occurring extremely rapidly. One epinephrine molecule can trigger the mobilization of thousands of glucose units, which then are ready to supply needed chemical energy.

The end product of glycogenolysis isn't actually glucose but glucose-1-phosphate. Cells that can do glycogenolysis also have an enzyme called *phosphoglucomutase,* which catalyzes the conversion of glucose-1-phosphate to its isomer, glucose-6-phosphate:

■ Epinephrine (adrenaline) and norepinephrine (noradrenaline)— see Special Topic 11.2, page 258— are called the "fight or flight" hormones because the brain causes their release in emergencies.

■ An estimated 30,000 molecules of glucose are released from glycogen for each molecule of epinephrine that initiates glycogenolysis.

Glucose-1-phosphate
(glucose-1-P or G1P)

Glucose-6-phosphate
(glucose-6-P or G6P)

■ The letter P is often used to represent the whole phosphate group in the structures of phosphate esters.

Glucose Is Trapped in the Muscle Cell When It Is in the Form of Glucose-6-phosphate. Glucose-6-phosphate, rather than glucose, is the form in which a glucose unit must be to enter glycolysis and produce ATP. Glucose-6-phosphate is also a derivative of glucose *that cannot migrate out of muscle cells.* Its glucose units, therefore, cannot be lost from muscle tissues needing them during physical work. Thus, glycogenolysis in muscle tissue is an important supplier of energy for muscular activities. When the supply of muscle glycogen is low, muscle cells take glucose from circulation, *trap it as glucose-6-phosphate,* and then convert this compound to glycogen.

■ Figure 19.3, page 429, shows how glucose units enter glycolysis.

Glucagon Activates Liver Glycogenolysis and Thus Affects the Blood Sugar Level. The α cells of the pancreas make a polypeptide hormone, *glucagon,* which helps to maintain a normal blood sugar level. When the blood sugar level decreases, the α cells release glucagon. Its target tissue is the liver, where it is a strong activator of glycogenolysis.

Unlike epinephrine, glucagon *inhibits* glycolysis, so one effect of glucagon is to keep the level of blood glucose up. Glucagon, also unlike epinephrine, does not cause

■ The glucagon molecule has 29 amino acid residues.

an increase in blood pressure or pulse rate, and it is longer-acting than epinephrine.

Growth requires energy, so the action of glucagon that helps to supply a source of energy—glucose—aids in the work of the *human growth hormone.* In some situations, such as a disfiguring condition known as *acromegaly,* there is an excessive secretion of human growth hormone that promotes too high a level of glucose in the blood. This is undesirable because a prolonged state of hyperglycemia from any cause can lead to some of the same blood capillary-related complications observed when diabetes is poorly controlled.

■ Acromegaly is sometimes call *giantism* because certain bone structures and visceral organs become enlarged.

Liver Cells Can Release Glucose to Circulation.   In the liver (but not in muscles), **glucose units** released as glucose-6-phosphate (or glucose-6-P) by glycogenolysis are **further** hydrolyzed to glucose.

$$\text{Glucose-6-P} + \text{H}_2\text{O} \xrightarrow[\text{(in the liver)}]{\text{Glucose-6-phosphatase}} \text{glucose} + \text{P}_i$$

The liver has the needed enzyme, as we said, but not the muscles. Glucose is thus freed in the liver to leave this organ and go into the blood. During periods of fasting, therefore, the overall process in the liver—glucose-1-phosphate to glucose-6-phosphate to glucose—is a major supplier of glucose for the blood. Glucagon, which triggers this, is thus an important regulator of the blood sugar level. Given the high daily demand of the brain for glucose, noted earlier, you can see how the brain depends on the liver during fasting to maintain its supply of chemical energy. We are beginning to see in chemical terms how vital the liver is to the performance of other organs. When circulating glucose is taken up by a brain cell, it is promptly trapped by being converted to glucose-6-phosphate.

Insulin Strongly Lowers the Blood Sugar Level.   The β cells of the pancreas make and release *insulin,* a polypeptide hormone. Its release is stimulated by an increase in the blood sugar level, which normally occurs after a carbohydrate-rich meal. As insulin moves into action, it finds its receptors at the cell membranes of muscle and adipose tissue. The insulin–receptor complexes make it possible for glucose molecules to move easily into the affected cells, and this activity decreases the blood sugar level.

■ If the receptors are continuously overloaded with work in someone who consumes a great deal of sugar, they can wear out, and the individual becomes diabetic.

Not all cells depend on insulin to take up glucose. Brain cells, red blood cells, and cells in the kidneys, the intestinal tract, and the lenses of the eyes take up glucose directly.

In one form of diabetes mellitus, type I diabetes, the β cells have been destroyed, and the pancreas is unable to release insulin. Such individuals must receive insulin, usually by intravenous injection. If more insulin is put into circulation than needed for the management of the blood sugar level, this level falls too low and *insulin shock* results.

■ Lifesaving first aid for someone in insulin shock is sugared fruit juice or candy to counter the hypoglycemia.

Somatostatin Inhibits Glucagon and Slows the Release of Insulin.   The hypothalamus, a specific region in the brain, makes *somatostatin,* another hormone that participates in the regulation of the blood sugar level. When the β cells of the pancreas secrete insulin, which helps to *lower* the blood sugar level, the α cells should not at the same time release glucagon, which helps to *raise* this level. Somatostatin acts at the pancreas to inhibit the release of glucagon as well as to slow down the release of insulin. It thus helps to prevent a wild swing in the blood sugar level that insulin alone might cause.

■ *Neo,* "new"; *neogenesis,* "new creation"; *gluconeogenesis,* "the synthesis of new glucose."

Gluconeogenesis Is a Source of Blood Glucose during Fasting.   The synthesis of glucose from smaller molecules is called **gluconeogenesis** (Figure 20.2). As we'd ex-

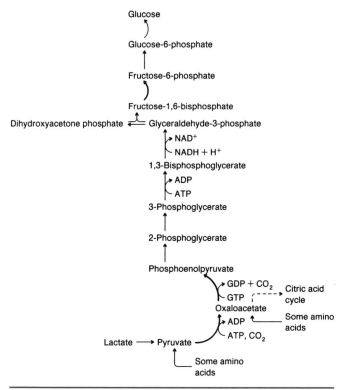

FIGURE 20.2

Gluconeogenesis. The straight arrows signify steps that are the reverse of corresponding steps in glycolysis. The heavy, curved arrows denote steps that are unique to gluconeogenesis.

pect, chemical energy is needed to take small molecules back up an energy hill to energy-richer carbohydrates. Guanosine triphosphate (GTP in the figure) is the source of the needed energy; it's a high-energy phosphate, like ATP. Gluconeogenesis is a vital means of supplying the body with glucose for maintaining the blood sugar level when this level drops and glycogen reserves are also low. Periods without food, known daily to millions of undernourished people in the world, quickly deplete the glycogen reserves. Vigorous exercise accompanied by anaerobic glycolysis is also a common cause of the loss of glycogen.

The end product of anaerobic glycolysis, as we learned in the previous chapter, is lactate. Thus, when extensive *anaerobic* glycolysis occurs, the lactate level in blood increases. Lactate is an example of a small species that can be made into glucose by gluconeogenesis. It still has useful chemical energy, and for roughly every six lactate ions, liver cells are able to convert five to glucose. The sixth is catabolized to supply the needed energy. Several amino acids can also be used as starting materials for gluconeogenesis.

Three steps in glycolysis cannot be directly reversed, those involving the curved arrows in heavy print in Figure 20.2. However, the liver and the kidneys have special enzymes that create bypasses, and they are integral parts of the enzyme team for gluconeogenesis. The other steps in gluconeogenesis are run as reverse shifts in equilibria that occur in the opposite direction in glycolysis.

20.3   GLUCOSE TOLERANCE

The ability of the body to tolerate swings in the blood sugar level is essential to health.

When glucose enters the bloodstream from the intestinal tract, some stays in circulation and some is removed by various mechanisms, as we have discussed. Muscle and liver cells, for example, trap glucose, and excess glucose is converted to other substances, like fatty acids and cholesterol. Glycolysis, which uses up glucose to make ATP, also tends to lower the blood sugar level. Many factors thus affect the blood sugar level, some tending to increase it and some doing the opposite (Figure 20.3). The ability of the body to keep its blood sugar level in the normal range and to respond quickly to widely changing conditions is called its **glucose tolerance.** Whenever hyperglycemia develops and tends to persist, something is wrong with the mechanisms for withdrawing glucose from circulation.

Glucose Tolerance Is Poor in Diabetes.   The subject of glucose tolerance is nowhere of greater concern than in connection with diabetes. **Diabetes** is defined clinically as a disease in which the blood sugar level persists in being much higher than warranted by the dietary and nutritional status of the individual. Invariably, a person with untreated diabetes has glucosuria, and the discovery of this condition often triggers the clinical investigations that are necessary to rule diabetes in or out.

There are two broad kinds of diabetes, type I and type II (see Special Topic 20.1). Type I diabetics are unable to manufacture insulin (at least not enough), and they need daily insulin therapy to manage their blood sugar levels. Rigorously maintaining a relatively even and normal blood sugar level is the best single strategy for the prevention of some of the vascular and neural problems that often complicate the lives of diabetics later. Type II diabetics are those who are largely able to manage their blood sugar levels by a good diet, weight control, exercise, and sometimes the use of oral medications.

In an individual with diabetes under poor control and so with sustained hyperglycemia, some blood glucose combines with hemoglobin to give *glycohemoglobin* (or glycosylated hemoglobin). The level of glycohemoglobin doesn't fluctuate as widely and quickly as the blood sugar level. Determination of the glycohemoglobin level,

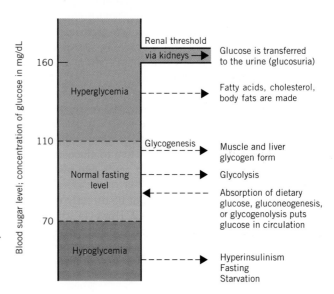

FIGURE 20.3
Factors that affect the blood sugar level.

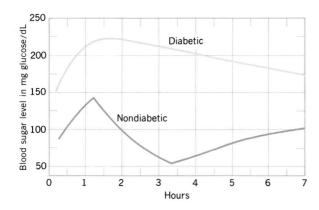

FIGURE 20.4
Glucose tolerance curves.

therefore, has become the best way to assess the average status of an individual's blood sugar level, better even than direct measurements of blood glucose. The initial discovery of poor glucose tolerance, however, is often made by using the *glucose tolerance test.*

The *Glucose Tolerance Test* Measures Glucose Tolerance.   In the **glucose tolerance test,** the individual is given a drink containing glucose, generally 75 g for an adult and 1.75 g/kg of body weight for children. Then the blood sugar level is checked at regular intervals and plotted (Figure 20.4).

The lower curve in Figure 20.4 is that of a person with normal glucose tolerance, and the upper curve is of one whose glucose tolerance is typical of a person with diabetes. In both, the blood sugar level increases sharply at first. A person with good glucose tolerance, however, soon manages the high level and brings it back down with the help of a normal flow of insulin and somatostatin. In the diabetic, however, the level comes down only very slowly and remains essentially in the hyperglycemic range throughout.

Notice in Figure 20.4 that the blood sugar level of a normal individual can sometimes drop to a mildly hypoglycemic level. This might happen, for example, after eating a carbohydrate-rich breakfast. When considerable glucose pours relatively quickly into the bloodstream, a bit more insulin than needed might be released. Now, an *overwithdrawal* of glucose from circulation might occur, which by midmorning, brings dizziness, sometimes fainting, and maybe even falling asleep in class! (Those who don't faint or fall asleep tend to become cranky.) The prevention isn't more sugared doughnuts but a balanced breakfast.

■ Unhappily, most people with midmorning sag start another round of coffee and sugared rolls. The glucose gives a short lift, but then an oversupply of insulin restores the mild hypoglycemia of the sag.

20.4   ABSORPTION, DISTRIBUTION, AND SYNTHESIS OF LIPIDS

Several lipoprotein complexes in the blood transport triacylglycerols, fatty acids, cholesterol, and other lipids.

The digestion of triacylglycerols produces a mixture of long-chain fatty acids and monoacylglycerols. As these migrate across the intestinal barrier, they are extensively reconstituted. Thus, what enters circulation consists mostly of triacylglycerols.

Lipids are insoluble in water, so they are carried in blood by proteins with which they form **lipoprotein complexes.** Defects in this system can cause heart disease, as discussed in Special Topic 20.2, a survey of recent developments in the relationship of dietary fat, cholesterol, and heart attacks. Lipoprotein complexes unload triacylglycerols at adipose tissue.

■ Because lipids are water insoluble, they attract the least amount of associated water in storage.

The name for this disorder is from the Greek *diabetes,* "to pass through a siphon," and *mellitus,* "honey sweet," meaning to pass urine that contains sugar. We'll call it diabetes for short. In severe, untreated diabetes, the victim's body wastes away despite efforts to satisfy a powerful thirst and hunger. To the ancients, it seemed as if the body were dissolving from within.

Between 1 and 6% of the U.S. population has diabetes, and almost as many others are believed to have this disease. It ranks third behind heart disease and cancer as the cause of death.

Between 10 and 25% of all cases of diabetes are of the severe, insulin-dependent variety in which the β-cells of the pancreas are unable to make and secrete insulin. This is **type I diabetes,** and insulin therapy is essential. It is also called **insulin-dependent diabetes mellitus** or **IDDM.** Most victims contract IDDM before the age of 40, often as adolescents, so IDDM has sometimes been called *juvenile-onset diabetes.*

The rest of all those with diabetes have a form called **type II diabetes,** or **non-insulin-dependent diabetes mellitus, NIDDM.** Most victims are able to manage their blood sugar levels by diet and exercise alone, without insulin injections. Their problem is not the lack of insulin but a breakdown in the machinery for taking advantage of it at insulin's target cells. Most who contract NIDDM do so when they are over 40, so NIDDM has been called *adult-onset diabetes.*

Type I Diabetes Develops in Six Stages. D. S. Eisenbarth, a diabetes specialist, divides the onset of type I diabetes into six stages. We'll review them as background for illustrations of equilibrium chemistry and factors that shift equilibria.

Stage 1 is thought to be an existing genetic defect most likely involving more than one gene. However, sets of identical twins are known in which only one twin becomes diabetic. In some way, the genetic problem must be related to a problem of the immune system.

Stage 2 is a triggering incident, like a viral infection. The mumps virus, for example, causes diabetes in some. Usually, the onset of virus-caused type I diabetes occurs slowly over a few years.

Stage 3 is the appearance in the blood of certain antibodies. (*Antibodies* are substances made by the immune system to counteract the effects of invading substances called *antigens* that are alien to the body.) Substances on the membranes of the pancreatic β cells have been altered (perhaps by the virus) so that the body's immune system sees them as foreign antigens and so makes antibodies against them. Type I diabetes is thus an autoimmune disease, *auto* because the body's immune system fails to recognize the proteins of its own body and sets out to destroy them and, therefore, itself.

Stage 4 is a period during which the pancreas loses its ability to secrete insulin. Stage 5 is diabetes and persistent hyperglycemia. Most of the pancreatic β cells have disappeared. Stage 6 is the period following complete destruction of β cells.

Immune-Suppressant Therapy Works If the Problem Is Caught in Its Early Stages. Cyclosporine is an agent used to suppress the rejection of transplanted organs, like kidney transplants. When used in the early stages of the onset of type I diabetes, cyclosporine prevents insulin dependence in a significant fraction of individuals tested.

Insulin Receptors Are a Problem in Type II Diabetes. The onset of type II diabetes is much slower than that of type I, and obesity, lack of exercise, and a sugar-rich diet are factors. Some scientists believe that sugar evokes such a continuous presence of insulin that the insulin receptor proteins of target cells literally wear out faster than they can be replaced. When the weight is reduced, particularly in connection with physical exercise, the relative numbers of receptor proteins rebound. Under the incessant demand to produce and secrete insulin, the β cells can eventually give out. In a veritable epidemic of NIDDM, over 60% of the older adult population of the Pacific islet of Nauru has become diabetic following a change in lifestyle from active fishing and farming to a very sedentary life. (The discovery of huge phosphate reserves on Nauru Island made the Nauruans one of the world's wealthiest peoples.)

The conventional wisdom that places insulin and its receptors at the heart of the NIDDM problem has been recently challenged by the discovery that the β cells secrete not only insulin but also another protein called amylin. Is it the relatively high level of amylin that suppresses the uptake of glucose by target cells? If so, would an amylin control drug be the answer to NIDDM? Needless to say, much research is in progress.

Glucosylation of Proteins May Cause the Long-Term Complications of Diabetes. The immediate complications of IDDM are an elevated blood sugar level, metabolic acidosis, and eventual death from coma and uremic poisoning. Insulin therapy corrects these immediate problems, but it deals less well with the longer-term complications.

The continuous presence of a high level of blood glucose in both IDDM and NIDDM shifts certain chemical equilibria in favor of glucosylated compounds. The aldehyde group of the open form of glucose, for example, can react with amino groups, like those on side chains of lysine residues, to form products called *Schiff bases.*

$$-CH{=}O + H_2N - \rightleftharpoons -CH{=}N - + H_2O$$

| Aldehyde group | Amino group | Schiff base system |

Hemoglobin, for example, gives this reaction, and a high level of glucose shifts this equilibrium to the right. The level of glucosylated hemoglobin thus increases. When the glucose level is brought down and kept within a normal range, the Schiff base level also declines.

The problem with the Schiff bases in the long term is that they undergo molecular rearrangements to more permanent products, called *Amadori compounds,* in which the C$=$N double bond has migrated to C$=$C positions. *After a time, the formation of the Amadori compounds is not reversible.*

When these reactions occur in the basement membrane of blood capillaries, they swell and thicken; the condition is called *microangiopathy.* (The basement membrane is the protein support structure that encases the single layer of cells of a capillary.) Microangiopathy is believed to lead to the other complications, most of which involve the vascular system or the neural networks: kidney problems, gangrene of the lower limbs, and blindness. Diabetes is the leading cause of new cases of blindness in the United States, and it is the second most common cause of blindness, overall.

Blindness from Diabetes May Also Reflect the Reduction of Glucose to Sorbitol. Glucose is reduced by the enzyme *aldose reductase* to sorbitol. It's a minor reaction in cells of the lens of the eye, but *an abundance of glucose shifts equilibria in favor of too much sorbitol.* Sorbitol, unlike glucose, tends to be trapped in lens cells, and as the sorbitol concentration rises so does the osmotic pressure in the fluid. *This draws water into the lens cells, which generates pressure and leads to cataracts.*

Blood Sugar Control Is Mandatory. The best single treatment of diabetes is any effort that keeps a strict control on the blood sugar level to avoid the episodes of upward surges followed by precipitous declines. A

report to this effect was issued in late 1993 by the Diabetes Control and Complications Trial Research Group, which involved hundreds of physicians and scientists from more than 30 institutions. Conventional therapy for IDDM patients has been one or two daily injections of insulin and daily self-monitoring of glucose in the blood or the urine. Over a 6-year period, the Research Group compared a group of such patients with another group given intensive therapy, which consisted of insulin injections three or more times a day with self-monitoring of blood glucose at least four times a day. The Research Group found that intensive therapy delays the onset and slows the progression of the vision-threatening complications of diabetes and other problems by a very significant factor.

A Number of Technologies Are in Use or Are Being Tested. People with IDDM gauge their insulin needs by blood tests for blood sugar levels. We mentioned an enzyme-based test in Section 17.4, but it requires a puncture to obtain a drop of blood. Another technology in a testing stage uses a hand-held, battery-driven source of infrared rays, which are focused onto the skin at the wrist or a fingertip. The meter converts the amount of light absorbed to a blood glucose level. With some diabetics, insulin pumps can be implanted much like heart pacemakers. These monitor the blood sugar level and release insulin according to need. The use of insulin nasal sprays immediately before a meal is another approach being tested. Several groups of scientists are working on insulin pills, a technology made difficult by the fact that insulin, like any protein, is digested.

When stripped of neighboring cells, β cells can be transplanted. They need not even be inserted into the receiver's pancreas, and they start to make insulin in a few weeks. Apparently the cells *adjacent to* the β cells are responsible for inducing the body's immune-centered rejection process, so without such cells the β cells are more safely transplanted.

Human β cells work best, of course, but those from pigs and cows also appear to be usable provided that they are encapsulated in very small plastic spheres. These spheres have microscopic holes large enough to let insulin molecules escape but not large enough to let antibodies inside. These techniques have cured type I diabetes in experimental animals. The first successful use of encapsulated human islet cells in a human patient was reported in 1994. Prior to treatment, the patient had experienced two to three hypoglycemic episodes per week. After treatment, the patient no longer needed insulin injections and had no hypoglycemic episodes in eight months.

Adipose Tissue Is the Chief Lipid Storage Depot.   There are two kinds of adipose tissue, brown and white. Both kinds are associated with internal organs, cushioning them against bumps and shocks and insulating them from swings in temperature. White adipose tissue stores energy as triacylglycerols chiefly in service to the energy budgets of other tissues. Brown adipose tissue serves to generate heat by the catabolism of fatty acids to keep the temperature of the body's core steady.

A 70-kg adult male has about 12 kg of triacylglycerol in storage. If he had to exist on no food, only water and a vitamin–mineral supplement, and if he needed 2500 kcal/day, this fat would supply his caloric needs for 43 days. Of course, during this time the body proteins would also be wasting away to serve gluconeogenesis, so metabolic acidosis would be a problem of growing urgency.

■ These data are for information; they're certainly not recommendations!

Fatty Acids Can Be Made from Acetyl CoA.   Acetyl CoA stands at a major metabolic crossroads, as we saw in Figure 20.1. It can be made from any monosaccharide in the diet, from virtually all amino acids, and from fatty acids. Once made, acetyl CoA can be shunted into the citric acid cycle, where its chemical energy can be used to make ATP; or its acetyl group can be made into other compounds, like fatty acids, that the body might need. We'll look in some detail at fatty acid synthesis because it provides an excellent example of a multienzyme complex in action.

Whenever acetyl coenzyme A molecules are made within mitochondria but aren't needed for the citric acid cycle and respiratory chain, they are exported to the cytosol. The enzymes for fatty acid synthesis are found there, not within the mitochondria, illustrating the general rule that *the body segregates catabolism from anabolism.*

■ Recall that the cytosol is the liquid part of the fluid outside of cellular organelles such as mitochondria and nuclei.

As might be expected, because fatty acid synthesis is in the direction of climbing an energy hill, the cell has to invest some energy of ATP to make fatty acids from smaller molecules. The first payment occurs in the first step in which the bicarbonate ion reacts with acetyl CoA.

■ Because the symbols ATP, ADP, and P_i are not given with their electrical charges, we can't provide an electrical balance to equations such as this.

$$\underset{\substack{\text{Acetyl}\\\text{coenzyme A}}}{\overset{\overset{\text{O}}{\|}}{CH_3CSCoA}} + HCO_3^- + ATP \longrightarrow \underset{\substack{\text{Malonyl}\\\text{coenzyme A}}}{\overset{\overset{\text{O}\quad\text{O}}{\|\quad\|}}{{}^-OCCH_2CSCoA}} + 2H^+ + ADP + P_i$$

The acetyl group is now activated for fatty acid synthesis.

The enzyme that now takes over is actually a huge complex of seven enzymes called *fatty acid synthase* (Figure 20.5). In the center of the complex is a molecular unit long enough to serve as a swinging arm carrier, like the boom of a construction crane. It's called the *acyl carrier protein,* or ACP, and it swings from active site to active site in the multienzyme complex. Thus, the arm brings what it carries over one enzyme after another. At each stop, a reaction is catalyzed that contributes to the lengthening of the chain of a fatty acid. Let's see how it works.

The malonyl unit in malonyl CoA, made according to the equation above, transfers to the swinging arm of ACP. In the meantime, a similar reaction occurs to another molecule of acetyl CoA at a different site on the synthase, called simply *E*.

Next, the acetyl group on *E* transfers to the malonyl group on ACP. As it does, carbon dioxide, the initial activator, is ejected; it has served its activating purpose. A four-carbon derivative of ACP, acetoacetyl ACP, forms. The *E* unit is vacated.

$$\underset{\text{Acetyl } E}{\overset{\overset{\text{O}}{\|}}{CH_3CS{-}E}} + \underset{\text{Malonyl ACP}}{\overset{\overset{\text{O}\quad\text{O}}{\|\quad\|}}{{}^-OCCH_2CS{-}ACP}} \longrightarrow \underset{\text{Acetoacetyl ACP}}{\overset{\overset{\text{O}\quad\text{O}}{\|\quad\|}}{CH_3CCH_2CS{-}ACP}} + CO_2 + E$$

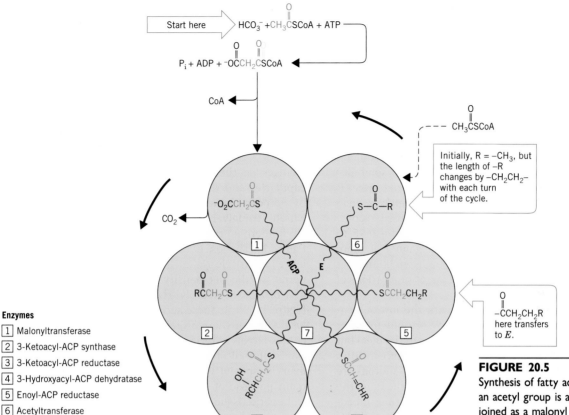

Enzymes

1. Malonyltransferase
2. 3-Ketoacyl-ACP synthase
3. 3-Ketoacyl-ACP reductase
4. 3-Hydroxyacyl-ACP dehydratase
5. Enoyl-ACP reductase
6. Acetyltransferase
7. Acyl carrier protein

FIGURE 20.5
Synthesis of fatty acids. At the top, an acetyl group is activated and joined as a malonyl unit to an arm of the acyl carrier protein, ACP. Another acetyl group transfers from acetyl CoA to site *E*. In a second transfer, this acetyl group is then joined to the malonyl unit as CO_2 splits back out. This gives a β-ketoacyl system whose keto group is reduced to CH_2 by the next series of steps. One turn of the cycle adds a CH_2CH_2 unit to the growing acyl chain.

Acetoacetyl ACP, now on the swinging boom, is moved over first one enzyme and then another until its keto group (CO) is reduced to CH_2. First, the keto group is reduced to a 2° alcohol, which is then dehydrated to introduce a double bond. Next, the double bond is hydrogenated (reduced) to the saturated system in butyryl ACP. The overall effect, as we said, is the reduction of the keto group to CH_2. Notice that NADPH, the reducing agent made by the pentose phosphate pathway of glucose catabolism, is used here, not NADH.

The butyryl group is now transferred to the vacant *E* unit of the synthase, the unit that initially held an acetyl group. This ends one complete turn of the cycle. To recapitulate, we have gone from two two-carbon acetyl units to one four-carbon butyryl unit. The steps now repeat so that the four-carbon unit is elongated to a six-carbon unit.

In the next turn, this six-carbon acyl group will be elongated to an eight-carbon group. And the process will repeat until the chain is as long as the system requires. Overall, the net equation for the synthesis of the palmitate ion from acetyl CoA is as follows, and you can see how much high-energy phosphate (ATP) and how much reducing agent (NADPH) are required per mole of palmitate ion.

■ Glucagon, epinephrine, and cyclic AMP—all stimulators of the use of glucose to make ATP—depress the synthesis of fatty acids in the liver. Insulin, however, promotes it.

$$8CH_3CSCoA + 7ATP + 14NADPH \longrightarrow$$

$$CH_3(CH_2)_{14}CO_2^- + 7ADP + 7P_i + 8CoA + 14NADP^+ + 6H_2O$$
Palmitate ion

SPECIAL TOPIC 20.2
LIPOPROTEIN COMPLEXES, "GOOD" AND "BAD" CHOLESTEROL, AND HEART DISEASE

Chylomicrons Carry Dietary Lipids and Cholesterol.
The solubility in water of a long-chain fatty acid is typically less than 10^{-6} mol/L, and triacylglycerols are comparably insoluble. However, when lipid molecules are carried as complexes with plasma albumin molecules, called **lipoprotein complexes,** the solubility functionally becomes about a thousandfold greater. There are several kinds of lipoprotein complexes, each with its own function, and they are classified according to their densities (d), which range from 0.95 to 1.21 g/cm^3. Cholesterol ($d = 1.05$ g/cm^3) is more dense than triacylglycerols ($d = 0.9$ g/cm^3), and the complexes of higher density have larger proportions of cholesterol.

The lipoprotein complexes with the lowest density are called *chylomicrons.* They are put together from exogenous lipid material (lipid from the diet) but include 2% protein or less. After their assembly within the cells of the intestinal membrane, chylomicrons are delivered to the lymph, which carries them to the bloodstream. When they enter the capillaries embedded in muscle and adipose tissue (fat tissue), chylomicrons encounter binding sites and are held up. An enzyme, *lipoprotein lipase,* now catalyzes the hydrolysis of the chylomicrons' triacylglycerols to fatty acids and monoacylglycerols, which are promptly absorbed by the nearby tissue. As the hydrolysis occurs, the chylomicrons shrink to *chylomicron remnants,* which still contain dietary cholesterol. The remnants break loose from the capillary surface, and circulate to the liver where they are absorbed (Figure 1). *The overall functions of the chylomicrons are thus to deliver exogenous fatty acids to muscle and adipose tissue and to carry dietary cholesterol to the liver.*

Lipoprotein Complexes Transport Endogenous Lipids. The liver is a site where breakdown products from carbohydrates are used to make fatty acids and, from them, triacylglycerols. Lipids made in the liver are called *endogenous* lipids ("generated from within"). Three similar lipoprotein complexes are used to transport endogenous lipids in the bloodstream: *very low density lipoproteins* (VLDL), *intermediate density lipoproteins* (IDL) and *low-density lipoproteins* (LDL). These three complexes also carry cholesterol. Another complex, *high-density lipoprotein* (HDL), carries back to the liver the cholesterol released from tissues and no longer needed by them.

VLDL Changes to IDL and Then to LDL As VLDL Moves from the Liver to Other Tissues. The liver packages triacylglycerols and cholesterol into VLDL and releases them into circulation. During circulation, the VLDL complex undergoes somewhat continuous changes as their triacylglycerols are hydrolyzed in capillaries and the hydrolysis products are taken up by adipose tissue and muscles. By these processes, which strongly resemble the changes that occur to chylomicrons, the VLDL particles change to IDL (see Figure 1). The loss of the lower density triacylglycerols leaves a particle richer in the higher density components, so the net density increases. With continued loss of triacylglycerols, the IDL change to LDL (see Figure 1). While these activities take place, the VLDLs lose some of their protein, and their cholesterol becomes largely esterified by fatty acids and changed to esters of cholesterol. Some LDL is reabsorbed by the liver at special *receptor proteins* for LDL, but the main purpose of LDL is to deliver cholesterol to extrahepatic tissue (tissue other than liver tissue) to be used to make cell membranes and, in specialized tissues, steroid hormones.

HDL Transports Cholesterol from Extrahepatic Tissue Back to the Liver. High density lipoprotein complexes are cholesterol scavengers. When cells of extrahepatic tissue break down for any reason, their cholesterol molecules are picked up by HDL particles and changed once again to cholesteryl esters. En route to the liver, HDL undergoes some changes of its own. There is evidence that HDL can transfer cholesteryl esters to VLDL. By one means or another, some of the HDL becomes more like LDL before entering liver cells by means of the LDL receptors (Figure 1). (There possibly are specific HDL receptors on liver cells, too.) When LDL receptors for cholesterol-bearing lipoprotein complexes are absent, defective, or in too few numbers, either at the liver or at extrahepatic tissue, there are serious consequences. American scientists J. L. Goldstein and M. S. Brown shared the 1985 Nobel prize in medicine for their work on LDL receptor proteins and how they help to control blood cholesterol levels.

Inadequate LDL Receptors Cause Serum Levels of LDL, the "Bad Cholesterol," to Increase. When the receptor proteins for LDL are reduced in number or are absent, there is little ability to remove cholesterol

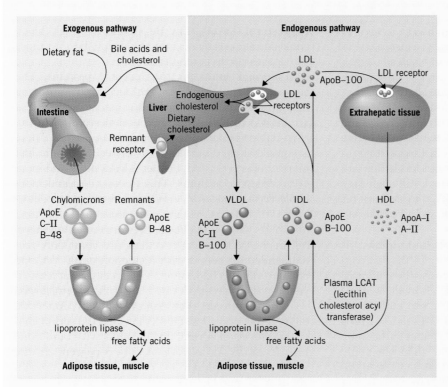

FIGURE I

Transport of cholesterol and triacylglycerols by lipoprotein complexes. The abbreviations ApoE, C-II, B-48, and so forth refer to specific proteins involved with the lipoprotein complexes. [After M. S. Brown and J. L. Goldstein *in* E. Brunwald, K. J. Isselbacher, R. G. Petersdort, J. B. Martin, and A. S. Fauci, Eds., *Harrison's Principles of Internal Medicine* (11th edition), p. 1652, McGraw-Hill (1987).]

from circulation. The level of cholesterol in the blood, therefore, becomes too high. The result is *atherosclerosis,* a disease in which several substances, including collagen, elastic fibers, and triacylglycerols, but chiefly cholesterol and its esters, form plaques in the arterial wall. Such plaques are the chief cause of heart attacks.

The ranges of serum cholesterol concentrations regarded as "desirable," "borderline high," and "risk" are given in Table 1. Some people have a genetic defect that bears specifically on the LDL receptors at the liver and causes elevated levels of LDL cholesterol. Two genes are involved. Those who carry two mutant genes have *familial hypercholesterolemia,* a genetically caused high level of cholesterol in the blood—

three to five times higher than average. Even on a zero-cholesterol diet, the victims have very high cholesterol levels. Their cholesterol along with other materials slowly comes out of the blood at valves and other sites and reduces the dimensions of the blood capillaries; this restricts blood flowage. Atherosclerosis has set in. Because elevated levels of LDL cholesterol are so often involved, LDL cholesterol is often called "bad cholesterol." Because the HDL complexes perform the desirable service of helping to carry cholesterol back to the liver, HDL cholesterol is sometimes called the "good cholesterol."

In atherosclerosis, the heart must work harder. Eventually arteries and capillaries in the heart itself become reduced so much in cross-sectional size that

TABLE 1   Serum Levels of Cholesterol in all Forms

Form	Desirable	Borderline High	Risk
Total cholesterol	< 200 mg/dL (< 5.18 mmol/L)	200–239 mg/dL (5.18–6.19 mmol/L)	> 239 mg/dL (> 6.20 mmol/L)
LDL	< 130 mg/dL	130–159 mg/dL	> 160 mg/dL
HDL	< 35 mg/dL (< 0.91 mmol/L)		

Source: New England Journal of Medicine, Sept. 3, 1992, page 718.

they no longer are able to bring sufficient oxygen to heart tissue. The plaques sometimes cause clots to form in capillaries of the heart and so block all oxygen delivery to affected heart tissue. Now a myocardial infarction ("heart attack") has occurred.

Victims of familial hypercholesterolemia generally have their first heart attacks as children and are dead by their early twenties. People with one defective gene and one normal gene for the LDL receptor proteins generally have blood cholesterol levels that are two or three times higher than normal. Although they number only about 0.5% of all adults, they account for 5% of all heart attacks among those younger than 60.

One Kind of Lipoprotein in HDL Is Undesirable. It isn't only the *relative concentration* of HDL that confers protection against atherosclerosis; the *composition* of the HDL is also a factor. HDL is made of about equal parts of lipid and protein. The two most abundant proteins are apolipoprotein A-I and A-II (or apoA-I and apoA-II). When *both* are found in the same HDL particles, the HDL is less "good" than when only apoA-I is present. It is the apoA-I that confers on the HDL its ability to protect against atherosclerosis, but this ability deteriorates when apoA-II is also present (as shown in 1993 by experiments in mice). *Thus, the ratio of HDL to LDL does not alone provide information concerning an individual's status regarding atherosclerosis. The composition of the HDL must also be considered.* Just how awaits further research.

Cholesterol Levels Can Be Reduced. High blood cholesterol levels occur in many people besides those with genes for hypercholesterolemia, even in people with normal genes for the receptor proteins. The causes of their high cholesterol levels have not been fully unraveled. Smoking, obesity, and lack of exercise are known to contribute to the cholesterol problem. High-cholesterol foods also appear to be factors. There is some evidence that as the liver receives more and more cholesterol from the diet it loses more and more of its receptor proteins. This forces more and more cholesterol to linger in circulation. Some people, however, are able to sustain high-cholesterol and high lipid diets with no ill effects. Eggs are 0.5% cholesterol (all in the yolk), so eggless diets are commonly recommended for people with high cholesterol. Yet, one 88-year-old man had eaten 25 eggs a day for at least 15 years without having elevated blood cholesterol levels!

The difference between a typically American high-fat diet and a low-fat diet lowered the serum cholesterol levels of several dozen adults with borderline high levels by an average of only 5 percent. In the same study, reported in 1993, lovastatin brought about a 27 percent reduction in serum cholesterol. Thus, the effect of diet on serum cholesterol levels among adults who are healthy, except for borderline high cholesterol levels, is relatively small. Serum cholesterol, of course, is not the only risk factor for heart disease, and the 1993 study in no way endorses a high-fat diet.

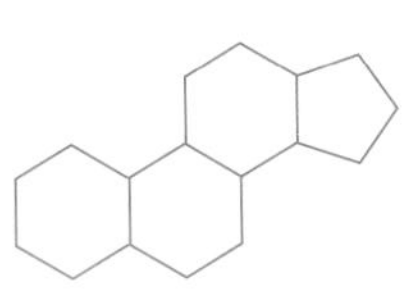

Cholesterol Is Also Made from Acetyl CoA. Every membrane of every cell in the body requires cholesterol. In addition, cholesterol is the raw material for making steroid hormones and bile salts. If insufficient cholesterol enters by the diet, the liver has enzyme systems to catalyze cholesterol synthesis from acetyl CoA. More than two dozen steps are involved.

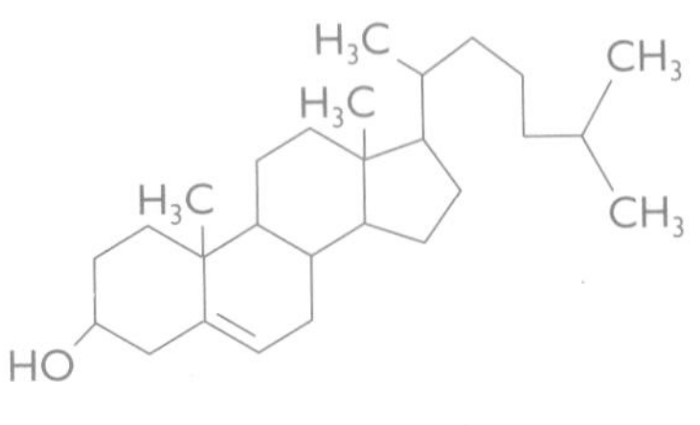

The synthesis of one of the enzymes needed to make cholesterol is deactivated by cholesterol itself, so cholesterol synthesis is under feedback control. Whenever the supply of cholesterol is sufficiently high, its further synthesis remains switched off. When the cholesterol supply falls enough, the cholesterol molecules that are deactivating the enzyme are removed for use elsewhere, and cholesterol synthesis restarts. Special Topic 20.2 told what can happen when the system isn't functioning normally.

One of the intermediates in cholesterol synthesis is the mevalonate ion. As this ion enters the next step in the synthesis, it must fit to a large enzyme. A medication used to inhibit cholesterol synthesis, *lovastatin,* is also able to bind to this enzyme. When lovastatin binds instead of mevalonate, it's like removing the enzyme itself, and cholesterol synthesis shuts down. You can see in the following structures how lovastatin can mimic mevalonate.

Steroid nucleus

Cholesterol

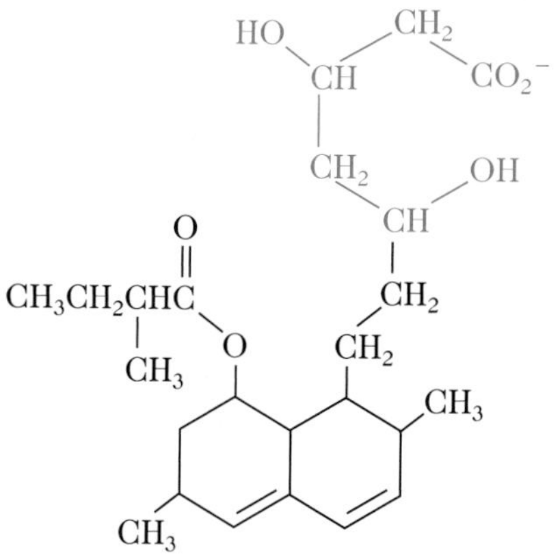

Mevalonate

Lovastatin (Mevacor)

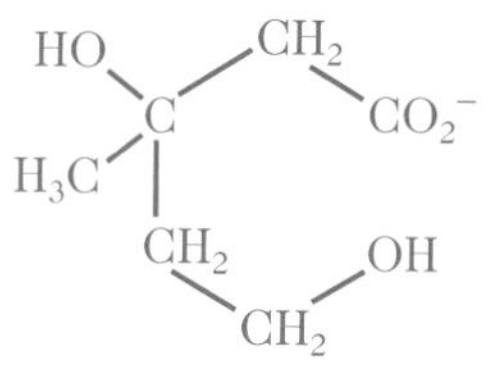

20.5 SYNTHESIS OF AMINO ACIDS

The body can manufacture a number of amino acids from intermediates in the catabolism of nonprotein substances.

Certain Amino Acids Must Be Provided by the Diet. We do not need all of the 20 amino acids in the diet, because we can make roughly half of them. Those that we *must* obtain by the diet are called the **essential amino acids,** named in the margin. *Essential* in this context refers *only* to a *dietary* need. In a larger context, the body must have all 20 amino acids to make polypeptides.

A number of food materials, particularly vegetables and grains, are deficient in one or more essential amino acids. Rice, for example, is low in lysine. Beans are low in valine. Corn carries too little tryptophan and lysine. Soybeans are low in methionine and cysteine. So is cassava, a root widely used in many third-world countries. If any one of these foods is the sole source of protein in the diet, severe malnutrition results.

Kwashiorkor is one of the world's widespread protein-deficiency diseases. While the first-born receives mother's milk, it gets perfectly balanced protein. There is none finer, although cow's milk and egg white are nearly as good. When the second-born arrives, the first goes onto a diet mostly of corn or cassava gruel with little or no milk, eggs, or meat. Now it gets too little of an essential amino acid to make its proteins, and the victim develops patchy skin and discolored hair, appears bloated, and falls behind developmentally. To obtain enough of all of the essential amino acids from cassava alone, the child would have to eat 27.5 lb/day, amounting to 16,400 kcal/day energy, an obvious impossibility. Contrast this with the 35 g of protein in human milk, which is all that is needed each day to provide all of the amino acids needed for growth.

Vegetarians, who are knowledgeable about what to eat and when to eat it, can combine two or more vegetables and achieve the right balance in essential amino acids. Thus, 43 g of a 1:1 mixture of rice (low in lysine) and beans (low in valine) is equal in value to the protein in 35 g of milk. The chief problem for vegetarians is vitamin B_{12}. No grain or vegetable has it.

The Nonessential Amino Acids Are Those the Body Needs but Can Make. The pathways for the synthesis of nonessential amino acids have several steps, and we won't examine any in detail. However, Figure 20.6 gives an overview showing in

Essential Amino Acids[a]

Isoleucine
Leucine
Lysine
Methionine
Phenylalanine
Threonine
Tryptophan
Valine

[a]Histidine is believed to be essential to infants.

■ A protein with all essential amino acids in the right proportions is called an **adequate protein.**

■ *Kwashiorkor* is from a Ghan dialect meaning "the condition the first-born enters when the second-born arrives."

general terms how some amino acids are made from the intermediates of glycolysis and the citric acid cycle.

Many of the syntheses outlined in Figure 20.6 depend on the availability of the glutamate ion, which is made from α-ketoglutarate by a reaction called **reductive amination**. The reaction uses the ammonium ion as a source of the amino group and

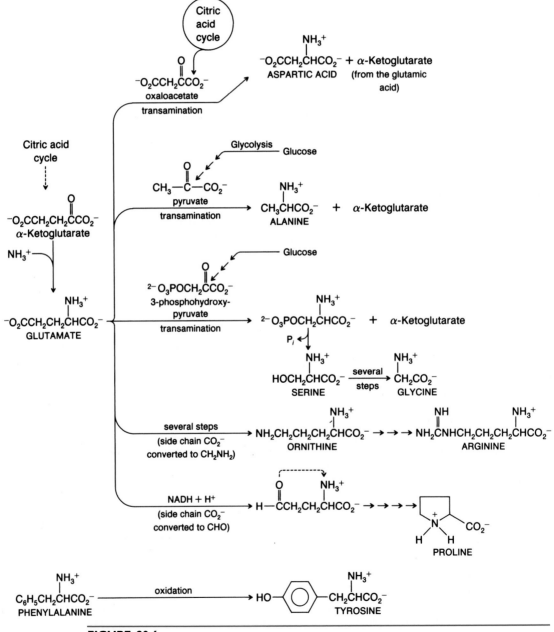

FIGURE 20.6
Biosynthesis of some nonessential amino acids.

NADPH/H$^+$ as the reducing agent. The overall result is:

$$^-O_2CCH_2CH_2\overset{\overset{\displaystyle O}{\|}}{C}CO_2^- + NH_4^+ + NADPH/H^+ \;\rightleftharpoons$$
α-Ketoglutarate

$$^-O_2CCH_2CH_2\overset{\overset{\displaystyle NH_3^+}{|}}{C}HCO_2^- + NADP^+ + H_2O$$
Glutamate

Glutamate now becomes the source of amino groups for still other amino acids that can be made by **transamination,** the transfer of an amino group from one molecule to another. We illustrate the general case; Figure 20.6 has several examples.

$$R\overset{\overset{\displaystyle O}{\|}}{C}CO_2^- + {}^-O_2CCH_2CH_2\overset{\overset{\displaystyle NH_3^+}{|}}{C}HCO_2^- \;\rightleftharpoons\; R\overset{\overset{\displaystyle NH_3^+}{|}}{C}HCO_2^- + {}^-O_2CCH_2CH_2\overset{\overset{\displaystyle O}{\|}}{C}CO_2^-$$
An α-keto acid   Glutamate   An α-amino acid   α-Ketoglutarate

The enzymes for transaminations, the *transaminases,* use the B vitamin pyridoxal to make their cofactors.

SUMMARY

Glycogen Metabolism The regulation of glycogenesis and glycogenolysis is a part of the machinery for glucose tolerance in the body. Hyperglycemia stimulates the secretion of insulin and somatostatin, and insulin helps cells of adipose tissue to take glucose from the blood. Somatostatin helps to suppress the release of glucagon (which otherwise stimulates glycogenolysis and leads to an increase in the blood sugar level).

When glucose is abundant, the body either replenishes its glycogen reserves or makes fat. In danger or other stresses, epinephrine stimulates the release of glucose from glycogen.

When glucose is in short supply, the body makes its own by gluconeogenesis from noncarbohydrate molecules, including several amino acids. In diabetes, some cells that are starved for glucose make their own, also. Such cells are unable to obtain glucose from circulation, so the blood sugar level is hyperglycemic to a glucosuric level. The glucose tolerance test is used to see how well the body handles an overload of glucose. In the management of diabetes, the rigorous maintenance of a steady blood sugar level in the normal range is vital.

Gluconeogenesis Most of the steps in gluconeogenesis are simply the reverse of steps in glycolysis, but there are a few that require rather elaborate bypasses. Special teams of enzymes and supplies of high-energy phosphates are used for these. Many amino acids can be used to make glucose by gluconeogenesis.

Lipid Absorption and Distribution As fatty acids and monoacylglycerols migrate out of the digestive tract they become reconstituted as triacylglycerols. The adipose tissue is the principal storage site, and fatty material comes and goes from this tissue according to the energy budget of the body.

Biosynthesis of Fatty Acids Fatty acids are made by a repetitive cycle of steps. The cycle begins by building one butyryl group from two acetyl groups. The four-carbon butyryl group is attached to an acyl carrier protein that acts as a swinging arm on the enzyme complex. This arm moves the growing fatty acyl unit over first one enzyme and then another as additional two-carbon units are added. The process consumes ATP and NADPH.

Biosynthesis of Cholesterol Cholesterol is made from acetyl groups by a long series of reactions. The synthesis of one of the enzymes is inhibited by excess cholesterol, which gives the system a mechanism for keeping its own cholesterol synthesis under control.

Amino Acid Synthesis Ten of the amino acids are essential amino acids, meaning they must be in the diet. Vegetables and grains are deficient in at least one essential amino acid, but when properly combined, they can furnish a balanced protein diet. Processes of transamination and reductive amination make nonessential amino acids from intermediates of glucose and fatty acid catabolism.

REVIEW EXERCISES

The answers to Review Exercises whose numbers are in color are found in Appendix V. The answers to the other Review Exercises are found in the Study Guide that accompanies this text. The more challenging questions are marked with asterisks.

Blood Sugar

20.1 What are the end products of the complete digestion of the carbohydrates in the diet?

20.2 Why can we treat the catabolism of carbohydrates as almost entirely that of glucose?

20.3 What is meant by *blood sugar level*? By *normal fasting level*?

20.4 What is the range of concentrations of glucose in mg/dL for the normal fasting level of whole blood?

*20.5 A level of 5.5 mmol/L for glucose in the blood corresponds to how many milligrams of glucose per deciliter?

20.6 What characterizes the following conditions?
(a) glucosuria   (b) hypoglycemia
(c) hyperglycemia   (d) glycogenolysis
(e) renal threshold   (f) glycogenesis

20.7 Explain how severe hypoglycemia can lead to disorders of the central nervous system.

Hormones and the Blood Sugar Level

20.8 When epinephrine is secreted, what soon happens to the blood sugar level?

20.9 At which one tissue is epinephrine the most effective?

20.10 What is the end product of glycogenolysis, and what does phosphoglucomutase do to it?

20.11 Why can liver glycogen but not muscle glycogen be used to resupply blood sugar?

20.12 What is glucagon? What does it do? What is its chief target tissue?

20.13 Which is probably better at increasing the blood sugar level, glucagon or epinephrine? Explain.

20.14 How does human growth hormone manage to promote the supply of the energy needed for growth?

20.15 What is insulin? Where is it released? What is its chief target tissue?

20.16 What triggers the release of insulin into circulation?

*20.17 If brain cells are not insulin-dependent cells, how can too much insulin cause insulin shock?

20.18 What is somatostatin? Where is it released? What kind of effect does it have on the pancreas?

Glucose Tolerance

20.19 What is meant by *glucose tolerance*?

20.20 What is the purpose of the glucose tolerance test?

20.21 Describe what happens when each of the following persons takes a glucose tolerance test.
(a) a nondiabetic individual   (b) a diabetic individual

20.22 Describe a circumstance in which hyperglycemia might arise in a nondiabetic individual.

Gluconeogenesis

20.23 In a period of prolonged fasting or starvation, what does the system do to try to maintain its blood sugar level?

*20.24 Amino acids are not excreted, and they are not stored in the same way that glucose residues are stored in a polysaccharide. What probably happens to the excess amino acids in a high-protein diet of an individual who does not exercise much?

20.25 The amino groups of amino acids can be replaced by keto groups. Which amino acids could give the following keto acids that participate in carbohydrate metabolism?
(a) pyruvic acid   (b) oxaloacetic acid

Absorption and Distribution of Lipids

20.26 What are the end products of the digestion of triacylglycerols?

20.27 What happens to the products of the digestion of triacylglycerols as they migrate out of the intestinal tract?

20.28 In what form are lipids carried in the blood?

20.29 Where and how are fatty acids chiefly stored in the body?

Biosynthesis of Fatty Acids

20.30 Where are the principal sites for each activity in a liver cell?
(a) fatty acid catabolism   (b) fatty acid anabolism

20.31 What metabolic pathway in the body is the chief supplier of NADPH for fatty acid synthesis?

20.32 How does cholesterol itself work to inhibit the activity of an enzyme needed to make cholesterol?

Biosynthesis of Amino Acids

20.33 Because the full complement of 20 amino acids is required to make all the body's proteins, why are fewer than half of this number considered as *essential* amino acids?

20.34 Alanine is not on the list of essential amino acids. Why?

20.35 Write the equation for the reductive amination that produces glutamate. (Use NADPH as the reducing agent.)

*20.36 Write the structure of the keto acid that forms when phenylalanine undergoes transamination with α-keto-glutarate.

*20.37 When valine and α-ketoglutarate undergo transamination, what new keto acid forms? Write its structure.

20.38 What vitamin is needed to make the transaminase enzymes?

Diabetes Mellitus (Special Topic 20.1)

20.39 What is the biochemical distinction between type I and type II diabetes?

20.40 Juvenile-onset diabetes is usually which type?

20.41 Adult-onset diabetes is usually which type?

20.42 Briefly state the six stages in the onset of type I diabetes.

20.43 Viruses that cause diabetes attack which target cells?

20.44 Type I diabetes is an autoimmune disease. What does this mean?

20.45 What are some explanations for the lack of glucose uptake in NIDDM?

20.46 Sustained hyperglycemia causes damage to which specific tissue, damage that might be responsible for other complications?

20.47 How is glucose involved in the formation of a Schiff base? What other kinds of compounds react with glucose in this way?

20.48 When the glucose level in blood drops, what happens to the level of glucosylated hemoglobin? Why?

20.49 What happens to the Schiff bases involving glucose if given enough time? Why is this serious?

20.50 Describe a theory that explains how the hydrogenation of glucose might contribute to blindness.

Lipoprotein Complexes and Cholesterol (Special Topic 20.2)

20.51 What are chylomicrons and what is their function?

20.52 What happens to chylomicrons as they move through capillaries of, say, adipose tissue?

20.53 What happens to chylomicron remnants when they reach the liver?

20.54 What are the two chief sources of cholesterol that the liver exports?

20.55 What do the following symbols stand for?
(a) VLDL   (b) IDL
(c) LDL   (d) HDL

20.56 The loss of what kind of substance from VLDL converts them into IDL?

20.57 What happens to cause the increase in density between VLDL and LDL?

20.58 What tissue can reabsorb IDL complexes?

20.59 What is the chief constituent of LDL?

20.60 In extrahepatic tissue, what two general uses await delivered cholesterol?

20.61 If the liver lacks the key receptor proteins, which specific lipoprotein complexes can't be reabsorbed?

20.62 Explain the relationship between the liver's receptor proteins for lipoprotein complexes and the control of the cholesterol level of the blood.

20.63 What is the chief job of the HDL?

*20.64 Why does HDL have a higher density than chylomicrons?

20.65 Which of the lipoprotein complexes is sometimes called "bad cholesterol?" Explain why it is so designated.

20.66 Which lipoprotein complex carries "good cholesterol?"

21

NUCLEIC ACIDS

Units of Heredity

Ribonucleic Acids

mRNA-Directed
  Polypeptide Synthesis

Viruses

Recombinant DNA
  Technology and
  Genetic Engineering

Hereditary Diseases

Bobcats have bobcat kittens,
not even tiger kittens. The
molecular basis for the reproduction of species is introduced in this chapter.

21.1   UNITS OF HEREDITY

Genetic information is carried by the sequence of side-chain bases on deoxyribonucleic acid, DNA.

Nearly every reaction in an organism requires an enzyme, and nearly all enzymes are proteins. The set of enzymes in one organism is not the same as in another, however, although many enzymes from different species are similar in structure and function. Important differences in enzymes also exist among individual members of a species. The question is, How does a member of any species obtain its own unique enzymes?

Nucleic Acids Carry Instructions for Making Enzymes.   In the reproductive process, each offspring receives from its parents not the enzymes themselves but the ability to make them uniquely. Duplicates of *enzymes* are not passed on, but rather duplicates of compounds bearing the *instructions* for assembling them. Instructions, of course, must be rich in information, and at the heart of the molecular basis of information are a family of polymers called the *nucleic acids.* Our purpose in this chapter is to study how nucleic acid molecules carry, duplicate, transmit, and use structure-encoded information.

Genes Are the Units of Heredity.   Each cell nucleus contains intertwined filaments of a *nucleoprotein* called **chromatin,** a complex of nucleic acid and protein. A chromatin filament is like a strand of pearls, each pearl made of proteins called *histones* around which are tightly coiled one of the kinds of nucleic acids, DNA for short. DNA also links the histone "pearls," and molecules of DNA have portions that constitute individual **genes,** the fundamental units of heredity.

■ The histones are not just spools for wrapping DNA strands; they contribute to the regulation of DNA activity.

■ Each "pearl" is called a *nucleosome.*

Prior to Cell Division, Genes Replicate.   When cell division begins, new nucleic acid and histones are made, so the chromatin strands thicken and become rodlike bodies called **chromosomes.** If all goes well, as it usually does, the new chromatin is an exact copy of the old. During the process, each gene undergoes **replication,** meaning that it is reproduced in duplicate. Thus, by the replication of DNA molecules, the genetic message of the first cell is made available to each of the two new cells.

■ There are 23 matched pairs of chromosomes in the human cell for a total of 46 chromosomes.

Nucleic Acids Are Polymers Made of Nucleotides.   The **nucleic acids** occur as two types of polymers given the symbols DNA and RNA. **DNA is deoxyribonucleic acid,** and **RNA is ribonucleic acid.** The monomers for these polymers are called **nucleotides.**

■ Every human cell has between 50,000 and 100,000 genes.

The nucleotides include phosphate ester groups, so they can be further hydrolyzed. The hydrolysis of a representative mixture of nucleotides produces three kinds of products: inorganic phosphate, a pentose sugar, and a group of heterocyclic amines called the **bases** (Figure 21.1). Usually the bases are referred to by their names or their single-letter symbols, as follows.

Bases from DNA		Bases from RNA	
Adenine	A	Adenine	A
Thymine	T	Uracil	U
Guanine	G	Guanine	G
Cytosine	C	Cytosine	C

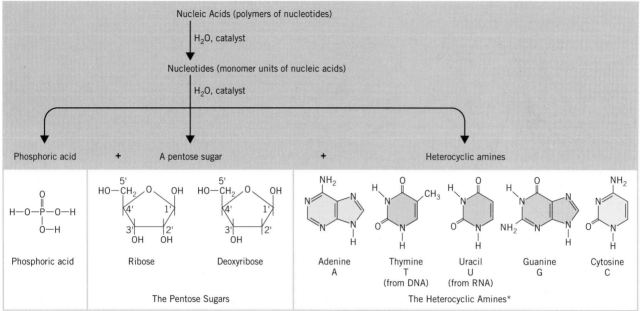

*These are the five principal heterocyclic amines obtainable from nucleic acids. Others, not shown, are known to be present. Although they differ slightly in structure, they are informationally equivalent to one or another of the five shown here.

FIGURE 21.1

Hydrolysis products of nucleic acids.

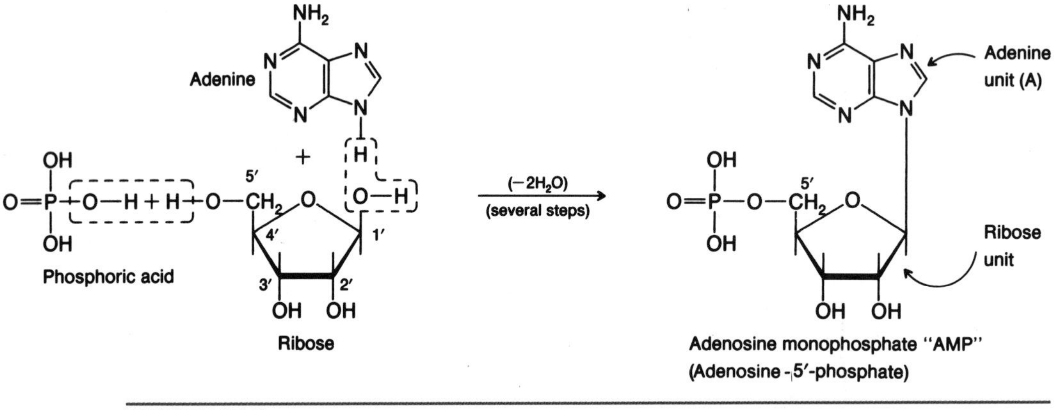

FIGURE 21.2

A typical nucleotide, AMP, and the smaller units from which it is assembled. The phosphate ester group forms by the splitting out of water between an OH group of phosphoric acid and an H atom at the 5′ OH group of the ribose unit. The splitting out of water between the 1′ OH group of the ribose unit and the H atom on a ring nitrogen of adenine (A) joins adenine to the sugar unit. Similar structures could be drawn with the other bases of Figure 21.1. In the formation of each such nucleotide, the H atom used in this splitting out of water is the one attached to the ring nitrogen drawn as the lowest in the structure of the base.

You can see that three bases, A, G, and C, are common to both DNA and RNA, and that one is different. Another difference between DNA and RNA is the sugar unit. In both, it is an aldopentose in its cyclic form, but the *R* in RNA stands for ribose and the *D* in DNA stands for deoxyribose.

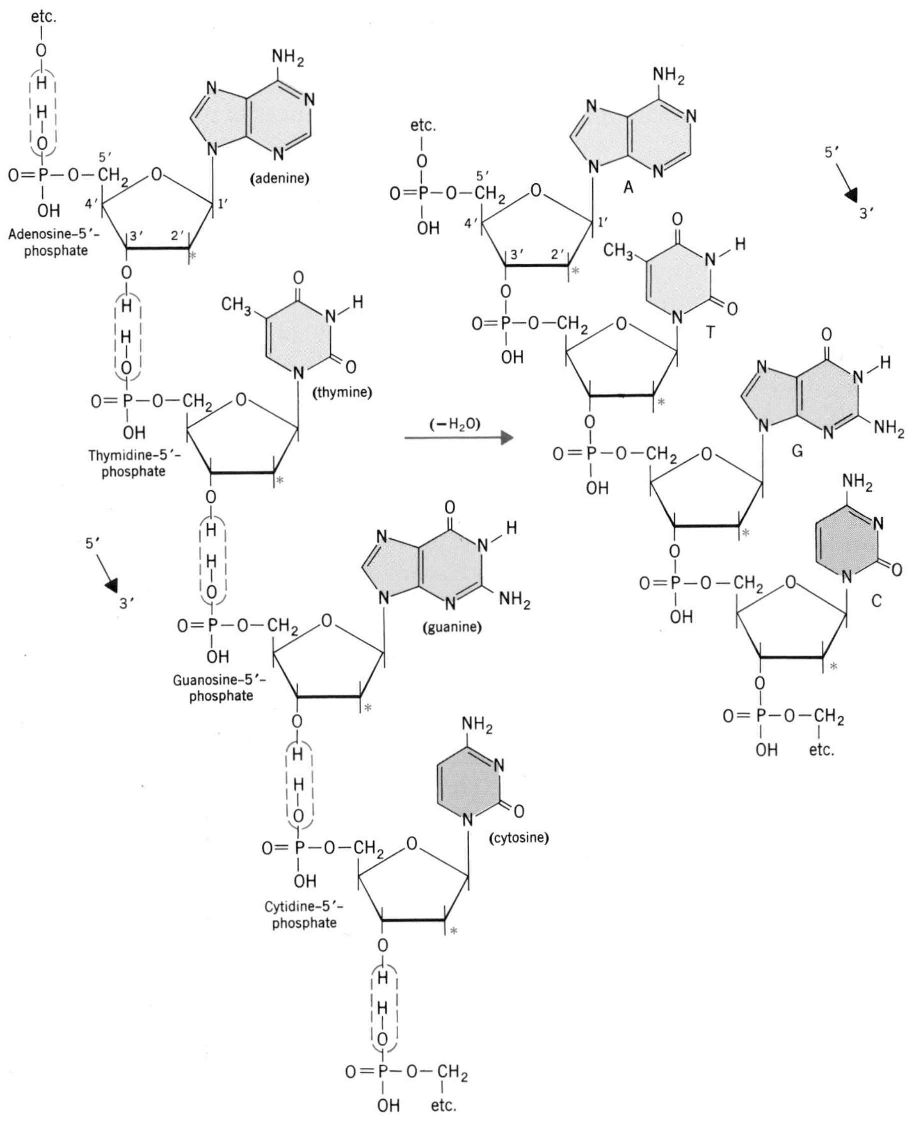

FIGURE 21.3
Relationship of a nucleic acid chain to its nucleotide monomers. On the right is a short section of a DNA strand. On the left are the nucleotide monomers from which it is made (after many steps). The colored asterisks by the pentose units identify the 2' positions of these rings where there would be another OH group if the nucleic acid were RNA (assuming that uracil also replaced thymine). The designation 5' ⟶ 3' means that the complete strand would have an unesterified OH group on C-5' of the first pentose unit and an unesterified C-3' on the other end, and that the sequence of bases is written from the 5' end to the 3' end. Thus, the sequence here is written as ATGC, not as CGTA.

With these hydrolysis products in mind, we can work backward and see how a nucleic acid polymer molecule is built up. First, we need nucleotides, the monomer molecules. A typical example is adenosine monophosphate (AMP), which is made from phosphoric acid, ribose, and its base, adenine (Figure 21.2). The other nucleotides are put together by using other bases and one or the other of the pentoses.

The Sequence of Bases Projecting from the Backbones of Nucleic Acids Is How Genetic Information Is Stored.   At least on paper, there is nothing more to putting nucleotides together into a nucleic acid than splitting out water between a phosphate unit of one nucleotide and an alcohol group on the pentose of the next (Figure 21.3). The result is a phosphodiester system linking sugar unit to sugar unit. Many steps are required *in vivo*, of course, each calling for its own enzyme, but when these are repeated thousands of times, a nucleic acid forms. Thus, the pattern for the backbone of a nucleic acid is alternating phosphate and pentose units.

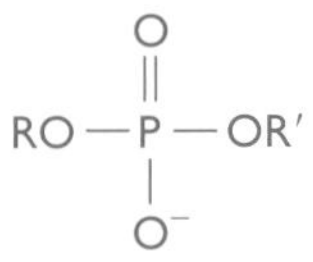

■ Some 20 enzymes are required, the chief being DNA polymerase, discovered in 1958 by Arthur Kornberg (Nobel prize, 1959).

Phospho-
diester
(anion form)

Each pentose unit of a nucleic acid backbone holds one of the four bases, *so the bases project at very regular intervals along the backbone.* The distinctiveness of a nucleic acid is the *sequence of these bases.* At the molecular level of life, the central genetic fact is that information is carried by unique sequences of bases in molecules of nucleic acids.

Pairs of Bases Are Attracted to Each Other by Hydrogen Bonds. The bases have functional groups so arranged geometrically that *base pairing occurs.* **Base pairing** means that unique pairs of bases are able to attract each other and "fit" together by means of hydrogen bonds (Figure 21.4). The geometries and relative locations of the functional groups on the bases are such that in DNA, G and C form a base pair and A and T form another. In RNA, G and C always pair, and U and T always pair. *The arrangements of hydrogen bond donor and acceptor sites allow only certain base pairs to exist.* Neither G nor C ever pairs with A, T, or U. They can't fit to each other.

DNA Occurs as Two Paired Strands Twisted into a Double Helix. In 1953, Francis Crick of England and James Watson of the United States proposed a **double helix** structure for DNA. Using X-ray data obtained by Rosalind Franklin, they deduced that the long DNA molecules occur in cell nuclei as pairs of complementary strands twisted into right-handed helices.

■ Crick and Watson shared the 1962 Nobel prize in medicine and physiology with Maurice Wilkins.

Being *complementary* means that whenever adenine (A) is on one strand, thymine (T) is opposite it on the other strand; whenever guanine (G) projects from one strand, cytosine (C) is opposite it on the other. Although hydrogen bonds determine which bases can pair, *hydrophobic interactions stabilize the double helix.*

■ Two irregular objects are *complementary* when one fits to the other, as your right hand would fit to its impression in clay.

A molecular model of the DNA double helix is shown in Figure 21.5*a*. The system resembles a spiral staircase in which the steps, which are perpendicular to the long axis of the spiral, consist of the base pairs. Hydrogen bonds between the pairs are centered around the long axis. Since Crick and Watson's work, a small number of variations in this structure have been found. Figure 21.5 shows what is now called DNA-B, the type Crick and Watson worked with and the most common kind.

■ *Hydrophobic interactions* are the water-avoiding responses of relatively nonpolar side chains to polar groups or water molecules (see page 357).

Figure 21.5*b*, a schematic representation, more clearly shows how the two entwined DNA strands, running in opposite directions, form something of a spiral staircase with the bases being like the steps.

Figure 21.5 shows only a short segment of a typical DNA double helix. What is not shown is that the helices are further twisted and coiled into superhelices, which is necessary if the cell's DNA is to fit into its nucleus. A typical human cell nucleus, for example, is only about 10^{-7} m across, but if all its DNA double helices were stretched out, they would measure more than 1 m, end to end. (The different types of DNA concern various ways of twisting and folding.)

FIGURE 21.4
Hydrogen bonding between base pairs. (a) Thymine (T) and adenine (A) form one base pair between which are two hydrogen bonds. (b) Cytosine (C) and guanine (G) form another base pair between which are three hydrogen bonds. Adenine can also base-pair to uracil (U).

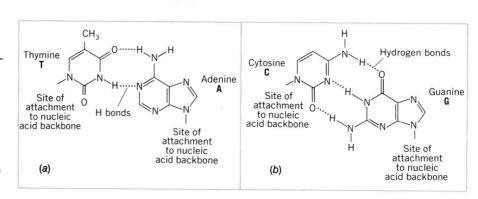

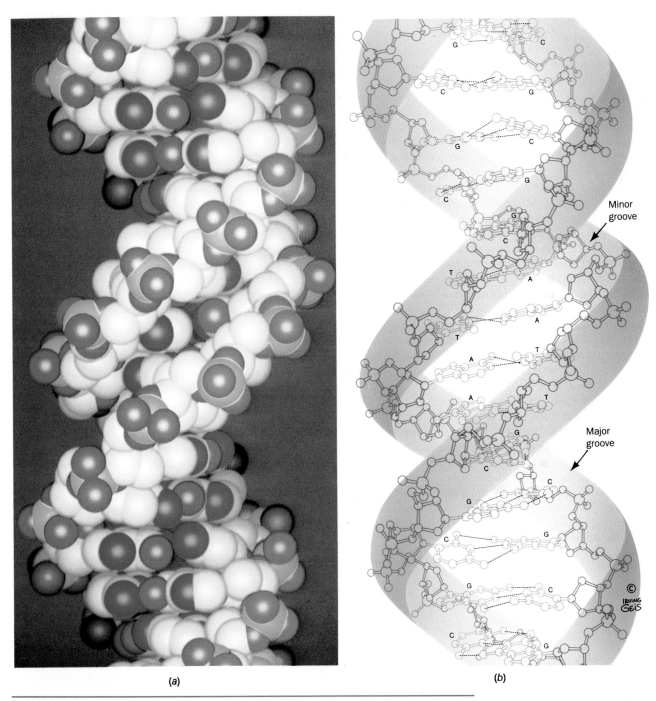

(a)   (b)

FIGURE 21.5
Structure of double-helix DNA—the native form (B-DNA)—seen here in (a) a computer-gen-
erated space-filling model and (b) a ball-and-stick drawing. (Drawing copyrighted © by Irving
Geis. Computer graphics courtesy of Robert Stodola, Fox Chase Cancer Center.)

In the entire set of human genes, the human *genome,* there are many regions consisting of nucleotide sequences repeated in tandem. These regions, called *minisatellites,* all have a common core sequence, but the *number* of repeated sequences in the minisatellites varies from individual to individual. These variations, which can be measured, are so considerable between individuals that they are the basis of a major new technique in discovering the truth concerning a suspect in certain crimes, particularly rape.

Suppose, for example, a sample of semen can be obtained from the sperm left by a rapist. The DNA is "amplified" or cloned by means of a special reaction called the *polymerase chain reaction* to convert it into a sample large enough for the analysis. (A share of the 1993 Nobel prize in chemistry went to American scientist Kary Mullis for discovering this reaction.) To apply the polymerase chain reaction, the DNA is first denatured by warming the sample, which causes the strands of duplex DNA to unwind and separate. (Denatured DNA renatures itself spontaneously if the temperature of the system is carefully controlled over a sufficient length of time.) A sample of the enzyme DNA polymerase is then added together with specially prepared, short DNA "primer" strands, and the mixture is incubated. During this period, cloning of the DNA occurs and new but identical copies of duplex DNA molecules form.

The sample of cloned DNA is next hydrolyzed with the use of a specific enzyme, called a *restriction enzyme.* The enzyme is able to catalyze the breakage (hydrolysis) of a DNA strand *only at specific sites,* which releases the core segments described earlier. The fragments of DNA can then be separated and made to bind (by base pairing) to short, radioactively labeled, specially made DNA that has been designed to be able to bind to core segments. Now the labeled DNA fragments are separated by a special technique. Because of their radiation, the fragments can affect photographic film, so the separated fragments are visualized as a series of 30 to 40 dark bands on a strip of film. Each band is caused by the molecules of a particular kind of DNA fragment. The resulting series of bands on film, called the individual's *DNA profile,* has been compared to a bar code used for pricing groceries at check out counters. Thanks to the polymerase chain reaction, enough DNA material can be made from the tiniest of crime scene samples to obtain a DNA profile. Because every cell in the body has an individual's entire genome, a single hair of a crime suspect can provide the same fundamental DNA profile as comes from a sample of the suspect's semen or blood.

If the two DNA profiles do not match, the prosecuting attorney looks for another suspect. The London Metropolitan Police Forensic Science Laboratory, for example, has cleared about 20% of rape suspects by failures of DNA profile matches.

When the DNA profile of a suspect matches that obtained from a crime scene (Figure 1), the jury has evidence for a conviction. It is evidence that juries, rightly instructed, do not necessarily take as final, however. The *strength* of DNA profile evidence is measured by the *match probability.* It bears on the question: How probable is it that an individual unrelated to the suspect or the crime will have a matching profile? In other words, what is the probability that an *unrelated* person, chosen at random from an appropriate

In DNA Replication, the Bases Guide the Formation of Complementary Strands. When DNA replicates, the cell makes an exact complementary strand for each of the original strands, and two identical double helices emerge (Figure 21.6). A number of enzymes are involved. The guarantee that each new strand is a complement to one of the old strands is the result of nothing more than the restrictions on base pairing, A to T and C to G.

In Higher Organisms, Sections of DNA Molecules Called *Exons* Collectively Carry the Message of One Gene. In single-celled species like bacteria, genes are *continuous, uninterrupted* sequences of nucleotide units. In higher organisms, however, one individual gene consists of a set of *separated* portions of a DNA strand. Virtually all single genes are divided or split up like this. The portions of a DNA chain that, taken together, constitute one gene are called **exons.** Separating the exons are other parts of the DNA chain called **introns,** which do not code for proteins. Only about 3% of all

■ The nucleotide monomer units are present in the fluid of the cell nucleus.

■ There are an estimated 3 billion base pairs in one nucleus of a human cell but only about 3% are a part of genes.

■ *Exon* refers to the part that is expressed, and *intron,* to the segments that *interrupt* the exons.

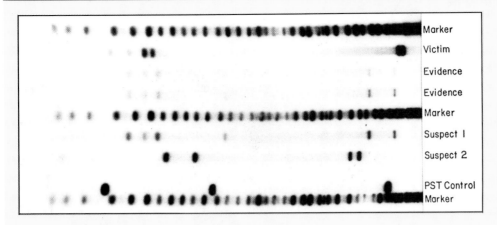

FIGURE I

DNA fingerprinting. The banding pattern of the DNA of suspect I matches that of the evidence. Neither the rape victim's DNA nor that of suspect 2 does.

population, will have a DNA profile matching a profile from the crime? A second and different question is: What is the probability that an *innocent* person will have a DNA profile that matches that of crime scene evidence? The first question concerns the probability of match; the second question is about the probability of guilt or innocence. Misrepresentations by prosecutors of data concerning match probability as being an answer to guilt probability have been the bases of challenges by defense attorneys.

The use of DNA profiling is going through rigorous court challenges. Does every individual actually have a completely unique DNA profile? (The DNA profiles of identical twins are the same.) Is the evidence from DNA typing truly like fingerprint evidence? (Members of a South American tribe founded by a few individuals and now excessively inbred have very similar DNA profiles.) Is what law enforcement officials call the *chain of evidence* complete and unbroken? (Have individuals signed off on the passage of physical evidence from one person to another in an unbroken series, including the chemists and computer special-

ists who do the final processing and then furnish the court with the DNA profile evidence?) Enough doubts have been raised over these questions by defense attorneys that some court systems will not admit DNA profiles as evidence unless or until these uncertainties are resolved.

The use of real fingerprints went through similar challenges. During the period of the challenges to DNA typing, recognizing that the technique could be an extraordinarily powerful tool to establish the truth, scientists and legal scholars have been at work to remove the doubts, improve the technology, and establish the reliability of every procedural step from the finding of raw evidence to the court appearance and its proper representation.

Suggested Readings
1. P. J. Neufeld and N. Colman, "When Science Takes the Witness Stand," *Scientific American,* May 1990, page 46.
2. D. J. Balding and P. Donnelly, "How Convincing is DNA Evidence?" *Nature,* March 24, 1994, page 285.

human DNA is exon DNA encoded for our proteins; roughly 97% of human DNA consists of introns. Once regarded as "junk DNA," intron DNA is presently the subject of intensive research to see what functions it has.

We'll see later in this chapter how the exons get their message together to give expression to a single gene. For the present, we can consider that a gene is a particular portion of a DNA strand minus all of the introns in this section. *A gene, in other words, is a specific series of bases strung in a definite sequence along a DNA backbone.* In humans, a single gene has between 900 and 3000 bases.

Base Sequences in Genes Can Be Determined. Although the genes of one human being are similar to those of another, there are many small variations. Each person, except for those who are identical twins, has unique genes just as each has unique fingerprints. These differences are behind a powerful tool for identifying criminals, genetic "fingerprinting," described in Special Topic 21.1.

■ British-born Richard J. Roberts and American-born Phillip A. Sharp shared the 1993 Nobel prize in physiology and medicine for discovering (in 1977) that genes are split and are interrupted by introns.

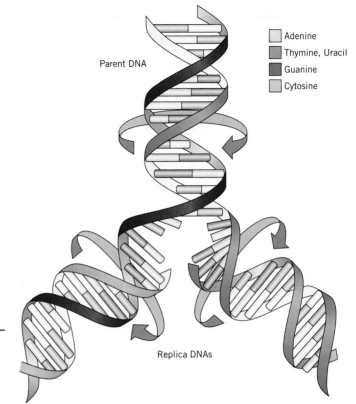

Parent DNA

Adenine
Thymine, Uracil
Guanine
Cytosine

Replica DNAs

FIGURE 21.6
Replication of DNA. The accuracy of the replication of DNA is related to the exclusive pairing of A with T and G with C.

21.2   RIBONUCLEIC ACIDS

The triplets of bases of the genetic code correlate with individual amino acids.

■ *Molecular biology* is the hybrid science that encompasses all of the chemistry of nucleic acids, genes, chromosomes, and genetic expressions in protein syntheses.

The information coded by the base sequence of an individual gene is translated into a side-chain sequence of a polypeptide with the aid of the other kind of nucleic acid, RNA. Genetic information flows from DNA to a specific polypeptide through four types of RNA (Figure 21.7).

Ribosomal RNA (rRNA) Is the Most Abundant RNA.   **Ribosomes** are small particles that occur by the thousands on the surfaces of membranelike tubules that twist and turn throughout a cell outside its nucleus. Each ribosome forms from two subunits, as shown in Figure 21.7, which come together to form a complex with messenger RNA, another type that we'll study soon.

■ More than 50 kinds of proteins exist in one ribosome.

Ribosomes contain both proteins and a type of RNA called **ribosomal RNA,** abbreviated **rRNA.** Except in a few viruses, rRNA is single stranded, but its molecules often have hairpin loops in which base pairing occurs. Ribosomes are the sites of polypeptide synthesis. Their rRNA does not itself direct this work. *Messenger RNA* does this, and it's made from the next kind of RNA in our study.

Heterogeneous Nuclear RNA (hnRNA) Is Complementary to DNA, Both Exons and Introns.   As indicated in Figure 21.7, when a cell uses a gene to direct the synthesis of a polypeptide, its first step is to use DNA to make **heterogeneous nuclear RNA,**

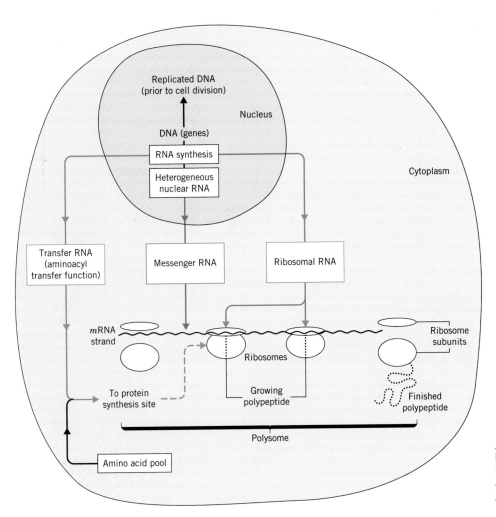

FIGURE 21.7

Relationships of nuclear DNA to the various RNAs and to the synthesis of polypeptides.

abbreviated **hnRNA.** A *single* strand of the DNA that bears the polypeptide's gene is used to guide the assemblage of a complementary molecule of hnRNA (Figure 21.8). Because it is an RNA, uracil is used instead of thymine, so when a DNA strand has an adenine side chain, then uracil not thymine takes the position opposite it on the complementary hnRNA. After making hnRNA, the next general step in gene-directed polypeptide synthesis is the processing of hnRNA to form still another kind of RNA, messenger RNA.

Messenger RNA (mRNA) Is Complementary to the Exons Making Up One Gene. Molecules of hnRNA have large sections complementary to the introns of the DNA, and these sections must be deleted. Special enzymes catalyze reactions that snip these pieces out and splice together only the units corresponding to the exons of the divided gene (Figure 21.9). The result is a shorter RNA molecule called **messenger RNA,** or **mRNA.** Because hnRNA is used to make mRNA, hnRNA is also called *precursor messenger RNA* or *pre-mRNA.*

In mRNA we have a sequence of bases complementary only to the gene's exons, so this mRNA now carries the unsplit genetic message. We have now moved the genetic message on DNA to a molecule of mRNA, and the name for this overall process is **transcription.**

■ Part of the grooming of hnRNA molecules installs at their 3′ ends a long poly(A) tail (about 200 adenosine units long), and at the other end a nucleotide triphosphate "cap."

■ A small family of nuclear ribonucleoproteins, snRNPs (or "snurps"), helps this resplicing process, the action occurring at a molecular complex called a *spliceosome.*

■ F. Jacob and J. Monod, French scientists (Nobel prizes in 1965), conceived the idea that a messenger RNA must exist.

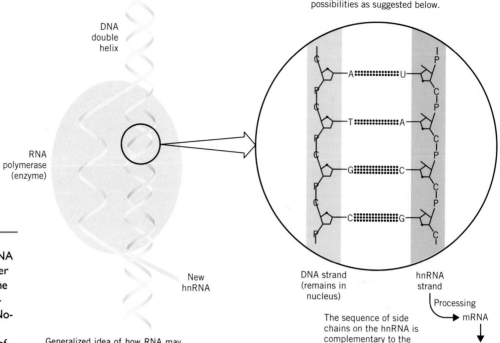

The order in which RNA subunits assemble is determined by pairing possibilities as suggested below.

DNA double helix

RNA polymerase (enzyme)

New hnRNA

Generalized idea of how RNA may be made using DNA as a "template"

DNA strand (remains in nucleus)

hnRNA strand

Processing → mRNA

To the ribosome

The sequence of side chains on the hnRNA is complementary to the corresponding sequence of the DNA strand.

FIGURE 21.8
DNA-directed synthesis of hnRNA in the nucleus of a cell in a higher organism. The shaded oval on the left represents a complex of enzymes that catalyze this step. (Notice that the direction of the hnRNA strand is opposite that of the DNA strand.)

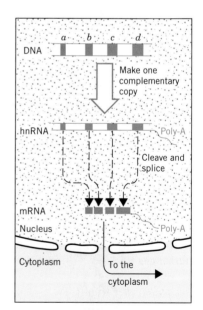

FIGURE 21.9
RNA made directly at a DNA strand is hnRNA. Only the segments made at sites *a*, *b*, *c*, and *d*—the exons—are needed to carry the genetic message to the cytoplasm. The hnRNA is processed, therefore, and its segments that matched the introns of the gene are snipped out. Then the segments that matched the exons are rejoined to make the mRNA strand.

Triplets of Bases on mRNA Are Genetic *Codons*. Each group of three adjacent bases on a molecule of mRNA constitutes a unit of genetic information called a **codon** (taken from the word code). Thus, a *sequence of codons on mRNA,* more than a sequence of individual bases, carries the genetic message. (We'll explain shortly why *three* bases per codon are necessary.)

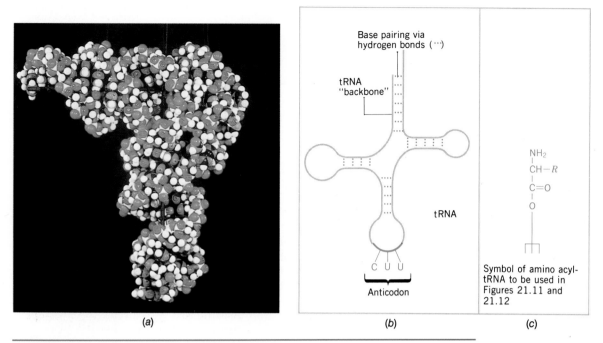

FIGURE 21.10

Transfer RNA (tRNA). *(a)* tRNA for phenylalanine. Its anticodon occurs at the tip of the base, and the place where the phenylalanyl residue can be attached is at the upper left point. *(b)* Highly schematic representation of the model to highlight the occurrence of double-stranded regions. *(c)* Symbol of the aminoacyl—tRNA unit that will be used in succeeding figures. (Molecular model courtesy of Academic Press/Molecular Design Incorporated.)

As indicated in Figure 21.7, once mRNA is made, it moves from the nucleus to the cytoplasm, where it attaches ribosomes. Many ribosomes can be strung like beads along one mRNA chain, and such a collection is called a *polysome* (short for polyribosome).

Ribosomes are traveling packages of enzymes intimately associated with rRNA. Each ribosome moves along its mRNA chain while the codons on the mRNA guide the synthesis of a polypeptide. To complete this system, we need a way to bring individual amino acids to the polysome's polypeptide assembly sites. For this, the cell uses still another type of RNA.

Transfer RNA (tRNA) Molecules Recognize Both Codons and Amino Acids. The substances that carry aminoacyl units to mRNA in the right order for a particular polypeptide are a collection of similar compounds called **transfer RNA or tRNA.** Their molecules are small, each typically having only 75 nucleotides, and are single stranded but with hairpin loops stabilized by base pairing (Figure 21.10).

tRNA Molecules Translate from the Genetic Language to the Polypeptide Language. We're now dealing with the molecular basis of information, so we can use language analogies. On a human level, we use language to convey information, and language involves words built from a common alphabet. We are aware that many languages exist among human societies and that the world knows several alphabets. To communicate between languages, we have to translate. The same need for translation occurs at the level of genes and polypeptides. *tRNA is the master translator in cells.*

TABLE 21.1   Codon Assignments

First	U	C	A	G	Third
	Second				
U	Phenylalanine	Serine	Tyrosine	Cysteine	U
	Phenylalanine	Serine	Tyrosine	Cysteine	C
	Leucine	Serine	CT[a]	CT	A
	Leucine	Serine	CT	Tryptophan	G
C	Leucine	Proline	Histidine	Arginine	U
	Leucine	Proline	Histidine	Arginine	C
	Leucine	Proline	Glutamine	Arginine	A
	Leucine	Proline	Glutamine	Arginine	G
A	Isoleucine	Threonine	Asparagine	Serine	U
	Isoleucine	Threonine	Asparagine	Serine	C
	Isoleucine	Threonine	Lysine	Arginine	A
	Methionine[b]	Threonine	Lysine	Arginine	G
G	Valine	Alanine	Aspartic acid	Glycine	U
	Valine	Alanine	Aspartic acid	Glycine	C
	Valine	Alanine	Glutamic acid	Glycine	A
	Valine	Alanine	Glutamic acid	Glycine	G

[a] The codon CT is a signal codon for chain termination.

[b] The codon for methionine, AUG, serves also as the codon for *N*-formylmethionine, the chain-initiating unit in polypeptide synthesis in bacteria and mitochondria.

tRNA is able to work with two "languages," the genetic and the polypeptide. The genetic language is expressed in an alphabet of 4 letters, the four bases, A, T (or U), G, and C. The polypeptide language has an alphabet of 20 letters, the side chains on the 20 amino acids. To translate from a 4-letter language to a 20-letter language requires that the 4 genetic letters be used in groups of a minimum of 3 letters. Then there are enough combinations of letters for the larger alphabet of the amino acids so that there can exist at least one genetic "word," built of 3 letters, for each of the 20 amino acids. This is exactly how tRNA is structured for its work with the genetic language. It is able to connect the three-letter codon "words" aligned on mRNA to a matching alignment of side chains of individual amino acid residues in a polypeptide being made.

One part of a tRNA molecule recognizes (i.e., forms base pairs with) a codon because it carries a triplet of bases complementary to the codon. A triplet on tRNA that is complementary to a codon on mRNA is called an **anticodon.** Each of the tRNAs carries a particular anticodon. In the tRNA molecule in Figure 21.10, the triplet CUU is its anticodon.

Another part of a tRNA molecule, an OH group at an end ribose unit, can attach a particular aminoacyl unit (by an ester bond). We can use the symbol tRNA-aa for this new compound, where we use "aa" for the aminoacyl group. Each amino acid has at least one tRNA designed to carry it as tRNA-aa. Most amino acids actually have two to six tRNA carriers, each with a unique anticodon. A given tRNA-aa molecule can be brought into alignment only with one codon of mRNA at a polysome.

As each molecule of tRNA-aa comes to mRNA, its aminoacyl unit is transferred to a growing polypeptide chain. *A unique series of codons can allow the polypeptide chain to grow only with an equally unique sequence of amino acid residues.* The pairing of the triplets of bases between the codons and anticodons permits only one sequence.

The Genetic Code Is the Correlation between Codons and Amino Acids. The **genetic code** consists of the known assignments of codons to amino acids (Table 21.1). Most amino acids are associated with more than one codon, but this apparently minimizes the harmful effects of genetic *mutations*. These, in molecular terms, are small changes in the structures of genes. Phenylalanine, for example, is coded either by UUU or by UUC. (Be sure that you can verify this using Table 21.1.) Alanine is coded by any one of four triplets: GCU, GCC, GCA, or GCG. Only two amino acids go with single codons, tryptophan (Trp) and methionine (Met).

■ UUU and UUC are called *synonyms* for Phe.

The Genetic Code Is Almost Universal for All Plants and Animals. A few single-celled species have been found with codon assignments not given in Table 21.1. Moreover, some of our genes occur in mitochondria, and many have unique codons. Apart from these exceptions, the genetic code of Table 21.1 is shared from the lowest to the highest forms of life in both the plant and animal kingdoms. We have a remarkable kinship with other forms of life.

■ Some (but not all) of the enzymes present in a mitochondrion are made *within* this organelle under the direction of mitochondrial nucleic acids. According to one theory, mitochondria evolved from single-celled organisms.

mRNA Codons Also Relate to DNA Triplets. It's important to remember that a codon cannot appear on a strand of mRNA unless a complementary triplet of bases was on an exon unit of the original DNA strand. For example, there could not be the UUC codon on mRNA unless the DNA strand had the triplet GAA, because G pairs with C and A of DNA pairs with U of mRNA.

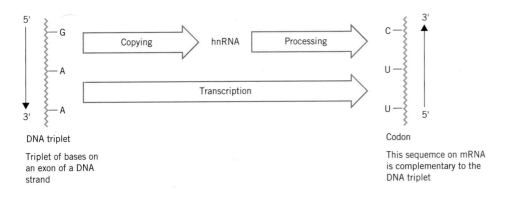

DNA triplet

Triplet of bases on an exon of a DNA strand

Codon

This sequemce on mRNA is complementary to the DNA triplet

The Direction in Which a Codon Triplet is Written Has Structural Meaning. As shown here, a DNA strand and the RNA strand made directly from it run in opposite directions. To avoid confusion in writing codons on a horizontal line, scientists use the following conventions. The 5′ end of a codon is written on the left end of the three-letter symbol, and the direction, left to right, is 5′ to 3′ (see also Figure 21.3). This is why the codon given above is written as UUC, not as CUU. Opposite this codon is the triplet GAA on the DNA strand, which is also written from the 5′ to 3′ end. To give another example, the complement to the mRNA codon, AAG, is the DNA triplet, CTT.

■ **PRACTICE EXERCISE 1** Using Table 21.1, what amino acids are specified by each of the following codons on an mRNA molecule?

(a) CCU   (b) AGA   (c) GAA   (d) AAG

■ **PRACTICE EXERCISE 2** What amino acids are specified by the following base triads on DNA?

(a) GGA   (b) TCA   (c) TTC   (d) GAT

21.3 mRNA-DIRECTED POLYPEPTIDE SYNTHESIS

tRNA molecules carry amino acyl groups to places at an mRNA strand where anticodons match codons.

In the previous section we saw how a particular genetic message can be transcribed from the exons of a divided gene to a series of codons on mRNA. We now learn in broad terms how the next general step in Figure 21.7 occurs, the mRNA-directed synthesis of a polypeptide.

Genetic *Translation* Follows Transcription.   The steps by which the sequence of codons on mRNA directs the formation of a matching sequence of amino acid residues in a polypeptide is called **translation.** As we have already mentioned, there is at least one kind of tRNA molecule for each of the 20 amino acids, and each aminoacyl unit is carried by its own tRNA molecule to the polypeptide assembly site at a ribosome. We will now use the very abbreviated forklike symbol for an amino acyl–tRNA combination shown in Figure 21.10c. For the rest of our study of translation, we will assume that all needed amino acyl–tRNA combinations have been made and are waiting like so many spare parts to be used at the assembly line.

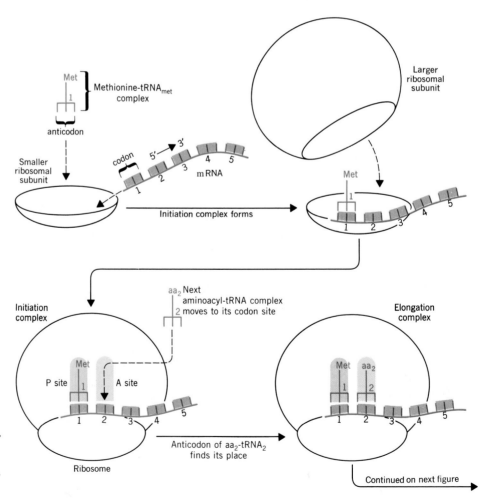

FIGURE 21.11

Formation of the elongation complex at the beginning of the synthesis of a polypeptide.

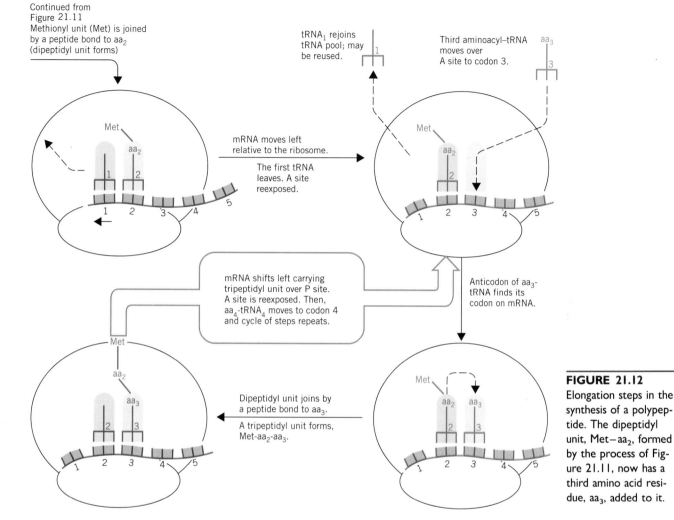

Continued from Figure 21.11 Methionyl unit (Met) is joined by a peptide bond to aa₂ (dipeptidyl unit forms)

mRNA moves left relative to the ribosome.

The first tRNA leaves. A site reexposed.

tRNA₁ rejoins tRNA pool; may be reused.

Third aminoacyl–tRNA moves over A site to codon 3.

Anticodon of aa₃-tRNA finds its codon on mRNA.

mRNA shifts left carrying tripeptidyl unit over P site. A site is reexposed. Then, aa₄-tRNA₄ moves to codon 4 and cycle of steps repeats.

Dipeptidyl unit joins by a peptide bond to aa₃.

A tripeptidyl unit forms, Met-aa₂-aa₃.

FIGURE 21.12
Elongation steps in the synthesis of a polypeptide. The dipeptidyl unit, Met–aa₂, formed by the process of Figure 21.11, now has a third amino acid residue, aa₃, added to it.

The cell begins a polypeptide with an N-terminal methionine residue. After the end of the synthesis, the aminoacyl group of methionine will be left in place only if the polypeptide is supposed to have it as its N-terminal unit. Otherwise, it will be removed, and the second aminoacyl group will be the final N-terminal unit.

The principal steps in making the polypeptide are as follows.

$$^+NH_3CHCO_2^-$$
$$|$$
$$CH_2CH_2SCH_3$$

Methionine, Met

1. *Formation of the elongation complex.* An *elongation complex* is made of several pieces: two subunits of the ribosome, the first amino acyl–tRNA unit (Met–tRNA₁), and the mRNA molecule, beginning at its first codon end (Figure 21.11). The complex forms as the Met–tRNA₁ comes to rest with its anticodon matched to the first codon and with the bulk of its system in contact with a portion of the ribosome's surface called the *P site*. The P site of the ribosome is a region with enzymes that catalyze the transfer of a growing polypeptide chain to a newly arrived aminoacyl unit.

    Now the second tRNA unit, tRNA₂, which holds the second aminoacyl group, aa₂, has to find the mRNA codon that matches its own anticodon, and it does this at another site on the ribosome called the *A site*. The elongation complex is now complete, and actual chain lengthening can start.

2. *Elongation of the polypeptide chain.* A series of repeating steps now occurs, illustrated in Figure 21.12. The methionine residue is transferred from its tRNA

■ The *P* in P site refers to the peptidyl transfer site. The A site is the aminoacyl binding site. (A third site, not shown, is the E site, which temporarily holds the departing tRNA.)

to the newly arrived aa_2. This makes the first peptide bond, and it takes place by acyl transfer much as we discussed when we studied the synthesis of amides in Section 13.5.

■ This kind of translocation of the growing polypeptide to the next amino acid is what is blocked by the toxin of the diphtheria bacillus.

$$CH_3SCH_2CH_2\overset{\overset{\displaystyle NH_2}{|}}{CH} \qquad \overset{\overset{\displaystyle NH_2}{|}}{CH}-R \qquad \xrightarrow{\;H^+\;} \qquad CH_3SCH_2CH_2\overset{\overset{\displaystyle NH_2}{|}}{CH}$$

Methionine unit on its tRNA   The next aminoacyl unit on tRNA $tRNA_1$ (to be recycled)   A dipeptidyl-tRNA unit

New peptide bond

■ Polypeptide synthesis can occur at several ribosomes moving along the mRNA strand at the same time.

The mRNA unit now shifts one codon over, leftward as we have drawn it. (It's actually a relative motion of the mRNA and the ribosome.) This movement positions what is now a *di*peptidyl–tRNA unit over the P site. Now the third amino acyl–tRNA moves in. Its anticodon finds the matching codon, the third codon of the mRNA strand. Now tRNA–aa_3 is over the recently vacated A site.

Further elongation occurs; the dipeptide unit transfers to the amino group of aa_3. Another peptide bond forms, and a *tri*peptidyl system has been made. The cycle of steps now takes place again. The mRNA chain first shifts relative to the ribosome so that the tripeptidyl–tRNA is placed over the P site. The fourth amino acyl group is carried by tRNA–aa_4 to the mRNA; the tripeptidyl unit transfers to it to make a *tetra*peptidyl unit, and so forth. The cycle of steps continues until a chain-terminating codon is reached.

■ In mammals, it takes only about one second to move each amino acid residue into place in a growing polypeptide.

3. **Termination of polypeptide synthesis.** Once a ribosome has moved to a chain-terminating codon (UAA, UAG, or UGA), the polypeptide synthesis is complete, and the polypeptide is released. The ribosome can be reused, and the polypeptide acquires its higher levels of structure.

Antibiotics Inhibit Polypeptide Synthesis in Bacteria. Several antibiotics kill bacteria by inhibiting their synthesis of polypeptides. Streptomycin, for example, inhibits the initiation of polypeptide synthesis. Chloramphenicol inhibits the ability to transfer newly arrived amino acyl units to the elongating strand. The tetracyclines inhibit the binding of tRNA–aa units when they arrive at the ribosome. Actinomycin binds tightly to DNA. Erythromycin, puromycin, and cycloheximide interfere with elongation.

21.4   VIRUSES

Viruses take over the genetic machinery of host cells to make more viruses.

■ A complete virus particle *outside* its host cell is called a *virion*.

Viruses are unique packages of (dead) chemicals at the borderline of living systems until they get inside their *host cells.* They then seem to be living things, because they reproduce.

Viruses Consist of Nucleic Acids and Proteins. Viruses are agents of infection made of nucleic acid molecules surrounded by overcoats of protein molecules. Unlike a cell, a virus has either RNA or DNA, but not both. Viruses must use host cells to reproduce because they can neither synthesize polypeptides nor generate their own energy for metabolism. The simplest virus has only 4 genes and the most complex has about 250.

Each kind of virus has something on its surface that is complementary to something on the surface of the host cell. The glycoproteins of membranes are involved together with a lock-and-key kind of recognition. Each virus, therefore, is unusually selective and has one particular kind of host cell. A virus that attacks, for example, the nerve cells in the spinal cord has no effect on heart muscle cells. A large number of viruses exist that do not affect any kind of human cell. Many viruses attack only plants.

The protein overcoat of some viruses includes an enzyme that catalyzes the breakdown of the cell membrane of the host cell. When such a virus particle sticks to the surface of a host cell, its overcoat catalyzes the opening of a hole into the host. Then the viral nucleic acid squirts into the cell, or the whole virus might move in.

Once a whole virus particle or part of one gets inside its host cell, one of two possible fates awaits it. It might become turned off and change into a *silent gene,* or it might take over the genetic machinery of the cell and reproduce so much of itself that it bursts the host cell walls. The new virus particles that spill out then infect neighboring host cells, and in this way the infection spreads. A virus that has become a silent gene might later be activated. Some cancer-causing agents, including ultraviolet light, are believed to initiate cancer by this mechanism.

RNA Viruses Either Carry or Make Enzymes for Synthesizing More RNA. Most viruses contain RNA, not DNA, so the manufacture of more of their RNA must somehow be managed without the direction of DNA. RNA viruses thus have a major problem if they are to infect a host cell. Host cells normally have no enzymes that can direct the synthesis of a copy of an RNA molecule *from the instructions of another RNA.* In healthy host cells, copies of RNA molecules are made by the direction of DNA, not RNA.

Two basic solutions to this problem occur involving two different enzymes. One is *RNA replicase,* an enzyme that can catalyze the manufacture of RNA *from the directions encoded on RNA.* Some viruses carry preformed RNA replicase. Others carry "instructions" directing the host cell to make it. Either way, once RNA replicase is inside the host cell, it handles the manufacture of the *mRNA* needed to make more viral RNA and protein, so new virus particles can form.

The second enzyme by which RNA is enabled to direct the synthesis of RNA is a DNA polymerase called *reverse transcriptase* carried by a family of viruses called *retroviruses.* Reverse transcriptase uses RNA information to make DNA and so aids the synthesis of *viral* DNA, which subsequently is used to direct the synthesis of more viral RNA. As we said, the work of reverse transcriptase is unusual because, normally, DNA information is used to make RNA, not the other way around.

■ David Baltimore and Howard Temin shared the 1975 Nobel prize in physiology and medicine (together with Renato Dulbecco) for the discovery of reverse transcriptase.

Cancer-Causing Viruses Transform Normal Genes in Host Cells. The retroviruses include the only known cancer-causing RNA viruses, technically termed the *oncogenic RNA viruses.* (Several DNA-based viruses also cause cancer.) They transform host cells so that they grow chaotically and continuously. They do this by changing normal genes in the host cell to *oncogenes,* genes that henceforth are able to continue the cancerous growth.

■ Oncogenic means cancer-inducing.

The Host Cell of the AIDS Virus Is Part of the Human Immune System. Acquired immunodeficiency syndrome, AIDS, is caused by a retrovirus, the human immuno-

AZT

deficiency virus or HIV. One reason why this virus is so dangerous is that its host cell, the T4 lymphocyte, is a vital part of the human immune system. By destroying T4 lymphocytes, HIV exposes the body to other infectious diseases, like pneumonia, or to certain rare types of cancer. One of the strategies that has been tried to retard the development of AIDS, if not cure it in those with this syndrome, is to offer the HIV virus a nucleotide that can bind to its reverse transcriptase but, once bound, will inhibit the further work of this enzyme. AZT, for example, a nucleotide with a modified sugar unit, has been used against AIDS. By itself, however, and not as part of a combination of drugs, AZT does not delay the onset of AIDS in patients known to be infected with HIV but who have not yet had the symptoms.

21.5   RECOMBINANT DNA TECHNOLOGY AND GENETIC ENGINEERING

Single-celled organisms can be made to manufacture the proteins of higher organisms.

■ The term *cloning* is used for the operation that places new genetic material into a cell where it becomes a part of the cell's gene pool. The new cells that follow this operation are called *clones*.

Human insulin and human growth hormone are now being manufactured by a technology that involves the production of *recombinant DNA*. With the aid of Figure 21.13, we'll learn how this technology works. It represents one of the important advances in scientific technology of this century. It has permitted the *cloning*—the synthesis of identical copies—of a number of genes. The use of recombinant DNA to make genes and the products of such genes is called **genetic engineering.**

Genes Alien to Bacteria Can Be Inserted into Bacterial Plasmids.   Bacteria generally make polypeptides using the same genetic code as humans. There are some differences in the machinery, however. An *Escherichia coli* bacterium, for example, has DNA not only in its single chromosome but also in large, circular, supercoiled DNA molecules called **plasmids.** Each plasmid carries just a few genes, but several copies of a plasmid can exist in one bacterial cell. Each plasmid can replicate independently of the chromosome.

The plasmids of *E. coli* can be removed and given new DNA material, such as a new gene, with base triplets for directing the synthesis of a particular polypeptide. It can be a gene completely alien to the bacteria, like the subunits of human insulin, or human growth hormone, or human interferon. The DNA of the plasmids is snipped open by special enzymes called *restriction enzymes* absorbed from the surrounding medium. This medium can also contain naked DNA molecules, such as those of the gene to be cloned. Then, with the aid of a DNA-knitting enzyme called *DNA ligase,* the new DNA combines with the open ends of the plasmid. This recloses the plasmid loops. The DNA of these altered plasmids is called **recombinant DNA.** The altered plasmids are then allowed to be reabsorbed by bacterial cells.

The remarkable feature of bacteria with recombinant DNA is that when they multiply, the plasmids in the offspring also have this new DNA. When these multiply, still more altered plasmids are made. Between their cell divisions, the bacteria manufacture the proteins for which they are genetically programmed, including the proteins specified by the recombinant DNA. In this way, bacteria can be tricked into making the *human* proteins we have mentioned.

■ Recombinant DNA technology isn't limited to using bacteria; yeast cells work, too.

Recombinant DNA Can Be Inserted into Cells of Higher Organisms.   Sometimes, before it will function, a desired polypeptide has to be "groomed" by a cell *after* it has

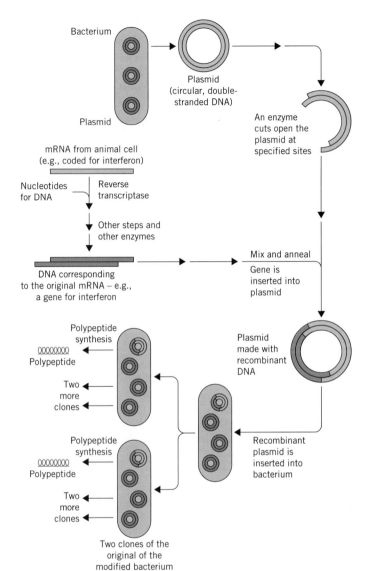

FIGURE 21.13

Recombinant DNA is made by inserting a DNA strand, coded for some protein not made by the bacteria, into the circular DNA of the bacterial plasmid.

been made by genetic translation. It might have to be attached to a carbohydrate molecule, for example. Bacteria lack the enzymes for such grooming work, so cells of higher organisms are used. When these cells are large enough, the new DNA can be inserted directly into them using glass pipets of extremely small diameters (0.1 μm). Although only a small fraction of such inserted DNA actually is taken up into the cell's chromosomes, it can be enough when amplified by successive cell divisions.

Another way to get new DNA inside cells is to let altered viruses carry it in. Retroviruses can be customized for this purpose, for example. Experimental therapy for cystic fibrosis is using this technology (see Section 21.6).

Genetic Engineering Offers Major Advances in Medicine. One of the hopes of genetic engineering research is to find ways to correct genetic faults. As we will study in the next section, a number of undesirable conditions are caused by flawed or absent genes. Dwarfism, for example, is caused by a lack of growth hormone, a relatively small polypeptide. In experiments with mice, genetic engineering has successfully introduced growth hormone into mice, with dramatic effects on mouse size.

In another medical application of genetic engineering, the smallpox vaccine is being remodeled to provide altered forms that might give immunity to many other diseases, ranging from malaria to influenza.

A polypeptide hormone made by the heart, which reduces blood pressure, can be manufactured by genetic engineering and used by victims of high blood pressure. The clot-dissolving enzyme called *tissue plasminogen activator* (TPA) has been genetically engineered for use in reducing the damage to heart tissue following a sudden heart attack (see Section 17.4).

The list of potential applications of genetic engineering to health problems grows yearly. It seems likely that kidney dialysis patients will need fewer blood transfusions if a blood cell-producing substance, erythropoietin, can be made by this technology. Hemophiliacs who lack a blood-clotting factor may have it available by genetic engineering. The synthesis by genetic engineering of several drugs, some that might be used against cancer, is being studied.

21.6 HEREDITARY DISEASES

About 4000 inherited disorders in humans are caused directly or indirectly by flawed genes.

■ The defective gene in cystic fibrosis makes a defective CFTR protein, after *cystic fibrosis transmembrane conductance regulator.*

In Cystic Fibrosis, a Defective Gene Makes a Defective Membrane Protein.   The victims of cystic fibrosis overproduce a thick mucus in the lungs and the digestive tract, which clogs these systems and which often leads to death in children. About one person in 20 carries the defective gene associated with this disease, and it hits about one in every 1000 newborns. The gene that is defective in cystic fibrosis normally directs the synthesis of a trans-membrane protein that regulates the movements of chloride ion through the cell membrane. These movements also affect the flow of water because water is vital for maintaining proper concentrations of solutes, like Cl^-; osmotic pressure relationships among internal fluids are always of high priority for health. When the cells of the lungs and airways carry the defective gene for cystic fibrosis, insufficient Cl^- ion, and so also insufficient water, moves out of such cells. When water is kept in diminished supply outside of the cells of the lungs and airways, the mucus thickens and does not flow properly. The thickened mucus makes breathing more difficult and provides a breeding ground for the bacteria that cause a certain form of pneumonia.

Gene Therapy for Cystic Fibrosis Is Being Tried.   In the early 1990s, teams of scientists began studies using *gene therapy* for cystic fibrosis. In the broadest terms, gene therapy seeks to replace defective genes by normal genes. If the correct gene (i.e., the correct DNA) for the transmembrane conductance regulator can somehow be inserted into cells of the lungs and airways of cystic fibrosis victims, perhaps the cells will be able to operate normally.

The general plan was to use an altered virus to piggyback correct DNA into the target cells. A common cold virus was used because its target cells were precisely those that needed fixing. The *exteriors* of the cold virus particles were left unchanged, because their surface proteins were needed for "finding" the host cells. The *interiors*, however, were denatured so that the virus could not cause an infection. The altered virus particles were treated so that they absorbed the correct DNA for the transmembrane protein, and then they were sprayed into the airways. The hope was that if corrected genes could become installed in only 10% of the airway cells, the defects in the movements of Cl^- and water would at least be rendered nonlethal. Early results with only three patients, reported in late 1993, made scientists optimistic. The flows of Cl^- and water in the three became normal and remained so for a few days without side effects.

Gene Therapy Has Worked against Enzyme Deficiencies. The very first effort to cure a human disease using gene therapy, begun in 1990, involved two little girls, ages 4 and 9. They were born with a defective ADA gene, which left them with almost no natural immunity. The children were injected with white blood cells that had been given the correct gene. Two years later, the girls had functioning immune systems and, instead of having to lead very isolated lives were in public school. Sickle-cell anemia is another disease caused by a defective gene, as we described in Special Topic 16.1.

PKU Is a Genetic Disease Treated Nutritionally. Phenylketonuria, or PKU disease, is a brain-damaging genetic disease in which abnormally high levels of the phenyl ketoacid called phenylpyruvic acid occur in the blood. This condition causes permanent brain damage in the newborn. Because of a defective gene, an enzyme needed to handle phenylalanine properly is not made, and this amino acid is increasingly converted to phenylpyruvic acid.

PKU can be detected by a simple blood test within 4 or 5 days of birth. If the diet is kept very low in phenylalanine, the infant can survive the critical danger period and experience no brain damage. The infant's diet should include no aspartame, a low-calorie sweetener (contained in NutraSweet), because it is hydrolyzed by the digestive processes to give phenylalanine.

Maintaining a low-phenylalanine diet, of course, is not the easiest and best solution, so genetic scientists are working to correct the fundamental gene defect.

The *Human Genome Project* Aims to Map All Human Genes and Determine Their DNA Sequences. The entire complement of genetic information of a species is called its **genome.** The *Human Genome Project,* formally launched in 1990, intends to discover the "map" of every human chromosome and to determine the sequence of bases in every human gene. It's perhaps the largest single project ever attempted in molecular biology, and its total cost may come within range of (but will probably be less than) what it cost to put an astronaut on the moon. If the genome is a library, if the chromosomes are book sections, and if the genes are individual books, then the Human Genome Project's goal in *mapping* chromosomes is to create the card catalog. In which section of the library is the gene for cystic fibrosis located? Answer: on chromosome 7. In which particular part ("shelf") of chromosome 7 is the gene responsible for cystic fibrosis located? Answer: in region q31. Now comes the *sequencing* question; it's not the same as the mapping question. What is the sequence of bases in the cystic fibrosis gene? The answer is known, but we can't give it here; the gene involves more than 6000 bases. What's wrong with this gene to make it cause cystic fibrosis? For about 70 percent of cystic fibrosis mutations, there is a deletion (an absence) of three bases in exon number 10. So much human suffering from so small a molecular defect!

Although an accurate genetic map will be a huge aid in medical diagnoses, the Human Genome Project is controversial. The ethical implications are immense. A small fraction (3%) of the project's \$3 billion budget, in fact, is targeted for studies of them. One question is, Exactly *how* will the knowledge be used? Will it be used, for

■ Additional positive evidence for the procedure was reported in late 1994.

$$C_6H_5CH_2CCO_2H$$
Phenylpyruvic acid

$$C_6H_5CH_2CHCO_2^-$$
$$NH_3^+$$
Phenylalanine

$$^+NH_3CHC-NHCHCOCH_3$$
$$CH_2 \quad CH_2$$
$$CO_2^- \quad C_6H_5$$
Aspartame

■ There is enough phenylalanine in just one slice of bread to be potentially dangerous to a PKU infant.

■ In early 1993, the location of the gene for Huntington's disease, a deadly neurological disorder (page 396), was found near the tip of chromosome 4.

example, to discriminate in employment against those deemed (by whom?) to have "bad" genes? [A reference: T. Wilkie, *Perilous Knowledge: The Human Genome Project and Its Implications,* Faber and Faber, London, 1993.]

SUMMARY

Hereditary Information   The genetic apparatus of a cell is mostly in its nucleus and consists of chromatin, a complex of DNA and proteins. Strands of DNA, a polymer, carry segments that are individual genes. Chromatin replicates prior to cell division, and the duplicates segregate as the cell divides. Each new cell thereby inherits exact copies of the chromatin of the parent cell.

DNA   Complete hydrolysis of DNA gives phosphoric acid, deoxyribose, and a set of four heterocyclic amines, the bases adenine (A), thymine (T), guanine (G), and cytosine (C). The molecular backbone of the DNA polymer is a series of deoxyribose units joined by phosphodiester groups. Attached to each deoxyribose is one of the four bases. The order in which triplets of bases occur is the cell's way of storing genetic information.

A gene in higher organisms consists of successive groups of triplets, the exons, separated by introns. Thus the gene is a divided system, not a continuous series of nucleotide units. DNA strands exist in cell nuclei as double helices, and the helices are held near each other by hydrogen bonds that extend from bases on one strand to bases on the other. Base A always pairs with T and C always pairs with G. Using this structure and the faithfulness of base pairing, Crick and Watson explained the accuracy of replication. After replication, each new double helix has one of the parent DNA strands and one new, complementary strand.

RNA   RNA is similar to DNA except that in RNA ribose replaces deoxyribose and uracil (U) replaces thymine (T). Four main types of RNA are involved in polypeptide synthesis. One is rRNA, which is in ribosomes. A ribosome contains both rRNA and proteins that have the enzyme activity needed during polypeptide synthesis.

mRNA is the carrier of the genetic message from the nucleus to the site where a polypeptide is assembled. mRNA results from chemical processing of the longer precursor RNA strand, symbolized as hnRNA (or pre-mRNA), which is made directly under the supervision of DNA.

tRNA molecules are the smallest RNAs, and their function is to convey aminoacyl units to the polypeptide assembly site. They recognize where they are to go by base pairing between an anticodon on tRNA and its complementary codon on mRNA. Both codon and anticodon consist of a triplet of bases.

Polypeptide Synthesis   Genetic information is first transcribed when DNA directs the synthesis of mRNA. Each base triplet on the exons of DNA specifies a codon on mRNA. The mRNA moves to the cytoplasm to form an elongation complex with subunits of a ribosome and the first and second tRNA—aa unit to become part of the developing polypeptide.

The ribosome then rolls down the mRNA as tRNA—aa units come to the mRNA codons during the moment when the latter are aligned over the proper enzyme site of a ribosome. Elongation of the polypeptide then proceeds to the end of the mRNA strand or to a chain-terminating codon. After chain termination, the polypeptide strand leaves, and it may be further modified to give it its final N-terminal amino acid residue.

Several antibiotics inhibit bacterial polypeptide synthesis, which causes the bacteria to die.

Viruses   Viruses are packages of DNA or RNA encapsulated by protein. Once they get inside their host cell, virus particles take over the cell's genetic machinery, make enough new virus particles to burst the cell, and then repeat this in neighboring cells. Viruses are implicated in human cancer. Some viruses make new RNA under the direction of existing RNA, using RNA replicase. Others, the retroviruses, use RNA and reverse transcriptase to make DNA, which then directs the synthesis of new RNA.

Recombinant DNA   Recombinant DNA is DNA made from bacterial plasmids and DNA obtained from another source and encoded to direct the synthesis of some desired polypeptide. The altered plasmids are reintroduced into the bacteria, where they become machinery for synthesizing the polypeptide (e.g., human insulin or growth hormone). Yeast cells can be used instead of bacteria for this technology.

Cells of higher organisms are also used as sites for inserting new DNA. In this kind of genetic engineering, microsyringes or tailored viruses have been used to insert the DNA.

Gene Therapy   Hereditary diseases stem from defects in DNA molecules that either prevent the synthesis of necessary enzymes or that make the enzymes in forms that won't work. Identification of the chromosomes bearing the defective genes has enabled the analysis of the genes themselves. In gene therapy, it is hoped that healthy genes can be substituted for those that are defective. Gene therapy for cystic fibrosis entails using a modified retrovirus to convey correct DNA material into the host cells without causing a viral infection.

REVIEW EXERCISES

The answers to Review Exercises whose numbers are in color are found in Appendix V. The answers to the other Review Exercises are found in the Study Guide that accompanies this text. The more challenging questions are marked with asterisks.

Hereditary Units

21.1   What is the relationship between a chromosome and chromatin?

21.2   The duplication of a gene occurs in what part of the cell?

21.3   In a broad, overall sense, what happens when DNA replicates?

Structural Features of Nucleic Acids

21.4   What is the general name for the chemicals that are most intimately involved in the storage and the transmission of genetic information?

21.5   The monomer units for the nucleic acids have what *general* name?

21.6   What are the names of the two sugars produced by the complete hydrolysis of all the nucleic acids in a cell?

21.7   What are the names and symbols of the four bases that are liberated by the complete hydrolysis of (a) DNA and (b) RNA?

21.8   How are all DNA molecules structurally alike?

21.9   How do different DNAs differ structurally?

21.10   How are all RNA molecules structurally alike?

21.11   What are the principal structural differences between DNA and RNA?

21.12   When DNA is hydrolyzed, the ratios of A to T and of G to C are each very close to 1:1, *regardless of the species investigated.* Explain.

21.13   What does base pairing mean, in general terms?

21.14   What is the chief stabilizing factor for the geometrical form taken by two molecules of DNA when they form a double helix?

21.15   If the AGGCTGA sequence appeared on a DNA strand, what would be the sequence on the DNA strand opposite it in a double helix?

21.16   The *accuracy* of replication is ensured by the operation of what factors?

21.17   What is the relationship between a single molecule of single-stranded DNA and a single gene?

21.18   Suppose that a certain DNA strand has the following groups of nucleotides, where each lowercase letter represents a group several side chains long.

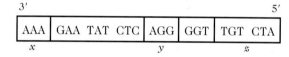

Which sections are most likely to be the introns? Why?

21.19   In general terms only, what particular contribution does a gene make to the structure of a polypeptide?

Ribonucleic Acids

21.20   What is the general composition of a ribosome, and what function does this particle have?

21.21   What is hnRNA, and what role does it have?

21.22   What is a codon, and what kind of nucleic acid is a continuous, uninterrupted series of codons?

21.23   What is an anticodon, and on what kind of RNA is it found?

21.24   Which triplet, ATA or CGC, cannot be a codon? Explain.

21.25   Which amino acids are specified by the following codons?

(a) UUU   (b) UCC   (c) ACA   (d) GAU

*__21.26__   What are the anticodons for the codons of Review Exercise 21.25?

*__21.27__   Suppose that sections *x*, *y*, and *z* of the following hypothetical DNA strand are the exons of one gene.

3′				5′
AAA	GAA TAT CTC	AGG	GGT	TGT CTA
x		*y*		*z*

What is the structure of each of the following substances made under its direction?
(a) the hnRNA
(b) the mRNA
(c) the tripeptide that is made using the given genetic information. (Use the three-letter symbol format for the tripeptide structure, referring as needed to Table 18.1 for these symbols.)

Polypeptide Synthesis

21.28   Use the identifying letters to arrange the following symbols or terms in the correct order in which they are synthesized in going from a gene to an enzyme. (Place the letter of the first material to be involved on the left.)

hnRNA	duplex DNA	polypeptide	mRNA
A	**B**	**C**	**D**
	<	<	<
	earliest		last to
	to be		appear
	involved		

21.29 What is meant by *translation,* as used in this chapter? And what is meant by *transcription?*

21.30 To make the pentapeptide Met-Ala-Trp-Ser-Tyr,
(a) What do the sequences of bases on the mRNA strand have to be?
(b) What is the anticodon on the first tRNA to move into place?

21.31 How do some of the antibiotics work at the molecular level?

21.32 The genetic code is the key to translating between what two "languages"?

21.33 What is meant by the statement that the genetic code is universal? And is the code strictly universal?

Viruses

21.34 What is a virus made of?

21.35 In general terms, how does a virus discriminate among all possible host cells and "find" just one kind of host cell?

21.36 In general terms, once a virus particle has joined to the membrane of its host cell, what must occur next if the viral infection is to advance?

21.37 In general terms, in the systems having it, what does RNA replicase do? What is true about a normal host cell that requires a virus to have RNA replicase?

21.38 In general terms and in connection with the work of (some) viruses, what does reverse transcriptase do?

21.39 What does the prefix *retro* signify in retrovirus?

21.40 What is meant by a *silent gene* and where does it come from?

21.41 Some viruses are called *oncogenic RNA viruses.* What does *oncogenic* mean here?

21.42 What is the full name of the HIV system?

21.43 What is the host cell of HIV and how does this fact make AIDS so dangerous?

Recombinant DNA

21.44 What is a plasmid, and what is it made of?

21.45 What is the name of the enzyme that can snip open the DNA of a plasmid?

21.46 Recombinant DNA is made from the DNA of two different kinds of sources. What are they?

21.47 Recombinant DNA technology is carried out to accomplish the synthesis of what kind of substance (in general terms)?

21.48 What does genetic engineering refer to?

Hereditary Diseases

21.49 At the molecular level of life, what kind of defect is the fundamental cause of a hereditary disease?

21.50 The defective gene in cystic fibrosis leads to an impairment of what specific activity of the cells of the affected tissues? Why does this activity result in a problem with mucus?

21.51 Only in the broadest terms, what is gene therapy meant to accomplish?

21.52 A cold virus instead of some other kind of virus is being used in gene therapy for cystic fibrosis. Why?

21.53 The alteration of the virus used in attempted gene therapy for cystic fibrosis is done to the virus interior, not to its exterior. Explain.

21.54 What is the molecular defect in PKU, and how does it cause the problems of the victims? How is it treated?

DNA Typing (Special Topic 21.1)

21.55 What fact about cells makes it possible to use cells from any part of the body of a suspect for DNA fingerprinting in a rape case for which a semen sample has been obtained?

21.56 What is meant by the *polymerase chain reaction,* and why is it used as part of the technology of DNA typing?

21.57 A restriction enzyme separates DNA molecules into pieces given what general name? How are they used to give the genetic "bar code"?

Additional Exercises

*21.58 Consider the following compound.

(a) Is it a mononucleotide, a dinucleotide, or one with even more nucleotides? How can you tell?
(b) Could it be obtained by the partial hydrolysis of DNA or RNA? How can you tell?
(c) Where is the 5′ end, at the bottom or top of the structure?
(d) In terms of the single-letter symbols for bases, how is the structure of this compound written?

22

RADIOACTIVITY AND NUCLEAR CHEMISTRY

Atomic Radiation

Ionizing Radiation— Dangers and Precautions

Units to Describe and Measure Radiation

Synthetic Radionuclides

Radiation Technology in the Food Industry

Radiation Technology in Medicine

Atomic Energy and Radionuclides

Radiation streaming toward the magnetic poles of our earth sometimes cause spectacular displays of northern lights. Radiation closer to home and hospital offers life-saving diagnostic and treatment tools, which must be handled carefully. In this chapter we'll learn about the kinds of radiation and how they can be dangerous or helpful.

22.1 ATOMIC RADIATION

Unstable atomic nuclei eject high-energy radiation as they change to more stable nuclei.

The nuclei of several isotopes are unstable. As they spontaneously change to more stable forms, they emit high-energy radiation and thus are said to be **radioactive.** Such isotopes are called **radionuclides,** and when carefully used, the potential benefits of their radiation outweigh the possible harm. In this chapter we study how radiation can be used wisely, how it is dangerous, and how radiation is detected and measured.

Decay of Radionuclides Produces Different Kinds of Radiation. Radioactivity was discovered in 1896 when a French physicist, A. H. Becquerel (1852–1908), happened to store some well-wrapped photographic plates in a drawer that contained samples of uranium ore. The film became fogged, meaning that when developed the picture was like a photograph of fog.

■ Becquerel shared the 1903 Nobel prize in physics with Pierre and Marie Curie.

Becquerel might have blamed the accident on faulty film or careless handling, but a mysterious radiation called X rays had recently been discovered by Wilhelm Roentgen (1845–1923), a German scientist. X rays had been found to be able to penetrate the packaging of unexposed film and ruin it. What fogged Becquerel's film, however, was soon shown to be a natural radiation resembling X rays. The same radiation was also found to be emitted by any compound of uranium as well as by uranium metal itself.

■ Roentgen won the 1901 Nobel prize in physics, the first to be awarded.

Several years later, two British scientists, Ernest Rutherford (1871–1937) and Frederick Soddy (1877–1956), explained radioactivity in terms of the *disintegration* or **radioactive decay** of unstable atomic nuclei. A decaying nucleus might eject a tiny particle into space, or it might emit a powerful radiation like an X ray but called a gamma ray, or it might do both. It depends on the radionuclide.

■ Nobel prizes in chemistry were awarded to both Rutherford (1908) and Soddy (1921).

The nuclei that remain after decay almost always are those of an entirely different element, so decay is usually accompanied by the **transmutation** of one isotope into another. The natural sources of radiation on our planet emit one or more of three kinds: *alpha radiation, beta radiation,* and *gamma radiation.* We receive another, called *cosmic radiation,* from the sun and outer space. It consists of gamma rays plus a mixture of particles—protons, neutrons, alpha and beta particles, and atomic nuclei of elements as high as atomic number 26 (iron).

Alpha Particles Are the Nuclei of Helium Atoms. One natural atomic radiation is called **alpha radiation.** It consists of particles called **alpha particles** that move with a

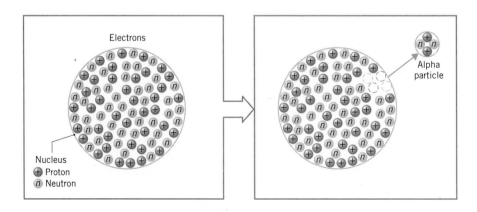

FIGURE 22.1
Emission of an alpha particle.

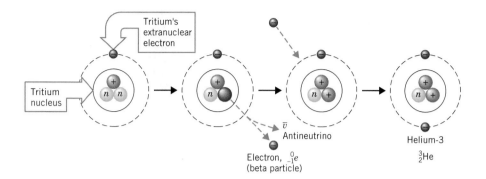

Tritium's extranuclear electron

Tritium nucleus

$\bar{v}$ Antineutrino

Electron, $_{-1}^{0}e$
(beta particle)

Helium-3
$_{2}^{3}\text{He}$

FIGURE 22.2
Emission of a beta particle.

velocity almost one-tenth the velocity of light as they leave the atom. Alpha particles are clusters of two protons and two neutrons, so they are actually the nuclei of helium atoms (Figure 22.1). They are the largest of the decay particles and have the greatest charge, so when alpha particles travel in air, they soon collide with molecules of nitrogen or oxygen and lose their energy (and charge). Alpha particles cannot penetrate even thin cardboard or the outer layer of dead cells on the skin, but they can cause a severe skin burn.

The most common isotope of uranium, uranium-238 or $_{92}^{238}\text{U}$, is an alpha emitter. When its nucleus ejects an alpha particle, it loses two protons, so the atomic number changes from 92 to 90. It also loses four units of mass number (two protons + two neutrons), so its mass number changes from 238 to 234. The result is that uranium-238 transmutes into an isotope of thorium, $_{90}^{234}\text{Th}$.

Beta Radiation Is a Stream of Electrons.   Another natural radiation, **beta radiation,** consists of a stream of particles called **beta particles,** which are actually electrons produced *within* the nucleus and then emitted (Figure 22.2). With less charge and a much smaller size, beta particles can penetrate matter, including air, more easily than alpha particles. Different sources emit beta particles with different energies, and those of lower energy cannot penetrate the skin. Those of the highest energy, however, can reach internal organs from outside the body.

As a nucleus emits a beta particle, a neutron changes into a proton (see Figure 22.2). Thus there is no loss in mass number, but the atomic number *increases* by 1 unit because of the new proton. For example, thorium-234, $_{90}^{234}\text{Th}$, is a beta emitter, and when it ejects a beta particle it changes to an isotope of protactinium, $_{91}^{234}\text{Pa}$.

Gamma Radiation Often Accompanies Other Radiation.   Atoms have different *nuclear* energy states just as they have different *electron* energy states. By ejecting small particles, the nuclei of radionuclides acquire a lower, more stable nuclear state. The energy lost by the nucleus is carried away by the particles, but often high-energy electromagnetic radiation is also emitted. This is **gamma radiation,** and it is like X rays or ultraviolet rays, but with more energy. Gamma radiation is very penetrating and very dangerous. It easily goes through the entire body.

The composition and symbols of the three kinds of radiation studied thus far are summarized in Table 22.1.

Nuclear Equations Display Balances of Both Mass Numbers and Atomic Numbers. In *chemical* reactions no changes in atomic nuclei occur, but *nuclear* reactions are nearly always accompanied by transmutations. An equation for a nuclear reaction, called a **nuclear equation,** is therefore different from chemical equations in important ways. In particular, it must describe changes in atomic numbers and mass numbers, as well as give the identities of the radionuclides.

■ Isotopes have special symbols that give the mass number as a left superscript and the atomic number as a left subscript. Thus in $_{92}^{238}\text{U}$, 238 is the mass number and 92 is the atomic number.

TABLE 22.1   **Radiation from Naturally Occurring Radionuclides**

Radiation	Composition	Mass Number	Electrical Charge	Symbols
Alpha	Helium nuclei	4	2+	$_{2}^{4}He$, α
Beta	Electrons	0	1−	$_{-1}^{0}e$, β
Gamma	X ray like	0	0	$_{0}^{0}\gamma$

In nuclear equations the alpha particle is symbolized as $_{2}^{4}He$, and although it is positively charged, the charge is omitted from the symbol. The particle soon picks up electrons anyway, taking them from the matter through which the particle moves. Thus, the alpha particle becomes a neutral atom of helium. The beta particle has the symbol $_{-1}^{0}e$ because its mass number is 0 and its charge is 1−. A photon of gamma radiation is symbolized simply by γ (sometimes, by $_{0}^{0}\gamma$). Gamma radiation has no associated mass number or charge.

A nuclear equation is *balanced* when the sums of the mass numbers on either side of the arrow are equal and when the sums of the atomic numbers are equal. The alpha decay of uranium-238, for example, is represented by the following equation:

$$_{92}^{238}U \longrightarrow\ _{90}^{234}Th + \ _{2}^{4}He$$

Notice that the sums of the atomic numbers agree: 92 = 90 + 2. The sums of the mass numbers also agree: 238 = 234 + 4. So the equation is balanced.

The beta decay of thorium-234, which also emits gamma radiation, is represented by the following nuclear equation.

$$_{90}^{234}Th \longrightarrow\ _{91}^{234}Pa + \ _{-1}^{0}e + \ _{0}^{0}\gamma$$

The sums of the atomic numbers agree: 90 = 91 + (−1). Likewise, the sums of the mass numbers agree. Thus the equation is balanced.

■ Losing one electron from the nucleus makes a proton out of a neutron, so no change in mass number occurs.

EXAMPLE 22.1   Balancing Nuclear Equations

Cesium-137, $_{55}^{137}Cs$, is one of the radioactive wastes that form during fission in a nuclear power plant or an atomic bomb explosion. This radionuclide decays by emitting both beta and gamma radiation. Write the nuclear equation for this decay.

ANALYSIS We start with an incomplete equation using the given information. Then we figure out any additional data we need.

SOLUTION The incomplete nuclear equation is

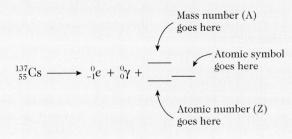

We have to figure out the atomic number, Z, first so that we can find the atomic symbol for the product in the table inside the front cover. To calculate Z we use the fact that the atomic number *(55)* on the left side of the equation must equal the sum of the atomic numbers on the right side.

$$55 = -1 + 0 + Z$$
$$Z = 56$$

In the periodic table we see that element 56 is Ba (barium), but which isotope of Ba? Recall that the sums of the mass numbers on either side of the equation must also be equal. Letting A equal the mass number of the barium isotope,

$$137 = 0 + 0 + A$$
$$A = 137$$

The balanced nuclear equation, therefore, is

$$^{137}_{55}\text{Cs} \longrightarrow {}^{0}_{-1}e + {}^{0}_{0}\gamma + {}^{137}_{56}\text{Ba}$$

■ It's proper to think of the electron as having an atomic number of -1.

EXAMPLE 22.2 Balancing Nuclear Equations

Until the 1950s, radium-226 was widely used as a source of radiation for cancer treatment. It is an alpha emitter and a gamma emitter. Write the equation for its decay.

ANALYSIS We have to look up the atomic number of radium, which turns out to be 88, so the symbol we'll use for this radionuclide is $^{226}_{88}\text{Ra}$. When one of its atoms loses an alpha particle, $^{4}_{2}\text{He}$, it loses 4 units in mass number—from 226 to 222. And it loses 2 units in atomic number—from 88 to 86. Thus, the new radionuclide has a mass number of 222 and an atomic number of 86. We have to look up the atomic symbol for element number 86, which turns out to be Rn, for radon.

SOLUTION Now we can assemble the nuclear equation.

$$^{226}_{88}\text{Ra} \longrightarrow {}^{222}_{86}\text{Rn} + {}^{4}_{2}\text{He} + {}^{0}_{0}\gamma$$

■ The radium used in cancer therapy was held in a thin, hollow gold or platinum needle to retain the alpha particles and all the decay products.

■ PRACTICE EXERCISE 1   Iodine-131 has long been used in treating cancer of the thyroid. This radionuclide emits beta and gamma rays. Write the nuclear equation for this decay.

■ PRACTICE EXERCISE 2   Plutonium-239 is a by-product of the operation of nuclear power plants. It can be isolated from used uranium fuel and made into fuel itself or into atomic bombs. A powerful alpha and gamma emitter, plutonium-239 is one of the most dangerous of all known substances. Write the equation for its decay.

A Short Half-Life Means a Rapid Decay.   Some radionuclides are much more stable than others, and we use the concept of *half-life* to describe the differences. The **half-life** of a radionuclide, symbolized as $t_{1/2}$, is the time it takes for half of the atoms in a sample of a single, pure isotope to decay. (The atoms that decay don't just vanish, of course. They change into different isotopes.) Table 22.2 gives the half-lives of several radionuclides.

The half-life of uranium-238 is 4.51×10^9 years, which means that if initially 100 g of this radionuclide is present, 50.0 g of uranium-238 would be left after 4.51×10^9

TABLE 22.2   Typical Half-Life Periods

Element	Isotope	Half-Life	Radiation or Mode of Decay
Naturally Occurring Radionuclides			
Potassium	$^{40}_{19}K$	1.3×10^9 y	beta, gamma
Tellurium	$^{123}_{52}Te$	1.2×10^{13} y	electron capture
Neodymium	$^{144}_{60}Nd$	5×10^{15} y	alpha
Samarium	$^{149}_{62}Sm$	4×10^{14} y	alpha
Rhenium	$^{187}_{75}Re$	7×10^{10} y	beta
Radon	$^{222}_{86}Rn$	3.82 d	alpha
Radium	$^{226}_{88}Ra$	1590 y	alpha, gamma
Thorium	$^{230}_{90}Th$	8×10^4 y	alpha, gamma
Uranium	$^{238}_{92}U$	4.51×10^9 y	alpha
Synthetic Radionuclides			
Tritium	$^{3}_{1}H$	12.26 y	beta
Oxygen	$^{15}_{8}O$	124 s	positron
Phosphorus	$^{32}_{15}P$	14.3 d	beta
Technetium	$^{99m}_{43}Tc$	6.02 h	gamma
Iodine	$^{131}_{53}I$	8.07 d	beta
Cesium	$^{137}_{55}Cs$	30 y	beta
Strontium	$^{90}_{38}Sr$	28.1 y	beta
Americium	$^{243}_{95}Am$	7.37×10^3 y	alpha

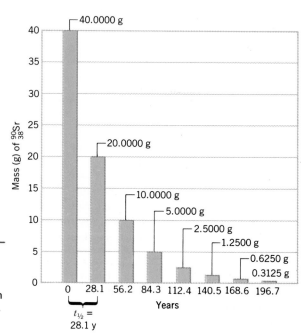

FIGURE 22.3
Each half-life period reduces the quantity of a radionuclide by a factor of 2. Shown here is the pattern for strontium-90, a radioactive pollutant with a half-life of 28.1 years.

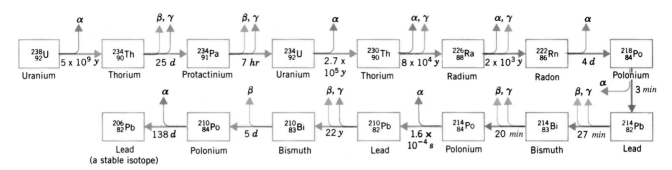

FIGURE 22.4

Uranium-238 radioactive disintegration series. The time given beneath the arrow is the half-life of the preceding isotope. *y*, year; *m*, month; *d*, day; *hr*, hour; *min*, minute; *s*, second.

years. Strontium-90, a by-product of nuclear power plants, is a beta emitter with a half-life of 28.1 years. Figure 22.3 shows graphically how an initial supply of 40 g is reduced successively by units of one-half for each half-life period. At the end of seven half-life periods (196.7 years, from 7 × 28.1), only 0.3 g of the initial 40 g of strontium-90 remains.

The shorter the half-life is, the larger is the number of decay events occurring per mole per second. Mole for mole, it's generally much safer to be near a radionuclide that has a long half-life and thus decays very slowly than to be near one that has a short half-life and decays very rapidly. The potential danger of a radionuclide, however, is a function of more factors than its half-life, as we will soon learn.

A Succession of Decays Occurs in a *Radioactive Disintegration Series.* The decay of one radionuclide sometimes produces not a *stable* isotope but only another radionuclide. It might, in turn, decay to still another radionuclide, with the process repeating until a stable isotope is finally reached. There are three such series still active in nature, called **radioactive disintegration series,** and uranium-238 is at the head of one that ends in a stable isotope of lead (Figure 22.4).

■ An extremely long half-life is typical of the radionuclides that head a radioactive disintegration series.

22.2 IONIZING RADIATION—DANGERS AND PRECAUTIONS

Atomic radiation creates unstable ions and radicals in tissue, which can lead to cancer, mutations, tumors, or birth defects.

The undesired effects of radiation are both *acute* and *latent.* Acute effects can be burns and any of the symptoms that are a part of radiation sickness, which we study in this section. Latent effects of radiation are those that do not show themselves until some time after exposure. Cancer, particularly leukemia, can be induced by radiation and is an example of a latent effect. Another example is the alteration of a gene.

Policy Makers Consider That No Safe *Threshold Exposure* to Radiation Exists. Any kind of radiation that penetrates the skin or enters the body on food or through the lungs is considered harmful, and *the damage can accumulate over a lifetime.*

Even the ultraviolet radiation in strong sunlight, which barely penetrates the skin, can alter the genetic molecules in skin cells so as to lead to skin cancer. No "tiny bit of exposure," no **threshold exposure,** is considered to exist for ionizing radiation below which no harm is possible. Cells, however, do have a significant capacity for the self-repair of radiation-caused damage, and some exposures carry very low risks. Thus, the apparent threshold exposure, surprisingly, appears to be high for transmittable genetic defects. A fetus, however, is much more susceptible to damage by radiation than are men and nonpregnant women. Nonetheless, the widespread and routine use of X rays for public health screenings has long been curtailed.

The latent effects of radiation correlate well with the accumulated dose regardless of how it is received, so medical personnel who work with radiation wear devices that automatically record their exposure. The exposure data are periodically logged into a permanent record book, and if the maximum permissible dose is attained, the worker must be transferred. Let's now see how the radiation has its effects and what are some steps for self-protection.

Unstable Ions and Radicals Are Produced in Tissue by Radiation. The different kinds of atomic radiation are dangerous because they can generate unstable, highly reactive particles as they travel through living tissues. For this reason, alpha and beta particles, as well as X rays and gamma rays, are called **ionizing radiation.** They can knock electrons from molecules as they strike them and so produce unstable polyatomic ions. Radiation, for example, can make ions from water molecules by the following reaction.

$$H-\overset{..}{\underset{..}{O}}-H \xrightarrow{\text{radiation}} \left[H-\overset{.}{\underset{..}{O}}-H\right]^+ + {}_{-1}^{0}e$$

The new cation, $\left[H-\overset{.}{\underset{..}{O}}-H\right]^+$, is unstable and one of its breakup paths is

$$\left[H-\overset{.}{\underset{..}{O}}-H\right]^+ \longrightarrow H^+ \ + \ :\overset{.}{\underset{..}{O}}-H$$
$$\qquad\qquad\qquad\qquad\text{Proton}\quad\text{Hydroxyl radical}$$

A proton forms, as does a hydroxyl radical, a kind of particle new to our study. It's a *neutral* particle with an odd number of electrons and without an octet for its oxygen atom. Any particle with an odd number of electrons is called a **radical,** and with few exceptions, radicals are *very* reactive species.

The new ions and radicals produced by ionizing radiation cause chemical reactions in the more stable substances around them, altering them in ways foreign to metabolism. If such chemical reactions occur in genes and chromosomes, the cell's genetic substances, subsequent reactions could lead to cancer, tumor growth, or a genetic mutation. If they happen in a sperm cell, an ovum, or a fetus, the result might be a birth defect. Relative to other tissues, the fetus is particularly sensitive to radiation, as we indicated.

Prolonged and repeated exposures to *low* levels of radiation are more likely to induce longer-term problems than bursts of high-level radiation. It depends on whether the injured cell is still able to duplicate itself by cell division. High-energy radiation bursts usually kill a cell outright, or at least render it reproductively dead. For this reason, high doses of radiation are used in cancer treatment. But low-level radiation that leaves a cell reproductively viable can alter the cell contents in ways that affect cell reproduction.

The Collection of Symptoms Caused by Ionizing Radiation Is Called *Radiation Sickness.* Molecules of hereditary materials in the cell's chromosomes are the primary sites of the most serious radiation damage. Damage to these molecules leads to all other problems, beginning with cell division. The first symptoms of exposure to radiation, therefore, usually occur in tissues whose cells divide most frequently, for

■ Sometimes the term used is *free radical.*

■ Technical terms:
Carcinogen: a cancer causer
Tumorigen: a tumor causer
Mutagen: a mutation causer
Teratogen: birth defect causer

example, the cells in bone marrow. These make white blood cells, so an early sign of radiation damage is a sharp decrease in the blood's white cell count. Cells in the intestinal tract also divide frequently, and even moderate exposure to X rays or gamma rays (as in therapy for cancer) may produce intestinal disorders.

Radiation sickness is the set of symptoms caused by nonlethal exposures to atomic radiation or X rays: nausea, vomiting, a drop in the white cell count, diarrhea, dehydration, prostration, hemorrhaging, and loss of hair. Many of these appear when strong bursts of radiation are used in an effort to halt the spread of cancer.

Protection from Ionizing Radiation Is Achieved by Deploying Shields, Using Short Exposure Times, and Moving Away from the Source. Shields have long been used to protect against ionizing radiation. (No doubt you have had a lead apron placed over your chest before a dental X ray.) Alpha and beta rays are the easiest to stop as the data in Table 22.3 show. Gamma radiation and X rays are stopped effectively only by particularly dense substances. Lead, a very dense metal but still fairly inexpensive, is the most common material used to shield against gamma or X rays. But notice in Table 22.3 that even 30 mm (3.0 cm, a little over an inch) of lead reduces the intensity of gamma radiation by only 10%. A vacuum is the least effective "shield," of course, and air isn't much better. Low-density materials, like cardboard, plastic, and aluminum, are poor shielders, but concrete works well if it is thick (and it's much cheaper than lead). Thus by a careful choice of a shielding material, protection can be obtained.

Another strategy to minimize exposure when ionizing radiation is used in medical diagnosis, such as in taking X rays, is to use fast film. With fast film, the *time* of exposure is kept as low as possible.

■ No technology in any medical field is entirely risk free. We take risks when we believe that the benefits outweigh them.

TABLE 22.3 Penetrating Abilities of Some Common Kinds of Radiation[a]

Type of Radiation	Common Sources	Approximate Energy When from These Sources	Approximate Depth of Penetration of Radiation into		
			Dry Air	Tissue	Lead
Alpha rays	Radium-226 Radon-222 Polonium-210	5 Mev	4 cm	0.05 mm	0
Beta rays	Tritium Strontium-90 Iodine-131 Carbon-14	0.01–0.2 Mev	0.3–6 cm[b]	0.06–4 mm[c]	0.005–0.3 mm
			Thickness to Reduce Initial Intensity by 10%		
Gamma rays	Cobalt-60 Cesium-137 Radium-226 decay products	1 MeV	400 cm	50 cm	30 cm
X rays Diagnostic		Up to 90 keV	120 m	15 cm	0.3 mm
Therapeutic		Up to 250 keV	240 m	30 cm	1.5 mm

[a]Data from J. B. Little, *New England Journal of Medicine*, Vol. 275, pages 929–938, 1966.

[b]The range of beta particles in air is about 30 cm/MeV.

[c]The protective layer of skin is about 0.07 mm thick. To penetrate it, alpha particles need about 7.5 MeV of energy, and beta particles, about 0.07 MeV.

TABLE 22.4   Average Radiation Doses Received Annually by the U.S. Population[a]

Types of Radiation	Dose (mrem)	Percent of Total
Natural Radiation—295 mrem, 82%		
Radon	200	55
Cosmic rays[b]	27	8
Rocks and soil	28	8
From inside the body	40	11
Artificial Radiation—65 mrem, 18%		
Medical X rays[c]	39	11
Nuclear medicine	14	4
Consumer products	10	3
Others	2	<1
Total[d]	360	

[a]Data from *Ionizing Radiation Exposure of the Population of the United States.* Report 93, 1987, National Council on Radiation Protection. These are averages. Individual exposures vary widely.

[b]Travelers in jet airplanes receive about 1 mrem per 1000 miles of travel.

[c]A normal chest X ray entails an exposure of 10–20 mrem.

[d]The federal standard for maximum safe occupational exposure in the United States is roughly 5000 mrem/y.

The least expensive self-protection step is to get as far from the source of radiation as you can. Radiation, like light from a bulb, moves in straight lines, spreading out from the source in all of the directions open to it. From any point on the surface of the source, the radiation forms a cone of rays. Fewer rays, therefore, strike a unit of surface area (opposite the cone's tip) the more distant is the surface from the source. The relationship between radiation intensity and distance from the source, given by the **inverse-square law** of radiation intensity, reflects the fact that the area of the base of a cone increases with the *square* of the distance from the tip.

INVERSE-SQUARE LAW   The intensity of radiation is inversely proportional to the square of the distance from the source.

$$\text{radiation intensity, } I \propto \frac{1}{d^2} \qquad (22.1)$$

This law is strictly true only in a vacuum, but it holds closely enough when the medium is air to make good estimates. If we move from location 1 to location 2, the variation of Equation 22.1 that we can use to compare the intensities, I_1 and I_2, at the two different places is given by the following equation.

$$\frac{I_1}{I_2} = \frac{d_2^{\,2}}{d_1^{\,2}} \qquad (22.2)$$

Suppose, for example, that you are 1.0 m from a radioactive source with an intensity of 30 units. If you move to a distance of 3.0 m, the exposure you would receive would not be 1/3rd but 1/9th as intense (the square of 1/3).

No Escape from the Natural Background Radiation Is Possible.   About 50 of the roughly 350 isotopes of all naturally occurring elements are radioactive. Some, like

Radon-222 is a naturally occurring radionuclide in the family of noble gases produced by the uranium-238 disintegration series. Chemically, it is as inert as the other noble gases, but radiologically it is a dangerous air pollutant. It is an alpha emitter and gamma emitter with a half-life of only 4 days. Produced in rocks and soil wherever uranium-238 is found, it migrates as a gas into the surrounding air. Basements not fully sealed act like fireplace chimneys to draw radon-222 into homes.

The first indication of how serious radon-222 pollution might be came when an engineer at a nuclear power plant in Pennsylvania set off radiation alarms just by his presence. The problem was traced to his home, where the radiation level in the basement was 2700 picocuries per liter of air. In the average home basement, the level is just 1 picocurie/L. (The picocurie is 10^{-12} curie.) The engineer had carried radon-222 and its radioactive decay products on his clothing into his workplace.

As a result of the incident, geologists went looking for unusual concentrations of uranium-238 nearby. They found that the Reading Prong, a formation of bedrock that cuts across Pennsylvania, New Jersey, New York State, and up into the New England states, is relatively rich in uranium-238.

Radon-222 enters the lungs with breathing, and some decays within the lungs. Several decay products after radon-222 in the radioactive disintegration series are not gases, like polonium-218 ($t_{1/2}$ 3 minutes, alpha emitter), lead-214 ($t_{1/2}$ 27 minutes, beta and gamma emitter), and polonium-214 ($t_{1/2}$ 1.6×10^{-4} seconds, alpha and gamma emitter).

Left in the lungs, alpha and gamma emitters can cause lung cancer. The health outcomes of uranium miners who worked for years in mines having high radon levels bear this out. Do exposures to the radon levels found in residences also contribute to the frequency of lung cancer? A Swedish study (1994) found that residential exposure to radon "is an important cause" of lung cancer in the general population. A Canadian study (1994), however, found no link between radon levels and lung cancer. Critics of both studies cite the extraordinary difficulty of obtaining reliable information concerning not only the historic radon levels in residences, but also the total hours of exposure of lung cancer victims to the radon. Reasonable assumptions, therefore, become embedded in any study, and these can be questioned. In the meanwhile, what do those of us in the general population do when experts disagree? We act prudently according to our own assumptions, biases, and "seats of the pants" estimations of risks. We do or do not smoke, for example, and we build mechanisms into new houses that reduce the interior radon levels. In the United States, the recommended upper limit on radon-222 concentration in home air is 4 picocuries/L (148 Bq/m^3). A conservative estimate puts the percentage of homes in the United States with levels above this at 1%. In Canada, the recommended level is 20 picocuries/L.

---

potassium-40 and carbon-14, are inevitably in the food we eat and the water we drink. Others, like radon, are in the air we breathe. Our bodies are exposed to radiation from cosmic ray showers and with every X ray we take. Radioactive materials are in the soils and rocks on which we walk and which we use to make building materials. On the average, the top 15 cm of soil on our planet has 1 g of radium per square mile. Shielding materials and distance, therefore, can never completely eliminate our exposure to the radiation, called **background radiation,** from all of these sources. Radon, a *chemically* inert but radioactive gas and a product of the uranium-238 disintegration series, makes the largest contribution to the background (see Special Topic 22.1).

Background radiation varies considerably from place to place, and only estimates (which vary widely with the estimator) are possible. Table 22.4 gives estimated average radiation dose equivalents from various sources for the U.S. population. The unit of *dose equivalent* used in the table, the millirem (mrem), will be discussed shortly. For comparisons, a dose of 500 rem (500,000 mrem) received by a large population would cause half to die in 30 days. In relationship to 500 rem, the intensity of background radiation is very small. At higher altitudes, the background radiation is greater because of cosmic rays, which have had less opportunity to be absorbed and destroyed by the thinner atmosphere at higher altitudes.

■ People became aware of the radon problem only in the early 1980s.

22.3 UNITS TO DESCRIBE AND MEASURE RADIATION

Units have been devised to describe the activity of a radioactive sample, radiation exposure, and radiation dose.

A number of units exist to describe radiation measurements. Each was invented to serve in answering one particular question about a given radiation or its source. If you note carefully what these questions are, the units will be easier to learn. The SI units that follow have been adopted in the United States by the National Council on Radiation Protection and Measurements (NCRP Report 82, 1985).

■ Most scientists, including those in health areas, now use the SI units, although older units still appear in the technical literature.

The *Becquerel* Describes How Active a Sample Is.   The **becquerel, Bq,** is the SI unit of activity and is used to answer the question, How *active* is a sample of a radionuclide? In terms of the number of disintegrations per second or dps,

$$1 \text{ Bq} = 1 \text{ dps}$$

The *curie*, the older unit of activity, is named after Marie Sklodowska Curie (1867–1934), a Polish scientist, who discovered radium. One **curie, Ci,** is the number of radioactive disintegrations that occur per second in a 1.0-g sample of radium, namely, 3.7×10^{10} dps:

$$1 \text{ Ci} = 3.7 \times 10^{10} \text{ dps}$$

■ Marie Curie is one of two scientists to win two Nobel prizes in a field of science, a share of the physics prize in 1903 and the chemistry prize in 1922.

Because 1 dps equals 1 Bq, 1 Ci equals 3.7×10^{10} Bq. The curie represents an intensely active rate, so fractions of the curie, such as the millicurie (mCi, 10^{-3} Ci), the microcurie (μCi, 10^{-6} Ci), and the picocurie (pCi, 10^{-12} Ci), are often used. Mole for mole, when the half-life is short, the radionuclide has a high activity.

Exposure to X Rays or Gamma Rays Is Described in Terms of the Quantity of Ions They Produce in Dry Air.   The older unit of X-ray or gamma ray exposure, the **roentgen,** is based on a specific total number of charges (2.1×10^9 units) created in 1 cm³ of dry air. The roentgen serves to answer the question, How *intense* is an exposure to X-ray or gamma-ray radiation? If members of a large population are exposed to 650 roentgens, half will die in 1 to 4 weeks. (The rest will have radiation sickness.)

The SI has no special unit for exposure to X rays or gamma rays, only a symbol, **X,** standing for the *ratio* of the total charge (of one sign) on the ions produced per 1 kilogram of dry air by the radiation.

The *Gray* Describes the *Absorbed Dose*.   The SI unit of *absorbed dose* is called the *gray*, named after a British radiologist, Harold Gray. It is used to answer the question, How much *energy* is *absorbed* by a unit mass of tissue or other materials? The **gray, Gy,** corresponds to the absorption of 1 joule (J) of energy per kilogram of tissue (4.184 J = 1 cal).

$$1 \text{ Gy} = 1 \text{ J/kg}$$

The older unit of absorbed dose is the **rad,** which is 1/100th of a gray.

$$1 \text{ rad} = 10^{-2} \text{ Gy}$$

■ *Rad* comes from *r*adiation *a*bsorbed *d*ose.

The roentgen and the rad are close enough in magnitude to be nearly equivalent from a health standpoint. Thus, one roentgen of the gamma radiation from a cobalt-60 source is the equivalent of 0.96 rad in muscle tissue and 0.92 rad in compact bone.

■ Cobalt-60 was once widely used for its gamma radiation in cancer therapy.

About 6 Gy or 600 rad of gamma radiation would be lethal to most people despite the fact that this corresponds to a very small quantity of energy. We must remember that it is not the quantity of energy that matters so much as the formation of unstable radicals and ions caused by this energy. A 6-Gy dose delivered to water breaks up only one molecule in every 36 million, but the radicals thus produced begin a cascade of harmful reactions in a cell.

The *Sievert*, the Unit of *Dose Equivalent*, Adjusts Absorbed Doses for Different Effects in Different Tissues.   A dose of 1 Gy of gamma radiation is not biologically the same as 1 Gy of beta radiation or of neutrons. The gray, therefore, does not serve as a good basis of comparison when working with *biological* effects. The **sievert (Sv)** is the SI unit that satisfies the need for a way to express dose equivalent that is additive for different kinds of radiation and different target tissues. If we let D stand for absorbed dose in grays, Q for *quality factor* (meaning relative effectiveness for causing harm in tissue), and N for any other modifying factors, then the dose equivalent, H, in sieverts is defined as follows.

■ The higher the value of Q, the more dangerous is the radiation.

$$H = DQN$$

In other words, we multiply the absorbed dose from some radiation, D, by a factor (Q) that takes into account biologically significant properties of the radiation, and by any other factor, N, bearing on the net effect. The value of the quality factor, Q, for alpha radiation is 20, but is only 1.0 for beta and gamma radiation. The much larger size of the alpha particle accounts for its extra danger. Its size and charge enable it to strike molecules with considerable (although undesirable) efficiency.

The older unit of dose equivalent is called the **rem**. One rem of any given radiation is the dose that has, in a human being, the effect of one roentgen. The sievert is 100 times larger than the rem but is still a quantity small in terms of energy and yet significant in terms of danger. Even millirem quantities of radiation should be avoided, and when this isn't possible the workers must wear monitoring devices that allow the day-to-day exposures to be determined.

■ *Rem* comes from *r*oentgen *e*quivalent for *m*an.

The *Electron-Volt* Describes the Energy of X Rays or Gamma Rays.   Those who work with X rays or gamma radiation use an old unit of energy, the **electron-volt, eV**, to describe the energies of such radiation.

■ In physics, the electron-volt is defined as the energy an electron receives when it is accelerated by a voltage of 1 volt.

$$1 \text{ eV} = 1.602 \times 10^{-19} \text{ J}$$

The 10^{-19} J term tells us that the electron-volt is an extremely small amount of energy. Multiples of the electron-volt are therefore very common, such as the kiloelectron-volt (1 keV = 10^3 eV), the megaelectron-volt (1 MeV = 10^6 eV), and even the gigaelectron-volt (1 GeV = 10^9 eV). X rays used for diagnosis are typically 100 keV or less. Beta radiation of 70 keV or more can penetrate the skin, but alpha particles (which are much larger than beta particles) need energies of more than 7 MeV to do this. Alpha radiation from radium-226 has an energy of 5 MeV. The cosmic radiation that enters our outer atmosphere has energies ranging from 200 MeV to 200 GeV.

■ Linear accelerators produce radiation for cancer treatment in the range 6 to 12 Mev.

Film Dosimeters, Scintillation Counters, and Geiger Counters Detect and Measure Radiation.   A *dosimeter* is a device for measuring exposure. The film badge is a common type. It contains photographic film, which becomes fogged by radiation, the measured degree of fogging being related to the exposure.

Ionizing radiation also affects substances called *phosphors*, salts with traces of rare earth metal ions that scintillate when struck by radiation. The scintillations—brief, sparklike flashes of light—can be translated into doses of radiation. Devices based on this technology are called *scintillation counters*.

■ Inside the screen of a color TV tube there is a coating that includes various phosphors; these glow with different colors when struck by the focused electron beam in the tube.

Geiger counters are devices used to measure beta and gamma radiation carrying enough energy to penetrate the thin window of a Geiger–Müller tube. Once in the tube, ionizing radiation activates an electrical circuit and a short pulse of electricity is sent to a counting device. The number of counts per second caused by a radioactive source is a measure of the source's activity, at least in beta and gamma radiation.

22.4   SYNTHETIC RADIONUCLIDES

Most radionuclides used in medicine are made by bombarding other atoms with high-energy particles.

Radioactive decay is nature's way of causing transmutations, but they can also be caused artificially by bombarding atoms with high-energy particles. Several hundred isotopes that do not occur naturally have been made this way. Several have been used successfully in medicine, both in diagnosis and in treatment.

Various bombarding particles, like alpha particles, neutrons, and protons, are used to make new isotopes. The first artificial transmutation, observed by Rutherford, was the alpha particle bombardment of nitrogen-14 to make oxygen-17 and gaseous protons. Alpha particles plowed right through the electron clouds of nitrogen-14 atoms and buried themselves in their nuclei. The new nuclei were those of fluorine-18, and they evidently had too much energy to exist for long. To rid themselves of excess energy, each ejected a proton, leaving behind an atom of oxygen-17. The equation is

$$\underset{\substack{\text{Alpha} \\ \text{particle}}}{^{4}_{2}\text{He}} + \underset{\substack{\text{Nitrogen} \\ \text{nucleus}}}{^{14}_{7}\text{N}} \longrightarrow \underset{\substack{\text{Fluorine} \\ \text{nucleus}}}{^{18}_{9}\text{F}} \longrightarrow \underset{\substack{\text{Oxygen} \\ \text{nucleus}}}{^{17}_{8}\text{O}} + \underset{\text{Proton}}{^{1}_{1}p}$$

Oxygen-17 is a rare but nonradioactive isotope of oxygen. Usually, transmutations caused by bombardments produce *radioactive* isotopes of other elements.

Certain isotopes of uranium in atomic reactors eject neutrons having sufficient energy to serve as bombarding particles. Because neutrons are electrically neutral, they aren't repelled either by the electrons surrounding an atom or by the nucleus itself. One important application of neutron bombardment is the synthesis of molybdenum-99 from molybdenum-98. Using $^{1}_{0}n$ as the symbol for the neutron, the equation is

$$^{98}_{42}\text{Mo} + ^{1}_{0}n \longrightarrow ^{99}_{42}\text{Mo} + \gamma$$

The decay of the product, molybdenum-99, provides an important application in medicine, as we'll see in Section 22.6.

22.5   RADIATION TECHNOLOGY IN THE FOOD INDUSTRY

Food irradiated with controlled doses of X rays, gamma rays, or electron beams is less likely to spoil.

When food products are passed through a beam of radiation, the effects depend on the energy of the beam. A low-dose beam—up to 100 kilorads—renders reproductively dead any insects that remain after harvest, and it inhibits the sprouting of potatoes

FIGURE 22.5
After 15 days of storage at 4 °C (38 °F), the unirradiated strawberries on the left became heavily covered by mold. Those on the right, however, had been protected by 200 kilorad of radiation.

and onions during storage. Low-energy beams also inactivate trichinae (*Trichinella spiralis,* a nematode worm) in pork, the parasite that causes trichinosis.

Medium-dose beams of radiation—100 to 1000 kilorads—significantly reduce the population of *Salmonella* bacteria in poultry, fish, and other meats, and also extend the shelf-lives of strawberries and other fruits that mold quickly (Figure 22.5).

High-dose beams—1000 to 10,000 kilorads—sterilize poultry, fish, and other meats and kill microorganisms and insects on seasonings and spices.

The irradiation of wheat and potatoes to control insects has been permitted in the United States more than 20 years, but no commercial operator is doing so. Some herbs and spices are now marketed after irradiation to reduce insects, bacteria, molds, and yeasts. The U.S. Food and Drug Administration (FDA) has approved the use of low-dose radiation to control trichinae in pork and to inhibit the spoilage of fruit and vegetables. None of these operations is as yet widespread, however, mostly because they are expensive.

Besides cost, customer confidence is another barrier to further use of irradiation. Any technology with the word *nuclear* or *radiation* in it makes people nervous. The legitimate question is, Does radiation produce substances—*radiolytic products*—that would not be present once foods have been otherwise processed, cooked, and digested? If so, Are they harmful at the levels at which they are present?

Food Irradiation Does Not Appear to Produce Any Unique Radiolytic Products.
Ionizing radiation causes chemical reactions in foods or insects or microorganisms in them. Because water is usually the most abundant substance present in a given food, the *primary* products of radiation result from the splitting of water into radicals and ions. Some recombine to form water and others combine to give hydrogen peroxide. Otherwise, the primary products of irradiation react with food molecules to give *secondary* products. Generally, however, these are the same as those produced by cooking, baking, or the subsequent digestion of food.

Specialists in food irradiation claim that their research has not yet turned up any radiolytic products uniquely caused by radiation. The *high-level* irradiation of certain foods gives traces of benzene, for example, and repeated environmental exposure to benzene is known to increase a person's likelihood of contracting leukemia. However,

■ The occurrence of trichinosis in the pork supply of the United States is small, and thorough cooking destroys it. (Pork is never served "rare.")

■ In the body, hydrogen peroxide, H_2O_2, quickly breaks down to water and oxygen.

■ A half a pound of the botulinum toxin would be enough to kill all of the people on earth.

small traces of benzene are naturally present in certain nonirradiated foods, like boiled eggs, at much higher levels than those produced by irradiation.

Still another safety issue concerns botulism, a particularly dangerous form of food poisoning caused by an odorless chemical, the botulinum toxin, produced by *Clostridium botulinum*. This bacterium is more resistant to radiation than the microorganisms causing food spoilage. Organisms that spoil foods give them dreadful odors which warn people, but the botulinum toxin cannot be detected by odor. High-dose food irradiation could thus prevent the kind of food spoilage associated with odors without necessarily destroying all botulinum bacteria.

 ## 22.6 RADIATION TECHNOLOGY IN MEDICINE

Both in diagnosis and in cancer treatment, ionizing radiation is used when its benefits are judged to outweigh its harm.

For medical uses, ionizing radiation is supplied as X rays, electron beams generated by special equipment, or by emissions from selected radionuclides. In *diagnostic* work, radiation is used to locate a cancer or tumor or to assess the function of some organ, like the thyroid gland. In *therapeutic* work, radiation is used to kill cancer cells or inhibit their growth. Generally, therapeutic doses are of much higher energy than those used for diagnosis. In this section we look briefly at some of the radioactive chemicals used in medicine, and we use Special Topics to discuss beam radiation technologies.

■ *Radiology:* the science of radioactive substances and of X rays
Radiologist: a specialist in radiology who also usually has a medical degree
Radiobiology: the science of the effects of radiation on living things

Both Chemical and Radiological Properties Are Important in the Selection of Radionuclides in Medicine. The *chemical* properties of radionuclides are identical to those of the stable isotopes of the same element. When radionuclides are selected for use in medicine, therefore, their chemical properties must be considered. A selected radionuclide must be *chemically* compatible with the living system when it is chosen for its ionizing radiation. Moreover, a radionuclide's *chemical* properties are what guide it naturally to the desired tissue. Iodine, for example, has only one stable isotope, iodine-127, which is used chemically by the thyroid gland to make the hormone thyroxin. This chemical property of iodine is held by all iodine isotopes and so will guide iodine-131, a beta emitter, to the thyroid gland. There, its radiation can be used to assess thyroid function or to treat thyroid cancer.

Minimizing Harm and Maximizing Benefit Guide the Selection of Radionuclides in Medicine. Exposing anyone to any radiation entails some risks, because prolonged exposure can produce cancer. No such exposure is permitted unless the expected benefit from finding and treating a dangerous disease is thought to be greater than the risk. To minimize the risks, the radiologist uses radionuclides that, as much as possible, have the following properties.

1. The radionuclide should have a half-life that is short. (Then it will decay *during* the diagnosis when the decay gives some benefit, and as little as possible of the radionuclide will decay later, when the radiation is of no benefit.)

2. The product of the decay of the radionuclide should have little if any radiation of its own and should be quickly eliminated. (Either the decay product should be a stable isotope or it should have a very long half-life.)

3. The half-life of the radionuclide must be long enough for it to be prepared and administered to the patient.

X rays are generated by bombarding a metal surface with high-energy electrons. These can penetrate the metal atom far enough to knock out one of its low-level electrons. This creates a "hole" in the electron configuration, and electrons at higher levels begin to drop down. In other words, the creation of this "hole" leads to electrons changing their energy levels. The difference between two of the lower levels corresponds to the energy of an X ray, which is emitted.

The refinement of X-ray techniques and the development of powerful computers made possible the generation of a diagnostic technology called computerized tomography, or CT for short. A. M. Cormack (United States) and G. N. Hounsfield (England) shared the 1979 Nobel prize in medicine for their work in the development of this technology. The instrument includes a large array of carefully positioned and focused X-ray generators. In the procedure called a CT scan, this array is rotated as a unit around the body or the head of the patient (Figure 1). Extremely brief pulses of X rays are sent in from all angles across one cross section of the patient (Figure 2).

The changes in the X rays that are caused by internal organs or by tumors are sent to a computer, which then processes the data and delivers a picture of the cross section. It's like getting a picture of the inside of a cherry pit without cutting open the cherry. The CT scan is widely used for locating tumors and cancers.

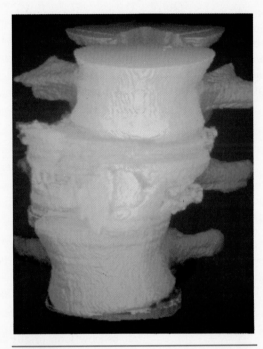

FIGURE 2
Three-dimensional image, based on 63 CT scans, of a section of the vertebrae of a man injured in a motorcycle accident. Both the compression and the twisting are clearly evident.

FIGURE 1
Instrument for the CT scan

4. If the radionuclide is to be used for diagnosis, it should decay by penetrating radiation entirely, which means gamma radiation. (Nonpenetrating radiation, like alpha and beta radiation, adds to the risk by causing internal damage without contributing to the detection of the radiation externally. For uses in *therapy,* as in cancer therapy, nonpenetrating radiation is preferred because a radionuclide well placed in cancerous tissue *should* cause damage to such tissue.)

5. The diseased tissue should concentrate the radionuclide, giving a "hot spot" where the diseased area exists, or it should do the opposite and reject the radionuclide, making the diseased area a "cold spot" insofar as external detectors are concerned.

SPECIAL TOPIC 22.3
POSITRON EMISSION TOMOGRAPHY—THE PET SCAN

A number of synthetic radionuclides emit positrons. These are particles that have the same small mass as an electron but carry one unit of *positive* charge. (They're sometimes called a positive electron.) A positron forms by the conversion of a proton into a neutron, as follows:

$$\underset{\substack{\text{Proton} \\ \text{(in atom's} \\ \text{nucleus)}}}{{}^{1}_{1}p} \longrightarrow \underset{\substack{\text{Neutron} \\ \text{(stays in} \\ \text{nucleus)}}}{{}^{1}_{0}n} + \underset{\substack{\text{Positron} \\ \text{(is emitted)}}}{{}^{0}_{1}e}$$

Positrons, when emitted, last for only a brief interval before they collide with an electron. The two particles annihilate each other, and in so doing their masses convert entirely into energy in the form of two photons of gamma radiation (511 keV). The gamma radiation formed in this way is called annihilation radiation.

$$\underset{\text{Electron}}{{}^{0}_{-1}e} + \underset{\text{Positron}}{{}^{0}_{1}e} \longrightarrow \underset{\substack{\text{Gamma} \\ \text{radiation}}}{2\,{}^{0}_{0}\gamma}$$

The two photons leave the collision site in almost exactly opposite directions.

To make a medically useful technology out of this property of the positron, a positron-emitting nuclide must be part of a molecule with a chemistry that will carry it into the particular tissue to be studied. Once the molecule gets in, the tissue now has a gamma radiator *on the inside*. Thus, instead of X rays being sent through the body, as in a CT scan (Special Topic 22.2), the radiation originates right within the site being monitored. The overall procedure is called positron emission tomography, or PET for short.

Three positron emitters are often used: oxygen-15, nitrogen-13, and carbon-11. Glucose, for example, can be made in which one carbon atom is carbon-11 instead of the usual carbon-12. Glucose can cross the blood–brain barrier and get inside brain cells. If some part of the brain is experiencing abnormal glucose metabolism, this will be reflected in the way in which positron-emitting glucose is handled, and gamma radiation detectors on the outside can pick up the differences (Figure 1). The use of the PET scan has led to the discovery that glucose metabolism in the brain is altered in schizophrenia and manic depression. PET scanning technology is able to identify extremely small regions in the brain that are in early stages of breakdown and that CT and MRI techniques miss. A brain involved with Alzheimer's disease gives a PET scan different from that of a normal brain (Figure 2). The PET scan is particularly useful in detecting abnormal brain function in infants, as in early stages of epilepsy.

Let us now look briefly at a few of the more important radionuclides employed in medicine.

Technetium-99m Is Widely Used in Medicine.
Technetium-99m is a radionuclide produced by the decay of molybdenum-99. You already know that gamma radiation often accompanies the emission of other types of radiation, but with molybdenum-99 the gamma radiation comes after a pause. Molybdenum-99 first emits a beta particle:

■ The synthesis of molybdenum-99 was described at the end of Section 22.4.

$$\underset{99}{^{42}}\text{Mo} \longrightarrow \underset{43}{^{99m}}\text{Tc} + {}^{0}_{-1}e$$

The other decay product is a metastable form of technetium-99, hence the *m* in 99*m*. *Metastable* functionally means "poised to move toward greater stability." Technetium-99m decays by emitting gamma radiation with an energy of 143 keV.

$$\underset{43}{^{99m}}\text{Tc} \longrightarrow \underset{43}{^{99}}\text{Tc} + \gamma$$

■ Technetium-99 decays to a stable isotope of ruthenium, $^{99}_{44}$Ru.

Technetium-99m fits almost ideally the criteria for a diagnostic radionuclide. Its half-life is short, 6.02 hours. Its decay product, technetium-99, has a very long half-life, 212,000 years, so it has too little activity to be of much concern. The half-life of technetium-99m, although short, is still long enough to allow time to prepare it and

PET technology is being used to study a number of neuropsychiatric disorders, including Parkinson's disease. When a drug labeled with carbon-11 is used, its molecules go to the parts of the brain that have nerve endings that release dopamine. A PET scan then discloses the dopamine-releasing potential of the patient. This potential becomes impaired in Parkinson's disease.

By labeling blood platelets with a positron emitter, scientists can follow the development of atherosclerosis in even the tiniest of human blood vessels. Blood flow in the heart can be monitored without having to insert a catheter.

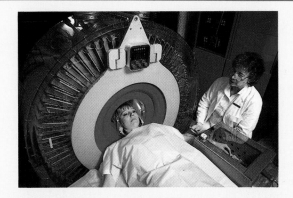

FIGURE 1
The patient is shown undergoing a brain scan performed by means of PET technology.

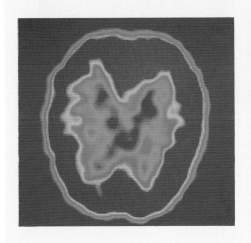 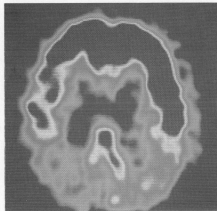

FIGURE 2
Left: Normal brain scan using PET technology. Right: PET scan of the brain of a patient with Alzheimer's disease.

administer it. It decays entirely by gamma radiation, which means that the maximum amount of the radiation gets to a detector to signal where the radionuclide is in the body. Finally, a variety of chemically combined forms of technetium-99m have been developed that permit either hot spots or cold spots to form.

One form of technetium-99m is the pertechnetate ion, TcO_4^-. It behaves in the body very much like a halide ion, so it tends to go where chloride ions, for example, go. TcO_4^- is eliminated by the kidneys, so it is used to assess kidney function. Other organs whose functions are also studied by technetium-99m are the liver, spleen, lungs, heart, brain, bones, and thyroid gland. Technetium-99m technology has received competition from the CT scan (see Special Topic 22.2), the PET scan (see Special Topic 22.3), and MRI imaging (see Special Topic 22.4).

Iodine-131 and Iodine-123. The thyroid gland, the only user of iodine in the body, takes iodide ion and, as we said, makes the hormone thyroxin. When either an underactive or an overactive thyroid is suspected, one technique is to let the patient drink a glass of flavored water that contains some radioactive iodine as I^-. The ability of the thyroid gland to concentrate iodide ion is so strong that if a small whole-body dose of iodine-131 is given, nearly 1000 times this much dose concentrates in a healthy thyroid. By placing radiation detection equipment near the thyroid gland, the

The CT scan subjects a patient to large numbers of short bursts of X rays. The PET scan exposes the patient to gamma radiation that is generated on the inside. Thus, both technologies carry the usual risks that attend ionizing radiation, and they are used when the potential benefits from correct diagnoses far outweigh such risks. MRI imaging technology operates without these dangers. At least, none has been discovered thus far. The principal developer of the first hardware for MRI imaging was Raymond Damadian.

MRI stands for magnetic resonance imaging. Atomic nuclei that have odd numbers of protons and neutrons behave as though they were tiny magnets, hence the *magnetic* part of MRI. The nucleus of ordinary hydrogen (which has no neutrons and one proton) constitutes the most abundant nuclear magnet in living systems, because hydrogen atoms are parts of water molecules and all biochemicals. Nuclear magnets spin about an axis much as the earth spins about its axis.

When molecules with spinning nuclear magnets are in a strong magnetic field and are simultaneously bathed with properly tuned radiofrequency radiation (which is of very low energy), the nuclear magnets flip their spins. (This is the *resonance* part of MRI.) As they resonate, they emit electromagnetic energy that is biologically harmless, being of very low energy, and this energy is picked up by detectors. The data are fed into computers, which produce an image much like that of a CT or a PET scan, but without having subjected the patient to ultrahigh energy electromagnetic radiation such as X rays and gamma rays. The MRI images are actually sophisticated plots of the distributions of the spinning nuclear magnets, of hydrogen atoms, for example.

MRI scanning has proved to be especially useful for studying soft tissue, the sort of tissue least well studied by X rays. Different soft tissues have different population densities of water molecules or of fat molecules (which are loaded with hydrogen atoms). And tumors and cancerous tissue have their own water inventories. Calcium ions do not produce any signals to confuse MRI scanning, so bone, which is rich in calcium, is transparent to MRI.

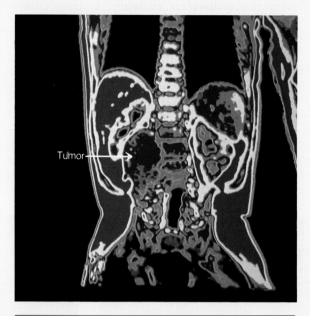

FIGURE I

An MRI scan of a 7-month-old child revealed a malignant tumor pushing its way into the spinal canal. (The tumor was treated in time.)

MRI technology has developed during the 1980s into a method superior to the CT scan for diagnosing tumors at the rear and base of the skull and equal to the CT scan for finding other brain tumors. MRI is now the preferred technology for assessing problems in joints (particularly the knees) and in the spinal cord, such as ruptured (herniated, "slipped") disks. Patients with heart pacemakers or embedded shrapnel or surgical clips present problems to the use of MRI because of the powerful magnets used.

CT scans are still better than MRI for the early detection of hemorrhages in the brain, so CT is the method of choice for finding them in potential stroke victims. CT scans are also still preferred for detecting tumors in the kidneys, lungs, pancreas, and spleen (Figure 1).

radiologist can tell how well this gland takes up iodide ion from circulation. For *diagnostic* purposes, iodine-123 is popular because it has a short half-life (13.3 hours) and it emits only gamma radiation (159 keV). Iodine-131 also has a short half-life (8 days), but it emits both beta particles (600 keV) and gamma radiation (mostly 360 keV). Certain types of thyroid cancer, therefore, have been treated with iodine-131.

The pertechnetate ion has about the same radius and charge density as the iodide ion. Cells of the thyroid gland, therefore, do not distinguish between these two ions at the point where they move inside thyroid cells. Thus, when the only intent is to get any kind of detectable radiation inside the thyroid to detect a tumor or cancer, TcO_4^- is preferred over iodine-123 because of the shorter half-life of technetium-99*m*.

Linear Accelerators Make High-Energy X Rays.   Special equipment called a *linear accelerator* is used to generate X rays with energies in the range 6 to 12 MeV for therapeutic uses. The equipment is easily mounted so that it can rotate about a central point, which makes it easy to position the patient and plan the treatment.

Other Medically Useful Radionuclides Are Available for Special Purposes.   Indium-111 ($t_{1/2}$ = 2.8 days; gamma emitter: 173 and 247 keV) has been found to be a good labeler of blood platelets.

Gallium-67 ($t_{1/2}$ = 78 hours; gamma emitter: 1.003 MeV) is used in the diagnosis of Hodgkin's disease, lymphomas, and bronchogenic carcinoma.

Phosphorus-32 ($t_{1/2}$ = 14.3 days; beta emitter: 1.71 MeV) in the form of the phosphate ion has been used to treat a form of leukemia, a cancer of the bone that affects white cells in the blood. Because the phosphate ion is part of the hydroxyapatite mineral in bone, this ion is a bone seeker.

22.7 ATOMIC ENERGY AND RADIONUCLIDES

Nuclear power plants generate atomic wastes that must be kept from human contact for several centuries.

All operating nuclear power plants in the United States today use fission to generate heat. **Fission** is the disintegration of a large atomic nucleus into small fragments following neutron capture, and it releases additional neutrons, radioactive isotopes, and enormous yields of heat. Unless the reactor is continuously cooled by a flowing coolant, usually water, the whole system will melt very quickly. The heat converts the coolant water into steam under high pressure, which drives electric turbines (Figure 22.6). In this way some of the energy of fission becomes electrical energy. What heat escapes is vented to the atmosphere or into a cooler river or lake.

Uranium-235 is the only naturally occurring radionuclide that spontaneously undergoes fission when it captures a slow-moving and relatively low-energy neutron. After neutron-capture, the nucleus spontaneously splits apart, which can occur in a number of ways giving different products. The equation for just one mode of splitting is

$$^{235}_{92}U + {}^1_0n \xrightarrow[\text{capture}]{\text{neutron}} {}^{236}_{92}U \xrightarrow{\text{fission}} {}^{139}_{56}Ba + {}^{94}_{36}Kr + 3{}^1_0n + \gamma + \text{heat}$$

More neutrons are released than are captured. One fission event, therefore, produces enough neutrons to initiate more than one new fission. In other words, a **nuclear chain reaction** can take place (Figure 22.7). It would almost instantly envelop all uranium-235 atoms present if the system were not designed to hold the ratio of neutrons produced to neutrons captured at 1:1, called the *critical* ratio. If the concentration of uranium-235 is too high and the ratio goes above 1:1, the system becomes supercritical and an atomic explosion will occur. To prevent this, the concentration of uranium-235 is kept so low in a nuclear reactor that if the system goes

■ Neutrons moving too rapidly are poorly captured.

■ Plutonium-239, made in reactors, also can fission, and it is used as a nuclear fuel and in making atomic bombs.

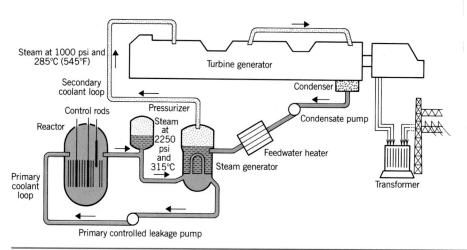

FIGURE 22.6

Pressurized water nuclear power station. Water in the primary coolant loop circulates around the reactor core and carrries away the heat of fission. This water is sealed under pressure, so its temperature can rise well above its normal boiling point. The water delivers its heat to the water in the secondary loop, which then turns to high-pressure steam and drives the electrical turbines. Maximum efficiency is reached by having as large a temperature drop as possible between the inlet steam temperature in this loop and the outlet water temperature. (Drawing from WASH 1261, U. S. Atomic Energy Commission, 1973.)

■ Steam explosions on April 26, 1986, tore apart one reactor of the Chernobyl energy facility in Russia, releasing a roughly estimated one million curies of pollutants.

supercritical, no atomic explosion is possible. Even so, the heat generated in a supercritical episode of a nuclear power plant might not be removable rapidly enough to prevent a meltdown of the whole system. A meltdown could generate a *steam* explosion that would rupture the reactor and expose the surrounding atmosphere to escaping radioactive pollutants. It is obviously essential that fission be under control at all times.

Nuclear reactors are designed to enable the control of nuclear fission at the critical ratio so that heat is generated without running any risk of an atomic explosion and is removed rapidly enough to prevent a meltdown. The nuclear "fuel" is uranium-235 at a concentration in nonfissionable solids of 3 to 5%. The mixture is formed into ceramic pellets and distributed among a large number of tubes with radii small enough to let a significant fraction of neutrons escape without causing fission. Nestled among the tubes are control rods that can be moved in or out. The control rods are made of materials that capture neutrons harmlessly. Circulating around and among the tubes is the coolant, which is heated by the atomic fission taking place.

■ The circulating water is also a good neutron moderator.

In *pressurized water reactors*, the coolant water is kept under pressure so that it can be heated above its normal boiling point. This water is in a closed, primary loop engineered so as to heat the water in a secondary loop (see Figure 22.6). The steam generated in the secondary loop drives the turbines.

Fission Products Are Potential Pollutants.   The new isotopes produced by fission are radioactive, and their decay leads to radioactive pollutants, which must be contained by the reactor and, later, by safe storage. They include strontium-90 (a bone-seeking element in the calcium family), iodine-131 (a thyroid gland seeker), and cesium-137 (a Group IA radionuclide that goes wherever Na^+ or K^+ can go). The U.S. government has set limits to the release of each radioactive isotope into the air and into the cooling water of nuclear power plants. Plants that operate in compliance with these standards

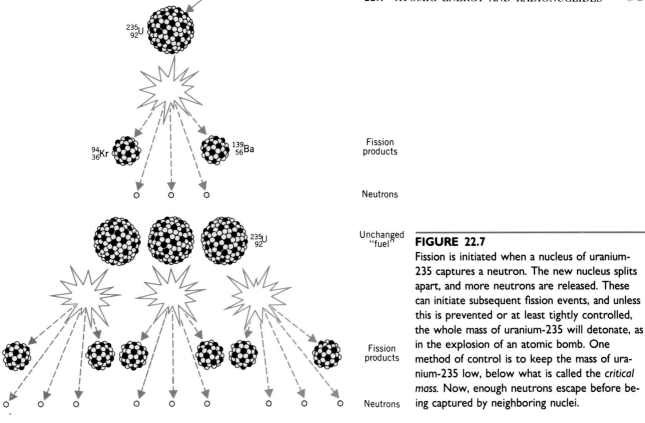

Fission products

Neutrons

Unchanged "fuel"

Fission products

Neutrons

FIGURE 22.7
Fission is initiated when a nucleus of uranium-235 captures a neutron. The new nucleus splits apart, and more neutrons are released. These can initiate subsequent fission events, and unless this is prevented or at least tightly controlled, the whole mass of uranium-235 will detonate, as in the explosion of an atomic bomb. One method of control is to keep the mass of uranium-235 low, below what is called the *critical mass*. Now, enough neutrons escape before being captured by neighboring nuclei.

expose people living near them to an extra dose of no more than 5% of the dose they normally receive from background radiation.

Wastes from Nuclear Power Plants Must Be Kept Apart from Human Contact for a Thousand Years.   One of the most vexing problems of nuclear energy has been the permanent storage of long-lived radioactive wastes. Most are now in temporary storage at nuclear power plants. Because several waste radionuclides have very long half-lives, the wastes must be sequestered from all human contact for at least a thousand years, and scientists are seeking deep geologic formations out of all contact with mining operations or underground water supplies into which these wastes can be placed. The sites must be marked and continuously maintained so that archeologists several centuries hence will not unknowingly venture into them.

Wastes Can Enter Food Chains.   Radioactive wastes, like iodine-131, enter food chains wherever they deposit. Milk from cows that have grazed on pastures contaminated by iodine-131 fallout carries this isotope into the human diet.
Alternatives to nuclear power pose several problems. Power plants that use petroleum or coal are huge emitters of carbon dioxide, a greenhouse gas. When coal and petroleum also contain sulfur impurities, the power plants emit sulfur dioxide, a contributor to acid rain. Moreover, what few people know is that deaths from the mining of coal (so far) greatly exceed deaths related to nuclear power, including the mining of nuclear fuels. Many people call for drastic reductions in the use of all currently exploited forms of energy—by the conservation of energy, by the use of wind energy and solar energy, and by other technologies.

SUMMARY

Atomic Radiation Radionuclides in nature emit alpha radiation (helium nuclei), beta radiation (electrons), and gamma radiation (high-energy X ray-like radiation). Radioactive decay is accompanied by transmutation. The penetrating abilities of the different kinds of radiation are a function of the sizes of the particles, their charges, and the energies with which they are emitted. Gamma radiation, which has no associated mass or charge, is the most penetrating.

Each decay can be described by a nuclear equation in which mass numbers and atomic numbers on either side of the arrow must balance. To describe how stable a radionuclide is we use its half-life, and the shorter this is, the more radioactive is the radionuclide.

The decay of one radionuclide doesn't always produce a stable nuclide. Uranium-238 is at the head of a radioactive disintegration series that involves several intermediate radionuclides until a stable isotope of lead forms.

Ionizing Radiation—Dangers and Precautions When radiation travels in matter, it creates unstable ions and radicals that have chemical properties dangerous to health. Intermittent exposure can lead to cancer, tumors, mutations, or birth defects. Intense exposures cause radiation sickness and death, but such exposures, focused on cancer tissue, are used in cancer therapy. Distance, fast film, and dense shielding material are the best strategies to guard against the hazards of ionizing radiation. According to the inverse-square law, the intensity of radiation falls off with the square of the distance from the source. Complete protection, however, is not possible because of the natural background radiation that now includes traces of radioactive pollutants.

Units of Radiation Measurement To describe activity we use the SI unit of the becquerel (Bq) or the curie (Ci). The common unit to describe the intensity of exposure to X rays or gamma radiation is the roentgen. The gray (Gy) or the rad is used to describe the absorbed dose, how much energy has been absorbed by a unit mass of tissue (or other matter). To put the damage that different kinds of radiation can cause when they have the same values of rads (or grays) on a comparable and additive basis, we use the sievert or the rem. Finally, to describe the energy possessed by a radiation, we use the electron-volt. Diagnostic X rays are on the order of 100 keV. The radiation used in cancer treatment is in the low megaelectron-volt range. To measure radiation, there are devices such as film badges, scintillation counters, and ionization counters (Geiger–Müller tubes).

Synthetic Radionuclides A number of synthetic radionuclides have been made by bombarding various isotopes with alpha radiation, neutrons, or accelerated protons. The target nucleus first accepts the mass, charge, and energy of the bombarding particle, and then it ejects something else to give the new nuclide.

Radiation Technology in the Food Industry X rays, gamma rays, and accelerated electrons can be used to inhibit or to kill insects, molds, and bacteria on seeds, potatoes, fruit, and meats. High doses fully sterilize the products. Low doses inhibit bacterial or mold growth. Radiolytic products, those formed by the chemical reactions induced by the radiation, are generally the same as the substances found naturally in food either before or after cooking and digestion. The resistance of the botulinum bacillus to radiation is one problem with the technology.

Radionuclides in Medicine For diagnostic uses, the radionuclide should have a short half-life but not so short that it decays before any benefit can be obtained. It should decay by gamma radiation only, and it should be chemically compatible with the organ or tissue so that either a hot spot or a cold spot appears. Its decay products should be as stable as possible and capable of being eliminated from the body.

Technetium-99*m* is almost ideal for diagnostic work, particularly for assessing the ability of an organ or tissue to function. Radionuclides of iodine (iodine-123 and iodine-131) are used in diagnosing or treating thyroid conditions. Gallium-67, indium-111, and phosphorus-32 are a few of the many other radionuclides used in diagnosis. Cobalt-60 has powerful gamma radiation and is used in cancer treatment. Linear accelerators also provide high-energy (6 to 12 MeV) radiation for cancer therapy.

Atomic Energy and Radioactive Pollutants The reactors of most nuclear power plants use uranium-235 as a fuel. When its atoms capture neutrons, they fission into smaller, usually radioactive atoms as neutrons are released that can cause additional fissions. The concentration of uranium-235 atoms is kept too small (3 to 5%) to make an atomic bomb type of explosion possible at a nuclear reactor. Circulating water keeps the reactor cool enough not to melt. The heat generated by fission converts water into high-pressure steam that drives electrical turbines.

Radioactive by-products of fission, such as iodine-131, strontium-80, and cesium-137, cannot be allowed to enter the food supply and must be contained. Some fission products have such long half-lives that radioactive wastes must be kept from human contact for a thousand years.

REVIEW EXERCISES

The answers to Review Exercises whose numbers are in color are found in Appendix V. The answers to the other Review Exercises are found in the Study Guide that accompanies this text. The more challenging questions are marked with asterisks.

Radioactivity and the Kinds of Radiation

22.1 Distinguish between *radioactive decay* and *transmutation*.

22.2 What are the names and symbols used in nuclear equations for the three types of naturally occurring atomic radiation?

22.3 The emission of which naturally occurring atomic radiation would *not* be accompanied by transmutation if it were the sole radiation from a radionuclide?

22.4 In balancing *chemical* equations we seek both a material balance and a charge balance.
(a) How do we check for material balance in a nuclear equation?
(b) Why do we ignore the question of charge balance by not showing electrical charges on reactants or products?
(c) In what other sense, however, do we consider charge balance?

22.5 The energy of an alpha particle is often higher than that of beta or gamma rays. Why is it, then, the least penetrating of the kinds of radiation?

22.6 The loss of an alpha particle changes the radionuclide's mass number by how many units? Its atomic number by how many units?

22.7 Why does the loss of a beta particle not change the radionuclide's mass number but *increases* its atomic number?

22.8 How many neutrons are in the nucleus of $^{232}_{92}U$?

22.9 What is the *one* most appropriate term to apply to a *radionuclide*: element, isotope, compound, mixture, atom, ion, or molecule? Explain

22.10 If electrons do not exist in the nucleus, how can one originate in a nucleus in beta decay?

Nuclear Equations

*22.11** Write the symbols of the missing particles in the following nuclear equations.
(a) $^{211}_{82}Pb \longrightarrow {}^{0}_{-1}e + \underline{\hspace{1cm}}$ (b) $^{220}_{86}Rn \longrightarrow {}^{4}_{2}He + \underline{\hspace{1cm}}$
(c) $^{140}_{56}Ba \longrightarrow {}^{0}_{-1}e + \underline{\hspace{1cm}}$

*22.12** Complete the following nuclear equations by writing the symbols of the missing particles.
(a) $^{149}_{62}Sm \longrightarrow {}^{4}_{2}He + \underline{\hspace{1cm}}$ (b) $^{245}_{96}Cm \longrightarrow {}^{4}_{2}He + $
(c) $^{22}_{9}F \longrightarrow {}^{0}_{-1}e + \underline{\hspace{1cm}}$

*22.13** Write a balanced nuclear equation for each of the following changes.
(a) alpha emission from neodymium-144
(b) beta emission from potassium-40
(c) beta emission from samarium-149
(d) alpha and gamma emission from californium-251

*22.14** Give the nuclear equation for each of the following radioactive decays.
(a) beta emission from rhenium-187
(b) alpha and gamma emission from plutonium-242
(c) beta emission from iodine-131
(d) alpha emission from americium-243

Half-Lives

22.15 Lead-214 is in the uranium-238 disintegration series. Its half-life is 26.8 minutes. Explain in your own words what being in this series means and what *half-life* means.

22.16 Which would be more dangerous to be near, a radionuclide that has a short half-life and decays by alpha emission only or a radionuclide that has the same half-life but decays by beta and gamma emission? Explain.

*22.17** A 12.00-ng sample of technetium-99*m* will still have how many nanograms of this radionuclide left after three half-life periods?

*22.18** If a patient is given 12.00 ng of iodine-123 (half-life 13.3 hours), how many nanograms of this radionuclide remain after 12 half-life periods (about a week)?

Dangers of Ionizing Radiation

22.19 We have ions in every fluid of the body. Why, then, is ionizing radiation dangerous?

22.20 What is a chemical *radical,* and why is it chemically reactive?

22.21 Ionizing radiation is a teratogenic agent. What does this mean?

22.22 Cesium-137 is a carcinogen. What does this mean?

22.23 What two properties of ionizing radiation are exploited in strategies for providing radiation protection?

22.24 Ionizing radiation is said to have no exposure threshold. What does this mean?

22.25 How is it that the same agent, radiation from a radionuclide, can be used both to cause cancer and to cure it?

22.26 The inverse-square law tells us that if we double the distance from a radioactive source, we will reduce the radiation intensity received by a factor of what number?

22.27 What general property of radiation is behind the inverse-square law?

22.28   List as many factors as you can that contribute to background radiation.

22.29   Why does a trip in a high-altitude jet plane increase a person's exposure to background radiation?

22.30   A radiologist discovered that at a distance of 1.80 m from a radioactive source, the intensity of radiation was 140 millirad. How far should the radiologist move away to reduce the exposure to 2 millirad?

22.31   Using a Geiger–Müller counter, a radiologist found that in a 20-minute period the dose from a radioactive source would measure 80 millirad at a distance of 10.0 m. How much dose would be received in the same time by moving to a distance of 2.00 m?

Units of Radiation Measurement

22.32   What SI unit is used to describe the activity of a radioactive sample?

22.33   A hospital purchased a sample of a radionuclide rated at 1.5 mCi. What does this rating mean?

22.34   What is the SI unit and the older unit that describe the intensity of an exposure to X rays?

22.35   What are the name and symbol of the SI unit used in describing how much energy a given mass of tissue receives from exposure to radiation? What are the name and symbol of the older, common unit?

22.36   How does the SI define the dose equivalent? What is meant by the *quality factor*? Which has the higher quality factor, alpha radiation or beta radiation? Explain.

22.37   Approximately how many rads would kill half of a large population within 4 weeks, assuming that each individual received this much? From a health protection standpoint, how do the roentgen and the rad compare in their potential danger?

22.38   We cannot add a 1-rad dose of gamma radiation to a 1-rad dose of neutron radiation given to the same organ and say that the total biologically effective dose to the organ is 2 rads. Why not?

22.39   How is the problem implied by the previous review exercise resolved?

22.40   In units of millirem, what is the average natural background radiation received by the U.S. population, exclusive of medical sources, radioactive pollutants, and fallout?

22.41   What is the name of the energy unit used to describe the energy associated with an X ray or a gamma ray?

22.42   In the unit traditionally used (Review Exercise 22.41), how much energy is associated with diagnostic X rays?

22.43   Why should diagnostic radiation ideally be of much lower energy than radiation used in therapy, in cancer treatment, for example?

22.44   In general terms, how does a film badge dosimeter work?

22.45   In your own words, how does a Geiger–Müller counter work? Why doesn't it detect alpha radiation?

Synthetic Radionuclides

*22.46   When manganese-55 is bombarded by protons, the neutron is one product. What else is produced? Write a nuclear equation.

*22.47   To make indium-111 for diagnostic work, silver-109 is bombarded with alpha particles. What forms if the nucleus of silver-109 captures one alpha particle? Write the nuclear equation for this capture.

*22.48   The compound nucleus that forms when silver-109 captures an alpha particle (previous review exercise) decays directly to indium-111 *plus* two other identical particles. What are they? Write the nuclear equation for this decay.

*22.49   To make gallium-67 for diagnostic work, zinc-66 is bombarded with accelerated protons. When a nucleus of zinc-66 captures a proton, the nucleus of what isotope forms? Write the nuclear equation.

*22.50   When fluorine-19 is bombarded by alpha particles, both a neutron and a nucleus of sodium-22 form. Write the nuclear equation, including the first nucleus that forms as an intermediate.

*22.51   When boron-10 is bombarded by alpha particles, nitrogen-13 forms and a neutron is released. Write the equation for this reaction.

*22.52   When nitrogen-14 is bombarded with deuterons, ^2_1H, oxygen-15, and a neutron form. Write the equation for this reaction.

Radiation Technology and the Food Industry

*22.53   The unit commonly used to describe how much radiation has been given to a food product is the *kilorad*, krad. Using conversion factors supplied in Section 22.3, calculate how much energy is in 1.00 krad in units of J/kg.

22.54   What is meant by the term *radiolytic product*?

22.55   What is meant by the term *unique* radiolytic product?

22.56   Are any radiolytic products known cancer-causing substances? Are they also found in nonirradiated foods? Give an example.

22.57   What nonradiation processes produce the same compounds as food irradiation?

22.58   What potential hazard in food is the most difficult to remove by radiation?

Medical Applications of Radiation

22.59   Why is it desirable to use a radionuclide of short half-life in diagnostic work, when we know that even small samples of such isotopes can be very active?

22.60 We know that gamma radiation is the most penetrating of all natural radiation. Why, then, is a diagnostic radionuclide that emits only gamma radiation preferable to one that gives, say, only alpha radiation?

22.61 Why is iodine-123 better for diagnostic work than iodine-131?

22.62 In what chemical form should phosphorus-32 be used to facilitate its seeking bone tissue? Explain.

Atomic Energy and Radionuclides

22.63 What is *fission?*

22.64 In general terms, how does fission differ from radioactive decay?

22.65 What fundamental aspect of fission makes it possible for it to proceed as a chain reaction?

22.66 Which naturally occurring radionuclide is able to undergo fission?

22.67 What concentration is the fissionable isotope in atomic reactors? What concentration is it in an atomic bomb?

22.68 The heat generated by fission in a power plant reactor is carried away in what way? And for what purpose?

22.69 Name three isotopes made in nuclear power plants that are particularly hazardous to health, and explain in what specific ways they endanger various parts of the body.

22.70 What fact about certain atomic wastes necessitates very long waste storage times?

Radon in the Environment (Special Topic 22.1)

22.71 How is radon-222 produced in the environment?

22.72 What kinds of radiation does it emit?

22.73 Besides its own radiation, what other factors make radon-222 in the lungs particularly hazardous?

22.74 What upper limit on the level of radon-222 in the home is recommended?

X Rays and CT Scans (Special Topic 22.2)

22.75 In general terms, how are X rays prepared?

22.76 How does the CT scanner differ from an ordinary X-ray machine?

Positron Emission Tomography—The PET Scan (Special Topic 22.3)

22.77 Compare the positron and electron in terms of mass and charge.

22.78 Describe how a positron forms in a positron-emitting radionuclide.

22.79 What property makes the lifetime of a positron extremely short?

22.80 When a radiologist uses the PET scan, what radiation is converted into an X ray-like picture?

22.81 In general terms, how can a positron-emitting radionuclide be gotten inside a tissue, like the brain?

Magnetic Resonance Imaging—The MRI Scan (Special Topic 22.4)

22.82 Why is the MRI less harmful a technique than the CT or PET scan?

22.83 How does MRI complement the use of X rays?

22.84 Why is bone transparent to the MRI?

Additional Exercises

22.85 Why is radiation sickness called an *acute* effect of radiation exposure? What is a particularly common *latent* effect of radiation exposure?

22.86 The hydroxyl radical is an electrically neutral species, yet it is dangerous in tissue. Explain.

I

MATHEMATICAL CONCEPTS

I.1 EXPONENTIALS

When numbers are either very large or very small, it's often more convenient to express them in what is called *exponential notation*. Several examples are given in Table I.1, which shows how multiples of 10, such as 10,000, and submultiples of 10, such as 0.0001, can be expressed in exponential notation.

Exponential notation expresses a number as the product of two numbers. The first is a digit between 1 and 10, and this is multiplied by the second, 10 raised to some whole-number power or exponent. For example, 55,000,000 is expressed in exponential notation as 5.5×10^7, in which 7 (meaning $+7$) is the exponent.

Exponents can be negative numbers, too. For example, 3.4×10^{-3} is a number with a negative exponent. Now let's learn how to move back and forth between the exponential and the expanded expressions.

Positive Exponents   A positive exponent is a number that tells how many times the number standing before the 10 has to be multiplied by 10 to give the same number in its expanded form. For example:

$$5.5 \times 10^7 = \underline{10 \times 10 \times 10 \times 10 \times 10 \times 10 \times 10}$$

$$10^7$$
$$6 \times 10^3 = 6 \times 10 \times 10 \times 10 = 6000$$
$$8.576 \times 10^2 = 8.576 \times 10 \times 10 = 857.6$$

The number before the 10 doesn't always have to be a number between 1 and 10. This is just a convention, which we sometimes might find useful to ignore; however, we can't ignore the rules of arithmetic in conversions from one form to another. For example,

$$0.00045 \times 10^5 = 0.00045 \times 10 \times 10 \times 10 \times 10 \times 10$$
$$= 45$$
$$87.5 \times 10^3 = 87.5 \times 10 \times 10 \times 10$$
$$= 87,500$$

TABLE I.1

NUMBER	EXPONENTIAL FORM
1	1×10^0
10	1×10^1
100	1×10^2
1000	1×10^3
10,000	1×10^4
100,000	1×10^5
1,000,000	1×10^6
0.1	1×10^{-1}
0.01	1×10^{-2}
0.001	1×10^{-3}
0.0001	1×10^{-4}
0.00001	1×10^{-5}
0.000001	1×10^{-6}

In most problem-solving situations you find that the problem is given the other way around. You encounter a large number, and (after you've learned the usefulness of exponential notation) you know that the next few minutes of your life could actually be easier if you could quickly restate the number in exponential form. This is very easy to do. Just count the number of places that you have to move the decimal point to the *left* to put it right after the first digit of the given number. For example, you might have to work with a number such as 1500 (as in 1500 mL). You'd have to move the decimal point three places leftward from where it is (or is understood to be) to put it immediately after the first digit.

$$\underset{3\ \ \ 2\ \ \ 1}{1\ 5\ 0\ 0}.$$

Each of these leftward moves counts as one unit for the exponent. Three leftward moves means an exponent of $+3$. Therefore 1500 can be rewritten as 1.500×10^3. A really large number and one that you'll certainly meet somewhere during the course is 602,000,000,000,000,000,000,000. It's called Avogadro's number, and you can see that manipulating it would be awkward. (How in the world does one even pronounce it?) In exponential notation, it's written simply as 6.02×10^{23}. Check it out. Do you have to move the decimal point leftward 23 places? (And now you could pronounce it: "six point oh two times ten to the twenty-third," but saying "Avogadro's number" is easier.) Do these exercises for practice.

■ **EXERCISE I.1** Expand each of these exponential numbers.

(a) 5.050×10^6 (b) 0.0000344×10^8
(c) 324.4×10^3

■ **EXERCISE I.2** Write each of these numbers in exponential form.

(a) 422,045   (b) 24,000,000,000,000,000,000
(c) 24.32

Answers to Exercises I.1 and I.2

I.1 (a) 5,050,000   (b) 3,440   (c) 324,400
I.2 (a) 4.22045×10^5 (b) 2.4×10^{19}
    (c) 2.432×10^1

Negative Exponents A negative exponent is a number that tells how many times the number standing before the 10 has to be *divided* by 10 to give the number in its expanded form. For example,

$$1 \times 10^{-4} = 1 \div 10 \div 10 \div 10 \div 10$$
$$= \frac{1}{10 \times 10 \times 10 \times 10} = \frac{1}{10,000} = \frac{1}{10^4}$$
$$= 0.0001$$
$$6 \times 10^{-3} = 6 \div 10 \div 10 \div 10$$
$$= \frac{6}{10 \times 10 \times 10} = \frac{6}{1000}$$
$$= 0.006$$
$$8.576 \times 10^{-2} = \frac{8.576}{10 \times 10} = \frac{8.576}{100} = 0.08576$$

You'll see negative exponents often when you study aqueous solutions that have very low concentrations.

Sometimes, you'll want to convert a very small number into its equivalent in exponential notation. This is also easy. This time we count *rightward* the number of times that you have to move the decimal point, one digit at a time, to place the decimal immediately to the right of the first nonzero digit in the number. For example, if the number is 0.00045, you have to move the decimal four times to the right to place it after the 4.

$$0 \cdot 0 \; 0 \; 0 \; 4 \; 5$$
$$1 \quad 2 \quad 3 \quad 4$$

Therefore we can write $0.00045 = 4.5 \times 10^{-4}$. Similarly, we can write $0.0012 = 1.2 \times 10^{-3}$. And $0.0000000000000011 = 1.1 \times 10^{-16}$. Now try these exercises.

■ **EXERCISE I.3** Write each number in expanded form.

(a) 4.3×10^{-2} (b) 5.6×10^{-10} (c) 0.00034×10^{-2}
(d) 4523.34×10^{-4}

■ **EXERCISE I.4** Write the following numbers in exponential forms.

(a) 0.115   (b) 0.00005000041
(c) 0.000000000000345

Answers to Exercises I.3 and I.4

I.3 (a) 0.043   (b) 0.00000000056   (c) 0.0000034
    (d) 0.452334
I.4 (b) 1.15×10^{-1} (b) 5.000041×10^{-5}
    (c) 3.45×10^{-13}

Now that we can write numbers in exponential notation, let's learn how to manipulate them.

How to Add and Subtract Numbers in Exponential Notation We'll not spend too much time on this, because it doesn't come up very often. The only rule is that when you add or subtract exponentials, all of the numbers must have the same exponents of 10. If they don't, we have to reexpress them to achieve this condition. Suppose you want to add 4.41×10^3 and 2.20×10^3. The result is simply 6.61×10^3.

$$(4.41 \times 10^3) + (2.20 \times 10^3) = [4.41 + 2.20] \times 10^3$$
$$= 6.61 \times 10^3$$

However, we can't add 4.41×10^3 to 2.20×10^4 without first making the exponents equal. We can do this in either of the following ways. In one way, we notice that $2.20 \times 10^4 = 2.20 \times 10 \times 10^3 = 22.0 \times 10^3$, so we have:

$$(4.41 \times 10^3) + (22.0 \times 10^3) = 26.41 \times 10^3$$
$$= 2.641 \times 10^4$$

Alternatively, we could notice that $4.41 \times 10^3 = 4.41 \times 10^{-1} \times 10^4 = 0.441 \times 10^4$, so we can do the addition as follows:

$$(0.441 \times 10^4) + (2.20 \times 10^4) = 2.641 \times 10^4$$

The result is the same both ways. The extension of this to subtraction should be obvious.[1]

[1]Whenever these operations are with pure numbers and not with physical quantities obtained by measurements, we are not concerned about the numbers of significant figures in the answers.

How To Multiply Numbers Written in Exponential Form
Use the following two steps to multiply numbers that are expressed in exponential forms.

1. Multiply the numbers in front of the 10s.
2. Add the exponents of the 10s algebraically.

EXAMPLE I.1

$$(2 \times 10^4) \times (3 \times 10^5) = 2 \times 3 \times 10^{(4+5)}$$
$$= 6 \times 10^9$$

Usually, the problem you want to solve involves very large or very small numbers that aren't yet stated in exponential form. When this happens, convert the given numbers into their exponential forms first, and then carry out the operation. The next example illustrates this and shows how exponentials can make a calculation easier.

EXAMPLE I.2

$$6576 \times 2000 = (6.576 \times 10^3) \times (2 \times 10^3)$$
$$= 13.152 \times 10^6$$
$$= 1.3152 \times 10^7$$

■ **EXERCISE I.5** Calculate the following products after you have converted large or small numbers to exponential forms.

(a) $6,000,000 \times 0.0000002$
(b) $10^6 \times 10^{-7} \times 10^8 \times 10^{-7}$
(c) $0.003 \times 0.002 \times 0.000001$
(d) $1,500 \times 3,000,000,000,000$

Answers

(a) 1.2 (b) 1 (c) 6×10^{-12} (d) 4.5×10^{15}

How to Divide Numbers Written in Exponential Form
To divide numbers expressed in exponential form, use the following two steps.

1. Divide the numbers that stand in front of the 10's.
2. *Subtract* the exponents of the 10's algebraically.

EXAMPLE I.3

$$(8 \times 10^4) \div (2 \times 10^3) = (8 \div 2) \times 10^{(4-3)}$$
$$= 4 \times 10^1$$

EXAMPLE I.4

$$(8 \times 10^4) \div (2 \times 10^{-3}) = (8 \div 2) \times 10^{[4-(-3)]}$$
$$= 4 \times 10^7$$

■ **EXERCISE I.6** Do the following calculations using exponential forms of the numbers.

(a) $6,000,000 \div 1500$ (b) $7460 \div 0.0005$

(c) $\dfrac{3\,000\,000 \times 6\,000\,000\,000}{20\,000}$ (d) $\dfrac{0.016 \times 0.0006}{0.000008}$

(e) $\dfrac{400 \times 500 \times 0.002 \times 500}{2\,500\,000}$

Answers

(a) 4×10^3 (b) 1.492×10^7 (c) 9×10^{11}
(d) 1.2   (e) 8×10^{-2}

The Pocket Calculator and Exponentials   The foregoing was meant to refresh your memory about exponentials, because you almost certainly studied them in any course in algebra or some earlier course. You probably own a good pocket calculator, at least one that can take numbers in exponential form. Go ahead and use it, but be sure that you understand exponentials well, first. Otherwise, there are many pitfalls.

Most pocket calculators have a key marked *EE* or *EXP*. This key is used to enter exponentials. Here is where an ability to *read* exponentials comes in handy. For example, the number 2.1×10^4 reads "two point one times ten to the fourth." The *EE* or *EXP* key on most calculators stands for "times ten to the" Therefore, to enter 2.1×10^4, punch the following keys.

$$\boxed{2}\ \boxed{\cdot}\ \boxed{1}\ \boxed{EE}\ \boxed{4}$$

Try this on your own calculator, and be sure to see that the display is correct. If it isn't, you may have a calculator that works differently than most, so recheck your operations of entering and then check the owner's manual.

To enter an exponential with a negative exponent, you have to use one more key, the $\boxed{+/-}$. This switches a positive number to its negative, and you *must* use it rather than the $\boxed{-}$ key in this situation. Thus, the number 2.1×10^{-5} enters as follows.

$$\boxed{2}\ \boxed{\cdot}\ \boxed{1}\ \boxed{EE}\ \boxed{+/-}\ \boxed{5}$$

Try it and check the display. To see what happens if you use the $\boxed{-}$ key instead of the $\boxed{+/-}$, clear the display and enter this number only using the $\boxed{-}$ key instead of the $\boxed{+/-}$ key.

I.2 CROSS-MULTIPLICATION

In this section we will learn how to solve for x in such expressions as

$$\frac{12}{x} = \frac{16}{25} \quad \text{or} \quad \frac{32.0}{11.2} = \frac{6.15x}{13.1}$$

The operation is called *cross-multiplication,* and its object is to get x to stand alone, all by itself, on one side of the = sign, and above any real or understood divisor line.

> To cross-multiply, move a number or a symbol both across the = sign and across a divisor line, and then multiply.

EXAMPLE I.5

PROBLEM: Solve for x in

$$\frac{25}{x} = 5$$

SOLUTION: Notice first the divisor lines; one of them is understood.

Divisor line $\longrightarrow \dfrac{25}{x} = 5 \longleftarrow$ The divisor line here is understood because

$$5 = \frac{5}{1}$$

Remember, we want x to stand alone, on top of a divisor line (even if this line is understood). To make this happen, we carry out cross-multiplication as indicated.

$$\frac{25}{x} \diagup 5$$

Notice that the arrows show moves that carry the quantities not only across the = sign but also across their respective divisor lines. *It is essential that both crossing-overs be done.* Now we have x standing alone above its (understood) divisor line.

$$\frac{25}{5} = x$$

Now we can do the arithmetic. $x = 5$.

EXAMPLE I.6

PROBLEM: Solve for x in:

$$\frac{25 \times 60}{12} = \frac{625}{x}$$

SOLUTION: To get x to stand alone, we carry out the following cross-multiplication.

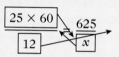

The result is

$$\begin{aligned} x &= \frac{625 \times 12}{25 \times 60} \\ &= 5 \end{aligned}$$

■ **EXERCISE I.7** Solve for x in the following.

(a) $\dfrac{12}{x} = \dfrac{16}{25}$ (b) $\dfrac{32.0}{11.2} = \dfrac{6.15x}{13.1}$

Answers

(a) $x = 18.75$ (b) $x = 6.085946574$

How to Do Chain Calculations with the Pocket Calculator Sometimes the steps in solving a problem lead to something like the following:

$$x = \frac{24.2 \times 30.2 \times 55.6}{2.30 \times 18.2 \times 4.44}$$

Many people will first calculate the value of the numerator and write it down. Then they'll compute the denominator and write it down. Finally, they'll divide the two results to get the final answer. There's no need to do this much work. All you have to do is enter the first number you see in the numerator, 24.2 in our example. Then use the ×̄ key for any number in the numerator and the ÷̄ key for any number in the denominator. *Each number in the denominator is entered with the ÷̄ key.* Any of the following sequences work. Try them.

$$24.2 \times 30.2 \times 55.6 \div 2.30 \div 18.2 \div 4.44 = 218.632...$$

Alternatively,

$$24.2 \div 2.30 \times 30.2 \div 18.2 \times 55.6 \div 4.44 = 218.632...$$

APPENDIX

II

SOME RULES FOR NAMING INORGANIC COMPOUNDS

Only rules considered sufficient to meet most of the needs of the users of this text are in this Appendix. The latest edition of the *Handbook of Chemistry and Physics*, published annually by the CRC Press, Boca Raton, Florida, under the general editorship of R. C. Weast, has a section on all of the rules. Virtually all college libraries have this reference.

I. **Binary Compounds**—those made from only two elements
   A. **One element is a metal and the other is a nonmetal**
      1. The name of the metal is written first in the name of the compound, and its symbol is placed first in the formula.
      2. The name ending of the nonmetal is changed to -*ide*. Thus the names of the simple ions of groups VIA and VIIA of the periodic table are

Group VIIA	Group VIA
Fluoride	Oxide
Chloride	Sulfide
Bromide	Selenide
Iodide	Telluride

   3. If the metal and the nonmetal each have just one oxidation number, a binary compound of the two is named simply by writing the name of the metal and then that of the nonmetal with its ending modified by -*ide*, as shown above. Greek prefixes such as mono-, di-, tri-, etc., are not necessary. Examples are

Some Compounds between Elements of Groups IA and VIIA		Some Compounds between Elements of Groups IA and VIA	
NaF	Sodium fluoride	Na_2O	Sodium oxide[a]
KCl	Potassium chloride	K_2S	Potassium sulfide
LiBr	Lithium bromide	Li_2O	Lithium oxide
RbI	Rubidium iodide	Cs_2S	Cesium sulfide
CsCl	Cesium chloride	Rb_2O	Rubidium oxide

[a]Not disodium oxide.

Some Compounds between Elements of Groups IIA and VIIA		Some Compounds between Elements of Groups IIA and VIA	
$BeCl_2$	Beryllium chloride	BeO	Beryllium oxide
$MgBr_2$	Magnesium bromide	MgS	Magnesium sulfide
CaF_2	Calcium fluoride	CaO	Calcium oxide
SrI_2	Strontium iodide	SrS	Strontium sulfide
$BaCl_2$	Barium chloride	BaO	Barium oxide

Some Compounds Between Elements of Groups IIIA and VIIA		Some Compounds Between Elements of Groups IIIA and VIA	
$AlCl_3$	Aluminum chloride	Al_2O_3	Aluminum oxide
AlF_3	Aluminum fluoride	Al_2S_3	Aluminum sulfide

4. If the metal has more than one oxidation number, but the nonmetal has just one, the formal name of the compound includes a roman numeral in parentheses following the name of the metal. This numeral stands for the oxidation number of the metal. Greek prefixes such as mono-, di, etc., are not needed. The following compounds of iron and copper illustrate this rule.

Compounds of Iron in Oxidation States of 2+ or 3+

	Formal Name	Common Name
$FeCl_2$	Iron(II) chloride[a]	Ferrous chloride
FeO	Iron(II) oxide	Ferrous oxide
Fe_2O_3	Iron(III) oxide	Ferric oxide
$FeCl_3$	Iron(III) chloride	Ferric chloride

[a]Pronounced "iron two chloride."

Compounds of Copper in Oxidation States of 1+ and 2+

	Formal Name	Common Name
Cu_2O	Copper(I) oxide	Cuprous oxide
$CuBr$	Copper(I) bromide	Cuprous bromide
$CuCl_2$	Copper(II) chloride	Cupric chloride
CuS	Copper(II) sulfide	Cupric sulfide

B. Molecular compounds of two elements.
Greek prefixes such as mono-, di-, etc., are used, sometimes for *both* elements.

Oxides of Nonmetals

Oxides of Carbon	Oxides of Sulfur
CO Carbon monoxide	SO_2 Sulfur dioxide
CO_2 Carbon dioxide	SO_3 Sulfur trioxide

Oxides of Nitrogen (older names in parentheses)

N_2O Dinitrogen monoxide (nitrous oxide)
NO Nitrogen oxide (nitric oxide)
N_2O_3 Dinitrogen trioxide
NO_2 Nitrogen dioxide
N_2O_4 Dinitrogen tetroxide
N_2O_5 Dinitrogen pentoxide

Oxides of Some Halogens

F_2O Difluorine monoxide
Cl_2O Dichlorine monoxide
Cl_2O_7 Dichlorine heptoxide

Some Halides of Carbon

CCl_4 Carbon tetrachloride
CBr_4 Carbon tetrabromide

Some Exceptions

H_2O Water NH_3 Ammonia CH_4 Methane

Li_2SO_4	Lithium sulfate	$MgSO_4$	Magnesium sulfate
Na_2SO_4	Sodium sulfate	$CaSO_4$	Calcium sulfate
K_2SO_4	Potassium sulfate	$Al_2(SO_4)_3$	Aluminum sulfate
$LiHCO_3$	Lithium hydrogen carbonate (lithium bicarbonate)[a]		
$NaHCO_3$	Sodium hydrogen carbonate (sodium bicarbonate)		
Li_2CO_3	Lithium carbonate	$KMnO_4$	Potassium permanganate
Na_2CO_3	Sodium carbonate	Na_2CrO_4	Sodium chromate
$CaCO_3$	Calcium carbonate	$Mg(NO_3)_2$	Magnesium nitrate
$Al_2(CO_3)_2$	Aluminum carbonate	$NaNO_2$	Sodium nitrite
$NaHSO_4$	Sodium hydrogen sulfate (sodium bisulfate)	Na_3PO_4	Sodium phosphate
NaH_2PO_4	Sodium dihydrogen phosphate	$Ca_3(PO_4)_2$	Calcium phosphate
K_2HPO_4	Potassium monohydrogen phosphate		
$MgHPO_4$	Magnesium monohydrogen phosphate		
$(NH_4)_2HPO_4$	Ammonium monohydrogen phosphate		

[a]*Bicarbonate* instead of *hydrogen carbonate* is used in this text because it is judged to be the more commonly used name for this ion, particularly among health scientists.

II. **Compounds of Three or More Elements**
   A. **A positive and a negative ion are combined.** The name of the positive ion is first followed by the name of the negative ion, just as with binary compounds between metals and nonmetals. Greek prefixes are not needed except where they occur in the name of an ion. (Older names are shown in parentheses.)

   B. **Molecular compounds of two or more elements.** Most are organic compounds, so their rules of nomenclature are given in the chapters on organic compounds.

III. **Important Inorganic Acids and Their Anions**

Formula	Name	Formula	Name
H_2CO_3	Carbonic acid	HCO_3^-	Hydrogen carbonate ion (bicarbonate ion)
		CO_3^{2-}	Carbonate ion
HNO_3	Nitric acid	NO_3^-	Nitrate ion
HNO_2	Nitrous acid	NO_2^-	Nitrite ion
H_2SO_4	Sulfuric acid	HSO_4^-	Hydrogen sulfate ion (bisulfate ion)
		SO_4^{2-}	Sulfate ion
H_2SO_3	Sulfurous acid	HSO_3^-	Hydrogen sulfite ion (bisulfite ion)
		SO_3^{2-}	Sulfite ion
H_3PO_4	Phosphoric acid (orthophosphoric acid)	$H_2PO_4^{2-}$	Dihydrogen phosphate ion
		HPO_4^{2-}	Monohydrogen phosphate ion
		PO_4^{3-}	Phosphate ion
$HClO_4$	Perchloric acid	ClO_4^-	Perchlorate ion
$HClO_3$	Chloric acid	ClO_3^-	Chlorate ion
$HClO_2$	Chlorous acid	ClO_2^-	Chlorite ion
$HClO$	Hypochlorous acid	ClO^-	Hypochlorite ion
HCl	Hydrochloric acid[a]	Cl^-	Chloride ion

[a]The name of the aqueous solution of gaseous HCl.

Some Generalizations about Names of Acids and Their Anions

1. The names of ions from acids whose names end in *-ic* all end in *-ate*.
2. When a nonmetal that forms an oxoacid whose name ends in *-ic* also forms an acid with one fewer oxygen atom, the name of the latter acid ends in *-ous*. (Compare nitric acid, HNO_3, and nitrous acid, HNO_2.)
3. When a nonmetal forms an oxoacid with one fewer oxygen atom than are in an *-ous* acid, then the prefix *hypo-* is used. (Compare chlor- ous acid, $HClO_2$, and hypochlorous acid, $HClO$.)
4. The binary hydrohalogen acids are called hydrogen halides when they occur as pure gases, but are called hydrohalic acids when they occur as aqueous solutions. Thus, hydrogen fluoride in water becomes hydrofluoric acid; hydrogen chloride in water becomes hydrochloric acid; and so on.

IUPAC NOMENCLATURE OF COMMON OXYGEN DERIVATIVES OF HYDROCARBONS

ALCOHOLS

Parent Compound   The parent compound is the longest continuous chain of carbons that includes the OH group. Replace the terminal -e of the name of the corresponding alkane by -ol. For the names of diols, triols, and so forth, retain the terminal -e but add *diol, -triol,* etc., according to the number of OH groups.

Numbering the Parent Chain   Number the parent chain from that end that gives the location of the OH group the lower number without regard to the numbers of any other attached hydrocarbon groups.

CH_3OH CH_3CH_2OH $CH_3CH_2CH_2OH$

Methanol   Ethanol   1-Propanol

$$\overset{OH}{\underset{|}{CH_3CHCH_3}} \qquad CH_3CH_2\overset{CH_3}{\underset{\underset{CH_3}{|}}{\overset{|}{C}}}CH_2OH$$

2-Propanol   2,2-Dimethyl-1-butanol

$$HOCH_2CH_2OH \qquad HOCH_2\overset{OH}{\underset{|}{CH}}CH_2OH$$

1,2-Ethanediol   1,2,3-Propanetriol

$$CH_3\overset{Cl}{\underset{|}{CH}}CH_2OH$$

2-Chloro-1-propanol   3-Methylcyclohexanol

ALDEHYDES

Parent Compound   The parent chain is the longest continuous chain of carbons *that includes the carbon atom of the aldehyde group.* Replace the terminal -e of the name of the corresponding alkane by *al.*

Numbering the Parent Chain   The carbon of the aldehyde group is given number 1. (This takes precedence over OH groups, double bonds, or alkyl groups.)

$$\overset{O}{\underset{||}{HCH}} \qquad \overset{O}{\underset{||}{CH_3CH}} \qquad \overset{H_3C}{\underset{|}{CH_3CH_2CH}}\overset{O}{\underset{||}{CH}} \qquad CH_3(CH_2)_4\overset{CH_3}{\underset{\underset{CH_3}{|}}{\overset{|}{C}}}CH_2\overset{O}{\underset{||}{CH}}$$

Methanal   Ethanal   2-Methylbutanal   3,3-Dimethyloctanal

KETONES

Parent Compound   The longest continuous chain of the parent must be selected so as to include the carbon atom of the keto group.

Numbering the Parent Chain   The parent chain is numbered from whichever end gives the number of the carbon of the keto group the lower number. This takes precedence over OH groups, alkene double bonds, and hydrocarbon groups. If numbering from each end gives the identical number to the carbon of the keto group, then the location numbers of other substituents are to be considered.

$$CH_3\overset{\overset{\displaystyle O}{\|}}{C}CH_3 \qquad CH_3\overset{\overset{\displaystyle O}{\|}}{C}CH_2CH_3 \qquad CH_3\overset{\overset{\displaystyle O}{\|}}{C}CH_2CH_2CH_3$$

2-Propanone   2-Butanone   2-Pentanone

$$CH_3CH_2CH_2\overset{\overset{\displaystyle O}{\|}}{C}CH_2\overset{\overset{\displaystyle CH_3}{|}}{\underset{\underset{\displaystyle C_6H_5}{|}}{C}}CH_3$$

2-Methyl-2-phenyl-4-heptanone

CARBOXYLIC ACIDS

Parent Compound   The parent chain is the longest continuous chain of carbons *that includes the carbon atom of the carboxyl group.* Replace the terminal *-e* of the name of the corresponding alkane by *-oic acid.*

Numbering the Parent Chain   The carbon of the carboxyl group is given number 1. (This takes precedence over OH groups, double bonds, or alkyl groups.)

HCO_2H CH_3CO_2H $CH_3CH_2CO_2H$

Methanoic acid   Ethanoic acid   Propanoic acid

$$CH_3\overset{\overset{\displaystyle CH_3}{|}}{C}HCH_2CO_2H$$

3-Methylbutanoic acid

CARBOXYLIC ACID DERIVATIVES

Parent Compound   First identify the parent carboxylic acid and name it. Then alter this name according to the rules for the various derivatives.

Carboxylic Acid Salts   Change *-ic acid* to *-ate* and precede this word by the name of the other ion (as a separate word).

$$CH_3CO_2^-Na^+ \qquad CH_3\overset{\overset{\displaystyle CH_3}{|}}{\underset{\underset{\displaystyle CH_3}{|}}{C}}CH_2CO_2^-K^+$$

Sodium ethanoate   Potassium 3,3-dimethylbutanoate

Esters   Change *-ic acid* to *-ate* and precede this word by the name of the hydrocarbon group (as a separate word).

$$CH_3CO_2CH_3 \qquad CH_3\overset{\overset{\displaystyle CH_3}{|}}{C}HCO_2CH_2CH_3$$

Methyl ethanoate   Ethyl 2-methylpropanoate

$$CH_3\overset{\overset{\displaystyle CH_3}{|}}{C}HCH_2CO_2CH_2CH_2CH_3$$

Propyl 3-methylbutanoate

Amides   Change the *-oic acid* part of the name of the parent acid to *-amide.* If a substituent is on the N atom instead of the carbon chain, locate it by the letter *N* instead of a number.

$$CH_3CONH_2 \quad CH_3\overset{\overset{\displaystyle CH_3}{|}}{C}HCH_2\overset{\overset{\displaystyle CH_3}{|}}{C}HCONH_2 \quad CH_3CH_2CONHCH_3$$

Ethanamide   2,4-Dimethylpentanamide   N-Methylpropanamide

Table IV.1 gives the recommended dietary allowances, the RDAs, of the Food and Nutrition Board, National Academy of Sciences/National Research Council (1989). In the judgment of this Board, these allowances are the intake levels of essential nutrients that are "adequate to meet the known nutritional needs of practically all healthy persons." A number of points and qualifications about the RDAs must be emphasized.

1. *The RDAs are not the same as the U.S. Recommended Daily Allowances (USRDAs).*

The USRDAs are set by the U.S. Food and Drug Administration, based on the RDAs, as standards for nutritional information on food labels.

2. *The RDAs are not the same as the Minimum Daily Requirements (MDRs) for any one individual.*

TABLE IV.1 Recommended Daily Dietary Allowances[a] of the Food and Nutrition Board, National Academy of Sciences/National Research Council, Revised 1989

Persons	Age (years)	Weight kg	Weight lb	Height cm	Height in.	Protein (g)	Fat-Soluble Vitamins A (μg)[b]	D (μg)[c]	E (mg)[d]	K (μg)
Infants	0.0–0.5	6	13	60	24	13	375	7.5	3	5
	0.5–1	9	20	71	28	14	375	10	4	10
Children	1–3	13	29	90	35	16	400	10	6	15
	4–6	20	44	112	44	24	500	10	7	20
	7–10	28	62	132	52	28	700	10	7	30
Males	11–14	45	99	157	62	45	1000	10	10	45
	15–18	66	145	176	69	59	1000	10	10	65
	19–24	72	160	177	70	58	1000	10	10	70
	25–50	79	174	176	70	63	1000	5	10	80
	51+	77	170	173	68	63	1000	5	10	80
Females	11–14	46	101	157	62	46	800	10	8	45
	15–18	55	120	163	64	44	800	10	8	55
	19–24	58	128	164	65	46	800	10	8	65
	25–50	63	138	163	64	50	800	5	8	65
	51+	65	143	160	63	50	800	5	8	65
Pregnant						60	800	10	10	65
Lactating	1–6 mo					65	1300	10	12	65
	7–12 mo					62	1200	10	11	65

[a]The allowances are intended to provide for individual variations among most normal persons as they live in the United States under usual environmental stresses. Diets should be based on a variety of common foods to provide other nutrients for which human requirements have been less well defined.
[b]Retinol equivalents. 1 retinol equivalent = 1 μg retinol or 6 μg β-carotene.
[c]As cholecalciferol. 10 μg cholecalciferol = 400 IU of vitamin D.

The MDRs are just that, minimums. They are set very close to the levels at which actual signs of deficiencies occur. The RDAs are two to six times the MDRs. Just as individuals differ greatly in height, weight, and appearance, they also differ greatly in specific biochemical needs. Therefore, an effort has been made to set the RDAs far enough above average requirements so that "practically all healthy people" will thrive. Most will receive more than they need; a few will not receive enough.

3. *The RDAs do not define therapeutic nutritional needs.*

People with chronic diseases such as prolonged infections or metabolic disorders, people who take certain medications on a continuing basis, and prematurely born infants all require special diets. Recently the importance of certain vitamins to pregnant women, particularly in the first few weeks of pregnancy, has been established by controlled studies.

The RDAs do cover people according to age, sex, and size, and they indicate special needs for pregnant and lactating women, but they do not include any other special needs. Therapeutic needs for water and salt increase during strenuous physical activity and prolonged exposure to high temperatures.

In some areas of the United States and in many parts of the world, intestinal parasites are common. These organisms rob the affected people of some of their food intake each day, and they also need special diets.

4. *The RDAs can (and ought to) be provided in the diet from a number of combinations and patterns of food.*

No single food contains all nutrients. People take dangerous risks with their health when they go on fad diets limited to one particular food, like brown rice, gelatin, yogurt, or liquid protein. The ancient wisdom of a varied diet that includes meat, fruit, vegetables, grains, nuts, pulses (e.g., beans), and dairy products may seem to be supported solely by cultural and esthetic factors. A varied diet, however, also assures us of getting any trace and needed nutrients that might not yet have been discovered.

TABLE IV.1   continued

Water-Soluble Vitamins							Minerals							
Ascorbic acid (mg)	Folate (µg)	Niacin[e] (mg)	Ribo-flavin (mg)	Thia-mine (mg)	Vita-min B_6 (mg)	Vita-min B_{12} (mg)	Calcium (mg)	Phos-phorus (mg)	Iodine (µg)	Iron (mg)	Mag-nesium (mg)	Sele-nium (µg)	Zinc (mg)
30	25	5	0.4	0.3	0.3	0.3	400	300	40	6	40	10	5
35	35	6	0.5	0.4	0.6	0.5	600	600	50	10	60	15	5
40	50	9	0.7	0.7	1.0	0.7	800	800	70	10	80	20	10
45	75	12	0.9	0.9	1.1	1.0	800	800	90	10	120	20	10
45	100	13	1.0	1.0	1.4	1.4	800	800	120	10	170	30	10
50	150	17	1.5	1.3	1.7	2.0	1200	1200	150	12	270	40	15
60	200	20	1.8	1.5	2.0	2.0	1200	1200	150	12	400	55	15
60	200	19	1.7	1.5	2.0	2.0	1200	1200	150	10	350	70	15
60	200	19	1.7	1.5	2.0	2.0	800	800	150	10	350	70	15
60	200	15	1.4	1.2	2.0	2.0	800	800	150	10	350	70	15
50	150	15	1.3	1.1	1.4	2.0	1200	1200	150	15	280	45	12
60	180	15	1.3	1.1	1.5	2.0	1200	1200	150	15	300	50	12
60	180	15	1.3	1.1	1.6	2.0	1200	1200	150	15	280	55	12
60	180	15	1.3	1.0	1.6	2.0	800	800	150	15	280	55	12
60	180	13	1.2	1.1	1.6	2.0	800	800	150	10	280	55	12
70	400	17	1.6	1.5	2.2	2.2	1200	1200	175	30	320	65	15
95	280	20	1.9	1.6	2.1	2.6	1200	1200	200	15	355	75	19
90	280	20	1.7	1.6	2.1	2.6	1200	1200	200	15	340	75	16

[d]α-Tocopherol equivalents. 1 mg (+)-α-tocopherol = 1 (+)-α-tocopherol equivalent.

[e]1 niacin equivalent (NE) equals 1 mg of niacin or 60 mg of dietary tryptophan.

Table IV.2 gives the chemical structures, good dietary sources, and deficiency diseases of the vitamins. The term *vitamin* applies to any compound or closely related group of compounds satisfying the following criteria.

1. It is organic rather than inorganic or an element.
2. It cannot be synthesized at all (or at least in sufficient amounts) by the body and it must be in the diet.
3. Its absence causes a specific *vitamin deficiency disease*.
4. Its presence is essential to normal growth and health.
5. It is present in foods in *small* concentrations, and it is not a carbohydrate, a hydrolyzable lipid, an amino acid, or a protein.

TABLE IV.2   Vitamins Important in Humans[a]

Vitamin Structure	Dietary Sources	Results of Deficiencies
Fat-Soluble Vitamins		
A_1 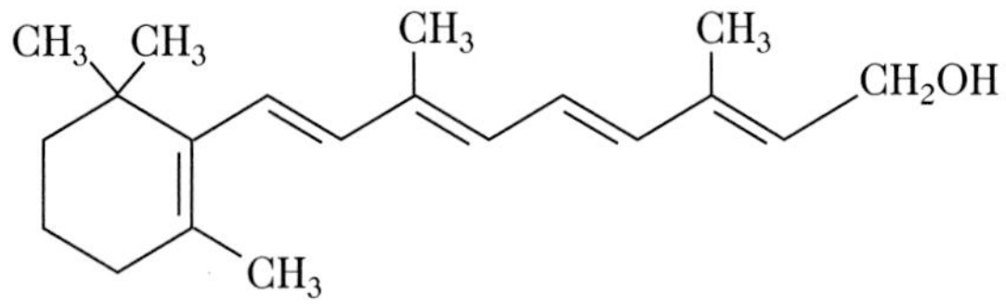 Retinol	Green and yellow vegetables, cod liver oil	Eye disease (xerophthalmia)
D_3 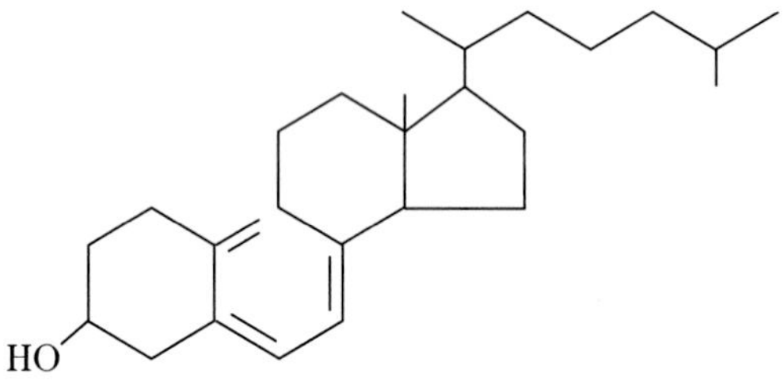 Cholecalciferol	Provitamins in the skin activated by sunlight Fish liver oils	Poor use of Ca^{2+} Poor bone formation Rickets
E 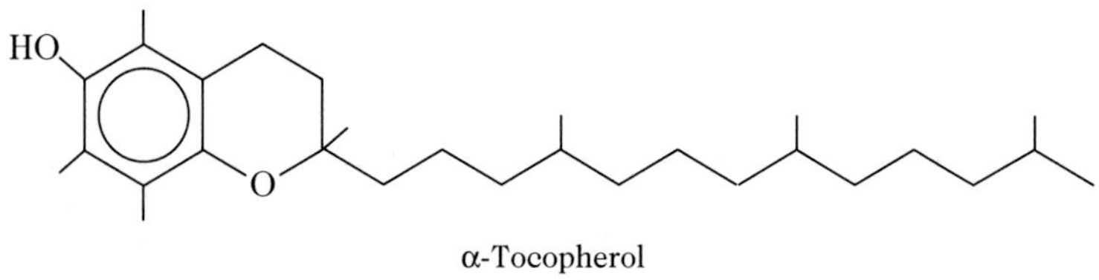 α-Tocopherol	Leafy vegetables, wheat germ oil, cottonseed oil.	Recent indications in coronary disease [detoxifies peroxides (ROOH); protects cell membranes]
K 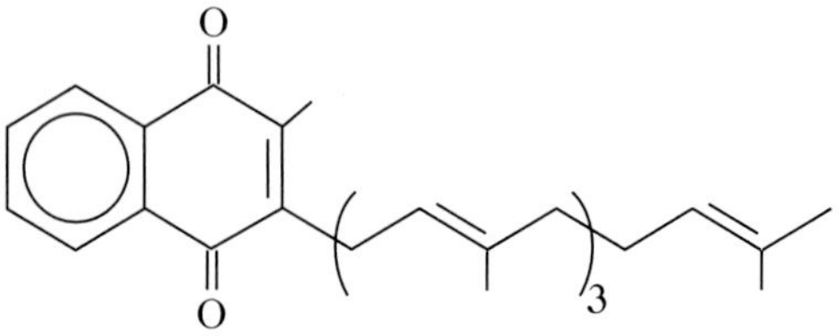 Phylloquinone	Spinach, cabbage Made in intestinal tract by microorganisms	Poor blood clotting

[a]More than one compound can sometimes provide the activity of a vitamin, but only one member of any such family is shown here.

TABLE IV.2 Vitamins Important in Humans[a]

Vitamin Structure	Dietary Sources	Results of Deficiencies
Water-Soluble Vitamins C 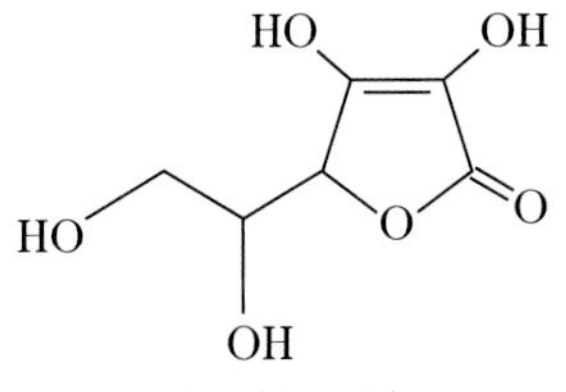 Asorbic acid	Citrus fruits, green vegetables, potatoes, berries	Scurvy
Biotin 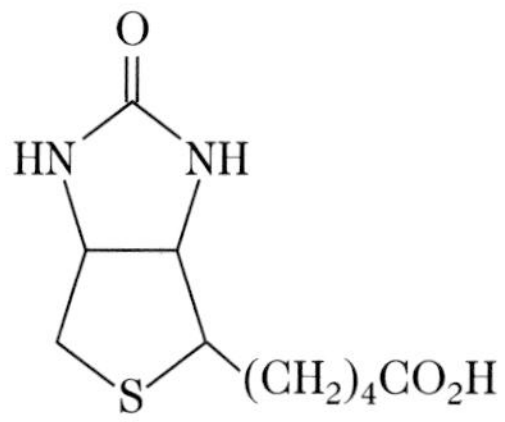	Liver, peas, lima beans	Skin disorders
Folate 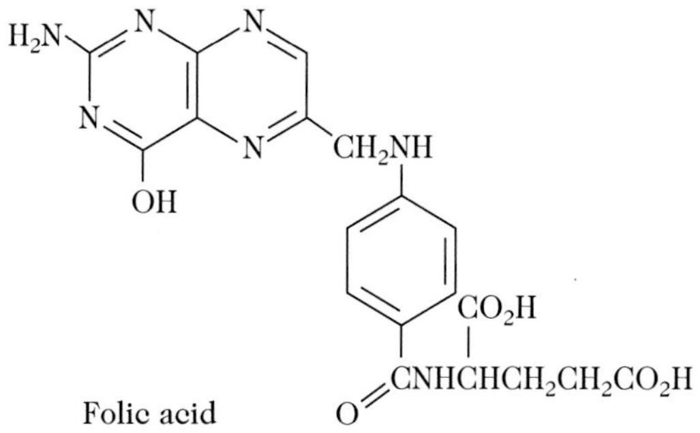 Folic acid	Animal organs, yeast, wheat germ, chicken, oysters	Gastrointestinal disorders
Pantothenic acid 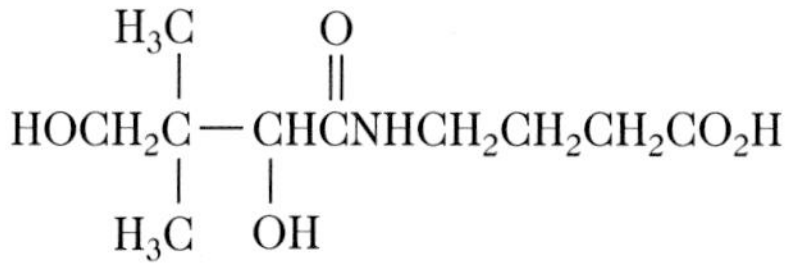	Egg yolks, kidneys, yeast, liver	Not clinically known for humans
Choline $HOCH_2CH_2\overset{+}{N}(CH_3)$	Peas, egg yolk, wheat germ, asparagus, spinach	Not clinically known for humans
Thiamine 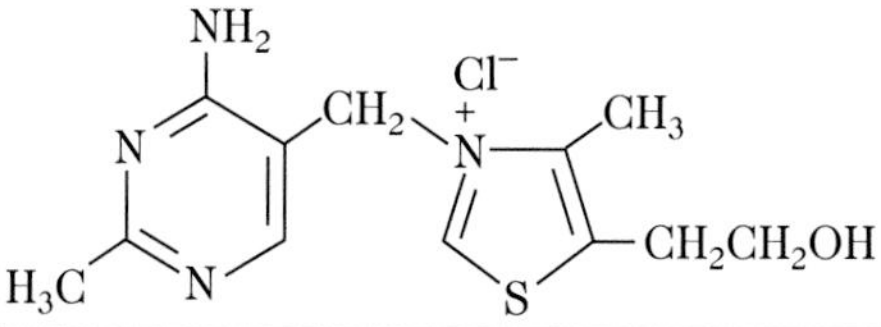	Cereal grains, nuts, liver, legumes, heart, kidney	Beriberi

TABLE IV.2   Vitamins Important in Humans[a]

Vitamin Structure	Dietary Sources	Results of Deficiencies
Riboflavin 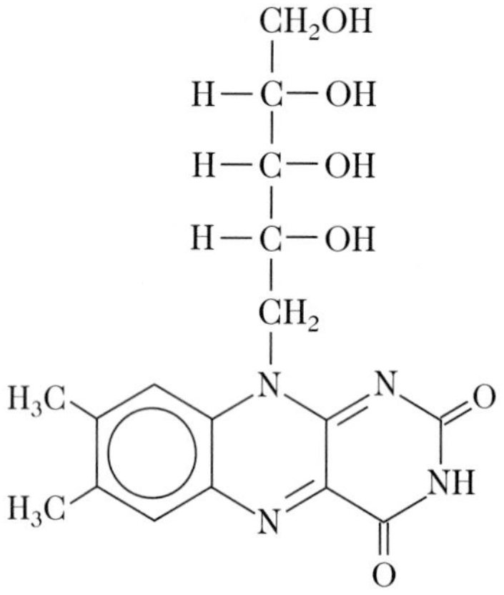	Milk, meat, cheese, eggs, peas, lima beans, wheat germ	Fissures at corners of mouth, reddening of tongue
B_6 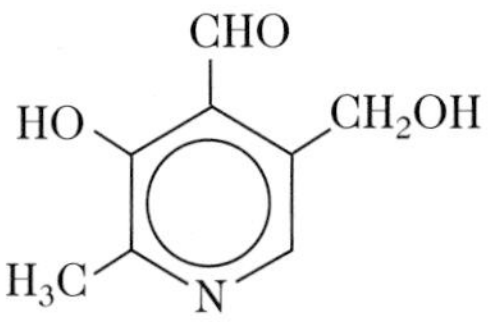	Cereals, liver, legumes, yeast, bananas	Skin disorders (in infants, convulsions)
Nicotinic acid 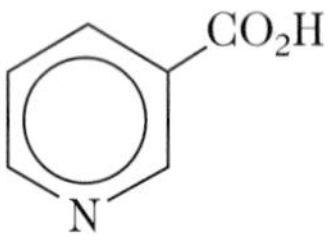	Meats, yeast, wheat germ	Pellagra
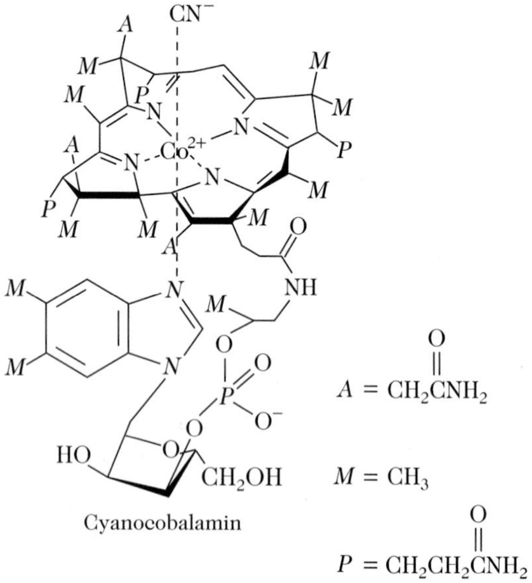	Eggs, meat, liver	Pernicious anemia

Riboflavin structure

Pyridoxal

Niacin

Cyanocobalamin

$A = CH_2CNH_2$ with $\overset{O}{\underset{\|}{}}$

$M = CH_3$

$P = CH_2CH_2CNH_2$ with $\overset{O}{\underset{\|}{}}$

APPENDIX

ANSWERS TO PRACTICE EXERCISES AND SELECTED REVIEW EXERCISES

CHAPTER 1

Practice Exercises, Chapter 1

1. 310 K
2. (a) 5.45×10^8 (b) 5.67×10^{12}
   (c) 6.454×10^3 (d) 2.5×10^1
   (e) 3.98×10^{-5} (f) 4.26×10^{-3}
   (g) 1.68×10^{-1} (h) 9.87×10^{-12}
3. (a) 10^{-6} (b) 10^{-9} (c) 10^{-6} (d) 10^3
4. (a) mL (b) μL (c) dL
   (d) mm (e) cm (f) kg
   (g) μg (h) mg
5. (a) Kilogram (b) Centimeter
   (c) Deciliter (d) Microgram
   (e) Milliliter (f) Milligram
   (g) Millimeter (h) Microliter
6. (a) 1.5 Mg (b) 3.45 μL (c) 3.6 mg
   (d) 6.2 mL (e) 1.68 kg (f) 5.4 dm
7. (a) 275 kg (b) 62.5 μL (c) 82 nm or 0.082 μm
8. (a) 95 (b) 11.36
   (c) 0.0263 (d) 1.3000
   (e) 16.1 (f) 3.8×10^2
   (g) 9.31 (h) 9.1×10^2
9. (a) $\dfrac{1\ g}{1000\ mg}$ or $\dfrac{1000\ mg}{1\ g}$

   (b) $\dfrac{1\ kg}{2.205\ lb}$ or $\dfrac{2.205\ lb}{1\ kg}$
10. 0.324 g of aspirin
11. (a) 324 mg of aspirin (b) 3.28×10^4 ft
    (c) 18.5 mL (d) 38.87 g
    (e) 4.78×10^3 μL
12. 40.0 °C
13. 59 °F (quite cool)
14. 20.7 mL
15. 32.1 g

Review Exercises, Chapter 1

1.5 The observation of a *chemical* property necessarily converts the substance into a different substance. The observation or measurement of a physical property does not do this.

1.6 A physical *quantity* gives a description of a physical property in terms of a number and a unit.

1.10 When we use a two-pan balance, we have both pans at almost identical locations so the gravitational attractions are the same for the objects on both pans when balanced. Therefore, their masses are the same, too.

1.12 A base unit

1.15 It can be defined (or *derived*) from a base quantity, length: volume = (length)^3.

1.18 The standard mass is a specific, one-of-its-kind object, stored at Sèvres, France, and is thus at risk of corrosion, fire, or theft. The standard meter is defined in terms of a characteristic of light, which is accessible to all.

1.19 (a) Meter (b) Inch
     (c) Ounce (d) Centimeter
     (e) Kilogram (f) Ton
     (g) Liter (h) Milliliter
     (i) Pound (j) Kilogram

1.21 1000 μg

1.25 (a) 100 (b) 180 (c) 100

1.27 The kelvin, by a factor of 9/5 (180/100)

1.29 This is the lowest degree of coldness obtainable.

1.31 41 °F (5 °C); too cold without a coat or heavy sweater for most people

1.33 20 °C

1.36 (a) 110 milliliters (b) 150 milligrams
(c) 16 kilometers (d) 50 micrograms
(e) 1.5 deciliters (f) 2.5 kilograms
(g) 75 microliters

1.38 (a) 1.30×10^{-1} L (b) 3.5685×10^3 m
(c) 4.2×10^{-6} g (d) 4.5×10^{-3} g

1.40 (a) 1.30 dL (b) 3.5685 km
(c) 4.2 μg (d) 4.5 mg

1.41 2.53×10^4 tickets

1.43 (a) A, B, D
(b) E, F, G
(c) C, H, I

1.45 (a) 2.00×10^5 (b) 2.000×10^5
(c) 2.0000×10^5 (d) 1.99899×10^5
(e) 1.9989891×10^5 (f) 1.99898909×10^5

1.47 (a) 1.5×10^{-4} (b) 3.5×10^4 (c) 2.50×10^{24}
(d) 6.7×10^1 (e) 2.8 (f) 4.502×10^1
(g) 3.0×10^1 (h) 1×10^{-1} (i) 2.00

1.49 (a) They are accurate; the average is 59.84. They are close to the true value.
(b) They are precise; they agree well with each other.

1.51 (a) $\dfrac{39.37 \text{ in.}}{1 \text{ m}}$ $\dfrac{1 \text{ m}}{39.37 \text{ in.}}$

(b) $\dfrac{1 \text{ L}}{1.057 \text{ quart}}$ $\dfrac{1.057 \text{ quart}}{1 \text{ L}}$

(c) $\dfrac{1 \text{ g}}{1000 \text{ mg}}$ $\dfrac{1000 \text{ mg}}{1 \text{ g}}$

(d) $\dfrac{1 \text{ kg}}{2.205 \text{ lb}}$ $\dfrac{2.205 \text{ lb}}{1 \text{ kg}}$

(e) $\dfrac{1 \text{ grain}}{0.0648 \text{ g}}$ $\dfrac{0.0648 \text{ g}}{1 \text{ grain}}$

1.53 (a) 250 mL (b) 1.6 km (c) 0.25 in. (d) 1 L

1.55 (a) 50.6 kg (b) 75.6 in.

1.57 10.0 liquid ounces

1.59 No. The vehicle has a mass of 2.04×10^3 kg, too much for the bridge.

1.61 3 tablets

1.63 7 mL

1.65 6195.7 m

1.67 0.500 g

1.69 All other forms of energy can be converted *quantitatively* into heat and so can be measured.

1.71 1.0 cal/g °C. Water's high specific heat means that relatively large quantities of heat can move into or out of a sample of water without a large change in the temperature of the sample.

1.74 675 g, 1.49 lb

1.76 0.911 g/mL, 274 mL

1.80 204 Calorie (unrounded)

1.81 114 days (unrounded)

1.83 Its density approaches that of water itself, so the urine contains little if any dissolved substances.

1.85 1.03

1.86 (a) 1.1×10^4 mg/6.0 pt; 3.9×10^3 mg/L
(b) 2.4×10^3 mg/6.0 pt; 8.5×10^2 mg/L

CHAPTER 2

Practice Exercises, Chapter 2

1. (a) 15 (b) 24 (c) 24

2. (a) 5 (b) 18 (c) 20

3. (a) 7 *p*, 7 *n*. Electron shells: 2 5
(b) 13 *p*, 14 *n*. Electron shells: 2 8 3
(c) 20 *p*, 20 *n*. Electron shells: 2 8 8 2

4. 52.5

5. (a) Sn (b) Cl (c) Rb (d) Mg (e) Ar

6. (a) 1 (b) 6 (c) 5 (d) 7

Review Exercises, Chapter 2

2.1 (a) Gas, liquid, and solid
(b) Elements, compounds, and mixtures

2.3 (a) Chemical (b) Physical
(c) Physical (d) Chemical

2.6 (a) Because both the elements sodium and chlorine "disappear" and become part of a new and different substance, sodium chloride.
(b) The reactants are sodium and chlorine; the product is sodium chloride.
(c) The ratio is a *fixed* ratio, 1:1

2.8 Law of definite proportions

2.10 Law of definite proportions

2.12 Their relative masses would be equal.

2.14 (a) 1+ (b) 1−
(c) Attract; they have opposite charges.
(d) *X*, 23; *Y*, 35

2.17 (a) Density $= 3.2 \times 10^{15}$ g/cm^3.
(b) 3.2×10^9 metric tons/cm^3 (or 3.2 billion metric tons per cubic centimeter)

2.19 1.000 g; almost identical, numerically, to the atomic mass of H

2.21 The protons. Whole numbers of protons, each with a whole-number charge of 1+ must result in a whole number for the nuclear charge.

2.27 Atomic number = 6. Mass number = 13. Electron shells: 2 4.

2.30 (a) 26 (b) No

2.32 (a) 2 8 2

(b) 2 8 8 1

(c) 2 7

(d) 2 1

2.34 13

2.36 Nonmetal

2.38 *M* and *X* are isotopes. *Q* and *Z* are isotopes.

2.40 (a) They have identical numbers of protons in their nuclei and identical numbers of electrons outside their nuclei.

(b) They have different numbers of neutrons in their nuclei.

2.42 *No* is the symbol of an element, nobelium, because the second letter is lower case. *NO* is the symbol of a compound, nitric oxide, made of the elements nitrogen (N) and oxygen (O).

2.49 Mass numbers are the sums of whole numbers—numbers of protons and neutrons per atom of an isotope. The atomic mass of an element (in u) is the average of the relative masses of the isotopes of the element, taking into account the percentages that the various isotopes have in naturally occurring samples of the element.

2.51 80

2.53 64.7

2.54 (a) Twice as heavy

(b) The mass of the pile of magnesium atoms would be twice as much as that of the pile of carbon atoms.

(c) A mass that is twice as much as that of the 2.0 g of carbon, 4.0 g of magnesium.

2.56 (a) 1.33 times as heavy   (b) 16.0 g of oxygen atoms

2.59 A representative element (one of the A elements)

2.61 (a) Group IA, alkali metal family

(b) Group VIIA, halogen family

(c) Group VIA, oxygen family

(d) Group IIA, alkaline earth metal family

2.63 (a) 8; Yes

(b) Nonmetal, because its outside level has over 4 electrons, as nearly all nonmetals do.

(c) Group VII, the halogen family

2.65 (a) 13, 14, 15, 16, 17

(b) 6, 14, 32

(c) 9, 17, 35

(d)

IIIA	IVA	VA	VIA	VIIA
5	6	7	8	9

(e) 2 (The element would be in Group IIA, so has two outside-shell electrons.)

(f) 6 (The element is in the same family as d, Group VIA, so it has six outside-shell electrons, just like d.)

(g) 9 (It stands high up in a group with more than 4 outside-level electrons.)

2.69 C: $1s^2 2s^2 2p_x^1 2p_y^1$ Na: $1s^2 2s^2 2p^6 3s^1$

CHAPTER 3

Practice Exercises, Chapter 3

1. (a) 2 8 8 1; 1+

(b) 2 8 6; 2−

(c) 2 8 4; no ion forms

2. (a) 2 8 8

(b) 2 8 8

3. (a) Cs^+ (b) F^- (c) P^{3-} (d) Sr^{2+}

4. (a) AgBr (b) Na_2O (c) Fe_2O_3 (d) $CuCl_2$

5. (a) Copper(II) sulfide; cupric sulfide

(b) Sodium fluoride

(c) Iron(II) iodide; ferrous iodide

(d) Zinc bromide

(e) Copper(I) oxide; cuprous oxide

6. (a) 3+ (b) 2+ (c) 2+

7. (a)

$$Cl - \underset{\underset{Cl}{|}}{\overset{\overset{Cl}{|}}{C}} - Cl$$

(b) H—S—H

(c)

$$Cl - \underset{\underset{Cl}{|}}{N} - Cl$$

(d)

$$H - \underset{\underset{Br}{|}}{\overset{\overset{Br}{|}}{C}} - Br$$

8. (a) $NaNO_3$ (b) KOH (c) $Ca(OH)_2$

(d) $MgCO_3$ (e) Na_2SO_4 (f) $(NH_4)_3PO_4$

9. (a) Lithium carbonate

(b) Sodium bicarbonate

(c) Potassium permanganate

(d) Sodium dihydrogen phosphate

(e) Ammonium monohydrogen phosphate

10. (a)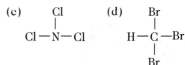

(b) $^{\delta+}H - F^{\delta-}$

(c) $^{\delta+}H - \overset{\delta-}{O} - \overset{\delta-}{O} - H^{\delta+}$

Review Exercises, Chapter 3

3.1 Unlike charges attract.

3.3 Molecular and ionic compounds

3.8 Group IIIA

3.11 (a) 2 8 8 (b) 2 8

3.12 Just two level 2 electrons; H^-

3.14 They differ by one electron. The sodium *atom* is electrically neutral; the sodium *ion* carries a charge of 1+.

3.16

Two sodium atoms   Oxygen atom

Two sodium ions   Oxide ion

3.19 (a) +1 (b) +3 (c) −2
(d) +2 (e) −1 (f) +2

3.21 Mn_2O_7

3.24 The sodium *ion*, Na^+

3.27 Sn^{2+} is the stannous ion and Sn^{4+} is the stannic ion.

3.28 (a) Lead(II)ion (b) Gold(III) ion
(c) Mercury(I) ion. The charge *per* Hg is 1+.
(d) Lead(IV) ion

3.30 (a) MgF_2 (b) Li_2O
(c) CuS (d) $FeCl_3$
(e) $NaBr$ (f) CaO

3.32 (a) Sodium fluoride
(b) Lithium oxide
(c) Copper(II) bromide, cupric bromide
(d) Magnesium chloride
(e) Zinc oxide
(f) Iron(III) bromide, ferric bromide

3.33 Both atoms and molecules are small particles and both are electrically neutral. A molecule, however, has two or more atomic nuclei and an atom has just one.

3.34 An ion and a molecule are both small particles, but the ion is electrically charged and a molecule is electrically neutral. (It is possible for ions to have more than one nucleus; molecules *always* do.)

3.35 *All* atoms in molecular elements have *identical* atomic numbers.

3.37 An *electrical* force of attraction. Its special name here is *covalent bond.*

3.41 The level 1 electron of the hydrogen atom pairs with an outer-level (level 3) electron of the chlorine atom as the two become shared between the two nuclei. This may be symbolized as follows:

$$H : \overset{..}{\underset{..}{Cl}} :$$

3.43 A

3.45 (a) 4 (b) IVA

3.47 S=C=S

3.49 H—O—Cl

3.52 (a) K_3PO_4 (b) Na_2CO_3
(c) $CaSO_4$ (d) NH_4CN
(e) $LiNO_2$ (f) $NaHSO_3$
(g) $CaCr_2O_7$ (h) $Mg(CH_3CO_2)_2$

3.54 (a) Sodium carbonate
(b) Ammonium nitrate
(c) Magnesium hydroxide
(d) Barium sulfate
(e) Potassium bicarbonate
(f) Calcium acetate
(g) Sodium nitrite
(h) Ammonium phosphate

3.56 (a) 14 (b) 22 (c) 15

3.58 49

3.60 *X* has the higher electronegativity *because* it has the larger nuclear charge. (Being in the same period, *X* and *Y* have the same inner-level electrons that screen their nuclei, so the atom with the higher nuclear charge has a higher *unscreened* ability to draw electrons of a covalent bond toward it.)

3.61 (a) Yes. (b) δ+ is near *X* and δ− is near *Y*.

3.64 No, now one of the two atoms has fully accepted an electron from the other and both atoms are now ions. Thus, the bond is an ionic bond.

3.67 (a) A 1s orbital of each H atom, one electron in each orbital
(b) A $2p_x$ orbital of each F atom, one electron in each orbital
(c) A 1s orbital of the H atom and a $2p_x$ orbital of the F atom, one electron in each orbital

3.69 The pairs of outer-level electrons are in four electron clouds that repel each other so that their axes are at angles of 109.5°.

CHAPTER 4[1]

Practice Exercises, Chapter 4

1. $3O_2 \rightarrow 2O_3$
2. $4Al + 3O_2 \rightarrow 2Al_2O_3$
3. (a) $2Ca + O_2 \rightarrow 2CaO$
(b) $2KOH + H_2SO_4 \rightarrow 2H_2O + K_2SO_4$
(c) $Cu(NO_3)_2 + Na_2S \rightarrow CuS + 2NaNO_3$

---

[1]If your calculated answers differ slightly from the answers given here, it may be caused by a conceptually unimportant difference in handling the calculation. All answers to computational problems given here were obtained by *chain* calculations. The formula masses needed for these calculations were first computed in the usual way from atomic masses rounded to the first decimal place (H to 1.01), and then the calculated formula masses were rounded to the number of significant figures allowed by the given data *before* they were used in the calculations.

(d) $2AgNO_3 + CaCl_2 \longrightarrow 2AgCl + Ca(NO_3)_2$

(e) $2Al + 3H_2SO_4 \longrightarrow Al_2(SO_4)_3 + 3H_2$

(f) $CH_4 + 2O_2 \longrightarrow 2H_2O + CO_2$

4. 8.68×10^{22} atoms of gold per ounce

5. (a)   180   (b)   58.3   (c)   858.6

6. 0.500 mol of H_2O

7. 4.20 mol of N_2 and 4.20 mol of O_2

8. 450 mol of H_2 and 150 mol of N_2

9. 408 g of NH_3

10. 0.0380 mol of aspirin

11. 11.5 g of O_2

12. 18.4 g of Na

13. (a)   2.45 g of H_2SO_4

(b)   9.00 g of $C_6H_{12}O_6$

14. 156 mL of 0.800 M Na_2CO_3 solution

15. 9.82 mL of 0.112 M H_2SO_4, when calculated step-by-step with rounding after each step
9.84 mL of 0.112 M H_2SO_4, when found by a chain calculation

Review Exercises, Chapter 4

4.2 Two molecules of nitrogen monoxide react with one molecule of oxygen to give two molecules of nitrogen dioxide.

4.4 (a) $N_2 + O_2 \longrightarrow 2NO$

(b) $MgO + 2HNO_3 \longrightarrow Mg(NO_3)_2 + H_2O$

(c) $CaBr_2 + 2AgNO_3 \longrightarrow Ca(NO_3)_2 + 2AgBr$

(d) $2HI + Mg(OH)_2 \longrightarrow MgI_2 + 2H_2O$

(e) $CaCO_3 + 2HBr \longrightarrow CaBr_2 + CO_2 + H_2O$

4.6 We need to know *numbers* because it is in a whole-number ratio of *numbers* of particles that substances interact, not in whole-number ratios of their masses. The problem is that the particles are too tiny to count directly (and so obtain them directly in whole-number ratios). Thus we must use an indirect method to count them, and Avogadro's number is the basis for solving this problem.

4.9 (a)   36.5   (b)   56.1   (c)   184.1

(d)   63.0   (e)   84.0   (f)   261.3

(g)   132.1   (h)   158.2   (i)   180.1

4.11 Formula masses vary from substance to substance, so the formula mass *taken in grams* must also vary accordingly. What is constant about 1 mol of all substances is that they all have identical numbers of formula units, Avogadro's number.

4.15 (a)   4.56 g HCL   (b)   7.01 g KOH

(c)   23.0 g $MgBr_2$ (d)   7.88 g HNO_3

(e)   10.5 g $NaHCO_3$ (f)   32.7 g $Ba(NO_3)_2$

(g)   16.5 g $(NH_4)_2HPO_4$ (h)   19.8 g $Ca(CH_3CO_2)_2$

(i)   22.5 g $C_6H_{12}O_6$

4.17 (a)   1.37 mol HCl   (b)   0.891 mol KOH

(c)   0.272 mol $MgBr_2$ (d)   0.794 mol HNO_3

(e)   0.595 mol $NaHCO_3$ (f)   0.191 mol $Ba(NO_3)_2$

(g)   0.379 mol $(NH_4)_2HPO_4$ (h)   0.316 mol $Ca(CH_3CO_2)_2$

(i)   0.277 mol $C_6H_{12}O_6$

4.19 2.15×10^{22} molecules N_2

4.21 6×10^{15} molecules O_3

4.24 $\dfrac{1\ \text{mol Ca(OH)}_2}{2\ \text{mol HCl}}$ $\dfrac{2\ \text{mol HCl}}{1\ \text{mol Ca(OH)}_2}$

4.26 (a)   0.417 mol O_2 (b)   0.278 mol Fe_2O_3

4.28 (a) $2Al_2O_3 \longrightarrow 4Al + 3O_2$

(b)   52.9 g Al   (c)   47.1 g O_2

(d)   100.0 g $(Al + O_2)$, which is identical to the initial mass of Al_2O_3 used (law of conservation of mass in chemical reactions)

4.30 (a)   0.740 mg H_2S (b)   5.38 mg Ag_2S

4.33 (a)   Saturated   (b)   Concentrated

4.35 The solubility in water of sodium hydroxide *increases* with temperature, so a saturated solution can become supersaturated by cooling the solution, provided that the excess solute does not precipitate. If the saturated solution is heated, it becomes unsaturated.

4.38 Molar concentration and molarity are exact synonyms. Each refers to a *concentration* in units of moles per liter. A *mole* is that mass of a substance that equals its formula mass in the unit of grams. A *molecule* is just one of the 6.02×10^{23} formula units in one mole of a molecular substance, so it is an exceedingly tiny *particle*.

4.40 (a)   0.0625 mol or 3.66 g NaCl

(b)   0.0250 mol or 4.50 g $C_6H_{12}O_6$

(c)   0.0250 mol or 2.45 g H_2SO_4

(d)   0.0625 mol or 6.63 g Na_2CO_3

4.42 66.7 mL

4.44 400 mL

4.46 0.300 mol Na_2CO_3

4.48 79.4 mL

4.50 22.8 mL

4.52 28.6 mL of 0.150 M Na_2SO_4 solution and 42.9 mL of 0.100 M $Ba(NO_3)_2$ solution

4.54 823 mL of 0.100 M HCl

CHAPTER 5

Practice Exercises, Chapter 5

1. 1.31×10^3 mL of helium

2. 547 mL

3. 484 atm

Review Exercises, Chapter 5

5.4 The column of air supported by the earth is shorter at the mountain top, so it exerts less weight on a unit of the earth's area.

5.8 0.329 atm

5.10 To a *mixture* of gases. Both the air we breathe in and what we breathe out are mixtures of gases of which one, oxygen, is an essential substance for life and another, carbon dioxide, is a waste product we must remove from the body.

5.12 601 mm Hg (This is what *partial pressure* means, the pressure that a component of a gas mixture would exert if it were all alone in the container and all other gases were removed.)

5.16 31 mm Hg

5.18 (a) T, V, n (b) V, n (c) T, n
(d) P, n (e) T, V (f) n

5.20 2.59 L

5.22 2.57 L

5.23 272 atm

5.24 740 mm Hg

5.26 257 mL

5.30 At STP

5.33 6.02×10^{23} molecules of H_2 (Avogadro's number)

5.35 64.9 mol; 2.08 kg O_2

5.37 1.3 mol O_2

5.39 (a) 4.00 mol NH_3
(b) 500 L NH_3
(c) 15.2 g H_2; 91.1 g NH_3

5.40 (a) 9.82×10^{-3} mol CO_2
(b) 9.82×10^{-3} mol $CaCO_3$
(c) 0.983 g $CaCO_3$
(d) 0.551 g CaO

5.42 It obeys all of the gas laws exactly.

5.46 (a) The Kelvin temperature; $PV \propto T$
(b) The Kelvin temperature is proportional to the average kinetic energy, or $T \propto KE$.

5.50 The distances between particles in liquids and solids are virtually zero, so physical properties of liquids and solids are much more sensitive to the chemical natures of the particles.

5.53 Electrical forces between the particles are weaker between nitrogen molecules and stronger between Na^+ and Cl^- ions.

5.58 Convection

5.62 (a) CO and O_2 (b) CO_2
(c) C (d) B
(e) Exothermic. The heat of reaction exceeds the energy of activation.
(f) E

5.63

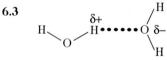

5.65 At a higher temperature, the frequency of all collisions increases and there is an increase in the *fraction* of successful collisions.

5.68 Increase concentrations.

5.71 (a) No effect (b) Reduces it
(c) No effect (d) Increases

5.73 Increase the temperature, increase the concentration, and use a catalyst.

5.74 21 L (The concentrations *in terms of relative volumes* are the same at all altitudes.)

5.78 Alcohol in the blood dilates blood capillaries, which would let the cold blood near the skin flood into and further chill the body's core.

5.79 43.9 g/mol. The formula mass is 43.9.

5.81 314 mL

Practice Exercises, Chapter 6

1. 10.2 g of 96% H_2SO_4

2. 1.25 g of glucose and 499 g of water (rounded from 498.75, and assuming that the density of water is 1.00 g/mL)

3. (a) 2.00 g of $KMnO_4$ (b) 0.25 g of NaOH

4. (a) 0.020 Osmol (b) 0.015 Osmol
(c) 0.100 Osmol (d) 0.150 Osmol

Review Exercises, Chapter 6

6.1 (a) Covalent bond
(b) It is a nonpolar molecule and there is no $\delta+$ or $\delta-$ on H.

6.3

6.5 (a) The boiling point of ammonia is much less than that of water. Ammonia molecules evidently attract each other less than do water molecules.
(b) Both $\delta+$ and $\delta-$ are larger in the water molecule than in the ammonia molecule.
(c) Oxygen is more electronegative than nitrogen.

6.13 The H ends of the polar O—H bonds in water molecules that carry $\delta+$ charges

6.16 Make sure that the solution has undissolved sodium nitrate in it.

6.17 $MgSO_4 \cdot 7H_2O(s) \xrightarrow{\text{heat}} MgSO_4(s) + 7H_2O(g)$

6.19 $Na_2SO_4(s) + 10H_2O \rightarrow Na_2SO_4 \cdot 10H_2O$

6.21 The pellets will absorb sufficient moisture from the air to form a film of concentrated NaOH solution on their surfaces.

6.22 $y = 3$. The formula is $X \cdot 3H_2O$.

6.25 (a) $NaCl(s) + heat \longrightarrow Na^+(aq) + Cl^-(aq)$

(b) $Na^+(aq) + Cl^-(aq) \longrightarrow NaCl(s) + heat$

(c) Forward. The forward reaction consumes energy as it occurs.

(d) Reverse (e) Become equal

(f) Lower (g) No

(h) The system would be required to absorb more heat.

(i) Shift. The equilibrium will shift to the right.

6.27 As equilibrium B

6.33 The air in contact with this blood under a total pressure of 1 atm has a partial pressure of O_2 equal to 80 mm Hg.

6.36 $\dfrac{0.915 \text{ g NaOH}}{100 \text{ g NaOH soln}}$ $\dfrac{100 \text{ g NaOH soln}}{0.915 \text{ g NaOH}}$

6.38 $\dfrac{30 \text{ mL alcohol}}{100 \text{ mL alcohol soln}}$ $\dfrac{100 \text{ mL alcohol soln}}{30 \text{ mL alcohol}}$

6.40 (a) 1.56 g NaI (b) 0.469 g NaBr

(c) 1.00 g glucose (d) 4.25 g H_2SO_4

6.42 (a) 7.50 g $Mg(NO_3)_2$ (b) 5.63 g NaBr

(c) 5.00 g KI (d) 0.938 g $Ca(NO_3)_2$

6.44 46.9 mL methyl alcohol

6.46 (a) 3.33 mL 3.00% KOH solution

(b) 33.3 mL 0.600% HCl solution

(c) 673 mL 1.00% NaCl solution

(d) 234 mL 3.00% KOH solution

6.48 30.8 g of the decahydrate (30.9 g by a chain calculation)

6.49 Solutions are always *homogeneous;* mixtures in general are often heterogeneous.

6.51 (a) Stirred suspensions

(b) Colloidal dispersions

(c) Colloidal dispersions

(d) Solutions

(e) Unstirred suspensions

6.56 Test for the Tyndall effect.

6.58 A colloidal dispersion of a solid in a liquid. Examples are jellies, paints, and starch in water.

6.60 $-1.86 \,°C$

6.61 The solute in the second solution breaks up (ionizes) into two ions per formula unit when it dissolves, so the effective concentration is 2.00 mol/1000 g H_2O.

6.65 This solution is 2.0 M in all solute particles, and osmolarity is a function of the molar concentrations of all osmotically active particles.

6.67 5.0% NaCl, which has 0.085 mol solute/100 g solution versus only 0.030 mol solute/100 g solution for 5% KI

6.70 (a) Hypertonic

(b) (1) Hemolysis (2) Crenation

6.72 0.33 Osmol/L (0.12 Osmol NaCl + 0.060 Osmol $NaHCO_3$ + 0.040 Osmol KCl + 0.11 Osmol glucose)/L

6.79 (a) The same (b) The same

(c) Less concentrated

6.81 1.64 mol/L; 3.28 Osmol/L

CHAPTER 7

Practice Exercises, Chapter 7

1. $HBr(g) + H_2O \longrightarrow H_3O^+(aq) + Br^-(aq)$
$HI(g) + H_2O \longrightarrow H_3O^+(aq) + I^-(aq)$

2. $HNO_3(aq) + KOH(aq) \longrightarrow KNO_3(aq) + H_2O$
$H^+(aq) + NO_3^-(aq) + K^+(aq) + OH^-(aq) \longrightarrow$
$\qquad\qquad\qquad K^+(aq) + NO_3^-(aq) + H_2O$
$H^+(aq) + OH^-(aq) \longrightarrow H_2O$

3. $Mg(OH)_2(s) + 2HCl(aq) \longrightarrow MgCl_2(aq) + 2H_2O$
$Mg(OH)_2(s) + 2H^+(aq) \longrightarrow Mg^{2+}(aq) + 2H_2O$

4. $2NaHCO_3(aq) + H_2SO_4(aq) \longrightarrow Na_2SO_4(aq) +$
$\qquad\qquad\qquad\qquad 2CO_2(g) + 2H_2O$
$2Na^+(aq) + 2HCO_3^-(aq) + 2H^+(aq) + SO_4^{2-}(aq)$
$\qquad \longrightarrow 2Na^+(aq) + SO_4^{2-}(aq) + 2CO_2(g) + 2H_2O$
$HCO_3^-(aq) + H^+(aq) \longrightarrow CO_2(g) + H_2O$

5. $K_2CO_3(aq) + H_2SO_4(aq)$
$\qquad\qquad \longrightarrow K_2SO_4(aq) + CO_2(g) + H_2O$
$2K^+(aq) + CO_3^{2-}(aq) + 2H^+(aq) + SO_4^{2-}(aq) \longrightarrow$
$\qquad\qquad 2K^+(aq) + SO_4^{2-}(aq) + CO_2(g) + H_2O$
$CO_3^{2-}(aq) + 2H^+(aq) \longrightarrow CO_2(g) + H_2O$

6. $MgCO_3(s) + 2HNO_3(aq) \longrightarrow Mg(NO_3)_2(aq) +$
$\qquad\qquad\qquad\qquad CO_2(g) + H_2O$
$MgCO_3(s) + 2H^+(aq) + 2NO_3^-(aq) \longrightarrow$
$\qquad\qquad Mg^{2+}(aq) + 2NO_3^-(aq) + CO_2(g) + H_2O$
$MgCO_3(s) + 2H^+(aq) \longrightarrow Mg^{2+}(aq) + CO_2(g) + H_2O$

7. (a) $NH_3(aq) + HBr(aq) \longrightarrow NH_4Br(aq)$
$NH_3(aq) + H^+(aq) \longrightarrow NH_4^+(aq)$

(b) $2NH^3(aq) + H_2SO_4(aq) \longrightarrow (NH_4)_2SO_4(aq)$
$NH_3(aq) + H^+(aq) \longrightarrow NH_4^+(aq)$

8. $Mg(s) + 2HCl(aq) \longrightarrow MgCl_2(aq) + H_2(g)$
$Mg(s) + 2H^+(aq) \longrightarrow Mg^{2+}(aq) + H_2(g)$

9. (a) Weak acid (b) Weak acid

(c) Weak acid

10. (a) Weak base (b) Weak base

(c) Strong base (d) Strong base

11. $Na_2S(aq) + Cu(NO_3)_2(aq) \longrightarrow CuS(s) + 2NaNO_3(aq)$
$S^{2-}(aq) + Cu^{2+}(aq) \longrightarrow CuS(s)$

12. The acetate ion combines with (neutralizes) the hydrogen ion. In the following equilibrium, the product is favored.
$$H^+(aq) + CH_3CO_2^-(aq) \rightleftharpoons CH_3CO_2H(aq)$$

13. (a)   AgCl precipitates: $Ag^+(aq) + Cl^-(aq) \rightarrow$
$$AgCl(s).$$

(b) CO_2 evolves. $CaCO_3(s) + 2H^+(aq) \rightarrow Ca^{2+}(aq)$
$$+ CO_2(g) + H_2O.$$

(c)   No reaction occurs.

Review Exercises, Chapter 7

7.2   (a)   No

(b)   Yes, any polyatomic ion such as $SO_4{}^{2-}$

(c)   No

(d)   No

7.4   (a)   Acidic   (b)   Neutral   (c)   Basic

7.6   They consist of oppositely charged ions, and between these are strong forces of attraction that set up rigid solids.

7.8   Aqueous solutions of electrolytes conduct electricity. Molten forms of electrolytes conduct electricity.

7.10   Negative

7.13   Ethyl alcohol does not furnish any ions of any kind in water.

7.15   It is a molecular compound. It does not consist of ions in the liquid state.

7.17   Salts do not supply a *common* ion in solution, whereas aqueous acids furnish hydrogen ions and aqueous bases supply hydroxide ions.

7.20   In a solution of a strong acid, the solute is fully ionized. In a solution of a weak acid, the percentage ionization is small.

7.24   Both H's in H_2SO_4 can react with a base, but only one H in CH_3CO_2H can neutralize a base.

7.25 $H_2A + H_2O \rightleftharpoons H_3O^+(aq) + HA^-(aq)$
$HA^-(aq) + H_2O \rightleftharpoons H_3O^+(aq) + A^{2-}(aq)$

7.26   With greater difficulty. The second H^+ ion has to pull away from a particle that has more opposite (negative) charge than the molecule from which the first H^+ ion came.

7.29 $CO_2(aq) + H_2O \rightleftharpoons H_2CO_3(aq)$

7.31   In the case of an ionic compound like KOH, both terms mean a high percentage dissociation into ions as it dissolves in water.

7.34   A solution of ammonia in water. Because it contains only a very low percentage of ammonium and hydroxide ions, it's inappropriate to call it ammonium hydroxide (although some chemical suppliers do).

7.37   (a)   Yes   (b)   No   (c)   No

7.39   (a) $2NaOH(aq) + H_2SO_4(aq) \rightarrow$
$$Na_2SO_4(aq) + 2H_2O$$
$OH^-(aq) + H^+(aq) \rightarrow H_2O$

(b) $K_2CO_3(aq) + 2HNO_3(aq) \rightarrow$
$$2KNO_3(aq) + CO_2(g) + H_2O$$
$CO_3{}^{2-}(aq) + 2H^+(aq) \rightarrow CO_2(g) + H_2O$

(c) $NaHCO_3(aq) + HBr(aq) \rightarrow$
$$NaBr(aq) + CO_2(g) + H_2O$$
$HCO_3{}^-(aq) + H^+(aq) \rightarrow CO_2(g) + H_2O$

(d) $CaCO_3(s) + 2HI(aq) \rightarrow$
$$CaI_2(aq) + CO_2(g) + H_2O$$
$CaCO_3(s) + 2H^+(aq) \rightarrow$
$$Ca^{2+}(aq) + CO^2(g) + H_2O$$

(e) $NH_3(aq) + HI(aq) \rightarrow NH_4I(aq)$
$NH_3(aq) + H^+(aq) \rightarrow NH_4{}^+(aq)$

(f) $Mg(OH)_2(s) + 2HBr(aq) \rightarrow MgBr_2(aq) + 2H_2O$
$Mg(OH)_2(s) + 2H^+(aq) \rightarrow Mg^{2+}(aq) + 2H_2O$

(g) $2Al(s) + 6HCl(aq) \rightarrow 2AlCl_3(aq) + 3H_2(g)$
$2Al(s) + 6H^+(aq) \rightarrow 2Al^{3+}(aq) + 3H_2(g)$

7.41 $MCO_3(s) + 2H^+(aq) \rightarrow M^{2+}(aq) + CO_2(g) + H_2O$

7.42 $M(OH)_2(s) + 2H^+(aq) \rightarrow M^{2+}(aq) + 2H_2O$

7.46   0.355 mol $KHCO_3$

7.48   7.61 g Na_2CO_3

7.50   5.10 g $NaHCO_3$

7.52   36.7 mL NaOH solution

7.54   (a) $CaCO_3(s) + 2HCl(aq) \rightarrow$
$$CaCl_2(aq) + CO_2(g) + H_2O$$
$CaCO_3(s) + 2H^+(aq) \rightarrow$
$$Ca^{2+}(aq) + CO_2(g) + H_2O$$

(b)   54.4 g $CaCO_3$; 235 mL HCl solution

7.56   (a) H_2SO_3 (b)   HBr

(c) H_3O^+ (d) CH_3CO_2H

7.58   (a) $NH_2{}^-$ (b) $NO_2{}^-$

(c) $SO_3{}^{2-}$ (d) $HSO_3{}^-$

7.60   (a) $NH_2{}^-$ (b) OH^- (c) S^{2-}

7.62   (a)   HCl   (b) H_2O (c) $HSO_4{}^-$

7.64   Potassium hydroxide, KOH
$KOH(aq) + HCl(aq) \rightarrow KCl(aq) + H_2O$
Potassium bicarbonate, $KHCO_3$
$KHCO_3(aq) + HCl(aq) \rightarrow KCl(aq) + CO_2(g) + H_2O$
Potassium carbonate, K_2CO_3
$K_2CO_3(aq) + 2HCl(aq) \rightarrow 2KCl(aq) +$
$$CO_2(g) + H_2O$$

7.66   Compounds c and d are insoluble in water.

7.68   Compounds a and d are insoluble in water.

7.70   (a) $Cl^-(aq) + Ag^+(aq) \rightarrow AgCl(s)$

(b)   No reaction

(c) $OH^-(aq) + H^+(aq) \rightarrow H_2O$

(d) $Pb^{2+}(aq) + 2Cl^-(aq) \rightarrow PbCl_2(s)$

(e)   No reaction

(f) $S^{2-}(aq) + Cu^{2+}(aq) \rightarrow CuS(s)$

(g) $SO_4{}^{2-}(aq) + Ba^{2+}(aq) \rightarrow BaSO_4(s)$

(h) $OH^-(aq) + H^+(aq) \rightarrow H_2O$

(i) $S^{2-}(aq) + Ni^{2+}(aq) \rightarrow NiS(s)$

(j) $Ag^+(aq) + Br^-(aq) \rightarrow AgBr(s)$

(k) $HCO_3{}^-(aq) + H^+(aq) \rightarrow CO_2(g) + H_2O$

(l) $Mg^{2+}(aq) + 2OH^-(aq) \rightarrow Mg(OH)_2(s)$

7.72 3.5–5.0 meq K^+/L

7.74 3.76 g or 3.76×10^3 mg of Cl^-

7.76 5.01 meq of K^+

7.79 Anion gap = 19 meq/L. This is above the normal range of 5–14 meq/L, so this anion gap suggests a disturbance in metabolism.

7.81 Ca^{2+}, Mg^{2+}, Fe^{3+} (Fe^{2+})

7.83 Hard water in which the chief anion is HCO_3^-. Its hardness is lost when such water is heated because HCO_3^- changes to CO_3^{2-}, which forms a carbonate precipitate with any of the hardness ions.

7.86 (a) A dilute (5% w/w) solution of NH_3 in water.

(b) $OH^-(aq)$

(c) It is present in the following equilibrium:
$$NH_3(aq) + H_2O \rightleftharpoons NH_4^+(aq) + OH^-(aq)$$

(d) $Ca^{2+}(aq) + 2OH^-(aq) \rightarrow Ca(OH)_2(s)$

7.88 Whichever aqueous solution is a nonconductor contains a molecular compound, unless the molecular compound reacts with water (ionizes) to give ions.

7.90 $NaHCO_3$

CHAPTER 8

Practice Exercises, Chapter 8

1. (a) 2.5×10^{-6} mol OH^-/L; basic

(b) 9.1×10^{-8} mol OH^-/L; acidic

(c) 1.1×10^{-7} mol OH^-/L; basic

2. (a) 1×10^{-7} to 1×10^{-8} mol H^+/L

(b) Slightly basic

(c) Acidosis

3. Use sodium bicarbonate solution. It neutralizes H^+ by
$$HCO_3^-(aq) + H^+(aq) \rightarrow CO_2(aq) + H_2O$$

4. (a) Basic (b) Basic

(c) Basic (d) Acidic

(e) Basic (f) Basic

5. Basic. The acetate ion, a proton acceptor or base, hydrolyzes.

6. Acidic. The hydrated Cu^{2+} ion hydrolyzes.

7. Yes. Decrease the pH. The NH_4^+ is a weak acid.

8. 0.105 M NaOH

9. 0.125 M H_2SO_4

Review Exercises, Chapter 8

8.3 Acid–base neutralization is heat producing or exothermic. This neutralization is the *reverse* reaction in the equilibrium

$$H_2O \rightleftharpoons H^+(aq) + OH^-(aq)$$

so the *forward* reaction must be heat consuming or endothermic. Therefore, the *addition* of heat shifts the equilibrium to the right in accordance with Le Châtelier's principle.

8.5 9.0×10^{-16}

8.7 Acidic

8.9 3.0

8.11 Because in water at pH = 7.00 (25 °C), $[H^+]$ = $[OH^-]$.

8.13 More acidic; acidosis

8.15 1×10^{-4} to 1×10^{-5} mol/L

8.17 Weak. At a pH of 5.72, the value of $[H^+]$ is between 1×10^{-5} to 1×10^{-6} mol/L, but the initial concentration of the acid is much higher (1×10^{-3} mol/L). Therefore, only a small percentage of the acid molecules can be ionized.

8.19 Acidosis, because 7.18 is less (so more acidic) than 7.35, the normal pH of blood

8.21 (a) Neutral (b) Acidic (c) Basic

(d) Acidic (e) Basic

8.24 Acidosis is a decrease in the pH of blood. Alkalosis is an increase in the pH of blood.

8.26 The pH will stay quite close to 7.45 even if small quantities of acid or base are added.

8.28 $H_2PO_4^-(aq) + OH^-(aq) \rightarrow HPO_4^{2-}(aq) + H_2O$

8.30 $CO_2(aq)$ and $HCO_3^-(aq)$

8.35 Involuntary hypoventilation forces the system to retain CO_2, which is equivalent to the retention of acid, causing acidosis.

8.36 (a) Alkalosis

(b) For each CO_2 lost at the lungs, one H^+ ion is permanently neutralized. The loss of H^+ ion, of course, means an increase in the value of the pH of the blood.

8.38 The molarity of the solution is accurately known.

8.40 The indicator is picked so that its color change occurs at or very close to the pH of the solution that forms when the calculated amount of base has been matched by that of the acid.

8.41 The titration of any strong acid (e.g., HCl) with any strong base (e.g., NaOH) produces a salt whose ions do not hydrolyze.

8.43 The titration of hydrochloric acid against aqueous ammonia produces NH_4Cl, an ammonium salt whose cation, NH_4^+, hydrolyzes to give a slightly acidic pH. (The anion, Cl^-, does not hydrolyze.)

8.45 (a) 0.3407 M HI

(b) 0.2000 M H_2SO_4

8.47 64.00

8.49 (a) 1.279 g HI

(b) 14.71 g H_2SO_4

(c) 59.62 g Na_2CO_3

8.50 (a) $Na_2CO_3(aq) + 2HCl(aq) \rightarrow 2NaCl(aq) + CO_2(g) + H_2O$

(b) $0.05206\ M\ Na_2CO_3$

(c) $5.518\ g\ Na_2CO_3/L$

8.52 5.7. Because of dissolved CO_2, some of which exists as carbonic acid.

8.55 The combustion of sulfur-containing fuels gives SO_2; some of this is changed by atmospheric reactions to SO_3.

8.58 In engines: $N_2(g) + O_2(g) \rightarrow 2NO(g)$
In air: $2NO(g) + O_2(g) \rightarrow 2NO_2(g)$

CHAPTER 9

Practice Exercises, Chapter 9

1. (a) $CH_3-CH_2-CH_3$ (b) 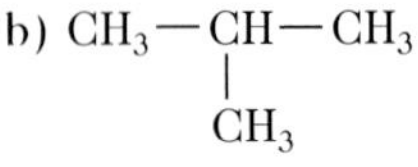 $CH_3-CH-CH_3$ with CH_3 branch

(c) 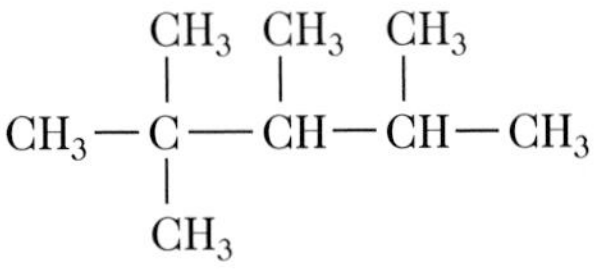 $CH_3-C(CH_3)_2-CH-CH-CH_3$ (with CH_3 groups)

2. (a) $CH_3CH_2CH_3$ (b) 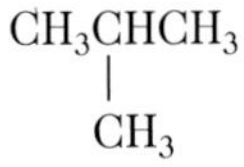 CH_3CHCH_3 with CH_3 branch

(c) 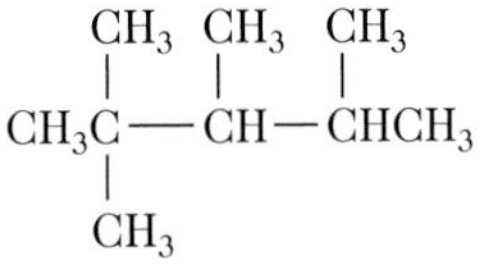 $CH_3C(CH_3)_2-CH-CHCH_3$ (with CH_3 groups)

3. (a), 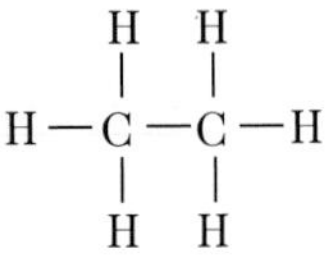 (b), 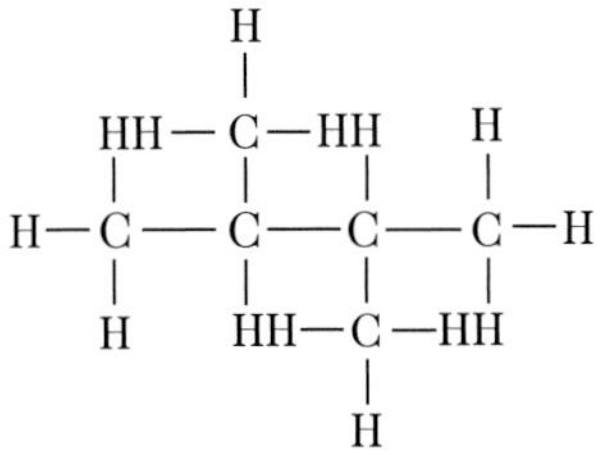 (c) 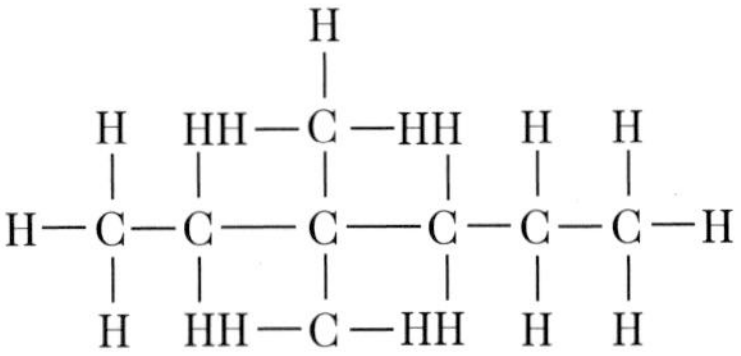 structures

4. Structures (b) and (c) violate the tetravalences of carbon at one point.

5. (a) Identical   (b) Isomers
(c) Identical   (d) Isomers
(e) Different in another way

Review Exercises, Chapter 9

9.1 Chemists had been unable to synthesize organic compounds from minerals. The vital force theory held that such a synthesis was inherently impossible without the presence of a "vital force."

9.4 Compounds b, d, and e are considered to be inorganic.

9.7 (a) Molecular; low melting and flammable

(b) Ionic; water soluble (and likely a carbonate or a bicarbonate)

(c) Molecular; no ionic compound is a gas (or a liquid) at room temperature.

(d) Ionic; high melting and nonflammable

(e) Molecular; most liquid organic compounds are insoluble in water but will burn.

(f) Molecular; no ionic compound is a liquid at room temperature.

9.9 Compounds a, d, and e are possible

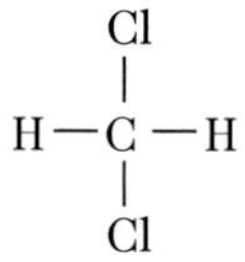

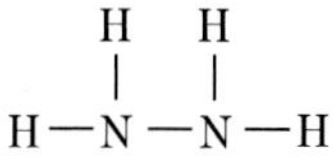

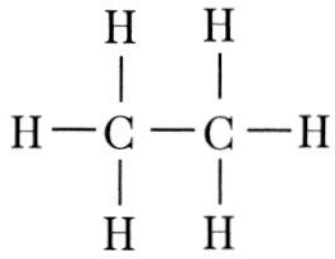

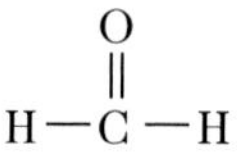

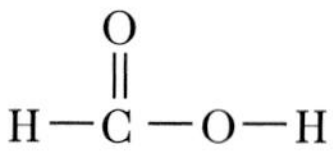

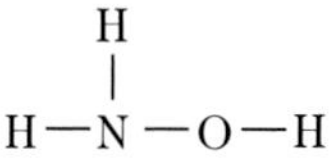

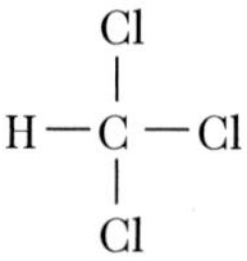

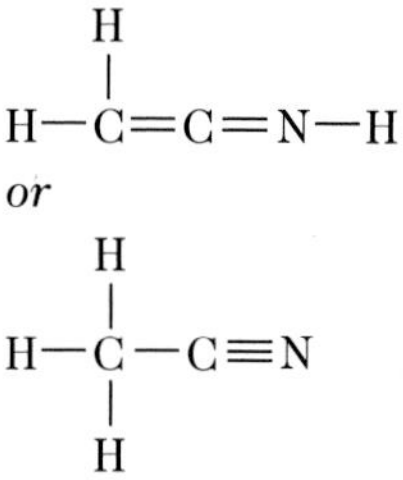

9.10 (a)–(l) structures

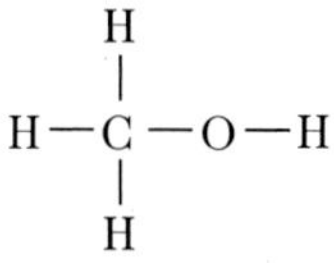

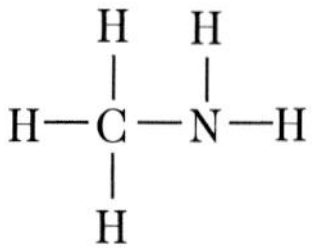

9.12 Identical compounds are a, b, e, and f. Isomers are d, g, h, i, j, k, and l. Structures that are neither identical nor isomers occur at c and m.

9.13 (a) Alkane   (b) Alcohol   (c) Aldehyde
(d) Ester   (e) Amine   (f) Ether

9.14 (a) Alcohol   (b) Alcohol
(c) Both thioalcohols   (d) First, alkene; second, cycloalkane
(e) Ketone   (f) Alkane
(g) Both amines   (h) First, carboxylic acid; second, alcohol + ketone

(i) First, ester; second, carboxylic acid. (If you called the first an aldehyde, instead of an ester, don't worry about it now. Why it is an ester will be explained later.)

(j) First, ester + alcohol; second, alcohol + carboxylic acid

(k) First, ether + ketone; second, ester

(l) Both alkanes

CHAPTER 10

Practice Exercises, Chapter 10

1. 2-Methyl-1-butanol (less polar)

2. (a) 3-Methylhexane
   (b) 4-Ethyl-2,3-dimethylheptane
   (c) 5-Ethyl-2,4,6-trimethyloctane

3. (a) Ethyl chloride
   (b) Butyl bromide
   (c) Isobutyl chloride
   (d) *t*-Butyl bromide

4. (a) 2-Methylpropene (or 2-methyl-1-propene)
   (b) 4-Isobutyl-3,6-dimethyl-3-heptene
   (c) 1-Chloropropene (or 1-chloro-1-propene)
   (d) 3-Bromopropene (or 3-bromo-1-propene)

5. *Cis–trans* isomerism is possible for a and b. For part (a), *cis* versus *trans* is based on the way the main chain passes through the double bond.

(a) 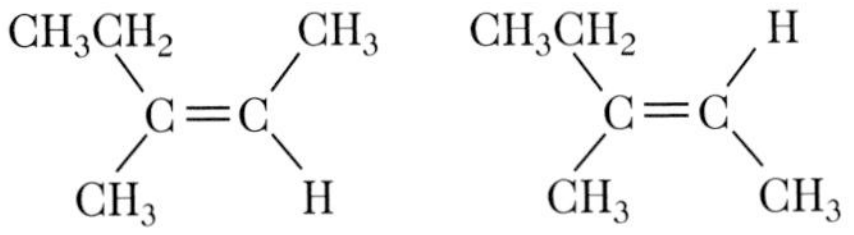 cis isomer   trans isomer

(b) 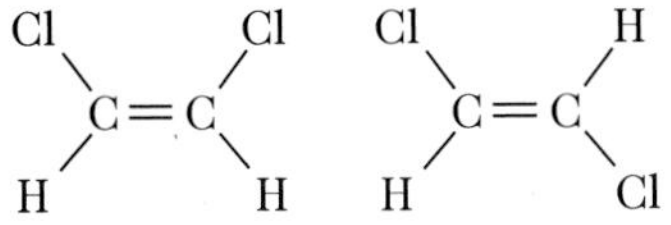 cis isomer   trans isomer

6. (a) $CH_3CH_2CH_3$ (b) No reaction
   (c) 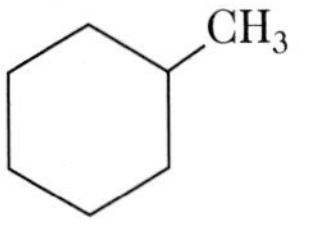 (d) $CH_3(CH_2)_{16}CO_2H$

7. (a) 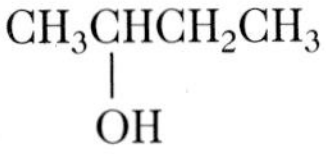 $CH_3CHCH_2CH_3$ (b) 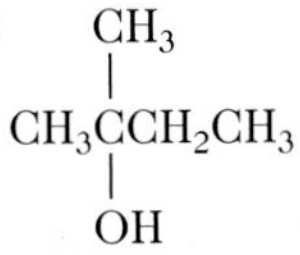 $CH_3CCH_2CH_3$
        $|$ $|$
        OH CH_3 ... OH

   (c) 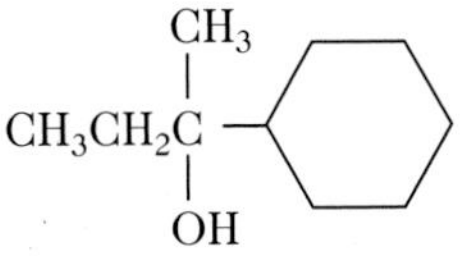 CH_3CH_2C —cyclohexane with CH₃ and OH

Review Exercises, Chapter 10

10.1 (a) Alkene (b) Alkane
     (c) Alkane (d) Alkyne

10.3 Compounds (b), (c), and (d) are saturated.

10.5

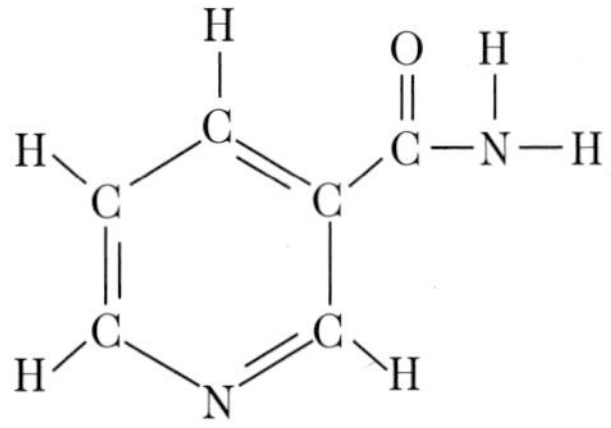

10.7 B. It consists of larger molecules.

10.9 Add a drop of the liquid to water. If it does not dissolve, it is hexane.

10.11

$CH_3CH_2CH_2CH_2CH_2CH_3$ Hexane

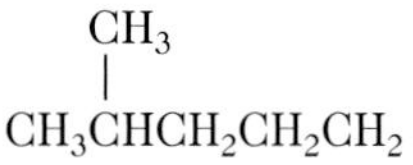

$$CH_3CHCH_2CH_2CH_2 \quad (CH_3) \quad \text{2-Methylpentane}$$

$$CH_3CH_2CHCH_2CH_3 \quad (CH_3) \quad \text{3-Methylpentane}$$

$$CH_3CCH_2CH_3 \quad (CH_3)(CH_3) \quad \text{2,2-Dimethylbutane}$$

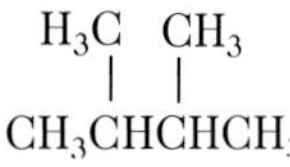

$$CH_3CHCHCH_3 \quad (H_3C)(CH_3) \quad \text{2,3-Dimethylbutane}$$

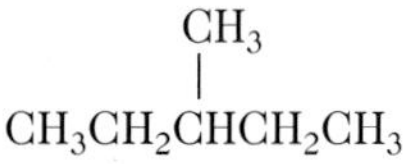

10.13 (a) 5-*sec*-Butyl-5-ethyl-2,3,3,9-tetramethyldecane
      (b) 7-*t*-Butyl-5-isobutyl-2-methyl-6-propyldecane

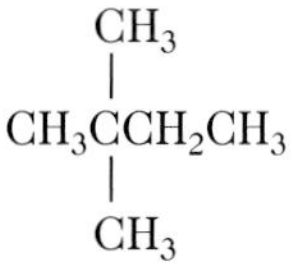

10.15 (a) $CH_3CH_2CH_2Cl$ (b) 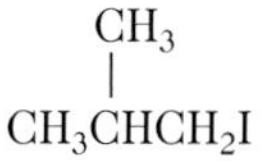 CH_3CHCH_2I (with CH₃)
      (c) 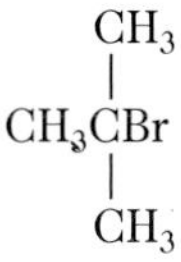 CH_3CBr (with CH₃ and CH₃)
      (d) CH_3CH_2Br

10.17 (a) 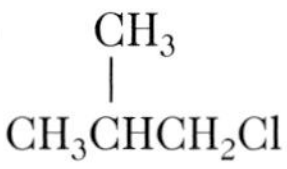 CH_3CHCH_2Cl (with CH₃)

      1-Chloro-2-methylpropane

      (b)

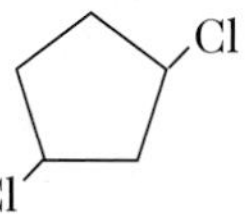

      1,3-Dichlorocyclopentane

      (c) 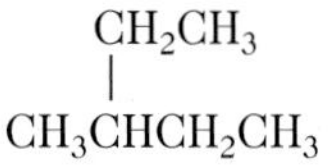 $CH_3CHCH_2CH_3$ (with CH₂CH₃)

      3-Methylpentane

      (d)

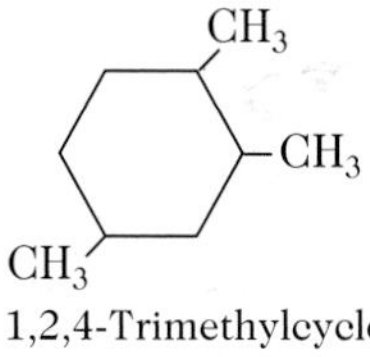

      1,2,4-Trimethylcyclohexane

10.18 $C_7H_{16} + 11O_2 \rightarrow 7CO_2 + 8H_2O$

10.20 No reaction occurs with any of the reactants.

10.21 (a) Identical (b) Identical
(c) Isomers (d) Identical
(e) Identical

10.23 (a)

$$F\overset{\displaystyle}{\underset{Br}{C}}=\overset{\displaystyle H}{\underset{Cl}{C}} \qquad F\overset{\displaystyle}{\underset{Br}{C}}=\overset{\displaystyle Cl}{\underset{H}{C}}$$

(b) No geometric isomers

(c)

$$H_3C\overset{\displaystyle}{\underset{H}{C}}=\overset{\displaystyle H}{\underset{CH=CH_2}{C}} \qquad H_3C\overset{\displaystyle}{\underset{H}{C}}=\overset{\displaystyle CH=CH_2}{\underset{H}{C}}$$

10.25 (a) 1-Octene
(b) 1-Bromo-3-methyl-1-butene
(c) 4-Methyl-2-propyl-1-hexene
(d) 4,4-Dimethyl-2-hexene

10.27 (a)

$$CH_3\underset{\displaystyle |}{\overset{CH_3}{C}}=CH_2 + H_2 \xrightarrow{\text{Catalyst}} CH_3\underset{\displaystyle |}{\overset{CH_3}{C}}HCH_3$$

(b)

$$CH_3\underset{\displaystyle |}{\overset{CH_3}{C}}=CH_2 + H_2O \xrightarrow{H^+} CH_3\underset{\displaystyle OH}{\overset{CH_3}{\underset{|}{C}}}CH_3$$

10.29 (a)

$$CH_3\underset{\displaystyle |}{\overset{CH_3}{C}}=CHCH_3 + H_2 \xrightarrow{\text{Catalyst}} CH_3\underset{\displaystyle |}{\overset{CH_3}{C}}HCH_2CH_3$$

(b)

$$CH_3\underset{\displaystyle |}{\overset{CH_3}{C}}=CHCH_3 + H_2O \xrightarrow{H^+} CH_3\underset{\displaystyle OH}{\overset{CH_3}{\underset{|}{C}}}HCH_2CH_3$$

10.31 57.5 g $KMnO_4$

10.33

(a) etc.—$CH_2\underset{CH_3}{\overset{CH_3}{C}}—CH_2\underset{CH_3}{\overset{CH_3}{C}}—CH_2\underset{CH_3}{\overset{CH_3}{C}}—CH_2\underset{CH_3}{\overset{CH_3}{C}}H$— etc.

(b)

$$-(CH_2\underset{CH_3}{\overset{CH_3}{C}})_n-$$

10.35 Dipentene has no benzene ring.

10.37 (a) $C_6H_6 + Cl_2 \xrightarrow{FeCl_3} C_6H_5Cl + HCl$

(b) No reaction

(c) No reaction

10.39 (a) $CH_3CH_2CH_2\underset{OH}{\overset{|}{C}}HCH_2CH_3$ (b) $CH_3\underset{|}{\overset{CH_3}{C}}HCH_2CH_3$

(c) $C_6H_5Br + HBr$ (d) No reaction

(e)

(f) $CH_3CH_2CH_2CH_2CH_3$

(g) No reaction

(h) $C_6H_5CH_2\underset{OH}{\overset{|}{C}}HC_6H_5$

(i) No reaction

(j) $C_5H_{12} + 8O_2 \longrightarrow 5CO_2 + 6H_2O$

10.53 $CH_3CH_2\overset{+}{C}HCH_3$

10.55

$$CH_3CH_2\underset{\displaystyle H}{\overset{CH_3}{\underset{|}{C}}}H-\overset{+}{\underset{\displaystyle H}{O}}: \qquad CH_3CH_2\underset{|}{\overset{CH_3}{C}}H-OH$$

The protonated alcohol   *sec*-Butyl alcohol

10.57 $O^* + O_2 + M \rightarrow O_3 + M + \text{heat}$
$O_3 + UV \text{ energy} \rightarrow O_2 + O^*$

10.59 UV photons break up CFC-11 molecules, releasing Cl atoms. The Cl atoms are able to react with both ozone and O atoms.

$$\begin{array}{l} Cl + O_3 \rightarrow ClO + O_2 \\ \underline{ClO + O \rightarrow Cl + O_2} \\ \text{Net} \quad O_3 + O \rightarrow 2O_2 \end{array}$$

10.60 It converts UV energy into heat.

10.61 Without the shield, UV photons would reach us with great frequency and lead to increased skin cancer cases and damage to agriculture.

CHAPTER 11

Practice Exercises, Chapter 11

1. (a) Alcohol (b) Phenol
(c) Carboxylic acid (d) Alkene + alcohol
(e) Alcohol (f) Alcohol

2. (a) Monohydric, secondary
(b) Monohydric, secondary
(c) Dihydric, unstable (two OH groups on the same carbon)
(d) Dihydric, both are secondary
(e) Monohydric, primary
(f) Monohydric, primary
(g) Monohydric, tertiary
(h) Monohydric, secondary
(i) Trihydric, unstable (three OH groups on the same carbon)

3. (a) $CH_3CH=CH_2$ (b) $CH_3CH=CH_2$
(b) $CH_2=\underset{|}{\overset{CH_3}{C}}CH_3$ (c)

4. (a)

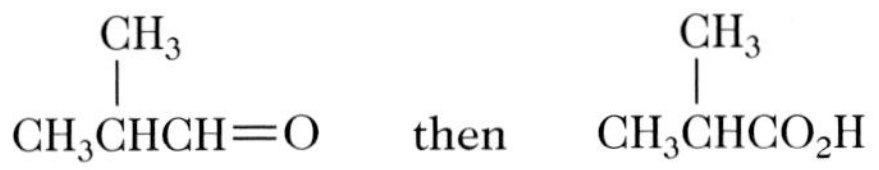

CH₃
|
CH₃CHCH=O   then   CH₃CHCO₂H (with CH₃ on middle carbon)

(b) C₆H₅CH=O   then   C₆H₅CO₂H

5. (a)

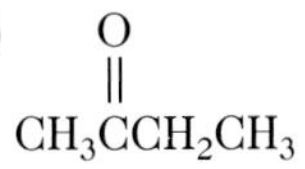

O
||
CH₃CCH₂CH₃

(b)

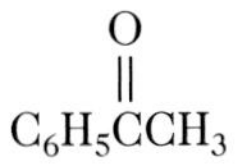

O
||
C₆H₅CCH₃

(c)

(cyclopentanone)

6. (a) CH₃OCH₃   (b) CH₃CH₂CH₂OCH₂CH₂CH₃

(c)

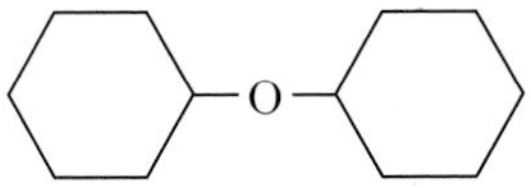

(dicyclohexyl ether)

7. (a) 2CH₃SH   (b)

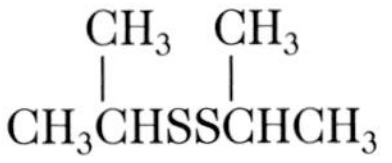

CH₃ CH₃
| |
CH₃CHSSCHCH₃

(c)

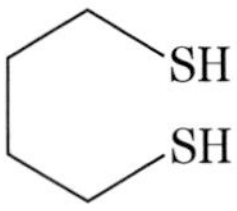

(cyclohexane with SH groups)

(d)

(dicyclopentyl disulfide)

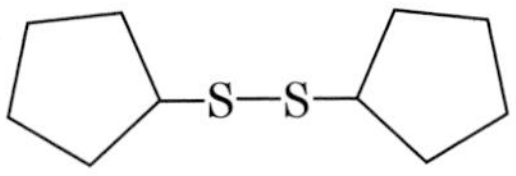

8. (a) C₆H₅NH₃⁺   (b) (CH₃)₃NH⁺   (c) ⁺NH₃CH₂CH₂NH₃⁺

9. (a)

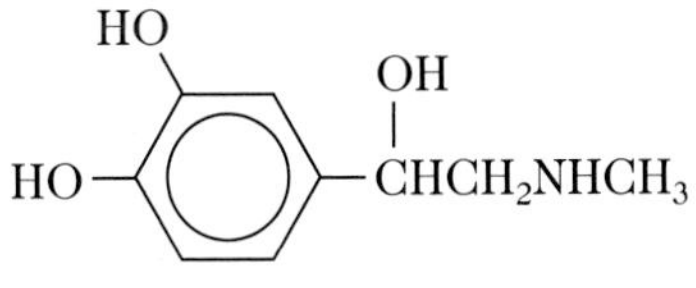

HO

HO—⟨ring⟩—CHCH₂NHCH₃
              |
              OH

(b)

OCH₃

CH₃O—⟨ring⟩—CH₂CH₂NH₂

OCH₃

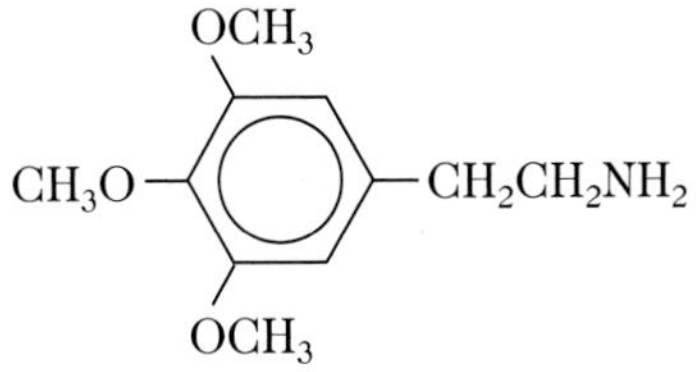

Review Exercises, Chapter 11

11.1 1, phenol; 2, 2° alcohol; 3, 2° alcohol

11.3 1, 1°alcohol; 2, ketone; 3, 2°alcohol; 4, ketone; 5, alkene

11.5 (a) amine, heterocyclic

(b) 1, amine; 2, ester; 3, amine (aromatic)

(c) 1, heterocyclic amine; 2, heterocyclic amine

(d) 1, 2° alcohol; 2, amine

11.9 (a) Ethyl alcohol   (b) Ethylene glycol

(c) Glycerol   (d) Isopropyl alcohol

11.11 HOCH₂CH₂OH, ethylene glycol (Remember, no carbon may hold two or more OH groups.)

11.13

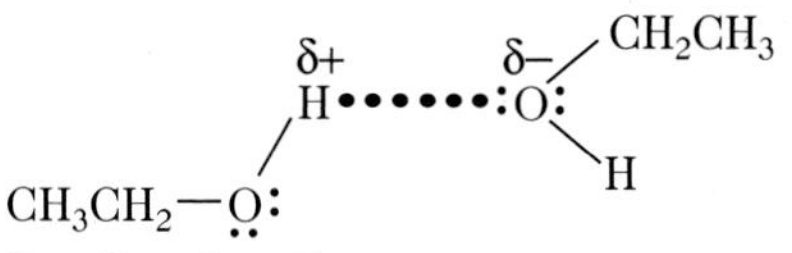

11.15 B < D < A < C

11.17 (a)

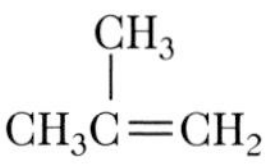

CH₃
|
CH₃C=CH₂

(b) CH₃CH=CHCH₃ +
mostly
CH₂=CHCH₂CH₃
some

(c)

CH₃

(methylcyclopentene) + (methylenecyclopentane)

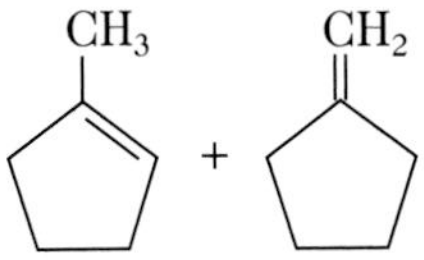

mostly   some

(d) C₆H₅CH=C(CH₃)₂

(e) 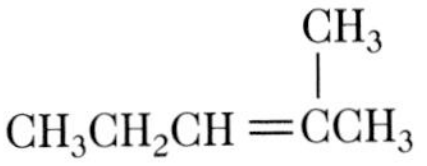 CH₃CH₂CH=CCH₃
              |
              CH₃
mostly

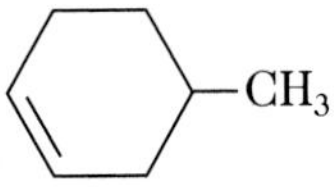

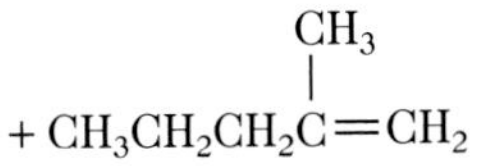

+ CH₃CH₂CH₂C=CH₂
              |
              CH₃
some

(f)

(methylcyclohexene)

11.19 (a) HOCH₂CH₂CH₃   or

CH₃CHCH₃
|
OH

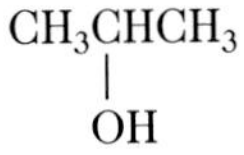

(b)

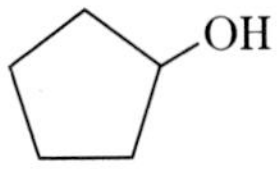

(cyclopentanol)

(c)

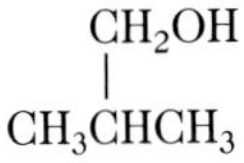

CH₂OH
|
CH₃CHCH₃

or

CH₃
|
CH₃CCH₃
|
OH

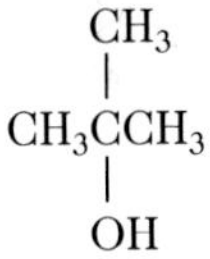

(d)

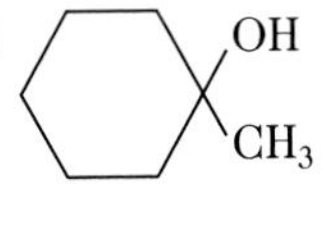

(1-methylcyclohexanol)   or   (2-methylcyclohexanol)

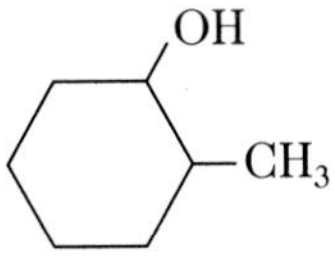

11.21 B. A is a 3° alcohol and so cannot be oxidized; B is a 2° alcohol.

11.23 A reacts with aqueous NaOH; B does not. B can be dehydrated to an alkene; A cannot. (Both react with oxidizing agents, but in different ways.)

11.25 (a) Two ether groups   (b) An ester

(c) A di-ester   (d) An ether

(e) Not a functional group we have studied (It's a peroxide.)

(f) An ether   (g) Ether + ketone

11.27 (a) CH₃CH₂CH₂OH   (b) 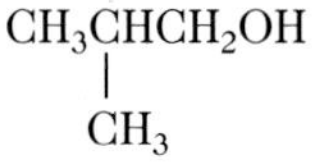 CH₃CHCH₂OH
                                        |
                                        CH₃

(c)

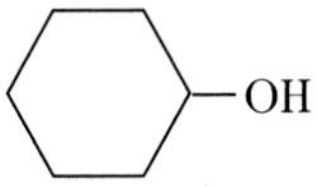

(cyclohexanol)

(d)

(methylcyclopentanol)

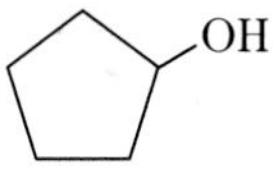

11.29 No reaction occurs. Ethers are stable in base.

11.30 (a) CH₃CH₂SSCH₂CH₃   (b) 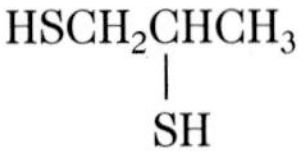 HSCH₂CHCH₃
                                           |
                                           SH

(c) 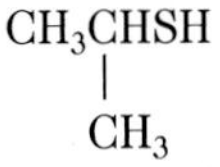 CH₃CHSH
       |
       CH₃

(d) HSCH₂CH₂CH₃

11.31 (a) CH₃CH₂CH₂SSCH₂CH₂CH₃   (b) (CH₃)₂CHCH₂SH

(c)

(dithiol cyclopentane)

S S
H H

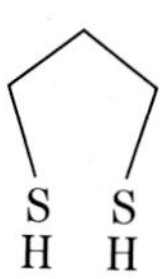

(d) (CH₃)₂CHSSCH(CH₃)₂

11.33 (a) Aliphatic amide + ether group

(b) Aliphatic amine + ester group

(c) Aliphatic, heterocyclic amide

(d) Aromatic *compound* overall because of the benzene ring, but the amine is an *aliphatic* amine (The amino group is not attached directly to the ring.)

11.35 (a) Isopropylpropylamine

(b) Ethylmethylpropylamine

(c) *t*-Butylisobutylamine

(d) Diisopropylamine

11.37 (a) $CH_3CH_2CH_2NH_3^+$ (b) $CH_3CH_2CH_2NH_2$

(c) No reaction (d) No reaction

11.39 **A** is the stronger base; it is an amine (plus a ketone). **B** is an amide.

11.41 (a)
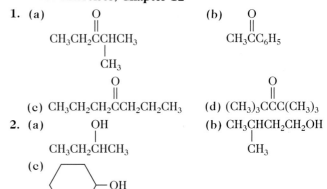
(b) $CH_3CH_2OCH_2CH_3$

(c)
$$CH_3\overset{\overset{\text{O}}{\|}}{C}CH_2CH_3$$

(d) cyclopentane with CH_3

(e) No reaction (f) No reaction

(g) No reaction (h) $CH_3CH_2\overset{\underset{\text{CH}_3}{|}}{C}HCH_3$

(i) cyclohexane ring with CH_3, OH, CH_3

(j) cyclopentanone with CH_3

(k) $CH_3CH_2CH_2NH_3^+Cl^-$ (l) $C_6H_5NH_2$

(m) $(CH_3)_2CHSSCH(CH_3)_2$

11.43 (a) 2.50 mol acetone

(b) 0.120 mol $KMnO_4$

(c) 22.1 g $KMnO_4$

(d) 12.2 g acetone; 12.1 g MnO_2

11.48 The partial pressure of ethyl alcohol in exhaled breath is proportional to the level of ethyl alcohol in the blood. The level of alcohol in the breath, measured by its oxidation in the breathalyzer, can thus be correlated to the level in the blood.

11.49 0%

11.56 The more water-soluble form is the protonated form of the free base, a salt with hydrogen chloride.

11.57 Cocaine has two ester groups that would be hydrolyzed in the digestive tract before being absorbed into the bloodstream.

11.58 The free base is molecular; the hydrogen chloride salt is ionic, and ionic compounds generally melt higher than molecular compounds and so are less volatile.

CHAPTER 12

Practice Exercises, Chapter 12

1. (a)
$$CH_3CH_2\overset{\overset{\text{O}}{\|}}{C}\overset{\underset{\text{CH}_3}{|}}{C}HCH_3$$

(b)
$$CH_3\overset{\overset{\text{O}}{\|}}{C}C_6H_5$$

(c) $CH_3CH_2CH_2\overset{\overset{\text{O}}{\|}}{C}CH_2CH_2CH_3$

(d) $(CH_3)_3C\overset{\overset{\text{O}}{\|}}{C}C(CH_3)_3$

2. (a)
$$CH_3CH_2\overset{\underset{\text{OH}}{|}}{C}HCH_3$$

(b) $CH_3\overset{\underset{\text{CH}_3}{|}}{C}HCH_2CH_2OH$

(c) cyclohexane with OH

3. (a) Not a hemiacetal (b) Not a hemiacetal

(c) $HO\overset{*}{C}H_2OCH_2CH_3$

(d) cyclohexane ring with $CH_3O\overset{*}{-}$ and HO

4. (a)
$$CH_3\overset{\underset{\text{OH}}{|}}{C}HOCH_3$$

(b) $CH_3CH_2CH_2\overset{\underset{\text{OH}}{|}}{C}HOCH_2CH_3$

(c)
$$C_6H_5\overset{\underset{\text{OH}}{|}}{C}HOCH_2CH_2CH_3$$

(d) $HOCH_2OCH_3$

5. (a)
$$CH_3CH_2\overset{\overset{\text{O}}{\|}}{C}H + HOCH_3$$

(b)
$$CH_3CH_2OH + H\overset{\overset{\text{O}}{\|}}{C}CH_2CH_3$$

(b) No reaction

6. (a)
$$2CH_3OH + H\overset{\overset{\text{O}}{\|}}{C}H$$

(c) $2CH_3OH + CH_3\overset{\underset{\text{H}_3\text{C}}{|}}{C}H\overset{\overset{\text{O}}{\|}}{C}CH_3$

Practice Exercises, Chapter 12

12.1 (a) Ketone (b) Aldehyde

(c) Ketone (d) Carboxylic acid

(e) Aldehyde (f) Ether + ketone

12.2

$CH_3CH_2\overset{\overset{\text{O}}{\|}}{C}H$	$CH_3\overset{\overset{\text{O}}{\|}}{C}CH_3$	$CH_3CH_2\overset{\overset{\text{O}}{\|}}{C}OH$	$CH_3\overset{\overset{\text{O}}{\|}}{C}OCH_3$
Aldehyde	Ketone	Carboxylic acid	Ester

12.5 (a) Benzaldehyde (b) Acetaldehyde

(c) Butyric acid (d) Propionaldehyde

12.7 Valeraldehyde

12.9 **B < A < D < C**

12.11 **B < A < D < C**

12.13
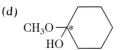

12.15 (a)

$$CH_3CH_2\overset{\displaystyle O}{\overset{\|}{C}}CH_3$$

(b)

$$\underset{\overset{|}{CH_3}}{CH_3CHCH_2CH_2CHO}$$

(c) H₃C

(d)

$$C_6H_5CH_2\underset{\overset{|}{OH}}{CHCHO}$$

12.17 C_3H_6O is $CH_3CH_2CH{=}O$; $C_3H_6O_2$ is $CH_3CH_2CO_2H$

12.19 A positive Tollens' test is given by (b).

12.26 (a)

$$CH_3\overset{\displaystyle O}{\overset{\|}{C}}CH_2CO_2^-$$

12.28 (a) $CH_3CH_2O^-$

(b) $CH_3CH_2O^- + H_2O \rightarrow CH_3CH_2OH + OH^-$

(c) Ethyl alcohol

12.30

$$\underset{\overset{|}{CH_2CH_2O^-}}{H_3\overset{+}{N}CHCO_2^-}$$

$$\underset{\overset{|}{CH_2CH_2OH}}{H_3\overset{+}{N}CHCO_2^-}$$

   A   **B**

12.32 (a)

$$CH_3\overset{\displaystyle O}{\overset{\|}{C}}CH_2CH_3$$

(b)

$$H\overset{\displaystyle O}{\overset{\|}{C}}\underset{\overset{|}{OCH_3}}{CHCH_2CH_3}$$

(c)

(d)

$$CH_3-\text{⬡}-\overset{\displaystyle O}{\overset{\|}{C}}H$$

12.34 (a) Acetal   (b) Hemiacetal

(c) Hemiacetal   (d) Something else (an ether and an alcohol)

12.36 (a)

$$CH_3\underset{\overset{|}{CH_3}}{\overset{\overset{\textstyle OCH_3}{|}}{C}}OCH_3$$

(b)

$$CH_3\underset{\overset{|}{CH_3}}{\overset{\overset{\textstyle OCH_2CH_3}{|}}{C}}OCH_2CH_3$$

12.37

$$\begin{array}{c} CH_3 \\ CH-OH \\ CH_2 \qquad CH{=}O \\ CH_2-CH_2 \end{array}$$

12.39

$$\begin{array}{c} OH \qquad CH_3 \\ CH_2 \qquad C{=}O \\ CH_2-CH_2 \end{array}$$

12.41 (a) $CH_3CH_2CHO + 2CH_3OH$ (b) No reaction

(c)

$$CH_3\overset{\displaystyle O}{\overset{\|}{C}}CH_3 + 2CH_3CH_2OH$$

(d)

$$\text{⬠}{=}O + 2CH_3OH$$

12.43 (a)

$$\underset{\overset{|}{CH_3}}{CH_3CHCH_2OH}$$

(b)

$$(CH_3)_2CH\overset{\displaystyle O}{\overset{\|}{C}}CH_3$$

(c) No reaction   (d) $CH_3CH_2CH_2CH_2CH_3$

(e)

$$\underset{\overset{|}{OH}}{CH_3CH_2CHOCH_3}$$

(f) $CH_3CH_2OH + Mtb^+$

(g)

$$\underset{\overset{|}{OCH_2CH_3}}{CH_3CHOCH_2CH_3}$$

(h) $CH_3CHO + 2CH_3OH$

(i)

$$\text{⬠}-CO_2H$$

(j) No reaction

(k) $C_6H_5CH_2NH_3^+Cl^-$

12.45

$$CH_3CH_2\overset{\displaystyle O}{\overset{\|}{C}}H$$

   A

$$CH_3CH_2CH_2OH$$

   B

$$CH_3CH{=}CH_2$$

   C

$$\underset{\overset{|}{OH}}{CH_3CHCH_3}$$

   D

$$CH_3\overset{\displaystyle O}{\overset{\|}{C}}CH_3$$

   E

12.47 (a) 0.173 mol butyraldehyde

(b) 1.23 mol CH_3OH

(c) Yes, 11.1 g CH_3OH needed but 39.4 g taken

(d) 3.11 g H_2O obtained

(e) To ensure that the equilibria involved in the reaction are all shifted as much as possible to the right, in favor of the products.

CHAPTER 13

Practice Exercises, Chapter 13

1. (a) $CH_3CH_2CO_2^-$ (b) $CH_3O-\text{⬡}-CO_2^-$

(c) $CH_3CH{=}CHCO_2^-$

2. (a) $CH_3O-\text{⬡}-CO_2H$ (b) $CH_3CH_2CO_2H$

(c) $CH_3CH{=}CHCO_2H$

3. (a)

$$CH_3\overset{\displaystyle O}{\overset{\|}{C}}OCH_3$$

(b)

$$CH_3\overset{\displaystyle O}{\overset{\|}{C}}OCH_2CH_2CH_3$$

(c)

$$CH_3\overset{\displaystyle O}{\overset{\|}{C}}O\underset{\overset{|}{CH_3}}{CHCH_3}$$

4. (a)

$$H\overset{\displaystyle O}{\overset{\|}{C}}OCH_2CH_3$$

(b)

$$CH_3CH_2\overset{\displaystyle O}{\overset{\|}{C}}OCH_2CH_3$$

(c)

$$C_6H_5\overset{\displaystyle O}{\overset{\|}{C}}OCH_2CH_3$$

5. (a) $CH_3OH + CH_3CO_2H$

(b) $(CH_3)_2CHOH + CH_3CH_2CO_2H$

(c) $CH_3CH_2CH_2OH + (CH_3)_2CHCO_2H$

6. (a) $C_6H_5OH + CH_3CO_2^-$

(b) 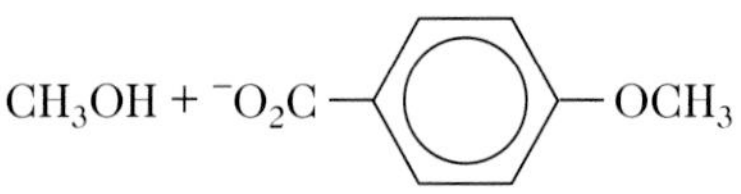 $CH_3OH + {}^-O_2C-$⬡$-OCH_3$

7. (a)

$(CH_3)_2CHCNHCH_3$ (with C=O)

(b)

$CH_3CNHC_6H_5$ (with C=O)

(c) No amide forms.   (d) No amide forms.

8. (a) $C_6H_5CO_2H + NH_2CH_3$

(b) No hydrolysis occurs.

(c) $C_6H_5NH_2 + HO_2CCH_3$

(d) $NH_2CH_2CH_2NH_2 + 2CH_3CO_2H$

Review Exercises, Chapter 13

13.4 (a) $CH_3CO_2^-Na^+$ (b) $CH_3CH_2CH_2CO_2H$

(c) $C_6H_5CO_2^-Na^+$ (d) $CH_3CH_2CH_2CO_2^-K^+$

13.5

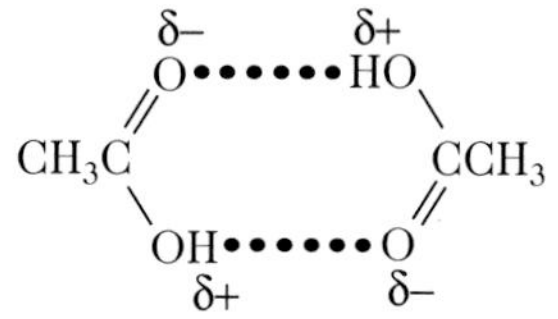

$$\overset{\delta-}{O}\cdots\cdots\overset{\delta+}{HO}$$

$CH_3C \qquad CCH_3$

$$\underset{\delta+}{OH}\cdots\cdots\underset{\delta-}{O}$$

13.7 $C < A < B$

13.9 Lactate ion

13.11 $B < A < C < D$

13.13 (a) $HO_2CCH_2CH_2CO_2H + 2OH^- \rightarrow$
$${}^-O_2CCH_2CH_2CO_2^- + 2H_2O$$

(b) $HOCH_2CH_2CH_2CO_2H + OH^- \rightarrow$
$$HOCH_2CH_2CH_2CO_2^- + H_2O$$

(c)

$HCCH_2CH_2CH_2CO_2H + OH^- \rightarrow$ (with C=O)
$$HCCH_2CH_2CH_2CO_2^- + H_2O \text{ (with C=O)}$$

(d) 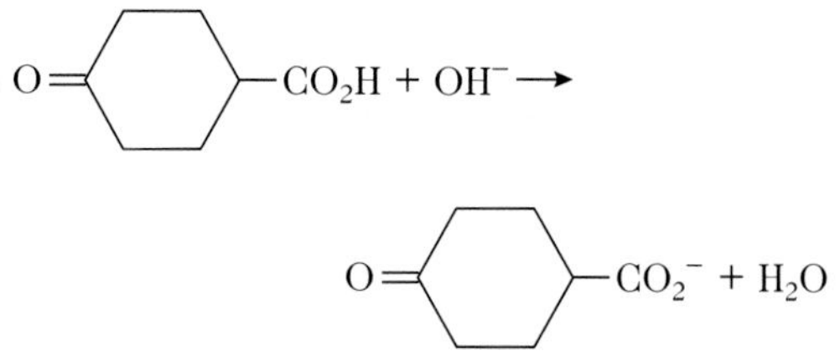 $O=$⬡$-CO_2H + OH^- \rightarrow$

$O=$⬡$-CO_2^- + H_2O$

13.14 **B.** The ionic compound, **A**, is very insoluble in a nonpolar solvent. **B** is not ionic.

13.16 (a) $CH_3CH_2CO_2^- + H^+ \rightarrow CH_3CH_2CO_2H$

(b) ${}^-O_2CCH_2CH_2CH_2CO_2^- + H^+ \rightarrow$
$$HO_2CH_2CH_2CH_2CO_2^-$$

(c) $NH_3 + H^+ \rightarrow NH_4^+$

13.18 (a)

$CH_3CH_2COCH_2CH_3 + H_2O$ (with C=O)

(b)

$(CH_3)_2CHCOCH_2CH_3 + H_2O$ (with C=O)

(c)

O_2N-⬡$-COCH_2CH_3 + H_2O$ (with C=O)

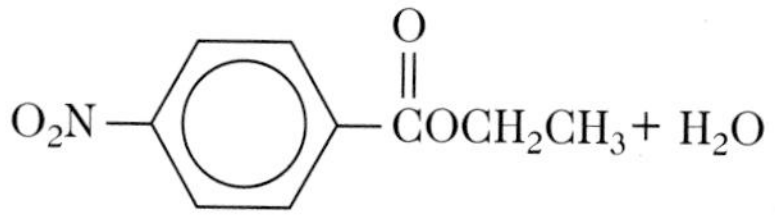

(d)

CH_3CH_2OC-⬡$-COCH_2CH_3$ (with two C=O)

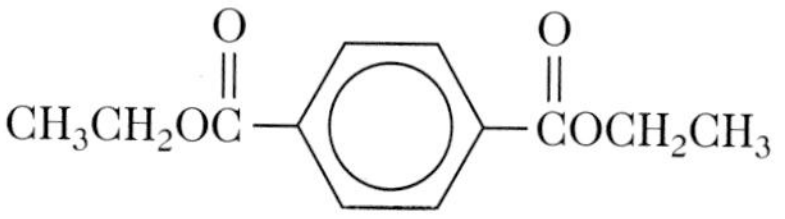

13.20 (a) $HCO_2CH(CH_3)_2$ (b) $C_6H_5CO_2CH_2CH_2CH_3$

13.22 $C < A < B < D$

13.24 (a)

$CH_3COCH_2CHCH_3 + H_2O \xrightarrow{H^+}$ (with C=O and CH₃ branch)

$CH_3COH + HOCH_2CHCH_3$ (with C=O and CH₃ branch)

(b)

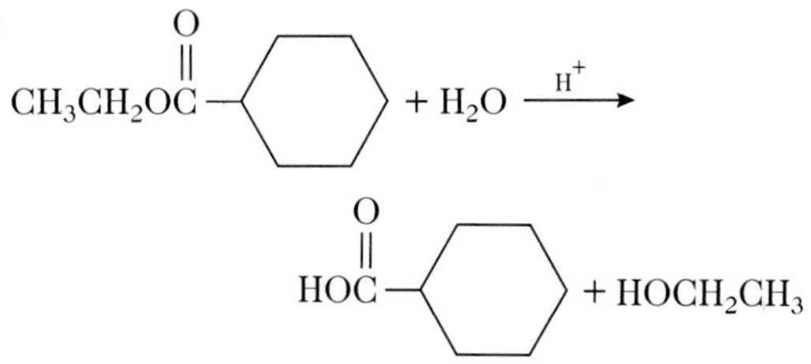

CH_3CH_2OC-⬡$+ H_2O \xrightarrow{H^+}$ (with C=O)

$HOC-$⬡$+ HOCH_2CH_3$ (with C=O)

(c) No reaction   (d) No reaction

13.26 $HOCH_2CHCH_2OH + CH_3(CH_2)_{12}CO_2H +$
$\qquad OH \qquad + CH_3(CH_2)_{10}CO_2H$
$$CH_3(CH_2)_{14}CO_2H$$

13.28 (a)

$CH_3CO^-Na^+ + HOCH_2CHCH_3$ (with C=O and CH₃ branch)

(b)

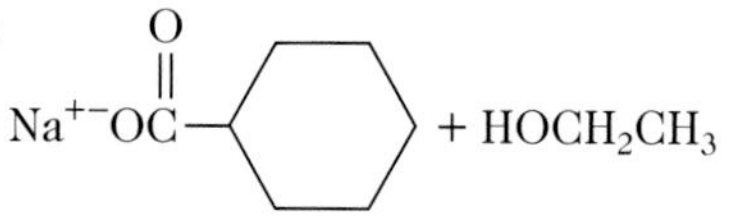

$Na^{+-}OC-$⬡$+ HOCH_2CH_3$ (with C=O)

(c) No reaction   (d) No reaction

13.30 $HOCH_2CHCH_2OH + CH_3(CH_2)_{12}CO_2^-Na^+ +$
$\qquad OH \qquad + CH_3(CH_2)_{10}CO_2^-Na^+$
$$CH_3(CH_2)_{14}CO_2^-Na^+$$

13.32 (a)

$$CH_3O-\overset{\overset{\displaystyle O}{\|}}{\underset{\underset{\displaystyle OH}{|}}{P}}-OH$$

(b)

$$CH_3CH_2O-\overset{\overset{\displaystyle O}{\|}}{\underset{\underset{\displaystyle OH}{|}}{P}}-O-\overset{\overset{\displaystyle O}{\|}}{\underset{\underset{\displaystyle OH}{|}}{P}}-OH$$

(c)

$$CH_3CH_2CH_2O-\overset{\overset{\displaystyle O}{\|}}{P}-O-\overset{\overset{\displaystyle O}{\|}}{P}-O-\overset{\overset{\displaystyle O}{\|}}{P}-OH$$
$$\qquad\quad\ \ \underset{OH}{|}\qquad\quad\ \underset{OH}{|}\qquad\quad\ \underset{OH}{|}$$

13.35 (a) (b) Two

$$NH_2CH_2\overset{\overset{\displaystyle O}{\|}}{C}NH\underset{\underset{\displaystyle CH_3}{|}}{CH}\overset{\overset{\displaystyle O}{\|}}{C}-$$

13.37 (a) $CH_3CONHC_6H_5$

(b) $CH_3CH_2CONHCH_3$

(c) $CH_3CH_2CH_2CON(CH_2CH_3)_2$

(d) $HCONH_2$

13.39 (a)

$$\underset{\underset{\displaystyle NH_2CHCH_2CO_2H}{}}{\overset{\overset{\displaystyle CH_3}{|}}{}}+NH_2CH_2CO_2H$$

(b)

$$2NH_3+HO_2C\underset{\underset{\displaystyle NH_2}{}}{CH_2}\overset{\overset{\displaystyle CH_3}{|}}{CH}CO_2H$$

(c) $CH_3\underset{\underset{\displaystyle NH_2}{|}}{CH}CH_2CH_2CH_2CO_2H$

(d) $2NH_3+(H_2CO_3)$.
The latter breaks up into CO_2+H_2O.

13.40 (a) $CH_3\underset{\underset{\displaystyle CH_3O}{|}}{CH}CO_2^-Na^+$ (b) $CH_3CH_2CO_2H+CH_3OH$

(c) CH_3CO_2H (d) ⬡

(e) $(CH_3)_2CHCH_2CO_2^-Na^++CH_3OH$

(f) $(CH_3)_2CH\overset{\overset{\displaystyle O}{\|}}{C}CH_3$ (g) $CH_3CH_2\overset{\overset{\displaystyle O}{\|}}{C}OCH_2CH_3$

(h) $CH_3CH_2CHO+2CH_3OH$

(i) No reaction (j)

$$C_6H_5\overset{\overset{\displaystyle O}{\|}}{C}OCH_2CH_3$$

(k) $CH_3\underset{\underset{\displaystyle Cl}{|}}{CH}CH_2CH_3$ (l) No reaction

13.42 (a) Water reacts quantitatively with alkenes, acetals or ketals, esters, and amides. The R groups can be alike or different. (H)R means that the group can be H or R.

$$\overset{\diagdown}{\diagup}C=C\overset{\diagup}{\diagdown}+H_2O\xrightarrow{H^+}-\underset{\underset{\displaystyle H}{|}}{C}-\underset{\underset{\displaystyle OH}{|}}{C}-$$
$$\text{Alkene}\qquad\qquad\qquad\text{Alcohol}$$

$$\underset{\underset{\displaystyle (H)R}{|}}{}(H)R\underset{}{}C(OR)_2+H_2O\xrightarrow{H^+}(H)R\overset{\overset{\displaystyle O}{\|}}{C}R(H)+2HOR$$
Acetal or   Aldehyde   Alcohol
ketal   or ketone

$$(H)R\overset{\overset{\displaystyle O}{\|}}{C}OR+H_2O\xrightarrow{H^+}(H)R\overset{\overset{\displaystyle O}{\|}}{C}OH+HOR$$
Ester   Carboxylic   Alcohol
               acid

$$(H)R\overset{\overset{\displaystyle O}{\|}}{C}NH_2+H_2O\xrightarrow{H^+}(H)R\overset{\overset{\displaystyle O}{\|}}{C}OH+NH_3$$
Amide   Carboxylic   Ammonia
             acid

(The H's on N of the amide can be replaced by one or two alkyl groups.)

(b) The groups that can be hydrogenated are alkenes, aldehydes and ketones, and disulfides.

$$\overset{\diagdown}{\diagup}C=C\overset{\diagup}{\diagdown}+H_2\xrightarrow{catalyst}-\underset{\underset{\displaystyle H}{|}}{C}-\underset{\underset{\displaystyle H}{|}}{C}-$$
Alkene   Alkane

$$(H)R\overset{\overset{\displaystyle O}{\|}}{C}R(H)+H_2\xrightarrow{catalyst}(H)R\underset{\underset{\displaystyle OH}{|}}{C}HR(H)$$
Aldehyde   Alcohol
or ketone

$$RSSR+H_2\xrightarrow{catalyst}2RSH$$
Disul-   Thioalcohol
fide

(c) Oxidizable groups in our study are 1° and 2° alcohols, aldehydes, and thioalcohols.

$$RCH_2OH+(O)\longrightarrow R\overset{\overset{\displaystyle O}{\|}}{C}H$$
1° Alcohol   Aldehyde

$$R\underset{\underset{\displaystyle R}{|}}{\overset{\overset{\displaystyle OH}{|}}{C}}HR+(O)\longrightarrow R\overset{\overset{\displaystyle O}{\|}}{C}R$$
2° Alcohol   Ketone

$$2RSH+(O)\longrightarrow RSSR+H_2O$$
Thioalcohol   Disulfide

$$R\overset{\overset{\displaystyle O}{\|}}{C}H+(O)\longrightarrow R\overset{\overset{\displaystyle O}{\|}}{C}OH$$
Aldehyde   Carboxylic acid

13.44 (a) 1.66 g benzoic acid (b) 28.3 mL 0.482 M HCl

13.45 (a) $CH_3CO_2H+CH_3OH$ (b)

$$CH_3\underset{\underset{\displaystyle }{}}{\overset{\overset{\displaystyle OH}{|}}{C}}HCH_2CH_3$$

(c) No reaction (d) $C_6H_5CO_2^-Na^+$
(Carbon dioxide and water are also products.)

(e) No reaction

(f) No reaction

(g) $CH_3CHO + 2HOCH_2CH_3$

(h) $CH_3CH_2CO_2H$

(i) No reaction

(j)
$$CH_3CH_2\overset{\overset{\displaystyle O}{\|}}{C}OCH_3$$

(k) $CH_3CH_2CO_2^- Na^+ + NH_3$

(l)
$$CH_3CH_2\overset{\overset{\displaystyle OCH_3}{|}}{C}HOCH_3$$

(m) $CH_3CH_2SSCH_2CH_3$

(n) $C_6H_5CO_2^- Na^+ + HOCH_2CH(CH_3)_2$

(o) $Cl^- {}^+NH_3CH_2CH_2CH(CH_3)_2$

(p) $CH_3CH_2CH_2CH_2OCH_3$

13.47 **B.** It is a carboxylic acid that will become an anion at the basic pH and so more soluble in water. (**A** is an ester and **C** is an amine.)

13.50 Two (or more) monomers are combined to give the polymer.

13.54 (a) Phenol group and carboxylic acid group

(b) Phenol group   (c) Carboxylic acid group

CHAPTER 14

Review Exercises, Chapter 14

14.5 (a) C (b) D

(c) A (d) B and D

14.7
$$HOCH_2\overset{\overset{\displaystyle O}{\|}}{C}H\overset{\overset{\displaystyle}{}}{\underset{\underset{\displaystyle OH}{|}}{C}}H$$

14.9 Polysaccharide

14.14

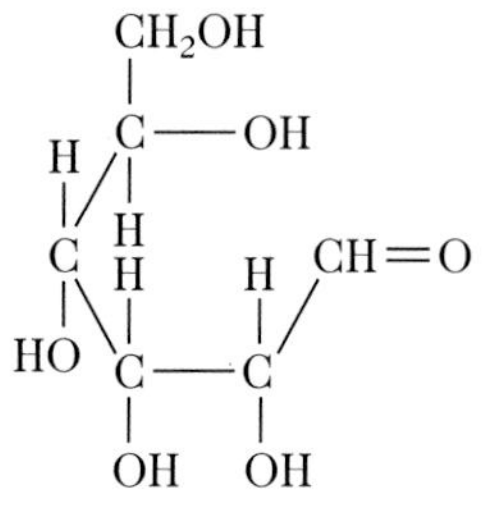

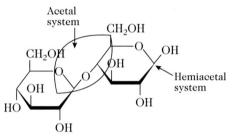

14.16

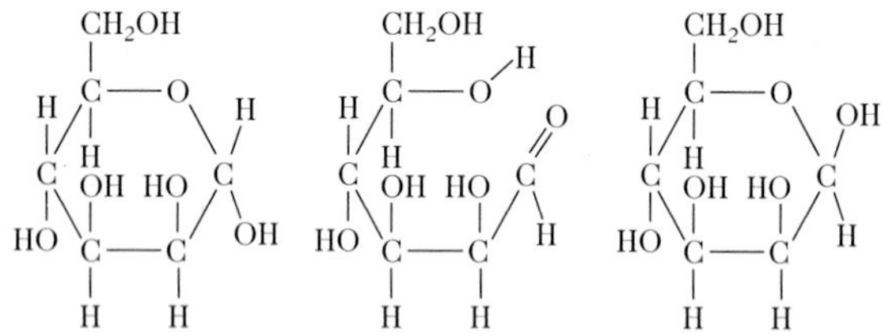

α-Mannose   Open form of mannose   β-Mannose

14.18 As molecules of the open-chain form are oxidized, the equilibrium continuously shifts to make more of the open chain form from the cyclic forms.

14.20

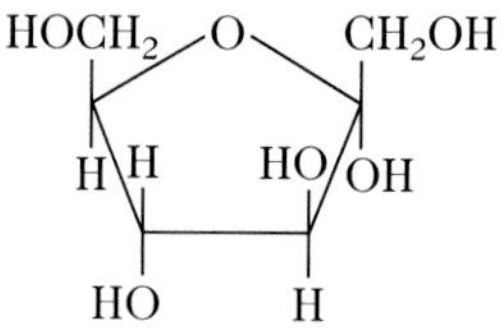

14.22

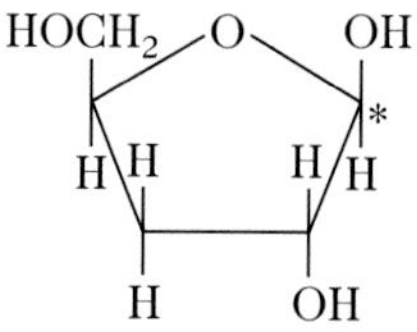

14.25 Yes, it has the potential aldehyde group (hemiacetal system).

14.27 (a) Dextrorotatory

(b) It will rotate the plane of plane-polarized light 26° to the left.

14.28 Their molecules are related as an object to its mirror image.

14.33 (a) Yes, see arrow.

(b) Yes, see enclosure.

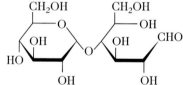

(c) Yes, it has the hemiacetal system so the open form of the corresponding ring (on the right) has an aldehyde group.

(d) Maltose has an α(1→4) bridge between the two rings.

(e) Two glucose molecules

14.35

14.39 They are polymers of α-glucose and have α(1→4) bridges.

14.40 Amylose is an entirely linear polymer of α-glucose in which all of the oxygen bridges are α(1→4), and amylopectin has branching in which α(1→6) bridges link amylose-like strands to other amylose strands.

14.45 Over a period of a week some of the sucrose (a nonreducing sugar) hydrolyzes to give glucose and fructose, both being reducing sugars that give a positive Tollens' test.

CHAPTER 15

Practice Exercises, Chapter 15

1. $CH_3(CH_2)_7$ $(CH_2)_7CO_2H$

$$C=C$$

H $\quad$ H

2. $CH_3(CH_2)_{26}CO_2(CH_2)_{25}CH_3$

3. I + 3NaOH $\longrightarrow$

$$\begin{array}{l} O \\ \| \end{array}$$
$CH_2OH + Na^{+-}OC(CH_2)_7CH=CH(CH_2)_7CH_3$

$|$
CHOH $\qquad$ O
$\qquad\qquad$ $\|$
$CH_2OH + Na^{+-}OC(CH_2)_{16}CH_3$

Glycerol $\qquad$ O
$\qquad\qquad\qquad$ $\|$
$\quad + Na^{+-}OC(CH_2)_7CH=CHCH_2CH=CH(CH_2)_4CH_3$

4. I + 3H$_2$ $\xrightarrow{\text{catalyst}}$
$$\begin{array}{l} O \\ \| \end{array}$$
$CH_2OC(CH_2)_{16}CH_3$
$|$ $\quad$ O
$\quad\quad$ $\|$
$CHOC(CH_2)_{16}CH_3$
$|$ $\quad$ O
$\quad\quad$ $\|$
$CH_2OC(CH_2)_{16}CH_3$

Review Exercises, Chapter 15

15.1 It is not obtainable from living plants or animals.

15.3 It is a soluble in water, and it isn't present in plant or animal sources.

15.7 (a) $CH_3(CH_2)_{14}CO_2H + NaOH \rightarrow$
$\qquad\qquad CH_3(CH_2)_{14}CO_2^- Na^+ + H_2O$
$\qquad\qquad\qquad\qquad\qquad\quad H^+$
(b) $CH_3(CH_2)_{14}CO_2H + CH_3OH \xrightarrow{\quad}$
$\qquad\qquad CH_3(CH_2)_{14}CO_2CH_3 + H_2O$

15.9 A. B is branched and has an uneven number of carbon atoms.

15.11
$$\begin{array}{l} O \\ \| \end{array}$$
$CH_2OC(CH_2)_7CH=CHCH_2CH=CHCH_2CH=CHCH_2CH_3$
$|$ $\quad$ O
$\quad\quad$ $\|$
$CHOC(CH_2)_7CH=CHCH_2CH=CH(CH_2)_4CH_3$
$|$ $\quad$ O
$\quad\quad$ $\|$
$CH_2OC(CH_2)_{14}CH_3$

15.13 HOCH$_2$CHCH$_2$OH
$\qquad\qquad |$
$\qquad\qquad$ OH
$\quad + HO_2C(CH_2)_7CH=CHCH_2CH=CH(CH_2)_4CH_3$
$\quad + HO_2C(CH_2)_{12}CH_3 + HO_2C(CH_2)_7CH=CH(CH_2)_7CH_3$

15.14 HOCH$_2$CHCH$_2$OH
$\qquad\qquad |$
$\qquad\qquad$ OH
$\quad + Na^{+-}O_2C(CH_2)_7CH=CHCH_2CH=CH(CH_2)_4CH_3$
$\quad + Na^{+-}O_2C(CH_2)_{12}CH_3 + Na^{+-}O_2C(CH_2)_7CH=CH(CH_2)_7CH_3$

15.16 More than one triacylglycerol structure is possible. The following is one, but another would have the oleic acid unit on the middle carbon of the glycerol unit.

$$\begin{array}{l} O \\ \| \end{array}$$
$CH_2OC(CH_2)_7CH=CH(CH_2)_7CH_3$
$|$ $\quad$ O
$\quad\quad$ $\|$
$CHOC(CH_2)_{10}CH_3$
$|$ $\quad$ O
$\quad\quad$ $\|$
$CH_2OC(CH_2)_{10}CH_3$

15.18 The triacylglycerol molecules that are present have several alkene units per molecule, so the substances are more "polyunsaturated" than the animal fats.

15.20 Butter melts on the tongue:, lard and tallow do not.

15.22 C. A is ruled out because *both* the acid and alcohol portions of the wax molecule are usually long-chain. B is ruled out because *both* of these portions are likely to have an even number of carbons.

15.24 The phosphate ester unit; the negative charge is on oxygen. A nitrogen atom carries the positive charge.

15.26 The OH (alcohol) groups on the sugar unit

15.31 (a)
$$\begin{array}{l} O \\ \| \end{array}$$
$CH_2OC(CH_2)_7(CH=CHCH_2)_3CH_3$
$|$ $\quad$ O
$\quad\quad$ $\|$
$CHOC(CH_2)_7CH=CH(CH_2)_7CH_3$
$|$ $\quad$ O
$\quad\quad$ $\|$
$CH_2OPOCH_2CH_2\overset{+}{N}(CH_3)_3$
$\qquad |$
$\qquad O^-$

(b) A glycerophospholipid, because it is based on glycerol, not sphingosine

(c) A lecithin, because its hydrolysis would give 2-(trimethylamino)ethanol

15.34 Vitamin D$_3$

15.36 Cholesterol

15.38 In various lipoprotein complexes

15.40 The hydrophobic tails intermesh with each other between the two layers of the bilayer.

15.42 The water-avoiding properties of the hydrophobic units and the water-attracting properties of the hydrophilic units

15.48 A synthetic detergent, because it works better in hard water

15.52 It's an *ester*, and digestive enzymes catalyze the hydrolysis of esters.

15.55 (a) Yes; it has both hydrophobic sections (hydrocarbon-like) and polar groups (OH).

(b) No, steroids have *three* six-membered rings plus a five-membered ring.

CHAPTER 16

Practice Exercises, Chapter 16

1. Glycine $^+NH_3CH_2CO_2^-$

   Alanine $^+NH_3CHCO_2^-$
   |
   CH_3

   Lysine $^+NH_3CHCO_2^-$
   |
   $CH_2CH_2CH_2CH_2NH_2$

   Glutamic acid $^+NH_3CHCO_2^-$
   |
   $CH_2CH_2CO_2H$

2. (a)
   $$^+NH_3CHCO^-$$
   |
   $CH_2CO_2^-$

   (b)
   $$^+NH_3CHCO^-$$
   |
   CH_2CONH_2

3.
   $$^+NH_3CHCO^- \qquad ^+NH_2$$
   |
   $CH_2CH_2CH_2NHCNH_2$

4. Hydrophilic; neutral (The side chain has an amide group, not an amino group.)

5.
   $$^+NH_3CHC—NHCHCO^- \qquad ^+NH_3CHC—NHCHCO^-$$
   | | | |
   CH_3 CH_2 CH_2 CH_3
              | |
              CH_2 CH_2
              | |
              CO_2H CO_2H

   Ala-Glu   Glu-Ala

Review Exercises, Chapter 16

16.1 **B.** Its NH_3^+ group is not on the same carbon that holds the CO_2^- group, the α-carbon.

16.3 $^+NH_3CH_2CO_2H$

16.6 **A.** It has amine-like groups that can both donate hydrogen bonds to water molecules and accept them. (**B** has an alkyl group side chain, which is hydrophobic.)

16.8 (a) At a pH of 1. In an acidic medium, all protons that can be accepted by groups of the amino acid will have been put into place.

(b) To the negatively charged electrode

(c) Hydrophilic

16.12 In the presence of additional acid, the following equilibrium shifts to the right in accordance with Le Châtelier's principle. This neutralizes the extra acid.

$$^+NH_3CH_2CO_2^- + H^+ \rightleftharpoons\ ^+NH_3CH_2CO_2H$$

In the presence of additional base, the following equilibrium shifts to the right in accordance with Le Châtelier's principle. This neutralizes the extra base.

$$^+NH_3CH_2CO_2^- + OH^- \rightleftharpoons NH_2CH_2CO_2^- + H_2O$$

16.13 A has an amide bond not to the amino group of the α-position of an amino acid unit but to an amino group of a side chain (that of lysine). **B** has a proper peptide bond.

16.15
$$^+NH_3CHC—NHCHCO^- \qquad ^+NH_3CHC——NHCHCO^-$$
   | | | |
   H $(CH_2)_2CO_2H$ $(CH_2)_2CO_2H$ H

   Gly-Glu   Glu-Gly

16.17 Gly-Cys-Ala Cys-Ala-Gly Ala-Gly-Cys
   Gly-Ala-Cys Cys-Gly-Ala Ala-Cys-Gly

16.19
$$^+NH_3CHC—NHCHC—NHCHC—NHCHC—NHCHCO^-$$
   | | | | |
   CH_3CHCH_3 $CH_2C_6H_5$ CH_3 H $CH_2CH(CH_3)_2$

16.21 (a) **A**

(b) **B.** It has only hydrophilic side chains. All those in **A** are hydrophobic.

16.23 Primary structure

16.25 Tertiary

16.27 It is a right-handed helix stabilized by hydrogen bonds between carbonyl oxygen atoms and H atoms on N atoms farther down the helix. The side chains project to the outside of the helix.

16.29 It aids in the hydroxylation of proline residues without which collagen is not adequately made.

16.31 No, they represents portions of the secondary structure of a polypeptide and often both features are present.

16.33 The force of attraction between a site bearing a full negative charge (e.g., a CO_2^- group on a glutamic acid or aspartic acid side chain) and a site with a full positive charge (e.g., an NH_3^+ group on a lysine side chain).

16.35 It consists of more than two polypeptides associated together in a specific way, each with primary, secondary, and tertiary structure.

16.37
$$^+NH_3CHCO^- + \ ^+NH_3CHCO^- + \ ^+NH_3CHCO^- +$$
   | | |
   CH_2OH CH_3 CH_3CHCH_3

$$^+NH_3CHCO^- + \ ^+NH_3CH_2CO_2^-$$
   |
   $(CH_2)_4NH_2$

16.39 *Digestion* is the hydrolysis of peptide bonds to give a mixture of amino acids. *Denaturation* is the disorganization of the overall shape of a protein without necessarily the breakup of peptide bonds.

16.41 A change in the value of something, like concentration, from one place to another.

16.46 Metabolic, energy-consuming reactions involving membrane components drive the transport of substances through membranes.

16.48 Fibrous proteins are insoluble in water; globular proteins are more soluble.

16.50 They both have strengthening functions in tissue; both are fibrous proteins. The action of hot water on collagen turns it to gelatin, but elastin is unaffected in this way.

16.52 Fibrin is the protein that forms a blood clot. Fibrinogen is changed to fibrin by the clotting mechanism.

16.59 (a) Yes. What is shown is a tripeptide written backward from the conventional way.

     (b) Phe

     (c) Phe−Ala−Cys (N terminus to C terminus)

     (d) No effect

CHAPTER 17

Review Exercises, Chapter 17

17.3 (a) An apoenzyme is the wholly polypeptide part of the enzyme.

     (b) A cofactor is a nonpolypeptide molecule or ion needed to make the complete enzyme.

     (c) A coenzyme is one kind of cofactor, an organic molecule.

17.5 It catalyzes the rapid reestablishment of the equilibrium after it has been disturbed.

17.7 Riboflavin

17.9 $H^+ + NADH + FAD \longrightarrow NAD^+ + FADH_2$
or $NADH/H^+ + FAD \longrightarrow NAD^+ + FADH_2$

17.11 (a) An oxidation

     (b) The transfer of a methyl group

     (c) A reaction with water

     (d) An oxidation−reduction equilibrium

17.13 Hydrolysis is a kind of reaction catalyzed by a hydrolase enzyme.

17.15 CK(MM) in skeletal muscle, CK(BB) in brain, and CK(MB) in heart muscle

17.17 By the necessity of the fitting of the substrate molecule to the surface of the enzyme much as a key must fit to a particular lock.

17.19 The inhibitor is a nonsubstrate molecule resembling the true substrate enough to enable the binding of the inhibitor to the enzyme. By thus occupying the active site, the enzyme's work is inhibited.

17.21 Because it shuts down a pathway when it is no longer needed but lets the pathway occur when it is needed.

17.23 A nonsubstrate that activates an enzyme

17.25 In the cytosol, 10^{-7} mol/L; in the fluid just outside the cell, 10^{-3} mol/L. Ca^{2+} ions that enter the cell are pumped back out after their work is complete. The calcium channels through the membrane are kept shut until an appropriate signal arrives.

17.27 No, most Ca^{2+} is held by calmodulin or troponin.

17.29 An enzyme might be activated or a muscle might be induced to contract. After such action, Ca^{2+} is pumped back out of the cell.

17.31 A proteolytic, blood clot-dissolving enzyme. It normally circulates in its inactive form, plasminogen.

17.33 *Antimetabolites* are compounds that interfere with the metabolism of disease-causing bacteria. *Antibiotics* are those antimetabolites that are made by microorganisms.

17.35 Viral hepatitis

17.37 The CK isoenzymes, GOT, and the LD isoenzymes

17.39 CK(MM)

17.41 Streptokinase and tissue plasminogen activator (TPA). TPA occurs in human blood and it can be manufactured by recombinant DNA technology.

17.43 (a) Adrenal glands (b) Nerve cells

17.45 They are primary chemical messengers.

17.47 When activated, adenylate cyclase catalyzes the formation of cyclic AMP (from ATP), which then activates an enzyme inside the target cell.

17.49 The cyclic AMP is hydrolyzed to AMP.

17.51 They are hydrocarbon-like and so slip through a hydrocarbon-like lipid bilayer.

17.53 It is hydrolyzed back to acetic acid and choline. The enzyme is choline acetyltransferase.

17.55 It blocks the receptor protein for acetylcholine.

17.57 They catalyze the deactivation of neurotransmitters such as norepinephrine and thus reduce the level of signal-sending activity that depends on such neurotransmitters.

17.59 They inhibit the reabsorption of norepinephrine by the presynaptic neuron and thus reduce the rate of its deactivation by the monoamine oxidases.

17.61 Dopamine

17.63 They accelerate the release of dopamine from the presynaptic neuron.

17.65 BDNF protects those cells that make dopamine from further degeneration, stimulates the cells to recover, and protects susceptible cells (at least in animal studies).

17.67 Enkephalin molecules enter pain-signaling neurons and inhibit the release of substance P, a neurotransmitter that helps to send pain signals. Thus enkephalin, like an opium drug, inhibits pain.

17.69 A substance made by a signal-receiving nerve cell that moves back to the signal-sending cell to strengthen the connection between the two cells

17.71 A disease-causing microorganism or virus

17.73 Cellular immunity, by attacking the helper T cells

17.75 The presence of glycoproteins with highly individual oligosaccharide units. This enables a lock-and-key kind of specificity between antibody and antigen.

17.77 It works to generate an antibody when introduced into the presence of type B blood.

17.79 A red cell with an oligosaccharide that functions as an antibody in the presence of the introduced type B blood

17.81 Inhibition by a nonsubstrate (or competitive inhibition)

17.83 165.1 mg of isoleucine

CHAPTER 18
Review Exercises, Chapter 18

18.2 Saliva, gastric juice, pancreatic juice, and intestinal juice

18.4 (a) Pepsin from its zymogen in gastric juice; trypsin, chymotrypsin, and elastase from zymogens in pancreatic juice

(b) Lipases provided in gastric juice, pancreatic juice, and intestinal juice

(c) Amylases in saliva, pancreatic juice, and intestinal juice

(d) Sucrase in intestinal juice

(e) Carboxypeptidase from its zymogen in pancreatic juice; aminopeptidase from its zymogen in intestinal juice

(f) Nucleases in pancreatic juice and intestinal juice

18.6 (a) Peptide (amide) bonds in proteins

(b) Acetal systems in carbohydrates

(c) Ester groups in triacylglycerols

18.8 They would catalyze the digestion of proteins that make up part of the pancreas to the serious harm of this organ.

18.10 (a) Lubricates the food

(b) Protects the stomach lining from gastric acid and pepsin

18.12 It helps to coagulate the protein in milk so that this protein stays longer in the stomach where it can be digested with the aid of pepsin.

18.14 Pancreatic juice delivers its zymogens and enzymes into the duodenum, whereas the enzymes of the intestinal juice work within cells of the intestinal wall.

18.16 They recombine to molecules of triacylglycerol during their migration from the intestinal tract toward the lymph ducts.

18.18 The concentration of soluble proteins is greater in blood.

18.22 Na^+ is in blood plasma and other extracellular fluids; K^+ is chiefly in intracellular fluids. The two ions help to maintain osmotic pressure relationships; are part of the regulatory mechanisms for acid–base balance; and participate in the smooth working of the muscles and the nervous system.

18.24 Blood pressure and osmotic pressure. The natural return of fluids to the blood on the venous side from the interstitial compartment is not balanced by the now reduced blood pressure on the venous side, so fluids return to the blood from which they left on the arterial side.

18.26 (a) Blood proteins leak out, which allows water to leave the blood and enter interstitial spaces throughout various tissues.

(b) Blood proteins are lost to the blood by being consumed, which also leads to loss of water from the blood and its appearance in interstitial compartments.

(c) Capillaries are blocked at the injured site reducing the return of blood in the veins, so fluids accumulate at the site.

18.28 Hemoglobin

18.30 $HHb + O_2 \rightleftharpoons HbO_2^- + H^+$

(a) To the left   (b) To the left

(c) To the right   (d) To the left

(e) To the left   (f) To the right

18.32 It generates H^+ needed to convert HCO_3^- to CO_2 and H_2O and to convert $HbCO_2^-$ to HHb and CO_2.

18.34 It helps to shift the following equilibrium to the left:

$$HHb + O_2 \rightleftharpoons HbO_2^- + H^+$$

18.36 As HCO_3^- in the serum and as $HbCO_2^-$ (carbaminohemoglobin) in red cells

18.38 The exchange of a chloride ion for a bicarbonate ion between a red blood cell and blood serum. This brings Cl^- inside the red cell when it is needed to help deoxygenate HbO_2^-.

18.40 Fetal hemoglobin can take oxygen from the oxyhemoglobin of the mother's blood and thus ensure that the fetus gets needed oxygen.

18.42 The pH of the blood decreases in both, but both pCO_2 and $[HCO_3^-]$ increase in respiratory acidosis and both decrease in metabolic acidosis.

18.44 Hypoventilation is observed in metabolic alkalosis, and isotonic ammonium chloride can be given to neutralize the excess base. Involuntary hypoventilation is observed in respiratory acidosis, and isotonic sodium bicarbonate might be given to neutralize excess acid.

18.46 In respiratory alkalosis. The involuntary loss of CO_2 reduces the level of this base neutralizer in the blood, allowing an increase in base.

18.48 In respiratory acidosis

18.50 (a) Respiratory alkalosis
(b) Metabolic alkalosis
(c) Respiratory acidosis
(d) Respiratory acidosis
(e) Metabolic acidosis
(f) Respiratory alkalosis
(g) Metabolic acidosis
(h) Metabolic alkalosis
(i) Respiratory acidosis
(j) Respiratory acidosis

18.52 CO_2 is removed at an excessive rate, which removes the blood's chief base neutralizer, so the blood becomes more alkaline and the pH of the blood rises (alkalosis).

18.54 The loss of acid with the loss of the stomach contents is followed by a loss of acid from the blood, which means an increase in the blood's pH and alkalosis.

18.56 Hypocapnia

18.58 The blood has become more concentrated in solutes.

18.60 Aldosterone is secreted from the adrenal cortex, and it instructs the kidneys to retain sodium ion in the blood.

18.62 The rate of diuresis increases.

18.64 The distention of the stomach caused by entering food causes the hormone gastrin to be secreted from cells of the gastric lining of the stomach. Gastrin stimulates the release of gastric juice.

18.66 A molecule of a monoacylglycerol retains a considerable hydrophobic unit (the fatty acyl "tail") and so enters the hydrophobic region of a membrane more easily than a molecule like glycerol with three hydrophilic OH groups.

18.68 The blood produced while at a high altitude has a higher concentration of hemoglobin. This aids in their ability to use oxygen during a race.

18.72 Oxygen uptake is reduced. In CO_2-enriched air, the concentration of $CO_2(aq)$ in the blood must increase (it's not being removed as gaseous CO_2), which shifts Equilibrium 18.3 to the left. This raises the level of H^+, which shifts Equilibrium 18.1 to the left.

CHAPTER 19

Review Exercises, Chapter 19

19.2

$$\text{Adenosine} - O - \overset{\displaystyle O}{\underset{\displaystyle O^-}{\overset{\displaystyle \|}{P}}} - O - \overset{\displaystyle O}{\underset{\displaystyle O^-}{\overset{\displaystyle \|}{P}}} - O - \overset{\displaystyle O}{\underset{\displaystyle O^-}{\overset{\displaystyle \|}{P}}} - O^-$$

19.4 The singly and doubly ionized forms of phosphoric acid, $H_2PO_4^- + HPO_4^{2-}$

19.6 (a) The aerobic synthesis of ATP
(b) The synthesis of ATP when a tissue operates anaerobically
(c) The supply of metabolites for the respiratory chain
(d) The supply of metabolites for the respiratory chain and for the citric acid cycle

19.8 An increase in its supply of ADP. The need to convert ADP back to ATP is met by metabolism, which requires oxygen.

19.10 (a) E < B < C < A < D
(b) B < C < A < D

19.12 A fatty acid

19.14 When a cell is temporarily low on oxygen for running the respiratory chain (as a source of ATP), the cell can continue (for a while) to make ATP anyway.

19.16 Glycolysis can operate and make ATP even when the oxygen supply is low, so a tissue in oxygen debt can continue to function.

19.18 It is reoxidized to pyruvate, which then undergoes oxidative decarboxylation to the acetyl group in acetyl CoA. This enters the citric acid cycle.

19.20 When the respiratory chain starts up, the citric acid cycle must start up to keep the chain going.

19.22 Two

19.24 ADP. The cycle helps the cell make ATP from ADP, so activation by ADP is logical.

19.26 An enzyme like isocitrate dehydrogenase, because the reaction is a dehydrogenation (not an addition of water to a double bond, catalyzed by an enzyme like aconitase)

19.28 Respiratory chain

19.30 It is a riboflavin-containing coenzyme that in its reduced form, $FADH_2$, passes electrons and H^+ into the respiratory chain.

19.32 The flow of protons across the inner mitochondrial membrane

19.34 Causes the release of ATP.

19.36 E < D < A < C < B

19.38 The citric acid cycle processes the acetyl units manufactured by the β-oxidation pathway and so fuels the respiratory chain.

19.40

$$CH_3CH_2CH_2\overset{\overset{\displaystyle O}{\|}}{C}SCoA + FAD \longrightarrow$$

$$CH_3CH\!=\!CH\overset{\overset{\displaystyle O}{\|}}{C}SCoA + FADH_2$$

$$CH_3CH\!=\!CH\overset{\overset{\displaystyle O}{\|}}{C}SCoA + H_2O \longrightarrow CH_3\overset{\overset{\displaystyle OH}{|}}{C}HCH_2\overset{\overset{\displaystyle O}{\|}}{C}SCoA$$

$$CH_3\overset{\overset{\displaystyle OH}{|}}{C}HCH_2\overset{\overset{\displaystyle O}{\|}}{C}SCoA + NAD^+ \longrightarrow$$

$$CH_3\overset{\overset{\displaystyle O}{\|}}{C}CH_2\overset{\overset{\displaystyle O}{\|}}{C}SCoA + NADH/H^+$$

$$CH_3\overset{\overset{\displaystyle O}{\|}}{C}CH_2\overset{\overset{\displaystyle O}{\|}}{C}SCoA + CoASH \longrightarrow 2CH_3\overset{\overset{\displaystyle O}{\|}}{C}SCoA$$

No more turns of the β-oxidation pathway are possible.

19.42 NADH passes its hydrogen into the respiratory chain and is changed back to NAD$^+$.

19.44 The entire collection of nitrogen compounds found anywhere in the body

19.46 Infancy

19.48

$$CH_3\overset{\overset{\displaystyle NH_3^+}{|}}{C}HCO_2^- + {}^-O_2CCH_2CH_2\overset{\overset{\displaystyle O}{\|}}{C}CO_2^- \longrightarrow$$

$$CH_3\overset{\overset{\displaystyle O}{\|}}{C}CO_2^- + {}^-O_2CCH_2CH_2\overset{\overset{\displaystyle NH_3^+}{|}}{C}HCO_2^-$$

$${}^-O_2CCH_2CH_2\overset{\overset{\displaystyle NH_3^+}{|}}{C}HCO_2^- + NAD^+ + H_2O \longrightarrow$$

$${}^-O_2CCH_2CH_2\overset{\overset{\displaystyle O}{\|}}{C}CO_2^- + NADH/H^+ + NH_4^+$$

19.50

19.52

$$C_6H_5CH_2\overset{\overset{\displaystyle O}{\|}}{C}CO_2H$$

19.54 Fatty acids

19.58 A proton or hydrogen ion, H$^+$. If the level of hydrogen ion increases, the problem is acidosis.

19.60 An above normal concentration of the ketone bodies in the blood

19.62 Enough acetone vapor in exhaled air to be detected by its odor

19.64 Metabolic acidosis brought on by an increase in the level of the ketone bodies in the blood.

19.66 Diuresis is accelerated to remove ketone bodies from the blood, and their removal requires the simultaneous removal of water, so the urine volume increases.

19.68 Amino acids are catabolized at a faster than normal rate to participate in gluconeogenesis, and their nitrogen is excreted largely as urea.

19.70 (a) $CH_3CH\!=\!O$

(b) Acetic acid, CH_3CO_2H (or acetate ion, $CH_3CO_2^-$)

(c) α-Ketoglutarate

CHAPTER 20

Review Exercises, Chapter 20

20.1 Glucose, fructose, and galactose

20.3 The concentration of reducing monosaccharides, chiefly glucose, in the blood is called the *blood sugar level*. The *normal fasting level* is the blood sugar level after several hours of fasting.

20.5 99 mg/dL

20.7 The lack of glucose means the lack of the one nutrient most needed by the brain.

20.9 Muscle tissue

20.11 Liver, but not muscles, has the enzyme glucose-6-phosphatase that catalyzes the hydrolysis of glucose-6-phosphate. This frees glucose for release from the liver to the bloodstream.

20.13 Glucagon, because it works better at the liver than epinephrine in initiating glycogenolysis, and when glycogenolysis occurs at the liver there is a mechanism for releasing glucose into circulation.

20.15 It is a polypeptide hormone released from the beta cells of the pancreas in response to an increase in the blood sugar level, and it acts most effectively at adipose tissue.

20.17 Too much insulin leads to a sharp decrease in the blood sugar level and, therefore, a decrease in the supply of glucose, the chief nutrient for the brain.

20.19 The body's ability to manage dietary glucose without letting the blood sugar level swing too widely from its normal fasting level

20.21 (a) The blood sugar level initially rises rapidly, but then drops sharply and slowly comes back to normal.

(b) The blood sugar level, already high to start, rises much higher and never sharply drops. It only very slowly comes back down.

20.23 It makes glucose out of smaller molecules obtained by the catabolism of fatty acids and amino acids.

20.25 (a) Alanine (b) Aspartic acid

20.27 They become reconstituted into triacylglycerols.

20.29 They are stored in adipose tissue as triacylglycerols.

20.31 The pentose phosphate pathway of glucose catabolism

20.33 The body can make several amino acids itself.

20.35

$$^{-}O_2CCH_2CH_2\overset{O}{\overset{\|}{C}}CO_2^{-} + NH_4^{+} + NADPH/H^{+} \rightleftharpoons {}^{-}O_2CCH_2CH_2\overset{NH_3^{+}}{\overset{|}{C}}HCO_2^{-} + NADP^{+} + H_2O$$

20.37

$$(CH_3)_2CH\overset{O}{\overset{\|}{C}}CO_2H$$

20.39 Type I diabetics cannot make insulin.

20.41 Type II

20.43 The beta-cells of the pancreas

20.45 One explanation: Insulin receptors "wear out." Another: The beta-cells of the pancreas "wear out." Still another: Amylin (a recently discovered protein component of what is released by the beta-cells) suppresses glucose uptake.

20.47 Its aldehyde group reacts with amino groups on proteins and genes to form C=N systems by means of which the glucose units are tied to other molecules.

20.49 They change to more permanent Amadori compounds, which constitute more permanent changes in cell molecules.

20.52 They unload some of their triacylglycerol.

20.54 Some cholesterol has originated in the diet and some has been synthesized in the liver.

20.56 Triacylglycerol

20.58 The liver

20.60 The synthesis of steroids and the fabrication of cell membranes

20.62 When the receptor proteins are reduced in number, the liver cannot remove cholesterol from the blood, so the blood cholesterol level increases.

20.64 The concentration of cholesterol, which has a higher density than triacylglycerols, is greater in HDL

20.66 HDL

CHAPTER 21

Practice Exercises, Chapter 21

1. (a) Proline (b) Arginine
   (c) Glutamic acid (d) Lysine
2. (a) Serine (b) CT (chain termination)
   (c) Glutamic acid (d) Isoleucine

Review Exercises, Chapter 21

21.1 *Chromatin* is a substance made up of nucleosomes—complexes of proteins (called histones) around which are wrapped coiled molecules of DNA—connected by DNA strands. *Chromosomes* are rodlike bodies made microscopically observable when chromatin thickens at the onset of the process that leads to cell division.

21.3 Duplicate copies of DNA are made.

21.5 Nucleotides

21.9 In the sequence of bases attached to the deoxyribose units of the main chain

21.11 DNA occurs as a double helix, and it alone has the base thymine (T). RNA alone has the base uracil (U). (The remaining three bases, A, G, and C, are the same in both DNA and RNA.) DNA molecules have one less OH group per pentose unit than RNA molecules.

21.12 A and T pair to each other, so they must be in a 1:1 ratio regardless of the species. Similarly, G and C pair to each other and must be in a 1:1 ratio.

21.14 Hydrophobic interactions stabilize the helices. (Hydrogen bonds hold two helices together.)

21.15 Given segment: $5' \rightarrow 3'$ AGGCTGA
Opposite segment: TCCGACT $3' \leftarrow 5'$

21.18 The introns are b, d, and f, because they are the longer segments.

21.22 A *codon* is a specific triplet of bases that corresponds to a specific amino acid residue in a polypeptide, and mRNA consists of a continuous sequence of codons.

21.25 (a) Phenylalanine (b) Serine
(c) Threonine (d) Aspartic acid

21.26 Writing them in the 5′ to 3′ direction:
(a) AAA (b) GGA
(c) UGU (d) AUC

21.27 (a) $5' \rightarrow 3'$ UUUCUUAUAGAGUCCCCAACAGAU
(b) $5' \rightarrow 3'$ UUUUCCACAGAU
(c) Phe-Ser-Thr-Asp

21.30 (a) A large number of sequences are possible because three of the specific amino acid residues are coded by more than one codon. The possibilities are indicated by:

Met	Ala	Trp	Ser	Tyr
AUG	GCU	UGG	UCU	UAU ($5' \rightarrow 3'$)
	GCC		UCC	UAC
	GCA		UCA	
	GCG		UCG	

(b) CAU ($5' \rightarrow 3'$) or if ($3' \rightarrow 5'$), then UAC

21.35 A molecular feature on the surface of a virus particle fits by a flexible lock-and-key mechanism to a specific glycoprotein on the membrane of one specific kind of host cell.

21.37 RNA replicase is able to direct the synthesis of RNA from the "directions" encoded on RNA. A normal host cell does not contain RNA replicase, so the virus either must bring it along or must direct its synthesis inside the host cell.

21.39 That the cell is able to move information backward, from RNA to DNA (Normally information always flows from DNA to RNA.)

21.41 Cancer-causing

21.48 The use of recombinant DNA to make genes and the products of genes

21.50 The synthesis of a transmembrane protein that lets chloride ion pass through the membranes of mucous cells in the lungs and the digestive tract. The reduced movement of Cl^- out of the cell means that less water is outside of the cell, and the mucus is thereby thickened and made more viscous. Breathing is thereby impaired.

21.53 The exterior surface carries the recognition molecules needed to find the appropriate host cells.

21.55 All of the cells of an individual have the entire genome.

21.58 (a) A dinucleotide. It has two side chain bases and two ribose units.

     (b) Of RNA, because the sugar units are those of ribose

     (c) At the top

     (d) AC

CHAPTER 22

Practice Exercises, Chapter 22

1. $^{131}_{53}I \longrightarrow ^{131}_{54}Xe + ^{0}_{-1}e + ^{0}_{0}\gamma$

2. $^{239}_{94}Pu \, ^{235}_{92}U + ^{4}_{2}He + ^{0}_{0}\gamma$

Review Exercises, Chapter 22

22.1 *Radioactive decay* is a change that occurs to radionuclides whereby they spontaneously emit radiation and change to other isotopes. In becoming other isotopes, the radionuclides undergo *transmutation* as well. Thus, radioactive decay is accompanied by transmutation.

22.3 Gamma radiation

22.5 It is the most massive of the particles and it carries the largest charge. Therefore, it collides very quickly with a molecule in the air or other matter that it enters and so travels the shorter distance (is least penetrating).

22.7 There is no change in mass number, because the mass number of the beta particle is 0. The atomic number increases by one because the loss of $^{0}_{-1}e$ has the effect of changing a neutron into a proton.

22.9 Isotope. A radionuclide is a radioactive *isotope* that consists of *atoms*.

22.11 (a) $^{211}_{83}Bi$ (b) $^{216}_{84}Po$ (c) $^{140}_{57}La$

22.13 (a) $^{144}_{60}Nd \longrightarrow ^{140}_{58}Ce + ^{4}_{2}He$

     (b) $^{40}_{19}K \longrightarrow ^{40}_{20}Ca + ^{0}_{-1}e$

     (c) $^{149}_{62}Sm \longrightarrow ^{149}_{63}Eu + ^{0}_{-1}e$

22.15 Lead-214 forms by successive decays of uranium-238. The quantity of lead-214 in the sample diminishes by half each 26.8 minutes.

22.17 1.500 ng

22.19 The ions that radiation produces are strange, unstable, highly reactive ions that initiate undesirable reactions in the body.

22.21 Radiation can cause birth defects.

22.23 Radiation intensity diminishes with the square of the distance; and radiation can be blocked by dense absorbing materials, like lead.

22.25 In low doses over a long period, radiation can initiate cancer. In well-focused, massive doses over a short period, radiation can kill cancer cells.

22.27 Radiation moves in straight lines and spreads out in the way that light from a light bulb spreads out.

22.29 Cosmic rays are more intense at higher altitudes.

22.31 2.0×10^3 millirad

22.33 The sample is undergoing 1.5×10^{-3} Ci $\times$ (3.7×10^{10} disintegrations/Ci) $= 5.6 \times 10^7$ disintegrations per second.

22.35 The gray, Gy. The older unit is the rad.

22.37 600 rad. They are roughly the same.

22.39 The rad doses are adjusted for the kinds of tissues and radiations and expressed as rems or sieverts.

22.41 The electron-volt

22.43 Radiation can cause cancer, so for diagnosis the lowest usable energies are in order.

22.46 Iron-55. $^{55}_{25}Mn + ^{1}_{1}p \longrightarrow ^{1}_{0}n + ^{55}_{26}Fe$

22.48 Neutrons. $^{113}_{49}In \longrightarrow ^{111}_{49}In + 2^{1}_{0}n$

22.50 $^{19}_{9}F + ^{4}_{2}He \longrightarrow ^{23}_{11}Na \longrightarrow ^{22}_{11}Na + ^{1}_{0}n$

22.52 $^{14}_{7}N + ^{2}_{1}H \longrightarrow ^{15}_{8}O + ^{1}_{0}n$

22.53 10 J/kg

22.55 A product produced by ionizing radiation that is not present in the food or produced either by cooking or digestion

22.57 Cooking and digestion

22.59 The shorter the half-life, the more active is the radionuclide and hence the smaller is the dose that is needed to get results.

22.61 It emits only gamma radiation, and it has a shorter half-life.

22.63 The splitting apart of a large nucleus, following neutron capture, into two smaller nuclei and neutrons

22.65 Each fission event produces more neutron initiators than were needed to cause the fission event.

22.67 About 3 to 5%. Essentially 100% in the bomb.

22.69 Strontium-90 is a bone seeker. Iodine-131 is taken up by the thyroid gland. Cesium-137 gets as widely distributed as the sodium ion.

22.71 As a product in the uranium-238 disintegration series

22.73 Its decay products, which are not gases but are also radioactive, stay in the lungs.

22.75 By bombarding a metal surface with high-energy electrons, which are able to expel low-level electrons from the metal atoms. As higher-level orbital electrons drop into the "holes," X rays are emitted.

22.77 Same masses, but the charge on the positron is 1+, not 1− as on the electron

22.79 In any collision with an electron it is annihilated, and electrons are abundant in matter.

22.82 No ionizing radiation or alien chemicals are used.

22.85 An *acute effect* is one produced at the time of the exposure or soon thereafter. Cancer (particularly leukemia) and gene alteration are *latent effects*.

GLOSSARY[1]

Absolute Zero The coldest temperature attainable; 0 K or -273.15 °C. (1.3)

Accuracy In science, the degree of conformity to some accepted standard or reference; freedom from error or mistake; correctness. (1.5)

Acetal Any organic compound in which two etherlike linkages extend from one CH unit. (12.4)

Achiral Not possessing chirality; that quality of a molecule (or other object) that allows it to be superimposed on its mirror image. (14.3)

Acid *Brønsted theory:* any substance that can donate a proton (H^+). (7.1, 7.2)

Acid–Base Balance The blood is in *acid–base balance* when its pH is in the range 7.35 to 7.45. (18.4)

Acid–Base Indicator (see *Indicator*)

Acid–Base Neutralization The reaction of an acid with a base. (7.1)

Acidic Solution A solution in which the molar concentration of hydronium ions is greater than that of hydroxide ions. (7.1, 8.1)

Acidity The molar concentration of hydronium ion in a solution or the pH of the solution. (8.1)

Acidosis A condition in which the pH of the blood is below normal. *Metabolic acidosis* is brought on by a defect in some metabolic pathway. *Respiratory acidosis* is caused by a defect in the respiratory centers or in the mechanisms of breathing. (8.1, 18.4, 19.7)

Acid Rain Rain made acidic by the presence of air pollutants such as oxides of sulfur and nitrogen. (Special Topic 8.1)

Active Transport The movement of a substance through a biological membrane against a concentration gradient and caused by energy-consuming chemical changes that involve parts of the membrane. (6.6, 16.8)

Activity Series A list of elements (or other substances) in the order of the ease with which they release electrons under standard conditions and become oxidized. (7.3)

Acyl Group

$$
\underset{\text{Acyl group}}{R-\overset{\displaystyle O}{\overset{\displaystyle \|}{C}}-} \qquad (13.5)
$$

Acyl Group Transfer Reaction Any reaction in which an acyl group transfers from a donor to an acceptor. (13.5)

Addition Reaction Any reaction in which two parts of a reactant molecule add to a double or a triple bond. (10.4)

Aerobic Sequence An oxygen-consuming sequence of catabolism that starts with glucose and proceeds through glycolysis, the citric acid cycle, and the respiratory chain. (19.1)

Agonist A compound whose molecules can bind to a receptor on a cell membrane and cause a response by the cell. (17.6)

Albumin One of a family of globular proteins that tend to dissolve in water, and that in blood contribute to the blood's colloidal osmotic pressure and aid in the transport of metal ions, fatty acids, cholesterol, triacylglycerols, and other water-insoluble substances. (16.9)

Alcohol Any organic compound whose molecules have the OH group attached to a saturated carbon; ROH. (11.1)

Alcohol Group The OH group when it is joined to a saturated carbon. (11.1)

Aldehyde An organic compound that has a carbonyl group joined to H on one side and C on the other. (12.1)

Aldehyde Group $-CH=O$ (12.1)

Aldohexose A monosaccharide whose molecules have six carbon atoms and an aldehyde group. (14.2)

Aldose A monosaccharide whose molecules have an aldehyde group. (14.2)

Aliphatic Compound Any organic compound whose molecules lack a benzene ring or a similar structural feature. (10.1, 10.6)

Alkali A strong base. (7.1)

Alkali Metals The elements of Group IA of the periodic table: lithium, sodium, potassium, rubidium, cesium, and francium. (2.5)

Alkaline Earth Metals The elements of Group IIA of the periodic table: beryllium, magnesium, calcium, strontium, barium, and radium. (2.5)

Alkaloid A physiologically active, heterocyclic amine isolated from plants. (11.6)

Alkalosis A condition in which the pH of the blood is above normal. *Metabolic alkalosis* is caused by a defect in metabolism. *Respiratory alkalosis* is caused by a defect in the respiratory centers of the brain or in the apparatus of breathing. (8.1, 18.4)

Alkane A saturated hydrocarbon; one that has only single bonds. A *normal* alkane is any whose molecules have straight chains. (10.1, 10.2)

---

[1]The entries in this Glossary include the terms that appear in boldface within the chapters, including the margin comments, as well as several additional entries. The numbers in parentheses following the definitions are the section numbers (or Special Topics) where the entry was introduced or discussed.

Alkene   A hydrocarbon whose molecules have one or more double bonds. (10.1, 10.3, 10.4)

Alkyl Group   A substituent group that is an alkane minus one H atom. (10.2)

Alkyne   A hydrocarbon whose molecules have triple bonds. (10.1)

Alpha Helix   (see *Helix*.)

Alpha (α) Particle   The nucleus of a helium atom; $^4_2$He. (22.1)

Alpha (α) Radiation   A stream of high-energy alpha particles. (22.1)

Amide   An organic compound whose molecules have a carbonyl-to-nitrogen single bond. (11.6, 13.5)

Amide Bond   The single bond that holds the carbonyl group to the nitrogen atom in an amide. (13.5)

Amine   An organic compound whose molecules have a trivalent nitrogen atom, as in RNH_2, RNHR, or R_3N. (11.6)

Amine Salt   An organic compound whose molecules have a positively charged, tetravalent, protonated nitrogen atom, as in RNH_3^+, $R_2NH_2^+$, or R_3NH^+. (11.6)

Amino Acid   Any organic compound whose molecules have both an amino group and a carboxyl group. (16.1)

Amino Acid Residue   A structural unit in a polypeptide,

$$-NH-\underset{\underset{R}{|}}{CH}-\overset{\overset{O}{\|}}{C}-$$

furnished by an α-amino acid, where R is the side-chain group of a particular amino acid. (16.1)

Aminoacyl Group

$$-NH_2-\underset{\underset{R}{|}}{CH}-\overset{\overset{O}{\|}}{C}-$$

where R is one of the amino acid side chains. (16.1)

Amphipathic Compound   A substance whose molecules have both hydrophilic and hydrophobic groups. (15.5)

Anaerobic Sequence   The oxygen-independent catabolism of glucose to lactate ion. (19.1, 19.2)

Anhydrous   Without water. (6.2)

Anion   A negatively charged ion. (7.1)

Anion Gap

$$\text{Anion gap} = \frac{\text{meq of Na}^+}{\text{L}} - \left(\frac{\text{meq of Cl}^-}{\text{L}} + \frac{\text{meq of HCO}_3^-}{\text{L}} \right)$$

(Special Topic 7.1)

Anode   The positive electrode to which negatively charged ions (anions) are attracted during electrolysis. (7.1)

Anoxia   A condition of a tissue in which it receives no oxygen. (18.4)

Antagonist   A compound that can bind to a membrane receptor but not cause any response by the cell. (17.6)

Antibiotics   Antimetabolites made by bacteria and fungi. (17.3)

Anticodon   A sequence of three adjacent side-chain bases on a molecule of tRNA that is complementary to a codon and that fits to its codon on an mRNA chain during polypeptide synthesis. (21.2)

Antimetabolite   A substance that inhibits the growth of bacteria. (17.3)

Apoenzyme   The wholly polypeptide part of an enzyme. (17.1)

Aromatic Compound   Any organic compound whose molecules have a benzene ring (or a feature very similar to this). (10.1, 10.6)

Atmosphere, Standard   (see *Standard Atmosphere*)

Atom   A small particle with one nucleus and zero charge; the smallest particle of a given element that bears the chemical properties of the element. (2.5)

Atomic Mass   The average mass, in atomic mass units (u), of the atoms of the isotopes of a given element as they occur naturally. (2.4)

Atomic Mass Number   (see *Mass Number*)

Atomic Mass Unit (u) $1.6605665 \times 10^{-24}$ g: a mass very close to that of a proton or a neutron. (2.2)

Atomic Number   The positive charge on an atom's nucleus; the number of protons in an atom's nucleus. (2.2)

Avogadro's Number 6.02×10^{23}: the number of formula units in one mole of any element or compound. (4.2)

Avogadro's Principle   Equal volumes of gases contain equal numbers of moles when they are compared at identical temperatures and pressures. (5.2)

Background Radiation   Cosmic rays plus the natural atomic radiation emitted by the traces of radioactive isotopes in soils and rocks plus any radiation that escapes from the operations of nuclear facilities. (22.2)

Balanced Equation   (see *Equation, Balanced*)

Barometer   An instrument for measuring atmospheric pressure. (5.1)

Basal Activities   The minimum activities of the body needed to maintain muscle tone, control body temperature, circulate the blood, handle wastes, breathe, and carry out other essential activities. (Special Topic 1.1)

Basal Metabolic Rate   The rate at which energy is expended to maintain basal activities. (Special Topic 1.1)

Basal Metabolism   The total of all of the chemical reactions that support basal activities. (Special Topic 1.1)

Base   *Brønsted theory:* a proton acceptor; a compound that neutralizes hydrogen ions. (7.1, 7.2)

Base, Heterocyclic   A heterocyclic amine obtained from the hydrolysis of nucleic acids: adenine, thymine, guanine, cytosine, or uracil. (21.1)

Base Pairing   In nucleic acid chemistry, the association by means of hydrogen bonds of two heterocyclic, side-chain bases: adenine with thymine (or uracil) and guanine with cytosine. (21.1)

Base Quantity   A fundamental quantity of physical measurement such as mass, length, and time; a quantity used to define derived quantities such as mass/volume for density. (1.3)

Base Unit   A fundamental unit of measurement for a base quantity, such as the kilogram for mass, the meter for length, the second for time, the kelvin for temperature degree, and the mole for quantity of chemical substance; a unit to which derived units of measurement are related. (1.3)

Basic Solution   A solution in which the molar concentration of hydroxide ions is greater than that of hydronium ions. (7.1, 8.1)

Becquerel (Bq)   The SI unit for the activity of a radioactive source; one nuclear disintegration (or other transformation) per second. (22.3)

Benedict's Reagent   The reagent that is used in Benedict's test. (12.3)

Benedict's Test   The use of Benedict's reagent [a solution of copper(II) sulfate, sodium citrate, and sodium carbonate] to detect the presence of any compound whose molecules have such easily oxidized functional groups as those present in monosaccharides. In a positive test the intensely blue color of the reagent disappears and a reddish precipitate of copper(I) oxide separates. (12.2)

Beta (β) Oxidation   The catabolism of a fatty acid by a series of repeating steps that produce acetyl units (in acetyl CoA); the fatty acid cycle of catabolism. (19.1, 19.5)

Beta (β) Particle   A high-energy electron emitted from a nucleus, $_{-1}^{0}e$. (22.1)

Beta (β) Radiation   A stream of high-energy electrons. (22.1)

Bile   A secretion of the gallbladder that empties into the upper intestine and furnishes bile salts; a route of excretion for cholesterol and bile pigments. (18.1)

Bile Salts   Steroid-based detergents in bile that emulsify fats and oils during digestion. (18.1)

Biochemistry   The study of the structures and properties of substances found in living systems. (14.1)

Blood Sugar   The carbohydrates—mostly glucose—that are present in blood. (14.2, 20.1)

Blood Sugar Level   The concentration of carbohydrate—mostly glucose—in the blood, usually stated in units of mg/dL. (20.2)

Boiling   The turbulent behavior in a liquid when its vapor pressure equals atmospheric pressure and when the liquid absorbs heat while experiencing no increase in temperature. (5.4)

Boiling Point, Normal   The temperature at which a substance boils when the atmospheric pressure is 760 mm Hg (1 atm). (5.4)

Bond, Chemical   A net electrical force of attraction that holds atomic nuclei near each other within compounds. (3.1)

Boron Family   The Group IIIA elements of the periodic table: boron, aluminum, gallium, indium, and thallium. (2.5)

Boyle's Law   (see *Pressure–Volume Law*)

Branched Chain   A sequence of atoms to which additional atoms are attached at points other than the ends. (9.2)

Brönsted Theory   An acid is a proton donor and a base is a proton acceptor. (7.1, 7.2)

Brownian Movement   The random, chaotic movements of particles in a colloidal dispersion that can be seen with a microscope. (6.5)

Buffer   A combination of solutes that holds the pH of a solution relatively constant even if small amounts of acids or bases are added. (8.3)

Butyl Group $CH_3CH_2CH_2CH_2$— (10.2)

sec-Butyl Group $CH_3CH_2CH(CH_3)$— (10.2)

t-Butyl Group $(CH_3)_3C$— (10.2)

Calorie   The amount of heat that raises the temperature of 1 g of water by 1 degree Celsius from 14.5 °C to 15.5 °C. (1.7)

Carbaminohemoglobin   Hemoglobin that carries chemically bound carbon dioxide. (18.3)

Carbohydrate   Any naturally occurring substance whose molecules are polyhydroxyaldehydes or polyhydroxyketones or can be hydrolyzed to such compounds. (14.2)

Carbonate Buffer   A mixture or a solution that includes bicarbonate ions and dissolved carbon dioxide in which the bicarbonate ion can neutralize added acid and carbon dioxide can neutralize added base. (8.3)

Carbon Family   The Group IVA elements in the periodic table: carbon, silicon, germanium, tin, and lead. (2.5)

Carbonyl Group   The atoms carbon and oxygen joined by a double bond, $C{=}O$. (12.1)

Carboxylic Acid   A compound whose molecules have the carboxyl group, CO_2H. (13.1)

Carcinogen   A chemical or physical agent that induces the onset of cancer or the formation of a tumor that might or might not become cancerous. (22.2)

Catabolism   The reactions of metabolism that break molecules down. (19.1)

Catalysis The phenomenon of an increase in the rate of a chemical reaction brought about by a relatively small amount of a chemical—the catalyst—that is not permanently changed by the reaction. (5.6)

Catalyst A substance that is able, in relatively low concentrations, to accelerate the rate of a chemical reaction without itself being permanently changed. (In living systems, catalysts are called enzymes.) (5.6)

Cathode The negative electrode to which positively charged ions—cations—are attracted during electrolysis. (7.1)

Cation A positively charged ion. (7.1)

Centimeter (cm) A length equal to one-hundredth of the meter.

$$1 \text{ cm} = 0.01 \text{ m} = 0.394 \text{ in. } (1.3)$$

Charles' Law (see *Temperature–Volume Law*)

Chemical Bond (see *Bond, Chemical*)

Chemical Energy The potential energy that substances have because their arrangements of electrons and atomic nuclei are not as stable as are alternative arrangements that become possible in chemical reactions. (1.7)

Chemical Equation A shorthand representation of a chemical reaction that uses formulas instead of names for reactants and products; that separates reactant formulas from product formulas by an arrow; that separates formulas on either side of the arrow by plus signs; and that expresses the mole proportions of the chemicals by simple numbers (coefficients) placed before the formulas. (4.1, 7.3)

Chemical Family A group of elements with similar chemical properties; usually a vertical group of elements in the periodic table. (2.5)

Chemical Property Any chemical reaction that a substance can undergo and the ability to undergo such a reaction. (1.2)

Chemical Reaction Any event in which substances change into different chemical substances. (1.2, 2.1)

Chemiosmotic Theory An explanation of how oxidative phosphorylation is related to a flow of protons in a proton gradient established by the respiratory chain, a gradient that extends across the inner membrane of a mitochondrion. (19.4)

Chemistry The study of the compositions and structures of substances and their ability to change into other substances. (1.1)

Chiral Having handedness in a molecular structure. (see also *Chirality*) (14.3)

Chirality The quality of handedness that a molecular structure has that prevents this structure from being superimposable on its mirror image. (14.3)

Chloride Shift An interchange of chloride ions and bicarbonate ions between a red blood cell and the surrounding blood serum. (18.3)

Chromatin Filaments of nucleoprotein made of histones and DNA in cell nuclei. (21.1)

Chromosomes Small threadlike bodies in a cell nucleus that carry genes in a linear array and that are microscopically visible during cell division. (21.1)

Citric Acid Cycle A series of reactions that dismantle acetyl units and send electrons (and protons) into the respiratory chain; a major source of metabolites for the respiratory chain. (19.1, 19.3)

Codon A sequence of three adjacent side-chain bases in a molecule of mRNA that codes for a specific amino acid residue when the mRNA participates in polypeptide synthesis. (21.1)

Coefficients Numbers placed before formulas in chemical equations to indicate the mole proportions of reactants and products. (4.1)

Coenzyme An organic compound needed to make a complete enzyme from an apoenzyme. (17.1)

Cofactor A nonprotein compound or ion that is an essential part of an enzyme. (17.1)

Collagen The fibrous protein of connective tissue that changes to gelatin in boiling water. (16.9)

Colligative Property A property of a solution that depends only on the concentrations of the solute and the solvent and not on their chemical identities (e.g., osmotic pressure). (6.6)

Colloidal Dispersion A relatively stable, uniform distribution in some dispersing medium of colloidal particles—those with at least one dimension between 1 and 1000 nm (6.5)

Colloidal Osmotic Pressure The contribution made to the osmotic pressure of a solution by substances colloidally dispersed in it. (6.6, 18.2)

Compound A substance made from the atoms of two or more elements that are present in a definite proportion by mass and by atoms. (2.1)

Concentration The quantity of some component of a mixture in a unit of volume or a unit of mass of the mixture. (4.6)

Condensed Structure (see *Structural formula*)

Constitutional Formula A formula that uses lines representing covalent bonds to connect the atomic symbols in the pattern that occurs in one molecule of a compound; a structural formula. (3.4, 9.2)

Constitutional Isomerism The existence of two or more compounds with identical molecular formulas but different atom-to-atom sequences. (9.3)

Constitutional Isomers Compounds with identical molecular formulas but different atom-to-atom sequences. (9.3)

Conversion Factor   A fraction that expresses a relationship between quantities that have different units, such as 2.54 cm/in. (1.7)

Cosmic Radiation   A stream of ionizing radiations from the sun and outer space that consists mostly of protons but also includes alpha particles, electrons, and the nuclei of atoms up to atomic number 28. (22.1)

Covalence   The number of covalent bonds that an atom can have in a molecule. (3.4)

Covalent Bond   The net force of attraction that arises as two atomic nuclei share a pair of electrons. One pair is shared in a single bond, two pairs in a double bond, and three pairs in a triple bond. (3.4)

Crenation   The shrinkage of red blood cells when they are in contact with a hypertonic solution. (6.7)

Curie (Ci)   A unit of activity of a radioactive source.

$$1 \text{ Ci} = 3.70 \times 10^{10} \text{ disintegrations/s}   (22.3)$$

Dalton's Law   (see *Law of Partial Pressures*)

Dalton's Theory   A theory that accounts for the laws of chemical combination by postulating that matter consists of indestructible atoms; that all atoms of the same element are identical in mass and other properties; that the atoms of different elements are different in mass and other properties; and that in the formation of a compound atoms join together in definite, whole-number ratios. (2.2)

Deamination   The removal of an amino group from an amino acid. (19.6)

Degree Celsius   One one-hundredth (1/100) of the interval on a thermometer between the freezing point and the boiling point of water. (1.3)

Degree Fahrenheit   One one-hundred-and-eightieth (1/80) of the interval on a thermometer between the freezing point and the boiling point of water. (1.3)

Deliquescence   The ability of a substance to attract water vapor to itself to form a concentrated solution. (6.2)

Denatured Protein   A protein whose molecules have suffered the loss of their native shape and form as well as their ability to function biologically. (16.2, 16.6)

Density   The ratio of the mass of an object to its volume; the mass per unit volume. Density = mass/volume (usually expressed in g/mL). (1.8)

Deoxyribonucleic Acid (DNA)   The chemical of a gene; one of a large number of polymers of deoxyribonucleotides and whose sequences of side-chain bases constitute the genetic messages of genes. (21.1)

Derived Quantity   A quantity based on a relationship that involves one or more base quantities of measurement such as volume (length^3) and density (mass/volume). (1.3)

Desiccant   A substance that combines with water vapor to form a hydrate and thereby reduces the concentration of water vapor in the air space around the substance. (6.2)

Detergent   A surface-active agent; a soap. (6.1, Special Topic 15.2)

D Family; L Family   The names of the two optically active families to which substances can belong when they are considered solely according to one kind of molecular chirality (molecular handedness) or the other. (14.3)

Diabetes Mellitus   A disease in which there is an insufficiency of effective insulin and an impairment of glucose tolerance. (20.3)

Dialysis   The passage through a dialyzing membrane of water and particles in solution, but not of particles that have colloidal size. (6.6)

Dialyzing Membrane   A membrane permeable to solvent and small ions or molecules but impermeable to colloidal-sized particles. (6.6)

Digestive Juice   A secretion into the digestive tract that consists of a dilute aqueous solution of digestive enzymes (or their zymogens) and inorganic ions. (18.1)

Dipeptide   A compound whose molecules have two α-amino acid residues joined by a peptide (amide) bond. (16.3)

Dipolar Ion   A molecule that carries one plus charge and one minus charge, such as an α-amino acid. (16.1)

Dipole, Electrical   A pair of equal but opposite (and usually partial) electrical charges separated by a small distance in a molecule. (3.5)

Diprotic Acid   An acid with two protons available per molecule to neutralize a base, e.g., H_2SO_4. (7.2)

Disaccharide   A carbohydrate that can be hydrolyzed into two monosaccharides. (14.2, 14.4)

Disulfide System   S—S as in R—S—S—R. (11.5)

DNA   (see *Deoxyribonucleic Acid; Double Helix DNA Model*)

Double Bond   A covalent bond in which two pairs of electrons are shared. (3.4)

Double Helix DNA Model   A spiral arrangement of two intertwining DNA molecules held together by hydrogen bonds between side-chain bases. (21.1)

Double Replacement   A reaction in which a compound is made by the exchange of partner ions between two salts. (7.5)

Dynamic Equilibrium   (see *Equilibrium, Dynamic*)

Edema   The swelling of tissue caused by the retention of water. (18.2)

Effector   A chemical other than a substrate that can activate an enzyme. (17.3)

Elastin   The fibrous protein of tendons and arteries. (16.9)

Electrical Balance   The condition of a net ionic equation wherein the algebraic sum of the positive and negative charges of the reactants equals that of the products. (7.3)

Electrical Dipole   (See *Dipole, Electrical.*)

Electrode   A metal object, usually a wire, suspended in an electrically conducting medium through which electricity passes to or from an external circuit. (7.1)

Electrolysis   A procedure in which an electrical current is passed through a solution that contains ions, or through a molten salt, for the purpose of bringing about a chemical change. (7.1)

Electrolyte   Any substance whose solution in water conducts electricity; or the solution itself of such a substance. (3.1, 7.1)

Electrolytes, Blood   The inorganic ionic substances dissolved in the blood. (18.2)

Electron   A subatomic particle that bears one unit of negative charge and has a mass that is 1/1836th the mass of a proton. (2.2)

Electron Cloud   A mental model that views the one or two rapidly moving electrons of an orbital as creating a cloudlike distribution of negative charge. (3.5)

Electron Configuration   The most stable arrangement (that is, the arrangement of lowest energy) of the electrons of an atom, ion, or molecule. (2.3)

Electron Density   The relative concentration of negative charge carried by electrons. (3.4)

Electronegativity   The ability of an atom joined to another by a covalent bond to attract the electrons of the bond toward itself. (3.5)

Electron Sharing   The joint attraction of two atomic nuclei toward a pair of electrons situated between the nuclei and between which, therefore, a covalent bond exists. (3.4)

Electron Shell   (see *Energy Level*)

Electron Volt (eV)   A very small unit of energy used to describe the energy of radiation. (22.3)

$$1 \text{ eV} = 1.6 \times 10^{-19} \text{ joule}$$
$$1 \text{ eV} = 3.8 \times 10^{-20} \text{ calorie}$$
$$10^3 \text{ eV} = 1 \text{ keV (1 kiloelectron volt)}$$
$$10^3 \text{ keV} = 1 \text{ MeV (1 megaelectron volt)}$$

Element   A substance that cannot be broken down into anything that is both stable and more simple; a substance in which all of the atoms have the same atomic number and the same electron configuration; one of the three broad kinds of matter, the others being compounds and mixtures. (2.4)

Emulsion   A colloidal dispersion of tiny microdroplets of one liquid in another liquid. (6.5)

Enantiomers   Isomers whose molecules are related as an object is related to its mirror image but that cannot be superimposed. (14.3)

Endothermic   Describing a change that needs a constant supply of heat energy to happen. (2.1)

End Point   The stage in a titration when the operation is stopped. (8.4)

Energy   A capacity to cause a change that can, in principle, be harnessed for useful work. (1.7)

Energy Level   A principal energy state in which electrons of an atom can be. (2.3)

Energy of Activation   The minimum energy that must be provided by the collision between reactant particles to initiate the rearrangement of electrons relative to nuclei that must happen if the reaction is to occur. (5.5)

Enzyme   A catalyst in a living system. (5.6, 17.1)

Enzyme–Substrate Complex   The temporary combination that an enzyme must form with its substrate before catalysis can occur. (17.2)

Equation, Balanced   A chemical equation in which all of the atoms represented in the formulas of the reactants are present in identical numbers among the products, and in which any net electrical charge provided by the reactants equals the same charge indicated by the products. (see also *Chemical Equation*) (4.1, 7.3)

Equilibrium, Dynamic   A situation in which two opposing events occur at identical rates so that no net change happens. (6.3)

Equivalence Point   The stage in a titration when the reactants have been mixed in the exact molar proportions represented by the balanced equation; in an acid–base titration, the stage when the moles of hydrogen ions furnished by the acid matches the moles of hydroxide ions (or other proton acceptor) supplied by the base. (8.4)

Equivalent (eq)   For an ion, usually its mass in grams divided by the amount of its electrical charge. (Special Topic 7.1)

Error   In a measurement, the difference between the measured value and the correct value of a physical quantity. (1.5)

Essential Amino Acid   An α-amino acid that the body cannot make from other amino acids and that must be supplied by the diet. (20.5)

Ester   A derivative of an acid and an alcohol that can be hydrolyzed to these parent compounds. Esters of carboxylic acids and phosphoric acid occur in living systems. (13.3)

Carboxylic acid ester   Phosphoric acid ester

Ether   An organic compound whose molecules have an oxygen attached by single bonds to separate carbon atoms neither of which is a carbonyl carbon atom: R—O—R′. (11.4)

Ethyl Group CH_3CH_2— (10.2)

Evaporation   The conversion of a substance from its liquid to its vapor state. (5.4)

Exact Number   A number having an infinite number of significant figures used in a physical quantity. (1.5)

Exon   A segment of a DNA strand that eventually becomes expressed as a corresponding sequence of aminoacyl residues in a polypeptide. (21.1)

Exothermic   Describing a change by which heat energy is released from the system. (2.1)

Extracellular Fluids   Body fluids that are outside of cells. (18.1)

Factor-Label Method   A strategy for solving computational problems that uses conversion factors and the cancellation of the units of physical quantities as an aid in working toward the solution. (1.6)

Family, Organic   Compounds whose molecules have the same functional group. (9.2)

Fatty Acid   Any carboxylic acid that can be obtained by the hydrolysis of animal fats or vegetable oils. (13.1, 15.1)

Fatty Acid Cycle   (see *Beta Oxidation*)

Feedback Inhibition   The competitive inhibition of an enzyme by a product of its own action. (17.3)

Fibrin   The fibrous protein of a blood clot that forms from fibrinogen during clotting. (16.9, 17.4)

Fibrinogen   A protein in blood that is changed to fibrin during clotting. (16.9, 17.4)

Fibrous Proteins   Water-insoluble proteins found in fibrous tissues. (16.9)

Fission   The splitting of the nucleus of a heavy atom approximately in half that is accompanied by the release of one or a few neutrons and energy. (22.7)

Formula, Chemical   A shorthand representation of a substance that uses atomic symbols and following subscripts to describe the elemental composition and the mole ratios in which the atoms of the elements are combined. (3.2)

Formula, Constitutional   (see Constitutional Formula)

Formula, Empirical   A chemical symbol for a compound that gives just the ratios of the atoms and not necessarily the composition of a complete molecule. (3.2)

Formula, Molecular   A chemical symbol for a substance that gives the composition of a complete molecule. (3.4)

Formula Mass   The sum of the atomic masses of the atoms represented in a chemical formula. (4.3)

Formula Unit   A small particle—an atom, a molecule, or a set of ions—that has the composition given by the chemical formula of the substance. (3.2)

Forward Reaction   In a chemical equilibrium, the reaction whereby substances to the left of the double arrows are changed to the products shown on the right-hand side of the arrows. (6.3)

Free Rotation   The absence of a barrier to the rotation of two groups with respect to each other when they are joined by a single, covalent bond. (9.2)

Functional Group   An atom or a group of atoms in a molecule that is responsible for the particular set of reactions that all compounds with this group have. (9.2)

Gamma radiation   A natural radiation similar to but more powerful than X rays. (22.1)

Gap Junctions   Tubules made of membrane-bound proteins that interconnect one cell to neighboring cells and through which materials can pass directly. (16.8)

Gas   Any substance that must be contained in a wholly closed space and whose shape and volume are determined entirely by the shape and volume of its container; a state of matter. (2.1)

Gas Constant, Universal *(R)*   The ratio of PV to nT for a gas, where P = the gas pressure, V = volume, n = the number of moles, and T = the Kelvin temperature. When P is in mm Hg and V is in mL,

$$R = 6.24 \times 10^4 \text{ mm Hg mL/mol K} \quad (5.2)$$

Gas Law, General

$$\frac{P_1V_1}{T_1} = \frac{P_2V_2}{T_2} \quad (5.2)$$

Gas Law, Universal, $PV = nRT$ (5.2)

Gas Tension   The partial pressure of a gas over its solution in some liquid when the system is in equilibrium. (6.3)

Gastric Juice   The digestive juice secreted into the stomach and that contains pepsinogen, hydrochloric acid, and gastric lipase. (18.1)

Gay–Lussac's Law   (see *Pressure–Temperature Law*)

Gel   A colloidal dispersion of a solid in a liquid that has adopted a semisolid form. (6.5)

Gene   A unit of heredity carried on a cell's chromosomes and consisting of DNA. (21.1)

Genetic Code   The set of correlations that specify which codons on mRNA chains are responsible for which amino acyl residues when the latter are steered into place during the mRNA-directed synthesis of polypeptides. (21.2)

Genetic Engineering   The use of recombinant DNA to manufacture substances or to repair genetic defects. (21.5)

Genome   The entire complement of genetic information of a species; all the genes of an individual. (21.7)

Geometric Isomerism   Isomerism caused by restricted rotation that gives different geometries to the same structural organization; cis–trans isomerism. (10.3)

Geometric Isomers Isomers whose molecules have identical atomic organizations but different geometries; cis–trans isomers. (10.3)

Globular Proteins Proteins that are soluble in water or in water that contains certain dissolved salts. (16.9)

Globulins Globular proteins in the blood (16.9)

Gluconeogenesis The synthesis of glucose from compounds with smaller molecules or ions. (19.6, 20.2)

Glucose Tolerance The ability of the body to manage the intake of dietary glucose while keeping the blood sugar level from fluctuating widely. (20.3)

Glucose Tolerance Test A series of measurements of the blood sugar level after the ingestion of a considerable amount of glucose; used to obtain information about an individual's glucose tolerance. (20.3)

Glucosuria The presence of glucose in urine. (20.2)

Glycerophospholipid A hydrolyzable lipid that has an ester linkage between glycerol and one phosphoric acid unit (this, in turn, forming another ester link to a small molecule). In *phosphatides,* the remaining two OH units of glycerol are esterified with fatty acids. In *plasmalogens,* one OH is esterified with a fatty acid and the other is joined by an ether link to a long-chain unsaturated alcohol. (15.3)

Glycogenesis The synthesis of glycogen. (20.2)

Glycogenolysis The breakdown of glycogen to glucose. (20.2)

Glycol A dihydric alcohol. (11.1)

Glycolipid A lipid whose molecules include a glucose unit, a galactose unit or some other carbohydrate unit. (15.3)

Glycolysis A series of chemical reactions that break down glucose or glucose units in glycogen until pyruvate remains (when the series is operated aerobically) or lactate forms (when the conditions are anaerobic). (19.1, 19.2)

Glycoprotein A protein, often membrane bound, that is joined to a carbohydrate unit. (16.8)

Gradient The presence of a change in value of some physical quantity with distance, as in a *concentration* gradient in which the concentration of a solute is different in different parts of the system. (16.8)

Gram (g) A mass equal to one-thousandth of the kilogram mass, the SI standard mass.

$$1 \text{ g} = 10^{-3} \text{ kg} = 10^3 \text{ mg}; \quad 1 \text{ lb} = 453.6\text{g} \quad (1.3)$$

Gray (Gy) The SI unit of absorbed dose of radiation equal to one joule of energy absorbed per kilogram of tissue. (22.3)

Group A vertical column in the periodic table; a family of elements. (2.5)

Half-Life The time needed for half of the atoms in a sample of a particular radioactive isotope to undergo radioactive decay. (22.1)

Halogens The elements of Group VIIA of the periodic table: fluorine, chlorine, bromine, iodine, and astatine. (2.5)

Hard Water Water that contains one or more of the metallic ions Mg^{2+}, Ca^{2+}, Fe^{2+}, and Fe^{3+}. The negative ions present are usually Cl^- and SO_4^{2-}. If HCO_3^- is the chief negative ion, the water is said to be *temporary hard water;* otherwise it is *permanent hard water.* (Special Topic 7.2)

Heat The form of energy that transfers between two objects in contact that have initially different temperatures. (1.7)

Heat of Fusion The quantity of heat that one gram of a substance absorbs when it changes from its solid to its liquid state at its melting point. (5.4)

Heat of Reaction The net energy difference between the reactants and the products of a reaction. (5.5)

Heat of Vaporization The Quantity of heat that one gram of a substance absorbs when it changes from its liquid to its gaseous state. (5.4)

α-Helix One kind of secondary structure of a polypeptide in which its molecules are coiled. (16.4)

Hemiacetal Any compound whose molecules have both an OH and an OR group coming to a CH unit. (12.4)

Hemiketal Any compound whose molecules have both an OH and an OR group coming to a carbon that otherwise bears no H atoms. (12.4)

Hemoglobin The oxygen-carrying protein in red blood cells. (16.5, 18.3)

Hemolysis The bursting of a red blood cell. (6.7)

Henry's Law (see *Pressure–Solubility Law*)

Heterocyclic Compound An organic compound with a ring in which an atom other than carbon takes up at least one position in the ring. (10.1)

Heterogenous Mixture A mixture in which the composition of one small portion is not identical with that of another. (6.5)

Heterogeneous Nuclear RNA (hnRNA) RNA made directly at the guidance of DNA and from which messenger RNA (mRNA) is made. (Formerly called primary transcript RNA, ptRNA.) (21.2)

Homeostasis The response of an organism to a stimulus such that the organism is restored to its prestimulated state. (17.3)

Homogeneous Mixture A mixture in which the composition and properties are uniform throughout. (6.5)

Hormone A primary chemical messenger made by an endocrine gland and carried by the bloodstream to a target organ where a particular chemical response is initiated. (17.5)

Hydrate A compound in which intact molecules of water are held in a definite molar proportion to the other components. (6.2)

Hydration   The association of water molecules with dissolved ions or polar molecules. (6.2)

Hydrocarbon   An organic compound that consists entirely of carbon and hydrogen. (10.5)

Hydrogen Bond   The force of attraction between a $\delta+$ on a hydrogen held by a covalent bond to oxygen or nitrogen (or fluorine) and a $\delta-$ charge on a nearby atom of oxygen or nitrogen (or fluorine). (6.1)

Hydrolase   An enzyme that catalyzes a hydrolysis reaction. (17.1)

Hydrolysis of Ions   Reactions in which ions (other than H^+ or OH^-) react with water and change the pH of a solution. (8.2)

Hydrolyzable Lipid   A lipid that can be hydrolyzed or saponified; formerly called a saponifiable lipid. (15.1)

Hydronium Ion H_3O^+ (7.1)

Hydrophilic Group   Any part of a molecular structure that attracts water molecules; a polar or ionic group such as OH, CO_2^-, NH_3^+, or NH_2. (15.5)

Hydrophobic Group   Any part of a molecular structure that has no attraction for water molecules; a nonpolar group such as any alkyl group. (15.5)

Hydrophobic Interaction   The water avoidance by nonpolar groups or side chains that is partly responsible for the shape adopted by a polypeptide or nucleic acid molecule in an aqueous environment. (16.1, 21.1)

Hydroxide Ion OH^- (7.1)

Hygroscopic   Describing a substance that can reduce the concentration of water vapor in the surrounding air by forming a hydrate. (6.2)

Hypercapnia   An elevated level of carbon dioxide in the blood as indicated by a partial pressure of CO_2 in venous blood above 50 mm Hg. (18.4)

Hyperglycemia   An elevated level of glucose in the blood—above 110 mg/dL in whole blood. (20.2)

Hyperthermia   A condition of a core body temperature above normal. (5.4, Special Topic 5.2)

Hypertonic   Having an osmotic pressure greater than some reference; having a total concentration of all solute particles higher than that of some reference. (6.7)

Hyperventilation   Breathing considerably faster and deeper than normal. (8.3)

Hypocapnia   A condition of a below-normal concentration of carbon dioxide in the blood as indicated by a partial pressure of CO_2 in venous blood of less than 35 mm Hg. (18.4)

Hypoglycemia   A low level of glucose in blood—below 65 mg/dL of whole blood. (20.2)

Hypothermia   A condition of a low body temperature. (5.4, Special Topic 5.2)

Hypotonic   Having an osmotic pressure less than some reference; having a total concentration of dissolved solute particles less than that of some reference. (6.7)

Hypoventilation   Breathing more slowly and less deeply than normal; shallow breathing. (8.3)

Hypoxia   A condition of a low supply of oxygen. (18.4)

Ideal Gas   A hypothetical gas that obeys the gas laws exactly. (5.3)

Indicator, Acid–Base   A dye that has one color in solution below a measured pH range and a different color above this range. (7.1, 8.1)

Induced-Fit Model   Many enzymes are induced by their substrate molecules to modify their shapes to accommodate the substrate. (17.2)

Inertia   The resistance of an object to a change in its position or its motion. (1.3)

Inhibitor   A substance that interacts with an enzyme to prevent its acting as a catalyst. (17.3)

Inner Transition Elements   The elements of the lanthanide and actinide series of the periodic table. (2.5)

Inorganic Compound   Any compound that is not an organic compound. (9.1)

Internal Environment   Everything enclosed within an organism. (18.1)

International System of Units (SI)   The successor to the metric system with new reference standards for the base units but with the same names for the units and the same decimal relationships. (1.3)

International Union of Pure and Applied Chemistry System (IUPAC System)   A set of systematic rules for naming compounds designed to give each compound one unique name and for which only one structure can be drawn. (10.2, Appendix III)

Interstitial Fluids   Fluids in tissues but not inside cells or the blood. (23.1)

Intestinal Juice   The digestive juice that empties into the duodenum from the intestinal mucosa and whose enzymes also work within the intestinal mucosa as molecules migrate through. (18.1)

Intracellular Fluids   Fluids inside cells. (18.1)

Intron   A segment of DNA strand that separates exons and that does not become expressed as a segment of a polypeptide. (21.1)

Inverse-Square Law   The intensity of radiation varies inversely with the square of the distance from its source. (22.2)

In Vitro   Occurring in laboratory vessels. (11.3)

In Vivo   Occurring within a living system. (11.3)

Ion   An electrically charged, atomic or molecular-sized particle; a particle that has one or a few atomic nuclei and either one or two (seldom, three) too many or too few electrons to render the particle electrically neutral. (3.1)

Ionic Bond   The force of attraction between oppositely charged ions in an ionic compound. (3.1)

Ionic Compound   A compound that consists of an orderly aggregation of oppositely charged ions that assemble in whatever ratio ensures overall electrical neutrality. (3.1)

Ionic Equation   A chemical equation that explicitly shows all of the particles—ions, atoms, or molecules—that are involved in a reaction, even if some are only spectator particles. (see also *Net Ionic Equation; Equation, Balanced*) (7.3)

Ionization   A change, usually involving solvent molecules, whereby molecules change into ions. (7.1)

Ionizing Radiation   Any radiation that can create ions from molecules within the medium it enters, such as alpha, beta, gamma, X, and cosmic radiation. (22.2)

Ion Product Constant of Water (K_w)   The product of the molar concentrations of hydrogen ions and hydroxide ions in water at a given temperature.

$$K_w = [H^+][OH^-]$$
$$= 1.0 \times 10^{-14} \text{ (at 25 °C)}   (1.1)$$

Isobutyl Group $(CH_3)_2CHCH_2—$ (10.6)

Isoelectric Molecule   A molecule that has an equal number of positive and negative sites. (16.1)

Isoelectric Point (pI)   The pH of a solution in which a specified amino acid or a protein is in an isoelectric condition; the pH at which there is no net migration of the amino acid or protein in an electrolysis experiment. (16.1)

Isoenzymes   Enzymes that have identical catalytic functions but that are made of slightly different polypeptides. (17.1)

Isohydric Shift   In actively metabolizing tissue, the use of a hydrogen ion, produced from newly released carbon dioxide to react with and liberate oxygen from oxyhemoglobin; in the lungs, the use of hydrogen ion, released when hemoglobin oxygenates, to combine with bicarbonate ion and liberate carbon dioxide for exhaling. (18.3)

Isomerase   An enzyme that catalyzes the conversion of a compound into one of its isomers. (17.1)

Isomerism   The phenomenon of the existence of two or more compounds with identical molecular formulas but different structures. (9.3)

Isomers   Compounds with identical molecular formulas but different structures. (9.3)

Isopropyl Group $(CH_3)_2CH—$ (10.2)

Isotonic   Having an osmotic pressure identical to that of a reference; having a concentration equivalent to that of the reference with respect to the ability to undergo osmosis. (6.7)

Isotope   A substance in which all of the atoms are identical in atomic number, mass number, and electron configuration. (2.4)

IUPAC System   (see *International Union of Pure and Applied Chemistry System*)

Joule (J)   The SI-derived unit of energy. (1.7)

K_w (see *Ion Product Constant of Water*)

Kelvin   The SI unit of temperature degree equal to 1/100th of the interval between the freezing point and the boiling point of water when measured under standard conditions. (1.3)

Kelvin Scale   The scale of absolute temperatures expressed in kelvins, beginning with 0 K for the coldest temperature attainable. (1.3)

Keratin   The fibrous protein of hair, fur, fingernails, and hooves. (16.9)

Ketal   A substances whose molecules have two OR groups joined to a carbon that also holds two hydrocarbon groups. (12.4)

Ketoacidosis   The acidosis caused by untreated ketonemia. (19.7)

Keto Group   The carbonyl group when it is joined on each side to carbon atoms. (12.1)

Ketohexose   A monosaccharide whose molecules contain six carbon atoms and have a keto group. (14.2)

Ketone   Any compound with a carbonyl group attached to two carbon atoms, as in $R_2C{=}O$. (12.1)

Ketone Bodies   Acetoacetate, β-hydroxybutyrate—or their parent acids—and acetone. (19.7)

Ketonemia   An elevated concentration of ketone bodies in the blood. (19.7)

Ketonuria   An elevated concentration of ketone bodies in the urine. (19.7)

Ketose   A monosaccharide whose molecules have a ketone group. (14.2)

Ketosis   The combination of ketonemia, ketonuria, and acetone breath. (19.7)

Kilocalorie (kcal)   The quantity of heat equal to 10^3 calories. (1.7)

Kilogram (kg)   The SI base unit of mass; 10^3 g; 2.205 lb. (1.3)

Kinase   An enzyme that catalyzes the transfer of a phosphate group. (17.1)

Kinetic Energy (KE)   The energy of an object by virtue of its motion.

$$\text{KE} = \tfrac{1}{2} \text{ mass} \times \text{velocity}^2   (1.7)$$

Kinetic Theory of Gases   A set of postulates about the nature of an ideal gas: that it consists of a large number of very small particles in constant, random motion; that in their collisions the particles lose no frictional energy; that between collisions the particles neither attract nor repel each other; and that the motions and collisions of the particles obey all the laws of physics. (5.3)

Krebs' Cycle   (see *Citric Acid Cycle*)

Law of Conservation of Mass   Matter is neither created nor destroyed in chemical reactions; the masses of all products equal the masses of all reactants. (2.1)

Law of Definite Proportions   The elements in a compound occur in definite proportions by mass. (2.1)

Law of Partial Pressures (Dalton's Law)   The total pressure of a mixture of gases is the sum of their individual partial pressures. (5.1)

Le Châtelier's Principle   If a system is in equilibrium and a change is made in its conditions, the system will change in whichever way most directly restores equilibrium. (6.3)

Length   The base quantity for expressing distances or how long something is. (1.3)

Ligase   An enzyme that catalyzes the formation of bonds at the expense of triphosphate energy. (17.1)

"Like Dissolves Like" Rule   Polar solvents dissolve polar or ionic solutes and nonpolar solvents dissolve nonpolar or weakly polar solutes. (10.1)

Lipid   A plant or animal product that tends to dissolve in such nonpolar solvents as ether, carbon tetrachloride, and benzene. (15.1)

Lipid Bilayer   The sheetlike array of two layers of lipid molecules, interspersed with molecules of cholesterol and proteins, that makes up the membranes of cells in animals. (15.5)

Lipoprotein Complex   A combination of lipid and protein molecules that serves as the vehicle for carrying the lipid in the bloodstream. (20.4)

Liquid   A state of matter in which a substance's volume, but not its shape, is independent of the shape of its container. (2.1)

Liter (L)   A volume equal to 1000 cm^3 or 1000 mL or 1.057 liquid quart. (1.3)

Lock-and-Key Theory   The specificity of an enzyme for its substrate is caused by the need for the substrate molecule to fit to the enzyme's surface much as a key fits to and turns only one tumbler lock. (17.2)

Lyase   An enzyme that catalyzes an elimination reaction to form a double bond. (17.1)

Macromolecule   Any molecule with a very high formula mass, generally several thousand or more. (10.5)

Markovnikov's Rule   In the addition of an unsymmetrical reactant to an unsymmetrical double bond of a simple alkene, the positive part of the reactant molecule (usually H$^+$) goes to the carbon that has the greater number of hydrogen atoms and the negative part goes to the other carbon of the double bond. (10.4)

Mass   A quantitative measure of inertia based on an artifact at Sèvres, France, called the standard kilogram mass; a measure of the quantity of matter in an object relative to this reference standard. (1.3)

Mass Number   The sum of the numbers of protons and neutrons in one atom of an isotope. (2.2)

Material Balance   The condition of a chemical equation in which all of the atoms present among the reactants are also found in the products. (7.3)

Matter   Anything that occupies space and has mass. (2.1)

Measurement   An operation that obtains a value for a physical quantity by the use of an instrument. (1.2)

Melting Point   The temperature at which a solid changes into its liquid form; the temperature at which equilibrium exists between the solid and liquid forms of a substance. (5.4)

Messenger RNA (mRNA)   RNA that carries the genetic code in the form of a specific series of codons for a specific polypeptide from the cell's nucleus to the cytoplasm. (21.2)

Metabolism   The sum total of all of the chemical reactions that occur in an organism. (5.4)

Metal   Any substance, usually an element, that is shiny, conducts electricity well, and (if a solid) can be hammered into sheets and drawn into wires. (2.4)

Metalloids   Elements that have some metallic and some nonmetallic properties. (2.5)

Meter (m)   The base unit of length in the International System of Units (SI).

$$1 \text{ m} = 100 \text{ cm} = 39.37 \text{ in.} = 3.280 \text{ ft}$$
$$= 1.093 \text{ yd}   (1.3)$$

Methyl Group   CH$_3$— (10.2)

Micelle   A globular arrangement of the molecules of an amphipathic compound in water in which their hydrophobic parts intermingle inside the globule and their hydrophilic parts are exposed to the water. (15.5)

Microgram (μg)   A mass equal to one-thousandth of a milligram.

$$1 \text{ μg} = 10^{-3} \text{ mg} = 10^{-6} \text{ g}   (1.3)$$

Microliter (μL)   A volume equal to one-thousandth of a milliliter.

$$1 \text{ μL} = 10^{-3} \text{ mL} = 10^{-6} \text{ L}   (1.3)$$

Milliequivalent (meq)   A quantity of substance equal to one-thousandth of an equivalent. (Special Topic 7.1)

Milligram (mg)   A mass equal to one-thousandth of a gram.

$$1 \text{ mg} = 10^{-3} \text{ g};   10^3 \text{ mg} = 1 \text{ g}   (1.3)$$

Milliliter (mL)   A volume equal to one-thousandth of a liter.

$$1 \text{ mL} = 10^{-3} \text{ L} = 1 \text{ cm}^3$$
$$1 \text{ liquid ounce} = 29.57 \text{ mL}$$
$$1 \text{ liquid quart} = 946.4 \text{ mL} \quad (1.3)$$

Millimeter (mm) A length equal to one-thousandth of a meter.

$$1 \text{ mm} = 10^{-3} \text{ m} = 0.0394 \text{ in.} \quad (1.3)$$

Millimeter of Mercury (mm Hg) A unit of pressure equal to 1/760th atm. (5.1)

Millimole (mmol) One-thousandth of a mole.

$$1 \text{ mmol} = 10^{-3} \text{ mol} \quad (4.4)$$

Mixture One of the three kinds of matter (together with elements and compounds); any substance made up of two or more elements or compounds combined physically in no particular proportion by mass and separable into its component parts by physical means. (2.1)

Molar concentration *(M)* A solution's concentration in units of moles of solute per liter of solution; molarity. (4.6)

Molarity (see *Molar Concentration*)

Molar Volume, Standard The volume occupied by one mole of a gas under standard conditions of temperature and pressure; 22.4 L at 273 K and 1 atm. (5.2)

Mole (mol) A mass of a compound or element that equals its formula mass in grams; Avogadro's number of a substance's formula units. (4.4)

Molecular Compound A compound whose smallest representative particle is a molecule; a covalent compound. (3.1)

Molecular Equation An equation that shows the complete formulas of all of the substances present in a mixture undergoing a reaction. (see also *Net Ionic Equation; Equation, Balanced*) (7.3)

Molecular Formula (see *Formula, Molecular*)

Molecular Mass The formula mass of a substance. (4.3)

Molecule An electrically neutral (but often polar) particle made up of the nuclei and electrons of two or more atoms and held together by covalent bonds; the smallest representative sample of a molecular compound. (3.1)

Monoamine Oxidase An enzyme that catalyzes the inactivation of neurotransmitters or other amino compounds of the nervous system. (17.6)

Monomer A compound that can be used to make a polymer. (10.5)

Monoprotic Acid An acid with one proton per molecule that can neutralize a base. (7.2)

Monosaccharide A carbohydrate that cannot be hydrolyzed. (14.2)

Mutagen A chemical or physical agent that can induce the mutation of a gene without preventing the gene from replicating. (22.2)

Myosins Proteins in contractile muscle. (16.9)

Native Protein A protein whose molecules are in the configuration and shape they normally have within a living system. (16.2)

Net Ionic Equation A chemical equation in which all spectator particles are omitted so that only the particles that participate directly are represented. (7.3)

Neurotransmitter A substance released by one nerve cell to carry a signal to the next nerve cell. (17.5)

Neutralization, Acid–Base A reaction between an acid and a base. (7.1)

Neutralizing Capacity The capacity of a solution or a substance to neutralize an acid or a base, expressed as a molar concentration. (8.4)

Neutral Solution A solution in which the molar concentration of hydronium ions exactly equals the molar concentration of hydroxide ions. (7.1, 8.1)

Neutron An electrically neutral subatomic particle with a mass of 1 u. (2.2)

Nitrogen Family The elements of Group VA of the periodic table: nitrogen, phosphorus, arsenic, antimony, and bismuth. (2.5)

Nitrogen Pool The sum total of all nitrogen compounds in the body. (19.6)

Noble Gases The elements of Group 0 of the periodic table: helium, neon, argon, krypton, xenon, and radon. (2.5)

Noble Gas Rule (see *Octet Rule*)

Nonelectrolyte Any substance that cannot furnish ions when dissolved in water or when melted. (7.1)

Nonfunctional Group A section of an organic molecule that remains unchanged during a chemical reaction at a functional group. (9.2)

Nonhydrolyzable Lipid Any lipid, such as the steroids, that cannot by hydrolyzed or similarly broken down by aqueous alkali. (15.1)

Nonmetal Any element that is not a metal. (see *Metal*) (2.4)

Normal Fasting Level The normal concentration of something in the blood, such as blood sugar, after about 4 hours without food. (20.2)

Normoglycemia The condition of the blood in which the level of glucose is normal for the nutritional status of the individual. (20.2)

Nuclear Chain Reaction The mechanism of nuclear fission by which one fission event makes enough fission initiators (neutrons) to cause more than one additional fission event. (22.7)

Nuclear Equation A representation of a nuclear transformation in which the chemical symbols of the reactants and products include mass numbers and atomic numbers. (2.1)

Nucleic Acid A polymer of nucleotides in which the repeating units are pentose phosphate diesters, each pentose unit bearing a side-chain base (one of five heterocyclic amines); polymeric compounds that are involved in the storage, transmission, and expression of genetic messages. (21.1)

Nucleotide A monomer of a nucleic acid that consists of a pentose phosphate ester in which the pentose unit carries one of five heterocyclic amines as a side-chain base. (21.1)

Nucleus In chemistry and physics, the subatomic particle that serves as the core of an atom and that is made up of protons and neutrons. (2.2) In biology, the organelle in a cell that houses DNA. (21.1)

Octet, Outer A condition of an atom or ion in which its highest occupied energy level has eight electrons—a condition of stability. (3.3)

Octet Rule (Noble Gas Rule) The atoms of a reactive element tend to undergo those chemical reactions that most directly give them the electron configuration of the noble gas that stands nearest the element in the periodic table (all but one of which have outer octets). (3.3)

Optical Isomer One of a set of compounds whose molecules differ only in their chiralities. (14.3)

Optically Active The ability of a substance to rotate the plane of polarization of plane-polarized light. (14.3)

Organic Compounds Compounds of carbon other than those related to carbonic acid and its salts, or to the oxides of carbon, or to the cyanides. (9.1)

Osmolarity The molar concentration of all osmotically active solute particles in a solution. (6.6)

Osmosis The passage of water only, without any solute, from a less concentrated solution (or pure water) to a more concentrated solution when the two solutions are separated by a semipermeable membrane. (6.6)

Osmotic Membrane A semipermeable membrane that permits only osmosis, not dialysis. (6.6)

Osmotic Pressure The pressure that would have to be applied to a solution to prevent osmosis if the solution were separated from water by an osmotic membrane. (6.6)

Outer Octet (see *Octet, Outer*)

Outside-Level Electrons Electrons occupying the highest energy level. (2.3, 3.3)

Oxidase (see *Oxidoreductase*)

Oxidation A reaction in which the oxidation number of one of the atoms of a reactant becomes more positive; in organic chemistry, the loss of hydrogen or the gain of oxygen. (3.1)

Oxidation Number For simple monatomic ions, the quantity and sign of the electrical charge on the ion. (3.1)

Oxidation–Reduction Reaction A reaction in which oxidation numbers change. (3.1)

Oxidative Deamination The change of an amino group to a keto group with loss of nitrogen. (19.6)

Oxidative Phosphorylation The synthesis of high-energy phosphates such as ATP from lower-energy phosphates and inorganic phosphate by the reactions that involve the respiratory chain. (19.4)

Oxidizing Agent A substance that can cause an oxidation. (3.1)

Oxidoreductase An enzyme that catalyzes the formation of an oxidation–reduction equilibrium. (17.1)

Oxygen Debt The condition in a tissue when anaerobic glycolysis has operated and lactate has been excessively produced. (19.2)

Oxygen Family The elements in Group VIA of the periodic table: oxygen, sulfur, selenium, tellurium, and polonium. (2.5)

Pancreatic Juice The digestive juice that empties into the duodenum from the pancreas. (18.1)

Partial Pressure The pressure contributed to the total pressure by an individual gas in a mixture of gases. (5.1)

Pentose Phosphate Pathway The synthesis of NADPH that uses chemical energy in glucose-6-phosphate and that involves pentoses as intermediates. (19.2)

Peptide Bond The amide linkage in a protein; a carbonyl-to-nitrogen bond. (16.3)

Peptide Unit

$$\overset{\displaystyle O}{\underset{\displaystyle R}{-NHCH\overset{\|}{C}-}} \qquad (16.1)$$

Percent (%) A measure of concentration.

Vol/vol (v/v) percent: The number of volumes of solute in 100 volumes of solution.

Wt/wt (w/w) percent: The number of grams of solute in 100 g of the solution.

Wt/vol (w/v) percent: The number of grams of solute in 100 mL of the solution.

Milligram percent: The number of milligrams of the solute in 100 mL of the solution. (6.4)

Period A horizontal row in the periodic table. (2.5)

Periodic Law Many properties of the elements are periodic functions of their atomic numbers. (2.5)

Periodic Table A display of the elements that emphasizes the family relationships. (2.5)

pH The negative power to which the base 10 must be raised to express the molar concentration of hydrogen ions in an aqueous solution.

$$[H^+] = 1 \times 10^{-pH}$$
$$-\log [H^+] = pH \quad (8.1)$$

Phenols Organic compounds whose molecules have an OH group attached to a benzene ring. (11.1, 11.3)

Phenyl Group The benzene ring minus one H atom; C_6H_5. (10.6)

Phosphate Buffer Usually a mixture or a solution that contains dihydrogen phosphate ions ($H_2PO_4^-$) and monohydrogen phosphate ions (HPO_4^{2-}) to neutralize H^+. (8.3)

Phosphatide A glycerophospholipid whose molecules are esters between glycerol, two fatty acids, phosphoric acid, and a small alcohol. (15.3)

Phosphoglyceride (see *Glycerophospholipid*)

Phospholipids Lipids such as the glycerophospholipids (phosphatides and plasmalogens) and the sphingomyelins whose molecules include phosphate ester units. (15.3)

Photosynthesis The synthesis in plants of complex compounds from carbon dioxide, water, and minerals with the aid of sunlight captured by the plant's green pigment, chlorophyll. (Special Topic 14.1)

Physical Change Any event in which chemical substances do not change into other chemical substances. (2.1)

Physical Property Any observable characteristic of a substance other than a chemical property, such as color, density, melting point, boiling point, temperature, and quantity. (1.2)

Physical Quantity A property of something to which we assign both a numerical value and a unit, such as mass, volume, or temperature;

physical quantity = number × unit (1.2)

Physiological Saline Solution A solution of sodium chloride with an osmotic pressure equal to that of blood. (6.7)

p*I* (see *Isoelectric Point*)

Plasmalogens Glycerophospholipids whose molecules include an unsaturated fatty alcohol unit. (15.3)

Plasmid A circular molecule of supercoiled DNA in a bacterial cell. (21.5)

β-Pleated Sheet A secondary structure for a polypeptide in which the molecules are aligned side by side in a sheet-like array with the sheet partially pleated. (16.4)

Polar Bond A bond at which we can write δ+ at one end and δ− at the other end, the end that has the more electronegative atom. (3.5)

Polar Molecule A molecule that has sites of partial positive and partial negative charge and a permanent electrical dipole. (3.5)

Polyatomic Ion A ion made from two or more atoms, such as OH^-, SO_4^{2-}, and CO_3^{2-}. (3.4)

Polymer Any substance with a very high formula mass whose molecules have a repeating structural unit. (10.5)

Polymerization A chemical reaction that makes a polymer from a monomer. (10.5)

Polypeptide A polymer with repeating α-aminoacid residues joined by peptide (amide) bonds. (16.1)

Polysaccharide A carbohydrate whose molecules are polymers of monosaccharides. (14.2, 14.5)

Potential Energy Stored or inactive energy. (1.7)

Precipitate A solid that separates from a solution as the result of a chemical reaction. (4.5)

Precipitation The formation and separation of a precipitate. (4.5)

Precision The fineness of a measurement or the degree to which successive measurements agree with each other when several are taken one after the other. (see also *Accuracy*) (1.5)

Pressure Force per unit area. (5.1)

Pressure–Solubility Law (Henry's Law) The concentration of a gas in a liquid at any given temperature is directly proportional to the partial pressure of the gas on the solution. (6.3)

Pressure–Temperature Law (Gay–Lussac's Law) The pressure of a gas is directly proportional to its Kelvin temperature when the gas volume is constant. (5.2)

Pressure–Volume Law (Boyle's Law) The volume of a gas is inversely proportional to its pressure when the temperature is constant. (5.2)

Primary Alcohol An alcohol in whose molecules an OH group is attached to a primary carbon, as in RCH_2OH. (11.1)

Primary Carbon In a molecule, a carbon atom that is joined directly to just one other carbon, such as the end carbons in $CH_3CH_2CH_3$. (10.2)

Primary Structure The sequence of amino acyl residues held together by peptide bonds in a polypeptide. (16.2, 16.3)

Primary Transcript RNA (ptRNA) [see *Heterogeneous Nuclear RNA (hnRNA)*]

Product A substance that forms in a chemical reaction. (2.1)

Property A characteristic of something by means of which we can identify it. (1.2)

Propyl Group $CH_3CH_2CH_2-$ (10.2)

Prosthetic Group A nonprotein molecule joined to a polypeptide to make a biologically active protein. (16.5)

Protein A naturally occurring polymeric substance made up wholly or mostly of polypeptide molecules. (16.1)

Proton A subatomic particle that bears one unit of positive charge and has a mass of 1 u. (2.2)

Proton-Pumping ATPase The enzyme on the matrix side of the inner mitochondrial membrane that catalyzes the formation of ATP from ADP and P_i under the influence of a flow of protons across this membrane. (19.4)

Quaternary Structure An aggregation of two or more polypeptide strands each with its own primary, secondary, and tertiary structure. (16.2, 16.6)

Rad One rad equals 100 ergs (1×10^{-5} J) of energy absorbed per gram of tissue as a result of ionizing radiation. (22.3)

Radiation In atomic physics, the emission of some ray such as an alpha, beta, or gamma ray; any of the rays themselves. (22.1)

Radiation Sickness The set of symptoms that develops following exposure to heavy doses of ionizing radiations. (22.1)

Radical A particle with an uneven number of electrons. (22.2)

Radioactive The property of unstable atomic nuclei whereby they emit alpha, beta, or gamma rays. (22.1)

Radioactive Decay The change occurring to a radioactive isotope by which it emits alpha rays, beta rays or gamma rays. (22.1)

Radioactive Disintegration Series A series of isotopes selected and arranged such that each isotope except the first is produced by the radioactive decay of the preceding isotope and the last isotope is nonradioactive. (22.1)

Radioactivity The ability to emit atomic radiations. (22.1)

Radionuclide A radioactive isotope. (22.1)

Reactant One of the substances that reacts in a chemical reaction. (2.1)

Reaction, Chemical An event in which chemical substances change into other substances. (1.2, 2.1)

Receptor Molecule A molecule of a protein built into a cell membrane that can accept a molecule of a hormone or a neurotransmitter. (16.8, 17.5)

Recombinant DNA DNA made by combining the natural DNA of plasmids in bacteria or the natural DNA in yeasts with DNA from external sources, such as the DNA for human insulin, and made as a step in a process that uses altered bacteria or yeasts to make specific proteins (e.g., human growth hormone and insulin). (21.5)

Redox Reaction Abbreviation of *reduction–oxidation;* a reaction in which oxidation numbers change. (3.1)

Reducing Agent A substance that can cause another to be reduced. (3.1)

Reducing Carbohydrate A carbohydrate that gives a positive Benedict's test. (14.2)

Reduction A reaction in which the oxidation number of an atom of one reactant becomes less positive or more negative; in organic chemistry, the gain of hydrogen or the loss of oxygen. (3.1)

Reductive Amination The conversion of a keto group to an amino group by the action of ammonia and a reducing agent. (20.5)

Reference Standard The physical embodiment of a base unit used to define the base unit in the International System of Units (SI). (1.3)

Rem One rem is the quantity of a radiation that produces the same effect in humans as one roentgen of X rays or gamma rays. (22.3)

Renal Threshold That concentration of a substance in blood above which it appears in the urine. (20.2)

Replication The reproductive duplication of a DNA double helix. (21.1)

Representative Element Any element in any A group of the periodic table; any element in Groups IA to VIIA and those in Group 0. (2.5)

Respiration The intake and chemical use of oxygen by the body and the release of carbon dioxide. (8.1)

Respiratory Chain The reactions that transfer electrons from intermediates made by other pathways to oxygen; the mechanism that creates a proton gradient across the inner membrane of a mitochondrion and that leads to ATP synthesis; the enzymes that handle these reactions. (19.1, 19.4)

Respiratory Enzymes The enzymes of the respiratory chain. (19.4)

Respiratory Gases Oxygen and carbon dioxide. (18.3)

Reverse Reaction The reaction that undoes the effect of the forward reaction of an equilibrium. (6.3)

Ribonucleic Acids (RNA) Polymers of nucleotides made using ribose that participate in the transcription and the translation of the genetic messages into polypeptides. [see also *Heterogeneous Nuclear RNA (hnRNA); Messenger RNA (mRNA); Ribosomal RNA (rRNA); Transfer RNA (tRNA)*] (21.1, 21.2)

Ribosomal RNA (rRNA) RNA that is incorporated into cytoplasmic bodies called ribosomes. (21.2)

Ribosome A granular complex of rRNA that becomes attached to a mRNA strand and that supplies some of the enzymes for mRNA-directed polypeptide synthesis. (21.2)

Ring Compound A compound whose molecules contain three or more atoms joined in a ring. (9.2)

RNA (see *Ribonucleic Acid*)

Roentgen One roentgen is the quantity of X rays or gamma radiation that generates ions with an aggregate of 2.1×10^9 units of charge in 1 mL of dry air at normal pressure and temperature. (22.3)

Saliva The digestive juice secreted in the mouth whose enzyme, amylase, catalyzes the partial digestion of starch. (18.1)

Salt Any crystalline compound that consists of oppositely charged ions (other than H^+, OH^-, or O^{2-}). (7.1, 7.5)

Salt Bridge A force of attraction between $(+)$ and $(-)$ sites on polypeptide molecules. (16.5)

Saponifiable Lipid (see *Hydrolyzable Lipid*).

Saponification The reaction of an ester with a base to give an alcohol and the salt of an acid. (13.3)

Saturated Compound A compound whose molecules have only single bonds. (10.1)

Scientific Notation The method of writing a number as the product of two numbers, one being 10^x, where x is some positive or negative whole number. (1.4)

Second (s) The SI unit of time; 1/60th minute (1.3)

Secondary Alcohol An alcohol in whose molecules an OH group is attached to a secondary carbon atom; R_2CHOH. (11.1)

Secondary Carbon Any carbon atom in an organic molecule that has two and only two bonds to other carbon atoms, such as the middle carbon atom in $CH_3CH_2CH_3$. (10.2)

Secondary Structure A shape, such as the α-helix or a unit in a β-pleated sheet, that all or a large part of a polypeptide molecule adopts under the influence of hydrogen bonds, salt bridges, and hydrophobic interactions after its peptide bonds have been made. (16.2, 16.4)

Semipermeable Descriptive of a membrane that permits only certain kinds of molecules to pass through and not others. (6.6)

Shock, Traumatic A medical emergency in which relatively large volumes of blood fluid leave the vascular compartment and enter the interstitial spaces. (18.2)

Side Chain A group attached to the α position of an α-amino acid. (16.1)

Sievert (Sv) The SI unit of radiation dose equivalent. (22.3)

Significant Figures The number of digits in a numerical measurement or in the result of a calculation that are known with certainty to be accurate plus one more digit. (1.5)

Simple Lipid (see *Triacylglycerol*)

Simple Salt A salt that consists of only one kind of cation and one kind of anion. (7.5)

Simple Sugar Any monosaccharide. (14.2)

Single Bond A covalent bond involving one shared pair of electrons. (3.4)

Soap A detergent that consists of the salts of long-chain fatty acids. (6.1, Special Topic 15.2)

Soft Water Water with little if any of the hardness ions— Mg^{2+}, Ca^{2+}, Fe^{2+}, or Fe^{3+}. (Special Topic 7.2)

Sol A colloidal dispersion of tiny particles of a solid in a liquid. (6.5)

Solid A state of matter in which the visible particles of the substance have both definite shapes and definite volumes. (2.1)

Solubility The extent to which a substance dissolves in a fixed volume or mass of a solvent at a given temperature. (4.5, 7.5)

Solute The component of a solution that is understood to be dissolved in or dispersed in a continuous solvent. (4.5)

Solution A homogeneous mixture of two or more substances that are at the smallest levels of their states of subdivision—at the ion, atom, or molecule level. (4.5)

Solution, Aqueous A solution in which water is the solvent. (4.5)

Solution, Concentrated A solution with a high ratio of solute to solvent. (4.5)

Solution, Dilute A solution with a low ratio of solute to solvent. (4.5)

Solution, Saturated A solution into which no more solute can be dissolved at the given temperature; a solution in which dynamic equilibrium exists between the dissolved and the undissolved solute. (4.5, 6.3)

Solution, Supersaturated An unstable solution that has a concentration of solute higher than that of the saturated solution. (4.5)

Solution, Unsaturated A solution into which more solute could be dissolved without changing the temperature. (4.5)

Solvent That component of a solution into which the solutes are considered to have dissolved; the component that is present as a continuous phase. (4.5)

Specific Gravity The ratio of the density of an object to the density of water. (Special Topic 1.2)

Specific Heat The amount of heat that one gram of a substance can absorb per degree Celsius increase in temperature:

$$\text{Specific heat} = \frac{\text{heat}}{g \; \Delta t}$$

where Δt = the change in temperature. (1.7)

Sphingolipid A lipid that, when hydrolyzed, gives sphingosine instead of glycerol, plus fatty acids, phosphoric acid, and a small alcohol or a monosaccharide; sphingomyelins and cerebrosides. (15.3)

Standard A physical description or embodiment of a base unit of measurement, such as the standard meter or the standard kilogram mass. (1.3)

Standard Atmosphere (atm) The pressure that supports a column of mercury 760 mm high when the mercury has a temperature of 0 °C. (5.1)

Standard Conditions of Temperature and Pressure (STP) 0 °C (or 273 K) and 1 atm (or 760 mm Hg). (5.2)

Standard Solution Any solution for which the concentration is accurately known. (8.4)

States of Matter The three possible physical conditions of aggregation of matter—solid, liquid, and gas. (2.1)

Steroids Nonhydrolyzable lipids such as cholesterol and several sex hormones whose molecules have the four fused rings of the steroid nucleus. (15.4)

Straight Chain A continuous, open sequence of covalently bound carbon atoms from which no additional carbon atoms are attached at interior locations of the sequence. (9.2)

Stress In equilibrium chemistry, anything that upsets an equilibrium. (6.3)

Strong Acid An acid with a high percentage ionization. Any species, molecule or ion, that has a strong tendency to donate a proton to some acceptor. (7.2)

Strong Base A metal hydroxide with a high percentage ionization in solution; any species, molecular or ionic, that binds an accepted proton strongly. (7.2) Any substance that has a high percentage ionization in solution. (7.1)

Structural Formula (see *Constitutional Formula*)

Structural Isomer (see *Constitutional Isomer*)

Structure Synonym for structural formula. (see *Constitutional Formula*)

Subatomic Particle An electron, a proton, or a neutron. The atomic nucleus as a whole is also a subatomic particle. (2.2)

Subscripts Numbers placed to the right and a half-space below the atomic symbols in a chemical formula. (3.2)

Substitution Reaction A reaction in which one atom or group replaces another atom or group in a molecule. (10.6)

Substrate The substance on which an enzyme performs its catalytic work. (14.3, 17.1)

Supersaturated Describing an unstable condition of a solution in which more solute is in solution than could be if there were equilibrium between the undissolved and dissolved states of the solute. (4.5)

Surface-Active Agent (see *Surfactant*)

Surface Tension The quality of a liquid's surface by which it behaves as if it were a thin, invisible, elastic membrane. (6.1)

Surfactant A substance, such as a detergent, that reduces the surface tension of water. (6.1, Special Topic 15.2)

Suspension A mixture in which the particles of at least one component have average diameters greater than 1000 nm. (6.5)

Target Cell A cell at which a hormone molecule finds a site where it can become attached and then cause some action that is associated with the hormone. (17.5)

Target Tissue The organ whose cells are recognizable by the molecules of a particular hormone. (17.5)

Temperature The measure of the hotness or coldness of an object. *Degrees* of temperature, such as those of the Celsius, Fahrenheit, or Kelvin scales, are intervals of equal separation on the thermometer. (1.3)

Temperature–Volume Law (Charles' Law) The volume of a gas is directly proportional to its Kelvin temperature when the pressure is kept constant. (5.2)

Teratogen A chemical or physical agent that can cause birth defects in a fetus other than inherited defects. (22.2)

Tertiary Alcohol An alcohol in whose molecules an OH group is held by a carbon from which three bonds extend to other carbon atoms; R_3COH. (11.1)

Tertiary Carbon A carbon in an organic molecule that has three and only three bonds to adjacent *carbon* atoms. (10.2)

Tertiary Structure The shape of a polypeptide molecule that arises from further folding or coiling of secondary structures. (16.2, 16.5)

Tetrahedral Descriptive of the geometry of bonds at a central atom in which the bonds project to the corners of a regular tetrahedron. (Special Topic 3.2)

Thioalcohol A compound whose molecules have the SH group attached to a saturated carbon atom; a mercaptan. (11.5)

Threshold Exposure The level of exposure to some toxic agent below which no harm is done. (22.2)

Time A period during which something endures, exists, or continues. (1.3)

Titration An experimental procedure for mixing two solutions using a buret in order to compare the concentration of one of the solutions with that of the other, the standard solution. (8.4)

Tollens' Test The use of Tollens' reagent [a slightly alkaline solution of the diammine complex of the silver ion, $Ag(NH_3)_2^+$, in water] to detect an easily oxidized group such as the aldehyde group. (12.2)

Torr A unit of pressure; 1 torr = 1 mm Hg; 1 atm = 760 torr. (5.1)

Transamination The transfer of an amino group from an amino acid to a receiver with a keto group such that the keto group changes to an amino group. (19.6)

Transcription The synthesis of messenger RNA under the direction of DNA. (21.2)

Transferase An enzyme that catalyzes the transfer of some group. (17.1)

Transfer RNA (tRNA) RNA that serves to carry an amino acyl group to a specific acceptor site of an mRNA molecule at a ribosome where the amino acyl group is placed into a growing polypeptide chain. (21.2)

Transition Elements The elements between those of Group IIA and Group IIA in the long periods of the periodic table; a metallic element other than one in Group IA or IIA or in the actinide or lanthanide families. (2.5)

Translation The synthesis of a polypeptide under the direction of messenger RNA. (21.3)

Transmutation The change of an isotope of one element into an isotope of a different element. (22.1, 22.4)

Triacylglycerol A lipid that can be hydrolyzed to glycerol and fatty acids; a triglyceride; sometimes, simply called a glyceride or a simple lipid. (15.1)

Tricarboxylic Acid Cycle   (see *Citric Acid Cycle*)

Triglyceride   (see *Triacylglycerol*)

Triple Bond A covalent bond involving the sharing of three pairs of electrons. (3.4)

Triple Helix The quaternary structure of tropocollagen in which three polypeptide chains are coiled together. (16.6)

Triprotic Acid An acid that can supply three protons per molecule. (7.2)

Tyndall Effect The scattering of light by colloidal-sized particles in a colloidal dispersion. (6.5)

Uncertainty The estimate of how finely a number can be read from a measuring instrument. (1.5)

Universal Gas Law   (see *Gas Law, Universal*)

Unsaturated Compound Any compound whose molecules have a double or a triple bond. (10.1)

Valence Shell The highest energy level of an atom that is occupied by electrons; the outside shell. (Special Topic 3.2)

Valence-Shell Electron-Pair Repulsion Theory (VSEPR) Bond angles at a central atom are caused by the repulsions of the electron clouds of valence-shell electron pairs. (Special Topic 3.2)

Vapor The gaseous form of a liquid. (5.4)

Vaporization   (see *Evaporation*)

Vapor Pressure The pressure exerted by a vapor that is in equilibrium with its liquid state at a given temperature. (5.4)

Vascular Compartment The entire network of blood vessels and their contents. (18.2)

Ventilation The movement of air into and out of the lungs by breathing. (8.3)

Virus One of a large number of substances that consist of nucleic acid surrounded by a protein overcoat and that can enter host cells, multiply, and destroy the host. (21.4)

Vitamin An organic substance that must be in the diet, whose absence causes a deficiency disease, that is present in foods in trace concentrations, and that isn't a carbohydrate, lipid, protein, or amino acid. (17.1)

Volume The capacity of an object to occupy space. (1.3)

Water of Hydration Water molecules held in a hydrate in some definite mole ratio to the rest of the compound. (6.2)

Wax A lipid whose molecules are esters of long-chain monohydric alcohols and long-chain fatty acids. (15.1)

Weak Acid An acid with a low percentage ionization in solution; any species, molecule or ion, that has a weak tendency to donate a proton and poorly serves as a proton donor. (7.2)

Weak Base A base with a low percentage ionization in solution; any species, molecule or ion, that weakly holds an accepted proton and poorly serves as a proton acceptor. (7.2)

Weak Electrolyte Any electrolyte that has a low percentage ionization in solution. (7.1)

Weight The gravitational force of attraction on an object as compared with that of some reference. (1.3)

Zwitterion   (see *Dipolar Ion*)

Zymogen A polypeptide that is changed into an enzyme by the loss of a few amino acid residues or by some other change in its structure; a proenzyme. (17.3)

PHOTO CREDITS

Chapter 1
Chapter 1 Opener: David Muench/AllStock, Inc.. Page 4: B. Daemmrich/The Image Works. Figure 1.1a: Michael Watson. Figure 1.1b: Courtesy Central Scientific Co.. Page 6: Ken Karp. Figure 1.2: Courtesy Des Poids et Mesures, France. Figure 1.3: Ken Karp. Page 23: FPG International.

Chapter 2
Chapter 2 Opener: Studio 7/The Stock Market. Page 34 (top):Culver Pictures, Inc.. Page 34 (bottom): Mark C. Burnett/Stock, Boston.

Chapter 3
Chapter 3 Opener: Dr. Dennis Kunkel/Phototake. Page 54: Courtesy Smithsonian Institution. Page 68: Courtesy Dryden & Palmer Rock Candy Co.. Page 70: Ken Karp. Page 71: Richard Megna/Fundamental Photographs.

Chapter 4
Chapter 4 Opener: Gary Gladstone/The Image Bank. Figure 4.2: Andy Washnik. Figure 4.3: Michael Watson.

Chapter 5
Chapter 5 Opener: Don Carroll/The Image Bank. Page 103: Eric Reynolds/Adventure Photo. Figure 5.5: Ben Rose/The Image Bank. Figure 5.9: Andy Washnik.

Chapter 6
Chapter 6 Opener: Tony Craddock/Tony Stone Images. Figure 6.7: OPC, Inc..

Chapter 7
Chapter 7 Opener: FPG International. Figure 7.3: Andy Washnik. Page 155: Robert Capece. Figure 7.4 and Figure 7.5: OPC, Inc.. Page 169: OPC, Inc.. Page 171: Courtesy of the Permutit Company, a division of Sybron Corp..

Chapter 8
Chapter 8 Opener: Oliver Streue/Tony Stone World Wide. Figure 8.2: OPC, Inc.. Figure 8.3: Courtesy Fisher Scientific. Page 189: Terraphotographics/BPS. Figure 8.5: Michael Watson.

Chapter 9
Chapter 9 Opener: Tripos Associates. Page 197, 201 and 204: Tripos Associates.

Chapter 10
Chapter 10 Opener: John Kelly/The Image Bank. Page 210: Tripos Associates. Page 213: Joseph P. Sinnot/ Fundamental Photographs. Page 216 - 217: Tripos Associates. Page 224 -225: Tripos Associates. Page 231: A. Baradshaw/Sipa Press. Page 233: Courtesy NASA. Figure 10.4: Russ Schleipman/Medichrome.

Chapter 11
Chapter 11 Opener: Jaime Villaseca/The Image Bank. Page 253: J. Pickerell/FPG International.

Chapter 12
Chapter 12 Opener: Louis Goldman/Photo Researchers. Page 275: Andy Washnik.

Chapter 13
Chapter 13 Opener: Jean-Francois Causse/Tony Stone World Wide. Page 296: Harry J. Przekop/Medichrome. Page 304: Lafoto/AllStock, Inc..

Chapter 14
Chapter 14 Opener: Tony Craddock/Tony Stone World Wide. Figure 14.1: Universal Press Syndicate. Page 323: Robert J. Capece.

Chapter 15
Chapter 15 Opener: COMSTOCK, Inc. Page 334: Tripos Associates. Page 337: Tripos Associates. Page 341-343: Courtesy of Richard Pastor/FDA.

Chapter 16
Chapter 16 Opener: Joe Van Os/The Image Bank. Page 366: Bill Longcore/Photo Researchers.

Chapter 17
Chapter 17 Opener: Jess Stock/Tony Stone World Wide. Figure 17.2: Courtesy of T.A. Steitz, Department of Molecular Biophysics and Biochemistry, Yale University, New Haven, CT. Page 388: Courtesy Boehmringer Mannheim Diagnostics.

Chapter 18
Chapter 18 Opener: Galen Rowell/Peter Arnold, Inc..

Chapter 19
Chapter 19 Opener: Al Tielemans/Duomo Photography, Inc.. Figure 19.2: Courtesy of Dr. Keith R. Porter.

Chapter 20
Chapter 20 Opener: David E. Myers/Tony Stone Images.

Chapter 21
Chapter 21 Opener: Art Wolfe/AllStock, Inc.. Page 477: Courtesy Lifecodes Corporation. Page 481: Courtesy of Academic Press/Molecular Design Incoparated.

Chapter 22
Chapter 22 Opener: David C. Fritts/AllStock, Inc. Figure 22.5: Courtesy of Council for Energy Awareness, Washington D.C.. Page 511: Courtesy GE Medicals Systems Group. Page 513 (top): Hank Morgan/Rainbow. Page 513 (center): Dan McCoy/Rainbow. Page 514: Howard Sochurek/Woodfin Camp & Associates.

Additional credits
Figure 16.2: Illustration Copyright by Irving Geis. Figure 16.4: After drawings by Jane Richardson, Duke University. Figure 16.5(b): Illustration Copyright by Irving Geis. Figure 16.6: Illustration Copyright by Irving Geis. Figure 21.5(b): Illustration Copyright by Irving Geis.

INDEX[1]

[1]Entries in italics refer to tables.